CRC Handbook
of
Clinical Chemistry

Volume I

Editor

Mario Werner, M.D.

Professor of Pathology
(Laboratory Medicine)
The George Washington University
Medical Center
Washington, D.C.

CRC Series in Clinical Laboratory Science

Editor-in-Chief

David Seligson, Sc.D., M.D.

Chairman and Professor
Department of Laboratory Medicine
School of Medicine
Yale University
New Haven, Connecticut

CRC

CRC Press, Inc.
Boca Raton, Florida

Library of Congress Cataloging in Publ. Data (Revised)
Main entry under title:

CRC handbook series in clinical laboratory
science.

Spine title: Handbook series in clinical
laboratory science.
Vols. for 1978 published in West Palm
Beach, Fla.; vols. for 1979-   published in
Boca Raton, Fla.
Includes bibliographies and indexes.
Contents: Section A. Nuclear medicine
v. -- Section B. Toxicology.   v. --[etc.]
-- Section G. Handbook of clinical chemistry.
--[etc.]
1. Medical laboratories. 2. Diagnosis,
Laboratory. I. Seligson, David. II. Chem-
ical Rubber Company. III. Title: Handbook
series in clinical laboratory science. [DNLM:
1. Chemistry, Clinical--Period. W1 HA 513n]
RB37.C18   616.07′56   76-27688
ISBN 0-8493-7080-9 (set)   AACR2

This book represents information obtained from authentic and highly regarded sources. Reprinted material is quoted with permission, and sources are indicated. A wide variety of references are listed. Every
reasonable effort has been made to give reliable data and information, but the author and the publisher
cannot assume responsibility for the validity of all materials or for the consequences of their use.

Direct all inquiries to CRC Press, Inc., 2000 Corporate Blvd., N.W., Boca Raton, Florida, 33431.

© 1982 by CRC Press, Inc.

International Standard Book Number 0-8493-7080-9 (Complete Set)
International Standard Book Number 0-8493-7081-7 (Volume I)

Library of Congress Card Number 76-27688
Printed in the United States

SERIES PREFACE

According to the *Webster — Merriam Dictionary,* 3rd ed., a handbook is

1a.   a book capable of being conveniently carried as a ready reference,
1b.   a concise reference book covering a particular field of knowledge,
2a.   a bookmaker's book of bets.

In order to accommodate the information in the field of clinical laboratory science we shall need more than 20 volumes. A series of volumes such as these, that will weigh in the order of 24 kg (53 pounds), will not fit the meaning of 1a: nor will 2a fit, although we are gambling that we can get the appropriate information together in a reasonable time. Definition 1b is more appropriate to our goal which is to accumulate, arrange, and present the qualitative and quantitative clinical laboratory data related to men and women (of all ages), so that anyone who needs to extract some information from the vast amount available will have it at his or her fingertips, provided he or she is within arm's length of these volumes.

To understand the biology of man, to know man, sick and well, requires knowledge of the body constituents and their interrelationships; the dynamic relationships of these constituents; and information about tools, chemicals and techniques used to get these data. Notation of the source of the data is important to scholars. It is our purpose to present this information in tabular and graphic form in a systematic way in order to help readers retrieve it.

If we are successful, we will make it possible for a user to find any information related to constituents of blood, body fluids, body tissues, and so on, whether they be organic or inorganic molecules, viruses, bacteria, or parasites. Users of these volumes should be able to find easily the composition of a buffer, the half-life of an isotope, or the life cycle of a parasite if that information has been used to study man or if the chemicals or particles are found in man.

We shall consider this *Series in Clinical Laboratory Science* successful if it meets most of our goals and if it provides the nidus for continuing growth. The editor of the series, the executive editor of each volume, the advisory boards, contributors, and publisher will make every effort to improve subsequent editions of this *Handbook.* The user will undoubtedly find errors of omission and commission or will have new information (all of which we hope will be called to our attention). All of us should strive to improve these volumes so that users and contributors the world over will find them useful in knowing more about man.

David Seligson
Editor-in-Chief

PREFACE

Some branches of laboratory medicine, such as microbiology, are defined according to pathology; others, such as hematology, are defined according to a functional system. Clinical chemistry, on the other hand, is defined by a technology, and so is capable of cutting across the conceptional structures and boundaries defined by pathology or organ systems. Such protean applicability is an obstacle to structuring a text which is comprehensive yet free of redundancy. To overcome this difficulty, the handbook has been divided into three parts which represent each a fundamental but conceptually different aspect of clinical chemistry: Elements of Methodology; Analytes; and Clinical Applications. This arrangement should allow the reader to find information intuitively at the most plausible place.

The material presented has been selected and edited to render it succinct and relevant from a practical viewpoint. Unavoidably, this necessitated omission of all information which may be of historical interest but has no longer practical use. It has been my good fortune to find support from the best talent in making the required critical judgments and decisions. The success of this editorial effort should be judged by the frequency this text is used as a convenient source of information.

Mario Werner

EDITOR-IN-CHIEF

Dr. David Seligson is Professor and Chairman of the Department of Laboratory Medicine (Clinical Pathology) and Professor of Pathology in the School of Medicine at Yale University. He is the Director of the Department of Clinical Laboratories of the Yale-New Haven Hospital. He has an Sc.D. in Biochemistry from Johns Hopkins University (1942) and an M.D. from the University of Utah (1946). He has been a President of the American Association of Clinical Chemists and the Academy of Clinical Laboratory Physicians and Scientists. He is a Fellow of the American Society of Clinical Pathologists, College of American Pathologists, American College of Physicians, and American Board of Clinical Chemistry.

He has been editor of *Standard Methods of the American Association of Clinical Chemists* (Volumes 2, 3, and 4) and he is a member of other editorial boards.

Dr. Seligson has an honorary Masters degree from Yale University and has received the following awards:

John G. Reinhold Award, Philadelphia Section of the American Association of Clinical Chemists, 1966

Fellowship of Medici Publici, University of Utah College of Medicine and Medical Alumni Association, 1966

Donald D. Van Slyke Award, American Association of Clinical Chemists 1971 Ames Award, American Association of Clinical Chemists, National Meeting, Seattle, Washington, August 1971

Gerald T. Evans Award, Academy of Clinical Laboratory Physicians and Scientists, June 1974

Honorary Member Association of Clinical Biochemists, London, April 1976

THE EDITOR

Mario Werner, M.D., Professor of Pathology (Laboratory Medicine) at The George Washington University Medical Center.

Dr. Werner obtained his medical training in Lausanne (Switzerland), Hamburg (Germany), Paris (France), and Zurich (Switzerland). He was awarded his medical license in 1956 and a Doctorate of Medical Science, University of Zurich, Switzerland, in 1960. After specializing in internal medicine at the University of Basle, Switzerland, he worked at the National Institute of Arthritis and Metabolic Diseases, Bethesda, Md. and at the Rockefeller University, New York, as a Fellow of the Swiss Academy of Medical Sciences. In 1964 he was called by the newly formed Medical School of Essen, West Germany to organize and head the Central Laboratory. Subsequently, he served as Assistant Professor of the University of California in San Francisco, as Associate Professor at Washington University, Saint Louis, Mo. and, since 1972, his present position.

Dr. Werner is a Fellow of the American Society of Clinical Pathology, the College of American Pathologists, and the National Academy of Clinical Biochemistry, and has held either elective office or committee appointments in these societies. Among other memberships, he belongs to the American Association for Clinical Chemistry, the Academy of Clinical Laboratory Physicians and Scientists, the Society for Experimental Biology and Medicine, the American Federation for Clinical Research, the German Society for Clinical Chemistry, and the International Society for Clinical Enzymology.

Dr. Werner has been a member of the Pathology, A Study Section of the National Institutes of Health, is a Director of the American Board of Clinical Chemistry and Chairman of its Examination Committee, has been a member of the Clinical Chemistry Panel of the Food and Drug Administration and is currently Chairman of the Committee on Clinical Pathology, College of American Pathologists.

Dr. Werner is the author of approximately 100 scientific papers and has edited four books. He sits on the editorial boards of *Clinical Chemistry, Clinica Chimica Acta, Clinical Biochemistry, Pathologist, Medical Laboratory Observer* (MLO) and *LAB* (Italy). His major current research interests relate to clinical enzymology, therapeutic drug monitoring, and the mathematical modeling of medical decision making.

ADVISORY BOARD

CONTRIBUTORS

Kaiser J. Aziz, Ph.D.
Chief
Clinical Chemistry Toxicology Branch
Division of In Vitro Diagnostic Device
  Standards
Bureau of Medical Devices
Food and Drug Administration
Silver Spring, Maryland

Bernard H. Armbrecht, Ph.D.
Bureau of Veterinary Medicine
Food and Drug Administration
Rockville, Maryland

William Beautyman, M.D.
Chairman
Department of Pathology
Berkshire Medical Center
Pittsfield, Massachusetts

George N. Bowers, Jr., M.D.
Director
Clinical Chemistry Laboratory
Department of Pathology
Hartford Hospital
Hartford, Connecticut

Joseph Chayen, D.Sc.
Head, Division of Cellular Biology
Head, WHO Collaborating Centre for
  Cytochemical Bioassays
Mathilda and Terence Kennedy
  Institute of Rheumatology
London, England

Gary S. Gross, D.V.M.
Hickory Ridge Animal Hospital
Columbia, Maryland

Guy T. Haven, M.D., Ph.D.
Professor
Department of Pathology
Nebraska Medical Center
Omaha, Nebraska

Robert E. Hill, Ph.D.
Associate Professor
Department of Pathology
McMaster University Medical Center
Hamilton, Ontario, Canada

Pierre W. Keitges, M.D.
Director of Laboratories
Physicians Reference Laboratory
and
Chairman, Department of Pathology
St. Joseph Hospital
Detroit, Michigan

Julius Kerkay, Ph.D.
Professor
Department of Chemistry
The Cleveland State University
Cleveland, Ohio

Ernest J. Kiser, Ph.D.
Clinical Chemist and Assistant
  Professor
Department of Pathology
University of Nebraska Medical Center
Omaha, Nebraska

Noel S. Lawson, M.D.
Associate Pathologist
St. John Hospital
and
Clinical Assistant Professor
Department of Pathology
Wayne State University
School of Medicine
Detroit, Michigan

Jacob B. Levine
Director
Laboratory Management Systems
Technicon Instruments Corporation
Tarrytown, New York

Anne McCann
Head, Information Services Group
Bureau of Foods
Food and Drug Administration
Washington, D.C.

Matthew J. McQueen, M.B., Ch.B.,
  Ph.D.
Head, Clinical Chemistry
Hamilton General Hospital
and
Associate Professor
Department of Pathology
McMaster University
Hamilton, Ontario, Canada

Robert S. Melville, Ph.D.
Director
Division of In Vitro Diagnostic Device
  Standards
Bureau of Medical Devices
Food and Drug Administration
Silver Spring, Maryland

Robert J. Mohrbacher, B.S.
Technical Manager, Clinical Chemistry
Department of Pathology
The George Washington University
  Hospital
Washington, D.C.

Herbert K. Naito, Ph.D.
Head, Lipid-Lipoprotein Laboratories
Department of Biochemistry
Division of Laboratory Medicine
Cleveland Clinic Foundation
and
Clinical Professor
Department of Chemistry
Cleveland State University
Cleveland, Ohio

Arthur W. Patterson, Jr., V.M.D.
Veterinary Medical Officer
Equine Specialist
Bureau of Veterinary Medicine
Food and Drug Administration
Rockville, Maryland

Albert F. Plant
Director
Chemical Expositions
American Chemical Society
Washington, D.C.

Edward R. Powsner, M.D.
Department of Pathology
St. John Hospital
Detroit, Michigan

Nadja N. Rehak, Ph.D.
Clinical Chemist
Clinical Chemistry Service
Clinical Pathology Department
National Institutes of Health
Bethesda, Maryland

Robert Rej, Ph.D.
Head, Clinical Chemistry Laboratories
Division of Laboratories and Research
New York State Department of Health
Albany, New York

Patrick A. Roche, Ph.D.
Clinical Chemist
Mt. Sinai Hospital
and
Clinical Assistant Professor
Department of Chemistry
Cleveland State University
Cleveland, Ohio

Jerald M. Rosenbaum, M.D.
Director and Attending Pathologist
Department of Pathology
Clinical Chemistry Service
Baystate Medical Center
Springfield, Massachusetts

John W. Ross, M.D.
Director of Laboratories
Kennestone Hospital
Marietta, Georgia

Robert D. Strickland, Ph.D.
Associate Professor
Department of Pathology
McMaster University
and
Clinical Chemist
Hamilton General Hospital
Hamilton, Ontario, Canada

Thomas O. Tiffany, Ph.D.
Research Director
Microchemical Division
Instrumentation Laboratory, Inc.
Spokane, Washington

William H. C. Walker, M.B.
Professor
Department of Pathology
McMaster University Medical Center
Hamilton, Ontario, Canada

TABLE OF CONTENTS

NOMENCLATURE, QUANTITIES, AND UNITS
Bernard H. Armbrecht, Editor
    Introduction 3
    Quantities and Units 5
    Reporting Laboratory Animal Data 23
    Clinical Laboratory Data for Equines 37
    Reporting Human Laboratory Data 43
    Chemical Nomenclature 59

PHYSICAL AND CHEMICAL DATA 67
Herbert K. Naito, Editor

REAGENTS, SPECIMEN COLLECTION, CALIBRATORS, REFERENCE MATERIAL AND STANDARDS 287
Pierre W. Keitges and Robert J. Mohrbacher, Editors

QUALITY ASSURANCE
Guy T. Haven and Noel S. Lawson, Editors
    Introduction 339
    Erroneous or Misleading Results Due to Preanalytic Factors 341
    Control Materials and Calibration Standards 359
    Analyte Stability in Control and Reference Materials 371
    Evaluation of Precision 391
    Implementation of an Internal Quality Control Program 423
    Regional Quality Control Programs 433
    External Quality Control Programs 439
    Vendors 449

APPARATUS AND INSTRUMENTATION
William H. C. Walker, Editor
    Introduction 453
    Analytical Balances 455
    Centrifuges and Centrifugation 463
    Temperature, Thermometers, and the Gallium Standard 469
    Automated Analyzers 477
    Continuous-Flow Analyzers 483
    Centrifugal Analyzers 489
    Calorimetry 503
    Mass Spectrometry 511
    Osmometry 523
    Scanning Microdensiometry 529
    Standards for Instrument Service 537

INDEX 541

HANDBOOK OF CLINICAL CHEMISTRY

Mario Werner, Editor

Volume I

Nomenclature, Quantities, and Units
Physical and Chemical Data
Reagents, Specimen Collection, Calibrators, Reference
Material and Standards
Quality Assurance
Apparatus and Instrumentation

Volume II

Optical Methods of Analysis
Electrochemistry
Ligand Assays
Separation Methods
Statistical and Mathematical Methods

Nomenclature, Quantities, and Units

INTRODUCTION

Bernard H. Armbrecht

Modern computer technology permits vast amounts of information to be stored and retrieved. When information is entered into data banks, it usually carries its existing nomenclature and unitage. Thus, biomedical numerical information obtained from foreign literature frequently is recorded in SI units while domestic data are stored with conventional units.

The Metric Conversion Act of 1975, P. L. 94-168, provides for the formation of the U.S. Metric Board (Suite 400, 1600 Wilson Boulevard, Arlington, Va. 22209) to coordinate the transition from conventional to metric units. The Board has issued planning guidelines.[1] The American National Metric Council (5410 Grosvenor Lane, Bethesda, Md. 20014) is a private organization which is coordinating the interests of the biomedical community.

This part of the handbook describes the modern metric system, or as it is formally designated, the International System of Units (SI). Other sections list biomedical quantities frequently measured in man and animals. These lists are not intended to be exhaustive but to illustrate the conversion from conventional to SI units.[2-4] Finally, a section describes the resources for obtaining information on nomenclature and general orientation.

REFERENCES

1. *Fed. Regist.*, 44 (222), 65940, 1979; 45(181); 61550-61560, 1980; 46(12), 5101-5103, 1981.
2. Lippert, H. and Lehmann, H. P., *SI Units in Medicine,* Urban and Schwarzenberg, Baltimore, 1978.
3. Kitchen, H. and Krehbiel, J. D., Eds., *Proc. 1st Inst. Symp. Equine Hematology,* The American Association of Equine Practitioners, Golden, Colo., 1975.
4. Mitruka, B. M. and Rawnsley, H. M., *Clinical Biochemical and Hematological Reference Values in Normal Experimental Animals,* Masson Publishing, New York, 1977.

QUANTITIES AND UNITS

Bernard H. Armbrecht

INTRODUCTION

Over 140 publications have appeared which are addressed to the SI units that comprise the expanded metric system. A considerable controversy has been raised among the various disciplines that utilize laboratory medical data. SI units are being adopted by more nations each year for reporting clinical laboratory results. In spite of the reluctance in several quarters to make the change from familiar traditional units, it is only a matter of time until the matter will be seriously addressed and an educational program instituted for the practitioner that uses this information. A first purpose is to outline the SI system of units.

A second purpose for compiling such information is related to the needs of regulatory agencies. Regulations for clinical investigators using animal test subjects have been issued under the general title, "Good Laboratory Practice for Nonclinical Laboratory Studies"[1] and the proposed establishment of regulations for clinical investigators using human test subjects was issued under the title, "Obligations of Clinical Investigators of Regulated Article".[2] Both documents are silent on standards for reporting clinical laboratory results. It is contemplated that these presentations will complement the documents listed above.

LEGISLATION

Metric Chronology

Units of weight, measure, and money have been in use since man engaged in commerce. In 1585, Simon Stévin pointed out the benefits of using a decimal notation for expressing numerical values less than unity. The decimal notation for units was proposed by Gabriel Mouton in 1670. Further opportunity was provided by the French Revolution to define units based on natural systems with a decimal notation. In 1791, the Académie des Sciences adopted the length of a metre as one forty-millionth part of the meridian passing through Paris.

An international committee, sponsored by the British Association for the Advancement of Science, was organized in 1862. In 1863, the committee recommended a system based on metre, gram, and second. However, in 1873, another committee of the British Association recommended a revision for the unit of length to the centimetre. This coherent system of base units gained wide acceptance and is called the CGS system.

At the metric convention (Convention du Métre), the nations in attendance at Paris signed an agreement on May 20, 1875 that created the laboratory, Bureau International des Poids et Mesures (BIPM), its executive committee, Comité International des Poids et Mesures (CIPM), and the Conférence Générale des Poids et Mesures (CGPM). The French government provided land and facilities for the bureau at Pavillion de Breteuil, Sèvres, France. There are eight consulting committees which act as advisory groups to CIPM. The official language of these organizations is French and when the names are translated to other languages, the French abbreviations (acronyms, Table 1) do not change.

In 1889, the first CGPM established the international metre (the distance between two marks on a bar of platinum-iridium alloy kept at BIPM) and the kilogram and arranged for national prototypes to be issued. At the third CGPM meeting in 1901,

Table 1
INTERNATIONAL BODIES ISSUING RECOMMENDATION AND THEIR ABBREVIATIONS

Name	Acronym
Bureau International des Poids et Mesures	BIPM
Committee on Data for Science and Technology (for the International Council of Scientific Unions)	CODATA
Comité International des Poids et Mesures	CIPM
Conférence Générale des Poids et Mesures	CGPM
International Commission on Radiation Units and Measurements	ICRU
International Committee for Standardization in Hematology (for International Society of Hematology)	ICSH
International Electrotechnical Commission	IEC
International Federation of of Clinical Chemistry	IFCC
International Standardization Organization	ISO
International Union of Biochemistry	IUB
International Union of Geodesy and Geophysics	IUGG
International Union of Pharmacology	IUPHAR
International Union of Physiological Sciences	IUPS
International Union of Pure and Applied Chemistry	IUPAC
International Union of Pure and Applied Physics	IUPAP
IUPAC-IUB Council on Biochemical Nomenclature	CBN
IUPAC Commission on Quantities and Units in Clinical Chemistry	CQUCC
IUPAP Commission on Symbols, Units, and Nomenclature	SUN Commission
National Committee for Clinical Laboratory Standards	NCCLS
World Association of Societies of Pathology (Anatomic and Clinical)	WASP

the litre was defined as the volume that pure water occupies at maximum density, normal atmosphere, having a mass of 1 kg at mean sea level as sanctioned in 1889, and measured with reference to the metre.

The Italian physicist Giovani Giorgi, in 1901, proposed another coherent system of units based on the metre, the kilogram, the second, and an electrical unit (finally in 1950, the ampere was selected). This system also gained widespread acceptance under the name MKS or "Giorgi System" which was extended after 1950 to the MKSA system. The MKSA system gained acceptance in the Metre Convention countries over the older CGS systems. The Conférence Générale des Poids et Mesures, in 1954, adopted the MKSA system and added other units.

At the 11th CGPM conference of 1960, the metre was redefined; the second was reconsidered; the status of the litre was considered; the Système International des Unités (acronym SI) was established with its first six base units, and two supplementary units were sanctioned (Table 2); the SI prefixes from tera through pico were authorized (Table 3); and SI derived units were published (Tables 4 and 5). Initially, degree kelvin was used, but since 1968, only the term kelvin remains. The 14th CGPM conference of 1971 approved the mole as a base unit; the SI prefix list was extended to include femto and atto. In 1975, the prefixes peta and exa were added. Thus, SI is a revised and expanded version of CGS/MKS/MKSA coherent systems as outlined above.

National Legislation

The U.S. Congress approved the metric system under an act of July 28, 1866. By the Metric Conversion Act of 1975, P.L. 94-168, the U.S. Congress implemented the metric system under a voluntary procedure together with a Metric Board to manage the transition. In October 1978, the U.S. General Accounting Office issued a report, raising many of the old issues that have persisted since 1866. Unfortunately, some of

Table 2
SI BASE UNITS AND SUPPLEMENTARY UNITS

Quantity	Unit	Symbol	Dimension	Symbol
SI base units				
Length	Metre[a]	m	L	*l*
Mass	Kilogram	kg	M	*m*
Time (interval)	Second	s	T	*t*
Electric current	Ampere	A	I	*I*
Thermodynamic temperature[b]	Kelvin	K	θ	*T*
Luminuous intensity	Candela	cd	J	*l*
Amount of substance	Mole	mol	N	*n*
SI supplementary units				
Plane	Radian	rad	$\propto$	*l*
Solid angle	Steradian	sr	Ω	*l*

[a] The recommended spelling is listed; spelling the unit as meter is discouraged. Analogously, litre is the recommended spelling of this unit.

[b] $T_0 = 273.15$ K by definition (based on the triple point of water), and the unit interval expressed in kelvins is identical with the unit interval expressed in degrees Celsius (symbol °C).

Table 3
SI PREFIXES

Factor	Prefix[a]	Symbol[b]
10^{18}	exa	E
10^{15}	peta	P
10^{12}	tera	T
10^{9}	giga	G
10^{6}	mega	M
10^{3}	kilo	k
10^{2}	hecto	h
10^{1}	deka	da
10^{-1}	deci	d
10^{-2}	centi	C
10^{-3}	milli	m
10^{-6}	micro	μ
10^{-9}	nano	n
10^{-12}	pico	p
10^{-15}	femto	f
10^{-18}	atto	a

[a] The use of the prefixes hecto, deka, deci, and centi should be avoided when possible; the remaining prefixes follow the ten to the third power rule.

[b] The use of double prefix symbols is not allowed (cf. Table 8 metric ton as an example).

the "so-called metric conversions" have resulted in decreasing the size of common units, e.g., ½ gal to 1.75 *l* or 1 lb to 450 g (actually, the value should be 454.5).

On January 1, 1978, the European Economic Community (EEC) countries implemented a directive for harmonizing legislation to implement SI. Recognition of various traditional units will be withdrawn, some immediately, and others after a 2-year-transition time interval. Legislation related to the remaining non-SI units will be published

Table 4
SI DERIVED UNITS

Quantity	SI unit	Symbol[a]
Area	Square metre	m^2
Volume	Cubic metre	m^3
Velocity[b]	Metre per second	m/s
Acceleration	Metre per second squared	m/s^2
Wave number	Unity per metre	m^{-1}
Density[c]	Kilogram per cubic metre	kg/m^3
Current density	Ampere per square metre	A/m^2
Magnetic field strength	Ampere per metre	A/m
Concentration[d]	Mole per cubic metre	mol/m^3
Specific volume	Cubic metre per kilogram	m^3/kg
Luminance	Candela per square metre	cd/m^2
Heat capacity[e]	Joule per kelvin	J/K
Specific heat capacity[e]	Joule per kilogram kelvin	$J/(kg \times K)$
Specific energy	Joule per kilogram	J/kg
Energy density	Joule per cubic metre	J/m^3
Molar energy	Joule per mole	J/mol
Molar heat capacity	Joule per mole kelvin	$J/(K \times mol)$
Irradiance	Watt per square metre	W/m^2
Thermal conductivity	Watt per metre kelvin	$W/(m \times K)$
Surface tension	Newton per metre	N/m
Moment of force	Newton metre	$N \times m$
Dynamic viscosity	Pascal second	$Pa \times s$

[a]   Formed by mathematical operations using no numeric values other than unity.

[b]   The term speed is sometimes used.

[c]   The term mass density may be used.

[d]   The International Committee of Weights and Measures has recognized that the litre needs to be retained for use with SI units ($1\ \ell = 1\ dm^3 = 1\ 10^{-3}m^3$). The term molar concentration is not acceptable because the term molar has a restricted definition relating to an extensive kind of quantity of the system divided by amount of substance of the system. An acceptable name is substance concentration as an alternate to a cumbersome systematic name.

[e]   Comparable terms for entropy can also be used.

at that time. The U.S. National Bureau of Standards issued guidelines for using the Metric System of Weights and Measures[3] that focused on the education of pupils and students. At the 1977 meeting of the Committee on Teaching of Chemistry, International Union of Pure and Applied Chemistry, held in Ljubljana, Yugoslavia, Brian T. Newbold was commissioned to survey the problems of conversion to SI units and its effects on teaching chemistry. The results of this inquiry were as follows.[4] Difficulties with SI units were more in the domain of the teacher than the pupil. Conversion to SI units in schools, on a world scale, is well underway. Problems encountered are of a relatively minor nature.

THE SI SYSTEM

Base Quantities and Units

A *quantity* is a measurable property, while a *unit* is a reference quantity. The *numerical value* is expressed as a ratio of these elements, namely,

$$Quantity = unit \times numerical\ value \tag{1}$$

Table 5
SI DERIVED UNITS WITH SPECIAL NAMES

Quantity	Name	Symbol	Other units[a]
Pressure	Pascal	Pa	N/m^2
Force	Newton	N	$(m \times kg)/s^2$
Frequency	Hertz	Hz	s^{-1}
Energy	Joule	J	$N \times m$
Power	Watt	W	J/s
Electric charge	Coulomb	C	$A \times s$
Electric potential	Volt	V	W/A
Capacitance	Farad	F	C/V
Electric resistance	Ohm	Ω	V/A
Conductance	Siemens	S	A/V
Magnetic flux	Weber	Wb	$V \times s$
Magnetic flux density	Telsa	T	Wb/m^2
Inductance	Henry	H	Wb/A
Luminous flux	Lumen	lm	$cd \times sr$
Illuminance	Lux	lx	lm/m^2
Radioactivity	Becquerel	Bq	s^{-1}
Absorbed dose[b]	Gray	Gy	J/kg
Dose equivalent[b]	Sievert	Sv	J/kg

[a] Expressed in terms of base units and other coherent derived units.

[b] Dose equivalent index, 16th CGpm (1979) Resoution 5. It should be noted that the quantity dose equivalent, *H*, is the product of the absorbed dose, *D*, of ionizing radiation, and the dimensionless factors *Q* (quality factor) and *N* (product of any other multiplying factors) stipulated by the International Commission on Radiological Protection. Thus, for a given irradiation, the numerical value in joules per kilogram of these two quantities *D* and *H* may differ, depending on the values of *Q* and *N*. To avoid any risk of confusion, the special names for the respective units should be used; i.e., *D* should be expressed in grays, and *H* should be expressed in sieverts.

Quantities are also characterized to have intensive or extensive properties. An *intensive quantity* is independent of sample size, while the *extensive quantity* is dependent on the sample size. The SI base quantities, units, symbols, and dimensions are listed in Table 2. The *dimension* of a quantity must be the same as its unit, that is, length is compared with length and not length with mass. This characteristic is described as *coherence* and any assembly of such units is called a "coherent system". The SI system is designed to satisfy these requirements.

The ranking use of dimensions, as given in Table 2, follows in descending order. Each quantity in a dimensional analysis has the dimensions listed: L^α, M^β, T^γ, I^δ, θ^ϵ, J^η, N^ξ, where coefficients ($\propto$, β, γ, ---) of the dimension not needed have a zero power (in essence, these unwanted dimension drop out of consideration). Positive or negative integers are used as needed. Both sides of an equation should have the same dimensions.

Definition of Base SI Units

The definitions of base units have changed over the years as measurement technology has improved. Except for the kilogram, the current definitions are based upon defined physical constants. New definitions are designed to fall within the variance of an older definition.

Metre

In 1791, the French Académie des Sciences adopted the length of the metre as one forty-millionth part of the meridian passing through Paris. In 1889, the first CGPM established the metre as the distance between two marks on a bar of platinum-iridium alloy.

In 1960, the 11th CGPM, Resolution 6, defined the metre as the length equal to 1 650 763,73 wavelengths in vacuum of the radiation corresponding to the transition between the levels $2p_{10}$ and $5d_s$ of the krypton 86 atom.

In a recent press announcement (C&EN October 22, 1979), the radiation of iodine was measured. The frequency is in the vicinity of 520 THz. Measurements were based upon the cesium 133 definition for a second. The Consultative Committee for the definition of the metre has proposed the definition as the distance traveled in 1/ 299 792 458 of a second by plane electromagnetic waves in a vacuum. This definition is about 10^3 times better than the krypton 86 based definition presently used. The proposal may be submitted to CGPM for a vote in 1983. (See Figure 1 for relative measurements.)

Kilogram

In 1889, the first CGPM defined kilogram as the mass of the international prototype held at Sèvres.

Second

In 1967, the 13th CGPM, Resolution 1, defined the second as the duration of 9 192 631 770 periods of the radiation corresponding to the transition between the two hyperfine levels of the ground state of cesium 133 atom.

Ampere

In 1948, the 9th CGPM, Resolution 2, defined the ampere as that constant electric current which, if maintained in two straight parallel conductors of infinite length, of negligible circular cross section, and placed 1 m apart in vacuum, would produce between these conductors a force equal to 2×10^{-7} N/m of length.

Kelvin

In 1967, the 13th CGPM, Resolution 4, defined the kelvin, unit of thermodynamic temperature, as the fraction 1/273, 16 of the triple point of water.

Candela

In 1967, the 13th CGPM, Resolution 5, defined candela as the luminous intensity in the perpendicular direction, of a surface of 1/600 000 m^2 of a black-body at the freezing temperature of platinum under a pressure of 101 325 Pa (N/m^2).

Because of the experimental difficulties in realizing a Planck radiator at high temperatures and the new possibilities offered by radiometry, i.e., the measurement of optical radiation power, the 16th CGPM adopted in 1979 the following new definition: "the candela is the luminous intensity, in a given direction, of a source that emits monochromatic radiation of frequency 540 × 10^{12} Hz and that has a radiant intensity in that direction of 1/683 W per sr" [16th CGPM (1979), Resolution 3].

Mole

In 1971, the 14th CGPM, Resolution 3, defined the mole as the amount of substance of a system that contains the same number of elementary entities as there are atoms in 0,012 kg of carbon 12.*

---

* When the mole is used, the elementary entities must be specified; these may be atoms, ions, molecules, electrons, particles, or specified groups of such particles.

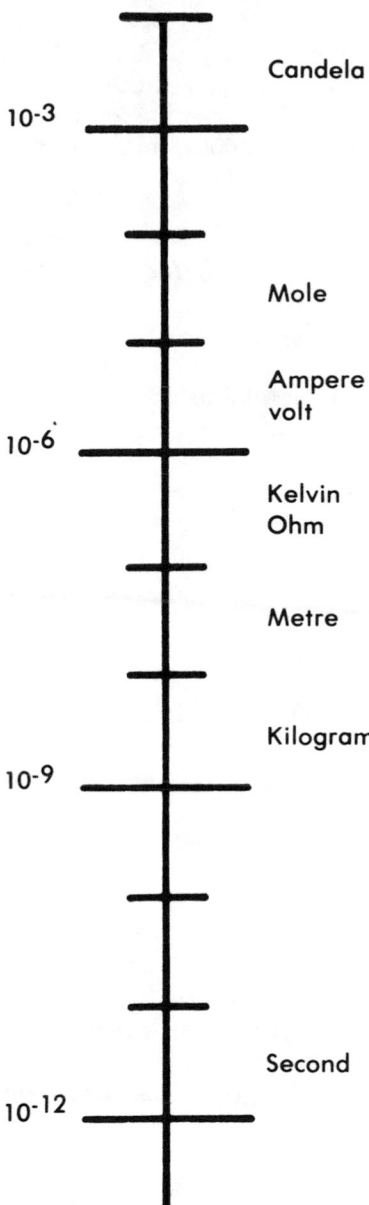

FIGURE 1. This figure compares the exactitude of measurement achievable for the SI system. These degrees differ, but all are adequate from a practical point of view. However, further advances of measurement technology are desirable from a scientific point of view.

Supplementary Units

With regard to the supplementary units (Table 2), CGPM did not state whether they were base or derived units. The International Standards Organization (ISO) has recommended the symbols shown. The dimension for these quantities is "dimensionless", or unity. As with base units, the supplementary units can be used to form derived units.

Radian

The radian is the plane angle with its vertex at the center of a circle that is subtended by an arc equal in length to the radius.

Steradian

The steradian is the solid angle with its vertex at the center of a sphere that is subtended by an area of the spherical surface equal to that of a square with sides equal in length to the radius.

SI-Derived Units

Derived units are formed by mathematical operations using no numerical coefficients. A base unit can be multiplied by itself or by combining two or more base units. Examples are illustrated in Tables 4 and 5.

Mathematical Operations

The International Standards Organization document ISO — 31.11[5] can also be consulted for a full discussion of theory and application of operations. Some of the operations that have ordinary biological applications are described below (see also Tables 6 and 7).

Multiplication

Multiplication is indicated among symbols by a small space, no space, a raised dot, or the sign (lower case, upright) ×. A dot on the line is used but this can cause confusion because it is also used as a decimal in U.S. and British documents. Care needs to be taken to avoid confusion with SI prefixes when other units are actually meant, for instance, moment of force is the newton metre and can be written as N × m, or N·m, or N m, or Nm. Written in the reverse order, m × N or m N would be comprehensible, but mN with the space omitted, would read millinewton (mN).

Division

Division can be shown by the use of a solidus (a diagonal stroke), by a horizontal line, the division sign (÷), or preferably, a negative exponent. The use of more than one solidus, as for example, mg/kg/day to express a drug dosage reading milligram of drug per kilogram of body weight per day is incorrect (unless parentheses are used) because the numerator cannot be identified. The use of a single line for writing the operation is easier to print or type and reserves the horizontal line for complex expressions. For example, specific heat capacity, joule per kilogram kelvin may be written $J/(kg \cdot K) = J\, kg^{-1}K^{-1} = J \times kg^{-1} \times K^{-1}$.

Exponents

Metre squared is written m^2; metre cubed is written m^3; and reciprocal metre is written m^{-1} or $1/m$. Fraction exponents indicate a root is taken.

Non-SI Units

Many traditional units are widely used by the scientific community and CGPM re-

Table 6
MATHEMATICAL SIGNS USED WITH QUANTITIES[5]

Term	Sign	Examples
Equal	=	$x = y$
Does not equal	≠	$x \neq y$
Is less than	<	$x < y$
Is more than	>	$x > y$
Is equal to or less than; is no more than	≤	$x \leq y$
Is equal to or more than; is no less than	≥	$x \geq y$
Is equivalent to	≙	$x \triangleq y$
Is identical to	≡	$x \equiv y$
Is proportional to, or is about equal to	∼	$x \sim y$
	≈	$x \simeq y$
Asymptotically equal	≃	$x \simeq y$
Approaches	→	$x \rightarrow y$
Infinity	∞	$1/0 = \infty$
Plus or minus	±	$\sqrt{4} = \pm 2$
Therefore	∴	
Since; because	∵	
Is a member of	∈	
Includes (as a member)	∋	
Is a subset of	⊂	
Includes (as a subset)	⊃	
Is the same or a subset of	⊆	

From ISO, ISO 31/II, 1978, International Standards Organization, Geneva, Switzerland. See Reference 17.

tained those listed in Table 8. The litre is one such unit. Initially, the volume of a litre* was defined with reference to the kilogram as described above. Presently, the litre is 1 dm³ exactly. The "old" litre contained 1 000,028 cm³ (c.c.) or 1 000 mℓ.

Retained for Use With SI
Table 8 lists these units. Some rules are included in the footnote.

Retained for Use With SI for a Limited Time
Table 9 lists these units.

Prefixes
Table 3 lists the prefixes. Some rules follow.

1. Spelling of a combined word where a double vowel results can be simplified by omitting one of the vowels, example, mega amperes can be written megamperes, but kilo amperes would be written kiloamperes.
2. It should be noted that the kilogram is formed by using a prefix with a gram. While this action does not conform to the rules, the name for this base unit is retained for traditional reasons. The kilogram should be used in the denominator of derived units.

* CIPM has recommended upper case L as the symbol for the litre, but CGPM has not acted on this proposal. For domestic usage, the symbol L is recommended (see Reference 12).

Table 7

OPERATORS USED WITH QUANTITIES AND THEIR SYMBOLS

Term	Symbol	Example	Alternatives
Logarithm to the base x	$\log_x$	$\log_x y$	$z = x^y$
Binary logarithm	lb	lb y	$\log_2 y$
Natural logarithm (Napierian Logarithm)	ln	ln y	$\log_e y$
Decadic logarithm (Briggsian Logarithm)	lg	lg y	$\log_{10} y$
Antilogarithm to base x	alog or $\log_x^{-1}$		
Exponent	exp	exp z	e^z
Decadic antilogarithm	alg	alg z	
Square root	$\sqrt{x}$	$\sqrt{x} \cdot \sqrt{x} = x$	$\sqrt{x}$, or $x^{1/2}$
Nth root	$\sqrt[n]{x}$		or $x^{1/n}$
Differential of x to y	dx/dy		
Partial differential	$\partial x/\partial y$		
Gradient; divergence	grad	$\mathrm{grad}_x q = \partial q/\partial x \cdot$ or ∇_q	
Limit of Q as q approaches zero	$\lim Q$ $q \rightarrow O$		
Integral of Q_q from q_1 to q_2	$\int_1^2 Q_q \cdot dq$		
Sum of q from q_1 to q_n	$\Sigma_1^n q$	$q_1 + q_2 + \ldots q_n \cdot$	
Product of q from q_1 to q_n	$\Pi_1^n q$	$q_1 \cdot q_2 \ldots q_n$	
Difference or interval between q_1 and q_2	$\Delta_1^2 q$	or $\delta_1^2 q$ for a small interval	
Quantity dose; quantity given, added, or introduced to system	$\Delta^x Q$	The superscript X indicates the input or output	
Quantity lost or removed from system	$-\Delta^x Q$		
Quantity increment in time t	$\Delta_t^x Q$	(Negative sign for a decrement)	
Limit of Q as q approaches zero	$\lim Q$ $q \rightarrow 0$		

From ISO, ISO 31/II, 1978, International Standards Organization, Geneva, Switzerland. See Reference 17.

3. Multiple use of prefixes is not allowed in the numerator and denominator except for Rule 2.
4. Very large or small numeric values can be reduced to a manageable size by selection of the appropriate prefix and the use of a factor in steps of 10^3.
5. Double usage of prefixes for a unit should not be practiced, for example, μkg should be formed from the gram and the appropriate prefix = mg.
6. SI base units should be used in the denominator of derived units.

PRACTICAL APPLICATIONS

1. World Health Assembly Actions

The World Health Assembly, by its resolution WHA29.65, instructed the Director-General to prepare a report on the use of SI units in medicine. The 30th World Health Assembly considered the Director-General's report and adapted its Resolution WHA30.39, May 1977; this text follows.

The 13th World Health Assembly,

Having considered the report of the Director-General submitted in accordance with resolution WHA29.65;

Table 8
NON-SI UNITS RETAINED FOR
GENERAL USE WITH SI

Name	Symbol	Value in SI
Time interval		
Minute	min	60 s
Hour	h	3 600 s[a]
Day	d	86 400 s
Litre	ℓ[b]	1 dm³ exactly
Metric ton (tonne)	t	10³ kg
Angular		
Degree	°	π/180 rad
Minute	′	π/10 800 rad
Second	″	π/648 000 rad

Note: Non-SI units of this class are referred to as noncoherent units.

[a] Numerical values are clustered in groups of three digits separated by a space. The use of a comma for this purpose is discouraged because the symbol is reserved for use as a decimal point. A dot on the line is acceptable for use in English language texts for the time being, but its continued use is discouraged.

[b] The alternative symbol L was recommended by CIPM to CGPM in 1978, and adopted by the 16th CGPM (1979, Resolution 6). In the U.S. the recommended symbol is L. Neither the word litre or its symbol should be used to express results of high precision. See footnote, Tabel 4 and Reference 12.

Table 9
NON-SI UNITS THAT ARE
PROVISIONALLY RETAINED FOR USE
WITH SI FOR A LIMITED TIME

Name	Symbol	Value in SI
Ångström	Å	10^{-10} m
Nautical mile	—	1 852 m
Knot	—	1 852/3 600 m/s
Hectare	ha	1 hm²
Bar	bar	10⁵Pa
Barn	b	100 fm²
Atmosphere, normal	atm	101 325 Pa
Are	a	1 dam²
Gal	Gal[b]	1 cm/s²
Rad	rad (rd)[c]	10^{-2}J/kg
Röntgen	R	2,58 × 10^{-4} C/kg

[a] CGPM has accepted these terms for use with SI for a limited time only. CGPM also discourages the continued use of the earlier CGS units that have names, i.e., erg, poise, dyne, gauss, stokes, oersted, maxwell, phot, and stilb.

[b] This term should not be confused with gallon.

[c] An alternate symbol that may be used to avoid confusion with radian (cf. Table 2).

Table 10
QUANTITIES AND UNITS FOR RADIATION

Quantity	SI Name	Conventional name	Conversion
Radioactivity	Bequerel	Curie	37 MBg = 1 mCi
Exposure	Coulomb per kilogram	Röntgen	0, 258μC/kg = 1 mR
Exposure rate	Coulomb per Kilogram second	Röntgen per Second	0, 258μC/(kg · s) = 1 m R/s
Absorbed dose	Gray	Rad	10 μGy = 1 mrad
Absorbed dose rate	Gray per second	Rad per second	10 μGy/s = 1 mrad/s
Dose equivalent[a]	Joule per kilogram	Rem	10 mSv = 1 rem

[a]   The gray also has comparable units but should not be used in this context. (Cf. Table 5, footnote b.)

Noting the wide endorsement that has been given by international scientific organizations to the Systéme International d'Unités (SI) developed by the Conférence Générale des Poids et Mesures, the intergovernmental body responsible for units of measurement;

Noting further that the change to the use of SI units in medicine has already taken place or is now under way in several countries;

Mindful, nevertheless, of the confusion that can arise if (few) new units of measurement are introduced without adequate preparation;

1.   *Recommends* the adoption of the SI by the entire scientific community, and particularly the medical community throughout the world;
2.   *Recommends* that, to minimize any confusion due to the simultaneous use of more than one system of units, the period of transition to the new system should not be unduly prolonged;
3.   *Recommends* that, in addition to the scale in kilopascals, the millimetre (or centimetre) of mercury and the centimetre of water be retained for the time being on the scales of instruments for the measurement of the pressure of body fluids, pending wider adoption of the use of the pascal in other fields;
4.   *Recommends* that, in making the change, institutions, scientific associations, and the like secure the best available advice and information, and give their personnel or members a course of intensive instruction in the theory and application of the SI prior to the time when the change takes effect;
5.   *Recommends* that all medical schools, and schools providing training in disciplines relating to medicine, include courses on the theory and use of the SI in their curricula;
6.   *Requests* the Director-General to assist the change by preparing a succinct, simple, and authoritative account of the SI that could be made available to Member States, medical associations, and medical journals.

In response to Recommendation 6, the World Health Orgaization issued "The SI for the Health Professions", in 1977 (ISBN 924 1540591). Now there is a handy pocket calculator that facilitates conversion from conventional to SI units and vice versa.

2. Radiation Units
Ionizing Radiation
In radiology, the International Commission on Radiation Units and Measurements (ICRU) followed the recommendations of CGPM in 1975 where the bequerel was adopted as a special name for the measurement of ionizing radiation. ICRU also recommended that the familiar units curie, röntgen, and rad be replaced with SI units over a period of 10 years. Old units should not be completely abandoned before 1985. Table 10 lists the quantities.

Nonionizing Radiation
Radiation can be expressed in three ways. First, electromagnetic radiation can be expressed as luminous flux, Φ_v in lumens, or as an amount of light Q_v in lumen seconds

(lm·s). The second way is to use radiant power (or radiant flux), Φ_e in watts or radiant energy Q_e in joules. For the third, the radiation appearing in the interval from 400 to 700 nm can be used, which is seen as "light" by eyes. The absorbed radiation energy is used to cause chemical reactions according to quanta (or photons) (an amount of substance) that involves Avagadro's number (N_A) but a convenient smaller constant called the Einstein equivalent (N_E). Unofficially, 1 Einstein = 1 mol and the entity is a photon.

The amount of photosynthic product is expressed by

$$N_E/S = \int_{400\ nm}^{200\ nm} \frac{\lambda \cdot t \cdot E_\lambda \cdot d\lambda}{N_A \cdot h \cdot c_0} \qquad (2)$$

Where S is the area of the absorbing surface (surface of leaves), λ wavelength, t is time interval of constant radiation, E_λ is the wavelength density function of incident spectral radiance, h is Plank's constant, and c_0 is the speed of light in a vacuum. The photochemical equivalent describes the amount of substance transformed by the quanta and is an expression of the photochemical efficienty. The SI unit is $mol \cdot m^{-2}$.

Radiation Passing a System

According to the Lambert law, monochromatic radiation passing through a homogeneous system decreases exponentially with path length. Beer's law states that this decrease is exponentially related to the concentration of an absorbing component in the system. These laws can be combined to give the following expression

$$\Phi = \Phi_0 \cdot e^{-\kappa lc(B)} \qquad (3)$$

which is expressed in terms of natural or Naperian terms and, when given in Briggsian terms,

$$\Phi = \Phi_0 \cdot 10^{-\epsilon lc(B)} \qquad (4)$$

The terms κ_B or ε_B are found in handbook tables; 1 is commonly 1 cm; the Φ_0 is measured as the incident beam or that passing through the reference system without component (B) present; Φ is measured directly; and c_B is calculated. A more detailed treatment is given under Optical Methods of Analysis.

3. Physiology Units

Several of the quantities that are likely to be used in clinical practice are given in Table 11.

Thermochemical Calorie

The thermochemical calorie or nutritional calorie is widely used in nutrition and dietary tables. The International Union of Nutritional Sciences is the authoritative body dealing with these matters. While the nutritional calorie has been defined in terms of the joule since 1935, the popular use of the calorie remains. The digestible energy obtained from foodstuffs and feedstuffs is usually less than the enthalpy measured by a combustion bomb calorimeter. Apparent digestability will vary when ruminant and monogastric animals or humans are considered for the energy obtained from fiber, lipid, carbohydrate, and protein.

Table 11
QUANTITIES AND UNITS FOR PHYSIOLOGY

Quantity	Symbol	SI name	Symbol	Conventional name	Symbol	Conversion
Force	F	Newton	N	Kilopond	kp	9,807
				Kilogram force	kg f	9,807
Work	W	joule	J	Kilopond metre	kp.m	9,807
				Kilogram force metre	kgf.m	9,807
Energy	E	joule	J	Calorie	cal$_{th}$	4,186
		kilojoule	kJ	Calorie[a]	k cal$_{th}$	4,186
Heat	Q_h	joule	J	Calorie	cal$_{th}$	4,186
Power	P	watt	W	Kilopond metre	kp·m/min	0,167
				Per minute		
Pressure	P	kilopascal	kPa	Millimetre mercury	mmHg	0,133 3
				Centimetre water	cmH$_2$O	0,098 07
Resistance	R_v	kilopascal	kPa·s/l	Millimetre mercury	mmHg	8
vascular[b]		second per litre	s/l	Minute per litre	min/l	
Resistance	R_p	kilopascal	kPa · s/l	Centimetre water	cmH$_2$O	0,098 07
pulmonary						
		second per litre	s/l	Second per litre	s/l	

[a]   The nutritional calorie is the thermochemical calorie which in turn is a kilocalorie or calorie (4 184 J). Also the term thermie, 1 Mcal or 4 184 MJ, can be confused with the therm, 1 000 BTU or 105,5 MJ.
[b]   Vascular resistance has been expressed in the GGS system as dyn·s/cm^5 that has a conversion factor of 0,1.

Pressure Measurement

The traditional pressure measurement unitage has been to express results in atmospheres, bars, millimetres, or inches for the height of a mercury column and analogously for water, dynes per centimetre squared (760 mm Hg = 1 012 630 dyn/cm^2), and pounds per square inch (1 atm = 14.7 lb/in.^{2}). The quantities derived from pressure, vascular and pulmonary resistance, need to be reported as given in Table 11 instead of the traditional methods listed which are not directly comparable .

Acoustical Phenomena
a. Hearing

Sound is that portion of the radiation spectrum which is detected by the ears of man and higher animals. Man can detect frequencies from 20 Hz to 20 kHz. In several species of animals, the range is extended up to about 100 kHz. The sound waves cause a vibration of the air particles with a definite frequency and amplitude at the vicinity of the typmpanic membrane (eardrum) external surface. An analogous vibration is induced in the eardrum which results from the pressure changes above and below atmospheric pressure found in the adjacent air particles. An anatomical description of the ear can be found in standard texts.

The loudest sound or pressure the average ear can tolerate is about 28 Pa (280 dyn/cm^2) over the hearing range. When measurements are taken at 1 kHz, the displacement of the eardrum is about 10 μm. Conversely, the faintest sound requires only about 20 μPa pressure (2 × 10^{-4} dyn/cm^2) and the corresponding amplitude is about 10 pm. Thus, the ear is a very sensitive organ.

Since the ear is sensitive to a large range of pressures, a logarithmic scale is used. The sound pressure level, L_p, is defined by

$$L_P = \ln (p/p_0) \tag{5}$$

Table 12
NOISE LEVELS FROM VARIOUS
SOURCES

Source	Noise level (dB)
Threshold of pain	120
Successive gun shots	100
Riveting, steel structure	95
Elevated train	90
Street traffic	70
Conversation	65
Whisper	20
Rustle of leaves	10
Threshold of hearing	0

Note: Static pressure, 1 dyn/cm^2 = 0,1 Pa; pressure in mm Hg × 0,133 3 = kPa; sound energy density, 1 erg/cm = 0,1 J/m^3; sound power, 1 erg/s = 1 × 10^{-7} W; and sound intensity, 1 erg/(s · cm^2) = 1 × 10^{-3} W/m^2.

and the unit is the decibel (dB), named in honor of Alexander Graham Bell. One dB is the sound pressure level when

$$20 \lg (p/p_o) = 1 \qquad (6)$$

Loudness level,

$$l_n = \ln (p\text{eff}/p_o\text{eff}) \qquad (7)$$

where p eff is the effective (root mean-square) sound pressure of a 1-kHz tone judged by a normal observer under standardized listening conditions as being equally loud and where

$$p^\circ = 20 \ \mu\text{Pa} \qquad (8)$$

the faintest sound recognized (about 1 × 10^{-16} W/cm^2, while 120 dB, the maximum tolerated level, is about 1 × 10^{-4} W/cm^2). Table 12 lists noise levels from various sources.

Loudness

The term loudness is reserved for the physiological subjective responses of a human observer. When loudness of a specific sound is increased, the response sensation also increases. The incremental scale follows, approximately, the decibel scale. Thus, the sensation produced by 40 dB sound at a specific frequency is about twice that for 20 dB and half that by 80 dB. The hearing capacity of a patient is evaluated by means of an audiogram where the threshold loudness level, expressed in dB, is plotted vs. frequency in Hz. Also, the upper threshold of the discomfort spectral curve is sometimes measured.

Individual differences exist among people in their capacity to hear sounds. The ear is, generally, most sensitive to tones of 2 to 3 kHz. Taking a 2.5 kHz tone, the hearing

Table 13
QUANTITIES AND UNITS FOR SOUND

Quantity	Symbol	Unit	Symbol	Definition
Perodic time	T	Second	s	time of one cycle
Frequency	f	Hertz	Hz	$1\ Hz = 1/s$
Frequency interval		Octave		Frequency interval for one octane is $f_2/f_1 = 2$ Octane $= lb(f_2/f_1)$
Wavelength	λ	Metre	m	
Velcoity of sound	c	Metre per second	m/s	
Circular frequency	ω	Reciprocal second	1/s	
Circular wave number	k	Reciprocal metre		$k = 2\Pi/\lambda = 2\Pi\sigma$ $\sigma = 1/\lambda$
Static pressure	ps	Pascal	Pa	Pressure in the absence of sound
Sound pressure	p	Bar	bar	Difference between instantaneous total pressure and ps
Sound particle displacement	ξ	Metre	m	Displacement of particle from position in absence of sound
Sound particle velocity	u	Metre per second	m/s	$u = \partial\xi/\partial t$
Sound particle acceleration	a	Metre per second	m/s	$a = \partial\mu/\partial t$
Sound power	P	Watt	W	Sound energy transferred in a certain time interval divided by that time interval
Sound intensity	I	Watt per square metre	W/m²	
Specific acoustic impedance	Zs	Pascal second per metre	Pa·s/m	Za $= Z$s$/S$ S is surface area
Acoustic impedance	Za	Pascal second per metre cubed	Pa·s/m³	Sound pressure at surface divided by volume flow rate
Mechanical impendance	Zm	Newton second per metre	N·s/m	Zm $= Z$sS
Reverberation time	T	Second	s	Time needed for sound energy density to decrease by 60 dB

From ISO, Quantities and units in acoustics, ISO 31/VII, 1978, International Standards Organization, Geneva, Switzerland (see Reference 6).

threshold at −5 dB can be recognized by 1% of the population, 50% at 8 dB, and 90% at 20 dB. The threshold of hearing at frequencies below 1 kHz and above 6 kHz diminishes significantly in normal individuals. Among people with hearing defects, the spectrum will depend, in part, as to the kind of injury found in the internal ear and cochlea.

Physical Measurement

Objective physical measurements are made with appropriate instruments. A full listing of quantities and units for acoustics is given in ISO 31/VII[6] and an abbreviated list is shown in Table 13.

REPORTING RESULTS

As laboratory analyses are conducted, a result is obtained for a measurable property of the component in question. The SI system only addresses quantities and units and remains silent on how the result is reported for use by a clinician. However, when information is exchanged among contemporaries, any misunderstanding can be clarified as needed. In order to convey the information contained in the results by an unambiguous manner in other situations, a number of attributes need to be described.

System

In clinical laboratory science, the sample is taken from a patient. It is sometimes convenient to describe the patient as the "super-system" and then taxonomic nomenclature of subsystems within the super-system is simplified to such systems as blood, tissue, urine, and so on. Subsystems in whole blood then could be erythrocytes, leukocytes, platelets, or a wide range of components in the plasma.

Component

The name of the component is the entity for which the analysis is conducted. The component name should conform with recognized nomenclature. A following section addresses this subject.

The name of the method needs to be identified. However, the method name should not supplant the component name.

Kind of Quantity

The various tables can be consulted for examples. The quantity refers to such matters as the amount of substance or substance concentration of the component that was measured.

Unit

Units can be represented by the symbols listed above in applicable tables or in another recognized compendium. Mixing unit names or their symbols belonging to different coherent systems such as the SI, CGS, or foot-pound-second (ft · lb · s or conventional) system is strongly discouraged by several recognized international nomenclature bodies and by CGPM itself.

The base unit should be used in the denominator when division is indicated. When results are reported according to SI, the units milliliter, gram, cubic centimetre, etc. taken from the CGS system should not be used, even though they might be constructed according to SI rules from SI base or supplementary units.

Numerical Value

The significant characters in a reported result should be extended only to that number of figures that are influenced by analytical method disturbances such as precision. For this, the papers of Kaiser[7] or Eisenhart[8] offer recommendations on the matter for statistical indication of uncertainty. The Committee on Data for Science and Technology (CODATA) has adopted the convention of printing the deviation value (1 SD plus other factors connected with the measurement process) under the final digits of the reported value.[9] The relationship of numerical value with a quantity and its unit is described by Equation 1.

In most applications, it is desirable to express the numerical value between 0.1 and 1 000[10]. The following examples are given.

Observed value	Expressed as
12 000 N	12 kN
0,000 3 s	0,3 ms
0,003 94 m	3,94 mm

REFERENCES

1. *Fed. Regist.*, 43(247), 59986, 1978.
2. *Fed. Regist.*, 43(153), 35210, 1978.
3. *Fed. Regist.*, 40(119), 25837, 1975.
4. **Newbold, B. T.**, SI units and the teaching of chemistry in schools, *Int. Newsl. Chem. Educ., IUPAC,* 10, 3, 1978.
5. ISO, Mathematical signs and symbols for use in the physical sciences and technology, ISO-31.11, 1978,* International Standards Organization, Geneva, Switzerland.
6. ISO, Quantities and units in acoustics, ISO 31/VII, 1978, International Standards Organization, Geneva, Switzerland.
7. **Kaiser, H.**, Quantitation in elemental analysis, *Anal. Chem.*, 42(4), 26A, 1970.
8. **Eisenhart, C.**, Expression of the uncertainties of final results, *Science,* 160, 1201, 1968.
9. Committee on Data for Science and Technology, Recommended consistent values of the fundamental physical constants 1973, *CODATA Bull.,* No. 11, December 1978.
10. ISO, Rules for the use of units of the international system of units ISO-R 1000, 1969, International Standards Organization, Geneva, Switzerland.
11. *Fed. Regist.*, 44(222), 65940, 1979.

GENERAL REFERENCES

12. **Goldman, D. T. and Bell, R. J.**, The international system of units (SI), *Natl. Bur. Stand. (U.S.) Spec. Publ.,* No. 330, 1981.
13. **Terrien, J. and de Boer, G.**, *Le Système International des Unités (SI),* 2nd ed., Bureau International des Poids et Mesures, Sèvres, France, 1973.
14. Commission on Quantities and Units, Quantities and units in clinical chemistry, *Pure Appl. Chem.,* 51, 2451, 1979.
15. Commission on Quantities and Units, List of quantities in clinical chemistry, *Pure Appl. Chem.,* 51, 2481, 1979.
16. WHO, *The SI for the Health Professions,* World Health Organization, Geneva, Switzerland, 1977.
17. ISO, Mathematic signs and symbols for use in the physical sciences and technology, ISO 31/11, 1978, International Standards Organization, Geneva, Switzerland.

---

*   The shortened references to ISO documents in this document are illustrated as ISO-31.11, 1978 following the title when necessary.

REPORTING LABORATORY ANIMAL DATA

Gary S. Gross

INTRODUCTION

This section presents a set of tabulated (Tables 1 to 6) clinical serum chemistry and hematologic reference values for the common laboratory animals. The data are presented in both the conventional metric units and the International System of Units (SI) nomenclature. The conventional data was extracted from the reference work by Mitruka and Rawnsley.[1] SI data was computed using conversion factors from the World Health Organization (WHO)[2] and through the use of an SI-COMPUTER that is designed solely for the conversion of "old" and SI units of measurement.

The presentation of any set of reference values on serum chemistry and hematology assumes the existence of a population of clinically "normal" animals. It is from this population that baseline physiological measurements can be accurately obtained. There are two sources of variation when defining a set of "normal" values for laboratory animals: the laboratory and biological variation.[3] The laboratory variation is due to the measuring process, i.e., laboratory procedure, and the biological variation is that which is derived from the individual animal in its own setting. In order to describe the inherent biological variation, the laboratory variation should be known and the laboratory variation should not contribute significantly to the total variation.[3]

NORMAL VALUES AND REFERENCE VALUES

Due to both laboratory and biological variation, the existence of a "normal" animal and, therefore, "normal" values, may be hypothetical. Consequently, "normal" animal and "normal" value are more accurately described as *reference individual* and *reference value*. A reference individual is an individual animal selected using defined criteria, specifically, knowledge of the health status of the individual and the husbandry practiced. The reference individual is one of a *reference population* from which a *reference sample group* may be selected to obtain a reference value — the reference value being an interval between and including two *reference limits*.[4] The reference interval values presented in this section are mean ± 1 SD. This presumes that the serum chemistry and hematology data are normally distributed, although this may not be the case.[5]

FACTORS AFFECTING THE REFERENCE POPULATION

The reference population consists of animals that are clinically healthy, drug free, of known nutritional status, maintained under defined conditions, and well-adapted to their environment. The environment should match as closely as possible the physiological needs of the animal. Unfortunately, the physiological status and maintenance requirements of some animals currently used for research may not be fully known. Age, weight, sex, strain, breed, and genetic factors must be as uniform as possible to further minimize biological and intraspecies variation.[6] However, inapparent or subclinical disease may still alter the serum chemistry and hematology of the reference group. In addition, the emotional and physiological state of the animal at the time of sampling may be altered by the methods of blood collection and physical or chemical restraint. This may result in further bias of the reference values.

Since there is laboratory and biological variation, each clinical laboratory should establish its own set of reference values. This is particularly true if data is to be used for diagnostic evaluation. The analytical data from clinical laboratories can only be comparable if the methods of blood collection, instruments, techniques, and controls are known to be equivalent.

LABORATORY ANIMAL DATA BANK

The *Laboratory Animal Data Bank* (LADB), sponsored by government health agencies and monitored by the National Library of Medicine, is a depository for control animal data from government sponsored research and private industry. This on-line computer system will be the single most accurate source for biological baseline data and reference information on many species and strains of animals commonly used in biomedical research. Eventually, data from tens of thousands of animals will provide data on clinical pathology, growth and lifespan, histopathology, environmental, and husbandry factors. Further information on the *Laboratory Animal Data Bank* can be obtained by writing: LADB, Specialized Information Services, National Library of Medicine, 8600 Rockville Pike, Bethesda, Md. 20014.

Table 1
HEMATOLOGIC REFERENCE VALUES FOR LABORATORY ANIMALS

Scientific name	Common name	Sex statistic	Platelet (10⁹/*l*)	RBC (10⁶/mm³)(10¹²/*l*)	PCV (—m*l*%)(*l*)	MCV (μ³)(f*l*)	Hb (g/d*l*)	Hb (mmol/*l*)	MCH (μμg)	MCH (fmol)	MCHC (%)	MCHC (mmol/*l*)
Mus musculus	Albino mouse	M x̄	232	9.30	41.5	49.0	11.3	7,01	12.2	0,76	27.2	16,88
		s	80	1.20	4.20	0.75	0.10	0,06	0.25	0,02	2.00	1,24
		F x̄	250	9.10	42.1	49.5	10.9	6,76	11.9	0,74	25.9	16,07
		s	77	1.12	1.20	1.25	0.10	0,06	0.42	0,03	1.80	1,12
Rattus norvegicus	Albino rat	M x̄	340	8.95	47.4	53.8	14.6	9,06	16.3	1,01	30.8	19,11
		s	70	0.40	1.50	2.00	0.58	0,36	1.00	0,06	2.30	1,42
		F x̄	319	7.98	44.1	58.2	13.8	8,56	17.3	1,07	31.3	19,42
		s	75	0.61	3.26	3.65	1.15	0,71	0.85	0,05	2.40	1,49
Mesocricetus auratus	Hamster (syrian)	M x̄	410	7.50	52.5	70.0	16.8	10,43	22.4	1,39	32.0	19,86
		s	75	1.40	2.30	3.19	1.20	0,74	1.27	0,08	2.23	1,38
		F x̄	360	6.96	49.0	70.0	16.0	9,92	23.0	1,43	32.6	20,23
		(s)	121	1.50	4.90	3.00	1.45	0,90	1.40	0,09	2.40	1,49
Cavia porcellus	Guinea pig	M x̄	500	5.60	42.0	77.0	14.4	8,93	25.7	1,59	34.3	21,29
		s	120	0.62	2.50	3.00	1.38	0,86	0.75	0,05	2.28	1,41
		F x̄	450	4.75	45.4	91.0	14.2	8,81	29.7	1,84	31.3	19,42
		s		1.20	2.25	2.45	1.42	0,88	0.80	0,05	1.55	0,96
Oryctolagus cuniculus	Rabbit	M x̄	480	6.70	41.5	62.5	13.9	8,62	20.7	1,28	33.5	20,79
		s	88	0.62	4.25	2.00	1.75	1,09	1.00	0,06	1.85	1,15
		F x̄	450	6.31	39.8	63.1	12.8	7,94	20.3	1,26	32.2	19,98
		s	90	0.60	4.40	1.92	1.50	0,93	1.60	0,10	1.74	1,08
Capra hircus	Goat	M x̄	46.5	18.0	30.0	19.3	10.5	6,52	5.83	0,36	31.8	19,73
		s	5.5	1.15	1.90	0.40	0.65	0,40	0.15	0,01	1.15	0,71
		F x̄	50	17.3	30.0	16.5	12.6	7,82	7.28	0,45	42.0	26,07
		s	5	1.20	2.20	0.38	0.81	0,50	0.20	0,01	1.20	0,74
Macaca mulatta	Rhesus monkey	M x̄	144	5.15	41.5	80.6	13.2	8,19	25.6	1,59	31.8	19,74
		s	15	0.91	4.20	9.98	1.35	0,84	3.17	0,20	1.97	1,22
		F x̄	130	5.19	41.8	83.5	13.4	8,32	25.8	1,60	32.0	19,86
		s	53	0.78	2.70	12.7	1.58	0,98	4.15	0,26	3.20	1,99

Table 1 (continued)
HEMATOLOGIC REFERENCE VALUES FOR LABORATORY ANIMALS

Scientific name	Common name	Sex statistic	Platelet (10⁹/l)	RBC (10⁶/mm³) (10¹²/l)	PCV (—ml%) (l)	MCV (μ^3) (fl)	Hb (g/dl)	Hb (mmol/l)	MCH ($\mu\mu$g)	MCH (fmol)	MCHC (%)	MCHC (mmol/l)
Canis familiaris	Dog (mongrel)	M $\bar{x}$	400	6.98	53.9	77.2	17.1	10,61	24.5	1,52	31.9	19,67
		s	105	0.68	2.70	2.25	0.90	0,56	1.00	0,06	1.50	0,93
		F $\bar{x}$	380	6.69	53.2	79.4	16.8	10,42	25.1	1,56	31.6	19,61
		s	90	0.50	3.10	3.50	1.00	0,62	1.42	0,09	1.00	0,62
Felis catus	Cat (mongrel)	M $\bar{x}$	235	7.40	41.4	56.6	10.7	6,64	14.4	0,89	25.8	16,01
		s	65	0.20	1.20	1.10	0.40	0,25	0.40	0,02	0.80	0,50
		F $\bar{x}$	220	7.10	39.6	56.6	10.3	6,39	14.5	0,90	26.0	16,14
		s	72	0.90	1.20	1.20	0.40	0,25	0.50	0,03	0.80	0,50

Note: RBC — erythrocytes; Hb — Hemoglobin; PCV — packed cell volume; MCV — mean corpuscular volume; MCH — mean corpuscular hemoglobin; MCHC — mean corpuscular hemoblobin concentration; M — male; F — female; $\bar{x}$ — mean; and s — standard deviation.

From Mitruka, B. M. and Rawnsley, H. M., *Clinical Biochemical and Hematological Reference Values in Normal Experimental Animals,* Masson Publishing, New York, 1977, 117. With permission.

Table 2
HEMATOLOGIC REFERENCE VALUES FOR LEUKOCYTES

Scientific name	Common name	Sex statistic	WBC (10⁹/l)	POLYS (10⁹/l)	POLYS (%)	Lymphs (10⁹/l)	Lymphs (%)	Monos (10⁹/l)	Monos (%)	Eos (10⁹/l)	Eos (%)	Baso (10⁹/l)	Baso (%)
Mus musculus	Albino mouse	M $\bar{x}$	14,2	2,16	17.4	11,2	72.6	0,45	2.35	0,38	2.09	0,09	0.52
		s	0,87	0,15	2.10	0,60	5.10	0,05	0.06	0,02	0.36	0,01	0.15
		F $\bar{x}$	12,9	2,20	17.1	9,28	71.9	0,40	2.04	0,31	2.41	0,06	0.49
		s	0,42	0,07	0.70	0,29	2.98	0,05	0.03	0,01	0.18	0,01	0.18

Differential Count

Species		WBC	POLYS	POLYS %	Lymps	Lymps %	Monos	Monos %	Eos	Eos %	Baso	Baso %
Ratus norvegicus — Albino rat	M x̄	9,92	2,42	24.4	7,48	75.4	0,02	0.23	0,03	0.35	0,02	0.20
	s	0,96	0,23	9.10	0,72	8.90	0,01	0.21	0,01	0.14	0,01	0.20
	F x̄	9,60	2,57	26.8	6,95	72.4	0,03	0.35	0,06	0.60	0,01	0.00
	s	1,50	0,40	11.4	1,08	11.1	0,01	0.73	0,01	0.68	0,01	0.20
Mesocricetus auratus — Hamster (syrian)	M x̄	7,62	1,68	22.1	5,60	73.5	0,19	2.50	0,07	0.90	0,08	1.00
	s	1,30	0,29	2.50	0,85	9.40	0,03	0.80	0,02	0.32	0,01	2.00
	F x̄	8,56	2,48	29.0	5,81	67.9	0,20	2.40	0,06	0.70	0,04	0.50
	s	1,54	0,30	3.12	0,70	8.52	0,02	1.00	0,01	0.24	0,01	0.70
Cavia porcellus — Guinea pig	M x̄	11,5	4,83	42.0	5,64	49.0	0,49	4.30	0,44	4.00	0,07	0.70
	s	3,00	1,26	7.00	1,47	6.75	0,12	0.50	0,11	1.50	0,02	0.50
	F x̄	10,8	3,36	31.1	6,85	63.4	0,19	1.80	0,38	3.50	0,20	0.20
	s	2,80	0,87	5.40	1,78	8.50	0,05	0.40	0,10	1.75	0,01	3.30
Oryctolagus cuniculus — Rabbit	M x̄	9,00	4,14	46.0	3,51	39.0	0,72	8.00	0,18	2.00	0,45	5.00
	s	1,75	0,82	4.00	0,68	5.50	0,14	2.00	0,04	0.75	0,08	1.25
	F x̄	7,90	3,43	43.4	3,30	41.8	0,71	9.00	0,16	2.00	0,34	4.30
	s	1,35	0,58	3.50	0,57	5.15	0,12	2.20	0,04	0.60	0,06	0.95
Capra hircus — Goat	M x̄	9,50	3,81	40.1	5,37	56.5	0,10	1.00	0,22	2.30	0,00	0.00
	s	3,35	0,90	5.25	1,28	5.12	0,02	0.20	0,05	0.44	0,01	0.10
	F x̄	7,42	2,71	36.6	4,30	58.0	0,15	2.00	0,26	3.50	0,00	0.00
	s	1,86	0,68	4.10	1,08	6.20	0,04	0.22	0,07	0.52	0,01	0.10
Macaca mulatta — Rhesus monkey	M x̄	10,6	3,98	37.6	6,23	58.8	0,17	1.60	0,19	1.80	0,02	0.20
	s	1,62	0,60	9.80	0,94	16.2	0,02	1.20	0,03	2.50	0,01	0.80
	F x̄	11,1	4,41	39.7	6,39	57.6	0,07	0.60	0,21	1.90	0,02	0.20
	s	2,50	0,99	10.7	0,60	1.44	0,02	1.20	0,04	2.90	0,01	0.60
Canis familiaris — Dog (mongrel)	M x̄	12,9	7,90	61.2	3,83	29.7	0,30	2.32	0,60	4.60	0,04	0.30
	s	2,40	1,45	14.7	0,72	6.20	0,06	0.10	0,12	2.70	0,01	0.10
	F x̄	12,1	7,57	62.6	3,50	28.9	0,30	2.47	0,60	4.94	0,04	0.30
	s	2,40	1,49	16.5	0,69	7.41	0,06	0.15	0,12	2.88	0,01	0.10
Felis catus — Cat (mongrel)	M x̄	17,4	10,1	58.3	5,57	32.0	0,49	2.80	1,20	6.90	0,00	0.00
	s	1,20	0,68	7.80	0,39	5.80	0,04	0.75	0,08	1.75	0,02	0.20
	F x̄	16,6	9,48	57.1	5,40	32.5	0,81	4.90	0,98	5.90	0,02	0.10
	s	1,40	0,80	8.10	0,46	4.60	0,07	1.84	0,09	1.90	0,01	0.20

Note: The quantity count $\times 10^3$mm^3 = count $10^9/l$. WBC — leukocytes; POLYS — polymorphonuclear neutrophil; Lymps — lymphocytes; Monos — monocytes; Eos — eosinophils; Baso — basophils; M — male; F — female; x̄ — mean; and s — standard deviation.

From Mitruka, B. M. and Rawnsley, H. M., *Clinical Biochemical and Hematological Reference Values in Normal Experimental Animals*, Masson Publishing, New York, 1977, 117. With permission.

Table 3
SUMMARY OF CHEMISTRY REFERENCE VALUES

Scientific name	Common name	Sex statistic	Na mEq (mmol)	K mEq (mmol)	Cl mEq (mmol)	CO₂ mEq (mmol)	Alb. (g/dl)	Alb. (g/l)	Pro. (g/dl)	Pro. (g/l)
Mus musculus	Albino mouse	M x̄	138	5,25	108	26,2	3.14	31,4	6.25	62,5
		s	2,90	0,13	0,60	2,10	0.36	3,6	0.75	7,5
		F x̄	134	5,40	107	24,8	2.92	29,2	6.10	61,0
		s	2,60	0,15	0,55	2,30	0.32	3,2	0.68	6,8
Rattus norvegicus	Albino rat	M x̄	147	5,82	102	24,0	3.73	37,3	7.61	76,1
		s	2,65	0,11	0,85	3,80	0.53	5,3	0.50	5,0
		F x̄	146	6,70	101	20,8	3.62	36,2	7.52	75,2
		s	2,50	0,12	0,90	3,60	0.52	5,2	0.32	3,2
Mesocricetus auratus	Hamster (syrian)	M x̄	128	4,66	96,7	37,3	3.23	32,3	6.94	69,4
		s	1,90	0,47	1,19	2,20	0.41	4,1	0.32	3,2
		F x̄	134	5,30	93,8	39,1	3.50	35,0	7.25	72,5
		s	2,30	0,50	1,20	2,30	0.34	3,4	0.48	4,8
Cavia porcellus	Guinea pig	M x̄	122	4,87	92,3	22,0	2.73	27,3	5.60	5,60
		s	0,98	0,84	1,04	4,00	0.30	3,0	0.28	2,8
		F x̄	125	5,06	96,5	20,9	2.42	24,2	4.80	48,0
		s	0,96	0,93	1,19	3,80	0.27	2,7	0.34	3,4
Oryctolagus cuniculus	Rabbit	M x̄	146	5,75	101	24,2	3.39	33,9	6.90	69,0
		s	1,15	0,20	1,45	3,15	0.29	2,9	0.36	3,6
		F x̄	141	6,40	105	22,8	3.04	30,4	6.70	67,0
		s	1,40	0,16	1,22	3,20	0.26	2,6	0.41	4,1
Capra hircus	Goat	M x̄	147	3,61	103	24,6	3.46	34,6	6.75	67,5
		s	3,52	0,18	0,52	2,10	0.41	4,1	0.35	3,5
		F x̄	149	2,95	106	26,1	3.35	33,5	6.90	69,0
		s	4,10	0,24	0,46	2,20	0.42	4,2	0.38	3,8
Macaca mulatta	Rhesus monkey	M x̄	150	5,10	112	29,9	3.50	35,0	7.20	72,0
		s	0,60	0,12	0,80	3,60	0.32	3,2	0.44	4,4
		F x̄	154	5,10	114	30,3	3.53	35,3	6.86	68,6
		s	0,80	0,12	0,70	3,40	0.33	3,3	0.56	5,6
Canis familiaris	Dog (mongrel)	M x̄	147	4,54	114	21,8	3.68	36,8	7.11	71,1
		s	2,20	1,10	1,15	3,60	0.22	2,2	0.20	2,0

Scientific name		Sex statistic															
Felis catus	Cat (mongrel)	F x̄	146	1,90	4,42	0,20	111	1,20	22,2	2,91	3.10	0.14	31,0	1,4	6.30	0.50	63,0	5,0

Felis catus — Cat (mongrel):

Sex statistic	Na							
F x̄	146	4,42	111	22,2	3.10	31,0	6.30	63,0
s	1,90	0,20	1,20	2,91	0.14	1,4	0.50	5,0
M x̄	150	4,25	120	20,4	2.81	28,1	6.10	61,0
s	1,15	0,24	1,10	2,40	0.21	2,1	0.32	3,2
F x̄	152	5,30	112	21,8	2.68	26,8	5.91	59,1
s	1,20	0,31	1,00	2,30	0.24	2,4	0.30	3,0

Note: Na — sodium; K — potassium; Cl — chloride; CO_2 — bicarbonate; Alb. — albumin; Pro. — protein; M — male; F — female; x̄ — mean; and s — standard deviation.

From Mitruka, B. M. and Rawnsley, H. M., *Clinical Biological and Hematological Reference Values in Normal Experimental Animals,* Masson Publishing, New York, 1977, 71. With permission.

Table 4
SUMMARY OF CHEMISTRY REFERENCE VALUES

Scientific name	Common name	Sex statistic	BIL (mg)	BIL (μmol)	CHO (mg)	CHO (mmol)	CR (mg)	CR (μmol)	GLC (mg)	GLC (mmol)	BUN (mg)	BUN (mmol)
Mus musculus	Albino mouse	M x̄	0.75	12,83	63.3	1,64	0.84	74,26	92.2	5,12	20.8	7,42
		s	0.05	0,86	11.8	0,31	0.19	16,80	10.5	0,58	5.86	2,09
		F x̄	0.70	11,97	65.5	1,69	0.67	59,23	85.0	4,72	17.9	6,39
		s	0.04	0,68	12.1	0,31	0.17	15,03	9.50	0,53	4.50	1,61
Rattus norvegicus	Albino rat	M x̄	0.35	5,99	28.3	0,73	0.46	40,66	78.0	4,33	15.5	5,53
		s	0.02	0,34	10.2	0,26	0.13	11,49	14.0	0,78	4.44	1,58
		F x̄	0.24	4,10	24.7	0,64	0.49	43,34	71.0	3,94	13.8	4,93
		s	0.07	1,20	9.62	0,25	0.12	10,61	16.0	0,89	4.15	1,48
Mesocricetus auratus	Hamster (syrian)	M x̄	0.42	7,18	54.8	1,42	1.05	92,82	73.4	4,07	23.4	8,35
		s	0.12	2,05	11.9	0,31	0.28	24,75	12.6	0,70	6.74	2,41
		F x̄	0.36	6,16	51.5	1,33	0.98	86,63	65.0	3,61	20.8	7,42
		s	0.11	1,88	11.0	0,28	0.30	26,52	10.5	0,58	5.64	2,01
Cavia porcellus	Guinea pig	M x̄	0.30	5,13	32.0	0,83	1.38	121,99	95.3	5,29	25.2	9,00
		s	0.08	1,37	10.5	0,39	0.39	34,48	11.9	0,66	6.37	2,27
		F x̄	0.32	5,47	26.8	0,69	1.40	123,76	89.0	4,94	21.5	7,67
		s	0.07	1,20	11.1	0,29	0.35	30,94	9.60	0,53	5.84	2,08

Table 4 (continued)
SUMMARY OF CHEMISTRY REFERENCE VALUES

Scientific name	Common name	Sex statistic	BIL (mg)	BIL (µmol)	CHO (mg)	CHO (mmol)	CR (mg)	CR (µmol)	GLC (mg)	GLC (mmol)	BUN (mg)	BUN (mmol)
Oryctolagus cuniculus	Rabbit	M x̄	0.32	5,47	26.7	0,69	1.59	140,56	135	7,49	19.2	6,85
		s	0.04	0,68	12.9	0,33	0.34	30,06	12.0	0,67	4.93	1,76
		F x̄	0.30	5,13	24.5	0,63	1.67	147,63	128	7,11	17.6	6,28
		s	0.04	0,68	11.2	0,29	0.38	33,59	14.0	0,78	4.36	1,56
Capra hircus	Goat	M x̄	0.05	0,86	70.0	1,81	1.36	120,22	83.5	4,64	20.5	7,32
		s	0.01	0,17	12.0	0,31	0.46	40,66	15.0	0,83	3.80	1,36
		F x̄	0.05	0,97	68.0	1,76	1.15	101,66	72.0	4,00	17.4	6,21
		s	0.01	0,17	14.0	0,36	0.42	37,13	16.5	0,92	3.62	1,29
Macaca mulatta	Rhesus monkey	M x̄	0.38	6,50	129	3,33	1.50	136,2	88.7	4,92	10.1	3,60
		s	0.18	3,08	26.6	0,69	0.09	7,96	10.6	0,59	1.90	0,68
		F x̄	0.57	9,75	150	3,88	1.28	113,15	77.0	4,27	12.8	4,57
		s	0.20	3,42	25.0	0,65	0.06	5,30	12.0	0,67	1.80	0,64
Canis familiaris	Dog (mongrel)	M x̄	0.25	4,28	211	5,46	1.35	119,34	132	7,32	15.0	5,35
		s	0.11	1,88	032	0,83	0.35	30,94	16.4	0,91	4.90	1,74
		F x̄	0.21	3,59	150	3,88	1.08	95,47	110	6,10	13.9	4,96
		s	0.10	1,71	17.0	0,44	0.15	13,26	12.5	0,69	3.20	1,14
Felis catus	Cat (mongrel)	M x̄	0.18	3,08	110	2,84	1.50	132,6	120	6,66	25.0	8,92
		s	0.05	0,86	28.0	0,72	0.50	44,2	14.0	0,78	5.00	1,78
		F x̄	0.15	2,57	101	2,61	1.40	123,76	114	6,33	27.5	9,81
		s	0.04	0,68	22.0	0,57	0.45	39,78	15.0	0,83	4.50	1,61

Note: BIL — bilirubin; CHO — cholesterol; CR — creatinine; GLC — glucose; BUN — blood urea nitrogen; M — male; F — female; x̄ — mean; and s — standard deviation.

From Mitruka, B. M. and Rawnsky, H. M., *Clinical Biochemical and Hematological Reference Values in Normal Experimental Animals*, Masson Publishing, New York, 1977, 117. With permission.

Table 5
SUMMARY OF CHEMISTRY REFERENCE VALUES

Scientific name	Common name	Sex statistic	PI (mg)	PI (mmol)	CA (mg)	CA (mmol)	MG (mg)	MG (mmol)	UA (mg)	UA (mmol)
Mus musculus	Albino mouse	M x̄	5.60	1,81	5.60	1,40	3.11	1,28	4.12	245,10
		s	1.61	0,52	0.40	0,10	0.37	0,15	1.10	65,43
		F x̄	6.55	2,11	7.40	1,85	1.38	0,57	3.90	232,01
		s	1.30	0,42	0.50	0,12	0.28	0,12	0.95	56,52
Rattus norvegicus	Albino rat	M x̄	7.56	2,44	12.2	3,04	3.12	1,28	1.99	118,39
		s	1.51	0,49	0.75	0,19	0.41	0,17	0.25	14,87
		F x̄	8.26	2,67	10.6	2,64	2.60	1,07	1.79	106,49
		s	1.14	0,37	0.89	0,22	0.21	0,04	0.24	14,28
Mesocricetus auratus	Hamster (syrian)	M x̄	5.29	1,71	9.52	2,38	2.54	1,04	4.58	272,46
		s	0.96	0,31	0.98	0,24	0.22	0,09	0.45	26,77
		F x̄	6.04	1,95	10.4	2,59	2.20	0,91	4.36	259,38
		s	1.10	0,36	0.92	0,23	0.14	0,06	0.50	29,75
Cavia porcullus	Guinea pig	M x̄	5.33	1,72	9.60	2,40	2.35	0,91	3.45	205,24
		s	1.15	0,37	0.63	0,16	0.25	0,10	0.40	23,80
		F x̄	5.30	1,71	10.7	2,67	2.46	1,01	3.38	201,08
		s	1.10	0,36	0.58	0,14	0.27	0,11	0.41	24,39
Oryctolagus cuniculus	Rabbit	M x̄	4.82	1,56	10.0	2,50	2.52	1,04	2.65	157,65
		s	1.05	0,34	1.11	0,28	0.24	0,10	0.88	52,35
		F x̄	5.06	1,63	9.50	2,37	3.20	1,32	2.62	155,86
		s	0.93	0,30	1.10	0,27	0.22	0,04	0.87	51,76
Capra hirucs	Goat	M x̄	10.9	3,52	10.3	2,57	2.50	1,03	0.67	39,86
		s	0.98	0,32	0.70	0,17	0.36	0,15	0.33	19,63
		F x̄	7.87	2,54	10.7	2,67	3.20	1,32	0.60	35,69
		s	1.42	0,46	0.62	0,15	0.35	0,14	0.30	17,85
Macaca mulatta	Rhesus monkey	M x̄	5.92	1,91	11.2	2,79	1.65	0,67	0.90	53,54
		s	1.05	0,34	0.66	0,16	0.32	0,13	0.11	6,54
		F x̄	5.79	1,93	10.8	2,69	1.82	0,75	1.29	76,74
		s	1.11	0,36	0.72	0,18	0.41	0,17	0.14	8,33
Canis familiaris	Dog (mongrel)	M x̄	4.40	1,42	10.2	2,54	2.10	0,86	0.55	32,72
		s	1.00	0,32	0.42	0,10	0.32	0,13	1.10	65,44
		F x̄	3.70	1,19	9.40	2,35	2.20	0,91	0.42	24,99
		s	0.50	0,16	0.50	0,12	0.28	0,12	0.10	5,95

Table 5 (continued)
SUMMARY OF CHEMISTRY REFERENCE VALUES

Scientific name	Common name	Sex statistic	PI (mg)	PI (mmol)	CA (mg)	CA (mmol)	MG (mg)	MG (mmol)	UA (mg)	UA (mmol)
Felis catus	Cat (mongrel)	M x̄	6.20	2,00	10.1	2,52	2.64	1,09	1.45	86,26
		s	1.07	0,35	0.85	0,21	0.25	6,10	0.22	13,09
		F x̄	6.40	2,07	11.2	2,79	2.54	1,04	1.30	77,34
		s	1.17	0,38	0.92	0,23	0.21	0,09	0.20	11,90

Note: PI — phosphorus; CA — calcium; MG — magnesium; UA — uric acid; M — male; F — female; x̄ — mean; and s — standard deviation.

From Mitruka, B. M. and Rawnsley, H. M., Clinical Biochemical and Hematological Reference Values in Normal Experimental Animals, Masson Publishing, New York, 1977, 117. With permission.

Table 6
SUMMARY OF SERUM ENZYME ACTIVITIES

Scientific name	Common name	Sex statistic	P'tase alk. (U/l)	P'tase alk. (nmol/s)	P'tase acid (U/l)	P'tase acid (nmol/s)	GPT (U/l)	GPT (nmols)	GOT (U/l)	GOT (nmols)	CPK (U/l)	CPK (nmols)	LDH (U/l)	LDH (nmols)
Mus musculus	Albino mouse	M x̄	21.6	360	14.7	245	13.5	225	36.2	604	3.70	61,7	149.0	2484
		s	4.56	76	3.50	58	5.30	88	6.13	101	1.45	24	19.0	317
		F x̄	16.8	280	11.8	197	12.6	210	34.5	590	2.50	41,7	119.0	1984
		s	5.37	89	2.70	45	4.40	73	6.10	102	1.52	25	21.0	350
Rattus norvegicus	Albino rat	M x̄	81.4	1357	39.0	650	25.2	420	62.5	1042	5.60	93	92.5	1542
		s	14.8	247	4.30	72	2.05	34	8.40	140	1.30	22	13.9	232
		F x̄	93.9	1565	37.5	625	22.5	375	64.0	1067	6.80	113	90.0	1500
		s	17.3	288	3.70	62	2.50	42	6.50	108	2.40	40	14.5	242
Mesocricetus auratus	Hamster (syrian)	M x̄	17.5	292	7.45	124	26.9	448	124.	2067	1.01	16,8	115.0	1917
		s	6.10	102	1.40	23	4.56	76	22.0	367	0.40	6,7	20.0	333
		F x̄	15.4	257	6.90	115	20.6	343	77.6	1294	0.85	14,2	110.0	1834
		s	4.20	70	1.56	26	3.95	66	14.5	242	0.32	5,3	27.0	450

Cavis porcellus	Guinea pig	M x̄	74.2	1237	32.2	537	44.6	744	48.2	804	0.95	15,8	46.9	782
		s	6.92	115	2.59	43	6.75	111	9.50	158	0.15	2,5	9.50	158
		F x̄	65.8	1097	28.7	478	38.8	647	45.5	759	1.10	18,3	52.1	869
		s	5.46	91	3.20	53	7.15	119	7.00	117	0.20	3,3	11.2	187
Oryctolagus cuniculus	Rabbit	M x̄	10.4	173	1.56	26	65.7	1095	72.3	1205	1.35	22,5	84.4	1407
		s	2.28	38	0.53	8,8	6.54	109	12.0	200	0.56	9,3	20.5	342
		F x̄	9.96	166	1.40	23	62.5	1042	68.1	1135	1.30	21,7	78.5	1309
		s	3.10	52	0.38	6,3	5.85	98	10.5	175	0.45	7,5	22.0	367
Capra hircus	Goat	M x̄	85.4	1424	37.4	624	22.1	368	67.5	1125	31.1	518	92.0	1534
		s	15.1	252	8.85	148	11.2	186	26.8	447	4.85	80,9	27.0	450
		F x̄	82.9	1382	35.9	599	24.5	408	66.5	1108	30.9	515	89.0	1484
		s	16.2	270	8.90	148	9.60	160	26.7	445	4.90	81,7	26.2	437
Macaca mulatta	Rhesus monkey	M x̄	9.60	160	31.5	525	24.5	408	27.9	465	11.5	192	175.0	2917
		s	5.10	85	3.32	55	9.50	158	6.60	110	1.40	23	70.0	1167
		F x̄	8.70	145	33.2	553	23.6	393	28.7	478	6.30	105	180.0	3000
		s	4.95	83	3.50	58	10.2	170	7.60	127	1.30	22	71.0	1184
Canis familiaris	Dog (mongrel)	M x̄	17.7	295	3.54	59	40.6	677	54.8	914	1.20	20	76.5	1275
		s	3.80	63	1.20	20	7.70	128	8.90	148	0.30	5	17.8	297
		F x̄	16.5	275	3.30	55	44.0	734	59.2	987	0.80	13	67.0	1117
		s	4.30	72	1.10	18	7.80	130	9.00	150	1.10	18	15.0	250
Felis catus	Cat (mongrel)	M x̄	14.3	238	2.67	45	19.6	327	19.0	317	1.70	28	67.5	1125
		s	3.54	59	1.15	19	4.95	83	4.80	80	1.10	18	10.6	177
		F x̄	10.4	173	2.48	41	18.5	308	17.0	283	1.90	32	87.4	1457
		s	3.20	53	1.27	21	3.60	60	4.90	82	1.30	22	11.5	192

Note: P'tase — phosphatase; GOT — aspartate aminotransferase; GTP — alanine aminotransferase; CPK — creatine kinase; LDH — lactate dehydrogenase; M — male; F — female; x̄ — mean; and s — standard deviation.

From Mitruka, B. M. and Rawnsley, H. M., *Clinical Biochemical and Hematological Reference Values in Normal Experimental Animals*, Masson Publishing, New York, 1977, 117. With permission.

REFERENCES

1. **Mitruka, B. M. and Rawnsley, H. M.**, *Clinical Biochemical and Hematological Reference Values in Normal Experimental Animals,* Masson Publishing, New York, 1977.
2. WHO, The SI for the Health Professions, World Health Organization, Geneva, 1977.
3. **Payne, B. J., Lewis, H. B., Murchison, T. E., and Hart, E. A.**, Hematology of laboratory animals; normal values, in *Handbook of Laboratory Animal Science,* Vol. 3, Melby, E. C., Jr. and Altman, N. H., Eds., CRC Press, Cleveland, 1974, 383.
4. Expert Panel on the Theory of Reference Values, Provisional recommendation on the theory of reference values, I. The concept of reference values, *Clin. Chim. Acta,* 87, 461F, 1978.
5. **Laird, C. W.**, Clinical pathology: blood chemistry, in *Handbook of Laboratory Animals,* Vol. 2, Melby, E. C. Jr. and Altman, N. H., Eds., CRC Press, Cleveland, 1974, 347.
6. **Duncan, J. R. and Prasse, K. W.**, Normal values, in *Veterinary Laboratory Medicine, Clinical Pathology,* Iowa State University Press, Ames, 1977, 185.
7. **Mitruka, B. M. and Rawnsley, H. M.**, Hematological values in experimental animals, in *Clinical Biochemical and Hematological Reference Values in Normal Experimental Animals,* Masson Publishing, New York, 1977, 71.
8. **Mitruka, B. M. and Rawnsley, H. M.**, Clinical biochemistry, in *Clinical Biochemical and Hematological Reference Values in Normal Experimental Animals,* Masson Publishing, New York, 1977, 117.

ADDITIONAL READING

Benjamin, M. M., Hematology, in *Outline of Veterinary Clinical Pathology,* 3rd ed., Iowa State University Press, Ames, 1978, 5.
Benjamn, M. M., Clinical chemistry, in *Outline of Veterinary Clinical Pathology,* 3rd ed., Iowa State University Press, Ames, 1978, 178.
T-W-Fiennes, R. N., Primates, in *The UFAW Handbook on the Care and Management of Laboratory Animals,* 5th ed., Longman Group, New York, 1976, 377.
Festing, M. F. W., Hamster, in *The UFAW Handbook on the Care and Management of Laboratory Animals,* 5th ed., Longman Group, New York, 1976, 248.
Festing, M. F. W., Guinea-pig, in *The UFAW Handbook on the Care and Management of Laboratory Animals,* 5th ed., Longman Group, New York, 1976, 228.
Adams, C. E., Rabbit, in *The UFAW Handbook on the Care and Management of Laboratory Animals,* 5th ed., Longman Group, New York, 1976, 172.
Sisk, D. B., Physiology, in *The Biology of the Laboratory Rabbit,* Wagner, J. E. and Manning, P. J., Eds., Academic Press, New York, 1976, 63.
Kozma, C., Macklin, W., Cummins, L. M., and Mauer, R., Anatomy, physiology and biochemistry, in *The Biology of the Laboratory Rabbit,* Weisbroth, S. H., Flatt, R. E., and Kraus, A. L., Eds., Academic Press, New York, 1974, 48.
Schalm, O. W., Jain, N. C., and Carroll, E. J., Normal values in blood morphology, in *Veterinary Hematology,* 3rd ed., Lea & Febiger, Philadelphia, 1975, 82.
Baker, H. J., Lindsey, J. R., and Weisbroth, S. H., *The Laboratory Rat, Biology and Diseases,* Vol. 1, Academic Press, New York, 1980.
Accelston, E., Normal haematological values in rats, mice and marmosets, in *Comparative Clinical Haematology,* Archer, R. K. and Jeffcott, L. B., Eds., Blackwell Scientific, London, 1977, 611.
Crispens, C. G., *Handbook on the Laboratory Mouse,* Charles C Thomas, Springfield, Ill., 1975, 95.
Medway, W., Prier, J. E., and Wilkinson, J. S., Blood chemistry, in *Textbook of Veterinary Clinical Pathology,* Williams & Wilkins, Baltimore, 1969, 18.
Medway, W., Prier, J. E. and Wildinson, J. S., Hematology, in *Textbook of Veterinary Clinical Pathology,* Williams & Wilkins, Baltimore, 1969, 205.
Siegmund, O. H., Ed., Hematology, in *The Merck Veterinary Manual,* 4th ed., Merck, Rahway, N.J., 1973, 1379.
Siegmund, O. H., Ed., Blood chemistry, in *The Merck Veterinary Manual,* 4th ed., Merck, Rahway, N.J., 1973, 1382.
McKelvie, D. H., Blood serum chemistry, in *The Beagle as an Experimental Dog,* Andersen, A. C., Ed., Iowa State University Press, Ames, 1970, 281.

Altman, P. L. and Dittmer, D. S., Eds, Blood and other body fluids, in *The Biology Data Book*, Vol. 3, 2nd ed., Federation of American Societies for Experimental Biology, Bethesda, Md., 1972, 1751.

McClure, H. M., Hematology and blood chemistry for the rhesus monkey, in *The Rhesus Monkey*, Bourne, G. H., Eds., Academic Press, New York, 1975, 409.

Schermer, S., *The Blood Morphology of Laboratory Animals*, 3rd ed., F. A. Davis, Philadelphia, 1967.

Anderson, A. E. and Schalm, O. W., Hematology, in *The Beagle as an Experimental Dog*, Andersen, A. C., Ed., Iowa State University Press, Ames, 1970, 261.

Loeb, W. F., Blood and blood forming organs, in *Feline Medicine and Surgery*, Catcott, E. J., Ed., American Veterinary Publications, Santa Barbara, Calif., 1975, 225.

Robinson, P. F., General aspects of physiology, in *The Golden Hamster, Its Biology and Use in Medical Research*, Hoffman, R. A., Robinson, P. F., and Magalhaes, H., Eds., Iowa State University Press, Ames, 1965.

Cornelius, E. C. and Kaneko, J. J., Eds., *Clinical Biochemistry of Domestic Animals*, Vol. 2, 2nd ed., Academic Press, New York, 1972.

Russell, E. S. and Bernstein, S. E., Blood and blood formation, in *Biology of the Laboratory Mouse*, Green, E. L. Ed., 2nd ed., Dover Publications, New York, 1968, 351.

Bernstein, S. E., Physiological characteristics, in *Biology of the Laboratory Mouse*, Green, E. L., Ed., 2nd ed., Dover Publications, New York, 1968, 337.

Bentinck-Smith, J., Roster of normal values, in *Current Veterinary Therapy, VI*, Kirk, R. W., Ed., W. B. Saunders, Philadelphia, 1977, 1355.

Schuchman, S. M., Individual care and treatment of rabbits, mice, guinea pigs, hampsters, and gerbils, in *Current Veterinary Therapy VI*, Kirk, R. W., Ed., W. B. Saunders, Philadelphia, 1977, 726.

Whitney, R. D., Johnson, D. J., and Cole, W. C., Physiological data, in *Laboratory Primate Handbook*, Academic Press, New York, 1973, 103.

Benjamin, M. M. and McKelvie, D. H., Clinical biochemistry, in *Pathology of Laboratory Animals*, Vol. 2, Benirschke, K., Garner, G. M., and Jones, I. C., Eds., Springer-Verlag, New York, 1978, 1749.

Loeb, W. F., Bannerman, R. M., Riniger, B. F., and Johnson, A. J., Hemotologic disorders, in *Pathology of Laboratory Animals*, Vol. 1, Benirschke, K., Garner, F. M., and Jones, T. C., Eds., Springer-Verlag, New York, 1978, 889.

New, A. E., Base-line blood determinations of the squirrel monkey (*Saimiri scirureus*), in *The Squirrel Monkey*, Rosenblum, L. A. and Cooper, R. W., Eds., Academic Press, New York, 1968, 417.

CLINICAL LABORATORY DATA FOR EQUINES

Arthur W. Patterson, Jr.

INTRODUCTION

The horse industry provides over $1 billion yearly tax revenue from an investment worth about $13 billion. Equine practitioners provide quality medical services to this industry based upon about 8 million individual animals used for personal pleasure riding, various kinds of animal herd control, draft, and racing events.

Over the years, the American Association of Equine Practitioners has sponsored symposia. The purposes of these conferences were: to assemble and publish current information on a specific subject area; to disseminate this knowledge to equine practitioners numbering well over 3300 world-wide; to identify problem areas that need further research; and to provide guidance and direction to these efforts. The first symposium was dedicated to the topical subject of equine nutrition. Subsequent subject areas addressed equine infectious diseases, pharmacology and therapeutics, colic, laminitis, and so on. A recent symposium of interest to the subject at hand addressed equine hematology.[1]

PATIENT CONDITIONS

It is imperative that the immediate prior history of the horse be documented when the diagnostic samples are collected. Considerable changes in blood parameters are found among animals at rest standing in a stall at sample collection, when previously the animal was in a sedentary resting condition at pasture or engaged in an active training program.[2-8]

Prior use of medication on either an acute or chronic basis can significantly alter certain blood parameters.[9-14] Among racing horses, phenylbutazone is used, but its prolonged use can produce hypoplastic anemia[15] along with lesions in the gastrointestinal tract, following oral administration.

It is desirable to document some additional information. This includes: the time of day, because diurnal effects may influence the observed result; the length of fasting and kind of feed last eaten; access to and use of drinking water; and the ambient temperature conditions over the past few days. Urinary constituents are affected by many factors described above for the blood parameters. Even more profound alterations are produced when diuretic agents, such as Lasix®, are administered.

SAMPLE COLLECTION AND HANDLING

Blood samples, usually taken by jugular venipuncture, are collected into clean sterile tubes and tested promptly, especially when enzyme assays are desired. Care needs to be taken when anticoagulants are used to avoid confusion of the results (sodium heparin for a sodium measurement) and influencing the constituent (interference) to be measured. Samples allowed to clot should have provisions to minimize hemolysis and remove fibrinogen.

The serum should be harvested within 1 to 2 hr after collection or when clot retraction takes place, and chilled. Samples destined for most of the low molecular weight constituent analyses can be frozen and stored at −20°C. Specific instructions, accompanying most modern automated instrumentation, should be consulted for the recommended collection and handling procedures.

REPORTING RESULTS

The problem of defining a "normal" horse is a complex and many exceptions need to be introduced that result from husbandry management, breeding, psychological factors, and subclinical disease conditions. For practical reasons, even when rigorously sorted by clinicians, the "normal" horse may, in reality, be an "unhealthy" horse. These animals have reached a condition of equilibrium with residual lingering lesions produced by, for example, parasites acquired at pasture when young, inadequate nutrition during growth, or virus infections that do not show clinical signs. In essence, using laboratory results for diagnostic purposes becomes a matter of recognizing an *abnormal sick* horse vs. a normally equilibrated "unhealthy" horse. The reference values, therefore, should be regarded as representing a population of normally equilibrated unhealthy horses comprising genetically stable breeds. The reader can consult recent reviews for more detailed information.[12,16,17] The data shown in Tables 1 to 3 should be regarded as a general guide,[18] more detailed findings are given in References 16, 17, 19, 22, and 23.

REFERENCE VALUES

Other monogastric and ruminant species are included to illustrate differences that must be considered when extrapolation among species is conducted. Reference values are not ideally distributed. In parametric terms, the clinician is familiar with the practical terms $\bar{x}$ (mean) and s or SD (standard deviation). A reference interval of $\bar{x} \pm 2\ s$ would include about 95% of all values from a population. A more conservative estimate might take $\bar{x} \pm 3\ s$, and outside of this range the result is considered abnormal. Some improvement can be developed when a natural log (ln) transformation of $\bar{x}$ and s is taken.[16] The imprecision of the method must also be taken into account[17] as well as how much this contributes to the range.[20] Nonparametric methodology for assessing laboratory results is another procedure that can be considered. More work needs to be done before its merits can be judged. In short, the theory of reference values is still in its formative stages.

Nomenclature and symbols follow the Recommendation 1978, made by the Commission on Quantity and Units in Clinical Chemistry, IUPAC, given in Reference 24.

Some interpretive comment on equine blood chemistry values has been provided by Coffman.[21] For lactate dehydrogenase (LDH), the examination of the isozymes present in the sample offers the usual tissue specificity to investigations of skeletal or cardiac muscle injury, liver disorders, and erythrocyte abnormalities. In liver injury, asparate amino transferase (GOT), alanine amino transferase (GPT), and alkaline phosphates levels are elevated. The latter enzyme levels in serum of foals can be twice the values found in adults. Sorbitol dehydrogenase is another useful indicator of liver injury. Creatine kinase (creatine phosphokinase) is a valuable indicator of muscle necrosis, sometimes found when horses in active training are excessively stressed. Elevated cholesterol levels are found in foals nursing high fat milk or in such disorders as hypothyroidism or diabetes mellitus.

Table 1
BLOOD CHEMISTRY VALUES IN DOMESTIC ANIMALS

System	Component	Kind of quantity	Units	Horse	Cow	Sheep	Pig
S,P,Hp	Alanine amino transferase	CATC	U/ℓ	58 — 94	20 — 34	68 — 90	8.2 — 22
S,Hp	Amylase	CATC	U/ℓ	75 — 100	—	—	—
S,P,Hp	Aspartate amino transferase	CATC	U/ℓ	1 — 6.7	4 — 11	10 — 12	9 — 17
S,P,Hp	Bilirubin			—	—	—	—
	Direct	CSUBSTC	mg/dℓ	0 0.4	0.04 0.44	0 0.27	0 0.3
	Total	CSUBSTC	mg/dℓ	0 — 2.0	0.1 — 0.42	0.1 — 0.42	0 — 0.6
S,Hp	Calcium II	CSUBSTC	mg/dℓ	11.2 — 13.6	9.7 — 12.4	11.5 — 12.8	7.1 — 11.6
B	Carbon dioxide	PART PRESS	mm Hg	38 — 46	35 — 44	37 — 46	—
B	Carbonate	CSUBSTC	meq/ℓ	24 — 32	21 — 32	21 — 28	—
S,P,Hp	Creatinine	CSUPSTC	mg/dℓ	1.2 — 1.9	1 — 2	1.2 — 1.9	1 — 2.7
S,Hp	Creatine kinase	CATC	U/ℓ	2.4 — 23	4.8 — 12	8 — 13	2.4 — 23
P,Hp	Fibrinogen	MASSC	mg/dℓ	200 — 400	200 — 500	—	—
S,P,Hp	Glucose	CSUBSTC	mg/dℓ	75 — 115	45 — 75	50 — 80	85 — 150
B	Hemoglobin	MASSC	g/dℓ	11 — 19	8 — 14	8 — 16	10 — 16
B	Hydrogen ion activity (pH)	log CSUBSTC	pH	7.32 — 7.44	7.31 — 7.53	7.32 — 7.54	—
S,Hp	Lactate dehydrogenase	CATC	U/ℓ	41 — 104	176 — 365	60 — 111	96 — 160
S,Hp	Magnesium II	CSUBSTC	mg/dℓ	2.2 — 2.8	1.8 — 2.3	2.2 — 2.8	2.7 — 3.7
S,Hp	Phosphatase alkaline	CATC	U/ℓ	97 — 269	0 — 332	46 — 263	80 — 269
S,Hp	Phosphate inorganic	CSUBSTC	mg/dℓ	3.1 — 5.6	5.6 — 6.5	5.0 — 7.3	5.3 — 9.6
S,Hp	Potassium ion	CSUBSTC	meq/ℓ	2.4 — 4.7	3.9 — 5.8	3.9 — 5.4	4.4 — 6.7
S	Protein	MASSC	g/dℓ	5.7 — 7.9	6.7 — 7.5	6 — 7.9	7.9 — 8.9
S,Hp	Sodium ion	CSUBSTC	meq/ℓ	132 — 146	132 — 152	139 — 152	135 — 150
S,Hp	Sorbitol dehydrogenase	CATC	U/ℓ	1.9 — 5.8	4.3 — 15	5.8 — 28	1 — 5.8
S	Thyroxin (T4)	CSUBSTC	μg/dℓ	0.9 — 2.8	—	—	—
S,P	Urea (BUN)	CSUBSTC	mg/dℓ	10 — 24	20 — 30	8 — 20	—

Note: Conventional reference values for enzyme activity need to be regarded as illustrative and are largely dependent on the specific method used for conducting the measurement. Kind of quantity — CATC: catalytic concentration; CSUBSTC: chemical substance concentration; MASSC: mass concentration; PART: partial; and PRESS: pressure. System — B: blood; d: day (as prefix); P: plasma; S: serum; and Hp: heparinized plasma. Units and values listed are conventional while in Table 2 the units and values follow the SI system.

From *Neb. Vet. Ext. Newsl.,* Department of Clinical Pathology, School of Veterinary Medicine, University of California, Davis, 6 (6), 1977. With permission.

Table 2
BLOOD CHEMISTRY VALUES IN DOMESTIC ANIMALS

System	Component	Kind of quantity	Units	Horse	Cow	Sheep	Pig
S,P,Hp	Alanine amino transferase	CATC	nkat/l	970 — 1570	330 — 570	1330 — 1500	140 — 370
S,Hp	Amylase	CATC	nkat/l	1250 — 1670	—	—	—
S,P,Hp	Aspartate amino transferase	CATC	nkat/l	17 — 110	67 — 180	170 — 200	150 — 280
S,P,Hp	Bilirubin			—	—	—	—
	Direct	CSUBSTC	μmol/l	0 6.8	0.7 7.5	0 4.6	0 5.1
	Total	CSUBSTC	μmol/l	0 — 34	1.7 — 7.2	1.7 — 7.2	0 — 10.3
S,Hp	Calcium II	SCUBSTC	mmol/l	2.8 — 3.4	2.4 — 3.0	2.9 — 3.2	1.8 — 2.9
B	Carbon dioxide	PART PRESS	kPa	5.1 — 6.1	4.7 — 5.9	4.9 — 6.1	—
B	Carbonate	CSUBSTC	mmol/l	24 — 32	21 — 32	21 — 28	—
S,P,Hp	Creatinine	CSUPSTC	μmol/l	91 — 145	76 — 150	92 — 145	76 — 200
S,Hp	Creatine kinase	CATC	nkat/l	38 — 380	80 — 200	133 — 220	40 — 380
P,Hp	Fibrinogen	MASSL	g/l	2 — 4	2 — 5	—	—
S,P,Hp	Glucose	CSUBSTC	mmol/l	4.3 — 6.4	2.5 — 4.2	2.8 — 4.4	4.7 — 8.3
B	Hemoglobin	MASSC	mmol/l	6.2 — 11.8	5.0 — 8.7	5.0 — 10.0	6.2 — 10.0
B	Hydrogen ion activity (pH)	log CSUBSTC	pH	7.32 — 7.44	7.31 — 6.53	7.32 — 7.54	—
S,Hp	Lactate dehydrogenase	CATC	nkat/l	680 — 1730	2930 — 6080	1000 — 1850	1600 — 2670
S,Hp	Magnesium II	CSUBSTC	mmol/l	.91 — 1.15	0.74 — 0.95	0.91 — 1.15	1.11 — 1.52
S,Hp	Phosphatase alkaline	CATC	nkat/l	1620 — 4500	0 — 5500	770 — 4380	1330 — 4500
S,Hp	Phosphate inorganic	CSUBSTC	mmol/l	1.0 — 1.8	1.8 — 2.1	1.6 — 2.4	1.7 — 3.1
S,Hp	Potassium ion	CSUBSTC	mmol/l	2.4 — 4.7	3.9 — 5.8	3.9 — 5.4	4.4 — 6.7
S	Protein	MASSC	mg/l	57 — 79	67 — 75	60 — 79	79 — 89
S,Hp	Sodium ion	CSUBSTC	mmol/l	132 — 146	132 — 152	139 — 152	135 — 150
S,Hp	Sorbitol dehydrogenase	CATC	nkat/l	32 — 97	72 — 250	97 — 470	17 — 97
S	Thyroxin (T4)	CSUBSTC	mmol/l	12 — 36	—	—	—
S,P	Urea (BUN)	CSUBSTC	mmol/l	1.7 — 4.0	3.3 — 5.0	1.3 — 3.3	—

Note: Kind of quantity — CATC: catalytic concentration; CSUBSTC: chemical substance concentration; MASSC: mass concentration; PART: partial; and PRESS: pressure. System — B: Blood; d: day (as prefix); P: plasma; S: serum; and Hp: heparinized plasma.

Table 3
HEMATOLOGIC VALUES FOR THE HORSE

Measurement	Units	Hot-blooded[a]		Cold-blooded[a]	
		$\overline{x}$	s	$\overline{x}$	s
RBC	$10^6/mm^3:10^{12}/l$	9.5	3	7.5	2
Hb	g/dl	15	4	11.5	2.5
	mmol/l	9.3	2.5	7.1	1.6
PCV	ml%:l	42	10	35	9
MCV	$\mu^3:fl$	46	12	—	—
MCHC	%	35	2	—	—
WBC	$10^9/l$	9.0	3.5	8.5	3.5
Polys	$10^9/l$	4.7	2.0	—	—
	%[b]	49	16	54	21
Lymphs	$10^9/l$	3.5	2.0	—	—
	%[b]	44	26	35	15
Monos	$10^9/l$	0.4	0.4	—	—
	%	4	3	5	5
Eos	$10^9/l$	0.38	0.54	—	—
	%	4	6	5	6
Basos	$10^9/l$	0.05	0.10	—	—
	%	0.5	2	0.5	2

Note: RBC: erythrocytes; WBC: white blood cells; Hb: hemoglobin; PCV: packed cell volume; MCV: mean corpuscular volume; MCH: mean corpuscular hemoglobin; MCHC: mean corpuscular hemoglobin concentration; M: male; F: female; $\overline{x}$: mean; s: standard deviation.

[a] Horses with 50% or more of Arabian ancestry are commonly referred to as hot-blooded. These are essentially the Arabians and the Thoroughbreds. The light and heavy draft breeds are essentially referred to as cold-blooded.
[b] Expressed as a percentage of leukocyte count.

From Schalm, O. W., *Veterinary Hematology,* Lea & Febiger, Philadelphia, 1965. With permission.

REFERENCES

1. **Kitchen, H. and Krehbiel, J. D.**, Eds., *Proc. 1st Int. Symp. Equine Hematology,* The American Association of Equine Practitioners, Golden, Colo., 1975.
2. **Snow, D. H.**, Biochemical changes in blood and muscle associated with exercise, in *Proc. 1st Int. Symp. Equine Hematology,* Kitchen, H. and Krehbiel, J. D., Eds., The American Association of Equine Practitioners, Golden, Colo., 1975.
3. **Gillespie, J. R.**, A review of hematological response to exercise, in *Proc. 1st Int. Symp. Equine Hematology,* Kitchen, H. and Krehbiel, J. D., Eds., The American Association of Equine Practitioners, Golden, Colo., 1975.
4. **Carlson, G. P.**, Hematologic alterations in endurance trained horses, in *Proc. 1st Int. Symp. Equine Hematology,* Kitchen, H. and Krehbiel, J. D., Eds., The American Association of Equine Practitioners, Golden, Colo., 1975.

5. **Gillespie, J. R., Kaufman, A., Steere, J., and White, L.,** Arterial blood gases and pH during long distance running in the horse, in *Proc. 1st Int. Symp. Equine Hematology,* Kitchen, H. and Krehbiel, J. D., Eds., The American Association of Equine Practitioners, Golden, Colo., 1975.
6. **Codazza, D.,** Observations on serum components in race horses after training, in *Proc. 1st Int. Symp. Equine Hematology,* Kitchen, H. and Krehbiel, J. D., Eds., The American Association of Equine Practitioners, Golden, Colo., 1975.
7. **Carlson, G. P.,** Fluid and electrolyte alterations in endurance trained horses, in *Proc. 1st Int. Symp. Equine Hematology,* Kitchen, H. and Krehbiel, J. D., Eds., The American Association of Equine Practitioners, Golden, Colo., 1975.
8. **Snow, D. H. and Munro, C. D.,** Changes in blood levels of several hormones following ACTH administration and during exercise, in *Proc. 1st Int. Symp. Equine Hematology,* Kitchen, H. and Krehbiel, J. D., Eds., The American Association of Equine Practitioners, Golden, Colo., 1975.
9. **Gabriel, K. L. and Martin, J. E.,** Phenylbutazone: short-term versus long-term administration to thoroughbred and standardbred horses, *J. Am. Vet. Med. Assoc.,* 140, 337, 1962.
10. **Elbert, E. F.,** Clinical use of phenylbutazone in large animals, *Vet. Med.,* 57, 33, 1962.
11. **Meaghes, D. M. and Tasker J. C.,** Effects of excitement and tranquilization on the equine hemogram, *Mod. Vet. Pract.,* 53, 41, 1972.
12. **Lumsden, J. H., Valli, V. E. O., and McSherry, B. J.,** The comparison of erthrocyte and leukocyte response to epinephrine and acepromazine maleat in standardbred horses, in *Proc. 1st Int. Symp. Equine Hematology,* Kitchen, H. and Krebhiel, J. D., Eds., The American Association of Equine Practitioners, Golden, Colo., 1975.
13. **Straub, R. and Gerber, H.,** Effects of prolonged use of corticoids, in *Proc. 1st Int. Symp. Equine Hematology,* Kitchen, H. and Krehbiel, J. D., Eds., The American Association of Equine Practitioners, Golden, Colo., 1975.
14. **Coffman, J. R.,** Clinical application of serum protein electrophoresis in the horse, *Proc. Am. Assoc. Equine Practitioners,* 265, 1968.
15. **Dunavant, M. L. and Murry, E. S.,** Clinical evidence of phenylbutazone induced hypoplastic anemia, in *Proc. 1st Int. Symp. Equine Hematology,* Kitchen, H. and Krehbiel, J. D., Eds., The American Association of Equine Practitioners, Golden, Colo., 1975.
16. **Gerber, H., Tschudi, P., and Straub, R.,** "Normal" values for different breeds of Horses, in *Proc. 1st Int. Symp. Equine Hematology,* Kitchen, H. and Krehbiel, J. D., Eds., The American Association of Equine Practitioners, Golden, Colo., 1975.
17. **Blackmore, D. J.,** A new concept of normal values, in *Proc. 1st Int. Symp. Equine Hematology,* Kitchen, H. and Krehbiel, J. D., Eds., The American Association of Equine Practitioners, Golden, Colo., 1975.
18. **Coffman, J.,** Blood chemistry values in domestic animals *Neb. Vet. Ext. Newsl.,* 6(6), 1977.
19. **Lumseden, J. H., Valli, V. E. O., and McSherry, J.,** Comparison on normal equine hematologic values utilizing the Coulter Model S and Model Z, in *Proc. 1st Int. Symp. Equine Hematology,* Kitchen, H. and Krehbiel, J. D., The American Association of Equine Practitioners, Golden, Colo., 1975.
20. **Young, D. S., Harris, E. K., and Cotlove, E.,** Results of a study designed to delineate long term analytical variation, *Clin. Chem.,* 17, 403, 1971.
21. **Coffman, J.,** Equine normal serum multiple analysis (SMA), *Neb. Vet. Ext. Newsl.,* 4(7), 1975.
22. **Catcott, E. J. and Smithcors, J. F., Eds.,** *Equine Medicine and Surgery,* American Veterinary Publications, Wheaton, Ill., 1972.
23. **Schalm, O. W.,** *Veterinary Hematology,* Lea & Febiger, Philadelphia, 1965.
24. Commission on Quantities and Units in Clinical Chemistry, Quantities and Units in Clinical Chemistry, *Pure and Applied Chemistry,* 51, 2451, 1979.

REPORTING HUMAN LABORATORY DATA

Edward R. Powsner

INTRODUCTION

Clinical laboratory reports must be comprehensible and unambiguous, as outlined below. To simplify interpretation, normal or reference values should be included in the report. Unless a standard or basal state of the patient and specimen are clearly implied, the report should include the details which may affect the results.

REFERENCE VALUES

Reference (formerly "normal") values for a number of the more commonly measured constituents of blood and urine are included in Table 1. Values are presented in both conventional metric and in the proposed SI units. Conventional data are based on those published by Conn,[1] Scully et al.,[2] and on the published and unpublished sources noted in the table. SI data are computed using conversion factors from Lehmann,[3] Lippert and Lehmann,[4] and WHO.[5] These reference values are representative of those which may be expected using one or more commonly used assays, but in many cases, e.g., enzyme measurements, results are strongly dependent on details of the method used. Where possible, reference values should be confirmed in each clinical laboratory by use of normal specimens or by analyses of "patient" specimens from among which a reasonably large proportion of normal samples may be expected. Where this is not possible, reference values must be those published for the specific method. Consistency of reference values can be controlled by regular use of commercial or other properly prepared specimens. Control values must also be interpreted with respect to the assay method used, as significantly different results may be obtained with different assays.[6]

PATIENT PREPARATION

Specimens for laboratory testing should be obtained from the properly prepared patient. Constituents of blood and urine are affected by one or more of the following elements that need to be documented:

1. Time of day
2. Time from, and nature of, last meal
3. State of hydration
4. Drugs administered
5. Position and activity of patient
6. Ambient temperature

As examples, plasma corticosteroids,[7] serum iron,[8] and plasma volume itself,[9] show diurnal fluctuation;[10-14] plasma glucose rises following a meal;[15-16] the concentration of serum albumin and blood hemoglobin are increased by dehydration. Among drugs, insulin lowers glucose, while cortisone increases it; diuretics affect blood pH, electrolytes, and the concentrations of fluid regulating hormones.[16,17] Blood and plasma volume are lower and hematocrit higher in the erect position.[18-21] Bed rest and cold weather also decrease plasma volume.[22] Conversely, plasma volume is increased by exercise, reclining, warm weather, and in pregnancy.[23] All such factors must be considered for their specific effects on any constituent measured. Where the effect of one variable is to be determined, others should be held constant. Largely for this reason, blood specimens are commonly obtained from the recumbent, fasting patient at a regular morning hour. Ambulatory patients can be instructed to report fasting and in the morning. When necessary to compare results to those for hospitalized patients, ambulatory patients should be asked to remain quietly in an anteroom for 30 to 60 min prior to testing. For clinical purposes the wait is safely omitted.

Urine constituents are affected by many of the same factors that affect those of blood. In addition, urine is affected by the length of time the urine has remained in the bladder and by the growth of bacteria both before and after voiding.

Fecal constituents including fat, hemoglobin, and urobilinoben vary widely from day-to-day. Where accuracy is important, specimens should be collected over a period of 3 or 4 days, pooled, and mixed before removing a sample for assay.[24]

SPECIMEN COLLECTION, TRANSPORTATION, AND STORAGE

Specimens should be collected carefully in sterile containers and tested promptly. More specifically, blood is best drawn into evacuated tubes by atraumatic venipuncture or when this is not feasible, from freely flowing capillary blood.[25,26] Capillary samples are not always equivalent to venous samples; platelet counts, for example, are lower in capillary samples.[27] Small bore needles, 22- to 25-gauge, are less traumatic to both patient and sample.[28] The tubes must be clean and free of significant contamination, especially when trace elements are measured. If an anticoagulant is used, it should be selected to minimize its effect on blood constituents.[16] As examples, sodium heparin is to be avoided in specimens for sodium measurement and oxylate avoided for calcium. If the sample is allowed to clot, the competing needs for rapid separation of cells from fluid and for adequate removal of fibrinogen and minimized hemolysis must be balanced in respect to the test to be performed. The effects of hemolysis in a number of constituents has been discussed by Caraway.[16] If cells are not separated, the effects of cell rupture must be considered and the handling procedures selected accordingly. Cell lysis is a required step in hemoglobin measurement, but even small numbers of damaged cells greatly increase serum potassium and serum lactate dehydrogenase.

After collection, specimens should be stored cool or frozen if not tested immediately. The stability of the constituent to be tested will dictate the permissible duration and the best temperature for storage.[16] When doubt exists about adequacy of specimen storage or transport, their effects should be tested.

Table 1

#	SYSTEM	CHEMICAL SUBSTANCE, BIOLOGIC ENTITY, PHYSIOLOGIC DESCRIPTOR, OR PHYSICAL QUANTITY	MEASURABLE PROPERTY (KIND OF QUANTITY) (see abbrev. 1st)	REFERENCE VALUES¹ AND UNITS CONVENTIONAL	SI (PROPOSED)
A01	S	ACETOACETATE	CSUBSTC	mg/dL	umol/L
A02	dU	ACETONE	AMCS	mg	umol
A03	U	--	CSUBSTC	mg/dL	umol/L
A04	S	ACETONE+ACETOACETATE etc.(KETONE BODIES)	ARB MASSC	0 arb un	0 arb un
A05	U	-- (KETONE BODIES)	MASSC	0.3-2.0 mg/dL	3-20 mg/L
A06	(B)Ercs	ACETYLCHOLINESTERASE	CATAMT	0.5-1.0 pH U	kat/mol
A07	S	--	CATC³	0.5-1.3 pH U	ukat/L
A08		ACID(H+) (see H19,etc.)			
A09	S	ACID PHOSPHATASE	CATC³	0.1-0.63 U/L	30-180 nkat/L
A10	S	ACID PHOSPHATASE, TARTRATE SENS	CATC³	U/L	ukat/L
A11		ACTH(see C48)			
A12		ADRENALINE(see E06,E07,E08)			
A13	S	ALANINE AMINOTRANSFERASE(SGPT; TRANSAMINASE)	CATC³	6-35 U/mL	50-280 nkat/L
A14	S	ALBUMIN	MASSC	3.5-5.0 g/dL	35-50 g/L
A15	Sf	--	MASSC	11-48 mg/dL	0.11-0.48 g/L
A16	(S)Prot	-- (A/G RATIO)	MASSFR	0.52-0.68	--
A17		-- (66500)	MASS/GLOB MASS	--	--
A18	S	-- (66500)	CSUBSTC	3.5-5.0 g/dL	510-740 umol/L
A19	Sf	ALDOSTERONE	CSUBSTC	mg/dL	umol/L
A20	dU	ALDOSTERONE PRODUCED	AMCS	3-20 ug/d	8-55 nmol
A21	dPt	ALIPH CARBOXYLATES(see F01,F02,F03)			
A22	S	ALKALINE PHOSPHATASE	CATC³	13-39 U/L	0.22-0.65 ukat/L
A23	S	AMINOACID NITROGEN(cf. N10)	CSUBSTC	mg/dL	mmol/L
A24	dU	DELTA-AMINOLEVULINATE	AMCS	1.3-7.0 mg/d	9.9-53 umol
A25	S	AMINOPEPTIDASE(CYTOSOL)	CATC³	4-25 U/mL	ukat/L
A26	dU	AMMONIA(NH3)	AMCS	mg/d	mmol
A27	B	-- (NH3)	CSUBSTC	80-110 ug/dL	47-64.6 umol/L
A28	P	-- (NH3)	CSUBSTC	56-122 ug/dL	33-71.6 umol/L
A29	Rect mucus	AMOEBA	ARB NUMC	arb un	arb un
A30	U(t'med)	ALPHA-AMYLASE	CATAMT	35-260 U/h	kat
A31	P	ANDROSTERONE	CATC⁵	U/mL	ukat/L
A32	S	ANION(A-)	CSUBSTC	meq/L	mmol/L
A33	S	ANTITHROMBIN III	CSUBSTC	ug/dL	nmol/L
A34	S	ALPHA-1-ANTITRYPSIN	ARB MASSC	mg/dL	mmol/L
A35	S	ARSENIC	ARB MASSC	arb un	arb un
A36	S	ARSENIC	AMCS	mg/dL	arb un
A37	dU	ARSENIC	AMCS	-100 ug/d	-1.3 nmol
A38	B	ARSENIC	CSUBSTC	3.5-7.2 ug/dL	0.47-0.96 mmol/L
A39	B	ASCORBATE(VITAMIN C)	CSUBSTC	0.4-1.5 mg/dL	20-85 umol/L

Table 1 (continued)

#	SYSTEM	CHEMICAL SUBSTANCE, BIOLOGIC ENTITY, PHYSIOLOGIC DESCRIPTOR, OR PHYSICAL QUANTITY	MEASURABLE PROPERTY (KIND OF QUANTITY) (see abbrev. list)	REFERENCE VALUES' AND UNITS — CONVENTIONAL	SI (PROPOSED)
A41	S	ASPARTATE AMINOTRANSFERASE(SGOT; TRANSAMINASE)	CATC'	15-40 U/mL	120-320 nkat/L
B01	U	BACTERIA	ARB NUMC	arb un	arb un
B02	U	BACTERIA, LIVING	NUMC	10^7/uL	10^9/L
B03	U	BARBITURATE(5,5-DIETHYLBARBITURIC ACID)	CSUBSTC	mg/dL	umol/L
B04	S	-- (5,5-DIETHYLBARBITURIC ACID)	CSUBSTC	0-0 mg/dL	0-0 umol/L
B05	aB	BASE(H+ BINDING, TITRATABLE)	CSUBSTC	meq/L	mmol/L
B06	aB	-- (H+ BINDING, BASE EXCESS)	CSUBSTC DIFF	meq/L	mmol/L
B07	(B)Ercs	BASOPHIL PUNCTATED ERYHTROCYTES(see E19)			
B08	B	BASOPHILS	NUMC	15-50 /uL	0.015-0.05 10^9/L
B09	(B)Lkcs	BASOPHILS	NUMFR	0-0.75 %	0-0.0075
B10	dU	N-BENZOYLGLYCINE(HIPPURIC ACID)	AMCS	g/d	mmol
B11	S	BICARBONATE(see H18)			
B12	S	BILIRUBIN, UNESTERIFIED(INDIRECT)	CSUBSTC	0.2-0.7 mg/dL	3-10 umol/l
B13	(S)Bili	BILIRUBIN ESTER	CSUBSTFR	--	--
B14	S	-- (DIRECT)	CSUBSTC	0.1-0.4 mg/dL	2-7 umol/l
B15	S	BILIRUBIN, TOTAL	CSUBSTC	0.3-1.1 mg/dL	5-19 umol/l
B16	Sf	BILIRUBIN	CSUBSTC	0 mg/dL	0 umol/l
B17	U	-- (GMELIN'S REACTION)	ARB CSUBSTC	0 arb un	0 arb un
B18	dF	BLOOD	VOL	mL	mL
B19	F	-- (BENZIDINE REACTION; cf. HO9)	ARB VOLFR	arb un	arb un
B20	Pt	BLOOD VOLUME	VOL	mL.(L)	L
B21	Pt	--	VOL/BODY MASS	85-90 mL/kg	85-90 mL/kg
B22	Pt	BLOOD, ARTERIAL, DIASTOLIC	PRESSURE	50-140 mmHg	7-19 kPa
B23	Pt	BLOOD, ARTERIAL, SYSTOLIC	PRESSURE	100-240 mmHg	13-32 kPa
B24	B	BLOOD COAGULUM LYSIS(CLOT LYSIS)	TIME	h	ks
B25	S	BROMIDE	CSUBSTC	0 mg/dL	0 mmol/L
B26	dF	BROMSULPHALEIN(BSP; see S21)			
C01	dF	CALCIUM II	AMCS	g/d	mmol
C02	dU	--	AMCS	-150 mg/d	-3.7 mmol
C03	U	--	CSUBSTC	mg/dL	mmol/L
C04	S	CALCIUM II ION(UNCHELATED)	CSUBSTC	2.1-2.6 meq/L	1.1-1.3 mmol/L
C05	S	CALCIUM II, CHELATED<1000	CSUBSTC	meq/L	mmol/L
C06	S	--	CSUBSTC	mg/dL	mmol/L
C07	S	CALCIUM II, PROTEIN BOUND	CSUBSTC	meq/L	mmol/L
C08	S	--	CSUBSTC	mg/dL	mmol/L
C09	S	CALCIUM II, TOTAL	CSUBSTC	4.2-5.2 meq/L	2.1-2.6 mmol/L
C10	Pt	CAPILLARY BLEEDING(BLEEDING TIME)	TIME	min	ks
C11		CARBAMIDE(see U05,etc.)			
C12	Gas(aB,eq)	CARBON DIOXIDE(pCO2)	PART PRESS	35-45 mmHg	4.7-6.0 kPa
C13	(aB)P	-- (pCO2)	CSUBSTC	mmol/L	mmol/L
C14	Expr gas	-- (pCO2)	VOLFR	--	--

Code	System	Analyte	Component	Conv. value	Conv. unit	SI value	SI unit
C15	(B)Hgb	CARBON MONOXIDE HBG(see C18)					
C16	B	CARBOXYHEMOGLOBIN(Fe)	CSUBSTC		mg/dL		umol/L
C17	P	CARBONATE+CARBON DIOXIDE(TOTAL CO2)	CSUBSTC	24-30	meq/L	24-30	mmol/L
C18	(B)Ercs	CARBOXYHEMOGLOBIN	MASSFR	0-5	%	0-0.05	—
C19	P	CAROTENOIDS(BETA-CAROTENE)	CSUBSTC	50-300	ug/dL	0.9-6	umol/L
C20	U	CASTS	ARB NUMC	arb un	arb un		10³
C21	U(timed)	--	NUM	10³/12h	10³/12h		arb un
C22	F	--	ARB CATC	arb un	arb un		arb un
C23		CATECHOLAMINES(see E07,E08)					
C24	S	CATION(B+)	CSUBSTC	0-5	meq/L	0-5	mmol/L
C25	Sf	CELLS(see L10)	NUMC		/uL		10⁶/L
C26	Pt	CEREBROSPINAL FLUID(CSF)	PRESSURE	70-180	mmH2O	0.69-1.77	kPa
C27	dU	CERULOPLASMIN(150000)	CSUBSTC	96-106	mg/dL	6.3-7.00	umol/L
C28	dU	CHLORIDE	AMCS	100-250	meq/d	100-250	mmol/d
C29	U	--					
C30	s	--		100-106	meq/L	100-106	mmol/L
C31	Sf	--		120-130	meq/L	120-130	mmol/L
C32	S	CHOLESTEROL, NONESTERIFIED	CSUBSTC		mg/dL		mmol/L
C33	S	CHOLESTEROL ESTERS	CSUBSTC	100-200	mg/dL	3-5	mmol/L
C34	S	CHOLESTEROL, TOTAL	CSUBSTC	120-220	mg/dL	3.1-5.7	mmol/L
C35	dU	CHORIONGONADOTROPIN	ARB MASSC	0	arb un	0	arb un
C36	s	CITRATE	CSUBSTC		mg/dL		umol/L
C37	B	COAGULATION(CLOTTING TIME)	TIME	5-15	min	0.3-0.90	ks
C38	P	COAG FACTORS(PROTHROMBIN TIME)	ARB TIME	12.0-14.0	s	12.0-14.0	s
C39	P	-- (PROTHROMBIN)	REL ARB MASSC				
C40	P	-- (PARTIAL THROMBOPLASTIN TIME)	ARB MASSC	35-45	s	35-45	s
C41	dU	COPPER	AMCS	0-30	ug/d	0-0.47	umol/L
C42	S	COPPER II	CSUBSTC	100-200	ug/dL	15.7-31.5	umol/L
C43	(B)Ercs	COPROPORPHYRIN	CSUBSTC		ug/dL		nmol
C44	dU	COPROPORPHYRINS I+III	AMCS	50-250	ug/d	76-382	nmol
C45	P,S	CORTICOSTEROIDS(HYDROYXSTEROIDS)	CSUBSTC	8-18	ug/dL	0.2-0.50	umol/L
C46	dU	--	AMCS		mg/d		mmol
C47	P	CORTICOSTERONE(COMPOUND B)	CSUBSTC		ug/dL		nmol/L
C48	P	CORTICOTROPIN(ACTH)	CSUBSTC	6-16	pg/mL		pmol/L
C49	P	CORTISOL(COMPOUND F)	CSUBSTC		ug/dL	200-440	nmol/L
C50	P	CORTISONE(COMPOUND E)	CSUBSTC		ug/dL		nmol/L
C51	dU	CREATINE	AMCS	0-40	mg/d	0-310	umol/L
C52	dU	--	AMCS	0-100	mg/d	0-763	umol
C53	U	--					
C54	s	--	AMCS	0.2-0.8	mg/dL	20-60	umol/L
C55	dU	CREATINE KINASE(CK; CPK)	CATC	0-50	U/L	0-0.84	ukat/L
C56	dU	CREATININE	AMCS/BODY MASS	15-25	mg/kg-d	130-220	umol/kg
C57	dU	--	CSUBSTC		mg/dL		mmol/L
C58	U	--					
C59	s	--	CSUBSTC	0.6-1.5	mg/dL	50-130	umol/L
C60	s	CRYOGLOBULIN	ARB MASSC		arb un		arb un
C61	P,S	CYANOCOBALAMIN(VITAMIN B12)	ABR CSUBSTC	200-800	pg/mL	100-600	pmol/L
C62	U	CYSTEINE	CSUBSTC		ug/dL		umol/L
C63	s	CYSTINE	CSUBSTC		arb u		arb u
DO1	s	11-DEOXYCORTICOSTERONE	CSUBSTC	0	arb u	0	arb u
DO2	U	11-DEOXYCORTISOL(SUBSTANCE S)	CSUBSTC	0	ug/dL	0	umol/L
DO3	P	2,5-DIHYDROXYPHENYLACETATE(see H16)	CSUBSTC		ng/mL		nmol/L

Table 1 (continued)

#	SYSTEM	CHEMICAL SUBSTANCE, BIOLOGIC ENTITY, PHYSIOLOGIC DESCRIPTOR, OR PHYSICAL QUANTITY	MEASURABLE PROPERTY (KIND OF QUANTITY) (see abbrev. 1st)	REFERENCE VALUES' AND UNITS CONVENTIONAL	SI (PROPOSED)
DO4	B,S	DILANTIN(see DO5)			
DO5		DIPHENYLHYDANTOIN	CSUBSTC	0 ug/dL	ug/dL
DO6		DIOXYGEN(see O10,O11,O12)			
EO1	Gastr secr	ELECTROLYTES(BASAL STATE)	CSUBSTC	17.3-27.6 meq/L	17.3-27.6 mmol/L
EO2	Gastr secr	-- (BASAL STATE)	AMCS	1.43-2.73 meq/h	1.43-2.73 mmol/L
EO3	Gastr secr	-- (AFTER STIMULATION)	AMCS	meq/h	mmol
EO4	B	EOSINOPHILS	NUMC	50-250 /uL	0.05-0.25 10^9/L
EO5	(B)Lkcs	--	NUMFR	1-3 %	0.01-0.03 --
EO6	dU	EPINEPHRINE	AMCS	0.5-20 ug/d	3-110 nmol
EO7	(dU)Ep+Nep	-- (CATECHOLAMINES)	CSUBSTFR	%	--
EO8	U	EPINEPHRINE+NOREPINEPHRINE(CATECHOLAMINES)	AMCS	ug/d	umol
EO9		EPITHELIAL CELLS	ARB NUMC	arb un	arb un
E10	(B)Ercs	ERYTHROCYTE DIAMETER, MEAN(MCD)	LENGTH	micron	um
E11		ERYTHROCYTE HEMOGLOBIN(see HO9,H10)			
E12	(B)Ercs	ERYTHROCYTE VOLUME, MEAN(MCV)	VOL	80-105 fL	80-105 fL
E13	(B)Ercs	ERYTHROCYTE LIFE, MEAN	TIME	d	d
E14	U	ERYTHROCYTES	ARB NUMC	0 arb un	0 arb un
E15	U(timed)	--	NUM	10^6/12h	10^6
E16	B	--	NUMC	4.6-6.2 10^6/uL	4.6-6.2 10^{12}/L
E17	Sf	ERYTHROCYTES	NUMC	/uL	10^6/L
E18	B	-- (PCV, HEMATOCRIT)	VOLFR	37-54 %	0.37-0.54 --
E19	(B)ERY	ERYTHROCYTES, BASOPHILIC STIPPLED	NUMFR	%	10^{-3}
E20	dU	ESTRADIOL	AMCS	ug/d	umol
E21	dU	--	AMCS	0-6 ug/d	0-0.02 umol
E22	dU	--	AMCS	0-14 ug/d	0-0.051 umol
E23	dU	ESTRIOL	AMCS	ug/d	umol
E24	dU	--	AMCS	1-11 ug/d	0.003-0.038 umol
E25	dU	--	AMCS	0-72 ug/d	0-0.25 umol
E26	dU	ESTROGEN	AMCS	ug/d	umol
E27	dU	-- (TOTAL)	AMCS	4-25 mg/d	0.01-0.090 umol
E28	dU	-- (TOTAL)	AMCS	5-100 mg/d	0.02-0.359 umol
E29	dU	ESTRONE	AMCS	3-8 ug/d	0.01-0.03 umol
E30	dU	--	AMCS	4-31 ug/d	0.01-0.11 umol
E31	B	ETHANOL	CSUBSTC	-5 mg/dL	-1 mmol/L
FO1	dF	FAT, TOTAL(see L14)			
FO2	(fPt)S	FATTY ACIDS, FREE(NEFA)	CSUBSTC	190-240 meq/L	190-240 mmol/L
FO3	(fPt)S	FATTY ACIDS, TOTAL	CSUBSTC	190-420 mmol/L	mmol/L
FO4	(FPT)S	-- (as OLEATE)	CSUBSTC	6.73-14.9 mg/dL	6.73-14.9 mmol/L
FO5	P	FIBRINOGEN(340000)	CSUBSTC	mg/dL	umol/L
FO6	P	--	MASSC	200-400 mg/dL	2-4 g/L
FO7	P	FIBRINOLYSIS	TIME	min	s

Code	System	Component	Kind-of-quantity	Value (conv.)	Unit	Value (SI)	Unit
F08	Water	FLUORIDE	CSUBSTC	1.5–3.0	mg/dL	0.015–0.030	umol/L
F09	(B)Hgb	FETAL HEMOGLOBIN	MASSFR		%		--
F10	(fPt)S	FOLATE(as PTEROYLGLUTAMIC ACID)	CSUBSTC	7–16	ng/mL	20–36	nmol/L
F11	U(timed)	FORMIMINOGLUTAMATE	AMCS		mg/8h		umol/d
F12	U	FREEXING-POINT DEPRESSION(OSMOLALITY)	TEMP DIFF		mosm/kg		mK
F13	S	FREEXING-POINT DEPRESSION (OSMOLALITY; cf. 123	TEMP DIFF		mosm/kg		mK
F14	S	FRUCTOSE-DIPHOSPHATE ALDOLASE	CATC3	0.8–3.0	U/dL	0.1–0.37	nkat/L
G01	P,S	GALACTOSE	CSUBSTC		mg/dL		mmol/L
G02	(B)Ercs	GALACTOSE-1-PHOSPHATE URIDYLTRANSFERASE	CATAMT		U		kat
G03	S	GLOBULIN	MASSC	2.3–3.5	g/dL	23–35	g/L
G04	Sf	--	ARB MASSC		arb un		arb un
G05	S	ALPHA-1-GLOBULIN	MASSC	0.2–0.4	g/dL	2–4	g/L
G06	(S)Prot	--	ARB MASSC	0.02–0.05		0.02–0.05	arb un
G07	S	--	MASSFR	0.2–0.4	g/dL	2–4	g/L
G08	S	ALPHA-2-GLOBULIN	MASSC	0.5–0.9	g/dL	5–9	g/L
G09	S	--	ARB MASSC	0.07–0.14		0.07–0.14	arb un
G10	(S)Prot	ALPHA-2-GLOBULIN	MASSFR	0.5–0.9	g/dL	5–9	g/L
G11	S	BETA-GLOBULIN	MASSC	0.6–1.1	g/dL		--
G12	(S)Prot	--	MASSFR	0.09–0.15	g/dL	0.09–0.15	arb un
G13	S	BETA-1-GLOBULIN	MASSC		g/dL		g/L
G14	S	--	ARB MASSC		g/dL		arb un
G15	(S)Prot	--	MASSFR		g/dL		g/L
G16	S	BETA-2-GLOBULIN	MASSC	0.7–1.7	g/dL	7–17	g/L
G17	(S)Prot	--	ARB MASSC		g/dL		arb un
G18	S	GAMMA-GLOBULIN	MASSC	0.7–1.7	g/dL	7–17	g/L
G19	(S)Prot	--	MASSFR	0.11–0.21	g/dL	0.11–0.21	arb un
G20	S	GLOMERULAR FILTRATE(CREATININE CLEAR.)	VOL RATE	104–125	mL/min	1.73–2.08	mL/s
G21	Pt	GLUCOSE	AMCS	0–300	mg/dL	0–2	mmol
G22	dU	--	ARB MASSC		arb un		arb un
G23	U	--	CSUBSTC	70–110	mg/dL	4–6.1	mmol/L
G24	P(fPt)	--	CSUBSTC	50–75	mg/dL	2.8–4.2	mmol/L
G25	P,S	--	CSUBSTC		mg/dL		mmol/L
G26	Sf	--	CSUBSTC		mg/dL		mmol/L
G27	U	--	ARB MASSC		arb un		arb un
G28	dU	GLUCOSE+REDUCING SUBSTANCES	CSUBSTC	0	mg/d	0	mmol
G29	U	-- (as GLUCOSE)	CATC3	–250	mg/d	–1.4	ukat/L
G30	dU	GLUCOSE-6-PHOSPHATE DEHYDROGENASE	ARB MASSC		arb un		arb un
G31	(B)Ercs	GONADOTROPINS, PITUITARY	MASSC		mU/mL		arb un
G32	dU	HAPTOGLOBIN	CSUBSTC	40–170	mg/dL	0.40–1.70	g/L
H01	S	--	MASSC		mg/dL		umol/L
H02	S	--	MASSC		mg/dL		g/L
H03	Pt	HEIGHT	LENGTH		ft		m
H04	Pt	HELMINTHES EGGS	ARB NUMC	0	arb un		arb un
H05	F	HEMIGLOBIN(see M16)	ARB MASSC		arb un		arb un
H06	(B)Hgb	HEMOGLOBIN(BENZIDINE REACTION)	ARB MASSC	0	arb un		arb un
H07	F	-- (BENZIDINE REACTION)	ARB MASSC		arb un		arb un
H08	U	-- (MONOMER. 16114). MEAN(MCH)	MASS	27–31	pg	17–19	fmol
H09	(B)Ercs	-- (MONOMER. 16114). MEAN(MCHC)	MASSC	32–36	g/dL	20–22	mmol/L
H10	(B)Ercs	-- (MONOMER. 16114)	MASSC	0–5.0	mg/dL	0–3.1	umol/L
H11	P	-- (MONOMER. 16114)	MASSC	12–16	g/dL	7.4–9.9	mmol/L
H12	B	-- (MONOMER. 16114)	MASSC		g/dL		mmol/L
H13	B	HEMOGLOBIN A2	MASSFR	1.5–3.0	%	0.015–0.030	arb un

Table 1 (continued)

#	SYSTEM	CHEMICAL SUBSTANCE, BIOLOGIC ENTITY, PHYSIOLOGIC DESCRIPTOR, OR PHYSICAL QUANTITY	MEASURABLE PROPERTY[1] (KIND OF QUANTITY) (see abbrev. list)	REFERENCE VALUES[2] AND UNITS	
				CONVENTIONAL	SI (PROPOSED)
H14	S	BETA-2-HEMOPEXIN	ARB MASSC	arb un	arb un
H15	U	HIPPURATE(see B10)			
H16	U	HOMOGENTISATE	ARB MASSC	0 arb un	arb un
H17		HYDROCORTISONE(see C49)			
H18	P	HYDROGEN CARBONATE ION(BICARBONATE; cf. C17)	CSUBSTC	meq/L	mmol/L
H19	dU	HYDROGEN ION	AMCS	meq/d	mmol
H20	aB	--	CSUBSTC	35.5-44.5 neq/L	35.5-44.5 nmol/L
H21	Gastr sec	--	AMCS	1-3 meq	1-3 mmol
H22	Gastr sec	--	CSUBSTC	20-30 meq/L	20-30 mmol/L
H23	aB	HYDROGEN ION ACTIVITY(pH)	LOG CSUBSTC	7.35-7.45 pH	
H24	U	-- (pH)	LOG CSUBSTC	4.6-8.0 pH	--
H25	Gastr secr	-- (pH)	LOG CSUBSTC	0.9-1.5 pH	--
H26	Sm	-- (pH)	pH	7.2-8.0 pH	--
H27	P	3-HYDROXYBUTYRATE(cf. A04)	CSUBSTC	mg/dL	umol/L
H28	S	2-HYDROXYBUTYRATE DEHYDROGENASE	CATC[3]	0-180 U/mL	kat/L
H29		HYDROXYSTEROIDS(see C45,C46,C50)			
H30	dU	5-HYDROXYINDOLYL ACETATE(SEROTONIN)	AMCS	2-9 mg/d	10-50 umol
H31	S	5-HYDROXYLINDOLYL ACETATE(SEROTONIN)	CSUBSTC	0.10-0.32 ug/mL	5-50 umol/L
H32	dU	4-HYDROXY-3-METHOXYMANDELATE(VMA)	AMCS	1-9 mg/d	umol
H33	dU	HYDROXYPROLINE	AMCS	mg/d	mmol
I01	S	IMMUNOGLOBULIN A (IgA)	MASSC	50-200 mg/mL	50-200 g/L
I02	S	-- (IgA)	ARB MASSC	IU/mL	
I03	S	IMMUNOGLOBULIN D (IgD)	MASSC	mg/mL	
I04	S	-- (IgD)	ARB MASSC	IU/mL	
I05	S	IMMUNOGLOBULIN E (IgE)	MASSC	mg/mL	
I06	S	-- (IgE)	ARB MASSC	IU/mL	
I07	S	IMMUNOGLOBULIN G (IgG)	MASSC	800-1200 mg/mL	800-1200 g/L
I08	S	-- (IgG)	ARB MASSC	IU/mL	
I09	Sf	-- (IgG)	MASSC	0-8.6 mg/mL	0-8.6 g/L
I10	S	IMMUNOGLOBULIN G1 (IgG 1)	MASSC	mg/mL	
I11	S	-- (IgG 1)	ARB MASSC	IU/mL	
I12	S	IMMUNOGLOBULIN G2 (IgG 2)	MASSC	mg/mL	
I13	S	-- (IgG 2)	ARB MASSC	IU/mL	
I14	S	IMMUNOGLOBULIN G3 (IgG 3)	MASSC	mg/mL	
I15	S	-- (IgG 3)	ARB MASSC	IU/mL	
I16	S	IMMUNOGLOBULIN G4 (IgG 4)	MASSC	mg/mL	
I17	S	-- (IgG 4)	ARB MASSC	IU/mL	
I18	S	IMMUNOGLOBULIN M (IgM)	MASSC	40-120 mg/mL	40-120 g/L
I19	S	-- (IgM)	ARB MASSC	IU/mL	
I20	P,S	INSULIN(5808; 41.67 U/ug)	CSUBSTC	6-26 uU/mL	40-190 pmol/L

Code	Spec	Name	Property	Conv. value	Conv. unit	SI value	SI unit
I21	S	IODINE, PROTEIN BOUND	CSUBSTC	3.5-8.0	ug/dL	0.28-0.63	umol/L
I22	S	IODINE, TOTAL	CSUBSTC		ug/dL		mmol/L
I23	S	ION(A-+B+) (OSMOLARITY; cf. F12)	CSUBSTC		mosm/L		umol
I24	dU	IRON II+III	AMCS		ug/dL		umol/L
I25	U	--	CSUBSTC		ug/dL		umol/L
I26	(fPt)S	-- (IN HEMOGLOBIN & TRANSFERRIN)	CSUBSTC		ug/dL		umol/L
I27	S	-- (IN HEMOGLOBIN & TRANSFERRIN)	CSUBSTC	50-175	ug/dL	9.0-31.3	umol/L
I28	(fPt)S	IRON III, TRANSFERRIN BOUND(cf. T11)	CSUBSTC		ug/dL		umol/L
I29	S	-- (cf. T11)	CSUBSTC		ug/dL		umol
K01	dU	17-KETO+KETOGENIC STEROIDS(as DHEA)	AMCS	1-26	mg/d	3-90	mmol/L
L01	B	LACTATE	CSUBSTC	6-16	mg/dL	0.7-1.8	mmol/L
L02	S	LACTATE DEHYDROGENASE	CATC³	150-450	U/mL	1.20-3.61	ukat/L
L03	dU	LEAD II	AMCS	-120	ug/d	-579	umol/L
L04	U	--	CSUBSTC	-8	ug/dL	-0.4	umol/L
L05	B	--	CSUBSTC	0-50	ug/dL	0-2.4	umol/L
L06	U	--	ARB NUMC		arb un		arb un
L07		LEUKOCYTES	NUMC				
L08	(B)Lkcs	--	NUMFR	5000-10000	/uL	5-10	10^9/L
L09	Sf	LEUKOCYTES TYPES(DIFFERENTIAL; see B08,etc.)	NUMC		%		10^6/L
L10	Sf	LEUKOCYTES	ARB NUMC	0-5	/uL	0-5	arb un
L11	Sm	--	NUM	0-	arb un		10^6
L12	U(timed)	LEUKOCYTES+EPITHELIAL CELLS	MASSFR		10^6/12h		g/kg
L13	F	LIPID	MASS		%		g
L14	dF	LIPIDS, TOTAL	MASSC	-6.0	g/d	-6.0	g/L
L15	(fPt)S	--	MASSC	450-1000	mg/dL	4.5-10	g/L
L16	S	LIPOPROTEINS, TOTAL	MASSFR		mg/dL		arb un
L17	S	ALPHA-1-LIPOPROTEIN	MASSC		%		arb un
L18	s	ALPHA-2-LIPOPROTEIN	MASSFR		mg/dL		arb un
L19	s	--	MASSC		%		arb un
L20	s	BETA-LIPOPROTEIN	MASSFR		mg/dL		arb un
L21	s	--	ARB TIME		%		s
L22	s	--	CSUBSTC		min		mmol/L
L23	Sm	LIQUIFICATION REACTION	NUMC	1500-3000	/uL	1.5-3	10^9/L
L24	P,S	LITHIUM	NUMFR	25-33	meq/L	0.25-0.33	--
L25	B	LYMPHOCYTES	ARB MASSC		/uL		arb un
L26	(B)Lkcs	--	AMCS		%		mmol
M01	S	ALPHA-2-MACROGLOBULIN	AMCS		arb un		mmol
M02	dF	MAGNESIUM II	CSUBSTC		mg/d		mmol/L
M03	dU	MAGNESIUM II	CSUBSTC		mg/d		mmol/L
M04	U	--	CSUBSTC		mg/dL		mmol/L
M05	U	MAGNESIUM II ION(NONCHELATED)	CSUBSTC		meq/L		mmol/L
M06	S	MAGNESIUM II, CHELATED<1000	MASS		meq/L		mmol/L
M07	S	MAGNESIUM II, PROTEIN BOUND	MASS	1.5-2	meq/L	0.75-1	mmol/L
M08	S	MAGNESIUM II, TOTAL	MASS		meq/L		kg
M09	Pt	MASS(WEIGHT)	ARB MASSC		kg.(lb)		g
M10	dF	MASS, DRY	CSUBSTC	23-32	g/d	23-32	g
M11	dF	MASS, TOTAL	AMCS	100-200	g/d	100-200	arb un
M12	U	MELANIN+MELANOGEN	MASSC		arb un		nmol/L
M13	U	MERCURY II	MASSFR	-10	ug/d	-50	nmol
M14	dU	--			ug/d		umol/L
M15	S	METHEMOGLOBIN(Fe)		0.03-0.13	g/dL	20-81	--
M16	(B)Hgb	--			%		

Table 1 (continued)

#	SYSTEM	CHEMICAL SUBSTANCE, BIOLOGIC ENTITY, PHYSIOLOGIC DESCRIPTOR, OR PHYSICAL QUANTITY	MEASURABLE PROPERTY (KIND OF QUANTITY)[1] (see abbrev. list)	REFERENCE VALUES[2] AND UNITS CONVENTIONAL	SI (PROPOSED)
M17	U	METHYLKETONES(KETONE BODIES; see AO4)			
M18	P	MOLALITY(see OO3)			
M19	U	-- (see OO4)			
M20	B	MONOCYTES	NUMC	280-500 /uL	0.28-0.5 10^9/L
M21	(B)Lkcs	--	NUMFR	3-7 %	0.03-0.07 --
M22	S	MUCOPROTEIN	MASSC	mg/dL	g/L
M23	Sputum	MYCOBACTERIUM SP	ARB NUMC	arb un	arb un
M24	P	MYOGLOBIN	MASSC	mg/dL	umol/L
M25	dU	--	AMCS	mg/d	umol
M26	U	--	MASSC	mg/dL	umol/L
N01	Uret secr	NEISSERIA SP	ARB NUMC	arb un	arb un
N02	B	NEUTROPHILS. NONSEGMENTED	NUMC	150-400 /uL	0.15-0.4 10^9/L
N03	(B)Lkcs	--	NUMFR	%	--
N04	B	NEUTROPHILS. SEGMENTED	NUMC	3000-6000 /uL	3-6 10^9/L
N05	(B)Lkcs	--	NUMFR	%	--
N06	P,S	NICKEL II	CSUBSTC	ug/dL	umol/L
N07	dF	NITROGEN	AMCS	-2.0 g/d	-140 mmol
N08	dU	--	AMCS	g/d	mol
N09	dU	NITROGEN, ALPHA AMINO	CSUBSTC	50-200 mg/d	4-10 mmol
N10	(fPt)S	--	CSUBSTC	3.0-5.5 mg/dL	2.1-3.9 mmol/L
N11	dU	NITROGEN, AMMONIA	CSUBSTC	20-70 mg/dL	20-70 mmol/L
N12	S	NITROGEN, NONPROTEIN(NPN)	CSUBSTC	mg/dL	11-25 mmol/L
N13	dU	NOREPINEPHRINE(NORADRENALINE)	AMCS	10-100 ug/d	59-591 nmol
O01	S	ORNITHINE CARBAMOYLTRANSFERASE	CATC[3]	U/L	kat/L
O02	S	ALPHA-1-OROSOMUCOID	ARB MASSC	arb un	arb un
O03	P	OSMOLALITY(cf. F12)	CSUBSTC	280-295 mosm/kg	280-295 mmol/kg
O04	U	-- (cf. F13)	CSUBSTC	38-1400 mosm/kg	38-1400 mmol/kg
O05	S	OSMOLARITY(see I23)			
O06	(B)Ercs	OSMOTIC PRESSURE REACTION(FRAGILITY)	ARB PRESSURE	0.40-0.45 arb un	arb un
O07	dU	OXALATE(as OXALIC ACID DIHYDRATE)	AMCS	mg/d	mmol
O08	dU	17-OXOSTEROID(see KO1)			
O09	B	OXYGEN(CAPACITY)	CSUBSTC	16-24 %	7.1-11 mmol/L
O10	aB	-- (CONTENT)	CSUBSTC	15-23 %	6.7-10 mmol/L
O11	vB	-- (CONTENT)	CSUBSTC	10-16 %	4.5-7.1 mmol/L
O12	Gas(aB eq)	-- (pO2)	PART PRESSURE	75-100 mmHg	10-13 kPa
O13	fPt	OXYGEN ABSORBED(BMR)	REL SUB RATE		--
O14	(aB)Hgb	OXYHEMOGLOBIN(OXYGEN SATURATION)	MASSFR	96-100 %	0.96-1.00 --
O15	(vB)Hgb	-- (OXYGEN SATURATION)	MASSFR	60-85 %	0.60-0.85 --
P01		pH(see H23,etc.).	LOG CSUBSTC		
P02	Pt	PHENOLSULFONPHTHALEIN TEST(PSP)	REL AMCS	%	--
P03	S	PHENYLALANINE	CSUBSTC	0-3 mg/dL	0-200 mmol/L

Reference table of quantities (component, specimen, kind-of-property, conventional value/unit, SI value/unit):

Code	Spec.	Component	Property	Conv. value	Conv. unit	SI value	SI unit
P04	U	PHENYLPYRUVATE	ARB MASSC	0	arb un	0	arb un
P05	dU	PHOSPHATE(P), INORGANIC	AMCS	0.9-1.3	g/d	30-42	mmol/L
P06	s	-- (P)	CSUBSTC	3.0-4.5	mg/dL	0.97-1.5	mmol/L
P07	s	-- (P), NONESTERIFIED, INORGANIC	CSUBSTC	9-16	mg/dL	3-5.2	mmol/L
P08	Pt	PHOSPHOLIPID(P)	MASSC				
P09	Pt	PLASMA VOLUME	VOL		mL.(L)		L
P10	Pt	--	VOLCONT	35-53	mL/kg	35-53	mL/kg
P11	B	PLATELETS	NUMC	150000-350000	/uL	150-350	10^9/L
P12	dU	PORPHOBILINOGEN	AMCS	0-0.2	mg/dL	0-9	umol/L
P13	U	--	CSUBSTC				umol/L
P14	dU	PORPHYRINS(cf. C44,U16)	ARB MASSC		arb un		arb un
P15	U	POTASSIUM ION	AMCS	25-100	meq/d	640-2560	mmol/L
P16	Ercs	--	CSUBSTC		meq/L		mmol/L
P17	P,S	--	CSUBSTC	3.5-5.0	meq/L	3.5-5.0	mmol/L
P18	dU	PREGNANEDIOL	AMCS		mg/d		umol
P19	dU	PREGNANETRIOL	AMCS	-2.5	mg/d	-7.4	umol
P20	s	PROGESTERONE	CSUBSTC		ug/dL		umol/L
P21	dU	PROTEIN	MASS	10-150	mg/d	10-150	mg
P22	ll	--	MASSC		mg/dL		mg
P23	s	--	MASSC	6.0-8.4	g/dL	60-84	g/L
P24	Sf(Lumb)	PROTEIN	MASSC	15-45	mg/dL	0.15-0.45	g/L
P25	Sf(Cist)	-- (cf. G04)	MASSC	15-25	mg/dL	0.15-0.25	g/L
P26	Sf(Vent)	--	MASSC	5-15	mg/dL	0.05-0.15	g/L
P27	U	--	ARB MASSC		arb un		arb un
P28	(B)Ercs	PROTEIN, BENCE JONES	ARB MASSC	0	arb un		arb un
P29		PROTEIN BOUND IODINE(PBI; see I21)					
P30	(B)Ercs	PROTOPORPHYRIN	CSUBSTC		ug/dL		umol/L
P31	Pt	PULSE	FREQUENCY	50-80	/min	0.8-1	Hz
P32	P	PYRUVATE	CSUBSTC	1.0-2.0	mg/dL	0.11-0.23	mmol/L
P33	B	--	CSUBSTC		mg/dL		mmol/L
P34	B	--	CSUBSTC	0-0.11	meq/L	0-0.11	mmol/L
P35	Pt	--	CSUBSTC				
R01	Pt	RECTUM	CELSIUS TEMP	36-38	°C.(°F)	36-38	°C
R02	Pt	--	THERMODYNAMIC	36-38	°C		K
R03	Pt	RESPIRATION	FREQUENCY	10-20	/min	0.2-0.3	Hz
R04	B	RETICULOCYTES	NUMC	25000-75000	/uL	25-75	10^9/L
R05	(B)Ercs	--	NUMFR	0.5-1.5	%	5-15	10^{-3}
S01	U	SACCHARIDE (see G29,G30)	CSUBSTC				
S02	P,S	SALICYLATE	CSUBSTC	0	mg/dL	0	mmol/L
S03	B	SEDIMENTATION REACTION(ESR)	LENGTH	0-20	mm	0-20	arb un
S04		SEROTONIN(see H30)					
S05		SIDEROPHYLIN(see T10)					
S06	dU	SODIUM ION	AMCS	130-260	meq/d	130-260	mmol
S07	U	--	CSUBSTC		meq/L		mmol/L
S08	P,S	--	CSUBSTC	135-145	meq/L	135-145	mmol/L
S09	P	SOMATOTROPINE(GROWTH HORMONE)	CSUBSTC		ng/mL		nmol/L
S10	U	SOLIDS(cf. S11,S12)	MASS DENS	30-70	g/L	30-70	g/L
S11	U	SPECIFIC GRAVITY	REL DENS	1.003-1.030	--	1.003-1.030	kg/L
S12	U	--	CSUBSTC	1.003-1.030	--		--
S13	S	SULFATE, INORGANIC(S)	CSUBSTC	0.8-1.2	mg/dL		mmol/L
S14	Gastr secr	SPECIFIC GRAVITY	MASS DENS	1.006-1.009	kg/L	1.006-1.009	kg/L

Table 1 (continued)

#	SYSTEM	CHEMICAL SUBSTANCE, BIOLOGIC ENTITY, PHYSIOLOGIC DESCRIPTOR, OR PHYSICAL QUANTITY	MEASURABLE PROPERTY[1] (KIND OF QUANTITY) (see abbrev. list)	REFERENCE VALUES[2] AND UNITS — CONVENTIONAL	SI (PROPOSED)
S15	Sm	SPERM	NUMC	60-150 10⁶/mL	60-150 10⁹/L
S16	(Sm)Sperm	SPERM, MOTILE	NUMFR	80- %	0.80-
S17	(Sm)Sperm	SPERM, NORMAL APPEARANCE	NUMFR	80-90 %	0.80-0.90
S18	P,S	SULFONAMIDE(SPECIFY COMPOUND)	CSUBSTC	0-0 mg/dL	mmol/L
S19	(B)Hgb	SULFHEMOGLOBIN	MASSFR	%	--
S20	B	-- (Fe)	MASSC	g/dL	umol/L
S21	Pt	SULFOBROMOPHTHALEIN TEST(BSP)	REL AMCS	%	--
T01	dU	TESTOSTERONE	AMCS	ug/d	umol
T02	S	--	CSUBSTC	ng/dL	nmol/L
T03	S	--			
T04	(fPt)S	THROMBOCYTES(see P11)	ARB MASSC	arb un	arb un
T05	S	THYMOL REACTION	MASSC		
T06	S	THYROID-STIMULATING HORMONE(TSH)	CSUBSTC	0-2.0 U/L	nmol/L
T07	S	THYROXINE(T4), TOTAL	CSUBSTC	4.4-9.9 ug/dL	57-130 nmol/L
T08	P,S	-- (T4), FREE	CSUBSTC	1.0-2.1 ng/dL	13-27 pmol/L
T09	S	TOCOPHEROLS(5.8-DIMETHYLTOCOL)	MASSC	mg/dL	umol/L
T10	S	TRANSFERRIN	CSUBSTC	mg/dL	g/L
T11	P,S	TRANSFERRIN, TOTAL(Fe BIND. CAP.; cf. I28)	CSUBSTC	250-410 ug/dL	45-73 umol/L
T12	P	TRANSFERRIN, UNSATURATED(Fe BIND. CAP.)	CSUBSTC	175-355 ug/dL	31.3-63.6 umol/L
T13	(fPt)S	TRIACYLGLYCEROL LIPASE(ATOXYL R RESISTANT)	CATC[3]	mU/mL	ukat/L
T14	(fPt)S	-- (QUININIUM RESISTANT)	CATC[3]		
T15	P,S	TRIGLYCERIDE(as GLYCEROL)	CSUBSTC	mg/dL	mmol/L
T16	P,S	TRIGLYCERIDES(as GLYCEROL TRIOLEATE)	CSUBSTC	40-150 mg/dL	0.45-1.69 mmol/L
T17	P	TRIIODOTHYRONINE(T3)	CSUBSTC	100-300 ng/dL	2-5 nmol/L
T18	(fPt)S	TRIIODOTHYRONINE REACTION	ARB MASSC	0.9-1.1 arb un	0.9-1.1 arb un
T19	F	TRYPSIN	CATC[3]	arb un	arb un
U01	dU	URATE(URIC ACID)	AMCS	mg/d	mmol
U02	(U)Stone	-- (URIC ACID)	ARB MASSC	arb un	arb un
U03	U	-- (URIC ACID)	CSUBSTC	mg/dL	mmol/L
U04	U	-- (URIC ACID)	CSUBSTC	3-7 mg/dL	200-400 umol/L
U05	B	UREA	CSUBSTC	21-43 mg/dL	3.5-7.2 mmol/L
U06	P,S	--	CSUBSTC	17-55 mg/dL	2.8-9.2 mmol/L
U07	dU	UREA NITROGEN(N)	AMCS	g/d	mmol
U08	U	-- (N)	CSUBSTC	g/dL	mol/L
U09	P,S	-- (BUN)	CSUBSTC	8-25 mg/dL	6-18 mmol/L
U10	U	UROBILIN	ARB MASSC	arb un	arb un
U11	dF	UROBILINOGEN	CSUBSTC	40-300 mg/d	70-500 umol
U12	(fPt)U	-- (EHRLICH'S REACTION)	ARB MASSC	arb un	arb un
U13	U(timed)	-- (EHRLICH'S REACTION)	ARB MASS	0-1.0 U/2h	arb un
U14	dU	--	AMCS	0-4.0 mg/d	0-6.7 umol

Code	Specimen	Analyte	Kind of quantity	Conventional ref.	Conv. unit	SI ref.	SI unit
U15	(B)Ercs	UROPORPHYRINS	CSUBSTC		ug/dL		nmol/L
U16	dU	UROPORPHYRINS I+III	CSUBSTC	10-30	ug/d	12-36	nmol
VO1	P,S	VITAMIN A(RETINOL)	AMCS	15-60	ug/dL	0.52-2.1	umol/L
VO2		VITAMIN B12 (see C61)	CSUBSTC				
VO3		VITAMIN C(see A40)					
VO4	(dPt)U	VOLUME	VOL	50-100	mL/d		L
VO5	Gastr secr	-- (FASTING STATE)	VOL	2-5	mL	0.050-0.100	L
VO6	Sm	--	VOL		ml	0.002-0.005	L
ZO1	dU	ZINC II	CSUBSTC	150-1200	ug/d	2.29-18	umol
ZO2	P,S	--	CSUBSTC	75-120	ug/dL	11-18	umol/L

LIST OF ABBREVIATIONS

1. KIND OF QUANTITY

AMCS........AMOUNT OF CHEMICAL SUBSTANCE
ARB.........ARBITRARY
CATAMT......CATALYTIC AMOUNT
CATC........CATALYTIC CONCENTRATION
CSUBSTC.....CHEMICAL SUBSTANCE CONCENTRATION
CSUBSTFR....CHEMICAL SUBSTANCE FRACTION
DENS........DENSITY
DIFF........DIFFERENCE
GLOB........GLOBULIN
LOG.........LOGARITHM
MASSC.......MASS CONCENTRATION
MASSFR......MASS FRACTION
MOLFR.......MOLE FRACTION
NUMC........NUMBER CONCENTRATION
NUMFR.......NUMBER FRACTION
PART........PARTIAL
REL.........RELATIVE
SUB.........SUBSTANCE
TEMP........TEMPERATURE
VOL.........VOLUME
VOLCONT.....VOLUME CONTENT
VOLFR.......VOLUME FRACTION

2. SYSTEM AND SYSTEM MODIFIER

a...........ARTERIAL
B...........BLOOD
d...........DAY, 24 h (as prefix)
Cist........CISTERNAL
eq..........EQUILIBRATED
Ercs(s).....ERYTHROCYTE(S)
f...........FASTING(as prefix)
F...........FECES
Hbg.........HEMOGLOBIN
Lkc(s)......LEUKOCYTE(S)
Lumb........LUMBAR
P...........PLASMA
Pt..........PATIENT
S...........SERUM
Sf..........SPINAL FLUID
U...........URINE
Vent........VENTRICULAR

3. OTHER

arb un......ARBITRARY UNIT
BIND........BINDING
CAP.........CAPICITY
CLEAR.......CLEARANCE

1   The measurable property (kind of quantity) listed corresponds to the proposed SI unit. For use with conventional units, substitute "mass" for "chemical substance".

2   Conventional reference values are taken directly or with modification from the sources cited.[1,2,29-33] Reference values in SI units are calculated from the corresponding conventional data and from published conversion factors.[3-5] The number of significant figures in each of the SI datum is taken from the corresponding conventional datum. In both groups, nonsignificant zeros are italicized.

3   For enzyme activity, the reference values are only illustrative. Results are very dependent on details of the method. For conventional units, values are from Conn[1] and Scully et al.[2] Although the proposed SI unit of activity, katal, is not widely accepted, reference values are included for completeness. Data in SI units are calculated using the conversion factors of Lippert and Lehmann.[4]

4   For hemoglobin and related compounds, the SI unit is based on the number of moles of monomer or, equivalently, on the number of moles of iron. The advantages of SI units for hemoglobin is somewhat lessened because hemoglobin does not always react as a multiple or submultiple of the number of molecules present.[34,35]

5   The structure and nomenclature usage follows NCCLS, C11-CR, Clinical Chemistry Committee Report on Quantities and Units: SI National Committee for Clinical Laboratory Standards, Villanova, Penna., 1981.

REFERENCES

1. Conn, R. B., Normal laboratory values of clinical importance, in *Textbook of Medicine*, Beeson, P. B. and McDermott, W., Eds., W. B. Saunders, Philadelphia, 1975, 1844.
2. Scully, R. E., McNeely, B. U., and Goldabini, J. J. ,Eds., Case records of the Massachusetts General Hospital, *N. Engl. J. Med.*, 298, 34, 1978.
3. Lehmann, H. P., SI units, *Crit. Rev. Clin. Lab. Sci.*, 147, 1979.
4. Lippert, H. and Lehmann, H. P., *SI Units in Medicine*, Urban and Schwarzenberg, Baltimore, 1978.
5. WHO, *The SI for the Health Professions*, 30th World Health Assembly, World Health Organization, Geneva, 1977.
6. Wilding, P., unpublished data, 1979.
7. Migeon, C. J., Tyler, F. H., Mahoney, J. P., Florentin, A. A., et al., Diurnal variations of plasma levels of 17-OH corticoids, *J. Clin. Endocrinol. Metabol.*, 16, 622, 1956.
8. Ramsay, W. N. M., Plasma Iron, in *Advances in Clinical Chemistry*, Vol. 1, Sobotka, H. and Steward, C. P., Eds., Academic Press, New York, 1958, 4.
9. Finlayson, D. C., Dagher, F. J., and Vandam, L. D., Diurnal variations in blood volume of man, *J. Surg. Res,.* 4, 286, 1964.
10. Hakulinen, A., Urinary excretion of vanilmandelic acid of children in normal and certain pathological conditions, *Acta Pediatr. Scand. Suppl.*, 212, 1, 1971.
11. Jarrett, R. J. and Keen, H., Further observations on the diurnal variation in oral glucose tolerance, *Br. Med. J.*, 4, 334, 1970.
12. Stamm, D., Tageschwankingen der Normalberiche diagnostisch wichtiger Blutbestandteile, *Verh. Dtsch. Ges. Inn. Med.*, 73, 982, 1967.
13. Studlar, M., Hammerl, H., Nebosis, G., and Pichler, O., Tagesprofile der freien Feltsauren, des freien Glycerins, der Triglyceride und des Blutzuckers bei Stoffwechselgesunden, *Klin. Wochenschr.*, 48, 238, 1970.
14. VonEuler, U. S., Hellner-Björkman, S., and Orwén, I., Diurnal variations in the excretion of free and conjugated nonadrenaline and adrenaline in urine from healthy subjects, *Acta Physiol. Scand.* 33(Suppl. 118), 10, 1955.
15. Annino, J. S. and Relman, A. S., The effect of eating on some of the clinically important chemical constituents of the blood, *Am. J. Clin. Pathol.*, 31, 155, 1959.
16. Caraway, W. T., Chemical and diagnostic specificity of laboratory tests, *Am. J. Clin. Pathol.*, 37, 445, 1962.
17. Young, D. S., Pestaner, L. C., and Gibberman, V., Effects of drugs on clinical laboratory tests, *Clin. Chem.*, 21, 1D, 1975.
18. Fawcett, J. K. and Wynn, V., Effects of posture on plasma volume and some blood constituents, *J. Clin. Pathol.*, 13, 304, 1960.
19. Thompson, W. O., Thompson, P. K., and Dailey, M. E., The effect of posture upon the composition and volume of the blood in man, *J. Clin. Invest.*, 5, 573, 1928.
20. Tombridge, T. L., Effect of posture on hematology results, *Am. J. Clin. Pathol.*, 49, 491, 1968.
21. Waterfield, R. L., The effect of posture on the volume of the leg, *J. Physiol.*, 72, 121, 1931.
22. Paloheimo, J. A., Seasonal variations of blood and plasma volumes in healthy men, *J. Clin. Lab. Invest.*, 15, 563, 1963.
23. Powsner, E. R. and Raeside, D. E., *Nuclear Medicine*, Grune & Stratton, New York, 1971, 324.
24. Anderson, E. T., Passovoy, M., and Trobaugh, F. E., Jr., Quantitation of gastrointestinal bleeding by use of a large volume scintillation detector, *J. Nucl. Med.*, 7, 612, 1966.
25. Kaplan, S. A., Yuceoglu, A. M., and Strauss, J., Chemical microanalyses: analysis of capillary and venous blood, *Pediatrics*, 24, 270, 1959.
26. Foster, G. L., Studies on carbohydrate metabolism. I. Some comparisons of blood sugar concentrations in venous blood and in finger blood, *J. Biol. Chem.*, 55, 291, 1923.
27. Feusner, J. H., Behrens, J. A., Detter, J. C., and Cullen, T. C., Platelet counts in capillary blood, *Am. J. Clin. Pathol.*, 72, 410, 1979.
28. Moss, G. and Staunton, C., Bloodflow, needle size and hemolyses — examining an old wives' tale, *N. Engl. J. Med.*, 282, 967, 1970.
29. Bartzack, C., unpublished data, 1976.
30. Borkowski, A. and Muquardt, C., Human chorionic gonadotropin in the plasma of normal, nonpregnant subjects, *N. Engl. J. Med.*, 301, 298, 1979.
31. McQueen, M. J., unpublished data, 1978.
32. Peart, W. S., Arterial hypertension, in *Textbook of Medicine*, Beeson, P. B. and McDermott, W., Eds., W. B. Saunders, Philadelphia, 1975, 981.
33. Stahl, T. J., Radioimmunoassay and the hormones of thyroid function, *Semin. Nucl. Med.*, 5, 221, 1975.

34. Dijkhuizen, P., Buursma, A., Fongers, T. M. E., Gerding, A. M., Oeseburg, B., and Zijlstra, W. G., The oxygen binding capacity of human haemoglobin, *Pflugers Arch.,* 369, 223, 1977.
35. ICSH, IFCC, WAPS, Recommendation for use of SI in clinical laboratory measurement, *Z. Klin. Chem. Klin. Biochem.,* 11, 93, 1973.

CHEMICAL NOMENCLATURE

Anne McCann

GENERAL REMARKS

A short discussion of nomenclature is included here to provide some assistance in resolving the problems research workers, chemists, and clinicians have in identifying chemical names. For instance, an unfamiliar name can often be found to have the same identity as another, more familiar name. If two names are recognized to be the same compound, a choice must be made between them. When using chemical names in scientific manuscripts intended for publication, one should consider alternative names carefully. Often a preferred name has been designated by a scientific body. These preferred names should be located and used when appropriate. The purpose of this section is to suggest how to find information about names from bibliographic sources.

Typically, the names with which one deals are surrogates, or merely fragments. The full names of most chemical compounds are long and complex, denoting what atoms are present in the molecule, how these atoms are arranged, and what structural features characterize the molecule as a member of one or more chemical classes. Knowing the full name of the compound, the chemist can draw a representation of its structure or, conversely, can devise its exact name from its known structure.

Because of the complexity of these names, widespread use of simpler names, usually known as common names, trivial names, nonproprietary names, or established names, has occurred. Common names, for instance, are often applied by industries which manufacture and use the chemical compounds and these may be the only names recognized by their users. Hence, such names appear widely in manufacturers' literature and product labels and, subsequently, find their way into professional journals, reviews, and books.

Many users of chemical names are not sufficiently familiar with chemical structures to be able to relate them to the names. However, these users should be aware that the only way to identify a chemical compound uniquely and to differentiate it from other chemical compounds is by this structure and the name that represents it. Published guides and information files have been developed which compile the information about chemical names. Through such tools, one can identify a chemical compound uniquely from a given name, differentiate it from any other chemical compound, and link it with synonyms, including systematic chemical names, trivial names, trade names, and with its Chemical Abstracts Service (CAS) Registry Number.

SYSTEMATIC NOMENCLATURE

Central to the process is systematic chemical nomenclature,[1-4] which is devised by chemists in order to define chemical compounds unambiguously. The International Union of Pure and Applied Chemistry (IUPAC) codifies the rules of chemical nomenclature world-wide. In the U.S., the American Chemical Society represents the views of American chemists, working through nomenclature committees in the Society's divisions or accepting the recommendations of independent experts.

The chemical name authority most widely used in the U.S. is the Chemical Abstracts Chemical Substance Index.[5] Chemical Abstracts has a continuing responsibility for developing and updating systematic nomenclature used in indexing its abstracts. At its own discretion, Chemical Abstracts revises its rules in the light of IUPAC recommendations as they become available.

For searchers of chemical names, Chemical Abstracts publishes two indexes, the Chemical Substance Index, which includes entries for chemical compounds and their derivatives, mixtures, enzymes, alloys, and other materials bearing CAS Registry Numbers, and the General Subject Index,[6] which covers classes of substances, biological substances, and incompletely defined materials. These indexes are cumulated over 5-year periods. Changes in names are introduced with each cumulation.

Each cumulative index has a companion volume, the Index Guide,[7] which is a comprehensive published source for chemical names. It contains the cross references extracted from the Chemical Substance and General Subject Indexes and, thus, links the commonly used names for chemical substances and terms used for general subjects to the proper Chemical Abstracts index headings. The Index Guide also provides valuable instructions for using the indexes, such as rules for selection of index names, to which is appended a current and historical bibliography of nomenclature of chemical substances.[8] The latest edition is the 1977 Index Guide, to accompany the 10th Cumulative Index (1977 to 1981). It is cumulated and supplemented annually and will be replaced by a Cumulated Index Guide at the end of the collective index period.

CAS REGISTRY SYSTEM

Since 1965, the CAS has maintained the CAS Registry System, a computer-based system that identifies chemical substances by their structural diagrams and assigns to each new substance a unique CAS Registry Number. This number has no chemical significance, but it is a permanent and concise identifier for the systematic name. The use of the CAS Registry Number is encouraged when the substance name is to appear in a technical publication. The CAS Registry Number is also often used in place of the index name in computer-processed index files.

ON-LINE DICTIONARIES

Information searchers today are finding that a quick and simple way to locate chemical names and synonyms is by using on-line computerized chemical dictionary files. Several of these are maintained as adjuncts to bibliographic data bases now commonly available.

One of these files. CHEMLINE,[9] is part of the National Library of Medicine's MEDLARS system. CHEMLINE contains records for more than 500,000 chemicals identified by CAS Registry Number in the TOXLINE data base. Records for chemical compounds include systematic chemical names used in the 8th and 9th Chemical Abstracts Collective Indexes, CAS Registry Numbers, molecular formulas, and numerous synonyms. All were extracted from the CAS Registry Nomenclature File, plus certain ring structure information from another CAS file.

Two other on-line chemical dictionaries also having records extracted from the CAS Registry File are Lockheed's CHEMNAME[10] and System Development Corporation's (SDC) CHEMDEX.[11] Since many medical and hospital libraries perform literature searches on the Lockheed and SDC computerized data bases, the on-line dictionaries are suggested as useful and accessible alternative sources of nomenclature information.

STANDARD NAMES AND DESIGNATIONS

Common and trivial names for chemical substances have tended to multiply, particularly for chemicals actively involved in commerce. However, professional and trade associations in several specialized areas have undertaken efforts to designate preferred names for chemical substances and to promote their uses.

Drugs and Cosmetics

The Food, Drug, and Cosmetic Act requires the use of an established name for any drug marketed in the U.S. Names of official (U.S. Pharmacopeia and National Formulary) drugs are recognized, and the authority to select new names has been delegated to the Commissioner of Food and Drugs. Drug names established since the law took effect are listed in the Code of Federal Regulations[12] with their chemical names but, unfortunately, their CAS Registry Numbers are not listed. Actual work on the standardization of nonproprietary names, however, is performed by the U.S. Adopted Names (USAN) Council. Drug manufacturers can propose nonproprietary names to the USAN Council. USAN's cumulative list, published annually, contains all the USAN names with their chemical names and CAS Registry Numbers.[13]

The Food, Drug and Cosmetic Act does not require established names for cosmetic ingredients. However, the cosmetic industry has undertaken a voluntary effort to standardize cosmetic ingredient names which appear on cosmetic package labels. By soliciting information from manufacturers, the Cosmetics, Toiletries and Fragrance Association was able to compile a dictionary of cosmetic ingredient descriptions, including chemical names, synonyms, and CAS Registry Numbers for their preferred names.[14]

Dyes

Color names are standardized in the Colour Index (CI), which is published by Great Britain's Society of Dyers and Colourists with technical assistance from the American Society of Textile Chemists and Colorists.[15] The Colour Index gives chemical names for colors and synonyms, including commercial names, but not CAS Registry Numbers. However, the Chemical Abstracts Index Guide includes as cross references CI names for most classes of dyes, thus locating their systematic names and CAS Registry Numbers.

Enzymes

Preferred names for enzymes are those recommended by the Nomenclature Committee of the International Union of Biochemistry (IUB).[16] These preferred names are assigned CAS Registry Numbers and indexed in the Chemical Abstracts Chemical Substance Index. The IUB Enzyme Commission class codes (E.C. codes) are found in the Chemical Abstracts Index Guide as references to the preferred names.

Biochemical Names

Extensive efforts have been made by the IUB to standardize the nomenclature for amino acids, vitamins, steroids, carbohydrates, and other classes of biochemical substances. Their nomenclature publications are listed in Reference 8. Some of them have been conveniently reprinted by the American Society of Biological Chemists.[17] Chemical Abstracts has applied systematic organic nomenclature to biochemical compounds, particularly specific derivatives of compounds having biologic activity.

Pesticides

Accepted common names must be used on the ingredient statements of pesticide product labels. Accepted names are generally those adopted by the American National Standards Institute or by the former Interdepartmental Committee on Pest Control.

The common or trivial name may be used alone if it is well known. If not, it must be followed by the chemical name. If no common name has been established, the chemical name alone is used. These names are compiled in an Environmental Protection Agency publication, Acceptable Common Names.[18] It includes chemical names, trade names, and CAS Registry Numbers for the common ingredient names, as well as a bibliography of publications useful in pesticide nomenclature and identification.

CONCLUSIONS

Although the emphasis in this discussion has been on nomenclature practices and on compendia of specialized and standardized names, the searcher should always be prepared to start his search in one of the basic chemical dictionaries that cover the full range of chemical names, such as the *Merck Index*,[19] the *Condensed Chemical Dictionary*,[20] and *Hackh's Chemical Dictionary*.[21] Much information can also be obtained from published data collections such as the Registry of Toxic Effects of Chemical Substances[22] and from the enormous Toxic Substances Control Act Chemical Inventory,[23] which brought many new chemicals into the CAS Registry System. Finally, trade name compendia, such as *Gardner's Handbook of Chemical Synonyms and Trade Names*,[24] can give needed support.

In the publication of papers requiring chemical names, writers and editors should insist upon the use of Chemical Abstracts systematized names or the standard names recommended by recognized professional bodies of the subject discipline area as appropriate. When standardized names are used, they should be attributed to the authorizing body. Authors should also use the CAS Registry Numbers in conjunction with both types of names. Practitioners should encourage more standardization efforts by professional and industrial organizations. This kind of standardization should be followed by interdisciplinary standardization, where appropriate.

REFERENCES

1. Fletcher, J. H., Dermer, O. C., and Fox, R. B., Eds., Nomenclature or Organic Compounds, Principles and Practice (*Adv. Chem. Ser.*, No. 126), American Chemical Society, Washington, D.C., 1974.
2. Weast, R. C., Ed., Nomenclature of inorganic chemistry, in *Handbook of Chemistry and Physics*, 60th ed., CRC Press, Boca Raton, Fla., 1979, B28.
3. Weast, R. C., Ed., Definitive rules for nomenclature of organic compounds, in *Handbook of Chemistry and Physics*, 60th ed., CRC Press, Boca Raton, Fla., 1979, C1.
4. Todd, R. C., Ed., Nomenclature of organic compounds, in *Pharmaceutical Handbook*, 18th ed., Pharmaceutical Society of Great Britain, Pharmaceutical Press, London, 1970, 442.
5. *Chemical Abstracts, 9th Collective Index*, Vols. 76-85, 1972-1976. *Chemical Substance Index*, 25 Vols., Chemical Abstracts Service, Columbus, Ohio.
6. *Chemical Abstracts, 9th Collective Index*, Vols. 76-85, 1972-1976. *General Subject Index*, 10 Vols., Chemical Abstracts Service, Columbus, Ohio.
7. *Chemical Abstracts, 1977 Index Guide*, Chemical Abstracts Service, Columbus, Ohio.
8. Selective Bibliography of Nomenclature of Chemical Substances, in *Chemical Abstracts, 1977 Index Guide*, 2101I.
9. CHEMLINE; Online Chemical Dictionary. MEDLARS subscription agreement handled by National Library of Medicine, Bethesda, Md.; billing handled by National Technical Information Services, Springfield, Va.
10. CHEMNAME; Online Chemical Dictionary. Subscription handled by Lockheed Dialog Information Retrieval Service, Palo Alto, Calif.

11. CHEMDEX; Online Chemical Dictionary. Subscription handled by System Development Corp., Santa Monica, Calif.
12. Drugs; Official Names, 21 *Code of Federal Regulations,* 299.20.
13. Griffiths, M. C., Ed., *USAN and the USP Dictionary of Drug Names 1982,* U.S. Pharmacopeial Convention, Rockville, Md: 1979.
14. Estrin, N. F., Ed., *CTFA Cosmetic Ingredient Dictionary,* 3rd ed., The Cosmetic, Toiletry and Fragrance Association, Washington, D.C., in press.
15. Colour Index, 3rd ed., Society of Dyers and Colourants, Bradford, England and American Society of Textile Chemists and Colorants, Research Triangle Park, N.C., 1971, (6 vols.).
16. International Union of Biochemistry, Nomenclature Committee, *Enzyme Nomenclature 1978,* Academic Press, New York, 1979.
17. *Collected Tentative Rules and Regulations of the Commission on Biochemical Nomenclature, IUPAC-IUB, and Related Documents,* 2nd ed., Reprinted by the American Society of Biological Chemists, Bethesda, Md., 1975.
18. Blalock, C. R., Shaughnessy, J. A., Johnson, D. E., and Caswell, R. L., Acceptable Common Names and Chemical Names for the Ingredient Statement on Pesticide Labels, 4th ed., Office of Pesticide Programs, U.S. Environmental Protection Agency, National Technical Information Services, Springfield, Va., 1979.
19. Windholz, M., Ed., *The Merck Index,* 9th ed., Merck & Co., Rahway, N.J., 1976.
20. Hawley, G. G., Ed., *The Condensed Chemical Dictionary,* 10th ed., Van Nostrand Reinhold, 1981.
21. Grant, J., Ed., *Hackh's Chemical Dictionary,* 4th ed., McGraw-Hill, New York, 1969.
22. Lewis, R. J. and Tatken, R. L., Ed., Registry of Toxic Effects of Chemical Substances, 1979 ed., Center for Disease Control, National Institute for Occupational Safety and Health, U.S. Public Health Service, Cincinnati, Ohio, 1980.
23. Toxic Substances Control Act (TSCA) Chemical Substance Inventory, Office of Toxic Substances, U.S. Environmental Protection Agency, Washington, 1979.
24. Gardner, W., Cooke, E. I., and Cooke, P. W. I., *Handbook of Chemical Synonyms and Trade Names,* 8th ed., CRC Press, Cleveland, 1978.

Physical and Chemical Data

PHYSICAL AND CHEMICAL DATA

Herbert K. Naito (Editor) with the assistance of the following contributors: Julius Kerkay, Patrick A. Roche, and Robert Rej

INTRODUCTION

This section contains data commonly and frequently used in the Clinical Laboratory. The choice has been highly selective to permit rapid retrieval, and it focuses on conversion factors; common buffers, indicators, and pH primary standards; boiling points of solvents; chemical and physical data on water, mercury, various liquids, amino acids, and fatty acids; flame emission spectra; transmission filters; as well as miscellaneous data pertinent to the clinical environment. On the other hand, most thermodynamic data have been omitted. In many instances, additional and more detailed information is found elsewhere in this handbook or by consulting the last table of this section. The latter provides a cross-referenced index of information compiled in five handbooks devoted primarily to chemical and physical data.

Table 1
PERIODIC TABLE OF THE ELEMENTS

KEY TO CHART

50	+2
Sn	+4
118.69	
18-18-4	

Atomic Number → ; Symbol → ; Atomic Weight → ; Oxidation States → ; Electron Configuration →

Transition Elements / Group 8

Z	Symbol	Oxidation States	Atomic Weight	Electron Configuration	Orbit
1	H	+1 −1	1.0079	1	K
2	He	0	4.00260	2	K
3	Li	+1	6.94	2-1	K-L
4	Be	+2	9.01218	2-2	K-L
5	B	+3	10.81	2-3	K-L
6	C	+2 +4 −4	12.011	2-4	K-L
7	N	+1 +2 +3 +4 +5 −2 −3	14.0067	2-5	K-L
8	O	−2	15.9994	2-6	K-L
9	F	−1	18.998403	2-7	K-L
10	Ne	0	20.179	2-8	K-L
11	Na	+1	22.98977	2-8-1	K-L-M
12	Mg	+2	24.305	2-8-2	K-L-M
13	Al	+3	26.98154	2-8-3	K-L-M
14	Si	+2 +4 −4	28.0855	2-8-4	K-L-M
15	P	+3 +5 −3	30.97376	2-8-5	K-L-M
16	S	+4 +6 −2	32.06	2-8-6	K-L-M
17	Cl	+1 +5 +7 −1	35.453	2-8-7	K-L-M
18	Ar	0	39.948	2-8-8	K-L-M
19	K	+1	39.0983	-8-8-1	-L-M-N
20	Ca	+2	40.08	-8-8-2	-L-M-N
21	Sc	+3	44.9559	-8-9-2	-L-M-N
22	Ti	+2 +3 +4	47.90	-8-10-2	-L-M-N
23	V	+2 +3 +4 +5	50.9415	-8-11-2	-L-M-N
24	Cr	+2 +3 +6	51.996	-8-13-1	-L-M-N
25	Mn	+2 +3 +4 +6 +7	54.9380	-8-13-2	-L-M-N
26	Fe	+2 +3	55.847	-8-14-2	-L-M-N
27	Co	+2 +3	58.9332	-8-15-2	-L-M-N
28	Ni	+2 +3	58.71	-8-16-2	-L-M-N
29	Cu	+1 +2	63.546	-8-18-1	-L-M-N
30	Zn	+2	65.38	-8-18-2	-L-M-N
31	Ga	+3	69.735	-8-18-3	-L-M-N
32	Ge	+2 +4	72.59	-8-18-4	-L-M-N
33	As	+3 +5 −3	74.9216	-8-18-5	-L-M-N
34	Se	+4 +6 −2	78.96	-8-18-6	-L-M-N
35	Br	+1 +5 −1	79.904	-8-18-7	-L-M-N
36	Kr	0	83.80	-8-18-8	-L-M-N
37	Rb	+1	85.4678	-18-8-1	-M-N-O
38	Sr	+2	87.62	-18-8-2	-M-N-O
39	Y	+3	88.9059	-18-9-2	-M-N-O
40	Zr	+4	91.22	-18-10-2	-M-N-O
41	Nb	+3 +5	92.9064	-18-12-1	-M-N-O
42	Mo	+6	95.94	-18-13-1	-M-N-O
43	Tc	+4 +6 +7	98.9062	-18-13-2	-M-N-O
44	Ru	+3	101.07	-18-15-1	-M-N-O
45	Rh	+3	102.9055	-18-16-1	-M-N-O
46	Pd	+2 +4	106.4	-18-18-0	-M-N-O
47	Ag	+1	107.868	-18-18-1	-M-N-O
48	Cd	+2	112.41	-18-18-2	-M-N-O
49	In	+3	114.82	-18-18-3	-M-N-O
50	Sn	+2 +4	118.69	-18-18-4	-M-N-O
51	Sb	+3 +5 −3	121.75	-18-18-5	-M-N-O
52	Te	+4 +6 −2	127.60	-18-18-6	-M-N-O
53	I	+1 +5 +7 −1	126.9045	-18-18-7	-M-N-O
54	Xe	0	131.30	-18-18-8	-M-N-O
55	Cs	+1	132.9054	-18-8-1	-N-O-P
56	Ba	+2	137.33	-18-8-2	-N-O-P
57*	La	+3	138.9055	-18-9-2	-N-O-P
72	Hf	+4	178.49	-32-10-2	-N-O-P
73	Ta	+5	180.9479	-32-11-2	-N-O-P
74	W	+6	183.85	-32-12-2	-N-O-P
75	Re	+4 +6 +7	186.207	-32-13-2	-N-O-P
76	Os	+3 +4 +6 +8	190.2	-32-14-2	-N-O-P
77	Ir	+3 +4	192.22	-32-15-2	-N-O-P
78	Pt	+2 +4	195.09	-32-16-2	-N-O-P
79	Au	+1 +3	196.9665	-32-18-1	-N-O-P
80	Hg	+1 +2	200.59	-32-18-2	-N-O-P
81	Tl	+1 +3	204.37	-32-18-3	-N-O-P
82	Pb	+2 +4	207.2	-32-18-4	-N-O-P
83	Bi	+3 +5	208.9804	-32-18-5	-N-O-P
84	Po	+2 +4	(209)	-32-18-6	-N-O-P
85	At	−1	(210)	-32-18-7	-N-O-P
86	Rn	0	(222)	-32-18-8	-N-O-P
87	Fr	+1	(223)	-18-8-1	O P Q
88	Ra	+2	226.0254	-18-8-2	O P Q
89**	Ac	+3	(227)	-18-9-2	O P Q
104	Rf	+4	(260)	-32-10-2	O P Q
105	Ha		(260)	-32-11-2	O P Q
106			(263)	-32-12-2	O P Q

*Lanthanides

Z	Symbol	Oxidation States	Atomic Weight	Electron Configuration	Orbit
58	Ce	+3 +4	140.12	-20-8-2	N O P
59	Pr	+3	140.9077	-21-8-2	N O P
60	Nd	+3	144.24	-22-8-2	N O P
61	Pm	+3	(145)	-23-8-2	N O P
62	Sm	+2 +3	150.4	-24-8-2	N O P
63	Eu	+2 +3	151.96	-25-8-2	N O P
64	Gd	+3	157.25	-25-9-2	N O P
65	Tb	+3	158.9254	-27-8-2	N O P
66	Dy	+3	162.50	-28-8-2	N O P
67	Ho	+3	164.9304	-29-8-2	N O P
68	Er	+3	167.26	-30-8-2	N O P
69	Tm	+3	168.9342	-31-8-2	N O P
70	Yb	+2 +3	173.04	-32-8-2	N O P
71	Lu	+3	174.967 ± 0.003	-32-9-2	N O P

**Actinides

Z	Symbol	Oxidation States	Atomic Weight	Electron Configuration	Orbit
90	Th	+4	232.0381	-18-10-2	O P Q
91	Pa	+5 +4	231.0359	-20-9-2	O P Q
92	U	+3 +4 +5 +6	238.029	-21-9-2	O P Q
93	Np	+3 +4 +5 +6	237.0482	-22-9-2	O P Q
94	Pu	+3 +4 +5 +6	(244)	-24-8-2	O P Q
95	Am	+3 +4 +5 +6	(243)	-25-8-2	O P Q
96	Cm	+3	(247)	-25-9-2	O P Q
97	Bk	+3 +4	(247)	-27-8-2	O P Q
98	Cf	+3	(251)	-28-8-2	O P Q
99	Es		(254)	-29-8-2	O P Q
100	Fm		(257)	-30-8-2	O P Q
101	Md		(258)	-31-8-2	O P Q
102	No		(259)	-32-8-2	O P Q
103	Lr		(260)	-32-9-2	O P Q

Numbers in parentheses are mass numbers of most stable isotope of that element

From Weast, R. C., Ed., *CRC Handbook of Chemistry and Physics*, 61st ed., CRC Press, Boca Raton, Fla., 1980. With permission.

Table 2
GREEK ALPHABET

Alpha	A	α	Nu	N	ν
Beta	B	β	Xi	Ξ	ξ
Gamma	Γ	γ	Omicron	O	o
Delta	Δ	δ	Pi	Π	π
Epsilon	E	ε	Rho	P	ϱ
Zeta	Z	ζ	Sigma	Σ	σ
Eta	H	η	Tau	T	τ
Theta	Θ	θ	Upsilon	Y	υ
Iota	I	ι	Phi	Φ	ϕ
Kappa	K	$\varkappa$	Chi	X	χ
Lambda	Λ	λ	Psi	Ψ	ψ
Mu	M	μ	Omega	Ω	ω

Table 3
NUMERICAL PREFIXES

½-Hemi	20-Elcosa	41-Hentetraconta
1-Mono	21-Henicosa	42-Dotetraconta
1½-Sesqui	22-Docosa	43-Tritetraconta
2-Di or Bi	23-Tricosa	44-Tetratetraconta
3-Tri	24-Tetracosa	45-Pentatetraconta
4-Tetra	25-Pentacosa	46-Hexatetraconta
5-Penta	26-Hexacosa	47-Heptatetraconta
6-Hexa	27-Heptacosa	48-Octatetraconta
7-Hepta	28-Octacosa	49-Nonatetraconta
8-Octa	29-Nonacosa	50-Pentaconta
9-Nona or Ennea	30-Triaconta	51-Henpentaconta
10-Deca or Deka	31-Hentriaconta	52-Dopentaconta
11-Undeca or Henadeca	32-Dotriaconta	53-Tripentaconta
12-Dodeca	33-Tritriaconta	54-Tetrapentaconta
13-Trideca	34-Tetratriaconta	55-Pentapentaconta
14-Tetradeca	35-Pentatriaconta	56-Hexapentaconta
15-Pentadeca	36-Hexatriaconta	57-Hentapentaconta
16-Hexadeca	37-Heptatriaconta	58-Octapentaconta
17-Heptadeca	38-Octatriaconta	59-Nonapentaconta
18-Octadeca	39-Nonatriaconta	60-Hexaconta
19-Nonadeca	40-Tetraconta	70-Heptaconta
		80-Octaconta
		90-Nonaconta

Table 4
CONVERSION FACTORS FOR VARIOUS UNITS

To convert	Into	Multiply by
Atmosphere (normal = 760 torr)	Bar	1.013 250
	Centimeter of mercury (0°C)	76.0
	Dyne per square centimeter	$1.013\ 250 \times 10^6$
	Foot of water (4°C)	33.90
	Foot of water (39.2°F)	33.899
	Gram per square centimeter	1 033.227
	Inch of mercury (0°C)	29.921 26
	Kilogram-force per square meter	$1.033\ 227 \times 10^4$
	Millimeter of mercury (0°C)*	760
	Newton per square meter	$1.013\ 250 \times 10^5$

Table 4 (continued)
CONVERSION FACTORS FOR VARIOUS UNITS

To convert	Into	Multiply by
Atmosphere	Pascal	$1.013\ 25 \times 10^5$
	Pound-force per square foot	2 116.22
	Pound-force per square inch	14.695 95
	Ton (short) per square foot	1.058 1
	Ton (short) per square inch	0.007 348 2
Atmosphere (technical = 1 kgf/cm²)	Pascal	$9.806\ 650 \times 10^4$
Cubic centimeter	Board foot	4.2377×10^{-4}
	Bucket (British, dry)	$5.499\ 3 \times 10^{-5}$
	Bushel (British)	$2.749\ 6 \times 10^{-5}$
	Cubic foot	$3.531\ 5 \times 10^{-5}$
	Cubic inch	0.061 023
	Cubic meter*	1×10^{-6}
	Cubic yard (U.S.)	$1.307\ 9 \times 10^{-6}$
	Drachm (British, fluid)	0.281 57
	Dram (U.S., fluid)	0.270 53
	Gallon (British)	$2.199\ 7 \times 10^{-4}$
	Gallon (U.S., liquid)	$2.641\ 7 \times 10^{-4}$
	Gill (British)	0.007 039 0
	Gill (U.S.)	0.008 453 5
	Liter*	0.001
	Milliliter*	1
	Minim (British)	16.894
	Minim (U.S.)	16.231
	Ounce (British, fluid)	0.035 196
	Ounce (U.S., fluid)	0.033 814
	Pint (U.S., dry)	0.001 816 2
	Pint (U.S., liquid)	0.002 113 4
	Quart (British, liquid)	$8.798\ 8 \times 10^{-4}$
	Quart (U.S., dry)	$9.080\ 8 \times 10^{-4}$
	Quart (U.S., liquid)	0.001 056 7
Curie	Disintegration per second*	3.70×10^{10}
Dalton	Gram	$1.649\ 8 \times 10^{-24}$
	Mass of atom of oxygen	1/16
Degree Celsius (°C)	Kelvin*	1
	See Table 2-6 for t°C	
Degree Fahrenheit (°F)	Kelvin*	5/9
Dyne per centimeter	Erg per square centimeter*	1
	Milligram per inch	2.590 1
	Milligram per millimeter	0.101 97
	Pascal*	10
Dram (apothecaries or troy)	Dram (avoirdupois)	2.194 286
	Grain	60
	Gram	3.887 935
	Ounce (avoirdupois)	0.137 142 9
	Ounce (troy)	0.125
	Pennyweight	2.5
	Pound (avoirdupois)	0.008 571 429
	Pound (troy)	0.010 416 67
	Scruple	3
Dram (U.S., fluid)	Cubic centimeter	3.696 691 195
	Cubic inch	0.225 570
	Gallon (U.S.)	$9.765\ 6 \times 10^{-4}$
	Gill (U.S.)	0.031 25
	Milliliter	3.696 61
	Minim (U.S.)	60

Table 4 (continued)
CONVERSION FACTORS FOR VARIOUS UNITS

To convert	Into	Multiply by
	Ounce (fluid)	0.125
	Pint (U.S., liquid)	0.007 812 5
	Quart (U.S., liquid)	0.003 906 25
Dyne per square centimeter	Atmosphere	$9.869\ 23 \times 10^{-7}$
	Bar*	1×10^{-6}
	Centimeter of water (4°C)	0.001 019 74
	Gram-force per square centimeter	0.001 019 716
	Inch of mercury (32°F)	$2.953\ 0 \times 10^{-5}$
	Inch of water (39.2°F)	$4.014\ 8 \times 10^{-4}$
	Kilogram-force per square meter	0.010 197 16
	Millimeter of mercury (0°C)	$7.500\ 617 \times 10^{-4}$
	Newton per square meter*	10
	Pound per square foot	0.002 088 6
	Pound per square inch	$1.450\ 4 \times 10^{-5}$
Electronvolt	Centimeter^{-1}	8 065.48
	Erg	$1.602\ 0 \times 10^{-12}$
	Joule	$1.602\ 19 \times 10^{-19}$
	Kilocalorie per mole	23.060 9
	Kilojoule per mole	9.648 46
	Mass unit	$1.073\ 7 \times 10^{-9}$
	Rydberg unit of energy	0.073 86
Foot (U.S.)	Centimeter	30.480 060 95
	Chain (Gunter's)	0.151 515
	Chain (Ramsden's)	0.01
	Fathom	1/6
	Foot (British)	1.000 002 8
	Furlong	0.001 515 15
	Hand (U.S.)	3
	Inch	12
	Link (Gunter's)	1.515 15
	Link (Ramsden's)	1
	Meter	0.304 8
	Mile (nautical)	$1.645\ 8 \times 10^{-4}$
	Mile (statute)	$1.893\ 939 \times 10^{-4}$
	Rod	0.060 606 1
	Yard	1/3
Foot-candle	Lumen per square foot*	1
	Lumen per square meter	10.763 910
	Lux	10.763 910
	Phot	0.001 076 4
Foot-lambert	Candela per square centimeter	$3.426\ 3 \times 10^{-4}$
	Candela per square foot	0.318 31
	Candela per square meter	3.426 3
	Lambert	$1.076\ 39 \times 10^{-3}$
	Meter-lambert	10.76
Foot-pound	Btu	0.001 285 4
	Calorie, gram	0.323 89
	Cubic foot-atmosphere	$4.725\ 3 \times 10^{-4}$
	Erg	$1.355\ 82 \times 10^{7}$
	Joule	1.355 82
Gallon (U.S., liquid)	Acre-foot	$3.068\ 89 \times 10^{-6}$
	Barrel (U.S., liquid)	0.031 746
	Cubic centimeter	3 785.434
	Cubic foot	0.133 680 5
	Cubic inch	231
	Cubic meter	0.003 785 434

Table 4 (continued)
CONVERSION FACTORS FOR VARIOUS UNITS

To convert	Into	Multiply by
	Cubic yard	0.004 951 1
	Gallon (British)	0.832 68
	Gill (U.S.)	32
	Liter	3.785 33
	Minim	61 440
	Ounce (U.S., fluid)	128
	Pint (U.S., liquid)	8
	Pound (avoirdupois) of water at 60°F	8.337 0
	Quart (U.S., liquid)	4
Gauss	Electrostatic cgs unit of magnetic flux density	$3.335\ 85 \times 10^{-4}$
Grain	Carat (metric)	0.324 0
	Dram (avoirdupois)	0.036 571 43
	Dram (troy)	0.016 667
	Dyne	63.545 3
	Gram*	0.064 798 91
	Ounce (avoirdupois)	0.002 285 7
	Ounce (troy)	0.002 083 3
	Pennyweight	0.041 666 7
	Pound (avoirdupois)	$1/7000 = 1.428\ 6 \times 10^{-4}$
	Pound (troy)	$1/5760 = 1.736\ 1 \times 10^{-4}$
	Scruple	0.05
Gram	Avogram	$6.022\ 8 \times 10^{23}$
	Carat (metric)	5
	Dram (avoirdupois)	0.564 383
	Dram (troy)	0.257 206
	Dyne	980.665
	Grain	15.4324
	Joule per centimeter	9.807×10^{-5}
	Milligram*	1 000
	Ounce (avoirdupois)	0.035 273 96
	Ounce (troy)	0.032 150 7
Inch (U.S.)	Angstrom unit*	2.540×10^{8}
	Centimeter*	2.540
	Chain (Gunter's)	0.001 262 63
	Chain (Ramsden's)	$8.333\ 333 \times 10^{-4}$
	Foot (U.S.)	1/12
	Inch (British)	1.000 002 8
	Link (Gunter's)	0.126 263
	Link (Ramsden's)	0.083 333 3
	Mil	1000
	Mile	$1.578\ 28 \times 10^{-5}$
	Millimeter*	25.40
	Pica (printers' type)	6
	Point (printers' type)	72
	Rod	0.005 050 51
	Yard	1/36
Inch of mercury at 32°F	Atmosphere	0.033 421 05
	Dyne per square centimeter	33 863.88
	Foot of water (39.2°F)	1.132 99
	Kilogram per square meter	345.315 5
	Pascal (N · m^{-2})	3 386.388
	Pound per square foot	70.727
	Pound per square inch	0.491 154 1
Inch of mercury at 60°F	Pascal (N · m^{-2})	3 376.85

Table 4 (continued)
CONVERSION FACTORS FOR VARIOUS UNITS

To convert	Into	Multiply by
Inch of water at 39.2°F (4°C)	Atmosphere	0.002 458 3
	Dyne per square centimeter	2 490.82
	Gram per square centimeter	2.539 9
	Inch of mercury at 32°F	0.073 554
	Kilogram per square meter	25.399
	Ounce per square inch	0.578 02
	Pascal $(N \cdot m^{-2})$	249.082
	Pound per square foot	5.202 2
	Pound per square inch	0.036 126
Inch of water at 60°F	Pascal	248.84
Inch per °F	Centimeter per °C	4.572 0
Joule	Btu (I.T.)	$9.478\ 172 \times 10^{-4}$
	Calorie	0.239 005 7
	Calorie (I.T.)	0.238 845 9
	Centigrade heat unit (chu)	$5.269\ 8 \times 10^{-4}$
	Cubic foot-atmosphere	$3.485\ 3 \times 10^{-4}$
	Cubic foot × (pound per square inch)	$5.121\ 960 \times 10^{-3}$
	Erg*	1×10^{7}
	Foot-pound	0.737 56
	Foot-poundal	23.730
	Gram-centimeter	10 197.2
	Horsepower-hour	$3.725\ 08 \times 10^{-7}$
	Joule (int)(1948)	0.999 835
	Kilogram-meter	0.101 972
	Kilowatt-hour	$2.777\ 78 \times 10^{-7}$
	Liter-atmosphere	0.009 869 233
	Watt-hour	$2.777\ 78 \times 10^{-4}$
	Watt-second*	1
Joule per ampere-hour	Joule per abcoulomb	0.002 777 8
	Joule per statcoulomb	$9.263\ 6 \times 10^{-14}$
Joule per centimeter	Dyne*	1×10^{7}
	Gram	1.020×10^{4}
	Joule per meter*	100
	Newton*	100
	Pound	22.48
	Poundal	723.3
Joule per coulomb	Joule per abcoulomb	10
Joule per coulomb per °F	Joule per coulomb per °C	1.8
Joule per °C	Btu (60°F) per °F	$5.267\ 9 \times 10^{-4}$
	Calorie per °C	0.239 005 7
Joule per electron	Joule per abcoulomb	$6.281\ 1 \times 10^{19}$
Lambert	Candela per square centimeter	0.318 310
	Candela per square inch	2.054
	Candela per square meter*	$(1/\pi) \times 10^{4}$
	Foot-lambert	929.03
	Lumen emitted per square centimeter of a perfectly diffusing surface*	1
	Lumen per square foot	929.03
Liter (before 1964)	Cubic centimeter	1 000.028
Liter (after 1964, 1 dm³)	Bushel (British)	0.027 496
	Bushel (U.S.)	0.028 377
	Cubic centimeter*	1 000
	Cubic foot	0.035 315
	Cubic inch	61.024
	Cubic meter*	1×10^{-3}

Table 4 (continued)
CONVERSION FACTORS FOR VARIOUS UNITS

To convert	Into	Multiply by
	Cubic yard	0.001 308
	Dram (U.S., fluid)	270.52
	Gallon (British)	0.219 970
	Gallon (U.S.)	0.264 170
	Gill (British)	7.039 033
	Gill (U.S.)	8.453 44
	Milliliter*	1 000
	Minim	16 230.6
	Ounce (British)	35.196
	Ounce (U.S., fluid)	33.813
	Peck (British)	0.109 984
	Peck (U.S.)	0.113 509
	Pint (British, liquid)	1.7598
	Pint (U.S., dry)	1.816 141
	Pint (U.S., liquid)	2.113 30
	Quart (British, liquid)	0.879 88
	Quart (U.S., dry)	0.908 071
	Quart (U.S., liquid)	1.056 789
Liter per minute	Cubic foot per hour	2.118 9
	Cubic foot per second	5.886×10^{-4}
	Gallon (U.S.) per second	0.004 403
Lumen	Candle power (spherical)	0.079 58
	Watt (maximum visible radiation)	0.001 47
Lumen per square centimeter	Lambert*	1
	Phot*	1
Lumen per square meter	Lumen per square foot (or foot-candle)	0.092 902
	Meter-candle*	1
	Phot*	1×10^{-4}
Lux	Foot-candle	0.092 902
	Lumen per square meter*	1
	Meter-candle*	1
	Milliphot	0.1
	Nox*	10^3
Mass of oxygen atom	Dalton	16
Mass unit	Electronvolt	$9.313 8 \times 10^8$
	Erg	$1.492 1 \times 10^{-5}$
Maxwell	Electrostatic cgs unit of magnetic flux	$3.335 6 \times 10^{-11}$
	Gauss-square centimeter*	1
	Kiloline	0.001
	Weber*	1×10^{-8}
Maxwell per square centimeter	Electromagnetic cgs unit of magnetic flux density	1
	Electrostatic cgs unit of magnetic flux density	$3.335 9 \times 10^{-11}$
	Maxwell per square inch	6.451 6
Meter	Angstrom unit*	1×10^{10}
	Cable length	0.004 556 6
	Chain (Gunter's)	0.049 709 6
	Chain (Ramsden's)	0.032 808 3
	Fathom	0.546 806
	Foot (British)	3.280 843
	Foot (U.S.)	3.280 833
	Furlong	0.004 970 96
	Inch (British)	39.370 113
	Inch (U.S.)	39.37

Table 4 (continued)
CONVERSION FACTORS FOR VARIOUS UNITS

To convert	Into	Multiply by
	Kilometer*	0.001
	Link (Gunter's)	4.970 960
	Link (Ramsden's)	3.280 83
	Micron (micrometer)*	1×10^{-6}
	Mile (nautical)	$5.395\ 93 \times 10^{-4}$
	Mile (statute)	$6.213\ 7 \times 10^{-4}$
	Nanometer (millimicron)*	1×10^{-9}
	Pied (French foot	3.078 34
	Rod (U.S.)	0.198 838
	Yard (British)	1.093 614
	Yard (U.S.)	1.093 611
Microfarad	Abfarad*	1×10^{-15}
	Farad*	1×10^{-6}
	Statfarad	$8.987\ 76 \times 10^{5}$
Micron (micrometer)	Angstrom unit*	1×10^{-4}
	Centimeter*	1×10^{-4}
	Inch	3.937×10^{-5}
	Mil	0.039 37
Milligram	Carat (metric)	0.005
	Dram (avoirdupois)	$5.643\ 83 \times 10^{-4}$
	Dram (troy)	$2.572\ 06 \times 10^{-4}$
	Grain	0.015 432
	Ounce (avoirdupois)	$3.527\ 396 \times 10^{-5}$
	Ounce (troy)	$3.215\ 074 \times 10^{-5}$
	Pennyweight	$6.430\ 15 \times 10^{-4}$
	Pound (avoirdupois)	$2.204\ 62 \times 10^{-6}$
	Pound (troy)	$2.679\ 23 \times 10^{-6}$
	Scruple	$7.716\ 18 \times 10^{-4}$
Millimeter of mercury at 0°C	Atmosphere	0.001 315 789 5
	Bar	0.001 333 22
	Dyne per square centimeter	1 333.224
	Foot of water at 39.2°F	0.044 604
	Gram per square centimeter	1.359 509 9
	Kilogram per square meter	13.595 099
	Pascal $(N \cdot m^{-2})$	133.322 4
	Pound per square foot	2.784 5
	Pound per square inch	0.019 336 78
Millimicron	See nanometer	
Nanometer (millimicron)	Angstrom unit*	10
	Centimeter*	1×10^{-7}
	Inch (U.S.)	3.937×10^{-8}
	Meter*	1×10^{-9}
Newton	Dyne*	1×10^{5}
Newton per square meter	See Pascal	
Ounce (troy or apothecaries)	Dram (avoirdupois)	17.554 28
	Dram (troy)	8
	Grain	480
	Gram	31.103 481
	Ounce (avoirdupois)	1.097 14
	Pennyweight	20
	Pound (avoirdupois)	0.068 571 43
	Pound (troy)	1/12
	Scruple	24
	Ton (short)	$3.428\ 57 \times 10^{-5}$
Ounce (U.S., fluid)	Cubic centimeter	29.573 7
	Cubic inch	1.804 69
	Dram (fluid)	8
	Gallon (U.S.)	1/128

Table 4 (continued)
CONVERSION FACTORS FOR VARIOUS UNITS

To convert	Into	Multiply by
	Gill (U.S.)	0.25
	Liter	0.029 573
	Minim (U.S.)	480
	Pint (U.S., liquid)	1/16
	Quart (U.S., liquid)	0.031 25
Part per million (ppm)	Grain per gallon (British)	0.070 155
	Grain per gallon (U.S.)	0.058 417
	Microgram per gram*	1
	Microgram per milliliter*	1
	Milligram per kilogram*	1
	Milligram per liter*	1
	Pound per million gallons (U.S.)	8.345 2
Pascal	Atmosphere (760 torr)	$9.869\ 233 \times 10^{-6}$
	Bar*	1×10^{-5}
	Barye*	10
	Centimeter of mercury (0°C)	$7.500\ 64 \times 10^{-4}$
	Centimeter of water (4°C)	0.010 197 44
	Dyne per square centimeter*	0.1
	Foot of water at 39.2°F	$3.345\ 623 \times 10^{-4}$
	Gram-force per centimeter	0.010 197 16
	Inch of mercury at 32°F	$2.952\ 998 \times 10^{-4}$
	Inch of mercury at 60°F	$2.961\ 340 \times 10^{-4}$
	Inch of water at 39.2°F	$4.014\ 742 \times 10^{-3}$
	Inch of water at 60°F	$4.018\ 6 \times 10^{-3}$
	Kip per square inch	$1.450\ 377 \times 10^{-7}$
	Millimeter of mercury (0°C)	$7.500\ 615 \times 10^{-3}$
	Poundal per square foot	0.671 969
	Pound-force per square foot	0.020 885 43
	Pound-force per square inch	$1.450\ 377 \times 10^{-4}$
	Torr	$7.500\ 638 \times 10^{-3}$
Pascal-second	Poise*	10
Square centimeter	Circular mil	$1.973\ 50 \times 10^{5}$
	Circular millimeter	127.32
	Square foot (U.S.)	0.001 076 387
	Square inch (U.S.)	0.154 999 69
	Square meter*	1×10^{-4}
	Square mil	$1.549\ 997 \times 10^{5}$
	Square rod	$3.953\ 67 \times 10^{-6}$
	Square yard	$1.196\ 0 \times 10^{-4}$
Volt	Abvolt*	1×10^{8}
	Statvolt	0.003 333 56
Volt (int) (1948)	Volt	1.000 330
Yard (U.S.)	Centimeter	91.44
	Chain (Gunter's)	0.045 454 5
	Chain (Ramsden's)	0.03
	Fathom	0.5
	Foot	3
	Furlong	1/220
	Inch	36
	Link (Gunter's)	4.545 45
	Meter	0.914 4
	Mile (statute)	$5.681\ 82 \times 10^{-4}$
	Rod	0.181 818

Note: Relations which are exact are indicated by an asterisk (*); I.T. — International Steam Table.

Adapted from Dean, J. A., Ed., *Lange's Handbook of Chemistry*, 12th ed., McGraw-Hill, New York, N.Y., 1979, 2-10. With permission.

Table 5
CONVERSION TABLE FOR UNITS OF MEASUREMENTS: SUMMARY OF U.S. AND BRITISH SYSTEMS

Where there is a difference between US and British units, the equivalents of the latter are given in *italics*

Length

Inches (in)	Feet (ft)	Yards (yd)	Rods (rd)	Miles (mi)
1	$^1/_{12} =$ 0.083 333	$^1/_{36} =$ 0.027 778	$^1/_{198} =$ 0.005 050 51	$^1/_{63360} =$ 0.000 015 782 8
12	1	$^1/_3 =$ 0.333 333	$^2/_{33} =$ 0.060 606 1	$^1/_{5280} =$ 0.000 189 394
36	3	1	$^2/_{11} =$ 0.181 818	$^1/_{1760} =$ 0.000 568 182
198	16.5	5.5	1	$^1/_{320} =$ 0.003 125
63,360	5280	1760	320	1

Area

Square inches (sq.in)	Square feet (sq.ft)	Square yards (sq.yd)	Square rods (sq.rd)	Acres (A.)	Square miles (sq.mi)
1	$^1/_{144} =$ 0.006 944 4	$^1/_{1296} =$ 0.000 771 6	$^1/_{39204} = 2.550$ 8 $\times 10^{-5}$	1.594 2 $\times 10^{-7}$	2.490 41 $\times 10^{-10}$
144	1	$^1/_9 =$ 0.111 1	$^4/_{1089} = 0.003$ 673 1	$^1/_{43560} = 2.295$ 68 $\times 10^{-5}$	3,587 01 $\times 10^{-8}$
1,296	9	1	$^4/_{121} = 0.033$ 057 85	$^1/_{4840} = 2.066$ 12 $\times 10^{-4}$	3.228 31 $\times 10^{-7}$
39,204	272.25	30.25	1	$^1/_{160} = 0.006$ 25	9.765 625 $\times 10^{-6}$
6,272,640	43,560	4 840	160	1	$^1/_{640} = 0.001$ 5625
4.0154 $\times 10^9$	27,878,400	3,097,600	102,400	640	1

Volume

Cubic inches (cu.in)	Cubic feet (cu.ft)	Cubic yards (cu.yd)
1	$^1/_{1728} = 0.000$ 578 704	$^1/_{46656} = 2.143$ 347 $\times 10^{-5}$
1,728	1	$^1/_{27} = 0.037$ 037 0
46,656	27	1

Capacity — Liquid Measure

Gills (gi)	Pints (pt)	Quarts (qt)	Gallons (gal)	Cubic inches (cu. in)	
1	$^1/_4 = 0.25$	$^1/_8 = 0.125$	$^1/_{32} = 0.031$ 25	7.218 75	*(8.669)*
4	1	$^1/_2 = 0.5$	$^1/_8 = 0.125$	28.875	*(34.678)*
8	2	1	$^1/_4 = 0.25$	57.75	*(69.355)*
32	8	4	1	231	*(277.421)*

Apothecaries' Fluid Measure

Minims (min or ♍)	Fluid drams (Brit.: fluid drachms) (fl.dr or ʒ fl.)	Fluid ounces (fl.oz or ʒ fl.)	Pints (pt)
1	$^1/_{60} =$ 0.016 667	$^1/_{480} = 0.002$ 083 3	$(^1/_{7680} = 0.000$ 130 21 $^1/_{9600} = 0.000$ 104 17*)*
60	1	$^1/_8 = 0.125$	$(^1/_{128} = 0.007$ 812 5 $^1/_{160} = 0.006$ 25*)*
480	8	1	$(^1/_{16} = 0.062$ 5 $^1/_{20} = 0.05$*)*
7680 *(9600)*	128 *(160)*	16 *(20)*	$(^1/_{16} = 0.062$ 5
			1

Dry Measure

Pints (pt)	Quarts (qt)	Pecks (pk)	Bushels (bu)	Cubic inches (cu. in)	
1	$^1/_2 =$ 0.5	$^1/_{16} = 0.062$ 5	$^1/_{64} = 0.015$ 625	33.600 3	
2	1	$^1/_8 = 0.125$	$^1/_{32} = 0.031$ 25	67.200 6	
16	8	1	$^1/_4 = 0.25$	537.605	*(554.6)*
64	32	4	1	2150.42	*(2219.3)*

Mass — Avoirdupois (commercial)

Grains (gr)	Drams (dr.av.)	Ounces (oz.av.)	Pounds (lb.av.)	Tons (short) (sh.tn)
1	0.036 571 43	0.002 285 7	$^1/_{7000} =$ 0.000 142 86	7.1429 $\times 10^{-7}$
27.343 75	1	$^1/_{16} =$ 0.062 5	$^1/_{256} =$ 0.003 906 25	1.9531 $\times 10^{-6}$
437.5	16	1	$^1/_{16} =$ 0.062 5	$^1/_{32000} = 0.000$ 031 25
7000	256	16	1	$^1/_{2000} = 0.0005$
1,400,000	512,000	32,000	2000	1

Mass — Troy Weight

Grains (gr)	Pennyweights (dwt)	Ounces (oz.t.)	Pounds (lb.t.)
1	$^1/_{24} =$ 0.041 667	$^1/_{480} =$ 0.002 0833	$^1/_{5760} = 0.000$ 173 611 1
24	1	$^1/_{20} =$ 0.05	$^1/_{240} = 0.004$ 166 7
480	20	1	$^1/_{12} = 0.083$ 333
5760	240	12	1

Mass — Apothecaries' Weight

Grains (gr)	Scruples (Ɔ or s.ap.)	Drams (Brit: drachms) (ʒ or dr.ap.)	Ounces (ʒ or oz.ap.)	Pounds (lb.ap.)
1	$^1/_{20} =$ 0.05	$^1/_{60} =$ 0.016 667	$^1/_{480} =$ 0.002 083 3	$^1/_{5760} = 0.000$ 173 611 1
20	1	$^1/_3 =$ 0.333 333	$^1/_{24} =$ 0.041 667	$^1/_{288} = 0.003$ 472 2
60	3	1	$^1/_8 = 0.125$	$^1/_{96} = 0.010$ 416 7
480	24	8	1	$^1/_{12} = 0.083$ 333
5760	288	96	12	1

Conversion Table for Apothecaries' and Metric Weights for Use in Prescribing[1]

Grains	Milligrams	Grains	Milligrams	Grains	Milligrams	Grains	Milligrams
10	600	$^1/_2$	30	$^1/_{20}$	3	$^1/_{150}$ }	0.4
$7^1/_2$	450	$^2/_5$	25	$^1/_{25}$	2.5	$^1/_{160}$ }	
5	300	$^1/_3$	20	$^1/_{30}$	2	$^1/_{200}$	0.3
4	250	$^1/_4$	15	$^1/_{40}$	1.5	$^1/_{210}$	0.25
3	200	$^1/_5$	12.5	$^1/_{50}$	1.25	$^1/_{300}$ }	0.2
$2^1/_2$	150	$^1/_6$	10	$^1/_{60}$	1	$^1/_{320}$ }	
2	125	$^1/_8$	7.5	$^1/_{75}$	0.8	$^1/_{480}$	0.125
$1^1/_2$	100	$^1/_{10}$	6	$^1/_{100}$	0.6	$^1/_{600}$	0.1
1	60	$^1/_{12}$	5	$^1/_{120}$ }	0.5		
$^3/_4$	50	$^1/_{15}$	4	$^1/_{130}$ }			

[1] Approved by the British Medical Association and the Pharmaceutical Society of Great Britain. Cf. Editorial, *Brit. med. J.*, **2**, 794 (1960).

From Diem, K., Ed., *Documenta Geigy Scientific Tables*, 6th ed., Geigy Pharmaceuticals, Ardsley, N.Y., 1962, 201. With permission.

Table 6
CONVERSION FACTORS FOR LENGTH, AREA, AND VOLUME UNITS IN THE U.S. AND BRITISH (IMPERIAL)

Imperial	Units of length				Units of area				Units of volume			
	Imp. comm.	Imp. scientific	United States	Standardized	Imp. comm.	Imp. scientific	United States	Standardized	Imp. comm.	Imp. scientific	United States	Standardized
commercial	1	1.000 000 866	0.999 997 133	0.999 999 133	1	1.000 001 732	0.999 994 266	0.999 998 266	1	1.000 002 598	0.999 991 399	0.999 997 399
scientific ...	0.999 999 133	1	0.999 996 267	0.999 998 267	0.999 998 266	1	0.999 992 534	0.999 996 534	0.999 997 399	1	0.999 988 801	0.999 994 801
United States .	1.000 002 866	1.000 003 732	1	1.000 002 000	1.000 005 732	1.000 007 464	1	1.000 004 000	1.000 008 598	1.000 011 196	1	1.000 006 000
Standardized .	1.000 000 866	1.000 001 732	0.999 998 000	1	1.000 001 732	1.000 003 464	0.999 996 000	1	1.000 002 598	1.000 005 196	0.999 994 000	1

From Diem, K., Ed., *Documenta Geigy Scientific Tables*, 6th ed., Geigy Pharmaceuticals, Ardsley, N.Y., 1962, 203. With permission.

Table 7
CONVERSION FACTORS FOR LENGTH UNITS

To convert from a unit in col. **A** into any unit under **C**, multiply by the factor in col. **B** and by the power of 10 given in the appropriate col. under **C**

To convert from a unit in column **A** into any unit under **D**, multiply by the factor given in the appropriate column under **D**

A	B	nm	μm	mm	cm	dm	m	km	XU	Å	AU	pc	ly	mi (US)	yd (US)	ft (US)	in (US)
Nanometer (millimicron) nm(mμ)	1	×1	10^{-3}	10^{-6}	10^{-7}	10^{-8}	10^{-9}	10^{-12}	9.97984×10^5	1.000000×10	6.688775×10^{-21}	3.242810×10^{-26}	1.057023×10^{-25}	6.213699×10^{-13}	1.093611×10^{-9}	3.28083×10^{-9}	3.937×10^{-8}
Micrometer (micron) μm (μ)	1	$\times 10^3$	1	10^{-3}	10^{-4}	10^{-5}	10^{-6}	10^{-9}	9.97984×10^8	1.000000×10^4	6.688775×10^{-18}	3.242810×10^{-23}	1.057023×10^{-22}	6.213699×10^{-10}	1.093611×10^{-6}	3.28083×10^{-6}	3.937×10^{-5}
Millimeter mm	1	$\times 10^6$	10^3	1	10^{-1}	10^{-2}	10^{-3}	10^{-6}	9.97984×10^{11}	1.000000×10^7	6.688775×10^{-15}	3.242810×10^{-20}	1.057023×10^{-19}	6.213699×10^{-7}	1.093611×10^{-3}	3.28083×10^{-3}	3.937×10^{-2}
Centimeter cm	1	$\times 10^7$	10^4	10	1	10^{-1}	10^{-2}	10^{-5}	9.97984×10^{12}	1.000000×10^8	6.688775×10^{-14}	3.242810×10^{-19}	1.057023×10^{-18}	6.213699×10^{-6}	1.093611×10^{-2}	3.28083×10^{-2}	3.937×10^{-1}
Decimeter dm	1	$\times 10^8$	10^5	10^2	10	1	10^{-1}	10^{-4}	9.97984×10^{13}	1.000000×10^9	6.688775×10^{-13}	3.242810×10^{-18}	1.057023×10^{-17}	6.213699×10^{-5}	1.093611×10^{-1}	3.28083×10^{-1}	3.937×1
Meter m	1	$\times 10^9$	10^6	10^3	10^2	10	1	10^{-3}	9.97984×10^{14}	1.000000×10^{10}	6.688775×10^{-12}	3.242810×10^{-17}	1.057023×10^{-16}	6.213699×10^{-4}	1.093611×1	3.28083×1	3.937×10
Decameter dkm	1	$\times 10^{10}$	10^7	10^4	10^3	10^2	10	10^{-2}	9.97984×10^{15}	1.000000×10^{11}	6.688775×10^{-11}	3.242810×10^{-16}	1.057023×10^{-15}	6.213699×10^{-3}	1.093611×10	3.28083×10	3.937×10^2
Hectometer hm	1	$\times 10^{11}$	10^8	10^5	10^4	10^3	10^2	10^{-1}	9.97984×10^{16}	1.000000×10^{12}	6.688775×10^{-10}	3.242810×10^{-15}	1.057023×10^{-14}	6.213699×10^{-2}	1.093611×10^2	3.28083×10^2	3.937×10^3
Kilometer km	1	$\times 10^{12}$	10^9	10^6	10^5	10^4	10^3	1	9.97984×10^{17}	1.000000×10^{13}	6.688775×10^{-9}	3.242810×10^{-14}	1.057023×10^{-13}	6.213699×10^{-1}	1.093611×10^3	3.28083×10^3	3.937×10^4
X-unit (Siegbahn unit) XU	1.00202×10^{-4}	10^{-4}	10^{-7}	10^{-10}	10^{-11}	10^{-12}	10^{-13}	10^{-16}	$\times 1$	1.00202×10^{-3}	6.70229×10^{-25}	3.24936×10^{-30}	1.05916×10^{-29}	6.22625×10^{-17}	1.09582×10^{-13}	3.28746×10^{-13}	3.94495×10^{-12}
Ångström Å	1.000000×10^{-1}	10^{-1}	10^{-4}	10^{-7}	10^{-8}	10^{-9}	10^{-10}	10^{-13}	9.97984×10^2	$\times 1$	6.688775×10^{-23}	3.242810×10^{-28}	1.057023×10^{-27}	6.213699×10^{-14}	1.093611×10^{-11}	3.280833×10^{-10}	3.937000×10^{-9}
Astronomical unit AU	1.495042×10^{20}	10^{20}	10^{17}	10^{14}	10^{13}	10^{12}	10^{11}	10^8	1.49203×10^{24}	1.495042×10^{21}	$\times 1$	4.848137×10^{-6}	1.580294×10^{-5}	9.289742×10^9	1.634995×10^{11}	4.904984×10^{11}	5.885980×10^{12}
Siriometer	1.495042×10^{26}	10^{26}	10^{23}	10^{20}	10^{19}	10^{18}	10^{17}	10^{14}	1.49203×10^{30}	1.495042×10^{27}	$\times 10^6$	4.848137×1	1.580294×10	9.289742×10^{15}	1.634995×10^{17}	4.904984×10^{17}	5.885980×10^{18}
Parsec pc	3.083745×10^{25}	10^{25}	10^{22}	10^{19}	10^{18}	10^{17}	10^{16}	10^{13}	3.07753×10^{29}	3.083745×10^{26}	2.062648×10^5	$\times 1$	3.259591×1	1.916147×10^{13}	3.372418×10^{16}	1.011726×10^{17}	1.214071×10^{17}
Distance of Sirius	1.541873×10^{26}	10^{26}	10^{23}	10^{20}	10^{19}	10^{18}	10^{17}	10^{14}	1.53876×10^{30}	1.541873×10^{27}	1.031324×10^6	$\times 1$	1.629795×10	9.580734×10^{13}	1.686209×10^{17}	5.058628×10^{17}	6.070353×10^{17}
Light year ly	9.460530×10^{24}	10^{24}	10^{21}	10^{18}	10^{17}	10^{16}	10^{15}	10^{12}	9.44146×10^{28}	9.460530×10^{25}	6.327936×10^4	3.067870×10^{-1}	$\times 1$	5.878489×10^{12}	1.034614×10^{16}	3.103842×10^{16}	3.724611×10^{17}
Mil (US)	2.540005×10^4	10	10^{-2}	10^{-3}	10^{-3}	10^{-4}	10^{-5}	10^{-8}	2.53488×10^8	2.540005×10^5	1.698952×10^{-16}	8.236753×10^{-22}	2.684844×10^{-21}	1.5782×10^{-8}	2.7×10^{-5}	8.3×10^{-5}	1×10^{-3}
Mil (imp.)	2.539996×10^4	10	10^{-2}	10^{-3}	10^{-3}	10^{-4}	10^{-5}	10^{-8}	2.53488×10^8	2.539996×10^5	1.698946×10^{-16}	8.236723×10^{-22}	2.684834×10^{-21}	1.578277×10^{-8}	2.777767×10^{-5}	8.333302×10^{-5}	9.999963×10^{-4}
Inch (US) in	2.540005×10^7	10^4	10	1	10^{-1}	10^{-1}	10^{-2}	10^{-5}	2.53488×10^{11}	2.540005×10^8	1.698952×10^{-13}	8.236753×10^{-19}	2.684844×10^{-18}	1.5782×10^{-5}	2.7×10^{-2}	8.3×10^{-2}	$\times 1$
Inch (imp.)	2.539996×10^7	10^4	10	1	10^{-1}	10^{-1}	10^{-2}	10^{-5}	2.53488×10^{11}	2.539996×10^8	1.698946×10^{-13}	8.236723×10^{-19}	2.684834×10^{-18}	1.578277×10^{-5}	2.777767×10^{-2}	8.333302×10^{-2}	9.999963×10^{-1}
Foot (US) ft	3.048006×10^8	10^8	10^5	10^2	10	1	10^{-1}	10^{-4}	3.04186×10^{12}	3.048006×10^9	2.038743×10^{-12}	9.884104×10^{-18}	3.221813×10^{-17}	1.893×10^{-4}	3.3×10^{-1}	$\times 1$	1.2×10
Foot (imp.)	3.047995×10^8	10^8	10^5	10^2	10	1	10^{-1}	10^{-4}	3.04185×10^{12}	3.047995×10^9	2.038735×10^{-12}	9.884067×10^{-18}	3.221801×10^{-17}	1.893932×10^{-4}	3.333321×10^{-1}	9.999963×10^{-1}	1.199996×10
Yard (US) yd	9.144018×10^8	10^8	10^5	10^2	10^2	10	10^{-1}	10^{-4}	9.12558×10^{12}	9.144018×10^9	6.116228×10^{-12}	2.965231×10^{-17}	9.665440×10^{-17}	5.681×10^{-4}	$\times 1$	3×1	3.6×10
Yard (imp.)	9.143984×10^8	10^8	10^5	10^2	10^2	10	10^{-1}	10^{-4}	9.12555×10^{12}	9.143984×10^9	6.116206×10^{-12}	2.965220×10^{-17}	9.665404×10^{-17}	5.681797×10^{-4}	9.999963×10^{-1}	2.999992×1	3.599987×10
Mile (US) mi	1.609347×10^{12}	10^{12}	10^9	10^6	10^5	10^4	10^3	1	1.60610×10^{16}	1.609347×10^{13}	1.076456×10^{-8}	5.218807×10^{-14}	1.701117×10^{-13}	$\times 1$	1.76×10^3	5.28×10^3	6.336×10^4
Mile (imp.)	1.609341×10^{12}	10^{12}	10^9	10^6	10^5	10^4	10^3	1	1.60610×10^{16}	1.609341×10^{13}	1.076452×10^{-8}	5.218787×10^{-14}	1.701111×10^{-13}	9.999963×10^{-1}	1.759993×10^3	5.279980×10^3	6.335976×10^4
Nautical mile (int.) n. mi	1.852×10^{12}	10^{12}	10^9	10^6	10^5	10^4	10^3	1	1.84827×10^{16}	1.852000×10^{13}	1.238761×10^{-8}	6.005684×10^{-14}	1.957607×10^{-13}	1.150777×1	2.025368×10^3	6.076103×10^3	7.291324×10^4
Nautical mile (imp.)	1.853181×10^{12}	10^{12}	10^9	10^6	10^5	10^4	10^3	1	1.84944×10^{16}	1.853181×10^{13}	1.239551×10^{-8}	6.009513×10^{-14}	1.958855×10^{-13}	1.151511×1	2.026659×10^3	6.079977×10^3	7.295973×10^4

Note: The basic SI unit of length is the meter.

From Diem, K., Ed., *Documenta Geigy Scientific Tables*, 6th ed., Geigy Pharmaceuticals, Ardsley, N.Y., 1962, 203. With permission.

Table 8
CONVERSION FACTORS FOR AREA UNITS

To convert from a unit in column **A** into any unit under **C**, multiply by the factor in column **B** and by the power of 10 given in the appropriate column under **C**

To convert from a unit in column **A** into any unit under **D**, multiply by the factor given in the appropriate column under **D**

A		B	C						D					
			μm^2	mm^2	cm^2	dm^2	m^2	km^2	circ.inch (US)	in^2 (US)	ft^2 (US)	yd^2 (US)	acre (US)	mi^2 (US)
Square micrometer	μm^2	1	1	10^{-6}	10^{-8}	10^{-10}	10^{-12}	10^{-18}	1.973517×10^{-9}	1.549997×10^{-9}	1.076387×10^{-11}	1.195985×10^{-12}	2.471044×10^{-16}	3.861006×10^{-19}
Square millimeter	mm^2	1	10^{6}	1	10^{-2}	10^{-4}	10^{-6}	10^{-12}	1.973517×10^{-3}	1.549997×10^{-3}	1.076387×10^{-5}	1.195985×10^{-6}	2.471044×10^{-10}	3.861006×10^{-13}
Square centimeter	cm^2	1	10^{8}	10^{2}	1	10^{-2}	10^{-4}	10^{-10}	1.973517×10^{-1}	1.549997×10^{-1}	1.076387×10^{-3}	1.195985×10^{-4}	2.471044×10^{-8}	3.861006×10^{-11}
Square decimeter	dm^2	1	10^{10}	10^{4}	10^{2}	1	10^{-2}	10^{-8}	1.973517×10	1.549997×10	1.076387×10^{-1}	1.195985×10^{-2}	2.471044×10^{-6}	3.861006×10^{-9}
Square meter	m^2	1	10^{12}	10^{6}	10^{4}	10^{2}	1	10^{-6}	1.973517×10^{3}	1.549997×10^{3}	1.076387×10	1.195985×1	2.471044×10^{-4}	3.861006×10^{-7}
Square decameter (are)	dam^2 (a)	1	10^{14}	10^{8}	10^{6}	10^{4}	10^{2}	10^{-4}	1.973517×10^{5}	1.549997×10^{5}	1.076387×10^{3}	1.195985×10^{2}	2.471044×10^{-2}	3.861006×10^{-5}
Square hectometer (hectare)	hm^2 (ha)	1	10^{16}	10^{10}	10^{8}	10^{6}	10^{4}	10^{-2}	1.973517×10^{7}	1.549997×10^{7}	1.076387×10^{5}	1.195985×10^{4}	2.471044×1	3.861006×10^{-3}
Square kilometer	km^2	1	10^{18}	10^{12}	10^{10}	10^{8}	10^{6}	1	1.973517×10^{9}	1.549997×10^{9}	1.076387×10^{7}	1.195985×10^{6}	2.471044×10^{2}	3.861006×10^{-1}
Circular inch (US)	circ.in	5.067 095	10^{8}	10^{2}	1	10^{-2}	10^{-4}	10^{-10}	1×1	7.853982×10^{-1}	5.454154×10^{-3}	6.060171×10^{-4}	1.252101×10^{-7}	1.956409×10^{-10}
Circular inch (imp.)	circ.in	5.067 057	10^{8}	10^{2}	1	10^{-2}	10^{-4}	10^{-10}	9.999925×10^{-1}	7.853923×10^{-1}	5.454113×10^{-3}	6.060126×10^{-4}	1.252092×10^{-7}	1.956394×10^{-10}
Square inch (US)	sq.in or in^2	6.451 626	10^{8}	10^{2}	1	10^{-2}	10^{-4}	10^{-10}	1.273240×1	1×1	6.94×10^{-3}	7.716049×10^{-4}	1.594225×10^{-7}	2.490977×10^{-10}
Square inch (imp.)	sq.in or in^2	6.451 578	10^{8}	10^{2}	1	10^{-2}	10^{-4}	10^{-10}	1.273230×1	9.999925×10^{-1}	6.944393×10^{-3}	7.715992×10^{-4}	1.594213×10^{-7}	2.490958×10^{-10}
Square foot (US)	sq.ft or ft^2	9.290 341	10^{10}	10^{4}	10^{2}	1	10^{-2}	10^{-8}	1.833465×10^{2}	1.44×10^{2}	1×1	1.1×10^{-1}	2.295684×10^{-5}	3.587006×10^{-8}
Square foot (imp.)	sq.ft or ft^2	9.290 272	10^{10}	10^{4}	10^{2}	1	10^{-2}	10^{-8}	1.833451×10^{2}	1.439989×10^{2}	9.999925×10^{-1}	1.111103×10^{-1}	2.295667×10^{-5}	3.586980×10^{-8}
Square yard (US)	sq.yd or yd^2	8.361 307	10^{11}	10^{5}	10^{3}	10	10^{-1}	10^{-7}	1.650118×10^{3}	1.296×10^{3}	9×1	1×1	2.066116×10^{-4}	3.228306×10^{-7}
Square yard (imp.)	sq.yd or yd^2	8.361 245	10^{11}	10^{5}	10^{3}	10	10^{-1}	10^{-7}	1.650106×10^{3}	1.295990×10^{3}	8.999933×1	9.999925×10^{-1}	2.066101×10^{-4}	3.228282×10^{-7}
Acre	A. or ac (US)	4.046 873	10^{15}	10^{9}	10^{7}	10^{5}	10^{3}	10^{-3}	7.986573×10^{6}	6.27264×10^{6}	4.356×10^{4}	4.84×10^{3}	1×1	1.5625×10^{-3}
Acre	A. or ac (imp.)	4.046 842	10^{15}	10^{9}	10^{7}	10^{5}	10^{3}	10^{-3}	7.986514×10^{6}	6.272593×10^{6}	4.355967×10^{4}	4.839964×10^{3}	9.999925×10^{-1}	1.562488×10^{-3}
Square mile (US)	sq.mi or mi^2	2.589 998	10^{18}	10^{12}	10^{10}	10^{8}	10^{6}	1	5.111407×10^{9}	4.014490×10^{9}	2.78784×10^{7}	3.0976×10^{6}	6.4×10^{2}	1×1
Square mile (imp.)	sq.mi or mi^2	2.589 979	10^{18}	10^{12}	10^{10}	10^{8}	10^{6}	1	5.111369×10^{9}	4.014460×10^{9}	2.787819×10^{7}	3.097577×10^{6}	6.399952×10^{2}	9.999925×10^{-1}

Note: The basic SI unit for area is square meter.

From Diem, K., Ed., *Documenta Geigy Scientific Tables*, 6th ed., Geigy Pharmaceuticals, Ardsley, N.Y., 1962, 204. With permission.

Table 9
CONVERSION FACTORS FOR UNITS OF VOLUME AND CAPACITY

To convert from a unit in column **A** into any unit under **C**, multiply by the factor in column **B** and by the power of 10 given in the appropriate column under **C**

To convert from a unit in column **A** into any unit under **D**, multiply by the factor given in the appropriate column under **D**

A		B	C					D							
			μm^3	mm^3	cm^3	dm^3	m^3	μl	ml	dl	l	hl	in^3 (US)	ft^3 (US)	yd^3 (US)
Cubic micrometer	μm^3	1	1	10^{-9}	10^{-12}	10^{-15}	10^{-18}	$9.999\,720\times10^{-10}$	$9.999\,720\times10^{-13}$	$9.999\,720\times10^{-15}$	$9.999\,720\times10^{-16}$	$9.999\,720\times10^{-18}$	$6.102\,338\times10^{-14}$	$3.531\,445\times10^{-17}$	$1.307\,943\times10^{-18}$
Cubic millimeter	mm^3	1	10^9	1	10^{-3}	10^{-6}	10^{-9}	$9.999\,720\times10^{-1}$	$9.999\,720\times10^{-4}$	$9.999\,720\times10^{-6}$	$9.999\,720\times10^{-7}$	$9.999\,720\times10^{-9}$	$6.102\,338\times10^{-5}$	$3.531\,445\times10^{-8}$	$1.307\,943\times10^{-9}$
Cubic centimeter	cm^3 (ccm)	1	10^{12}	10^3	1	10^{-3}	10^{-6}	$9.999\,720\times10^2$	$9.999\,720\times10^{-1}$	$9.999\,720\times10^{-3}$	$9.999\,720\times10^{-4}$	$9.999\,720\times10^{-6}$	$6.102\,338\times10^{-2}$	$3.531\,445\times10^{-5}$	$1.307\,943\times10^{-6}$
Cubic decimeter	dm^3	1	10^{15}	10^6	10^3	1	10^{-3}	$9.999\,720\times10^5$	$9.999\,720\times10^2$	$9.999\,720\times1$	$9.999\,720\times10^{-1}$	$9.999\,720\times10^{-3}$	$6.102\,338\times10$	$3.531\,445\times10^{-2}$	$1.307\,943\times10^{-3}$
Cubic meter (stere)	m^3	1	10^{18}	10^9	10^6	10^3	1	$9.999\,720\times10^8$	$9.999\,720\times10^5$	$9.999\,720\times10^3$	$9.999\,720\times10^2$	$9.999\,720\times1$	$6.102\,338\times10^4$	$3.531\,445\times10$	$1.307\,943\times1$
Microliter	μl	1.000 028	10^9	1	10^{-3}	10^{-6}	10^{-9}	$\times1$	$\times10^{-3}$	$\times10^{-5}$	$\times10^{-6}$	$\times10^{-8}$	$6.102\,509\times10^{-5}$	$3.531\,544\times10^{-8}$	$1.307\,979\times10^{-9}$
Milliliter	ml	1.000 028	10^{12}	10^3	1	10^{-3}	10^{-6}	$\times10^3$	$\times1$	$\times10^{-2}$	$\times10^{-3}$	$\times10^{-5}$	$6.102\,509\times10^{-2}$	$3.531\,544\times10^{-5}$	$1.307\,979\times10^{-6}$
Deciliter	dl	1.000 028	10^{14}	10^5	10^2	10^{-1}	10^{-4}	$\times10^5$	$\times10^2$	$\times1$	$\times10^{-1}$	$\times10^{-3}$	$6.102\,509\times1$	$3.531\,544\times10^{-3}$	$1.307\,979\times10^{-4}$
Liter	l	1.000 028	10^{15}	10^6	10^3	1	10^{-3}	$\times10^6$	$\times10^3$	$\times10$	$\times1$	$\times10^{-2}$	$6.102\,509\times10$	$3.531\,544\times10^{-2}$	$1.307\,979\times10^{-3}$
Hectoliter	hl	1.000 028	10^{17}	10^8	10^5	10^2	10^{-1}	$\times10^8$	$\times10^5$	$\times10^3$	$\times10^2$	$\times1$	$6.102\,509\times10^3$	$3.531\,544\times1$	$1.307\,979\times10^{-1}$
Cubic inch (US)	cu.in or in^3	1.638 716	10^{13}	10^4	10	10^{-2}	10^{-5}	$1.638\,670\times10^4$	$1.638\,670\times10$	$1.638\,670\times10^{-1}$	$1.638\,670\times10^{-2}$	$1.638\,670\times10^{-4}$	$\times1$	$5.787\,03\times10^{-4}$	$2.143\,347\times10^{-5}$
Cubic inch (imp.)		1.638 698	10^{13}	10^4	10	10^{-2}	10^{-5}	$1.638\,652\times10^4$	$1.638\,652\times10$	$1.638\,652\times10^{-1}$	$1.638\,652\times10^{-2}$	$1.638\,652\times10^{-4}$	$9.999\,888\times10^{-1}$	$5.786\,972\times10^{-4}$	$2.143\,323\times10^{-5}$
Cubic foot (US)	cu.ft or ft^3	2.831 702	10^{16}	10^7	10^4	10	10^{-2}	$2.831\,622\times10^7$	$2.831\,622\times10^4$	$2.831\,622\times10^2$	$2.831\,622\times10$	$2.831\,622\times10^{-1}$	1.728×10^3	$\times1$	3.703×10^{-2}
Cubic foot (imp.)		2.831 670	10^{16}	10^7	10^4	10	10^{-2}	$2.831\,591\times10^7$	$2.831\,591\times10^4$	$2.831\,591\times10^2$	$2.831\,591\times10$	$2.831\,591\times10^{-1}$	$1.727\,981\times10^3$	$9.999\,888\times10^{-1}$	$3.703\,662\times10^{-2}$
Cubic yard (US)	cu.yd or yd^3	7.645 594	10^{17}	10^8	10^5	10^2	10^{-1}	$7.645\,380\times10^8$	$7.645\,380\times10^5$	$7.645\,380\times10^3$	$7.645\,380\times10^2$	$7.645\,380\times1$	$4.665\,6\times10^4$	2.7×10	$\times1$
Cubic yard (imp.)		7.645 509	10^{17}	10^8	10^5	10^2	10^{-1}	$7.645\,295\times10^8$	$7.645\,295\times10^5$	$7.645\,295\times10^3$	$7.645\,295\times10^2$	$7.645\,295\times1$	$4.665\,548\times10^4$	$2.699\,970\times10$	$9.999\,888\times10^{-1}$
Register ton (US)	reg.tn	2.831 702	10^{18}	10^9	10^6	10^3	1	$2.831\,622\times10^9$	$2.831\,622\times10^6$	$2.831\,622\times10^4$	$2.831\,622\times10^3$	$2.831\,622\times10$	1.728×10^5	1×10^2	3.703×1
Register ton (imp.)		2.831 670	10^{18}	10^9	10^6	10^3	1	$2.831\,591\times10^9$	$2.831\,591\times10^6$	$2.831\,591\times10^4$	$2.831\,591\times10^3$	$2.831\,591\times10$	$1.727\,981\times10^5$	$9.999\,888\times10$	$3.703\,662\times1$

Note: The basic SI unit for volume is cubic meter.

From Diem, K., Ed., *Documenta Geigy Scientific Tables*, 6th ed., Geigy Pharmaceuticals, Ardsley, N.Y., 1962, 205. With permission.

Table 10
CONVERSION FACTORS FOR U.S. DRY CAPACITY MEASURES

To convert from a unit in column **A** into any unit under **B**, multiply by the factor given in the appropriate column under **B**

A		B							C see footnote
		m^3	l	in^3 (US)	dry pt (US)	dry qt (US)	pk	bu	Imperial
Cubic meter	m^3	$\times 1$	$9.999\,720 \times 10^2$	$6.102\,338 \times 10^4$	$1.816\,155 \times 10^3$	$9.080\,775 \times 10^2$	$1.135\,097 \times 10^2$	$2.837\,742 \times 10$	—
Liter	l	$1.000\,028 \times 10^{-3}$	$\times 1$	$6.102\,509 \times 10$	$1.816\,206 \times$	$9.081\,030 \times 10^{-1}$	$1.135\,129 \times 10^{-1}$	$2.837\,822 \times 10^{-2}$	—
Cubic inch .....(US) cu.in or in^3		$1.638\,716 \times 10^{-5}$	$1.638\,670 \times 10^{-2}$	$1 \quad \times 1$	$2.976\,163 \times 10^{-2}$	$1.488\,081 \times 10^{-2}$	$1.860\,102 \times 10^{-3}$	$4.650\,254 \times 10^{-4}$	$1.000\,011\,20$
Dry pint .....(US) dry pt		$5.506\,138 \times 10^{-4}$	$5.505\,984 \times 10^{-1}$	$3.360\,031 \times 10$	$1 \quad \times 1$	$5 \quad \times 10^{-1}$	$6.25 \quad \times 10^{-2}$	1.5625×10^{-2}	$0.968\,944\,72$
Dry quart .....(US) dry qt		$1.101\,228 \times 10^{-3}$	$1.101\,197 \times 1$	$6.720\,062 \times 10$	$2 \quad \times 1$	$1 \quad \times 1$	$1.25 \quad \times 10^{-1}$	$3.125 \quad \times 10^{-2}$	$0.968\,944\,72$
Peck .....(US) pk		$8.809\,820 \times 10^{-3}$	$8.809\,574 \times 1$	$5.376\,05 \times 10^2$	$1.6 \quad \times 10$	$8 \quad \times 1$	$1 \quad \times 1$	$2.5 \quad \times 10^{-1}$	$0.968\,944\,72$
Bushel .....(US) bu		$3.523\,928 \times 10^{-2}$	$3.523\,829 \times 10$	$2.150\,42 \times 10^3$	$6.4 \quad \times 10$	$3.2 \quad \times 10$	$4 \quad \times 1$	$1 \quad \times 1$	$0.968\,944\,72$

To convert a US (or imperial) unit under **A** into the corresponding imperial (or US) unit, multiply (or divide) by the appropriate factor under **C**. The same applies to any of the equivalents given under **B**; for an example see the footnote on the following page.

From Diem, K., Ed., *Documenta Geigy Scientific Tables*, 6th ed., Geigy Pharmaceuticals, Ardsley, N.Y., 1962, 205. With permission.

Table 11
CONVERSION FACTORS FOR U.S. LIQUID CAPACITY MEASURE

A			To convert from a unit in column **A** into any unit under **B**, multiply by the factor given in the appropriate column under **B**									**C** see footnote
			B									
		m^3	l	in^3 (US)	min (US)	fl.dr. (US)	fl.oz (US)	gi (US)	liq.pt (US)	liq.qt (US)	gal (US)	Imperial
Cubic meter	m^3	**1** × 1	9.999 720 × 10^2	6.102 338 × 10^4	1.623 063 × 10^7	2.705 106 × 10^5	3.381 382 × 10^4	8.453 455 × 10^3	2.113 364 × 10^3	1.056 682 × 10^3	2.641 705 × 10^2	—
Liter	l	1.000 0**28** × 10^{-3}	**1** × 1	6.102 509 × 10	1.623 109 × 10^4	2.705 181 × 10^2	3.381 477 × 10	8.453 692 × 1	2.113 423 × 1	1.056 711 × 1	2.641 779 × 10^{-1}	—
Cubic inch (US)	cu.in or in^3	1.638 716 × 10^{-5}	1.638 670 × 10^{-2}	**1** × 1	2.659 740 × 10^2	4.432 900 × 1	5.541 126 × 10^{-1}	1.385 281 × 10^{-1}	3.463 203 × 10^{-2}	1.731 602 × 10^{-2}	4.329 004 × 10^{-3}	1.000 011 20
Minim (US)	min	6.161 189 × 10^{-8}	6.161 016 × 10^{-5}	3.759 766 × 10^{-3}	**1** × 1	1.6̇ × 10^{-2}	2.083̇ × 10^{-3}	5.208 3̇ × 10^{-4}	1.302 083̇ × 10^{-4}	6.510 416̇ × 10^{-5}	1.627 604 × 10^{-5}	0.960 754 24
Fluid dram (US)	fl.dr	3.696 713 × 10^{-6}	3.696 610 × 10^{-3}	2.255 859 × 10^{-1}	**6** × 10	**1** × 1	**1.25** × 10^{-1}	**3.125** × 10^{-2}	7.812 **5** × 10^{-3}	3.906 **25** × 10^{-3}	9.765 6**25** × 10^{-4}	0.960 754 24
Fluid ounce (US)	fl.oz	2.957 371 × 10^{-5}	2.957 288 × 10^{-2}	1.804 688 × 1	**4.8** × 10^2	**8** × 1	**1** × 1	**2.5** × 10^{-1}	**6.25** × 10^{-2}	**3.125** × 10^{-2}	7.812 **5** × 10^{-3}	0.960 754 24
Gill (US)	gi	1.182 948 × 10^{-4}	1.182 915 × 10^{-1}	7.218 **75** × 1	**1.92** × 10^3	**3.2** × 10	**4** × 1	**1** × 1	**2.5** × 10^{-1}	**1.25** × 10^{-1}	**3.125** × 10^{-2}	0.832 679 13
Liquid pint (US)	liq.pt	4.731 793 × 10^{-4}	4.731 661 × 10^{-1}	2.887 **5** × 10	**7.68** × 10^3	**1.28** × 10^2	**1.6** × 10	**4** × 1	**1** × 1	**5** × 10^{-1}	**1.25** × 10^{-1}	0.832 679 13
Liquid quart (US)	liq.qt	9.463 586 × 10^{-4}	9.463 321 × 10^{-1}	**5.775** × 10	**1.536** × 10^4	**2.56** × 10^2	**3.2** × 10	**8** × 1	**2** × 1	**1** × 1	**2.5** × 10^{-1}	0.832 679 13
Gallon (US)	gal	3.785 434 × 10^{-3}	3.785 329 × 1	**2.31** × 10^2	**6.144** × 10^4	**1.024** × 10^3	**1.28** × 10^2	**3.2** × 10	**8** × 1	**4** × 1	**1** × 1	0.832 679 13

To convert a US (for imperial) unit under **A** into the corresponding imperial (or US) unit, multiply (or divide) by the appropriate factor under **C**. The same applies to any of the equivalents given under **B**, for example: 1 fl.dr (US) = 0.960 754 24 fl.dr (imp.); 1 fl.dr (imp.) = (3.696 610 × 10^{-3}) × 0.960 754 24 l = 3.551 534 × 10^{-3} l.

From Diem, K., Ed., *Documenta Geigy Scientific Tables,* 6th ed., Geigy Pharmaceuticals, Ardsley, N.Y., 1962, 206. With permission.

Table 12
CONVERSION FACTORS FOR BRITISH (IMPERIAL) CAPACITY MEASURES

To convert from a unit in column **A** into any unit under **B**, multiply by the factor given in the appropriate column under **B**

A	m³	l	in³ (imp.)	min (imp.)	fl.dr (imp.)	fl.oz (imp.)	gi (imp.)	pt (imp.)	qt (imp.)	gal (imp.)	pk (imp.)	bu (imp.)	C — US liquid	C — US dry
Cubic meter .. m³	1	$9.999\,720\times10^{2}$	$6.102\,406\times10^{4}$	$1.689\,364\times10^{7}$	$2.815\,606\times10^{5}$	$3.519\,508\times10^{4}$	$7.039\,015\times10^{3}$	$1.759\,754\times10^{3}$	$8.798\,769\times10^{2}$	$2.199\,692\times10^{2}$	$1.099\,846\times10^{2}$	$2.749\,615\times10$	—	—
Liter l	$1.000\,028\times10^{-3}$	1	$6.102\,577\times10$	$1.689\,411\times10^{4}$	$2.815\,685\times10^{2}$	$3.519\,606\times10$	$7.039\,213\times1$	$1.759\,803\times1$	$8.799\,016\times10^{-1}$	$2.199\,754\times10^{-1}$	$1.099\,877\times10^{-1}$	$2.749\,692\times10^{-2}$	—	—
Cubic inch cu.in or in³	$1.638\,098\times10^{-5}$	$1.638\,652\times10^{-2}$	1	$2.768\,357\times10^{2}$	$4.613\,928\times1$	$5.767\,410\times10^{-1}$	$1.153\,482\times10^{-1}$	$2.883\,705\times10^{-2}$	$1.441\,852\times10^{-2}$	$3.604\,631\times10^{-3}$	$1.802\,316\times10^{-3}$	$4.505\,789\times10^{-4}$	$0.999\,988\,80$	$0.999\,988\,80$
Minim ... min	$5.919\,388\times10^{-8}$	$5.919\,223\times10^{-5}$	$3.612\,251\times10^{-3}$	1	$1.\dot{6}\times10^{-2}$	2.083×10^{-3}	$4.1\dot{6}\times10^{-4}$	$1.041\,\dot{6}\times10^{-4}$	$5.208\,\dot{3}\times10^{-5}$	$1.302\,083\times10^{-5}$	$6.510\,416\times10^{-6}$	$1.627\,604\times10^{-6}$	$1.040\,84891$	$1.040\,84891$
Fluid drachm f.dr	$3.551\,633\times10^{-6}$	$3.551\,534\times10^{-3}$	$2.167\,351\times10^{-1}$	6×10	1	1.25×10^{-1}	2.5×10^{-2}	6.25×10^{-3}	3.125×10^{-3}	7.8125×10^{-4}	$3.906\,25\times10^{-4}$	$9.765\,625\times10^{-5}$	$1.040\,84891$	$1.040\,84891$
Fluid ounce fl.oz	$2.841\,306\times10^{-5}$	$2.841\,227\times10^{-2}$	$1.733\,881\times10$	4.8×10^{2}	8	1	2×10^{-1}	5×10^{-2}	2.5×10^{-2}	6.25×10^{-3}	3.125×10^{-3}	7.8125×10^{-4}	$1.040\,84891$	$1.040\,84891$
Gill gi	$1.420\,653\times10^{-4}$	$1.420\,613\times10^{-1}$	$8.669\,403\times10$	2.4×10^{3}	4×10	5	1	2.5×10^{-1}	1.25×10^{-1}	3.125×10^{-2}	1.5625×10^{-2}	$3.906\,25\times10^{-3}$	$1.200\,94280$	$1.200\,94280$
Pint pt	$5.682\,613\times10^{-4}$	$5.682\,454\times10^{-1}$	$3.467\,761\times10$	9.6×10^{3}	1.6×10^{2}	2×10	4	1	5×10^{-1}	1.25×10^{-1}	6.25×10^{-2}	1.5625×10^{-2}	$1.200\,94280$	$1.200\,94280$
Quart qt	$1.136\,523\times10^{-3}$	$1.136\,491\times1$	$6.935\,522\times10$	1.92×10^{4}	3.2×10^{2}	4×10	8	2	1	2.5×10^{-1}	1.25×10^{-1}	3.125×10^{-2}	$1.200\,94280$	$1.200\,94280$
Gallon gal	$4.546\,090\times10^{-3}$	$4.545\,963\times1$	$2.774\,209\times10^{2}$	7.68×10^{4}	1.28×10^{3}	1.6×10^{2}	3.2×10	8	4	1	5×10^{-1}	1.25×10^{-1}	$1.200\,94280$	$1.200\,94280$
Peck pk	$9.092\,181\times10^{-3}$	$9.091\,926\times1$	$5.548\,418\times10^{2}$	1.536×10^{5}	2.56×10^{3}	3.2×10^{2}	6.4×10	1.6×10	8	2	1	2.5×10^{-1}	—	$1.032\,05062$
Bushel bu	$3.636\,872\times10^{-2}$	$3.636\,770\times10$	$2.219\,367\times10^{3}$	6.144×10^{5}	1.024×10^{4}	1.28×10^{3}	2.56×10^{2}	6.4×10	3.2×10	8	4	1	—	$1.032\,05062$

To convert a US (or imperial) unit into the corresponding imperial (or US) unit, multiply (or divide) by the appropriate factor under **C**. The same applies to any of the equivalents given under **B**, for example:
1 fl.dr (US) = 0.960 754 24 fl.dr (imp.); 1 fl.dr (imp.) = $(3.696\,610 \times 10^{-3}) \times 0.960\,754\,24$ l = $3.551\,534 \times 10^{-3}$ l.

From Diem, K., Ed., *Documenta Geigy Scientific Tables*, 6th ed., Geigy Pharmaceuticals, Ardsley, N.Y., 1962, 206. With permission.

Table 13
CONVERSION FACTORS FOR UNITS OF MASS

To convert from a unit in column **A** into any unit under **C**, multiply by the factor in column **B** and by the power of 10 given in the appropriate column under **C**. To convert from a unit in column **A** into any unit under **D**, multiply by the factor given in the appropriate column under **D**.

A (unit)		B	C — μg	mg	g	kg	t	D — gr	dr.av.	oz.av.	lb.av.	ctl (imp.), sh.cwt (US)	cwt (imp.), l.cwt (US)	sh.tn (US)	tn (imp.), l.tn (US)
Microgram (gamma) *(The term gamma should no longer be used.)*	μg (γ)	$1\times$	1	10^{-3}	10^{-6}	10^{-9}	10^{-12}	$1.543\,236\times10^{-5}$	$5.643\,833\times10^{-7}$	$3.527\,396\times10^{-8}$	$2.204\,622\times10^{-9}$	$2.204\,622\times10^{-11}$	$1.968\,413\times10^{-11}$	$1.102\,311\times10^{-12}$	$9.842\,064\times10^{-13}$
Milligram	mg	$1\times$	10^{3}	1	10^{-3}	10^{-6}	10^{-9}	$1.543\,236\times10^{-2}$	$5.643\,833\times10^{-4}$	$3.527\,396\times10^{-5}$	$2.204\,622\times10^{-6}$	$2.204\,622\times10^{-8}$	$1.968\,413\times10^{-8}$	$1.102\,311\times10^{-9}$	$9.842\,064\times10^{-10}$
Gram	g	$1\times$	10^{6}	10^{3}	1	10^{-3}	10^{-6}	$1.543\,236\times10$	$5.643\,833\times10^{-1}$	$3.527\,396\times10^{-2}$	$2.204\,622\times10^{-3}$	$2.204\,622\times10^{-5}$	$1.968\,413\times10^{-5}$	$1.102\,311\times10^{-6}$	$9.842\,064\times10^{-7}$
Kilogram	kg	$1\times$	10^{9}	10^{6}	10^{3}	1	10^{-3}	$1.543\,236\times10^{4}$	$5.643\,833\times10^{2}$	$3.527\,396\times10$	$2.204\,622$	$2.204\,622\times10^{-2}$	$1.968\,413\times10^{-2}$	$1.102\,311\times10^{-3}$	$9.842\,064\times10^{-4}$
Metric ton	t	$1\times$	10^{12}	10^{9}	10^{6}	10^{3}	1	$1.543\,236\times10^{7}$	$5.643\,833\times10^{5}$	$3.527\,396\times10^{4}$	$2.204\,622\times10^{3}$	$2.204\,622\times10$	$1.968\,413\times10$	$1.102\,311$	$9.842\,064\times10^{-1}$
Grain, avoirdupois, ap. or troy (US/imp.)	gr	$6.479\,892\times$	10^{4}	10	10^{-2}	10^{-5}	10^{-8}	1	$3.657\,143\times10^{-2}$	$2.285\,714\times10^{-3}$	$1.428\,571\times10^{-4}$	$1.428\,571\times10^{-6}$	$1.275\,511\times10^{-6}$	$7.142\,857\times10^{-8}$	$6.377\,551\times10^{-8}$
Dram, avoirdupois (US/imp.)	dr.av.	$1.771\,845\times$	10^{6}	10^{3}	1	10^{-3}	10^{-6}	$2.734\,375\times10$	1	6.25×10^{-2}	$3.906\,25\times10^{-3}$	$3.906\,25\times10^{-5}$	$3.487\,723\times10^{-5}$	$1.953\,125\times10^{-6}$	$1.743\,862\times10^{-6}$
Ounce, avoirdupois (US/imp.)	oz.av.	$2.834\,953\times$	10^{7}	10^{4}	10	10^{-2}	10^{-5}	4.375×10^{2}	1.6×10	1	6.25×10^{-2}	6.25×10^{-4}	$5.580\,357\times10^{-4}$	3.125×10^{-5}	$2.790\,179\times10^{-5}$
Pound, avoirdupois (US/imp.)	lb.av.	$4.535\,924\times$	10^{8}	10^{5}	10^{2}	10^{-1}	10^{-4}	7×10^{3}	2.56×10^{2}	1.6×10	1	1×10^{-2}	$8.928\,571\times10^{-3}$	5×10^{-4}	$4.464\,286\times10^{-4}$
Cental (imp.) ctl / Hundredweight, short (US) sh.cwt	ctl / sh.cwt	$4.535\,924\times$	10^{10}	10^{7}	10^{4}	10	10^{-2}	7×10^{5}	2.56×10^{4}	1.6×10^{3}	1×10^{2}	1	$8.928\,571\times10^{-1}$	5×10^{-2}	$4.464\,286\times10^{-2}$
Hundredweight (imp.) cwt / Hundredweight, long (US) l.cwt	cwt / l.cwt	$5.080\,235\times$	10^{10}	10^{7}	10^{4}	10	10^{-2}	7.84×10^{5}	2.8672×10^{4}	1.792×10^{3}	1.12×10^{2}	1.12	1	5.6×10^{-2}	5×10^{-2}
Ton, short (US) sh.tn	sh.tn	$9.071\,849\times$	10^{11}	10^{8}	10^{5}	10^{2}	10^{-1}	1.4×10^{7}	5.12×10^{5}	3.2×10^{4}	2×10^{3}	2×10	$1.785\,714\times10$	1	$8.928\,571\times10^{-1}$
Ton (imp.) tn / Ton, long (US) l.tn	tn / l.tn	$1.016\,047\times$	10^{12}	10^{9}	10^{6}	10^{3}	1	1.568×10^{7}	5.7344×10^{5}	3.584×10^{4}	2.24×10^{3}	2.24×10	2×10	1.12	1

Note: The basic SI unit for mass is kilogram.

From Diem, K., Ed., *Documenta Geigy Scientific Tables*, 6th ed., Geigy Pharmaceuticals, Ardsley, N.Y., 1962, 207. With permission.

Table 14
CONVERSION FACTORS FOR UNITS OF TROY WEIGHT

To convert from a unit in column **A** into any unit under **B**, multiply by the factor given in the appropriate column under **B**

A		g	c	gr	s.ap	dwt	dr.ap.	oz.ap, oz.t.	lb.ap., lb.t.	lb.av.
Gram	g	1 × 1	5 × 1	1.543 236×10	7.716 179×10⁻¹	6.430 149×10⁻¹	2.572 059×10⁻¹	3.215 074×10⁻²	2.679 229×10⁻³	2.204 622×10⁻³
Carat	c	2 ×10⁻¹	1 × 1	3.086 471×10	—	1.286 030×10⁻¹	—	6.430 148×10⁻²	5.358 457×10⁻³	4.409 245×10⁻³
Grain, avoirdupois, apoth. or troy (US/imp.)	gr	6.479 892×10⁻²	3.239 946×10⁻¹	1 × 1	5 ×10⁻²	4.16 ×10⁻²	1.6 ×10⁻²	2.083 ×10⁻³	1.736 1 ×10⁻⁴	1.428 571×10⁻⁴
Scruple, apothecary (US/imp.)	s.ap.	1.295 978× 1	—	2 ×10	1 × 1	—	3.3 ×10⁻¹	4.16 ×10⁻²	3.472 ×10⁻³	2.857 143×10⁻³
Pennyweight, troy (US/imp.)	dwt	1.555 174× 1	7.775 870× 1	2.4 ×10	—	1 × 1	—	5 ×10⁻²	4.16 ×10⁻³	3.428 571×10⁻³
Dram, apothecary (US) / Drachm, apothecary (imp.)	dr.ap.	3.887 935× 1	—	6 ×10	3 × 1	—	1 × 1	1.25 ×10⁻¹	1.041 6 ×10⁻²	8.571 429×10⁻³
Ounce, apothecary or troy (US/imp.)	oz.ap. or oz.t.	3.110 348×10	1.555 174×10²	4.8 ×10²	2.4 ×10	2 ×10	8 × 1	1 × 1	8.3 ×10⁻²	6.857 143×10⁻²
Pound, apothecary or troy (US)	lb.ap. or lb.t.	3.732 418×10²	1.866 209×10³	5.76 ×10³	2.88 ×10²	2.4 ×10²	9.6 ×10	1.2 ×10	1 × 1	8.228 571×10⁻¹
Pound, avoirdupois (US/imp.)	lb.av.	4.535 924×10²	2.267 962×10³	7 ×10³	3.5 ×10²	2.916 ×10²	1.16 ×10²	1.458 3 ×10	1.215 27 × 1	1 × 1

From Diem, K., Ed., *Documenta Geigy Scientific Tables*, 6th ed., Geigy Pharmaceuticals, Ardsley, N.Y., 1962, 208. With permission.

Table 15
CONVERSION OF THERMOMETER SCALES

Celsius to Fahrenheit
$F = (C \times 9/5) + 32$

Fahrenheit to Celsius
$C = (F - 32) \times 5/9$

Celsius to Kelvin
$K = C + 273.1$

Kelvin to Celsius
$C = K - 273.1$

Celsius to Réaumur
$Re = C \times 4/5$

Réaumur to Celsius
$C = R\acute{e} \times 5/4$

Fahrenheight to Réaumur
$R\acute{e} = (F - 32) \times 4/9$

Réaumur to Fahrenheit
$F = (R\acute{e} \times 9/4) + 32$

Fahrenheit to Rankine
$R = F + 459.58$

Note: SI usage for degrees Kelvin is Kelvins, abbreviated as K.

Table 16
CONVERSION FORMULAS FOR SOLUTIONS HAVING CONCENTRATIONS EXPRESSED IN VARIOUS WAYS

A = Weight per cent of solute
B = Molecular weight of solvent
E = Molecular weight of solute
F = Grams of solute per liter of solution
G = Molality
M = Molarity
N = Mole fraction
R = Density of solution grams per cc

Concentration of solute— SOUGHT	Concentration of solute—GIVEN				
	A	N	G	M	F
A	—	$\dfrac{100N \times E}{N \times E + (1 - N)B}$	$\dfrac{100G \times E}{1000 + G \times E}$	$\dfrac{M \times E}{10R}$	$\dfrac{F}{10R}$
N	$\dfrac{\frac{A}{E}}{\frac{A}{E} + \frac{100 - A}{B}}$	—	$\dfrac{B \times G}{B \times G + 1000}$	$\dfrac{B \times M}{M(B - E) + 1000R}$	$\dfrac{B \times F}{F(B - E) + 1000R \times E}$
G	$\dfrac{1000A}{E(100 - A)}$	$\dfrac{1000N}{B - N \times B}$	—	$\dfrac{1000M}{1000R - (M \times E)}$	$\dfrac{1000F}{E(1000R - F)}$
M	$\dfrac{10R \times A}{E}$	$\dfrac{1000R \times N}{N \times E + (1 - N)B}$	$\dfrac{1000R \times G}{1000 + E \times G}$	—	$\dfrac{F}{E}$
F	$10AR$	$\dfrac{1000R \times N \times E}{N \times E + (1 - N)B}$	$\dfrac{1000R \times G \times E}{1000 + G \times E}$	$M \times E$	—

From Weast, R. C., Ed., *CRC Handbook of Chemistry and Physics,* 61st ed., CRC Press, Boca Raton, Fla., 1980, D-154. With permission.

Table 17
FORMULAS FOR CALCULATING TITRATION DATA, pH VS. ML OF REAGENT

A Substance Titrated V_0 ml. of solution M_0 its molarity	B Initial $[H^+]$ or $[OH^-]$	C Intermediate Points 10, 50, 99, etc., per cent neutralized, V_1 ml. of reagent of M_1 molarity added	D Equivalence Point	E Excess of Reagent V_1 volume reagent M_1 its molarity V_T total volume
(1) Strong Acid	$[H^+] = M_0$	$[H^+] = \dfrac{V_0M_0 - V_1M_1}{V_0 + V_1}$	$\sqrt{K_w}$	$[OH^-] = \dfrac{V_1M_1}{V_T}$
(2) Strong Base	$[OH^-] = M_0$	$[OH^-] = \dfrac{V_0M_0 - V_1M_1}{V_0 + V_1}$	$\sqrt{K_w}$	$[H^+] = \dfrac{V_1M_1}{V_T}$
(3) Weak Acid ($K_a = 10^{-5}$ to 10^{-8})	$[H^+] = \sqrt{M_0K_a}$	$[H^+] = \dfrac{[\text{Acid}]}{[\text{Salt}]} K_a$	$[OH^-] = \sqrt{\dfrac{K_w}{K_a}c}$	$[OH^-] = \dfrac{V_1M_1}{V_T}$ (Value in column D to be added)
(4) Weak Base ($K_b = 10^{-5}$ to 10^{-8})	$[OH^-] = \sqrt{M_0K_b}$	$[OH^-] = \dfrac{[\text{Base}]}{[\text{Salt}]} K_b$	$[H^+] = \sqrt{\dfrac{K_w}{K_b}c}$	$[H^+] = \dfrac{V_1M_1}{V_T}$ (Value in column D to be added)
(5) Salt of a Very Weak Acid (e.g. KCN)	$[OH^-] = \sqrt{\dfrac{K_w}{K_a}c}$	$[H^+] = \dfrac{[\text{Acid}]}{[\text{Salt}]} K_a$	$[H^+] = \sqrt{[\text{Acid}]K_a}$	$[H^+] = \dfrac{V_1M_1}{V_T}$ (Correct for value in column D)
(6) Salt of a Very Weak Base	$[H^+] = \sqrt{\dfrac{K_w}{K_b}c}$	$[OH^-] = \dfrac{[\text{Base}]}{[\text{Salt}]} K_b$	$[OH^-] = \sqrt{[\text{Base}]K_b}$	$[OH^-] = \dfrac{V_1M_1}{V_T}$ (Add to $[OH^-]$ found in column D)

From Weast, R. C., Ed., *CRC Handbook of Chemistry and Physics,* 61st ed., CRC Press, Boca Raton, Fla., 1980, D-154. With permission.

Table 18
CONVERSION OF CONCENTRATION UNITS

a) *Conversion of mg/100 ml into mmol/l and vice versa*

$$mmol/l = \frac{10 \times mg/100 \; ml}{\text{molecular weight}} = \frac{10,000 \times g/100 \; ml}{\text{molecular weight}}$$

$$mg/100 \; ml = \frac{mmol \times \text{molecular weight}}{10}$$

b) *Conversion of ml of gas/100 ml into mmol/l and vice versa*
(at $0°C$; 2.24 = millimolar normal volume of an ideal gas)

$$mmol/l = \frac{ml/100 \; ml}{2.24}$$

$$ml/100 \; ml = 2.24 \times mmol/l$$

c) *Conversion of mg/100 ml into mEq/l* and vice versa*

$$mEq/l = \frac{10 \times mg/100 \; ml \times \text{valency}}{\text{molecular weight}}$$

$$mg/100 \; ml = \frac{mEq/l \times \text{molecular weight}}{10 \times \text{valency}}$$

Note: In clinical chemistry, concentration data for fluids should be related to the unit liter, those for solids to the unit kilogram. The use of ambiguous units like g%, mg%, p.p.m., etc. should not be used since they do not make it clear whether the data relate to mass or to volume.

Table 19
RELATION OF CONCENTRATION UNITS

1 μg/mL = 1 mg/liter = 1 ppm (w/v)

1 meq/liter = (weight in grams per equivalent/1000) in 1 liter

1 mg atom per liter = (atomic weight in grams/1000) in 1 liter

1 μg/mL = (1 meq/liter) (1000/weight in grams per equivalent)

Table 20
TEMPERATURE VARIATION OF MOLARITY AND NORMALITY OF AQUEOUS SOLUTIONS

Temperature (°C)	Factor
14	1.0010
15	1.0009
16	1.0007
17	1.0006
18	1.0004
19	1.0002
20	1.0000
21	0.9998
22	0.9996
23	0.9994
24	0.9991
25	0.9989
26	0.9986
27	0.9983

Note: The above conversion factors for temperatures deviating from the normal temperature of 20°C are for 0.1-N solutions and assume a coefficient of expansion for glass of 0.000027[1].

Adapted from Kolthoff, I. M., *Die Massanalyse,* (Part 2), Springer, Berlin, 1928, 30. With permission.

Table 21
AQUEOUS SOLUTIONS — CONVERSION FACTORS FOR ELECTROLYTES (I)

This table gives (*a*) the molecular weight and solubilities (in hot and cold water) of the electrolytes;
(*b*) for a given weight of electrolyte, the millimoles of undissociated solute, the milligram equivalents (mEq) and weights of the cation and anion, and the milliosmoles of solute on the assumption of complete dissociation.

	Electrolyte (data for 1 gram unless otherwise stated)		Molecular weight	Undiss. solute mmol	Solubility[1] g per 1000 g water cold	hot	Cation mEq	mg		Anion mEq	mg		Milli-osmoles*
	Calcium, Ca												
1	acetate	Ca($C_2H_3O_2$)$_2$ + H_2O	176.19	5.68	436[20]	343[100]	11.35	228	Ca++	11.35	670	$C_2H_3O_2$-	17.03
2		Ca($C_2H_3O_2$)$_2$ + 2 H_2O	194.20	5.15	347[0]	335[50]	10.30	206	Ca++	10.30	608	$C_2H_3O_2$-	15.45
3	chloride	$CaCl_2$ + 2 H_2O	147.03	6.80	977[0]	3260[50]	13.60	273	Ca++	13.60	482	Cl-	20.40
4		$CaCl_2$ + 6 H_2O	219.09	4.56	2790[0]	5360[20]	9.13	183	Ca++	9.13	324	Cl-	13.69
5	citrate	Ca$_3$($C_6H_5O_7$)$_2$ + 4 H_2O	570.52	1.75	8.5[18]	9.6[23]	10.52	211	Ca++	10.52	663	$C_6H_5O_7$---	8.76
6	D-gluconate	Ca($C_6H_{11}O_7$)$_2$ + H_2O	448.40	2.23	33[15]		4.46	89	Ca++	4.46	870	$C_6H_{11}O_7$-	6.69
7	lactate	Ca($C_3H_5O_3$)$_2$ + 5 H_2O	308.31	3.24	31[0]	79[30]	6.49	130	Ca++	6.49	578	$C_3H_5O_3$-	9.73
8	levulinate	Ca($C_5H_7O_3$)$_2$ + 2 H_2O	306.33	3.27	400		6.53	131	Ca++	6.53	752	$C_5H_7O_3$-	9.79
9	oxide (lime)**	CaO	56.08	17.83	1.31[10]d	0.78[0]d	35.66	715	Ca++				
10	phosphate, dibasic	$CaHPO_4$ + 2 H_2O	172.10	5.81	0.2[25]	0.75[100]	11.62	233	Ca++	11.62	558/180	HPO_4-- / P	11.62
11	thiosulfate	CaS_2O_3 + 6 H_2O	260.31	3.84	1000[2]	d	7.68	154	Ca++	7.68	431/246	S_2O_3-- / S	7.68
	Chlorine, Cl												
12	Ammonium chloride	NH_4Cl	53.50	18.69	297[0]	758[100]	18.69	337	NH_4+	18.69	663	Cl-	37.38
13	Hydrochloric acid (10% solution)												
	1 gram	(0.1 g HCl)	36.465	2.74	} 823[0]	} 561[60] {	2.74	2.8	H+	2.74	97.2	Cl-	5.48
	1 milliliter	(0.1047 g HCl)	36.465	2.87			2.87	2.9	H+	2.87	101.8	Cl-	5.74
	See also Calcium (3,4), Magnesium (14,15), Potassium (22) and Sodium (35)												
	Magnesium, Mg												
14	chloride	$MgCl_2$	95.23	10.50	542.5[20]	727[100]	21.00	255	Mg++	21.0	745	Cl-	31.50
15		$MgCl_2$ + 6 H_2O	203.33	4.92	1670	3670	9.84	120	Mg++	9.84	349	Cl-	14.75
16	hydroxide	Mg(OH)$_2$	58.34	17.14	0.009[18]	0.04[100]	34.28	417	Mg++				
17	oxide (magnesia)**	MgO	40.32	24.80	0.0062	0.086[30]	49.60	603	Mg++				
18	sulfate (Epsom salts)	$MgSO_4$ + 7 H_2O	246.50	4.06	710[20]	910[40]	8.11	98.7	Mg++	8.11	390/130	SO_4-- / S	8.11
	Phosphorus, P												
	See Calcium (10), Potassium (26, 27) and Sodium (30, 31, 40–42)												
	Potassium, K												
19	acetate	K($C_2H_3O_2$)	98.15	10.19	2530[20]	4920[62]	10.19	398	K+	10.19	602	$C_2H_3O_2$-	20.38
20	bicarbonate	$KHCO_3$	100.12	9.99	224	600[60]	9.99	391	K+	9.99	610	HCO_3-	19.98
21	bromide	KBr	119.02	8.40	534.8[0]	1020[100]	8.40	329	K+	8.40	672	Br-	16.81
22	chloride	KCl	74.56	13.41	347[20]	567[100]	13.41	525	K+	13.41	476	Cl-	26.83
23	citrate	K$_3$($C_6H_5O_7$) + H_2O	324.42	3.08	1670[15]	1997[31]	9.25	362	K+	9.25	583	$C_6H_5O_7$---	12.33
24	D-gluconate	K($C_6H_{11}O_7$)	234.25	4.27			4.27	167	K+	4.27	833	$C_6H_{11}O_7$-	8.54
25	oxide**	K_2O	94.20	10.62	d	d	21.23	830	K+				
26	phosphate, monobasic	KH_2PO_4	136.09	7.35	330[15]	v.s.	7.35 7.35 }	287 7.4 }	K+ H+ }	14.70	705/228	HPO_4-- / P	22.04
27	phosphate, dibasic	K_2HPO_4	174.18	5.74	1670[20]	v.s.	11.48	449	K+	11.48	551/178	HPO_4-- / P	17.22
	Sodium, Na												
28	acetate	Na($C_2H_3O_2$) + 3 H_2O	136.09	7.35	762[0]	1388[50]	7.35	169	Na+	7.35	434	$C_2H_3O_2$-	14.70
29	acid citrate	Na$_2$H($C_6H_5O_7$) + 1½ H_2O	263.12	3.80	v.s.	v.s.	7.60 7.60 } 3.80 }	175 175 3.83 }	Na+ Na+ H+ }	11.4	719	$C_6H_5O_7$---	15.20
30	acid phosphate	NaH_2PO_4 + H_2O	138.00	7.25	599[0]	427[100]	7.25 7.25 }	167 7.3 }	Na+ H+ }	14.49	696/224	HPO_4-- / P	21.74
31		NaH_2PO_4 + 2 H_2O	156.01	6.41	710[0]	3900[83]	6.41 6.41 }	147 6.5 }	Na+ H+ }	12.82	615/199	HPO_4-- / P	19.23
32	aminosalicylate	Na($C_7H_6O_3N$) + 2 H_2O	211.16	4.74			4.74	109	Na+	4.74	720	$C_7H_6O_3N$-	9.47
33	bicarbonate***	$NaHCO_3$	84.01	11.90	69[0]	164[60]	11.90	274	Na+	11.90	726	HCO_3-	23.80
34	bromide	NaBr	102.91	9.72	795[0]	1210[100]	9.72	224	Na+	9.72	777	Br-	19.43
35	chloride (common salt)	NaCl	58.45	17.11	357[0]	391[100]	17.11	393	Na+	17.11	607	Cl-	34.22
36	citrate	Na$_3$($C_6H_5O_7$) + 2 H_2O	294.11	3.40	720[25]	1670[100]	10.19	235	Na+	10.19	643	$C_6H_5O_7$---	13.60
37		Na$_3$($C_6H_5O_7$) + 5½ H_2O	357.17	2.80	926[25]	2500[100]	8.40	193	Na+	8.40	529	$C_6H_5O_7$---	11.20
38	lactate***	Na($C_3H_5O_3$)	112.06	8.92	v.s.		8.92	205	Na+	8.92	795	$C_3H_5O_3$-	17.85
39	oxide**	Na_2O	61.98	16.13	d	d	32.26	742	Na+				
40	phosphate	Na_2HPO_4	141.97	7.04			14.08	324	Na+	14.08	676/218	HPO_4-- / P	21.13
41		Na_2HPO_4 + 2 H_2O	178.00	5.62	1000[60]	1170[80]	11.24	258	Na+	11.24	539/174	HPO_4-- / P	16.85
42		Na_2HPO_4 + 12 H_2O	358.16	2.79	41.5	874[34]	5.58	128	Na+	5.58	268/86.5	HPO_4-- / P	8.38
43	salicylate	Na($C_7H_5O_3$)	160.11	6.25	1110[15]	1250[25]	6.25	144	Na+	6.25	856	$C_7H_5O_3$-	12.49
44	sulfate (anhydrous)	Na_2SO_4	142.05	7.04	47.6[0]	427[100]	14.08	324	Na+	14.08	676/226	SO_4-- / S	21.12
45	sulfate (Glauber's salt)	Na_2SO_4 + 10 H_2O	322.21	3.10	110[0]	927[30]	6.21	143	Na+	6.21	298/100	SO_4-- / S	9.31
46	thiosulfate	$Na_2S_2O_3$	158.11	6.32	500	2310[100]	12.65	291	Na+	12.65	709/406	S_2O_3-- / S	18.97
	Sulfur, S												
	See Calcium (11), Magnesium (18) and Sodium (44–46)												

* On the assumption of complete dissociation.
** The oxides have been included in view of the continuing use of the older nutritional tables.
*** The sodium content of 1 gram sodium bicarbonate corresponds to that of 1.33 grams sodium lactate. The sodium content of 1 gram sodium

lactate corresponds to that of 0.75 gram sodium bicarbonate.

1) Data from Hodgman et al. (Eds.), *Handbook of Chemistry and Physics*, 42nd ed., Cleveland, 1960. The index figures are the temperatures in °C; v.s. = very soluble; d = decomposes.

From Diem, K., Ed., *Documenta Geigy Scientific Tables*, 6th ed., Geigy Pharmaceuticals, Ardsley, N.Y., 1962, 328. With permission.

Table 22
AQUEOUS SOLUTIONS — CONVERSION FACTORS FOR ELECTROLYTES
(II)

For a given number of milliosmoles of solute on the assumption of complete dissociation, this table gives the corresponding weight and number of milli-moles (mmol) of the undissociated solute and the corresponding milligram equivalents (mEq) and weights of the cation and anion.

	Electrolyte (data for 10 millimoles of solute unless otherwise stated)		Undissociated solute		Cation			Anion		
			g	mmol	mEq	mg		mEq	mg	
	Calcium, Ca									
1	acetate	$Ca(C_2H_3O_2)_2 + H_2O$	0.587	$3^1/_3$	$6^2/_3$	134	Ca^{++}	$6^2/_3$	394	$C_2H_3O_2^-$
2		$Ca(C_2H_3O_2)_2 + 2H_2O$	0.647	$3^1/_3$	$6^2/_3$	134	Ca^{++}	$6^2/_3$	394	$C_2H_3O_2^-$
3	chloride	$CaCl_2 + 2H_2O$	0.490	$3^1/_3$	$6^2/_3$	134	Ca^{++}	$6^2/_3$	236	Cl^-
4		$CaCl_2 + 6H_2O$	0.730	$3^1/_3$	$6^2/_3$	134	Ca^{++}	$6^2/_3$	236	Cl^-
5	citrate	$Ca_3(C_6H_5O_7)_2 + 4H_2O$	1.141	2	12	241	Ca^{++}	12	756	$C_6H_5O_7^{---}$
6	D-gluconate	$Ca(C_6H_{11}O_7)_2 + H_2O$	1.495	$3^1/_3$	$6^2/_3$	134	Ca^{++}	$6^2/_3$	1301	$C_6H_{11}O_7^-$
7	lactate	$Ca(C_3H_5O_3)_2 + 5H_2O$	1.028	$3^1/_3$	$6^2/_3$	134	Ca^{++}	$6^2/_3$	594	$C_3H_5O_3^-$
8	levulinate	$Ca(C_5H_7O_3)_2 + 2H_2O$	1.021	$3^1/_3$	$6^2/_3$	134	Ca^{++}	$6^2/_3$	767	$C_5H_7O_3^-$
10	phosphate, dibasic	$CaHPO_4 + 2H_2O$	0.861	5	10	200	Ca^{++}	10	480 / 155	HPO_4^{--} / P
11	thiosulfate	$CaS_2O_3 + 6H_2O$	1.302	5	10	200	Ca^{++}	10	561 / 321	$S_2O_3^{--}$ / S
	Chlorine, Cl									
12	Ammonium chloride	NH_4Cl	0.268	5	5	90	NH_4^+	5	177	Cl^-
13	Hydrochloric acid (10% solution)									
	1 gram	(0.1 g HCl/g)	1.823	5	5	5	H^+	5	177	Cl^-
	1 milliliter	(0.1047 g HCl/ml)	1.741	5	5	5	H^+	5	177	Cl^-
	See also Calcium (3,4), Magnesium (14,15), Potassium (22) and Sodium (35)									
	Magnesium, Mg									
14	chloride	$MgCl_2$	0.317	$3^1/_3$	$6^2/_3$	81	Mg^{++}	$6^2/_3$	236	Cl^-
15		$MgCl_2 + 6H_2O$	0.678	$3^1/_3$	$6^2/_3$	81	Mg^{++}	$6^2/_3$	236	Cl^-
18	sulfate	$MgSO_4 + 7H_2O$	1.233	5	10	122	Mg^{++}	10	480 / 160	SO_4^{--} / S
	Phosphorus, P									
	See Calcium (10), Potassium (26, 27) and Sodium (30, 31, 40–42)									
	Potassium, K									
19	acetate	$K(C_2H_3O_2)$	0.491	5	5	196	K^+	5	295	$C_2H_3O_2^-$
20	bicarbonate	$KHCO_3$	0.501	5	5	196	K^+	5	305	HCO_3^-
21	bromide	KBr	0.595	5	5	196	K^+	5	400	Br^-
22	chloride	KCl	0.373	5	5	196	K^+	5	177	Cl^-
23	citrate	$K_3(C_6H_5O_7) + H_2O$	0.811	$2^1/_2$	$7^1/_2$	293	K^+	$7^1/_2$	473	$C_6H_5O_7^{---}$
24	D-gluconate	$K(C_6H_{11}O_7)$	1.171	5	5	196	K^+	5	976	$C_6H_{11}O_7^-$
26	phosphate, monobasic	KH_2PO_4	0.454	$3^1/_3$	$3^1/_3$	130	K^+	$6^2/_3$	320 / 103	HPO_4^{--} / P
27	phosphate, dibasic	K_2HPO_4	0.581	$3^1/_3$	$6^2/_3$	261	K^+	$6^2/_3$	320 / 103	HPO_4^{--} / P
	Sodium, Na									
28	acetate	$Na(C_2H_3O_2) + 3H_2O$	0.681	5	5	115	Na^+	5	295	$C_2H_3O_2^-$
29	acid citrate	$Na_2H(C_6H_5O_7) + 1\frac{1}{2}H_2O$	0.658	$2^1/_2$	5	115	Na^+	$7^1/_2$	473	$C_6H_5O_7^{---}$
30	acid phosphate	$NaH_2PO_4 + H_2O$	0.460	$3^1/_3$	$3^1/_3$	77	Na^+	$6^2/_3$	320	HPO_4^{--}
31		$NaH_2PO_4 + 2H_2O$	0.520	$3^1/_3$	$3^1/_3$	77	Na^+	$6^2/_3$	320 / 103	HPO_4^{--} / P
32	aminosalicylate	$Na(C_7H_6O_3N) + 2H_2O$	1.056	5	5	115	Na^+	5	761	$C_7H_6O_3N^-$
33	bicarbonate	$NaHCO_3$	0.420	5	5	115	Na^+	5	305	HCO_3^-
34	bromide	$NaBr$	0.515	5	5	115	Na^+	5	400	Br^-
35	chloride	$NaCl$	0.292	5	5	115	Na^+	5	177	Cl^-
36	citrate	$Na_3(C_6H_5O_7) + 2H_2O$	0.735	$2^1/_2$	$7^1/_2$	173	Na^+	$7^1/_2$	473	$C_6H_5O_7^{---}$
37		$Na_3(C_6H_5O_7) + 5\frac{1}{2}H_2O$	0.893	$2^1/_2$	$7^1/_2$	173	Na^+	$7^1/_2$	473	$C_6H_5O_7^{---}$
38	lactate	$Na(C_3H_5O_3)$	0.560	5	5	115	Na^+	5	445	$C_3H_5O_3^-$
40	phosphate	Na_2HPO_4	0.473	$3^1/_3$	$6^2/_3$	153	Na^+	$6^2/_3$	320 / 103	HPO_4^{--} / P
41		$Na_2HPO_4 + 2H_2O$	0.593	$3^1/_3$	$6^2/_3$	153	Na^+	$6^2/_3$	320 / 103	HPO_4^{--} / P
42		$Na_2HPO_4 + 12H_2O$	1.194	$3^1/_3$	$6^2/_3$	153	Na^+	$6^2/_3$	320 / 103	HPO_4^{--} / P
43	salicylate	$Na(C_7H_5O_3)$	0.801	5	5	115	Na^+	5	686	$C_7H_5O_3^-$
44	sulfate (anhydrous)	Na_2SO_4	0.474	$3^1/_3$	$6^2/_3$	153	Na^+	$6^2/_3$	320 / 107	SO_4^{--} / S
45	sulfate	$Na_2SO_4 + 10H_2O$	1.074	$3^1/_3$	$6^2/_3$	153	Na^+	$6^2/_3$	320 / 107	SO_4^{--} / S
46	thiosulfate	$Na_2S_2O_3$	0.527	$3^1/_3$	$6^2/_3$	153	Na^+	$6^2/_3$	374 / 214	$S_2O_3^{--}$ / S
	Sulfur, S									
	See Calcium (11), Magnesium (18) and Sodium (44–46)									

From Diem, K., Ed., *Documenta Geigy Scientific Tables*, 6th ed., Geigy Pharmaceuticals, Ardsley, N.Y., 1962, 329. With permission.

Table 23
AQUEOUS SOLUTIONS — CONVERSION FACTORS FOR ELECTROLYTES
(III)

Inorganic ions   Left-hand column: **Given:** weight of the inorganic ions. **Required:** corresponding weight of the salt.
Right-hand column: **Given:** milliequivalents of the ions. **Required:** corresponding weight of the salt.

1	**1 gram = 49.90 mEq Calcium (Ca^{++}) corresponds to:**				**1 mEq = 20.04 mg Calcium (Ca^{++}) corresponds to:**		
2	4.396 g	Calcium acetate	Ca(C$_2$H$_3$O$_2$)$_2$ + H$_2$O		88.10 mg	Calcium acetate	Ca(C$_2$H$_3$O$_2$)$_2$ + H$_2$O
3	4.845 g	dihydrate	Ca(C$_2$H$_3$O$_2$)$_2$ + 2 H$_2$O		97.10 mg	dihydrate	Ca(C$_2$H$_3$O$_2$)$_2$ + 2 H$_2$O
4	3.668 g	Calcium chloride	CaCl$_2$ + 2 H$_2$O		73.51 mg	Calcium chloride	CaCl$_2$ + 2 H$_2$O
5	5.466 g	hexahydrate	CaCl$_2$ + 6 H$_2$O		109.54 mg	hexahydrate	CaCl$_2$ + 6 H$_2$O
6	4.745 g	Calcium citrate	Ca$_3$(C$_6$H$_5$O$_7$)$_2$ + 4 H$_2$O		95.08 mg	Calcium citrate	Ca$_3$(C$_6$H$_5$O$_7$)$_2$ + 4 H$_2$O
7	11.188 g	Calcium D-gluconate	Ca(C$_6$H$_{11}$O$_7$)$_2$ + H$_2$O		224.20 mg	Calcium D-gluconate	Ca(C$_6$H$_{11}$O$_7$)$_2$ + H$_2$O
8	7.692 g	Calcium lactate	Ca(C$_3$H$_5$O$_3$)$_2$ + 5 H$_2$O		154.15 mg	Calcium lactate	Ca(C$_3$H$_5$O$_3$)$_2$ + 5 H$_2$O
9	7.643 g	Calcium levulinate	Ca(C$_5$H$_7$O$_3$)$_2$ + 2 H$_2$O		153.16 mg	Calcium levulinate	Ca(C$_5$H$_7$O$_3$)$_2$ + 2 H$_2$O
11	4.294 g	Calcium phosphate, dibasic	CaHPO$_4$ + 2 H$_2$O		86.05 mg	Calcium phosphate, dibasic	CaHPO$_4$ + 2 H$_2$O
12	6.495 g	Calcium thiosulfate	CaS$_2$O$_3$ + 6 H$_2$O		130.15 mg	Calcium thiosulfate	CaS$_2$O$_3$ + 6H$_2$O

1 gram Carbon dioxide, CO$_2$, corresponds to 1.387 g = 22.72 mEq bicarbonate ions (HCO$_3^-$)

**1 vol% Carbon dioxide, CO$_2$, at 0°C and 760 mm Hg corresponds to 27.40 mg/l = 0.449 mEq/l bicarbonate ions (HCO$_3^-$)*

1 mEq = 61.02 mg Bicarbonate ions (HCO$_3^-$) corresponds to 44.01 mg carbon dioxide (CO$_2$)

1 mEq/l = 61.02 mg/l Bicarbonate ions (HCO$_3^-$) corresponds at 0°C and 760 mm Hg to 2.23 vol% carbon dioxide (CO$_2$)*

13	**1 gram = 28.20 mEq Chloride (Cl$^-$) corresponds to:**				**1 mEq = 35.457 mg Chloride (Cl$^-$) corresponds to:**		
14	1.509 g	Ammonium chloride	NH$_4$Cl		53.50 mg	Ammonium chloride	NH$_4$Cl
4	2.073 g	Calcium chloride	CaCl$_2$ + 2 H$_2$O		73.51 mg	Calcium chloride	CaCl$_2$ + 2 H$_2$O
5	3.090 g	hexahydrate	CaCl$_2$ + 6 H$_2$O		109.54 mg	hexahydrate	CaCl$_2$ + 6 H$_2$O
15	10.28 g or 9.551 ml Hydrochloric acid, 10%				364.7 mg or 348.28μl Hydrochloric acid, 10%		
17	1.343 g	Magnesium chloride	MgCl$_2$		47.62 mg	Magnesium chloride	MgCl$_2$
18	2.867 g	hexahydrate	MgCl$_2$ + 6 H$_2$O		101.67 mg	hexahydrate	MgCl$_2$ + 6 H$_2$O
27	2.103 g	Potassium chloride	KCl		74.56 mg	Potassium chloride	KCl
41	1.648 g	Sodium chloride	NaCl		58.45 mg	Sodium chloride	NaCl
16	**1 gram = 82.24 mEq Magnesium (Mg^{++}) corresponds to:**				**1 mEq = 12.16 mg Magnesium (Mg^{++}) corresponds to:**		
17	3.916 g	Magnesium chloride	MgCl$_2$		47.62 mg	Magnesium chloride	MgCl$_2$
18	8.361 g	hexahydrate	MgCl$_2$ + 6 H$_2$O		101.67 mg	hexahydrate	MgCl$_2$ + 6 H$_2$O
21	10.136 g	Magnesium sulfate	MgSO$_4$ + 7 H$_2$O		123.25 mg	Magnesium sulfate	MgSO$_4$ + 7 H$_2$O

22	**1 gram Phosphorus, P, corresponds to:**		
11	5.556 g	Calcium phosphate, dibasic	CaHPO$_4$ + 2 H$_2$O
31	4.394 g	Potassium phosphate, monobasic	KH$_2$PO$_4$
32	5.623 g	Potassium phosphate, dibasic	K$_2$HPO$_4$
36	4.455 g	Sodium acid phosphate	NaH$_2$PO$_4$ + H$_2$O
37	5.037 g	dihydrate	NaH$_2$PO$_4$ + 2 H$_2$O
46	4.583 g	Sodium phosphate	Na$_2$HPO$_4$
47	5.747 g	dihydrate	Na$_2$HPO$_4$ + 2 H$_2$O
48	11.563 g	dodecahydrate	Na$_2$HPO$_4$ + 12 H$_2$O

At pH 4.3 one gram Phosphorus (P) corresponds to 32.28 mEq H$_2$PO$_4^-$ ions, and 1 mEq of H$_2$PO$_4^-$ ions corresponds to 30.98 mg phosphorus (with only a small error these figures can be used for urine).
At pH 9.6 one gram Phosphorus (P) corresponds to 64.56 mEq HPO$_4^{--}$ ions, and 1 mEq of HPO$_4^{--}$ ions corresponds to 15.49 mg phosphorus.
At pH 7.4 1 gram Phosphorus (P) corresponds to 58.1 mEq of phosphate ions, and 1 mEq of Phosphate ions corresponds to 17.21 mg phosphorus (ca. 20°/₀ H$_2$PO$_4^-$ ions and ca. 80°/₀ HPO$_4^{--}$ ions).

23	**1 gram = 25.58 mEq Potassium (K$^+$) corresponds to:**				**1 mEq = 39.10 mg Potassium (K$^+$) corresponds to:**		
24	2.510 g	Potassium acetate	K(C$_2$H$_3$O$_2$)		98.15 mg	Potassium acetate	K(C$_2$H$_3$O$_2$)
25	2.561 g	Potassium bicarbonate	KHCO$_3$		100.12 mg	Potassium bicarbonate	KHCO$_3$
26	3.044 g	Potassium bromide	KBr		119.02 mg	Potassium bromide	KBr
27	1.907 g	Potassium chloride	KCl		74.56 mg	Potassium chloride	KCl
28	2.766 g	Potassium citrate	K$_3$(C$_6$H$_5$O$_7$) + H$_2$O		108.14 mg	Potassium citrate	K$_3$(C$_6$H$_5$O$_7$) + H$_2$O
29	5.991 g	Potassium D-gluconate	K(C$_6$H$_{11}$O$_7$)		234.25 mg	Potassium D-gluconate	K(C$_6$H$_{11}$O$_7$)
31	3.481 g	Potassium phosphate, monobasic	KH$_2$PO$_4$		136.09 mg	Potassium phosphate, monobasic	KH$_2$PO$_4$
32	2.227 g	Potassium phosphate, dibasic	K$_2$HPO$_4$		87.09 mg	Potassium phosphate, dibasic	K$_2$HPO$_4$

33	**1 gram = 43.48 mEq Sodium (Na$^+$) corresponds to:**				**1 mEq = 22.991 mg Sodium (Na$^+$) corresponds to:**		
34	5.919 g	Sodium acetate	Na(C$_2$H$_3$O$_2$) + 3 H$_2$O		136.09 mg	Sodium acetate	Na(C$_2$H$_3$O$_2$) + 3 H$_2$O
35	5.722 g	Sodium acid citrate	Na$_2$H(C$_6$H$_5$O$_7$) + 1½ H$_2$O		131.55 mg	Sodium acid citrate	Na$_2$H(C$_6$H$_5$O$_7$) + 1¼ H$_2$O
36	6.002 g	Sodium acid phosphate	NaH$_2$PO$_4$ + H$_2$O		138.00 mg	Sodium acid phosphate	NaH$_2$PO$_4$ + H$_2$O
37	6.786 g	dihydrate	NaH$_2$PO$_4$ + 2 H$_2$O		156.01 mg	dihydrate	NaH$_2$PO$_4$ + 2 H$_2$O
38	9.184 g	Sodium aminosalicylate	Na(C$_7$H$_6$O$_3$N) + 2 H$_2$O		211.16 mg	Sodium aminosalicylate	Na(C$_7$H$_6$O$_3$N) + 2 H$_2$O
39	3.654 g	Sodium bicarbonate	NaHCO$_3$		84.01 mg	Sodium bicarbonate	NaHCO$_3$
41	2.542 g	Sodium chloride	NaCl		58.45 mg	Sodium chloride	NaCl
42	4.264 g	Sodium citrate	Na$_3$(C$_6$H$_5$O$_7$) + 2 H$_2$O		98.03 mg	Sodium citrate	Na$_3$(C$_6$H$_5$O$_7$) + 2 H$_2$O
43	5.178 g	Sodium citrate	Na$_3$(C$_6$H$_5$O$_7$) + 5½ H$_2$O		119.05 mg	Sodium citrate	Na$_3$(C$_6$H$_5$O$_7$) + 5½ H$_2$O
44	4.874 g	Sodium lactate	Na(C$_3$H$_5$O$_3$)		112.06 mg	Sodium lactate	Na(C$_3$H$_5$O$_3$)
46	3.088 g	Sodium phosphate	Na$_2$HPO$_4$		70.98 mg	Sodium phosphate	Na$_2$HPO$_4$
47	3.871 g	dihydrate	Na$_2$HPO$_4$ + 2 H$_2$O		89.00 mg	dihydrate	Na$_2$HPO$_4$ + 2 H$_2$O
48	7.789 g	dodecahydrate	Na$_2$HPO$_4$ + 12 H$_2$O		179.08 mg	dodecahydrate	Na$_2$HPO$_4$ + 12 H$_2$O
49	6.964 g	Sodium salicylate	Na(C$_7$H$_5$O$_3$)		160.11 mg	Sodium salicylate	Na(C$_7$H$_5$O$_3$)
50	3.089 g	Sodium sulfate (anhydrous)	Na$_2$SO$_4$		71.02 mg	Sodium sulfate (anhydrous)	Na$_2$SO$_4$
51	7.007 g	Sodium sulfate	Na$_2$SO$_4$ + 10 H$_2$O		161.10 mg	Sodium sulfate	Na$_2$SO$_4$ + 10 H$_2$O
52	3.439 g	Sodium thiosulfate	Na$_2$S$_2$O$_3$		79.05 mg	Sodium thiosulfate	Na$_2$S$_2$O$_3$

53	**1 gram Sulfur (S) corresponds to:**		
12	4.059 g	Calcium thiosulfate	CaS$_2$O$_3$ + 6 H$_2$O
21	7.687 g	Magnesium sulfate	MgSO$_4$ + 7 H$_2$O
50	4.430 g	Sodium sulfate (anhydrous)	Na$_2$SO$_4$
51	10.048 g	Sodium sulfate	Na$_2$SO$_4$ + 10 H$_2$O
52	2.465 g	Sodium thiosulfate	Na$_2$S$_2$O$_3$

1 gram Sulfur (S) corresponds to 62.37 mEq SO$_4^{--}$ ions and 1 mEq SO$_4^{--}$ ions corresponds to 16.03 mg sulfur.

At pH 7.4 and 38°C and with an albumin/globulin ratio of 1.8, **1 gram of serum proteins** corresponds to 0.241 basic mEq of ionized serum proteins, and **1 basic mEq** of ionized serum proteins corresponds to 4.15 grams of serum proteins[1].

* The conversion factors (0.449 and 2.23) given here for vol% CO$_2$ into mmol CO$_2$/l and mEq CO$_2$/l (bicarbonate-CO$_2$) are derived from the molar volume of this gas (22.257 liters at 0°C and 760 mm Hg). The conversion factor 2.24 often used in medical literature is mistakenly based on the molar volume of ideal gases (22.412 liters). For practical purposes the difference between these two factors is negligible. The same factor, rounded off to 2.226, is used in the section on blood gases in Synopsis of Blood, page 571.

1) After Van Slyke et al., *J. biol. Chem.*, **79**, 768 (1928).

From Diem, K., Ed., *Documenta Geigy Scientific Tables*, 6th ed., Geigy Pharmaceuticals, Ardsley, N.Y., 1962, 330. With permission.

Table 24
COMMON CHEMICAL CONVERSION FACTORS USED FOR SOME COMMON LABORATORY TESTS

For converting	Factor	log₁₀	For converting	Factor	log₁₀
Acetone into acetoacetic acid	1.758	0.245 50	Acetoacetic acid into acetone	0.5689	0.7550—1
Acetone into β-hydroxybutyric acid	1.792	0.2533	β-Hydroxybutyric acid into acetone	0.5579	0.7466—1
Ca into CaO	1.399	0.1458	CaO into Ca	0.7147	0.8541—1
Cl into NaCl	1.649	0.2172	NaCl into Cl	0.6066	0.7829—1
K into K_2O	1.205	0.0810	K_2O into K	0.8302	0.9192—1
Mg into MgO	1.658	0.2195	MgO into Mg	0.6032	0.7805—1
Na into NaCl	2.542	0.4052	NaCl into Na	0.3934	0.5948—1
Na into Na_2O	1.348	0.1297	Na_2O into Na	0.7419	0.8704—1
P into P_2O_5	2.291	0.3600	P_2O_5 into P	0.4364	0.6399—1
P into H_3PO_4	3.164	0.5002	H_3PO_4 into P	0.3161	0.4998—1
S into SO_3	2.497	0.3974	SO_3 into S	0.4005	0.6026—1
S into H_2SO_4	3.059	0.4857	H_2SO_4 into S	0.3269	0.5144—1
Protein-N into protein	6.25	0.7959	Protein into protein-N	0.16	0.2041—1
Ammonia-N into ammonia	1.216	0.0849	Ammonia into ammonia-N	0.8224	0.9151—1
Creatine-N into creatine	3.121	0.4943	Creatine into creatine-N	0.3204	0.5057—1
Creatinine-N into creatinine	2.692	0.4301	Creatinine into creatinine-N	0.3715	0.5700—1
Urea-N into urea	2.144	0.3312	Urea into urea-N	0.4665	0.6689—1
Uric acid-N into uric acid	3.001	0.4772	Uric acid into uric acid-N	0.3333	0.5228—1
Lipid-P into phosphatides	23.5	1.3711			
Lipid-P into lecithin	25	1.3979	Lecithin into lipid-P	0.040	0.6021—2

From Diem, K., Ed., *Documenta Geigy Scientific Tables,* 6th ed., Geigy Pharmaceuticals, Ardsley, N.Y., 1962, 312. With permission.

Table 25
VALUES OF 2.3026 *RT/F* AT SEVERAL TEMPERATURES (IN MILLIVOLTS)

$t°C$	Value	$t°C$	Value	$t°C$	Value	$t°C$	Value
0	54.197	25	59.157	50	64.118	80	70.070
5	55.189	30	60.149	55	65.110	85	71.062
10	56.181	35	61.141	60	66.102	90	72.054
15	57.173	38	61.737	65	67.094	95	73.046
18	57.767	40	62.133	70	68.086	100	74.038
20	58.165	45	63.126	75	69.078		

From Dean, J. A., Ed., *Lange's Handbook of Chemistry,* 12th ed., McGraw-Hill, New York, 1979, 5-70. With permission.

pH MEASUREMENTS

Basic Concept

The pH value is defined for an aqueous solution in an operational (arbitrary but reproducible) manner according to the Bates-Guggenheim convention:

$$pH_x = pH_s + \frac{E_r - E_s}{2.3026RT/F}$$

where R is the gas constant per mole. T is the temperature on the absolute scale, and F is the faraday. The pH_x of the unknown medium is calculated from that of an accepted standard (pH_s) and the measured difference in the emf (E) of the electrode combination when the standard solution is removed from the cell and replaced by the unknown. The partition "A" indicates a liquid junction. Electrodes as fabricated exhibit variations in the reproducibility of the reference electrode, in the liquid-junction potential, and, with glass electrodes, in the asymmetry potential. These differences are all eliminated in the standardizing procedure with standard reference pH buffers (see Bates, R. G., *Determination of pH, Theory and Practice,* Wiley, New York, 1964).

Electrode reversible to hydrogen ions	Standard reference buffer or unknown solution	Salt bridge (KCl, 3.5 *M* or saturated)	Reference electrode
		"A"	

An electrometric pH-measurement system consists of (1) pH-responsibe electrode, (2) reference electrode, and (3) potential-measuring device — some form of high-impedance electronic voltmeter for glass-electrode combinations and this or a potentiometer arrangement for other pH-responsive electrodes. Electronic pH meters are simply voltmeters with scale divisions in pH units which are equivalent to the values of $2.3026RT/F$ (in mV) per pH unit. There is no compensation incorporated in the meter for the changes in pH of the test solution as a function of temperature. Reliability of an indicator-reference electrode combination must be ascertained by standardization of the pH meter with one standard buffer and checking the pH response by immersing the combination in a second and different reference buffer.

The temperature compensator on a pH meter varies the instrument definition of a pH unit from 54.20 mV at 0° to perhaps 66.10 mV at 60°C. This permits one to measure the pH of the sample (and reference buffer standard) at its actual temperature and thus avoid error due to dissociation equilibria and to junction potentials which have significant temperature coefficients.

To prepare the standard pH buffer solutions recommended by the National Bureau of Standards, the indicated weights of the pure materials in Table 26 should be dissolved in water of specific conductivity not greater than 5 micromhos. The tartrate, phthalate, and phosphates can be dried for 2 hr at 110°C before use. Potassium tetroxalate and calcium hydroxide need not be dried. Fresh-looking crystals of borax should be used. Before use, excess solid potassium hydrogen tartrate and calcium hydroxide must be removed. Buffer solutions pH 6 or above should be stored in plastic containers and should be protected from carbon dioxide with soda-lime traps. The solutions should be replaced within 2 to 3 weeks, or sooner if formative of mold is noticed. A crystal of thymol may be added as preservative.

The following tables give approximate pH values for a number of substances such as acids, bases, foods, biological fluids, etc. All values are rounded off to the nearest tenth and are based on measurements made at 25°C. A few buffer systems with their pH values are also given.

Table 26
PRIMARY STANDARD SOLUTIONS FOR THE DETERMINATION OF pH

NATIONAL BUREAU OF STANDARDS REFERENCE pH BUFFER SOLUTIONS

Temperature, °C	Secondary standard, 0.05 MK tetroxalate	KH tartrate (satd. at 25°C)	0.05 M KH Phthalate	Phosphate 0.025 M KH_2PO_4, 0.025 M Na_2HPO_4	Phosphate 0.0087 M KH_2PO_4, 0.0302 M Na_2HPO_4	0.01 M $Na_2B_4O_7$, 10 H_2O	0.025 M $NaHCO_3$, 0.025 M Na_2CO_3	Secondary standard, $Ca(OH)_2$ (satd. at 25°C)
0	1.666		4.003	6.984	7.534	9.464	10.317	13.423
5	1.668		3.999	6.951	7.500	9.395	10.245	13.207
10	1.670		3.998	6.923	7.472	9.332	10.179	13.003
15	1.672		3.999	6.900	7.448	9.276	10.118	12.810
20	1.675		4.002	6.881	7.429	9.225	10.062	12.627
25	1.679	3.557	4.008	6.865	7.413	9.180	10.012	12.454
30	1.683	3.552	4.015	6.853	7.400	9.139	9.966	12.289
35	1.688	3.549	4.024	6.844	7.389	9.102	9.925	12.133
38	1.691	3.548	4.030	6.840	7.384	9.081		12.043
40	1.694	3.547	4.035	6.838	7.380	9.068	9.889	11.934
45	1.700	3.547	4.047	6.834	7.373	9.038		11.841
50	1.707	3.549	4.060	6.833	7.367	9.011	9.828	11.705
55	1.715	3.554	4.075	6.834		8.985		11.574
60	1.723	3.560	4.091	6.836		8.962		11.449
70	1.743	3.580	4.126	6.845		8.921		
80	1.766	3.609	4.164	6.859		8.885		
90	1.792	3.650	4.205	6.877		8.850		
95	1.806	3.674	4.227	6.886		8.833		
Dilution value, $\Delta pH_{1/2}$	+0.186	+0.049	+0.052	+0.080	+0.07	+0.01	0.079	−0.28
Buffer value, β	0.070	0.027	0.016	0.029	0.016	0.020	0.029	0.09
Composition (air weight per liter of Buffer Solution)	12.61 g	Saturate @ 25°C (~0.034 M)	10.12 g	3.39 g, 3.53 g	1.17 g, 4.30 g	3.80 g	2.10 g, 2.65 g	Saturated @ 25°C

Adapted from Dean, J. A., Ed., *Lange's Handbook of Chemistry,* 12th ed., McGraw-Hill, New York, 1979, 5-74. With permission.

Table 27
COMMON pH BUFFER SOLUTIONS

Unless otherwise stated, stock solutions and buffers should be prepared and made up with distilled water free from carbon dioxide. Standard reagents should be used. The strength of solutions made up with reagents of doubtful purity or degree of hydration must be checked by titration. The amounts x and y of the stock solutions required to yield a desired pH value are given in the table on the opposite page. In the table below, the buffers are arranged in the alphabetical order of their chemical names.

No.	Buffer	pH range	Stock solutions A	Stock solutions B	Composition of the buffer
1	WALPOLE's acetate[1,2]	3.6–5.6	0.2 molar acetic acid (12.0 g/l)	0.2 molar sodium acetate (16.4 g $C_2H_3O_2Na$ or 27.2 g $C_2H_3O_2Na \cdot 3H_2O$ per liter)	x ml A + y ml B made up to 100 ml
2	GOMORI's aconitate[2]	2.5–5.7	0.5 molar aconitic acid (87.1 g/l)	0.2-N NaOH	10 ml A + x ml B made up to 100 ml
3	MICHAELIS's barbital sodium[3]	6.8–9.6	0.1 molar barbital sodium (20.6 g/l)	0.1-N HCl	x ml A + (100–x) ml B
4	MICHAELIS's barbital sodium–acetate[4]	2.6–9.4	$1/7$ molar sodium acetate in $1/7$ molar barbital sodium (19.43 g $C_2H_3O_2 \cdot 3H_2O$ + 29.43 g barbital sodium in 1 liter)	0.1-N HCl	50 ml A + x ml B + 20 ml 8.5% NaCl solution made up to 250 ml
5	CLARK and LUBS's borate[5]	7.8–10.0	0.1 molar boric acid in 0.1 molar KCl (6.2 g H_3BO_3 + 7.46 g KCl per liter)	0.1-N NaOH	50 ml A + x ml B made up to 100 ml
6	KOLTHOFF's borax–phosphate[6,7]	5.8–9.2	0.05 molar borax (19.1 g/l)	0.1 molar monopotassium phosphate (13.6 g KH_2PO_4 per liter)	x ml A + (100–x) ml B
7	KOLTHOFF's borax–succinic acid[7]	3.0–5.8	0.05 molar borax (19.1 g/l)	0.05 molar succinic acid (5.9 g/l)	x ml A + (100–x) ml B
8	PLUMEL's cacodylate[2,8]	5.0–7.4	0.2 molar sodium cacodylate (42.8 g $Na[CH_3]_2AsO_2 \cdot 3H_2O$ per liter)	0.2-N HCl	25 ml A + x ml B made up to 100 ml
9	DELORY and KING's carbonate–bicarbonate[2,9]	9.2–10.7	0.2 molar anhydrous sodium carbonate (21.2 g/l)	0.2 molar sodium bicarbonate (16.8 g/l)	x ml A + y ml B made up to 100 ml
10	SØRENSEN's citrate I[10,11]	2.2–4.8	0.1 molar disodium citrate (21.0 g citric acid [1H_2O] diss. in 200 ml 1-N NaOH and made up to 1 liter)	0.1-N HCl	x ml A + (100–x) ml B
11	SØRENSEN's citrate II[10,11]	5.0–6.8	As No. 10	0.1-N NaOH	x ml A + (100–x) ml B
12	TEORELL and STENHAGEN's citrate–phosphate–borate[12]	2.0–12.0	To citric acid and phosphoric acid solutions (ca. 100 ml), each corr. to 100 ml 1-N NaOH, add 3.54 g cryst. orthoboric acid and 343 ml 1-N NaOH, and make up the mixture to 1 liter	0.1-N HCl	20 ml A + x ml B made up to 100 ml
13	McILVAINE's citric acid–phosphate[13]	2.2–8.0	0.1 molar citric acid (21.0 g $C_6H_8O_7 \cdot 1H_2O$ per liter)	0.2 molar disodium phosphate (35.6 g $Na_2HPO_4 \cdot 2H_2O$ per liter)	x ml A + (100–x) ml B
14	STAFFORD, WATSON and RAND's dimethylglutarate[14]	3.2–7.6	0.1 molar ββ-dimethylglutaric acid (16.02 g/l)	0.2-N NaOH	(a) 100 ml A + x ml B made up to 1000 ml; (b) 100 ml A + x ml B + 5.845 g NaCl made up to 1000 ml ($\triangle$ 0.1 molar NaCl)
15	SØRENSEN's glycine I[10,11]	1.2–3.6	0.1 molar glycine in 0.1-N NaCl (7.5 g glycine + 5.85 g NaCl in 1 liter)	0.1-N HCl	x ml A + (100–x) ml B
16	SØRENSEN's glycine II[10,11]	8.4–13.0	As No. 15	0.1-N NaOH	x ml A + (100–x) ml B
17	SØRENSEN's phosphate[11,15]	5.0–8.2	$1/15$ molar monopotassium phosphate (9.08 g KH_2PO_4 per liter)	$1/15$ molar disodium phosphate (11.88 g $Na_2HPO_4 \cdot 2H_2O$ per liter)	x ml A + (100–x) ml B
18	CLARK and LUBS's phthalate I[5]	2.2–3.8	0.1 molar potassium biphthalate (20.4 g $KHC_8H_4O_4$ per liter)	0.1-N HCl	50 ml A + x ml B made up to 100 ml
19	CLARK and LUBS's phthalate II[5]	4.0–6.2	As No. 18	0.1-N NaOH	50 ml A + x ml B made up to 100 ml
20	SMITH and SMITH's piperazine[16]	4.8–7.0 / 8.8–11.0	Molar piperazine dihydrochloride (159.1 g/l)	1-N NaOH	1000 ml A + x ml B
21	SMITH and SMITH's piperazine (sea-water)[16]	5.4–9.8	0.01 molar piperazine dihydrochloride in filtered sea water (pH 8.0) (1.591 g/l)	1-N NaOH	1000 ml A + x ml B
22	SMITH and SMITH's piperazine–glycylglycine[16]	4.4–10.8	0.01 molar piperazine dihydrochloride in 0.01 molar glycylglycine (1.591 g piperazine·2HCl + 1.321 g glycyl-glycine in 1 liter)	1-N NaOH	1000 ml A + x ml B
23	CLARK and LUB's potassium chloride–hydrochloric acid[5]	1.0–2.2	0.2-N KCl (14.9 g/l)	0.2-N HCl	25 ml A + x ml B made up to 100 ml
24	GOMORI's succinate[2]	3.8–6.0	0.2 molar succinic acid (23.62 g/l)	0.2-N NaOH	25 ml A + x ml B made up to 100 ml
25	GOMORI's tris[2,17]	7.2–9.0	0.2 molar tris (24.2 g tris[hydroxy-methyl]aminomethane per liter)	0.2-N HCl	25 ml A + x ml B made up to 100 ml
26	GOMORI's tris–maleate[2,17]	5.2–8.6	0.2 molar tris acid maleate (24.2 g tris-[hydroxymethyl]aminomethane + 23.2 g maleic acid or 19.6 g maleic anhydride in 1 liter)	0.2-N NaOH	25 ml A + x ml B made up to 100 ml

1) WALPOLE, G. S., *J. chem. Soc.*, 105, 2501 (1914). 2) GOMORI, G., in COLOWICK and KAPLAN (Eds.), *Methods in Enzymology*, vol. I, New York, 1955, page 138. 3) MICHAELIS, L., *J. biol. Chem.*, 87, 33 (1930). 4) MICHAELIS, L., *Biochem. Z.*, 234, 139 (1931). 5) CLARK and LUBS, *J. Bact.*, 2, 1 (1917). 6) KOLTHOFF, I. M., *Säure-Basen-Indicatoren*, Berlin, 1932, page 257. 7) KOLTHOFF, I. M., *J. biol. Chem.*, 63, 135 (1925). 8) PLUMEL, M., *Bull. Soc. chim. biol. (Paris)*, 30, 129 (1948). 9) DELORY and KING, *Biochem. J.*, 39, 245 (1945). 10) SØRENSEN, S. P. L., *Biochem. Z.*, 21, 131 (1909). 11) SØRENSEN, S. P. L., *Ergebn. Physiol.*, 12, 393 (1912). 12) TEORELL and STENHAGEN, *Biochem. Z.*, 299, 416 (1938). 13) McILVAINE, T. C., *J. biol. Chem.*, 49, 183 (1921). 14) STAFFORD et al., *Biochim. biophys. Acta*, 18, 319 (1955); KREBS and HEMS, personal communication (1957). 15) SØRENSEN, S. P. L., *Biochem. Z.*, 22, 352 (1909). 16) SMITH and SMITH, *Biol. Bull.*, 96, 233 (1949). 17) GOMORI, G., *Proc. Soc. exp. Biol. (N. Y.)*, 68, 354 (1948).

From Diem, K., Ed., *Documenta Geigy Scientific Tables*, 6th ed., Geigy Pharmaceuticals, Ardsley, N.Y., 1962, 314. With permission.

Table 28
COMMON pH BUFFER SOLUTIONS

pH	24 (23°C) x	25 (23°C) x	23 (23°C) x	22 (25°C) x	21 (25°C) x	20 (25°C) x	19 (20°C) x	18 (20°C) x	17 (18°C) x	15 (18°C) x	14b (21°C) x	14a (21°C) x	13 (23°C) x	12 (20°C) x	11 (18°C) x	10 (°C) x	8 (23°C) x	7 (18°C) x	6 (18°C) x	5 (20°C) x	4 (23°C) x	2 (23°C) x	1 (23°C) x	1 y	pH
1.0			48.50																			—	46.3	3.7	1.0
.2			32.25																			12.3	44.0	6.0	.2
.4			13.15																						.4
.6			8.40																						.6
.8																									.8
2.0			5.30																			16.0	41.0	9.0	2.0
.2			3.35																			20.0	36.5	13.5	.2
.4																						24.0	30.5	19.5	.4
.6																						28.0	25.5	24.5	.6
.8																						32.0	20.0	30.0	.8
13.0											7.5														13.0

pH variation negligible over the normal working temperature range.

Note: This table gives the quantities (x or x,y) of stock solutions required to make up any of the numbered buffers listed on the page opposite to a desired pH value at the temperature given.

From Diem, K., Ed., *Documenta Geigy Scientific Tables*, 6th ed., Geigy Pharmaceuticals, Ardsley, N.Y., 1962, 315. With permission.

Table 29
COMPOSITION AND pH VALUES OF BUFFER SOLUTIONS FROM pH 1.00 TO 13.00

25 mℓ 0.2MKCl + ×mℓ 0.2 MHCl, diluted to 100 mℓ			50 mℓ 0.1MKH Phthalate + ×mℓ 0.1M HCl diluted to 100 mℓ			50 mℓ 0.1 MKH Phthalate + ×mℓ 0.1 M NaOH, diluted to 100 mℓ		
pH	×	β	pH	×	β	pH	×	β
1.00	67.0	0.31	2.20	49.5		4.20	3.0	0.017
1.20	42.5	0.34	2.40	42.2	0.036	4.40	6.6	0.020
1.40	26.6	0.19	2.60	35.4	0.033	4.60	11.1	0.025
1.60	16.2	0.077	2.80	28.9	0.032	4.80	16.5	0.029
1.80	10.2	0.049	3.00	22.3	0.030	5.00	22.6	0.031
2.00	6.5	0.030	3.20	15.7	0.026	5.20	28.8	0.030
2.20	3.9	0.022	3.40	10.4	0.023	5.40	34.1	0.025
			3.60	6.3	0.018	5.60	38.8	0.020
			3.80	2.9	0.015	5.80	42.3	0.015

50 mℓ 0.1 MKH$_2$PO$_4$ + ×mℓ 0.1 MNaOH, diluted to 100 mℓ			50 mℓ of a mixture 0.1 $M$with respect to both KCl and H$_3$BO$_3$ + ×mℓ 0.1 MNaOH, diluted to 100 mℓ			50 mℓ 0.05 MNaHCO$_3$ + ×mℓ 0.1 MNaOH, diluted to 100 mℓ ΔpH/Δt ≃ −0.009 I = 0.001(25 + 2×)		
pH	×	β	pH	×	β	pH	×	β
5.80	3.6		8.00	3.9		10.40	16.5	0.013
6.00	5.6	0.010	8.20	6.0	0.011	10.60	19.1	0.012
6.20	8.1	0.015	8.40	8.6	0.015	10.80	21.2	0.009
6.40	11.6	0.021	8.60	11.8	0.018	11.00	22.7	
6.60	16.4	0.027	8.80	15.8	0.022			
6.80	22.4	0.033	9.00	20.8	0.027			
7.00	29.1	0.031	9.20	26.4	0.029			
7.20	34.7	0.025	9.40	32.1	0.027			
7.40	39.1	0.020	9.60	36.9	0.022			
7.60	42.4	0.013	9.80	40.6	0.016			
7.80	44.5	0.009	10.00	43.7	0.014			
8.00	46.1		10.20	46.2				

50 mℓ 0.05 MNa$_2$HPO$_4$ + ×mℓ 0.1 MNaOH, diluted to 100 mℓ ΔpH/Δt ≃ −0.025 I = 0.001(77 + 2×)			25 mℓ 0.2 MKCl + × mℓ 0.2 MNaOH, diluted to 100 mℓ ΔpH/Δt ≃ −0.033 I = 0.001(50 + 2×)		
pH	×	β	pH	×	β
11.00	4.1	0.009	12.00	6.0	0.028
11.20	6.3	0.012	12.20	10.2	0.048
11.40	9.1	0.017	12.40	16.2	0.076
11.60	13.5	0.026	12.60	25.6	0.12
11.80	19.4	0.034	12.80	41.2	0.21
11.90	23.0	0.037	13.00	66.0	0.30

Note: Values based on the Convential activity pH scale as defined by the National Bureau of Standards and pertain to a temperature of 25°C. Buffer value is denoted by column headed β.

Adapted from Dean, J. A., Ed., *Lange's Handbook of Chemistry*, 12th ed., McGraw-Hill, New York, N.Y., 1979, 5-77. With permission.

Table 30
pH VALUES OF SECONDARY BRITISH STANDARDS

Solution	12°C	25°C	38°C
Potassium tetroxalate 0.1 M	—	1.48	1.50
HCl 0.01 M KCl 0.09 M	—	2.07	2.08
Acetic acid 0.1 M, sodium acetate 0.1 M	4.65	4.64	4.65
Acetic acid 0.01 M, sodium acetate 0.01 M	4.71	4.70	4.72
KH_2PO_4 0.025 M, Na_2HPO_4 0.025 M	—	6.85	6.84
Sodium tetraborate decahydrate (borax) 0.05 M	—	9.18	9.07
$NaHCO_3$ 0.025 M, Na_2CO_3 0.025 M	—	10.00	—

From Dean, J. A., Ed., *Lange's Handbook of Chemistry,* 12th ed., McGraw-Hill, New York, 1979. With permission.

Table 31
APPROXIMATE pH VALUES OF COMMON REAGENT SOLUTIONS AT OR NEAR ROOM TEMPERATURE

Solution	Molarity	pH
Ammonia water	0.1	11.3
Ammonium chloride	0.1	4.6
Ammonium dihydrogen phosphate	0.1	4.0
Ammonium oxalate	0.1	6.4
Ammonium sulphate	0.1	5.5
Barbital sodium	0.1	9.4
Benzoic acid	Saturated	2.8
Boric acid	0.1	5.3
Calcium hydroxide	Saturated	12.4
Citric acid	0.1	2.1
Diammonium hydrogen phosphate	0.1	7.9
Disodium hydrogen phosphate	0.1	9.2
Hydrochloric acid	0.1	1.1
Oxalic acid	0.1	1.3
Potassium acetate	0.1	9.7
Potassium aluminium sulphate	0.1	4.2
Potassium bicarbonate	0.1	8.2
Potassium carbonate	0.1	11.5
Potassium dihydrogen phosphate	0.1	4.5
Salicylic acid	Saturated	2.4
Sodium acetate	0.1	8.9
Sodium benzoate	0.1	8.0
Sodium bicarbonate	0.1	8.3
Sodium bisulphate	0.1	1.4
Sodium carbonate	0.1	11.5
Sodium dihydrogen phosphate	0.1	4.5
Sodium hydroxide	0.1	12.9
Sodium tetraborate decahydrate	0.1	9.4
Succinic acid	0.1	2.7
Tartaric acid	0.1	2.0
Trichloracetic acid	0.1	1.2

From Dean, J. A., Ed., *Lange's Handbook of Chemistry,* 12th ed., McGraw-Hill, New York, 1979. With permission.

Table 32
APPROXIMATE pH VALUES OF SOME ACIDS, BASES, AND BIOLOGIC MATERIAL

Acids

Hydrochloric, N	0.1	Oxalic, 0.1N	1.6
Hydrochloric, 0.1N	1.1	Tartaric, 0.1N	2.2
Hydrochloric, 0.01N	2.0	Malic, 0.1N	2.2
Sulfuric, N	0.3	Citric, 0.1N	2.2
Sulfuric, 0.1N	1.2	Formic, 0.1N	2.3
Sulfuric, 0.01N	2.1	Lactic, 0.1N	2.4
Orthophosphoric, 0.1N	1.5	Acetic, N	2.4
Sulfurous, 0.1N	1.5	Acetic, 0.1N	2.9

Acetic, 0.01N	3.4
Benzoic, 0.01N	3.1
Alum, 0.1N	3.2
Carbonic (saturated)	3.8
Hydrogen sulfide, 0.1N	4.1
Arsenious (saturated)	5.0
Hydrocyanic, 0.1N	5.1
Boric, 0.1N	5.2

Bases

Sodium hydroxide, N	14.0	Lime (saturated)	12.4
Sodium hydroxide, 0.1N	13.0	Trisodium phosphate, 0.1N	12.0
Sodium hydroxide, 0.01N	12.0	Sodium carbonate, 0.1N	11.6
Potassium hydroxide, N	14.0	Ammonia, N	11.6
Potassium hydroxide, 0.1N	13.0	Ammonia, 0.1N	11.1
Potassium hydroxide, 0.01N	12.0	Ammonia, 0.01N	10.6
Sodium metasilicate, 0.1 N	12.6	Potassium cyanide, 0.1 N	11.0

Magnesia (saturated)	10.5
Sodium sesquicarbonate, 0.1M	10.1
Ferrous hydroxide (saturated)	9.5
Calcium carbonate (saturated)	9.4
Borax, 0.1N	9.2
Sodium bicarbonate, 0.1N	8.4

Biologic Materials

Blood, plasma, human	7.3—7.5	Gastric contents, human	1.0—3.0	Milk, human	6.6—7.6
Spinal fluid, human	7.3—7.5	Duodenal contents, human	4.8—8.2	Bile, human	6.8—7.0
Blood, whole, dog	6.9—7.2	Feces, human	4.6—8.4		
Saliva, human	6.5—7.5	Urine, human	4.8—8.4		

From Weast, R. C., Ed., CRC Handbook of Chemistry and Physics, 61st ed., CRC Press, Boca Raton, Fla., 1980, D-149. With permission.

Table 33
ACID-BASE RATIOS AT A GIVEN pH

Fraction base or acid (%)	pH	Fraction base or acid (%)	pH
5 or 95	$pK'_a \pm 1.25$	30 or 70	$pK'_a \pm 0.37$
10 or 90	$pK'_a \pm 0.95$	35 or 65	$pK'_a \pm 0.27$
15 or 85	$pK'_a \pm 0.75$	40 or 60	$pK'_a \pm 0.18$
20 or 80	$pK'_a \pm 0.60$	45 or 55	$pK'_a \pm 0.09$
25 or 75	$pK'_a \pm 0.48$	50 or 50	$pK'_a \pm 0$

Note: The above relationships are useful to have readily available to estimate the ratio of acid to base at a given pH or to estimate the buffer ration of acid required to give a given pH.

From Jencks, W.P. and Regenstein, J., in *CRC Handbook of Biochemistry and Molecular Biology*, 3rd ed., Fasman, G.D., Ed., CRC Press, Boca Raton, Fla., 1976, 306.

Table 34
IONIZATION CONSTANTS (K) FOR COMMON ACIDS AND BASES IN WATER[a]

Acid or base	K	pK_a	Acid or base	K	pK_a
Acetic acid	1.75×10^{-5}	4.76	Imidazole	1.01×10^{-7}	6.95
Acetoacetic acid	2.62×10^{-4}	3.58	Isocitric acid	5.13×10^{-4}	3.29
		(18°C)		1.99×10^{-5}	4.70
Ammonia	5.6×10^{-10}	9.25		3.98×10^{-7}	6.40
Barbituric acid	9.8×10^{-5}	4.01	Malic acid	3.9×10^{-4}	3.40
Benzoic acid	6.46×10^{-5}	4.19		7.8×10^{-6}	5.11
Boric acid[b]	6.4×10^{-10}	9.19	*p*-Nitrophenol	7×10^{-8}	7.15
Carbonic acid	4.47×10^{-7}	6.35	Oxalacetic acid	2.75×10^{-3}	2.56
	4.68×10^{-11}	10.34		4.27×10^{-5}	4.37
Citric acid	7.4×10^{-4}	3.13	Oxalic acid	6.5×10^{-2}	1.19
	1.7×10^{-5}	4.77		6.1×10^{-5}	4.21
	4.0×10^{-7}	6.40	Phosphoric acid	7.5×10^{-3}	2.12
Diethylbarbituric acid (Veronal)	3.7×10^{-8}	7.43		6.2×10^{-8}	7.21
				4.8×10^{-13}	12.32
Ethylenediamine	1.4×10^{-7}	6.85	Phosphorous acid	5×10^{-2}	1.30
	1.12×10^{-10}	9.93		2.6×10^{-7}	6.59
Ethylenediamine tetraacetate	1.00×10^{-2}	2.00	Pyruvic acid	3.23×10^{-3}	2.49
	2.16×10^{-3}	2.67	Succinic acid	6.2×10^{-5}	4.21
	6.92×10^{-7}	6.16		2.3×10^{-6}	5.64
	5.50×10^{-11}	10.26	Sulfuric acid	$\gg 1$	—
Formic acid	1.76×10^{-4}	3.75		1.2×10^{-2}	1.92
Glycine	4.5×10^{-3}	2.35	Tartaric acid	1.1×10^{-3}	2.96
	1.7×10^{-10}	9.77		6.9×10^{-5}	4.16
Glycylglycine	7.24×10^{-4}	3.14	Triethanolamine	1.26×10^{-8}	7.90
	5.62×10^{-9}	8.25	Tris(hydroxymethyl)-amino methane	8.32×10^{-9}	8.08
Hydroxylamine	9.1×10^{9}	8.04			

[a] Temperature at or near room temperature (25°C) unless otherwise indicated.
[b] Boric acid acts as a monotropic acid in aqueous solution.

Adapted from Tietz, N. W., Ed., *Fundamentals of Clinical Chemistry*, W. B. Saunders, Philadelphia, 1970, 925.

Table 35
pH INDICATOR METHODS

pH RANGES OF INDICATORS

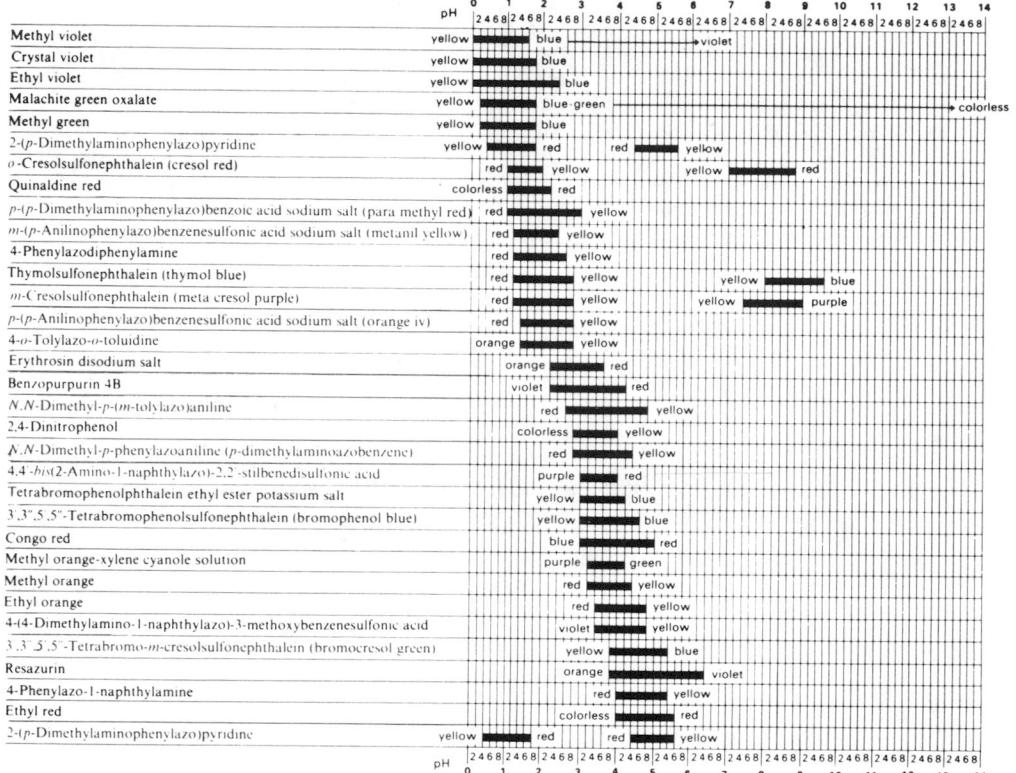

From Fasman, G. D., Ed., *CRC Handbook of Biochemistry and Molecular Biology*, Vol. 1, 3rd ed., CRC Press, Boca Raton, Fla., 1975, 388. With permission.

Acid-base indicators have the property of altering the color of a solution in the region 1 to 2 pH units as color of a solution in the region 1 to 2 pH units as the pH changes. They are therefore useful for pH measurements, although in general the accuracy is inferior to that obtainable by electrometric procedures. A list of suitable indicators, their pH ranges, and color changes is shown in Tables 35 and 36.

Fluorescent indicators are substances which show definite changes in fluorescence with change in pH. Some fluorescent materials are not suitable for indicators since their change in fluorescence is too gradual. Fluorescent indicators find greatest utility in the titration of opaque, highly turbid or deeply colored solutions. A long wavelength ultraviolet ("black light") lamp in a dimly lighted room provides the best environment for titrations involving fluorescent indicators, although bright daylight is sometimes sufficient to evoke a response in the bright green, yellow, and orange fluorescent indicators. Titrations are carried out in nonfluorescent glassware. One should check the glassware prior to use to make certain that it does not fluoresce due to the wavelengths of light involved in the titration. The meniscus of the liquid in the burette can be followed when a few particles of an insoluble fluorescent solid are dropped onto its surface.

Table 36
INDICATORS FOR VOLUMETRIC WORK AND pH DETERMINATIONS

Indicator		Acid color	pH range	Basic color	Preparation
Methyl violet 6B	Tetra and pentamethylated p-rosaniline hydrochloride	Y	0.1–1.5	B	pH: 0.25% water
Metacresol purple (acid range)	m-Cresolsulfonphthalein	R	0.5–2.5	Y	pH: 0.10 g. in 13.6 ml. 0.02 N NaOH, diluted to 250 ml. with water
Metanil yellow	4-Phenylamino-azobenzene-3′-sulfonic acid	R	1.2–2.3	Y	pH: 0.25% in ethanol
p-Xylenol blue (acid range)	1,4-Dimethyl-5-hydroxybenzene-sulfonphthalein	R	1.2–2.8	Y	pH: 0.04% in ethanol
Thymol blue (acid range)	Thymolsulfonphthalein	R	1.2–2.8	Y	pH: 0.1 g. in 10.75 ml. 0.02 N NaOH, diluted to 250 ml. with water
Tropaeolin OO	Sodium p-diphenylaminoazo-benzenesulfonate	R	1.4–2.6	Y	pH: 0.1% in water Vol.: 1% in water
Quinaldine red	2-(p-Dimethylaminostyryl)-quinoline ethiodide	C	1.4–3.2	R	Vol.: 0.1% in ethanol
Benzopurpurine 4B	Ditolyl-diazo-bis-α-naphthyl-amine-4-sulfonic acid	B-V	1.3–4.0	R	pH, vol.: 0.1% in water
Methyl violet 6B	Tetra and pentamethylated p-rosaniline hydrochloride	B	1.5–3.2	V	pH, vol.: 0.25% in water
2,4-Dinitrophenol		C	2.6–4.0	Y	pH, vol.: 0.1 g. in 5 ml. ethanol, diluted to 100 ml. with water
Methyl yellow	p-Dimethylaminoazobenzene	R	2.9–4.0	Y	pH, vol.: 0.05% in ethanol
Bromphenol blue	Tetrabromophenolsulfon-phthalein	Y	3.0–4.6	B	pH: 0.1 g. in 7.45 ml. 0.02 N NaOH, diluted to 250 ml. with water
Tetrabromophenol blue	Tetrabromophenol-tetrabromo-sulfonphthalein	Y	3.0–4.6	B	pH: 0.1 g. in 5.00 ml. 0.02 N NaOH, diluted to 250 ml. with water
Direct purple	Disodium 4,4′-bis(2-amino-1-naphthylazo)-2,2′-stilbene-disulfonate	B-P	3.0–4.6	R	Vol.: 0.1 g. in 7.35 ml. 0.02 N NaOH, diluted to 100 ml. with water
Congo red	Diphenyl-diazo-bis-1-naphthyl-amine-4-sodium sulfonate	B	3.0–5.2	R	pH: 0.1% in water
Methyl orange	4′-Dimethylaminoazobenzene-4-sodium sulfonate	R	3.1–4.4	Y	Vol.: 0.1% in water
Brom-chlorphenol blue	Dibromodichlorophenolsulfon-phthalein	Y	3.2–4.8	B	pH: 0.1 g. in 8.6 ml. 0.02 N NaOH, diluted to 250 ml. with water Vol.: 0.04% in ethanol
p-Ethoxychrysoidine	4′-Ethoxy-2,4-diaminoazo-benzene	R	3.5–5.5	Y	Vol.: 0.1% in ethanol
α-Naphthyl red		R	3.7–5.0	Y	Vol.: 0.1% in ethanol
Sodium alizarinsulfonate	Dihydroxyanthraquinone sodium sulfonate	Y	3.7–5.2	V	pH, vol.: 1% in water
Bromcresol green	Tetrabromo-m-cresolsulfon-phthalein	Y	3.8–5.4	B	pH: 0.10 g. in 7.15 ml. 0.02 N NaOH, diluted to 250 ml. with water
2,5-Dinitrophenol		C	4.0–5.8	Y	pH, vol.: 0.10 g. in 20 ml. ethanol, then dilute to 100 ml. with water
Methyl red	4′-Dimethylaminoazobenzene-2-carboxylic acid	R	4.2–6.2	Y	pH: 0.10 g. in 18.6 ml. 0.02 N NaOH, diluted to 250 ml. with water Vol.: 0.1% in ethanol
Lacmoid		R	4.4–6.2	B	Vol.: 0.5% in ethanol
Azolitmin		R	4.5–8.3	B	Vol.: 0.5% in water
Litmus		R	4.5–8.3	B	Vol.: 0.5% in water
Cochineal	Complex hydroxyanthraquinone derivative	R	4.8–6.2	V	Vol.: Triturate 1 g. with 20 ml. ethanol and 60 ml. water, let stand 4 days, and filter
Hematoxylin		Y	5.0–6.0	V	Vol.: 0.5% in ethanol.
Chlorphenol red	Dichlorophenolsulfonphthalein	Y	5.0–6.6	R	pH: 0.1 g. in 11.8 ml. 0.02 N NaOH, diluted to 250 ml. with water Vol.: 0.04% in ethanol
Bromcresol purple	Dibromo-o-cresolsulfonphthalein	Y	5.2–6.8	Pu	pH: 0.1 g. in 9.25 ml. 0.02 N NaOH, diluted to 250 ml. with water Vol.: 0.02% in ethanol

Table 36 (continued)
INDICATORS FOR VOLUMETRIC WORK AND pH DETERMINATIONS

Indicator	Chemical name	Acid color	pH range	Basic color	Preparation
Bromphenol red	Dibromophenolsulfonphthalein	Y	5.2–7.0	R	pH: 0.1 g. in 9.75 ml. 0.02 N NaOH, diluted to 250 ml. with water Vol.: 0.04% in ethanol
Alizarin	1,2-Dihydroxyanthraquinone	Y	5.5–6.8	R	Vol.: 0.1% in ethanol
Dibromophenoltetrabromo-phenolsulfonphthalein		Y	5.6–7.2	Pu	pH: 0.1 g. in 1.21 ml. 0.1 N NaOH, diluted to 250 ml. with water
p-Nitrophenol		C	5.6–7.6	Y	pH, vol.: 0.25% in water
Bromothymol blue	Dibromothymolsulfonphthalein	Y	6.0–7.6	B	pH: 0.1 g. in 8 ml. 0.02 N NaOH, diluted to 250 ml. with water Vol.: 0.1% in 50% ethanol
Indo-oxine	5,8-Quinolinequinone-8-hydroxy-5-quinolyl-5-imide	R	6.0–8.0	B	Vol.: 0.05% in ethanol
Cucumin		Y	6.0–8.0	Br-R	Vol: saturated aq. soln.
Quinoline blue	Cyanine	C	6.6–8.6	B	Vol.: 1% in ethanol
Phenol red	Phenolsulfonphthalein	Y	6.8–8.4	R	pH: 0.1 g. in 14.20 ml. 0.02 N NaOH, diluted to 250 ml. with water Vol.: 0.1% in ethanol
Neutral red	2-Methyl-3-amino-6-dimethyl-aminophenazine	R	6.8–8.0	Y	pH, vol.: 0.1 g. in 70 ml. ethanol, diluted to 100 ml. with water
Rosolic acid aurin; corallin		Y	6.8–8.2	R	pH, vol.: 1% in 50% ethanol
Cresol red	o-Cresolsulfonphthalein	Y	7.2–8.8	R	pH: 0.1 g. in 13.1 ml. 0.02 N NaOH, diluted to 250 ml. with water Vol.: 0.1% in ethanol
α-Naphtholphthalein		P	7.3–8.7	G	pH, vol.: 0.1% in 50% ethanol
Metacresol purple (alkaline range)	m-Cresolsulfonphthalein	Y	7.4–9.0	P	pH: 0.1 g. in 13.1 ml. 0.02 N NaOH, diluted to 250 ml. with water Vol.: 0.1% in ethanol
Ethylbis-2,4-dinitrophenylacetate		C	7.5–9.1	B	Vol.: saturated soln. in equal volumes of acetone and ethanol
Tropaeolin OOO No. 1	Sodium α-naphtholazobenzene-sulfonate	Y	7.6–8.9	R	Vol.: 0.1% in water
Thymol blue (alkaline range)	Thymolsulfonphthalein	Y	8.0–9.6	B	pH: 0.1 g. in 10.75 ml. 0.02 N NaOH, diluted to 250 ml. with water Vol.: 0.1% in ethanol
p-Xylenol blue	1,4-Dimethyl-5-hydroxybenzene-sulfonphthalein	Y	8.0–9.6	B	pH, vol.: 0.04% in ethanol
o-Cresolphthalein		C	8.2–9.8	R	pH, vol.: 0.04% in ethanol
α-Naphtholbenzein		Y	8.5–9.8	G	pH, vol.: 1% in ethanol
Phenolphthalein	3,3-Bis(p-hydroxyphenyl)-phthalide	C	8.2–10	R	Vol.: 1% in ethanol
Thymolphthalein		C	9.3–10.5	B	pH, vol.: 0.1% in ethanol
Nile blue A	Aminonaphthodiethylamino-phenoxazine sulfate	B	10–11	P	Vol.: 0.1% in water
Alizarin yellow GG	3-Carboxy-4-hydroxy-3'-nitro-azobenzene	Y	10–12	L	pH, vol.: 0.1% in 50% ethanol
Alizarin yellow R	3-Carboxy-4-hydroxy-4'-nitro-azobenzene sodium salt	Y	10.2–12.0	R	pH, vol.: 0.1% in water
Poirrer's blue C4B		B	11–13	R	pH: 0.2% in water
Tropaeolin O	p-Benzenesulfonic acid-azo-resorcinol	Y	11–13	O	pH: 0.1% in water
Nitramine	Picrylnitromethylamine	C	10.8–13	Br	pH: 0.1% in 70% ethanol
1,3,5-Trinitrobenzene		C	11.5–14	O	pH: 0.1% in ethanol
Indigo carmine	Sodium indigodisulfonate	B	11.6–14	Y	pH: 0.25% in 50% ethanol

Table 36 (continued)
INDICATORS FOR VOLUMETRIC WORK AND pH DETERMINATIONS

MIXED INDICATORS

Composition	Solvent	Transition pH	Acid color	Transition color	Basic color
Dimethyl yellow, 0.05% + Methylene blue, 0.05%	alc.	3.2	Blue–violet	—	Green
Methyl orange, 0.02% + Xylene cyanole FF, 0.28%	50% alc.	3.9	Red	Gray	Green
Methyl yellow, 0.08% + Methylene blue, 0.004%	alc.	3.9	Pink	Straw–pink	Yellow–green
Methyl orange, 0.1% + Indigocarmine, 0.25%	aq.	4.1	Violet	Gray	Yellow–green
Bromcresol green, 0.1% + Methyl orange, 0.02%	aq.	4.3	Orange	Light green	Dark green
Bromcresol green, 0.075% + Methyl red, 0.05%	alc.	5.1	Wine–red	—	Green
Methyl red, 0.1% + Methylene blue, 0.05%	alc.	5.4	Red–violet	Dirty blue	Green
Bromcresol green, 0.05% + Chlorphenol red, 0.05%	aq.	6.1	Yellow–green	—	Blue–violet
Bromcresol purple, 0.05% + Bromthymol blue, 0.05%	aq.	6.7	Yellow	Violet	Violet–blue
Neutral red, 0.05% + Methylene blue, 0.05%	alc.	7.0	Violet–blue	Violet–blue	Green
Bromthymol blue, 0.05% + Phenol red, 0.05%	aq.	7.5	Yellow	Violet	Dark violet
Cresol red, 0.025% + Thymol blue, 0.15%	aq.	8.3	Yellow	Rose	Violet
Phenolphthalein, 0.033% + Methyl green, 0.067%	alc.	8.9	Green	Gray–blue	Violet
Phenolphthalein, 0.075% + Thymol blue, 0.025%	50% alc.	9.0	Yellow	Green	Violet
Phenolphthalein, 0.067% + Naphtholphthalein, 0.033%	50% alc.	9.6	Pale rose	—	Violet
Phenolphthalein, 0.033% + Nile blue, 0.133%	alc.	10.0	Blue	Violet	Red
Alizarin yellow, 0.033% + Nile blue, 0.133%	alc.	10.8	Green	—	Red–brown

Note: The indicator colors are abbreviated as follows: B, blue; Br, brown; C, colorless; G, green; L, lilac; O, orange; P, pink; Pu, purple; R, red; V, violet; and Y, yellow.

From Fasman, G. D., Ed., *CRC Handbook of Biochemistry and Molecular Biology*, Vol. 1, 3rd ed., CRC Press, Boca Raton, Fla., 1975, 380. With permission.

Table 37
FLUORESCENT INDICATORS

pH 0 to 2

Indicator	C.I.	From pH	To pH
Benzoflavine	—	0.3, yellow fl.	1.7, green fl.
3,6-Dioxyphthalimide	—	0, blue fl.	2.4, green fl.
Eosine YS	768	0, yellow colored	3.0, yellow fl.
Erythrosine	772	0, yellow colored	3.6, yellow fl.
Esculin	—	1.5, colorless	2, blue fl.
4-Ethoxyacridone	—	1.2, green fl.	3.2, blue fl.
3,6-Tetramethyldiaminooxanthone	—	1.2, green fl.	3.4, blue fl.

pH 2 to 4

Indicator	C.I.	From pH	To pH
Chromotropic acid	—	3.5, colorless	4.5, blue fl.
Fluorescein	766	4, colorless	4.5, green fl.
Magdala Red	—	3.0, purple colored	4.0, fl.
α-Naphthylamine	—	3.4, colorless	4.8, blue fl.
β-Naphthylamine	—	2.8, colorless	4.4, violet fl.
Phloxine	774	3.4, colorless	5.0, bright yellow fl.
Salicylic acid	—	2.5, colorless	3.5, blue fl.

In Table 37 the indicators are arranged by approximate pH range covered. In the case of some of the dyestuffs the end point may vary slightly with the source or manufacturer.

Table 37 (continued)
FLUORESCENT INDICATORS

pH 4 to 6

Acridine	788	4.9, green fl.	5.1, violet colored
Dichlorofluorescein	—	4.0, colorless	5.0, green fl.
3,6-Dioxyxanthone	—	5.4, colorless	7.6, blue-violet fl.
Erythrosine	772	4.0, colorless	4.5, yellow-green fl.
β-Methylesculetin	—	4.0, colorless	6.2, blue fl.
Neville-Winther acid	—	6.0, colorless	6.5, blue fl.
Resorufin	—	4.4, yellow fl.	6.4, weak orange fl.
Quininic acid	—	4.0, yellow colored	5.0, blue fl.
Quinine [first end point]	—	5.0, blue fl.	6.1, violet fl.

pH 6 to 8

Acid R Phosphine	—	(claimed for range pH 6.0–7.0)	
Brilliant Diazol Yellow	—	6.5, colorless	7.5, violet fl.
Cleves acid	—	6.5, colorless	7.5, green fl.
Coumaric acid	—	7.2, colorless	9.0, green fl.
3,6-Dioxyphthalic dinitrile	—	5.8, blue fl.	8.2, green fl.
Magnesium 8-hydroxyquinolinate	—	6.5, colorless	7.5, golden fl.
β-Methylumbelliferone	—	7.0, colorless	7.5, blue fl.
1-Naphthol-4-sulfonic acid	—	6.0, colorless	6.5, blue fl.
Orcinaurine	—	6.5, colorless	8.0, green fl.
Patent Phosphine	789	(for the range pH 6.0–7.0, green-yellow fl.)	
Thioflavine	816	(for the region pH 6.5–7.0, yellow fl.)	
Umbelliferone	—	6.5, colorless	7.6, blue fl.

pH 8 to 10

Indicator	C.I.	From pH	To pH
Acridine Orange	788	8.4, orange colored	10.4, green fl.
Ethoxyphenylnaphthostilbazonium chloride	—	9, green fl.	11, non-fl.
G Salt	—	9.0, dull blue fl.	9.5, bright blue fl.
Naphthazol derivatives	—	8.2, colorless	10.0, yellow or green fl.
α-Naphthionic acid	—	9, blue fl.	11, green fl.
2-Naphthol-3,6-disulfonic acid	—	9.5, dark blue fl.	Light blue fl. at higher pH
β-Naphthol	—	8.6, colorless	Blue fl. at higher pH
α-Naphtholsulfonic acid	—	8.0, dark blue fl.	9.0, bright violet fl.
1,4-Naphtholsulfonic acid	—	8.2, dark blue fl.	Light blue fl. at higher pH
Orcinsulfonphthalein	—	8.6, yellow colored	10.0 fl.
Quinine [second end point]	—	9.5, violet fl.	10.0, colorless
R-Salt	—	9.0, dull blue fl.	9.5, bright blue fl.
Sodium 1-naphthol-2-sulfonate	—	9.0, dark blue fl.	10.0, bright violet fl.

pH 10 to 12

Coumarin	—	9.8, deep green fl.	12, light green fl.
Eosine BN	771	10.5, colorless	14.0, yellow fl.
Papaverine (permanganate oxidized)	—	9.5, yellow fl.	11.0, blue fl.
Schaffers Salt	—	5.0, violet fl.	11.0, green-blue fl.
SS-Acid (sodium salt)	—	10.0, violet fl.	12.0, yellow colored

pH 12 to 14

Cotarnine	—	12.0, yellow fl.	13.0, white fl.
α-Naphthionic acid	—	12, blue fl.	13, green fl.
β-Naphthionic acid	—	12, blue fl.	13, violet fl.

From DeMent, J., in Weast, R. C., Ed., *CRC Handbook of Chemistry and Physics*, 61st ed., CRC Press, Boca Raton, Fla., 1981, D-152. With permission.

Table 38
SELECTED LIST OF OXIDATION-REDUCTION INDICATORS

Name	Reduction potential (30°C) in Volts at		Suitable pH range	Color change upon oxidation
	pH = 0	pH = 7		
Bis(5-bromo-1,10-phenanthroline) ruthenium(II) dinitrate	1.41[a]			Red to faint blue
Tris(5-nitro-1,10-phenanthroline) iron(II) sulfate	1.25[a]			Red to faint blue
Iron(II)-2,2′,2″-tripyridine sulfate	1.25[a]			Pink to faint blue
Tris(4,7-diphenyl-1,10-phenanthroline) iron(II) disulfate	1.13 (4.6 $M H_2SO_4$)[a]			Red to faint blue
	0.87(1.0 $M H_2SO_4$)[a]			
o m′-Diphenylaminedicarboxylic acid	1.12			Colorless to blue-violet
Setopaline	1.06 (*trans*)[b]			Yellow to orange
p-Nitrodiphenylamine	1.06			Colorless to violet
Tris(1,10-phenanthroline)-iron(II) sulfate	1.06 (1.00 $M H_2SO_4$)[a]			Red to faint blue
	1.00 (3.0 $M H_2SO_4$)[a]			
	0.89 (6.0 $M H_2SO_4$)[a]			
Setoglaucine O	1.01 (*trans*)[b]			Yellow-green to yellow-red
Xylene cyanole FF	1.00 (*trans*)[b]			Yellow-green to pink
Erioglaucine A	1.00 (*trans*)[b]			Green-yellow to bluish red
Eriogreen	0.99 (*trans*)[b]			Green-yellow to orange
Tris(2,2′-bipyridine)-iron(II) hydrochloride	0.97[a]			Red to faint blue
2-Carboxydiphenylamine [N-phenylanthranilic acid]	0.94			Colorless to pink
Benzidine dihydrochloride	0.92			Colorless to blue
o-Toluidine	0.87			Colorless to blue
Bis(1,10-phenanthroline-osmium(II) perchlorate	0.859 (0.1 $M H_2SO_4$)			Green to pink
Diphenylamine-4-sulfonate-(Na salt)	0.85			Colorless to violet
3,3′-Dimethoxybenzidine dihydrochloride [o-dianisidine]	0.85			Colorless to red
Ferrocyphen	0.81			Yellow to violet
4′-Ethoxy-2,4-diaminoazobenzene	0.76			Red to pale yellow
N,N-Diphenylbenzidine	0.76			Colorless to violet
Diphenylamine	0.76			Colorless to violet
N,N-Dimethyl-p-phenylenediamine	0.76			Colorless to red
Variamine blue B hydrochloride	0.712[c]	0.310	1.5—6.3	Colorless to blue
N-Phenyl-1,2,4-benzenetriamine	0.70			Colorless to red

Table 38 (continued)
SELECTED LIST OF OXIDATION-REDUCTION INDICATORS

Name	Reduction potential (30°C) in Volts at		Suitable pH range	Color change upon oxidation
	pH = 0	pH = 7		
Bindschedler's green	0.680[c]	0.224	2—9.5	Colorless to blue
2,6-Dichloroindophenol (Na salt)	0.668[c]	0.217	6.3—11.4	Colorless to blue
2,6-Dibromophenolindophenol	0.668[c]	0.216	7.0—12.3	Colorless to blue
Brilliant cresyl blue [3-amino-9-dimethyl-amino-10-methylphenoxy-azine chloride]	0.583	0.047	0—11	Colorless to blue
Iron(II)-tetrapyridine chloride	0.59			Red to faint blue
Thionine [Lauth's violet]	0.563[c]	0.064	1—13	Colorless to violet
Starch (soluble potato, I₃; present)	0.54			Colorless to blue
Gallocyanine (25°C)	0.532[c]	0.021		Colorless to violet-blue
Methylene blue		0.011	1—13	Colorless to blue
Nile blue A [aminonaphthodiethylamino-phenoxazine sulfate]	0.406[c]	−0.119	1.4—12.3	Colorless to blue
Indigo-5,5',7,7'-tetrasulfonic acid (Na salt)	0.365[c]	−0.046	<9	Colorless to blue
Indigo-5,5',7-trisulfonic acid (Na salt)	0.332[c]	−0.081	<9	Colorless to blue
Indigo-5,5'-disulfonic acid (Na salt)	0.291[c]	−0.125	<9	Colorless to blue
Phenosafranine	0.280[c]	−0.252	1—11	Colorless to violet-blue
Indigo-5-monosulfonic acid (Na salt)	0.262[c]	−0.157	<9	Colorless to blue
Safranine T	0.24[c]	−0.289	1—12	Colorless to violet-blue
Bis(dimethylglyoximato)-iron(II) chloride	0.155		6—10	Red to colorless
Induline scarlet	0.047[c]	−0.299	3—8.6	Colorless to red
Neutral red		−0.323	2—11	Colorless to red-violet

[a] Transition point is at higher potential than the tabulated formal potential because the molar absorptivity of the reduced form is very much greater than that of the oxidized form.

[b] Trans = first noticeable color transition; often 60 mV less than E°.

[c] Values of E° are obtained by extrapolation from measurements in weakly acid or weakly alkaline systems.

From Dean, J. A., Ed., *Lange's Handbook of Chemistry*, 12th ed., McGraw-Hill, New York, 1979, 6-20. With permission.

PREPARATION OF BUFFERS FOR USE IN ENZYME STUDIES*

The buffers described in this section are suitable for use either in enzymatic or histochemical studies. The accuracy of the tables is within ±0.05 pH at 23°. In most cases the pH values will not be off by more than ±0.12 pH even at 37° and at molarities slightly different from those given (usually 0.05 M).

The methods of preparation described are not necessarily identical with those of the original authors. The titration curves of the majority of the buffers recommended have been redetermined by the writer. The buffers are arranged in the order of ascending pH range. For more complete data on phosphate and acetate buffers over a wide range of concentrations, see Vol. 1[10].*

* From Gomori, in *Methods in Enzymology,* Vol. 1, Colowick and Kaplan, Eds., Academic Press, New York, 1955, 138. With permission.

Table 39
HYDROCHLORIC ACID-POTASSIUM CHLORIDE BUFFER[a]

x	pH
97.0	1.0
78.0	1.1
64.5	1.2
51.0	1.3
41.5	1.4
33.3	1.5
26.3	1.6
20.6	1.7
16.6	1.8
13.2	1.9
10.6	2.0
8.4	2.1
6.7	2.2

[a]   Stock solutions

Note: A: 0.2 M solution of KCl (14.91 g in 1,000 ml). B: 0.2 M HCl 50 ml of A + x ml of B, diluted to a total of 200 ml.

From Fasman, G. D., Ed., *CRC Handbook of Biochemistry and Molecular Biology, Physical and Chemical Data,* Vol. 1, 3rd ed., CRC Press, Boca Raton, Fla., 1975, 370. With permission.

Table 41
PHTHALATE-HYDROCHLORIC ACID BUFFER[a]

x	pH	x	pH
46.7	2.2	14.7	3.2
39.6	2.4	9.9	3.4
33.0	2.6	6.0	3.6
26.4	2.8	2.63	3.8
20.3	3.0		

[a]   Stock solutions.

Note: A:0.2 M solution of potassium acid phthalate (40.84 g in 1,000 ml). B:0.2 M HCl 50 ml of A + x ml of B, diluted to a total of 200 ml.

From Fasman, G. D., Ed., *CRC Handbook of Biochemistry and Molecular Biology, Physical and Chemical Data,* Vol. 1, 3rd ed., CRC Press, Boca Raton, Fla., 1975, 371. With permission.

Table 40
GLYCINE-HCl BUFFER[a]

x	pH	x	pH
5.0	3.6	16.8	2.8
6.4	3.4	24.2	2.6
8.2	3.2	32.4	2.4
11.4	3.0	44.0	2.2

Note: A: 0.2 M solution of glycine (15.01 g in 1,000 ml). B: 0.2 M HCl 50 ml of A + x ml of B, diluted to a total of 200 ml.

[a]   Stock solutions.

From Fasman, G. D., Ed., *CRC Handbook of Biochemistry and Molecular Biology, Physical and Chemical Data,* Vol. 1, 3rd ed., CRC Press, Boca Raton, Fla., 1975, 370. With permission.

Table 42
ACONITATE BUFFER[a]

x	pH	x	pH
15.0	2.5	83.0	4.3
21.0	2.7	90.0	4.5
28.0	2.9	97.0	4.7
36.0	3.1	103.0	4.9
44.0	3.3	108.0	5.1
52.0	3.5	113.0	5.3
60.0	3.7	119.0	5.5
68.0	3.9	126.0	5.7
76.0	4.1		

[a]   Stock solutions.

Note: A: 0.5 M solution of aconitic acid (87.05 g in 1,000 ml). B: 0.2 M NaOH 20 ml of A + x ml of B, diluted to a total of 200 ml.

From Fasman, G. D., Ed., *CRC Handbook of Biochemistry and Molecular Biology, Physical and Chemical Data,* Vol. 1, 3rd ed., CRC Press, Boca Raton, Fla., 1975, 371. With permission.

Table 43
CITRATE BUFFER[a]

x	y	pH
46.5	3.5	3.0
43.7	6.3	3.2
40.0	10.0	3.4
37.0	13.0	3.6
35.0	15.0	3.8
33.0	17.0	4.0
31.5	18.5	4.2
28.0	22.0	4.4
25.5	24.5	4.6
23.0	27.0	4.8
20.5	29.5	5.0
18.0	32.0	5.2
16.0	34.0	5.4
13.7	36.3	5.6
11.8	38.2	5.8
9.5	41.5	6.0
7.2	42.8	6.2

[a] Stock solutions.

Note: A:0.1 M solution of citric acid (21.01 g in 1,000 ml). B:0.1 M solution of sodium citrate (29.41 g $C_6H_5O_7Na_3 \cdot 2H_2O$ in 1,000 ml; the use of the salt with 5½ H_2O is not recommended.) x ml of A + y ml of B, diluted to a total of 100 ml.

From Fasman, G. D., Ed., *CRC Handbook of Biochemistry and Molecular Biology, Physical and Chemical Data,* Vol. 1, 3rd ed., CRC Press, Boca Raton, Fla., 1975, 371. With permission.

Table 44
ACETATE BUFFER[a]

x	y	pH
46.3	3.7	3.6
44.0	6.0	3.8
41.0	9.0	4.0
36.8	13.2	4.2
30.5	19.5	4.4
25.5	24.5	4.6
20.0	30.0	4.8
14.8	35.2	5.0
10.5	39.5	5.2
8.8	41.2	5.4
4.8	45.2	5.6

[a] Stock solutions.

Note: A:0.2 M solution of acetic acid (11.55 ml in 1,000 ml). B:0.2 M solution of sodium acetate (16.4 g of $C_2H_3O_2Na$ or 27.2 g of $C_2H_3O_2Na \cdot 3H_2O$ in 1,000 ml) x ml of A + y ml of B, diluted to a total of 100 ml.

From Fasman, G. D., Ed., *CRC Handbook of Biochemistry and Molecular Biology, Physical and Chemical Data,* Vol. 1, 3rd ed., CRC Press, Boca Raton, Fla., 1975, 372. With permission.

Table 45
TRIS(HYDROXYMETHYL)-AMINO METHANE-MALEATE (TRIS-MALEATE) BUFFER[a,b]

x	pH	x	pH
7.0	5.2	48.0	7.0
10.8	5.4	51.0	7.2
15.5	5.6	54.0	7.4
20.5	5.8	58.0	7.6
26.0	6.0	63.5	7.8
31.5	6.2	69.0	8.0
37.0	6.4	75.0	8.2
42.5	6.6	81.0	8.4
45.0	6.8	86.5	8.6

Note: A: 0.2 M solution of Tris acid maleate (24.2 g of tris(hydroxymethyl)amino-methane + 23.2 g of maleic acid or 19.6 g of maleic anhydride in 1,000 ml). B: 0.2 M NaOH 50 ml of A + x ml of B, diluted to a total of 200 ml.

[a]   Stock solutions.
[b]   A buffer-grade Tris can be obtained from the Sigma Chemical Co., St. Louis, Mo., or From Matheson Coleman & Bell, East Rutherford, N.J.

From Fasman, G. D., Ed., *CRC Handbook of Biochemistry and Molecular Biology, Physical and Chemical Data*, Vol. 1, 3rd ed., CRC Press, Boca Raton, Fla., 1975, 374. With permission.

Table 47
PHTHALATE-SODIUM HYDROXIDE BUFFER[a]

x	pH	x	pH
3.7	4.2	30.0	5.2
7.5	4.4	35.5	5.4
12.2	46	39.8	5.6
17.7	4.8	43.0	5.8
23.9	5.0	45.5	6.0

Note: A:0.2 M solution of potassium acid phthalate (40.84 g in 100 ml). B:0.2 M NaOH 50 ml of A + x ml of B, diluted to a total of 200 ml.

[a]   Stock solutions.

From Fasman, G. D., Ed., *CRC Handbook of Biochemistry and Molecular Biology, Physical and Chemical Data*, Vol. 1, 3rd ed., CRC Press, Boca Raton, Fla., 1975, 373. With permission.

Table 46
SUCCINATE BUFFER[a]

x	pH	x	pH
7.5	3.8	26.7	5.0
10.0	4.0	30.3	5.2
13.3	4.2	34.2	5.4
16.7	4.4	37.5	5.6
20.0	4.6	40.7	5.8
23.5	4.8	43.5	6.0

Note: A:0.2 M solution of succinic acid (23.6 g in 1,000 ml). B:0.2 M NaOH 25 ml of A + x ml of B, diluted to a total of 100 ml.

[a]   Stock solutions.

From Fasman, G. D., Ed., *CRC Handbook of Biochemistry and Molecular Biology, Physical and Chemical Data*, Vol. 1, 3rd ed., CRC Press, Boca Raton, Fla., 1975, 373. With permission.

Table 48
MALEATE BUFFER[a]

x	pH	x	pH
7.2	5.2	33.0	6.2
10.5	5.4	38.0	6.4
15.3	5.6	41.6	6.6
20.8	5.8	44.4	6.8
26.9	6.0		

Note: A:0.2 M solution of acid sodium maleate (8 g of NaOH + 23.2 g of maleic acid or 19.6 g of maleic anhydride in 1,000 ml). B:0.2 M NaOH 50 ml of A + x ml of B, diluted to a total of 200 ml.

[a]   Stock solutions.

From Fasman, G. D., Ed., *CRC Handbook of Biochemistry and Molecular Biology, Physical and Chemical Data*, Vol. 1, 3rd ed., CRC Press, Boca Raton, Fla., 1975, 373. With permission.

Table 49
CACODYLATE BUFFER[a]

x	pH	x	pH
2.7	7.4	29.6	6.0
4.2	7.2	34.8	5.8
6.3	7.0	39.2	5.6
9.3	6.8	43.0	5.4
13.3	6.6	45.0	5.2
18.3	6.4	47.0	5.0
23.8	6.2		

Note: A:0.2 M solution of sodium cacodylate (42.8 g of $Na(CH_3)_2AsO_2 \cdot 3H_2O$ in 1,000 mℓ). B:0.2 M HCl 50 mℓ of A + x mℓ of B, diluted to a total of 200 mℓ.

[a]   Stock solutions.

From Fasman, G. D., Ed., *CRC Handbook of Biochemistry and Molecular Biology, Physical and Chemical Data,* Vol. 1, 3rd ed., CRC Press, Boca Raton, Fla., 1975, 373. With permission.

Table 50
BARBITOL BUFFER[a,b]

x	pH
1.5	9.2
2.5	9.0
4.0	8.8
6.0	8.6
9.0	8.4
12.7	8.2
17.5	8.0
22.5	7.8
27.5	7.6
32.5	7.4
39.0	7.2
43.0	7.0
45.0	6.8

Note: A:0.2 M solution of sodium barbital (veronal) (41.2 g in 1,000 mℓ). B:0.2 M HCl 50 mℓ of A + x mℓ of B, diluted to a total of 200 mℓ.

[a]   Stock solutions.
[b]   Solutions more concentrated than 0.05 M may crystallize on standing, especially in the cold.

From Fasman, G. D., Ed., *CRC Handbook of Biochemistry and Molecular Biology, Physical and Chemical Data,* Vol. 1, 3rd ed., CRC Press, Boca Raton, Fla., 1975, 375. With permission.

Table 51
CARBONATE-BICARBONATE BUFFER[a]

x	y	pH
4.0	46.0	9.2
7.5	42.5	9.3
9.5	40.5	9.4
13.0	37.0	9.5
16.0	34.0	9.6
19.5	30.5	9.7
22.0	28.0	9.8
25.0	25.0	9.9
27.5	22.5	10.0
30.0	20.0	10.1
33.0	17.0	10.2
35.5	14.5	10.3
38.5	11.5	10.4
40.5	9.5	10.5
42.5	7.5	10.6
45.0	5.0	10.7

Note: A:0.2 M solution of anhydrous sodium carbonate (21.2 g in 1,000 mℓ). B:0.2 M solution of sodium bicarbonate (16.8 g in 1,000 mℓ) x mℓ of A + y mℓ of B, diluted to a total of 200.

[a]   Stock solutions.

From Fasman, G. D., Ed., *CRC Handbook of Biochemistry and Molecular Biology, Physical and Chemical Data*, Vol. 1, 3rd ed., CRC Press, Boca Raton, Fla., 1975, 377. With permission.

Table 52
TRIS(HYDROXYMETHYL) -AMINOMETHANE (TRIS) BUFFER[a,b]

x	pH
5.0	9.0
8.1	8.8
12.2	8.6
16.5	8.4
21.9	8.2
26.8	8.0
32.5	7.8
38.4	7.6
41.4	7.4
44.2	7.2

Note: A:0.2 M solution of tris(hydroxymethyl)aminomethane (24.2 g in 1,000 mℓ). B:0.2 M HCl 50 mℓ of A + x mℓ of B, diluted to a total of 200 mℓ.

[a]   Stock solutions.
[b]   A buffer-grade Tris can be obtained from the Sigma Chemical Co., St. Louis, Mo., or from Matheson Coleman & Bell, East Rutherford, N.J.

From Fasman, G. D., Ed., *CRC Handbook of Biochemistry and Molecular Biology, Physical and Chemical Data*, Vol. 1, 3rd ed., CRC Press, Boca Raton, Fla., 1975, 375. With permission.

Table 53
BORAX-NaOH BUFFER[a]

x	pH
0.0	9.28
7.0	9.35
11.0	9.4
17.6	9.5
23.0	9.6
29.0	9.7
34.0	9.8
38.6	9.9
43.0	10.0
46.0	10.1

Note: A:0.05 *M* solution of borax (19.05 g in 1,000 ml; 0.02 *M* in terms of sodium borate). B:0.2 *M* NaOH 50 ml of A + x ml of B, diluted to a total of 200 ml.

[a] Stock solutions.

From Fasman, G. D., Ed., *CRC Handbook of Biochemistry and Molecular Biology, Physical and Chemical Data,* Vol. 1, 3rd ed., CRC Press, Boca Raton, Fla., 1975, 376. With permission.

Table 54
BORIC ACID-BORAX BUFFER[a]

x	pH	x	pH
2.0	7.6	22.5	8.7
3.1	7.8	30.0	8.8
4.9	8.0	42.5	8.9
7.3	8.2	59.0	9.0
11.5	8.4	83.0	9.1
17.5	8.6	115.0	9.2

Note: A:0.2 *M* solution of boric acid (12.4 g in 1,000 ml). B:0.05 *M* solution of borax (19.05 g in 1,000 ml; 0.2 *M* in terms of sodium borate) 50 ml of A + x ml of B, diluted to a total of 200 ml.

[a] Stock solutions.

From Fasman, G. D., Ed., *CRC Handbook of Biochemistry and Molecular Biology, Physical and Chemical Data,* Vol. 1, 3rd ed., CRC Press, Boca Raton, Fla., 1975, 376. With permission.

Table 55
CITRATE-PHOSPHATE BUFFER[a]

x	y	pH
44.6	5.4	2.6
42.2	7.8	2.8
39.8	10.2	3.0
37.7	12.3	3.2
35.9	14.1	3.4
33.9	16.1	3.6
32.3	17.7	3.8
30.7	19.3	4.0
29.4	20.6	4.2
27.8	22.2	4.4
26.7	23.3	4.6
25.2	24.8	4.8
24.3	25.7	5.0
23.3	26.7	5.2
22.2	27.8	5.4
21.0	29.0	5.6
19.7	30.3	5.8
17.9	32.1	6.0
16.9	33.1	6.2
15.4	34.6	6.4
13.6	36.4	6.6
9.1	40.9	6.8
6.5	43.6	7.0

Note: A:0.1 *M* solution of citric acid (19.21 g in 1,000 ml). B:0.2 *M* solution of dibasic sodium phosphate (53.65 g of $Na_2-HPO_4 \cdot 7H_2O$ or 71.7 g of $Na_2HPO_4 \cdot 12H_2O$ in 1,000 ml x ml of A + y ml of B, diluted to a total of 100 ml.

[a] Stock solutions.

From Fasman, G. D., Ed., *CRC Handbook of Biochemistry and Molecular Biology, Physical and Chemical Data,* Vol. 1, 3rd ed., CRC Press, Boca Raton, Fla., 1975, 372. With permission.

Table 56
GLYCINE-NaOH BUFFER[a]

x	pH	x	pH
4.0	8.6	22.4	9.6
6.0	8.8	27.2	9.8
8.8	9.0	32.0	10.0
12.0	9.2	38.6	10.4
16.8	9.4	45.5	10.6

Note: A:0.2 M solution of glycine (15.01 g in 1,000 ml). B:0.2 M NaOH 50 ml of A + x ml of B, diluted to a total of 200 ml.

[a]   Stock solutions.

From Fasman, G. D., Ed., *CRC Handbook of Biochemistry and Molecular Biology, Physical and Chemical Data,* Vol. 1, 3rd ed., CRC Press, Boca Raton, Fla., 1975, 370. With permission.

Table 57
PHOSPHATE BUFFER[a]

x	y	pH	x	y	pH
93.5	6.5	5.7	45.0	55.0	6.9
92.0	8.0	5.8	39.0	61.0	7.0
90.0	10.0	5.9	33.0	67.0	7.1
87.7	12.3	6.0	28.0	72.0	7.2
85.0	15.0	6.1	23.0	77.0	7.3
81.5	18.5	6.2	19.0	81.0	7.4
77.5	22.5	6.3	16.0	84.0	7.5
73.5	26.5	6.4	13.0	87.0	7.6
68.5	31.5	6.5	10.5	90.5	7.7
62.5	37.5	6.6	8.5	91.5	7.8
56.5	43.5	6.7	7.0	93.0	7.9
51.0	49.0	6.8	5.3	94.7	8.0

Note: A:0.2 M solution of monobasic sodium phosphate (27.8 g in 1,000 ml). B:0.2 M solution of dibasic sodium phosphate (53.65 g of $Na_2HPO_4 \cdot 7H_2O$ or 71.7 g of $Na_2HPO_4 \cdot 12H_2O$ in 1,000 ml) x ml of A + y ml of B, diluted to a total of 200 m$_l$.

[a]   Stock solutions.

From Fasman, G. D., Ed., *CRC Handbook of Biochemistry and Molecular Biology, Physical and Chemical Data,* Vol. 1, 3rd ed., CRC Press, Boca Raton, Fla., 1975, 374. With permission.

AMINE BUFFERS USEFUL FOR BIOLOGICAL RESEARCH

All of these amines are highly polar, water-soluble substances. Their advantages and disadvantages must be determined empirically for each biological reaction system. For best buffering performance they should be used at pH's close to the pKa, preferable within ±0.5 pH units of the pKa and never more than ±1.0 unit from the pKa. Note that the pKa's, and therefore the pH's of buffered solutions, change with temperature in the manner indicated.

Table 58

AMINE BUFFERS USEFUL FOR BIOLOGICAL RESEARCH

Chemical name	Trivial name or acronym	Structure	pKa at 20°C	ΔpKa/°C
2-(N-Morpholino)ethanesulfonic acid	MES		6.15	-0.011
Bis(2-Hydroxyethyl)imino-tris-(hydroxymethyl)methane	Bistris	$(HOCH_2CH_2)_2=N-C\equiv(CH_2OH)_3$	6.5	—
N-(2-Acetamido)iminodiacetic acid	ADA[a]		6.6	-0.011
Piperazine-N,N'-bis(2-ethanesulfonic acid)	PIPES		6.8	-0.0085
1,3-Bis[tris(hydroxymethyl)-methylamino]propane	Bistrispropane	$(HOCH_2)_3=C-NH(CH_2)_3NH-C\equiv(CH_2OH)_3$	6.8(9.0)	—
N-(Acetamido)-2-aminoethanesulfonic acid	ACES	$H_2NCOCH_2\ N^+H_2CH_2CH_2SO_3^-$	6.9	-0.020
3-(N-Morpholino)propanesulfonic acid	MOPS		7.15	-0.013
N,N'-Bis(2-Hydroxyethyl)-2-amino-ethanesulfonic acid	BES	$(HOCH_2CH_2)_2=N^+HCH_2CH_2SO_3^-$	7.15	-0.016
N-Tris(hydroxymethyl)methyl-2-amino-ethanesulfonic acid	TES	$(HOCH_2)_3=C-N^+H_2CH_2CH_2SO_3^-$	7.5	-0.020
N-2-Hydroxyethylpiperazine-N'-ethanesulfonic acid	HEPES[b]		7.55	-0.014
N-2-Hydroxyethylpiperazine-N'-propanesulfonic acid	HEPPS[b]		8.1	-0.015
N-Tris(hydroxymethyl)methylglycine	Tricine[a]	$(HOCH_2)_3\equiv C-N^+H_2CH_2COO^-$	8.15	-0.021
Tris(hydroxymethyl)aminomethane	Tris	$(HOCH_2)_3\equiv CNH_2$	8.3	-0.031
N,N-Bis(2-hydroxyethyl)glycine	Bicine[a]	$(HOCH_2CH_2)_2=N^+HCH_2COO^-$	8.35	-0.018
Glycylglycine	Glycylglycine[a]	$H_3N^+CH_2CONHCH_2COO^-$	8.4	-0.028
N-Tris(hydroxymethyl)methyl-3-amino-propanesulfonic acid	TAPS	$(HOCH_2)_3=C-N^+H_2(CH_2)_3SO_3^-$	8.55	-0.027
1,3-Bis[tris(hydroxymethyl)-methylamino] propane	Bistrispropane	$(HOCH_2)_3=C-NH(CH_2)_3NH-C\equiv(CH_2OH)_3$	9.0(6.8)	—
Glycine	Glycine[a]	$H_3N^+CH_2COO^-$	9.9	—

Note: For further information on these and other buffers, see Good and Izawa, in *Methods in Enzymology, Part B*, Vol. 24, Pietro, Ed., Academic Press, New York, 1972, 53.

[a] These substances may bind certain di- and polyvalent cations and therefore they may sometimes be useful for providing constant, low level concentrations of free heavy metal ions (Heavy metal buffering).

[b] These substances interfere with and preclude the Folin protein assay.

From Fasman, G. D., Ed., *CRC Handbook of Biochemistry and Molecular Biology, Physical and Chemical Data*, Vol. 1, 3rd ed., CRC Press, Boca Raton, Fla., 1975, 368. With permission.

Table 59
BOILING POINTS OF WATER

A. Barometric Pressures at Various Temperatures

Temp. °C.	0.0° mm of Hg	0.2° mm of Hg	0.4° mm of Hg	0.6° mm of Hg	0.8° mm of Hg
80	355.40	358.28	361.19	364.11	367.06
81	370.03	373.01	376.02	379.05	382.09
82	385.16	388.25	391.36	394.49	397.64
83	400.81	404.00	407.22	410.45	413.71
84	416.99	420.29	423.61	426.95	430.32
85	433.71	437.12	440.55	444.01	447.49
86	450.99	454.51	458.06	461.63	465.22
87	468.84	472.48	476.14	479.83	483.54
88	487.28	491.04	494.82	498.63	502.46
89	506.32	510.20	514.11	518.04	521.99
90	525.97	529.98	534.01	538.07	542.15
91	546.26	550.40	554.56	558.75	562.96
92	567.20	571.47	575.76	580.08	584.43
93	588.80	593.20	597.63	602.09	606.57
94	611.08	615.62	620.19	624.79	629.41
95	634.06	638.74	643.45	648.19	652.96
96	657.75	662.58	667.43	672.32	677.23
97	682.18	687.15	692.15	697.19	702.25
98	707.35	712.47	717.63	722.81	728.03
99	733.28	738.56	743.87	749.22	754.59
100	760.00	765.44	770.91	776.42	781.95

B. Boiling Points of Water at Various Pressures

Pressure, atm.	Boiling Point, °C.	Pressure, atm.	Boiling Point, °C.	Pressure, atm.	Boiling Point, °C.	Pressure, atm.	Boiling Point, °C.
0.5	80.9	7	164.2	14	194.1	21	213.9
1	100.0	8	169.6	15	197.4	22	216.2
2	119.6	9	174.5	16	200.4	23	218.5
3	132.9	10	179.0	17	203.4	24	220.8
4	142.9	11	183.2	18	206.1	25	222.9
5	151.1	12	187.1	19	208.8	26	225.0
6	158.1	13	190.7	20	211.4	27	227.0

From Dean, J. A., Ed., *Lange's Handbook of Chemistry*, 12th ed., McGraw-Hill, New York, 1979, 10-54. With permission.

Table 60
DENSITY OF WATER

Maximum density (4°C, 760 mm Hg, free of air)² .... ρ_m(H₂O) = 1 g/ml = (0.999 972 ± 0.000 003) g/cm³

Density between 0 and 40°C (760 mm Hg, free of air)³:

°C	g/ml	°C	g/ml	°C	g/ml	°C	g/ml
0	0.999 8676	4	1.0	8	0.999 8765	12	0.999 5261
1	0.999 9265	5	0.999 9919	9	0.999 8092	13	0.999 4059
2	0.999 9678	6	0.999 9683	10	0.999 7281	14	0.999 2732
3	0.999 9922	7	0.999 9297	11	0.999 6336		

°C	.0	.1	.2	.3	.4	.5	.6	.7	.8	.9
15	0.999 1286	0.999 1134	0.999 0982	0.999 0828	0.999 0674	0.999 0518	0.999 0360	0.999 0202	0.999 0043	0.998 9882
16	0.998 9721	0.998 9558	0.998 9394	0.998 9229	0.998 9062	0.998 8895	0.998 8726	0.998 8557	0.998 8386	0.998 8214
17	0.998 8041	0.998 7867	0.998 7691	0.998 7515	0.998 7337	0.998 7158	0.998 6979	0.998 6798	0.998 6616	0.998 6433
18	0.998 6248	0.998 6063	0.998 5877	0.998 5689	0.998 5501	0.998 5311	0.998 5120	0.998 4928	0.998 4735	0.998 4541
19	0.998 4346	0.998 4150	0.998 3953	0.998 3754	0.998 3555	0.998 3355	0.998 3153	0.998 2950	0.998 2747	0.998 2542
20	0.998 2336	0.998 2130	0.998 1922	0.998 1713	0.998 1503	0.998 1292	0.998 1080	0.998 0867	0.998 0653	0.998 0438
21	0.998 0221	0.998 0004	0.997 9786	0.997 9567	0.997 9346	0.997 9125	0.997 8903	0.997 8679	0.997 8455	0.997 8230
22	0.997 8003	0.997 7776	0.997 7547	0.997 7318	0.997 7088	0.997 6856	0.997 6624	0.997 6390	0.997 6156	0.997 5921
23	0.997 5684	0.997 5447	0.997 5208	0.997 4969	0.997 4729	0.997 4487	0.997 4245	0.997 4002	0.997 3758	0.997 3512
24	0.997 3266	0.997 3019	0.997 2771	0.997 2522	0.997 2272	0.997 2021	0.997 1769	0.997 1516	0.997 1262	0.997 1007
25	0.997 0751	0.997 0494	0.997 0237	0.996 9978	0.996 9718	0.996 9458	0.996 9196	0.996 8934	0.996 8671	0.996 8406

°C	g/ml	°C	g/ml	°C	g/ml	°C	g/ml
26	0.996 8141	30	0.995 6783	34	0.994 4030	38	0.992 9970
27	0.996 5437	31	0.995 3722	35	0.994 0635	39	0.992 6260
28	0.996 2642	32	0.995 0575	36	0.993 7159	40	0.992 2473
29	0.995 9757	33	0.994 7344	37	0.993 3604		

From Diem, K., Ed., *Documenta Geigy Scientific Tables*, 6th ed., Geigy Pharmaceuticals, Ardsley, N.Y., 1962, 245. With permission.

Table 61
REFRACTIVE INDEX, VISCOSITY, DIELECTRIC CONSTANT, AND SURFACE TENSION OF WATER FROM 0 TO 100°C

Temp., °C	Refractive Index (n_D)	Viscosity, mN·s·m⁻²	Dielectric Constant (ε)	Surface Tension, dyn·cm⁻¹
0	1.333 95	1.770 2	87.74	75.83
5	1.333 88	1.510 8	85.76	75.09
10	1.333 69	1.303 9	83.83	74.36
15	1.333 39	1.137 4	81.95	73.62
20	1.333 00	1.001 9	80.10	72.88
21	1.332 90	0.976 4	79.73	72.73
22	1.332 80	0.953 2	79.38	72.58
23	1.332 71	0.931 0	79.02	72.43
24	1.332 61	0.910 0	78.65	72.29
25	1.332 50	0.890 3	78.30	72.14
26	1.332 40	0.870 3	77.94	71.99
27	1.332 29	0.851 2	77.60	71.84
28	1.332 17	0.832 8	77.24	71.69
29	1.332 06	0.814 5	76.90	71.55
30	1.331 94	0.797 3	76.55	71.40
35	1.331 31	0.719 0	74.83	70.66
40	1.330 61	0.652 6	73.15	69.92
45	1.329 85	0.597 2	71.51	69.18
50	1.329 04	0.546 8	69.91	68.45
55	1.328 17	0.504 2	68.35	67.71
60	1.327 25	0.466 9	66.82	66.97
65		0.434 1	65.32	66.23
70		0.405 0	63.86	65.49
75		0.379 2	62.43	64.75
80		0.356 0	61.03	64.01
85		0.335 2	59.66	63.28
90		0.316 5	58.32	62.54
95		0.299 5	57.01	61.80
100		0.284 0	55.72	61.80

From Dean, J. A., Ed., *Lange's Handbook of Chemistry*, 12th ed., McGraw-Hill, New York, 1979, 10-99. With permission.

Table 62
IONIC PRODUCT CONSTANT OF WATER

Temp., °C	pK_w	Temp., °C	pK_w	Pressure, atm	pK_w
0	14.944	60	13.017	1	13.997
5	14.734	65	12.908	250	13.907
10	14.535	70	12.800	500	13.824
15	14.346	75	12.699	750	13.747
20	14.167	80	12.598	1000	13.667
25	13.997	85	12.510	1250	13.585
30	13.833	90	12.422	1500	13.524
35	13.680	95	12.341	1750	13.449
40	13.535	100	12.259	2000	13.394
45	13.396	110	12.126		
50	13.262	120	12.002		
55	13.137	130	11.907		

Note: This table gives value of pK_w on a molal scale, where K_w is the ionic activity product constant of water. Taking pK_w = 13.997 at 25°C and 1 atm, values of pK_w are given at pressures up to 2000 atm.

From Dean, J. A., Ed., *Lange's Handbook of Chemistry*, 12th ed., McGraw-Hill, New York, 1979, 5-7. With permission.

Table 63
VOLUME PROPERTIES OF WATER AT 1 atm

	ρ, kg m^{-3},	$10^6\ \alpha$, K^{-1},	$10^6\ \kappa T$/bar^{-1}		ρ, kg m^{-3},	$10^6\ \alpha$, K^{-1},	$10^6\ \kappa T$/bar^{-1}
	Equation 1	Equation 1	Equation 2		Equation 1	Equation 1	Equation 2
−30	983.857	−1400.0	80.79	38	992.9683	369.79	44.3051
−25	989.588	−955.9	70.94	39	992.5973	377.59	44.2697
−20	993.550	−660.6	64.25	40	992.2187	385.30	44.2391
−15	996.286	−450.3	59.44	41	991.8327	392.91	44.2131
−10	998.120	−292.4	55.83	42	991.4394	400.43	44.1917
−9	998.398	−265.3	55.22	43	991.0388	407.85	44.1747
−8	998.650	−239.5	54.64	44	990.6310	415.19	44.1620
−7	998.877	−214.8	54.08	45	990.2162	422.45	44.1536
−6	999.080	−191.2	53.56	46	989.7944	429.63	44.1494
−5	999.259	−168.6	53.06	47	989.3657	436.73	44.1494
−4	999.417	−146.9	52.58	48	988.9303	443.75	44.1533
−3	999.553	−126.0	52.12	49	988.4881	450.71	44.1613
−2	999.669	−106.0	51.69	50	988.0393	457.59	44.1732
−1	999.765	−86.7	51.28	51	987.5839	464.40	44.189
0	999.8425	−68.05	50.8850	52	987.1220	471.15	44.209
1	999.9015	−50.09	50.5091	53	986.6537	477.84	44.232
2	999.9429	−32.74	50.1505	54	986.1791	484.47	44.259
3	999.9672	−15.97	49.8081	55	985.6982	491.04	44.290
4	999.9750	0.27	49.4812	56	985.2111	497.55	44.324
5	999.9668	16.00	49.1692	57	984.7178	504.01	44.362
6	999.9432	31.24	48.8712	58	984.2185	510.41	44.403
7	999.9045	46.04	48.5868	59	983.7132	516.76	44.448
8	999.8512	60.41	48.3152	60	983.2018	523.07	44.496
9	999.7838	74.38	48.0560	61	982.6846	529.32	44.548
10	999.7026	87.97	47.8086	62	982.1615	535.53	44.603
11	999.6081	101.20	47.5726	63	981.6327	541.70	44.662
12	999.5004	114.08	47.3474	64	981.0981	547.82	44.723
13	999.3801	126.65	47.1327	65	980.5578	553.90	44.788
14	999.2474	138.90	46.9280	66	980.0118	559.94	44.857
15	999.1026	150.87	46.7331	67	979.4603	565.95	44.928
16	998.9460	162.55	46.5475	68	978.9032	571.91	45.003
17	998.7779	173.98	46.3708	69	978.3406	577.84	45.081
18	998.5986	185.15	46.2029	70	977.7726	583.74	45.162
19	998.4082	196.08	46.0433	71	977.1991	589.60	45.246
20	998.2071	206.78	45.8918	72	976.6203	595.43	45.333
21	997.9955	217.26	45.7482	73	976.0361	601.23	45.424
22	997.7735	227.54	45.6122	74	975.4466	607.00	45.517
23	997.5415	237.62	45.4835	75	974.8519	612.75	45.614
24	997.2995	247.50	45.3619	76	974.2520	618.46	45.714
25	997.0479	257.21	45.2472	77	973.6468	624.15	45.817
26	996.7867	266.73	45.1392	78	973.0366	629.82	45.922
27	996.5162	276.10	45.0378	79	972.4212	635.46	46.031
28	996.2365	285.30	44.9427	80	971.8007	641.08	46.143
29	995.9478	294.34	44.8537	81	971.1752	646.67	46.258
30	995.6502	303.24	44.7707	82	970.5446	652.25	46.376
31	995.3440	312.00	44.6935	83	969.9091	657.81	46.497
32	995.0292	320.63	44.6221	84	969.2686	663.34	46.621
33	994.7060	329.12	44.5561	85	968.6232	668.86	46.748
34	994.3745	337.48	44.4956	86	967.9729	674.37	46.878
35	994.0349	345.73	44.4404	87	967.3177	679.85	47.011
36	993.6872	353.86	44.3903	88	966.6576	685.33	47.148
37	993.3316	361.88	44.3452	89	965.9927	690.78	47.287

Table 63 (continued)
VOLUME PROPERTIES OF WATER AT 1 atm

	ρ, kg m^{-3},	10^6 α, K^{-1},	10^6 κT/bar^{-1}	
	Equation 1	Equation 1	Equation 2	Equation 3
90	965.3230	696.23	47.429	47.428
91	964.6486	701.66	47.574	47.574
92	963.9693	707.08	47.722	47.722
93	963.2854	712.49	47.874	47.873
94	962.5967	717.89	48.028	48.028
95	961.9033	723.28	48.185	48.185
96	961.2052	728.67	48.346	48.346
97	960.5025	734.04	48.509	48.510
98	959.7951	739.41	48.676	48.677
99	959.0831	744.78	48.846	48.847
100	958.3665	750.14	49.019	49.020
101	957.645	755.5		49.20
102	956.920	760.8		49.38
103	956.189	766.2		49.56
104	955.454	771.5		49.74
105	954.715	776.9		49.93
106	953.971	782.2		50.13
107	953.222	787.6		50.32
108	952.469	792.9		50.52
109	951.712	798.3		50.72
110	950.950	803.6		50.93
115	947.073	830.4		52.01
120	943.085	857.4		53.17
125	938.987	884.7		54.43
130	934.778	912.3		55.79
135	930.459	940.3		57.24
140	926.029	968.9		58.80
145	921.487	998.0		60.47
150	916.832	1027.8		62.25

Equations:

$$\rho/\text{kg m}^{-3} = (999.83952 + 16.945176\ t - 7.9870401 \times 10^{-3}\ t^2 - 46.170461 \times 10^{-6}\ t^3 + 105.56302 \times 10^{-9}\ t^4 - 280.54253 \times 10^{-12}\ t^5)/(1 + 16.879850 \times 10^{-3}\ t) \tag{1}$$

$$10^6\ \kappa_T/\text{bar}^{-1} = (50.88496 + 0.6163813\ t + 1.459187 \times 10^{-3}\ t^2 + 20.08438 \times 10^{-6}\ t^3 - 58.47727 \times 10^{-9}\ t^4 + 410.4110 \times 10^{-12}\ t^5)/(1 + 19.67348 \times 10^{-3}\ t) \tag{2}$$

$$10^6\ \kappa_T/\text{bar}^{-1} = (50.884917 + 0.62590623\ t + 1.3848668 \times 10^{-3}\ t^2 + 21.603427 \times 10^{-6}\ t^3 - 72.087667 \times 10^{-9}\ t^4 + 465.45054 \times 10^{-12}\ t^5)/(1 + 19.859983 \times 10^{-3}\ t) \tag{3}$$

$$\kappa_S = (\partial \ln \rho/\partial P)_S = \frac{1}{\rho U^2} \tag{4}$$

* Density ρ, thermal expansivity $\alpha = -\ (\partial \ln \rho/\partial T)_p$, and isothermal compressibility $\kappa T = (\partial \ln \rho/\partial p)T$. For purposes of this table, ordinary water is that with a maximum density of 999.972 kg m^{-3}. Equation 4 for the compressibility should be used for temperatures $0 \leqslant t \leqslant 100°$C, and Equation 3 for 100 $\leqslant + \leqslant 150°$C. The liquid is metastable below $0°$C and above $100°$C. Values below $0°$C were obtained by extrapolation, and no claim is made for their accuracy.

Table 64

DENSITY AND VOLUME OF MERCURY BASED ON THE DENSITY OF MERCURY AT 0°C BY THIESEN AND SCHEEL (SELECTED FROM SMITHSONIAN TABLES)

Temp. °C.	Mass in gr. per ml.	Vol. of 1 gr. in ml.	Temp. °C.	Mass in gr. per ml.	Vol. of 1 gr. in ml.	Temp. °C.	Mass in gr. per ml.	Vol. of 1 gr in ml.	Temp. °C.	Mass in gr. per ml.	Vol. of 1 gr. in ml.
−10	13.6202	0.0734205	11	5684	7011	30°	13.5217	0.0739552	150	2330	5688
−9	6177	4338	12	5659	7145	31	5193	9686	160	2093	7044
−8	6152	4472	13	5634	7278	32	5168	9820	170	1856	8402
−7	6128	4606	14	5610	7412	33	5144	9953	180	1620	9764
−6	6103	4739	15	13.5585	0.0737546	34	5119	40087	190	13.1384	0.0761128
−5	13.6078	0.0734873	16	5561	7680	35	13.5095	0.0740221	200	1148	2495
−4	6053	5006	17	5536	7813	36	5070	0354	210	0913	3865
−3	6029	5140	18	5512	7947	37	5046	0488	220	0678	5239
−2	6004	5273	19	5487	8081	38	5021	0622	230	0443	6616
−1	5979	5407	20	13.5462	0.0738215	39	4997	0756	240	13.0209	0.0767996
0	13.5955	0.0735540	21	5438	8348	40	13.4973	0.0740891	250	12.9975	9381
1	5930	5674	22	5413	8482	50	4729	2229	260	9741	70769
2	5906	5808	23	5389	8616	60	4486	3569	270	9507	2161
3	5881	5941	24	5364	8750	70	4244	4910	280	9273	3558
4	5856	6075	25	13.5340	0.0738883	80	4003	6252	290	12.9039	0.0774958
5	13.5832	0.0736209	26	5315	9017	90	13.3762	0.0747594	300	8806	6364
6	5807	6342	27	5291	9151	100	3522	8939	310	8572	7774
7	5782	6476	28	5266	9285	110	3283	50285	320	8339	9189
8	5758	6610	29	5242	9419	120	3044	1633	330	8105	80609
9	5733	6744	30	13.5217	0.0739552	130	2805	2982	340	12.7872	0.0782033
10	13.5708	0.0736877				140	13.2567	0.0754334	350	7638	3464
									360	7405	4900

From Weast, R. C., Ed., *CRC Handbook of Chemistry and Physics,* 61st ed., CRC Press, Boca Raton, Fla., 1980, F-7. With permission.

Table 65
DENSITY OF VARIOUS LIQUIDS

(Selected from Smithsonian Tables.)

Liquid	Grams per cu. cm	Pounds per cu. ft.	Temp °C	Liquid	Grams per cu. cm	Pounds per cu. ft.	Temp. °C
Acetone	0.792	49.4	20°	Milk	1.028–1.035	64.2–64.6	15
Alcohol, ethyl	0.791	49.4	20	Naphtha, petroleum ether	0.665	41.5	0
Alcohol, methyl	0.810	50.5	0	Naphtha, wood	0.848–0.810	52.9–50.5	
Benzene	0.899	56.1	0	Oils:			
Carbolic acid	0.950–0.965	59.2–60.2	15	castor	0.969	60.5	15
Carbon disulfide	1.293	80.7	0	cocoanut	0.925	57.7	15
Carbon tetrachloride	1.595	99.6	20	cotton seed	0.926	57.8	16
Chloroform	1.489	93.0	20	creosote	1.040–1.100	64.9–68.6	15
Ether	0.736	45.9	0	linseed, boiled	0.942	58.8	15
Gasoline	0.66–0.69	41.0–43.0	0	olive	0.918	57.3	15
Glycerin	1.260	78.6	0	Sea water	1.025	63.99	15
Kerosene	0.82	51.2		Turpentine (spirits)	0.87	54.3	15
Mercury	13.6	849.0		Water	1.00	62.43	4

From Weast, R. C., Ed., *CRC Handbook of Chemistry and Physics*, 61st ed., CRC Press, Boca Raton, Fla., 1980, F-3. With permission.

Table 66
DENSITY OF ETHYL ALCOHOL IN GRAMS PER CUBIC CENTIMETER COMPUTED FROM MENDELEEFF'S FORMULA

(Selected from Smithsonian Tables.)

Temp. °C	0	1	2	3	4	5	6	7	8	9
0	.80625	.80541	.80457	.80374	.80290	.80207	.80123	.80039	.79956	.79872
10	.79788	.79704	.79620	.79535	.79451	.79367	.79283	.79198	.79114	.79029
20	.78945	.78860	.78775	.78691	.78606	.78522	.78437	.78352	.78267	.78182
30	.78097	.78012	.77927	.77841	.77756	.77671	.77585	.77500	.77414	.77329

From Weast, R. C., Ed., *CRC Handbook of Chemistry and Physics*, 61st ed., CRC Press, Boca Raton, Fla., 1980, F-3. With permission.

Table 67

SOLUBILITIES OF INORGANIC COMPOUNDS AND METAL SALTS OF ORGANIC ACIDS IN WATER AT VARIOUS TEMPERATURES

Substance	Formula	0°	10°	20°	30°	40°	60°	80°	90°	100°
Aluminum chloride	$AlCl_3$	43.9	44.9	45.8	46.6	47.3	48.1	48.6		49.0
fluoride	AlF_3	0.56	0.56	0.67	0.78	0.91	1.1	1.32		1.72
nitrate	$Al(NO_3)_3$	60.0	66.7	73.9	81.8	88.7	106	132	153	160
perchlorate	$Al(ClO_4)_3$	122	128	133						182
sulfate	$Al_2(SO_4)_3$	31.2	33.5	36.4	40.4	45.8	59.2	73.0	80.8	89.0
thallium(I) sulfate	$Al_2Tl_2(SO_4)_4$	3.15	4.60	6.39	9.37	14.39	35.35			
Ammonium aluminum sulfate	$NH_4Al(SO_4)_2$	2.10	5.00	7.74	10.9	14.9	26.7			
azide	NH_4N_3	16.0		25.3		37.1				
bromide	NH_4Br	60.5	68.1	76.4	83.2	91.2	108	125	135	145
chloride	NH_4Cl	29.4	33.2	37.2	41.4	45.8	55.3	65.6	71.2	77.3
chloroiridate(IV)	$(NH_4)_2IrCl_6$	0.56	0.71	0.95	1.20	1.56	2.45	4.38		
chloroplatinate(IV)	$(NH_4)_2PtCl_6$	0.289	0.374	0.499	0.637	0.815	1.44	2.16	2.61	3.36
chromate	$(NH_4)_2CrO_4$	25.0	29.2	34.0	39.3	45.3	59.0	76.1		
chromium(III) sulfate	$(NH_4)Cr(SO_4)_2$	3.95			18.8	32.6				
cobalt(II) sulfate	$(NH_4)_2Co(SO_4)_2$	6.0	9.5	13.0	17.0	22.0	33.5	49.0	58.0	75.1
dichromate	$(NH_4)_2Cr_2O_7$	18.2	25.5	35.6	46.5	58.5	86.0	115		156
dihydrogen arsenate	$NH_4H_2AsO_4$	33.7		48.7		63.8	83.0	107	122	
dihydrogen phosphate	$NH_4H_2PO_4$	22.7	29.5	37.4	46.4	56.7	82.5	118		173
dithionate	$(NH_4)_2S_2O_6$	133	151	166	179					
formate	NH_4CHO_2	102		143		204	311	533		
hydrogen carbonate	NH_4HCO_3	11.9	16.1	21.7	28.4	36.6	59.2	109	170	354
hydrogen phosphate	$(NH_4)_2HPO_4$	42.9	62.9	68.9	75.1	81.8	97.2			
hydrogen tartrate	$NH_4C_4H_5O_6$	1.00	1.88	2.70						
iodide	NH_4I	155	163	172	182	191	209	229		250
iron(II) sulfate	$(NH_4)_2Fe(SO_4)_2$	12.5	17.2	26.4	33	46				
Ammonium magnesium sulfate	$(NH_4)_2Mg(SO_4)_2$	11.8	14.6	18.0	21.7	25.8	35.1	48.3		65.7
nickel sulfate	$(NH_4)_2Ni(SO_4)_2$	1.00	4.00	6.50	9.20	12.0	17.0			
nitrate	NH_4NO_3	118	150	192	242	297	421	580	740	871
oxalate	$(NH_4)_2C_2O_4$	2.2	3.21	4.45	6.09	8.18	14.0	22.4	27.9	34.7
perchlorate	NH_4ClO_4	12.0	16.4	21.7	27.7	34.6	49.9	68.9		
selenite	$(NH_4)_2SeO_3$	96	105	115	126	143	192			

Substance	Formula	Solubility values (low → high temperature)
sulfate	$(NH_4)_2SO_4$	70.6, 73.0, 75.4, 78.0, 81, 88, 95, 103
sulfite	$(NH_4)_2SO_3$	47.9, 54.0, 60.8, 68.8, 78.4, 104, 144, 150, 153
tartrate	$(NH_4)_2C_4H_4O_6$	45.0, 55.0, 63.0, 70.5, 76.5, 86.9
thioantimonate(V)	$(NH_4)_3SbS_4$	71.2, 91.2, 120
thiocyanate	NH_4SCN	120, 144, 170, 208, 234, 346
vanadate	NH_4VO_3	0.48, 0.84, 1.32, 2.42
zinc sulfate	$(NH_4)_2Zn(SO_4)_2$	7.0, 9.5, 12.5, 16.0, 20.0, 30.0, 46.6, 58.0, 72.4
Antimony(III) chloride	$SbCl_3$	602, 910, 1087, 1368
fluoride	SbF_3	385, 444, 562
Arsenic hydride (760 mm), cc	AsH_3	42, 30, 28 [completely miscible at 72°]
oxide (pent-)	As_2O_5	59.5, 62.1, 65.8, 69.8, 71.2, 73.0, 75.1, 76.7
oxide (tri-)	As_2O_3	1.20, 1.49, 1.82, 2.31, 2.93, 4.31, 6.11, 8.2
Barium acetate	$Ba(C_2H_3O_2)_2 \cdot 3H_2O$	58.8, 62, 72, 75, 78.5, 75.0, 74.0, 74.8
azide	$Ba(N_3)_2$	12.5, 16.1, 17.4[17°]
bromate	$Ba(BrO_3)_2 \cdot H_2O$	0.29, 0.44, 0.65, 0.95, 1.31, 2.27, 3.52, 5.39
bromide	$BaBr_2 \cdot 2H_2O$	98, 101, 104, 109, 114, 123, 135, 149
n-butyrate	$Ba(C_4H_7O_2)_2$	37.0, 36.1, 35.4, 34.9, 35.2, 37.2, 41.7, 45.5, 48.1[95°]
caproate	$Ba(C_6H_{11}O_2)_2 \cdot 3.5H_2O$	11.71, 8.38, 6.89, 5.87, 5.79, 8.39, 14.71, 19.28
chlorate	$Ba(ClO_3)_2 \cdot H_2O$	20.3, 26.9, 33.9, 41.6, 49.7, 66.7*, 84.8, 105
chloride	$BaCl_2 \cdot 2H_2O$	31.2, 33.5, 35.8, 38.1, 40.8, 46.2, 52.5, 59.4
chlorite	$Ba(ClO_2)_2$	43.9, 44.6, 45.4, 47.9, 53.8, 55.8, 66.6, 80.8
fluoride	BaF_2	0.159, 0.160, 0.162
formate	$Ba(CHO_2)_2$	26.2, 28.0, 29.9, 31.9, 34.0, 38.6, 44.2, 51.3
hydroxide	$Ba(OH)_2$	1.67, 2.48, 3.89, 5.59, 8.22, 20.94, 101.4
iodate	$Ba(IO_3)_2$	0.035, 0.046, 0.057
iodide	$BaI_2 \cdot 2H_2O$	182, 201, 223, 250, 264, 291, 301
nitrate	$Ba(NO_3)_2$	4.95, 6.67, 9.02, 11.48, 14.1, 20.4, 27.2, 34.4
nitrite	$Ba(NO_2)_2 \cdot H_2O$	50.3, 60, 72.8, 102, 151, 222, 261, 325
perchlorate	$Ba(ClO_4)_2 \cdot 3H_2O$	239, 336, 416, 495, 575, 653
propionate	$Ba(C_3H_5O_2)_2 \cdot H_2O$	57.2, 56.8, 57.5, 59.0, 62.0, 67.8, 73.0, 82.7
isosuccinate	$BaC_4H_4O_4$	0.421, 0.432, 0.418, 0.393, 0.366, 0.306, 0.237
sulfamate	$Ba(SO_3NH_2)_2$	18.3, 22.3, 26.8, 32.5, 38.5, 49.6, 61.5, 73.5
sulfide	BaS	2.88, 4.89, 7.86, 10.38, 14.89, 27.69, 49.91, 60.29
tartrate	$Ba(C_2H_2O_3)_2$	0.021, 0.024, 0.028, 0.032, 0.035, 0.044, 0.053
Beryllium nitrate	$Be(NO_3)_2$	97, 102, 108, 113, 125, 178
sulfate	$BeSO_4$	37.0, 37.6, 39.1, 41.4, 45.8, 53.1, 67.2, 82.8
Boric acid	H_3BO_3	2.67, 3.73, 5.04, 6.72, 8.72, 14.81, 23.62, 40.25
Cadmium bromide	$CdBr_2$	56.3, 75.4, 98.8, 129, 152, 153, 156, 160
chlorate	$Cd(ClO_3)_2$	299, 308, 322, 348, 376, 455
chloride	$CdCl_2 \cdot 2.5H_2O$	90, 100, 113, 132, 135
chloride	$CdCl_2 \cdot H_2O$	135, 135, 136, 140, 147

Table 67 (continued)
SOLUBILITIES OF INORGANIC COMPOUNDS AND METAL SALTS OF ORGANIC ACIDS IN WATER AT VARIOUS TEMPERATURES

Substance	Formula	0°	10°	20°	30°	40°	60°	80°	90°	100°
formate	Cd(CHO₂)₂	8.3	11.1	14.4	18.6	25.3	59.5	80.5	85.2	94.6
iodide	CdI₂	78.7		84.7	87.9	92.1	100	111		125
nitrate	Cd(NO₃)₂	122	136	150	167	194	310	713		272
perchlorate	Cd(ClO₄)₂·6H₂O		180	188	195	203	221	243		
selenate	CdSeO₄	72.5	68.4	64.0	58.9	55.0	44.2	32.5	27.2	22.0
sulfate	CdSO₄	75.4	76.0	76.6		78.5	81.8	66.7	63.1	60.8
Calcium acetate	Ca(OAc)₂·2H₂O	37.4	36.0	34.7	33.8	33.2	32.7	33.5	31.1	29.7
benzoate	Ca(OBz)₂·3H₂O	2.32	2.45	2.72	3.02	3.42	4.71	6.87	8.55	8.70
bromide	CaBr₂·6H₂O	125	132	143	$185^{34°}$	213	278	295		$312^{105°}$
butyrate	Ca(C₄H₇O₂)₂	20.31	19.15	18.20	17.25	16.40	15.15	14.95		15.85
cacodylate	Ca(C₂H₆AsO₂)₂·9H₂O	48	52	59	71					
chloride	CaCl₂·6H₂O	59.5	64.7	74.5	100	128	137	147	154	159
chromate	CaCrO₄	4.5		2.25	1.83	1.49	0.83			
(mn)	CaCrO₄·2H₂O	17.3		16.6	16.1					
formate	Ca(CHO₂)₂	16.15		16.60		17.05	17.50	17.95	36.80	18.40
gluconate	Ca(C₆H₁₁O₇)₂·H₂O			3.72		5.29		12.11		$57.29^{98°}$
hydrogen carbonate	Ca(HCO₃)₂	16.15		16.60		17.05	17.50	17.95		18.40
hydroxide	Ca(OH)₂	0.189	0.182	0.173	0.160	0.141	0.121		0.086	0.076
Calcium iodate	Ca(IO₃)₂·6H₂O	0.090		0.24	0.38	0.52	0.65	0.66	0.67	
iodide	CaI₂	64.6	66.0	67.6	69.0	70.8	74	78		81
lactate	Ca(C₃H₅O₃)₂·5H₂O	3.1		$5.4^{15°}$	7.9					
levulinate	Ca(C₁₀H₁₄O₆)·2H₂O	38.1		$45.1^{16°}$	55.0	$70.3^{45°}$	$88.7^{55°}$			
malonate	Ca(C₃H₂O₄)	0.29	0.33	0.36	0.40	0.42	0.46	0.48		
nitrate	Ca(NO₃)₂·4H₂O	102	115	129	152	191		358		363
nitrite	Ca(NO₂)₂·4H₂O	63.9		$84.5^{18°}$	104		134	151	166	178
propionate	Ca(C₃H₅O₂)₂·H₂O	42.80		39.85			38.25	39.85	42.15	48.44
selenate	CaSeO₄·2H₂O	9.73	9.77	9.22	8.79	7.14				
succinate	Ca(C₃H₂O₂)₂·3H₂O	1.127	1.22	1.28		1.18	0.89	0.68		0.66
sulfamate	Ca(SO₃NH₂)₂	56.5	62.8	72.3	84.5	100.1	150.0	215.2	$242^{95°}$	
sulfate	CaSO₄·½H₂O			0.32	$0.29^{25°}$	$0.263^{5°}$	$0.21^{45°}$	$0.145^{65°}$	$0.127^{5°}$	0.071
sulfate	CaSO₄·2H₂O	0.223	0.244	$0.255^{18°}$	0.264	0.265	$0.244^{65°}$	$0.234^{75°}$		0.205
tartrate	CaC₄H₄O₆·4H₂O	0.026	0.029	0.034	0.046	0.063	0.091	0.130		
uranyl carbonate	Ca₂UO₂(CO₃)₃·10H₂O	0.1		$0.43^{23°}$		0.8	$1.5^{55°}$			
valerate	Ca(C₅H₉O₂)₂	9.82	9.25	8.80	8.40	8.05	7.78	7.95	8.20	8.78
isovalerate	Ca(C₅H₉O₂)₂·3H₂O	26.05	22.70	21.80	21.68	22.00	18.38	16.88	16.65	16.55

The following is a solubility table (grams per 100 g water unless otherwise noted). The column temperature headings do not appear on this page; values are given left-to-right in order of increasing temperature.

Compound									
Carbon disulfide CS_2	0.204	0.194	0.179	0.155	0.111				
oxide sulfide (STP) mL/100 mL COS	133.3	83.6	56.1	40.3					
tetrafluoride (STP) mL/100 g CF_4		0.595	0.490	0.415	0.366				
Cerium(III) ammonium nitrate $Ce(NH_4)_2(NO_3)_5$		242	276	318	376	681			
(IV) ammonium nitrate $Ce(NH_4)_2(NO_3)_6$			135	150	169	213			
(III) ammonium sulfate $Ce(NH_4)_4(SO_4)_2$	39.5	37.2	35.2	33.2	32.6	13.7			
(III) selenate $Ce_2(SeO_3)_3$	21.4							2.1	
(III) sulfate $Ce_2(SO_4)_3 \cdot 9H_2O$	18.8	9.84	7.24	5.63	3.87	2.02	1.33		
$Ce_2(SO_4)_3 \cdot 8H_2O$	0.21	9.43	7.10	5.70	4.04	2.00	4.6	5.40	
Cesium aluminum sulfate $Cs_2Al_2(SO_4)_4$		0.30	0.40	0.61	0.85			10.5	22.7
bromate $CsBrO_3$	2.46	3.8	6.2	9.5	13.8	26.2	45.0	58.0	79.0
chlorate $CsClO_3$			3.66^{25}	4.53	5.30^{35}	8.9	19.5	27.7	37.9
chloride $CsCl$	161	175	187	197	208	230	250	260	271
chloroaurate(III) $CsAuCl_4$	0.0047	0.0064	0.0087	0.0119	0.0158	0.0290	0.0525	0.0675	0.0914
chloroplatinate(IV) Cs_2PtCl_6									
formate $CsCHO_2$	335	381	450	533	694				
iodide CsI	44.1	58.5	76.5	96	124^{45}	150	190	205	
nitrate $CsNO_3$	9.33	14.9	23.0	33.9	47.2	83.8	134	163	197
perchlorate $CsClO_4$	0.8	1.0	1.6	2.6	4.0	7.3	14.4	20.5	30.0
sulfate Cs_2SO_4	167	173	179	184	190	200	210	215	220
Chlorine dioxide ClO_2	2.76	6.00	8.70^{15}	130^{25}	152^{35}				
Chromium(III) nitrate $Cr(NO_3)_3$	108	124^{15}	130^{25}	167.2	172.5	183.9	191.6		206.8
(VI) oxide CrO_3	164.8								
(III) perchlorate $Cr(ClO_4)_3$	104	123	130	112	128	163			257
Cobalt(II) bromide $CoBr_2$	91.9	162	180	180	195	214	227	241	
chlorate $Co(ClO_3)_2$	135	162				316			
chloride $CoCl_2$	43.5	47.7	52.9	59.7	69.5	93.8	97.6	101	106
iodate $Co(IO_3)_2$		0.24	1.02	0.90	0.88	0.82	0.73		0.70
nitrate $Co(NO_3)_2$	84.0	89.6	97.4	111	125	174	204	300	
nitrite $Co(NO_2)_2$	0.076	0.24	0.40	0.40	0.61	0.85			
sulfate $CoSO_4$	25.5	30.5	36.1	42.0	48.8	55.0	53.8	45.3	38.9
$CoSO_4 \cdot 7H_2O$	44.8	56.3	65.4	73.0	88.1	101	76.5	76.5	
Copper(II) ammonium chloride $CuCl_2 \cdot 2NH_4Cl$	28.2	32.0^{12}	35.0	38.3	43.8	56.6	76.5	86.1	
ammonium sulfate $CuSO_4 \cdot (NH_4)_2SO_4$	11.5	15.1	19.4	24.4	30.5	46.3	69.7		
bromide $CuBr_2$	107	116	126	128	131^{50}	174	204		107
chloride $CuCl_2$	68.6	70.9	73.0	77.3	87.6	96.5	104	108	120

Table 67 (continued)
SOLUBILITIES OF INORGANIC COMPOUNDS AND METAL SALTS OF ORGANIC ACIDS IN WATER AT VARIOUS TEMPERATURES

Substance	Formula	0°	10°	20°	30°	40°	60°	80°	90°	100°
fluorosilicate	$CuSiF_6$	73.5	76.5	81.6	$84.1^{25°}$	$91.2^{50°}$		$93.2^{75°}$		
nitrate	$Cu(NO_3)_2$	83.5	100	125	156	163	182	208	222	247
potassium sulfate	$CuSO_4 \cdot K_2SO_4$	5.1	7.2	10.0	13.6	18.2				
selenate	$CuSeO_4$	12.04	14.53	17.51	21.04	25.22	36.50	53.68		
sulfate	$CuSO_4 \cdot 5H_2O$	23.1	27.5	32.0	37.8	44.6	61.8	83.8		114
tartrate	$CuC_4H_4O_6 \cdot 3H_2O$		$0.020^{15°}$	0.042	0.089	0.142	0.197	0.144		
Gadolinium bromate	$Gd(BrO_3)_3 \cdot 9H_2O$	50.2	70.1	95.6	126	166				
sulfate	$Gd_2(SO_4)_3$	3.98	3.30	2.60	2.32					
Germanium(IV) oxide	GeO_2		0.49	0.43	0.50	0.61				
Holmium sulfate	$Ho_2(SO_4)_3 \cdot 8H_2O$			8.18	$6.71^{25°}$	4.52				
Hydrazinium (1+) nitrate	$N_2H_5NO_3$		175	266	402	607	2127			
(2+) sulfate	$N_2H_6SO_4$			2.87	3.89	4.15	9.08	14.39		
(1+) sulfate	$(N_2H_5)_2SO_4$				221	300	554			
Hydrogen bromide	HBr	221.2	210.3	$204.0^{15°}$		$171.5^{50°}$		$150.5^{75°}$		130.0
chloride	HCl	82.3	$78.0^{8°}$	71.9	67.3	63.3	56.1			
selenide, mL at STP	H_2Se	386	351	289						
Iodine	I_2	0.014	0.020	0.029	0.039	0.052	0.100	0.225	0.315	0.445
Iridium(IV) ammonium chloride	$(NH_4)_2IrCl_6$	0.556	0.706	0.77	1.21	1.57	2.46	4.38	dec	
sodium chloride	Na_2IrCl_6		$34.46^{15°}$		56.17	96.00	191.2	279.3		
Iron(II) ammonium sulfate	$FeSO_4 \cdot (NH_4)_2SO_4 \cdot 6H_2O$	17.23	31.0	36.47	45.0					
(II) bromide	$FeBr_2$	101	109	117	124	133	144	168	176	184
(II) chloride	$FeCl_2$	49.7	59.0	62.5	66.7	70.0	78.3	88.7	92.3	94.9
(III) chloride	$FeCl_3 \cdot 6H_2O$	74.4		91.8	106.8					
(II) fluoro-silicate	$FeSiF_6 \cdot 6H_2O$	72.1	74.4		$77.0^{25°}$		$83.7^{50°}$	$88.1^{75°}$		$100.1^{106°}$
(II) nitrate	$Fe(NO_3)_2 \cdot 6H_2O$	113	134				266			
(III) nitrate	$Fe(NO_3)_3 \cdot 9H_2O$	112.0		137.7		175.0				
(III) perchlorate	$Fe(ClO_4)_3$	289		368	422	478	772			
(II) sulfate	$FeSO_4 \cdot 7H_2O$	28.8	40.0	48.0	60.0	73.3	100.7	79.9	68.3	57.8
Lanthanum bromate	$La(BrO_3)_3$	98	120	149	200	168	247			
nitrate	$La(NO_3)_3$	100		136						
selenate	$La_2(SeO_3)_3$	50.5	45	45	45	45	18.5	5.4	2.2	

Compound	Formula									
sulfate	$La_2(SO_4)_3$	3.00	2.72	2.33	1.90	1.67	1.26	0.91	0.79	0.68
Lead(II) acetate	$Pb(C_2H_3O_2)_2$	19.8	29.5	44.3	69.8	116				
bromide	$PbBr_2$	0.45	0.63	0.86	1.12	1.50	2.29	3.23	3.86	4.55
chloride	$PbCl_2$	0.67	0.82	1.00	1.20	1.42	1.94	2.54	2.88	3.20
fluorosilicate	$PbSiF_6$	190		222		403		428		463
iodide	PbI_2	0.044	0.056	0.069	0.090	0.124	0.193	0.294		0.42
nitrate	$Pb(NO_3)_2$	37.5	46.2	54.3	63.4	72.1	91.6	111		133
Lithium acetate	$LiC_2H_3O_2$	31.2	35.1	40.8	50.6	68.6				
ammonium sulfate	$LiNH_4SO_4$		55.2		55.9	56.1	56.5			
azide	LiN_3	61.3	64.2	67.2	71.2	75.4	86.6			100
benzoate	$LiC_7H_5O_2$	38.9	41.6	44.7	53.8					
borate (meta-)	$LiBO_2$	0.90	1.3	2.7	5.7	10.9				
bromate	$LiBrO_3$	154	166	179	198	221	269	308	329	355
bromide	$LiBr$	143	147	160	183	211	223	245		266
carbonate	Li_2CO_3	1.54	1.43	1.33	1.26	1.17	1.01	0.85		0.72
chlorate	$LiClO_3$	241	283	372	488	604	777			
chloride	$LiCl$	69.2	74.5	83.5	86.2	89.8	98.4	112	121	128
chloroaurate(III)	$LiAuCl_4$	105	113	136	167	206	324	599		
cyanoplatinate(II)	$Li_2Pt(CN)_4$			141	153	160	178	216	239	
formate	$LiCHO_2$	32.3	35.7	39.3	44.1	49.5	64.7	92.7	116	138
hydrogen phosphite	Li_2HPO_3	9.97			7.61	7.11	6.03			4.43
hydroxide	$LiOH$	11.91	12.11	12.35	12.70	13.22	14.63	16.56		19.12
iodide	LiI	151	157	165	171	179	202	435	440	481
molybdate	Li_2MoO_4	82.6		79.5	79.4	78.0				73.9
nitrate	$LiNO_3$	53.4	60.8	70.1	138	152	175	233	272	
nitrite	$LiNO_2$	70.9	82.5	96.8	114	133	177			324
perchlorate	$LiClO_4$	42.7	49.0	56.1	63.6	72.3	92.3	128	151	
phosphate (meta-)	$LiPO_3$	0.101	$0.058^{25°}$			0.048				
selenite	Li_2SeO_3	25.0	23.3	21.5	19.6	17.9	14.7	11.9	11.1	
sulfate	Li_2SO_4	36.1	35.5	34.8	34.2	33.7	32.6	31.4	30.9	9.9
tartrate (*d*-)	$Li_2C_4H_4O_6$	42.0	31.8	27.1	26.6	27.2	29.5			
thiocyanate	$LiSCN$			114	131	153				
vanadate	Li_3VO_4	2.50		4.82	6.28	4.38	2.67			
Magnesium acetate	$Mg(C_2H_3O_2)_2$	56.7	59.7	53.4	68.6	75.7	118			
bromide	$MgBr_2$	98	99	101	104	106	112			125
chlorate	$Mg(ClO_3)_2$	114	123	135	155	178	242		268	
chloride	$MgCl_2$	52.9	53.6	54.6	55.8	57.5	61.0	66.1	69.5	73.3
fluorosilicate	$MgSiF_6$	26.3	30.8	30.8		34.9	44.4			
formate	$Mg(CHO_2)_2$	14.0	14.2	14.4	14.9	15.9	17.9	20.5	22.2	23.9
iodate	$Mg(IO_3)_2$		7.2	8.6	10.0	11.7	15.2	15.5	15.6	
iodide	MgI_2	120	140	140		173		186		

Table 67 (continued)
SOLUBILITIES OF INORGANIC COMPOUNDS AND METAL SALTS OF ORGANIC ACIDS IN WATER AT VARIOUS TEMPERATURES

Substance	Formula	0°	10°	20°	30°	40°	60°	80°	90°	100°
Magnesium nitrate	Mg(NO₃)₂	62.1	66.0	69.5	73.6	78.9	78.9	91.6	106	
selenate	MgSeO₄	20.0	30.4	38.3	44.3	48.6	55.8			
sulfate	MgSO₄	22.0	28.2	33.7	38.9	44.5	54.6	55.8	52.9	50.4
sulfite	MgSO₃	0.339	0.446	0.573	0.751	0.959	0.779	0.642	0.622	
tartrate	MgC₄H₄O₆	0.54	0.78	1.06		1.02				
Manganese bromide	MnBr₂	127	136	147	157	169	197	225	226	228
chloride	MnCl₂	63.4	68.1	73.9	80.8	88.5	109	113	114	115
fluoride	MnF₂			1.06		0.67	0.44			0.48
nitrate	Mn(NO₃)₂	102	118	139	206					
oxalate	MnC₂O₄	0.020	0.024	0.028	0.033					
sulfate	MnSO₄	52.9	59.7	62.9	62.9	60.0	53.6	45.6	40.9	35.3
Mercury(II) bromide	HgBr₂	0.30	0.40	0.56	0.66	0.91	1.68	2.77		4.9
(II) chloride	HgCl₂	3.63	4.82	6.57	8.34	10.2	16.3	30.0		61.3
(I) perchlorate	Hg₂(ClO₄)₂	282	325	367	407	455	499	541		580
Molybdenum trioxide	MoO₃			0.134	0.285	0.454	1.08	1.74		
Neodymium bromate	Nd(BrO₃)₃	43.9	59.2	75.6	95.2	116				
chloride	NdCl₃	127	96.7	98.0	99.6	102	105			
nitrate	Nd(NO₃)₃	127	133	142	145	159	211			
selenate	Nd₂(SeO₃)₃	46.2	44.6	41.8	39.9	39.9	43.9	7.0	3.3	
sulfate	Nd₂(SO₄)₃	13.0	9.7	7.1	5.3	4.1	2.8	2.2	1.2	
Nickel bromide	NiBr₂	113	122	131	138	144	153	154		155
chlorate	Ni(ClO₃)₂	111	120	133	155	181	221	308		
chloride	NiCl₂	53.4	56.3	60.8	70.6	73.2	81.2	86.6		87.6
fluoride	NiF₂		2.55	2.56			2.56		2.59	
iodate	Ni(IO₃)₂				1.15		1.06		1.00	
	Ni(IO₃)₂·4H₂O	0.74		1.09	1.43					
iodide	NiI₂	124	135	148	161	174	184	187	188	
nitrate	Ni(NO₃)₂	79.2		94.2	105	119	158	187	188	
perchlorate	Ni(ClO₄)₂	105	107	110	113	117				
Nickel sulfate	NiSO₄·6H₂O	(pale blue)		40.1	43.6	47.6	55.6	64.5	70.1	76.7
		(green)		44.4	46.6	49.2				
	NiSO₄·7H₂O	26.2	32.4	37.7	43.4	50.4				

Compound	Formula									
Osmium tetroxide	OsO_4	5.26	5.75	6.43	14.23	21.52	44.32	84.5	120	
Oxalic acid	$H_2C_2O_4$	3.54	6.08	9.52						
Potassium acetate	$KC_2H_3O_2$	216	233	256	283	324	350	381	398	
aluminum sulfate	$KAl(SO_4)_2$	3.00	3.99	5.90	8.39	11.7	24.8	71.0	109	
azide	KN_3	41.4	46.2	50.8	55.8	61.0				106
benzoate	$KC_7H_5O_2$		65.8	70.7	76.7	82.1				
bromate	$KBrO_3$	3.09	4.72	6.91	9.64	13.1	22.7	34.1		49.9
bromide	KBr	53.6	59.5	65.3	70.7	75.4	85.5	94.9	99.2	104
cadmium bromide	$KCdBr_3$	116	133	150	170	191	233	276	298	325
cadmium chloride	$KCdCl_3$	26.6	32.3	38.9	45.6	53.1	67.5	83.5		101
carbonate	K_2CO_3	105	108	111	114	117	127	140	148	156
chlorate	$KClO_3$	3.3	5.2	7.3	10.1	13.9	23.8	37.6	46.0	56.3
chloride	KCl	28.0	31.2	34.2	37.2	40.1	45.8	51.3	53.9	56.3
chloroaurate(III)	$KAuCl_4$		38.3	61.8	94.9	145	405			
chloroplatinate(IV)	K_2PtCl_6	0.48	0.60	0.78	1.00	1.36	2.45	3.71		5.03
chromate	K_2CrO_4	56.3	60.0	63.7	66.7	67.8	70.1	74.5		
citrate	$K_3C_6H_5O_7$		153	172	194					
cobalt(II) sulfate	$K_2Co(SO_4)_2$	8.5	11.7	15.5	19.3	23.3	32.5	47.7		
copper(II) sulfate	$K_2Cu(SO_4)_2$	5.1	7.2	10.0	13.6	18.2				
cyanoplatinate(II)	$K_2Pt(CN)_4$	11.6	19.8	33.9	52.0	78.3	139	177	194	
dichromate	$K_2Cr_2O_7$	4.7	7.0	12.3	18.1	26.3	45.6	73.0		
dihydrogen phosphate	KH_2PO_4	14.8	18.3	22.6	28.0	33.5	50.2	70.4	83.5	
dithionate	$K_2S_2O_6$	2.6	4.2	6.6	9.3					
ferricyanide	$K_3Fe(CN)_6$	30.2	38	46	53	59.3	70			91
ferrocyanide	$K_4Fe(CN)_6$	14.3	21.1	28.2	35.1	41.4	54.8	66.9	71.5	74.2
fluoride	KF	44.7	53.5	94.9	108	138	142	150		
fluorogermanate(IV)	K_2GeF_6	0.25	0.36	0.50	0.66	0.96				
fluorosilicate	K_2SiF_6	0.077	0.102	0.151	0.202	0.253				
fluorotitanate(IV)	K_2TiF_6	0.55	0.91	1.28						
formate	$KCHO_2$		313	337	361	398	471	580	658	
hydrogen carbonate	$KHCO_3$	22.5	27.4	33.7	39.9	47.5	65.6			
Potassium hydrogen fluoride	KHF_2	24.5	30.1	39.2	46.8	56.5	78.8	114		
hydrogen selenite	$KH_3(SeO_3)_2$	115	162	215	300	408	900			
hydrogen sulfate	$KHSO_4$	36.2	48.6	54.3	61.0	76.4	96.1			122
hydrogen tartrate	$KC_4H_5O_6$	0.231	0.358	0.523	0.762					
hydroxide	KOH	95.7	103	112	126	134	154			178
iodate	KIO_3	4.60	6.27	8.08	10.3	12.6	18.3	24.8		32.3
iodide	KI	128	136	144	153	162	176	192	198	206
iron(II) sulfate	$K_2Fe(SO_4)_2$	19.6	24.5	32.1	39.1	44.9	57.2	63.4		
magnesium sulfate	$K_2Mg(SO_4)_2$	14.0	19.5	25.0	30.4	36.6	50.2			

Table 67 (continued)
SOLUBILITIES OF INORGANIC COMPOUNDS AND METAL SALTS OF ORGANIC ACIDS IN WATER AT VARIOUS TEMPERATURES

Substance	Formula	0°	10°	20°	30°	40°	60°	80°	90°	100°
nickel sulfate	$K_2Ni(SO_4)_2$	3.37	4.50	5.94	7.72	9.85	15.4	23.0	27.8	33.4
nitrate	KNO_3	13.9	21.2	31.6	45.3	61.3	106	167	203	245
nitrite	KNO_2	279	292	306	320	329	348	376	390	410
oxalate	$K_2C_2O_4$	25.5	31.9	36.4	39.9	43.8	53.2	63.6	69.2	75.3
perchlorate	$KClO_4$	0.76	1.06	1.68	2.56	3.73	7.3	13.4	17.7	22.3
periodate	KIO_4	0.17	0.28	0.42	0.65	1.0	2.1	4.4	5.9	
permanganate	$KMnO_4$	2.83	4.31	6.34	9.03	12.6	22.1			
peroxodisulfate	$K_2S_2O_8$	1.65	2.67	4.70	7.75	11.0				
perrhenate	$KReO_4$	0.34	0.63	0.99	1.47	2.2	4.58	8.7		
phosphate	K_3PO_4		81.5	92.3	108	133				
salicylate	$KC_7H_5O_3$	21.2	32.4	47.1	61.3	78.6	116	156		
selenate	K_2SeO_4	107	109	111	113	115	119	121		122
selenite	K_2SeO_3	169	186	203	217	217	220			217
sulfate	K_2SO_4	7.4	9.3	11.1	13.0	14.8	18.2	21.4	22.9	24.1
sulfite	K_2SO_3	106	106	106	107	107	108			112
tellurate	K_2TeO_4	8.8		27.5	50.4					
thioantimonate(V)	K_3SbS_4	306	320		302	315		381		
thiocyanate	$KSCN$	177	198	224	255	289	372	492	571	675
thiosulfate	$K_2S_2O_3$	96		155	175	205	238	293	312	
zinc sulfate	$K_2Zn(SO_4)_2 \cdot 6H_2O$	13.0		25.9	35.0	44.9	72.1			
Praseodymium bromate	$Pr(BrO_3)_3$	55.9	73.0	91.8	114	144				
nitrate	$Pr(NO_3)_3$			112	162	178				
selenate	$Pr_2(SeO_3)_3$	36.2			32.4	31.2	30.4	5.43	3.6	
sulfate	$Pr_2(SO_4)_3$	19.8	15.6	12.6	9.89	2.56	5.04	3.5	1.1	0.91
Rubidium aluminum sulfate	$Rb_2Al_2(SO_4)_4$	0.72	1.05	1.50	2.20	3.25	7.40	21.6		
bromate	$RbBrO_3$				3.6	5.1				
bromide	$RbBr$	90	99	108	119	132	158			
chlorate	$RbClO_3$	2.1	3.4	5.4	8.0	11.6	22	38	49	63
chloride	$RbCl$	77	84	91	98	104	115	127	133	143
chloroaurate(III)	$RbAuCl_4$		4.8	9.9	15.5	21.5	36.2	54.6	65.8	79.2
chloroplatinate(IV)	Rb_2PtCl_6	0.014	0.020	0.028	0.040	0.056	0.090	0.182	0.247	0.333
chromate	Rb_2CrO_4	62.0	67.5	73.6	78.9	85.6	95.7			

Compound	Formula									
cobalt sulfate	Rb₂Co(SO₄)₂	5.10	7.47	10.8	14.5	18.2	30.2	44.9	55.0	70.1
dichromate (mn)	Rb₂Cr₂O₇			5.9	10.0	15.2	32.3			
(tric)				5.8	9.5	14.8	32.4			
formate	RbCHO₂		443	554	614	694	900			
iron(III) sulfate	RbFe(SO₄)₂·12H₂O		8.0	20	35	52				
nitrate	RbNO₃	19.5	33.0	52.9	81.2	117	200	310	374	452
perchlorate	RbClO₄	1.09	1.19	1.55	2.20	3.26	6.27	11.0	15.5	22.0
salicylate	RbC₇H₅O₃		187	212	238	268	324			
sulfate	Rb₂SO₄	37.5	42.6	48.1	53.6	58.5	67.5	75.1	78.6	81.8
Samarium bromate	Sm(BrO₃)₃	34.2	47.6	62.5	79.0	98.5				
chloride	SmCl₃		92.4	93.4	94.6	96.9				
Selenic acid	H₂SeO₄	426		567	1328					
Selenious acid	H₂SeO₃	90.1	122.2	166.7	235.6	344.4	383.1	383.1	385.4	
Selenium dioxide	SeO₂		222	257	291	335	440			
Silver acetate	AgC₂H₃O₂	0.73	0.89	1.05	1.23	1.43	1.93	2.59		
bromate	AgBrO₃		0.11	0.16	0.23	0.32	0.57	0.94	1.33	
chlorate	AgClO₃		10.4	15.3	20.9	26.8				
fluoride	AgF	85.9	120	172	190	203				
nitrate	AgNO₃	122	167	216	265	311	440	585	652	733
nitrite	AgNO₂	0.16	0.22	0.34	0.51	0.73	1.39			
perchlorate	AgClO₄	455	484	525	594	635				793
sulfamate	AgNH₂SO₃	2.30	4.82	7.53	10.3	15.3	28.5			
sulfate	Ag₂SO₄	0.57	0.70	0.80	0.89	0.98	1.15	1.30	1.36	1.41
Sodium acetate	NaC₂H₃O₂	36.2	40.8	46.4	54.6	65.6	139	153	161	170
aluminum sulfate	Na₂Al₂(SO₄)₄	37.4	39.3	39.7	41.7	43.8				
azide	NaN₃	38.9	39.9	40.8						
benzoate	NaC₇H₅O₂	62.6	62.8	62.8	62.9	63.1	64.5	68.6	70.6	55.3
borate (penta-)	Na₂B₁₀O₁₆	6.4	8.6	12.0	16.4	22.0	37.9	63.4	83.5	73.3
borate (tetra-)	Na₂B₄O₇	1.11	1.60	2.56	3.86	6.67	19.0	31.4	41.0	108
bromate	NaBrO₃	24.2	30.3	36.4	42.6	48.8	62.6	75.7		52.5
bromide	NaBr	80.2	85.2	90.8	98.4	107	118	120	121	90.8
carbonate	Na₂CO₃	7.00	12.5	21.5	39.7	49.0	46.0	43.9	43.9	121
chlorate	NaClO₃	79.6	87.6	95.9	105	115	137	167	184	204
chloride	NaCl	35.7	35.8	35.9	36.1	36.4	37.1	38.0	38.5	39.2
chloroaurate(III)	NaAuCl₄		139	151	178	227	900			
chloroiridate(IV)	Na₂IrCl₆		31.6	39.3	56.2					
chromate	Na₂CrO₄	31.7	50.1	84.0	88.0	96.1	115	125		126
cyanide	NaCN	40.8	48.1	58.7	71.2	96.0	192	279		
dichromate	Na₂Cr₂O₇	163	172	183	198	215	269	376	405	415
diethyl barbiturate	NaC₈H₁₁N₂O₃		12.7	21.5	24.7				48.0	
dihydrogen phosphate (ortho-)	NaH₂PO₄	56.5	69.8	86.9	107	133	172	211	234	

Table 67 (continued)
SOLUBILITIES OF INORGANIC COMPOUNDS AND METAL SALTS OF ORGANIC ACIDS IN WATER AT VARIOUS TEMPERATURES

Substance	Formula	0°	10°	20°	30°	40°	60°	80°	90°	100°
dihydrogen phosphate (pyro-)	$Na_2H_2P_2O_7$	4.47	6.95	12.0	17.1	18.4				
dithionate	$Na_2S_2O_6$	6.3	11.1	15.1	19.6	24.7	36.1	49.3	56.3	64.7
dodecanesulfonate	$NaC_{12}H_{25}SO_3$			0.13	0.25	6.54				
dodecanoate	$NaC_{12}H_{23}O_2$				4.58	22.7	105	170		
EDTA (Y)*	$Na_2H_2Y \cdot 2H_2O$	10.6		11.1	12.8	14.2	17.0	22.2	24.3	27.09*
ferrocyanide	$Na_4Fe(CN)_6$	11.2	14.8	18.8	23.8	29.9	43.7	62.1		
fluoride	NaF	3.66		4.06	4.22	4.40	4.68	4.89		5.08
fluoroberyllate	Na_2BeF_4	1.33		1.44		1.92	2.24	2.62	2.73	
fluorogermanate	Na_2GeF_6	1.52	1.68		2.25	2.83		3.36		
fluorosilicate	Na_2SiF_6	4.35	5.7	7.2	8.6	10.3	14.3	18.7	21.5	24.5
formate	$NaCHO_2$	43.9	62.5	81.2	102	108	122	138	147	160
germanate	Na_2GeO_3	14.4	18.8	23.8	28.7	37.2	65.0	116		
hydrogen arsenate	Na_2HAsO_4	5.9	13.0	33.9	49.3	69.5	144	186	188	198
hydrogen carbonate	$NaHCO_3$	7.0	8.1	9.6	11.1	12.7	16.0			
hydrogen phosphate	Na_2HPO_4	1.68	3.53	7.83	22.0	55.3	82.8	92.3	102	104
hydrogen phosphite	Na_2HPO_3	418	424	429	566					
hydrogen succinate	$NaC_4H_5O_4$	17.5	25.3	34.8	47.7	61.6	74.5	90.1		
hydroxide	$NaOH$	46.0	98	109	119	129	174			
hydroxostannate(IV)	$Na_2Sn(OH)_6$	29.4	36.4	43.7	42.7	38.9				
hypochlorite	$NaClO$			53.4	100	110				
iodate	$NaIO_3$	2.48	4.59	8.08	10.7	13.3	19.8	26.6	29.5	33.0
iodide	NaI	159	167	178	191	205	257	295		302
molybdate	Na_2MoO_4	44.1	64.7	65.3	66.9	68.6	71.8			
nitrate	$NaNO_3$	73.0	80.8	87.6	94.9	102	122	148		180
nitrite	$NaNO_2$	71.2	75.1	80.8	87.6	94.9	111	133		160
oxalate	$Na_2C_2O_4$	2.69	3.05	3.41	3.81	4.18	4.93	5.71		6.50
perchlorate	$NaClO_4$	167	183	201	222	245	288	306		329
periodate	$NaIO_4$	1.83	5.6	10.3	19.9	30.4				
phosphate	Na_3PO_4	4.5	8.2	12.1	16.3	20.2	29.9	60.0	68.1	77.0
potassium tartrate	$NaKC_4H_4O_6$	31.9	46.6	67.8	102	117	130	144		
salicylate	$NaC_7H_5O_3$		44.7	95.3						
selenate	Na_2SeO_4	13.3	25.2	26.9	77.0	81.8	78.6	74.8	73.0	72.7

Compound	Formula									
selenite	Na_2SeO_3	78.6	81.2	86.2	94.2	96.5	91.6	86.6	84.5	82.5
sulfate	Na_2SO_4	4.9	9.1	19.5	40.8	48.8	45.3	43.7	42.7	42.5
	$Na_2SO_4 \cdot 7H_2O$	19.5	30.0	44.1						
sulfide	Na_2S	9.6	12.1	15.7	20.5	26.6	39.1	55.0	65.3	
sulfite	Na_2SO_3	14.4	19.5	26.3	35.5	37.2	32.6	29.4	27.9	
thioantimonate(V)	Na_3SbS_4	13.4	20.0	27.9	37.2	49.3	53.8	88.3		
thiocyanate	$NaSCN$	111	134	164	176	192	210	218		
thiosulfate	$Na_2S_2O_3 \cdot 5H_2O$	50.2	59.7	70.1	83.2	104	90.8			97.2
tungstate	Na_2WO_4	71.5	73.0		77.6		90.8			
vanadate	$NaVO_3$	19.3	22.5	26.3	33.0		40.8			
Strontium acetate	$Sr(C_2H_3O_2)_2$	37.0	42.9	41.1	39.5	38.3	36.8	36.1	36.2	36.4
bromide	$SrBr_2$	85.2	93.4	102	112	123	150	182		223
chloride	$SrCl_2$	43.5	47.7	52.9	58.7	65.3	81.8	90.5		101
chromate	$SrCrO_4$	0.085	0.090					0.058		
Strontium fluoride	SrF_2	0.0113	0.0113	0.0117	0.0119					
formate	$Sr(CHO_2)_2$	9.1	10.6	12.7	15.2	17.8	25.0	31.9	32.9	34.4
hydroxide	$Sr(OH)_2$	0.91	1.25	1.77	2.64	3.95	8.42	20.2	44.5	91.2
iodide	SrI_2	165	178	192	218		270		365	383
nitrate	$Sr(NO_3)_2$	39.5	52.9	69.5	88.7	89.4	93.4	96.9	98.4	
nitrite	$Sr(NO_2)_2$	65	72	79	97			130	134	
oxide	SrO	1.03	1.05	3.40	9.15			13.13		12.15
sulfate	$SrSO_4$	0.0129	0.0132	0.0138	0.0141	0.0131	0.0116	0.0016	0.0115	
Sulfamic acid	H_2NSO_3H	14.7	18.6	21.3	26.1	29.5	37.1	47.1		
Telluric acid	H_2TeO_4	16.2	33.8	41.6	50.0	57.2	77.5	106		155
Terbium bromate	$Tb(BrO_3)_3 \cdot 9H_2O$	66.4	89.7	117	152	198				
Thallium(I) azide	TlN_3	0.171	0.236	0.364						
bromide	$TlBr$	0.022	0.032	0.048	0.068	0.097	0.177			
carbonate	Tl_2CO_3	2.00	3.92	5.3	12.2	12.7^{50}				
chlorate	$TlClO_3$	0.21	0.33	0.52	0.80	1.20	1.80			
chloride	$TlCl$	0.25	0.42	0.57	0.83					
hydroxide	$TlOH$	25.4	29.6	35.0	40.4	49.4	73.3	106	126	150
iodide	TlI	0.002	0.006	0.015	0.035	0.070	0.120			
nitrate	$TlNO_3$	3.90	6.22	9.55	14.3	21.0	46.1	110	200	414
nitrite	$TlNO_2$	17.9	28.9	40.3	53.2	83.6	216	1150	750	
perchlorate	$TlClO_4$	6.00	8.04	13.1	19.7	28.3	50.8	81.5		
picrate	$TlOC_6H_2(NO_2)_3$	0.135	0.40	0.57	0.83	1.73				
selenate	Tl_2SeO_4	2.17	2.80				8.50			10.8
sulfate	Tl_2SO_4	2.73	3.70	4.87	6.16	7.53	11.0	14.6	16.5	18.4

* Properly called dihydrogen ethylenediaminetetraacetate ($Na_2H_2EDTA \cdot 2H_2O$).

Table 67 (Continued)
SOLUBILITIES OF INORGANIC COMPOUNDS AND METAL SALTS OF ORGANIC ACIDS IN WATER AT VARIOUS TEMPERATURES

Substance	Formula	0°	10°	20°	30°	40°	60°	80°	90°	100°
Thorium nitrate	$Th(NO_3)_4$	186	187	191						
sulfate	$Th(SO_4)_2 \cdot 4H_2O$				1.99	4.04	1.63			
	$Th(SO_4)_2 \cdot 9H_2O$					3.00				
Tin(II) iodide	SnI_2	0.74	0.99	1.38	1.17	1.42	2.11	3.04	3.58	4.20
Uranium(IV) sulfate	$U(SO_4)_2 \cdot 4H_2O$			0.99	10.1	9.0	7.7			
	$U(SO_4)_2 \cdot 8H_2O$			11.9	17.9	29.2	55.8			
Uranyl nitrate	$UO_2(NO_3)_2$	98	107	122	141	167	317	388	426	474
oxalate	$UO_2C_2O_4$		0.45	0.50	0.61	0.80	1.22	1.94		3.16
Ytterbium sulfate	$Yb_2(SO_4)_3$	44.2	37.5		22.2	17.2	10.4	6.4	5.8	4.7
Yttrium bromide	YBr_3	63.9		75.1		87.3	101	116	123	
chloride	YCl_3	77.3	78.1	78.8	79.6	80.8				
nitrate	$Y(NO_3)_3$	93.1	106	123	143	163	200			
sulfate	$Y_2(SO_4)_3$	8.05	7.67	7.30	6.78	6.09	4.44	2.89	2.2	
Zinc bromide	$ZnBr_2$	389		446	528	591	618	645		672
chlorate	$Zn(ClO_3)_2$	145	152	200	209	223				
chloride	$ZnCl_2$	342	363	395	437	452	488	541		614
formate	$Zn(CHO_2)_2$	3.70	4.30	5.20	6.10	7.40	11.8	21.2	28.8	38.0
iodide	ZnI_2	430		432		445	467	490		510
nitrate	$Zn(NO_3)_2$	98			138	211				
sulfate (rh)	$ZnSO_4$	41.6	47.2	53.8	61.3	70.5	75.4	71.1		60.5
sulfate (mn)			54.4	60.0	65.5					
tartrate	$ZnC_4H_4O_6$			0.022	0.041	0.060	0.104	0.059		

Note: Solubilities are expressed as the number of grams of substance of stated molecular formula which when dissolved in 100 g of water make a saturated solution at the temperature stated (°C).

From Dean, J. A., Ed., *Lange's Handbook of Chemistry,* 12th ed., McGraw-Hill, New York, 1979, 10-2. With permission.

Table 68
MISCIBILITY OF ORGANIC SOLVENT PAIRS

Compound number	Compounds	Acetone	Acetyl acetone	2-Amino-2-methyl-1-propanol	Aniline	Benzaldehyde	Benzene	Benzin	Benzyl alcohol	Butyl acetate	Butyl alcohol	n-Butyl ether	Capryl alcohol	Carbon tetrachloride	Diacetone alcohol	Diethanolamine	Diethyl cellosolve	Diethyl ether	Dimethylaniline	Ethyl alcohol	Ethyl benzoate	Ethylene glycol	2-Ethylhexanol	Formamide	Furfuryl alcohol	Glycerol	Hydroxyethyl-ethylenediamine	Isoamyl alcohol	Methyl isobutyl ketone	Nitromethane	Dibutoxytetra-ethylene glycol	Pyridine	Triethanolamine	Trimethylene glycol
1	Acetone	:	M	M	:	M	M	M	M	M	M	M	M	M	M	M	M	M	M	M	M	M	M	M	M	I	M	M	M	M	M	M	:	M
2	Acetyl acetone	M	:	R	:	M	M	M	M	M	M	M	M	M	M	R	M	M	M	M	M	M	M	M	M	Is	R	M	M	M	M	M	:	M
3	Adiponitrile	M	M	M	M	M	M	:	M	M	M	I	M	I	M	:	M	M	M	M	M	M	I	M	M	I	M	L	M	M	M	M	M	M
4	2-Amino-2-methyl-1-propanol	R	M	:	M	M	M	:	M	M	M	M	:	M	R	:	M	M	M	M	M	M	M	M	M	I	R	M	M	:	M	M	M	M
5	Benzaldehyde	M	R	M	:	M	M	L	M	M	M	Is	M	M	M	M	M	M	M	M	M	Is	M	M	M	Is	R	M	M	M	M	M	M	M
6	Benzene	M	M	L	M	M	:	M	M	M	M	M	M	M	M	I	M	M	M	M	M	L	M	I	M	L	L	M	M	L	M	M	:	I
7	Benzin	M	M	I	:	M	M	:	L	M	M	M	M	M	L	:	M	M	M	M	M	I	M	:	I	I	Is	M	M	:	M	M	:	I
8	Benzonitrile	M	M	L	M	M	M	M	M	M	M	M	M	M	M	M	M	M	M	M	M	L	M	M	M	L	M	M	M	M	M	M	M	I
9	Benzothiazole	M	M	M	:	M	M	:	M	M	M	M	:	M	M	I	M	M	M	M	M	M	M	M	M	L	M	M	M	:	M	M	M	I
10	Benzyl alcohol	M	M	M	M	M	M	:	M	M	M	M	M	M	M	M	M	M	M	M	M	M	M	M	M	M	M	M	M	M	M	M	M	M
11	Benzyl mercaptan	M	M	I	:	M	M	:	M	M	M	M	M	M	M	I	M	M	M	M	M	I	M	:	I	I	L	M	M	:	M	M	:	M
12	Butyl acetate	M	M	L	:	M	M	M	M	M	M	M	M	M	M	I	M	M	M	M	M	L	M	M	M	I	L	M	M	M	M	M	R	M
13	Butyl alcohol	M	M	M	M	M	M	M	M	M	M	M	M	M	M	Is	M	M	M	M	M	M	M	M	M	M	L	M	M	M	M	M	M	M
14	n-Butyl ether	M	M	I	:	M	M	M	M	M	M	:	M	M	Is	I	M	M	Is	M	M	I	M	:	I	I	L	M	M	I	M	M	:	M
15	Capryl alcohol	M	M	Is	:	M	M	M	M	M	M	M	M	M	M	I	M	M	M	M	M	M	M	:	M	L	R	M	M	:	M	M	:	Is
16	Carbon tetrachloride	M	M	M	M	M	M	M	M	M	M	M	M	:	Is	I	M	M	M	M	M	I	M	:	M	L	R	M	M	I	M	M	:	M
17	Diacetone alcohol	M	M	R	:	M	M	:	M	M	M	M	M	Is	M	M	M	M	Is	M	M	M	M	M	M	M	M	M	M	:	M	M	M	M
18	Diethanolamine	M	R	M	M	M	I	:	M	I	M	I	:	I	M	M	M	M	M	M	I	M	I	M	M	M	M	M	I	M	I	M	M	M
19	Diethyl Cellosolve	M	M	M	:	M	M	M	M	M	M	M	M	M	M	M	M	M	M	M	M	M	M	M	M	M	M	M	M	M	M	M	M	M
20	Diethyl ether	M	M	L	M	M	M	M	M	M	M	M	M	M	M	I	M	:	M	M	M	L	M	:	I	L	L	M	M	Is	M	M	:	M
21	Dimethylaniline	M	M	M	:	M	M	M	M	M	M	M	M	M	M	I	M	M	M	M	M	I	M	:	M	L	L	M	M	M	M	M	M	M
22	Di-N-propylaniline	M	M	I	:	M	M	:	M	M	M	M	M	M	M	Is	M	M	M	M	M	I	M	:	I	L	L	M	M	:	M	M	:	M
23	Ethyl alcohol	M	M	M	M	M	M	M	M	M	M	M	M	M	M	M	M	M	M	M	M	M	M	M	M	M	M	M	M	M	M	M	M	M
24	Ethyl benzoate	M	M	R	:	M	M	M	M	M	M	M	M	M	M	I	M	M	M	M	M	I	M	:	M	I	R	M	M	:	M	M	M	M
25	Ethyl isothiocyanate	M	M	M	M	M	M	M	M	M	M	M	M	M	M	M	M	M	M	M	M	M	M	M	M	Is	M	M	M	M	M	M	M	Is
26	Ethyl thiocyanate	M	M	M	:	M	M	M	M	M	M	M	M	M	M	M	M	M	M	M	M	M	M	M	M	I	M	M	M	M	M	M	M	M
27	Ethylene glycol	M	M	M	:	M	M	M	M	M	M	M	:	M	M	M	M	M	M	M	M	M	M	M	M	M	M	M	L	M	M	M	M	M
28	2-Ethylhexanol	M	M	M	M	M	I	I	M	M	M	M	M	M	M	M	M	M	M	M	M	M	M	M	M	I	M	M	M	I	M	M	M	M

Table 68 (continued)
MISCIBILITY OF ORGANIC SOLVENT PAIRS

Compound number	Compounds	Acetone	Acetyl acetone	2-Amino-2-methyl-1-propanol	Aniline	Benzaldehyde	Benzene	Benzin	Benzyl alcohol	Butyl acetate	Butyl alcohol	n-Butyl ether	Capryl alcohol	Carbon tetrachloride	Diacetone alcohol	Diethanolamine	Diethyl cellosolve	Diethyl ether	Dimethylaniline	Ethyl alcohol	Ethyl benzoate	Ethylene glycol	2-Ethylhexanol	Formamide	Furfuryl alcohol	Glycerol	Hydroxyethyl-ethylenediamine	Isoamyl alcohol	Methyl isobutyl ketone	Nitromethane	Dibutoxytetra-ethylene glycol	Pyridine	Triethanolamine	Trimethylene glycol
29	Formamide	M	M	M	.	M	M	I	M	M	M	I	I	I	M	M	M	I	I	M	M	M	M		M	M	M	M	M	M	M	M	.	M
30	Furfuryl alcohol	M	R	M	.	M	M	Is	M	M	M	I	I	I	M	M	M	I	I	M	M	M	M	M		.	M	I	I	I	I	M	.	M
31	Glycerol	I	I	M	.	I	I	I	I	I	I	I	M	I	I	M	I	I	I	M	I	M	I	M	.		M	I	I	I	I	M	.	M
32	Hydroxyethyl-ethylenediamine	M	M	M	.	M	M	I	M	M	I	M	I	I	R	M	M	I	I	M	I	M	I	M	M	M		M	M	M	M	M	M	M
33	Isoamyl alcohol	M	M	I	.	M	M	M	M	M	M	M	M	M	R	I	M	M	M	M	M	M	M	M	I	I	I		M	I	M	M	M	I
34	Isoamyl sulfide	M	M	M	M	M	M	.	M	M	M	M	.	M	M	.	M	M	M	M	M	M	M	M	M	I	M	.	M	.	M	M	M	M
35	Isobutyl mercaptan	M	M	I	.	.	M	.	M	M	M	M	.	M	.	.	M	M	M	M	M	I	M	I	L	I	L	.	M	.	M	M	.	L
36	Methyl disulfide	M	M	M	M	M	M	.	M	M	M	M	.	M	M	.	M	M	M	M	M	M	M	M	I	I	I	M	M	.	M	M	R	R
37	Methyl isobutyl ketone	M	M	M	M	M	M	Is	M	M	M	M	Is	M	M	.	M	M	M	M	M	M	M	Is	I	I	I	M	M	M	M	M	I	R
38	Nitromethane	M	M	M	.	M	M	I	M	M	M	M	Is	M	M	I	M	M	M	M	M	M	M	Is	M	I	M	M	M		M	M	.	I
39	Dibutoxytetra-ethylene glycol	M	M	M	.	M	M	M	M	M	M	M	M	M	M	.	M	M	M	M	M	M	M	M	M	I	M	M	.	M		M	.	M
40	Pyridine	M	M	M	.	M	M	M	M	M	M	M	.	M	M	M	M	M	M	M	M	M	M	M	M	M	M	M	M	M	M		M	M
41	Tri-n-butylamine	M	M	I	I	.	I	I	I	M	M	M	M	I	M	.	M	M	M	M	M	I	M	M	M	I	L	I	I	I	I	M	I	.
42	Trimethylene glycol	M	M	M	M	M	I	I	M	M	M	I	M	I	M	M	M	M	I	M	Is	M	M	M	M	M	M	I	I	I	M	M	I	.

Note: The classifications were made by shaking together 5 ml of each of the solvents listed in a test tube for 1 minute, then allowing the mixture to settle. If no interfacial meniscus was observed, the solvent pair was considered miscible. If such a meniscus was present, the solvent pair was regarded as immiscible. The classification of immiscible is a qualitative one since solvent pairs may exhibit some degree of partial miscibility while existing as separate phases. Solvent pairs possessing a pronounced degree of partial miscibility are designated by the symbol Is. The designation R indicates that the two solvents reacted.

Reprinted with permission from Drury, J. S., Ind. Eng. Chem., 44(11), November 1952. Copyright by the American Chemical Society.

Table 69

SOLUBILITY CHART OF VARIOUS COMPOUNDS

No.	Compound	Al	NH₄	Sb	Ba	Bi	Cd	Ca	Cr	Co	Cu	Au′	Au‴	H	Fe″	Fe‴
1	Acetates —($C_3H_5O_2$)	W $Al(-)_3$	W $NH_4(-)$	....	W $Ba(-)_2$	W $Bi(-)_3$	W $Cd(-)_2$	W $Ca(-)_2$	W $Cr(-)_3$	W $Co(-)_2$	W $Cu(-)_2$	....	....	W $C_3H_5O_2$	W $Fe(-)_2$	W $Fe(-)_3$
2	Arsenate —(AsO_4)	a $Al(-)$	W $(NH_4)_3(-)$	A $Sb(-)$	W $Ba_3(-)_2$	W $Bi(-)$	W $Cd_3(-)_2$	W $Ca_3(-)_2$	....	A $Co_3(-)_2$	A $Cu_3(-)_2$	....	?	W H_3AsO_4	A $Fe_3(-)_2$	A $Fe(-)$
3	Arsenite —(AsO_2)	....	W NH_4AsO_2	A $Sb(-)$	....	....	....	....	....	....	A $CuH(-)$	....	....	W H_3AsO_3	....	....
4	Benzoate —($C_7H_5O_2$)	....	W $NH_4(-)$	....	W $Ba(-)_2$	A $Bi(-)_3$	W $Cd(-)_2$	W $Ca(-)_2$	....	W $CoH_2(-)_4$	W $Cu(-)_2$	....	....	W $C_7H_5O_2$	W $Fe(-)_2$	A $Fe(-)_3$
5	Bromide	W $AlBr_3$	W NH_4Br	d $SbBr_3$	W $BaBr_2$	d $BiBr_3$	W $CdBr_2$	W $CaBr_2$	W(I)* $CrBr_3$	W $CoBr_2$	W $CuBr_2$	W $AuBr$	W $AuBr_3$	W HBr	W $FeBr_2$	W $FeBr_3$
6	Carbonate	....	W $(NH_4)_2CO_3$	....	A $BaCO_3$	A $Bi_2(CO_3)_3$	A $CdCO_3$	A $CaCO_3$	W Cr_2CO_3	A $CoCO_3$	....	....	....		W $FeCO_3$	....
7	Chlorate —(ClO_3)	W $Al(-)_3$	W $NH_4(-)$	....	W $Ba(-)_2$	W $Bi(-)_3$	W $Cd(-)_2$	W $Ca(-)_2$	....	W $Co(-)_2$	W $Cu(-)_2$	....	....	W $HClO_3$	W $Fe(-)_2$	W $Fe(-)_3$
8	Chloride	W $AlCl_3$	W NH_4Cl	W $SbCl_3$	W $BaCl_2$	d $BiCl_3$	W $CdCl_2$	W $CaCl_2$	I $CrCl_3$	W $CoCl_2$	W $CuCl_2$	W $AuCl$	W $AuCl_3$	W HCl	W $FeCl_2$	W $FeCl_3$
9	Chromate —(CrO_4)	....	W $(NH_4)_2(-)$	....	A $Ba(-)$	A $Bi(-)$	A $Cd(-)$	W $Ca(-)_2$	....	A $Co(-)$	....	....	....	W $HCrO_2$	W $Fe(-)_2$	A $Fe(-)_3$
10	Citrate —($C_6H_5O_7$)	W $Al(-)$	W $(NH_4)_3(-)$	....	W $Ba_3(-)_2$	A $Bi(-)$	W $Cd(-)_2$	W $Ca_3(-)_2$	A $Cr(-)$	W $Co(-)_2$	A $Cu(-)_2$	....	....	W $C_6H_5O_7$	W $Fe(-)_2$	W $Fe(-)$
11	Cyanide	W $AlCN$	W NH_4CN	....	W $Ba(CN)_2$	W $Bi(CN)_3$	W $Cd(CN)_2$	W $Ca(CN)_2$	A $Cr(CN)_2$	A $Co(CN)_2$	W $Cu(CN)_2$	W $AuCN$	W $Au(CN)_3$	W HCN	a $Fe(CN)_2$	a $Fe(-)_3$
12	Ferricy′de —($Fe(CN)_6$)	....	W $(NH_4)_3(-)$	....	W $Ba(-)_2$	A $Bi(-)_3$	W $Cd_3(-)_2$	W $Ca_3(-)_2$	A $Cr(-)$	I $Co_3(-)_2$	I $Cu_3(-)_2$	....	....	W $H_4(-)$	I $Fe(-)$	A $Fe(-)$
13	Ferrocy′de —($Fe(CN)_6$)	W $Al_4(-)_3$	W $(NH_4)_4(-)$	....	W $Ba_2(-)$	A $Bi_4(-)_3$	A $Cd_2(-)$	W $Ca_2(-)$	....	I $Co_2(-)$	I $Cu_2(-)$	....	....	W $H_4(-)$	W $Fe_2(-)$	a $Fe_4(-)_3$
14	Fluoride	W AlF_3	W NH_4F	W SbF_3	W BaF_2	W BiF_3	W CdF_2	A CaF_2	W(a)* CrF_2	A CoF_2	W CuF_2	....	....	W HF	W FeF_2	a FeF_3
15	Formate —(CHO_2)	W $Al(-)_3$	W $NH_4(-)$	....	W $Ba(-)_2$	A $Bi(-)_3$	W $Cd(-)_2$	W $Ca(-)_2$	A $Cr(-)_3$	W $Co(-)_2$	W $Cu(-)_2$	....	....	W CHO_2	W $Fe(-)_2$	W $Fe(-)_3$
16	Hydroxide	A $Al(OH)_3$	W NH_4OH	d	W $Ba(OH)_2$	A $Bi(OH)_3$	A $Cd(OH)_2$	W $Ca(OH)_2$	A $Cr(OH)_3$	A $Co(OH)_2$	A $Cu(OH)_2$	W $AuOH$	A $Au(OH)_3$	H_2O	A $Fe(OH)_2$	A $Fe(OH)_3$
17	Iodide	W AlI_3	W NH_4I	d SbI_3	W BaI_2	A BiI_3	W CdI_2	W CaI_2	W CrI_2	W CoI_2	a CuI	a AuI	a AuI_3	W HI	W FeI_2	W FeI_3
18	Nitrate	W $Al(NO_3)_3$	W NH_4NO_3	....	W $Ba(NO_3)_2$	W $Bi(NO_3)_3$	W $Cd(NO_3)_2$	W $Ca(NO_3)_2$	W $Cr(NO_3)_3$	W $Co(NO_3)_2$	W $Cu(NO_3)_2$	....	....	W HNO_3	W $Fe(NO_3)_2$	W $Fe(NO_3)_4$
19	Oxalate —(C_2O_4)	A $Al_2(-)_3$	W $(NH_4)_2(-)$	....	W $Ba(-)$	A $Bi_2(-)_3$	A $Cd(-)$	A $Ca(-)$	a $Cr(-)$	A $Co(-)$	A $Cu(-)$	A Au_2O	A Au_2O_3	W $C_2H_2O_4$	A $Fe(-)$	a $Fe_2(-)_3$
20	Oxide	a Al_2O_3	....	W Sb_2O_3	W BaO	A Bi_2O_3	A CdO	W CaO	a Cr_2O_3	A CoO	A CuO	A Au_2O	A Au_2O_3	W H_2O	A FeO	A Fe_2O_3
21	Phosphate	A $AlPO_4$	W $NH_4H_2PO_4$	....	A $Ba_3(PO_4)_2$	A $BiPO_4$	A $Cd_3(PO_4)_2$	A $Ca_3(PO_4)_2$	W $Cr_3(PO_4)_2$	A $Co_3(PO_4)_2$	A $Cu_3(PO_4)_2$	....	....	W H_3PO_4	A $Fe_3(PO_4)_2$	W $FePO_4$
22	Silicate, —(SiO_3)	I $Al_2(-)_3$	....	....	W $Ba(-)$	....	A $Cd(-)$	W $Ca(-)$	A $Cr_2(-)_3$	A Co_2SiO_4	A $Cu(-)$	....	....	I H_2SiO_3	I $Fe(-)$	W $FePO_4$

Table 69 (continued)

SOLUBILITY CHART OF VARIOUS COMPOUNDS

No.		Al	NH₄	Sb	Ba	Bi	Cd	Ca	Cr	Co	Cu	Au′	Au‴	H	Fe″	Fe‴
23	Sulfate	$Al_2(SO_4)_3$ W	$(NH_4)_2SO_4$ W	$Sb_2(SO_4)_3$ A	$BaSO_4$ a	$Bi_2(SO_4)_3$ d	$CdSO_4$ W	$CaSO_4$ W	$Cr_2(SO_4)_3$ W(I)*	$CoSO_4$ W	$CuSO_4$ W	...	...	H_2SO_4 W	$FeSO_4$ W	$Fe_2(SO_4)_3$ W
24	Sulfide	Al_2S_3 d	$(NH_4)_2S$ W	Sb_2S_3 A	BaS d	Bi_2S_3 A	CdS A	CaS W	Cr_2S_3 d	CoS A	CuS A	Au_2S I	Au_2S_3 I	H_2S W	FeS A	Fe_2S_3 d
25	Tartrate —(C₄H₄O₆)	$Al_2(—)_3$ W	$(NH_4)_2(—)$ W	$Sb_2(—)_3$ A	$Ba(—)_2$ W	$Bi_2(—)_3$ W	$Cd(—)$ W	$Ca(—)$ W	...	$Co(—)$ W	$Cu(—)$ d	...	...	$C_4H_6O_6$ W	$Fe(—)$ W	$Fe_2(—)_3$ W
26	Thiocy'te	...	NH_4CNS W	...	...	...	...	$Ca(CNS)_2$ W	...	$Co(CNS)_2$ W	$CuCNS$	...	...	$CNSH$ W	$Fe(CNS)_2$ W	$Fe(CNS)_3$ W

No.		Pb	Mg	Mn	Hg′	Hg″	Ni	K	Ag	Na	Sn″″	Sn″	Sr	Zn	Pt
1	Acetate —(C₂H₃O₂)	$Pb(—)_2$ W	$Mg(—)_2$ W	$Mn(—)_2$ W	$Hg(—)$ W	$Hg(—)_2$ W	$Ni(—)_2$ W	$K(—)$ W	$Ag(—)$ a	$Na(—)$ W	$Sn(—)_4$ W	$Sn(—)_2$ d	$Sr(—)_2$ W	$Zn(—)_2$ W	...
2	Arsenate —(AsO₄)	$PbH(—)$ A	$Mg_3(—)$ A	$MnH(—)$ W	$Hg_3(—)$ A	$Hg_3(—)_2$ A	$Ni_3(—)_2$ A	$K_3(—)$ W	$Ag_3(—)$ A	$Na_3(—)$ W	$Sn(—)_4$ W	$Sn_3(—)_2$ A	$SrH(—)$ W	$Zn_3(—)_2$ A	...
3	Arsenite —(AsO₃)	...	$Mg_3(—)_2$ W	$Mn(—)_2$ W	$Hg_3(—)$ A	$Hg(—)_2$ W	$Ni(—)_2$ W	K_3AsO_3 W	$Ag_3(—)$ A	$Na_3H(—)$ W	...	$Sn(—)_2$ W	$Sr_3(—)_2$ W	$Zn(—)_2$ W	...
4	Benzoate —(C₇H₅O₂)	$Pb(—)_2$ W	$Mg(—)_2$ W	$Mn_3H_6(—)_4$	$Hg_3(—)$	$Hg(—)_2$ W	$Ni(—)_2$ W	$K(—)$ W	$Ag(—)$ A	$Na(—)$ W	...	...	$Sr(—)_2$ W	$Zn(—)_2$ W	...
5	Bromide	$PbBr_2$ W	$MgBr_2$ W	$MnBr_2$ W	$HgBr$ A	$HgBr_2$ W	$NiBr_2$ W	KBr W	$AgBr$ a	$NaBr$ W	$SnBr_4$ W	$SnBr_2$ W	$SrBr_2$ W	$ZnBr_2$ W	$PtBr_4$ W
6	Carbonate	$PbCO_3$ A	$MgCO_3$ A	$MnCO_3$ A	Hg_2CO_3 A	...	$NiCO_3$ A	K_2CO_3 W	Ag_2CO_3 A	Na_2CO_3 W	...	...	$SrCO_3$ A	$ZnCO_3$ W	...
7	Chlorate —(ClO₃)	$Pb(—)_2$ W	$Mg(—)_2$ W	$Mn(—)_2$ W	$Hg(—)$ A	$Hg(—)_2$ W	$Ni(—)_2$ W	$K(—)$ W	$Ag(—)$ W	$Na(—)$ W	$Sn(—)_4$ W	$Sn(—)_2$ W	$Sr(—)_2$ W	$Zn(—)_2$ W	...
8	Chloride	$PbCl_2$ W	$MgCl_2$ W	$MnCl_2$	$HgCl$ a	$HgCl_2$ W	$NiCl_2$ W	KCl W	$AgCl$ A	$NaCl$ W	$SnCl_4$ W	$SnCl_2$ W	$SrCl_2$ W	$ZnCl_2$ W	$PtCl_4$ W
9	Chromate —(CrO₄)	$Pb(—)$ A	$Mg(—)$ W	...	$Hg(—)$ W	$Hg(—)$ W	$Ni(—)$ W	$K_2(—)$ W	$Ag_2(—)$ A	$Na_2(—)$ W	$Sn(—)_2$ W	$Sn(—)$ A	$Sr(—)$ W	$Zn(—)$ W	...
10	Citrate —(C₆H₅O₇)	$Pb_3(—)_2$ W	$Mg_3(—)_2$ W	$MnH(—)$ W	$Hg_3(—)$ W	$Hg(—)$ W	$Ni_3(—)_2$ W	$K_3(—)$ W	$Ag_3(—)$ A	$Na_3(—)$ W	$Sn_3(—)_2$ W	$Sn(—)$ A	$SrH(—)$ A	$Zn_3(—)_2$ W	...

No.		Pb	Mg	Mn	Hg′	Hg″	Ni	K	Ag	Na	Sn″″	Sr	Zn	Pt
11	Cyanide	$Pb(CN)_2$ W	$Mg(CN)_2$ W	...	$HgCN$ W	$Hg(CN)_2$ W	$Ni(CN)_2$ a	KCN W	$AgCN$ I	$NaCN$ W	...	$Sr(CN)_2$ W	$Zn(CN)_2$ A	$Pt(CN)_2$ I
12	Ferricy'de —Fe(CN)₆	$Pb(CN)$ W	$Mg_3(—)_2$ W	...	...	$Hg_3(—)_2$ A	$Ni_3(—)_2$ I	$K_3(—)$ W	$Ag_3(—)$ I	$Na_3(—)$ W	...	$Sr_3(—)_2$ W	$Zn_3(—)_2$ A	...
13	Ferrocy'de —Fe(CN)₆	$Pb_2(—)$ a	$Mg_2(—)$ W	$Mn_2(—)$ A	$Hg_2(—)$ d	$Hg_2(—)$ I	$Ni_2(—)$ I	$K_4(—)$ W	$Ag_4(—)$ I	$Na_4(—)$ W	Sn_4F_4	$Sr_2(—)$ W	$Zn_2(—)$ I	...
14	Fluoride	PbF_2	MgF_2	MnF_2 A	HgF	HgF_2 d	NiF_2	KF	AgF	NaF	SnF_4 W	SrF_2 W	ZnF_2 W	PtF_4 W

Anion	Pb	Mg	Mn	Hg₂	Hg	Ni	K	Ag	Na	Sn(IV)	Sn(II)	Sr	Zn	Pt
15 Formate –(CHO₂)	W $Pb(-)_2$	W $Mg(-)_2$	W $Mn(-)_2$	W $Hg(-)$	W $Hg(-)_2$	W $Ni(-)_2$	W $K(-)$	W $Ag(-)$	W $Na(-)$	......	......	W $Sr(-)_2$	W $Zn(-)_2$	......
16 Hydroxide	W $Pb(OH)_2$	A $Mg(OH)_2$	A $Mn(OH)_2$	......	A $Hg(OH)_2$	A $Ni(OH)_2$	W KOH	......	W $NaOH$	W $Sn(OH)_4$	A $Sn(OH)_2$	W $Sr(OH)_2$	A $Zn(OH)_2$	A $Pt(OH)_4$
17 Iodide	w PbI_2	W MgI_2	W MnI_2	A HgI	w HgI_2	W NiI_2	W KI	I AgI	W NaI	d SnI_4	d SnI_2	W SrI_2	W ZnI_2	I PtI_4
18 Nitrate	W $Pb(NO_3)_2$	W $Mg(NO_3)_2$	W $Mn(NO_3)_2$	W $HgNO_3$	W $Hg(NO_3)_2$	W $Ni(NO_3)_2$	W KNO_3	W $AgNO_3$	W $NaNO_3$	......	W $Sn(NO_3)_2$	W $Sr(NO_3)_2$	W $Zn(NO_3)_2$	W $Pt(NO_3)_4$
19 Oxalate –(C₂O₄)	A $Pb(-)$	A $Mg(-)$	A $Mn(-)$	a $Hg(-)$	a $Hg(-)$	A $Ni(-)$	W $K_2(-)$	a $Ag_2(-)$	W $Na_2(-)$	......	A $Sn(-)$	A $Sr(-)$	A $Zn(-)$	......
20 Oxide	A PbO	A MgO	A MnO	A Hg_2O	w HgO	A NiO	W K_2O	A Ag_2O	d Na_2O	A SnO_2	A SnO	W SrO	A ZnO	A PtO
21 Phosphate	A $Pb_3(PO_4)_2$	W $Mg_3(PO_4)_2$	W $Mn_3(PO_4)_2$	A Hg_3PO_4	A $Hg_3(PO_4)_2$	A $Ni_3(PO_4)_2$	W K_3PO_4	A Ag_3PO_4	W Na_3PO_4	......	A $Sn_3(PO_4)_2$	A $Sr_3(PO_4)_2$	A $Zn_3(PO_4)_2$	......
22 Silicate –(SiO₃)	A $Pb(-)$	A $Mg(-)$	A $Mn(-)$	......	......	......	W $K_2(-)$	......	W $Na_2(-)$	......	A $Sn(-)$	A $Sr(-)$	A $Zn(-)$	......
23 Sulfate	A $PbSO_4$	W $MgSO_4$	W $MnSO_4$	W $HgSO_4$	d $HgSO_4$	W $NiSO_4$	W K_2SO_4	A Ag_2SO_4	W Na_2SO_4	W $Sn(SO_4)_2$	W $SnSO_4$	W $SrSO_4$	W $ZnSO_4$	W $Pt(SO_4)_2$
24 Sulfide	A PbS	d MgS	A MnS	I Hg_2S	I HgS	A NiS	W K_2S	A Ag_2S	W Na_2S	A SnS_2	A SnS	W SrS	A ZnS	I PtS
25 Tartrate –(C₄H₄O₆)	A $Pb(-)$	W $Mg(-)$	W $Mn(-)$	I $Hg(-)$	......	A $Ni(-)$	W $K_2(-)$	A $Ag(-)$	W $Na_2(-)$	......	A $Sn(-)$	A $Sr(-)$	A $Zn(-)$	......
26 Thiocy'te	W $Pb(CNS)_2$	W $Mg(CNS)_2$	A $Mn(CNS)_2$	$Hg_2(CNS)_2$	W $Hg(CNS)_2$	......	W $KCNS$	I $AgCNS$	W $NaCNS$	......	......	$Sr(CNS)_2$	W $Zn(CNS)_2$	......

Note: Abbreviations: W, soluble in water; A, insoluble in water but soluble in acids; w, sparingly soluble in water but soluble in acids; a, insoluble in water and only sparingly soluble in acids; I, insoluble in both water and acids; d, decomposes in water. *Certain salts occur in two modifications.

From Weast, R. C., Ed., *CRC Handbook of Chemistry and Physics*, 61st ed., CRC Press, Boca Raton, Fla., 1980, D-145, D-146. With permission.

Table 70

SOLUBILITIES OF THE AMINO ACIDS IN GRAMS PER 100 GRAMS OF WATER

Amino acid	Temperature, °C.					Ref. No.
	0°	25°	50°	75°	100°	
DL-Alanine	12.11	16.72	23.09	31.89	44.04	1
L-Alanine	12.73	16.65	21.79	28.51	37.30	1
DL-Aspartic acid	0.262	0.778	2.000	4.456	8.594	1
L-Aspartic acid	0.209	0.500	1.199	2.875	6.893	1
L-Cystine‡ × 10²	0.502	1.096	2.394	5.229	11.42	2
Diiodo-DL-tyrosine × 10	0.149	0.340	0.773	—	—	3
Diiodo-L-tyrosine × 10	0.204	0.617	1.862	5.62	17.00	1
DL-Glutamic acid	0.855	2.054	4.934	11.86	28.49	1
L-Glutamic acid	0.341	0.864	2.186	5.532	14.00	1
Glycine	14.18	24.99	39.10	54.39	67.17	1
L-Histidine	—	4.19	—	—	—	4
Hydroxy-L-Proline	28.86	36.11	45.18	51.67*	—	5
DL-Isoleucine	1.826	2.229	3.034	4.607	7.802	1
L-Isoleucine	3.791	4.117	4.818	6.076	8.255	2
DL-Leucine	0.797	0.991	1.406	2.276	4.206	1
L-Leucine	2.270	2.426†	2.887†	3.823	5.338	1
DL-Methionine	1.818	3.381	6.070	10.52	17.60	2
DL-Phenylalanine	0.997	1.411	2.187	3.708	6.886	1
L-Phenylalanine	1.983	2.965	4.431	6.624	9.900	2
L-Proline × 10⁻¹	12.74	16.23	20.67	23.90*	—	3
DL-Serine	2.204	5.023	10.34	19.21	32.24	2
L-Tryptophan	0.823	1.136	1.706	2.795	4.987	2
DL-Tyrosine × 10	0.147	0.351	0.836	—	—	3
L-Tyrosine × 10	0.196	0.453	1.052	2.438	5.650	1
D-Tyrosine × 10	0.196	0.453	1.052	—	—	3
DL-Valine	5.98	7.09	9.11	12.61	18.81	1
L-Valine	8.34	8.85	9.62	10.24*	—	6

*   Value at 65°.

†   Dunn and Stoddard (7) report 2.19 g. at 25° for L-Leucine rendered methionine-free by repeated recrystallization from 6 N HCl. Hlynka (8) found 2.20 g. at 25° and 2.66 g. at 50° for L-Leucine rendered methionine-free [by S. W. Fox (9)] by fractional crystallization of the formyl derivative and identical values for D-Leucine obtained by resolution of the DL form.

‡   The following values were found by Loring and du Vigneaud (10): DL-Cystine (0.0049 g), D-Cystine (0.0108 g), and meso-cystine (0.0056 g) at 25°.

REFERENCES

1. Dalton, J. B., and Schmidt, C. L. A., J. Biol. Chem., 103, 549 (1933).
2. Dalton, J. B., and Schmidt, C. L. A., J. Biol. Chem., 109, 241 (1935).
3. Winnek, P. S., and Schmidt, C. L. A., J. Gen. Physiol., 18, 889 (1934-35).
4. Dunn, M. S., Frieden, E. H., and Brown, H. V., unpublished data.
5. Tomiyama, T., and Schmidt, C. L. A., J. Gen. Physiol., 19, 379 (1935-36).
6. Dalton, J. B., and Schmidt, C. L. A., J. Gen. Physiol., 19, 767 (1935-36).
7. Dunn, M. S., and Stoddard, M. P., unpublished data.
8. Hlynka, I., Thesis (1939), California Institute of Technology, Pasadena, California.
9. Fox. S. W., Science, 84, 163 (1936).
10. Loring, H. S., and du Vigneaud, V., J. Biol Chem., 107, 270 (1934).

From Weast, R. C., Ed., CRC Handbook of Chemistry and Physics, 61st ed., CRC Press, Boca Raton, Fla., 1980, C-714. With permission.

Table 71

SOLUBILITIES OF THE AMINO ACIDS IN GRAMS PER 100 GRAMS OF WATER-ETHANOL MIXTURES

DL-Alanine

Per cent ethanol by volume	Temp. °C	Grams amino acid per 100 grams solvent	Ref. No.
24.93	0.00	3.84	1
50.10	0.00	1.16	1
74.50	0.00	0.305	1
95.14	0.00	0.0167	1
10	25	12.25	2
24.93	24.97	7.09	1
50.10	24.97	2.52	1
74.20	24.97	0.573	1
95.14	25.09	0.0329	1
25.28	45.16	10.6	1
50.10	44.96	4.25	1
74.20	44.98	0.949	1
95.14	45.19	0.0545	1
24.93	64.96	15.9	1
50.10	64.94	6.68	1
74.20	64.94	1.48	1
95.09	65.15	0.0851	1

DL-Aspartic acid

Per cent ethanol by volume	Temp. °C	Grams amino acid per 100 grams solvent	Ref. No.
24.93	0.03	0.0703	1
50.10	0.03	0.0267	1
74.20	0.02	0.0111	1
24.55	25.06	0.266	1
50.25	25.06	0.0992	1
74.28	25.14	0.0317	1
95.14	25.07	0.0020	1
24.74	45.25	0.680	1
50.18	45.25	0.255	1
74.28	45.27	0.0608	1
95.14	45.21	0.0042	1
24.93	64.91	1.53	1
50.10	64.91	0.588	1
74.20	65.07	0.132	1
95.14	65.00	0.0129	1

L-Aspartic acid

Per cent ethanol by volume	Temp. °C	Grams amino acid per 100 grams solvent	Ref. No.
20	25	0.204	3
50	25	0.0633	3
70	25	0.0224	3
90	25	0.0034	3

L-Glutamic acid

Per cent ethanol by volume	Temp. °C	Grams amino acid per 100 grams solvent	Ref. No.
24.74	0.01	0.0855	1
50.18	0.01	0.0371	1
74.28	0.03	0.0163	1
24.56	25.05	0.292	1
50.25	25.07	0.131	1
74.35	25.07	0.0370	1
95.14	25.04	0.0044	1
24.55	45.01	0.811	1
50.18	45.27	0.378	1
74.35	44.93	0.0885	1
95.14	45.20	0.0127	1

Glycine

Per cent ethanol by volume	Temp. °C	Grams amino acid per 100 grams solvent	Ref. No.
24.93	0.02	3.95	1
50.10	0.02	1.03	1
74.50	0.02	0.200	1
95.09	0.01	0.0080	2
10	25	17.13	1
24.93	24.97	8.72	1
50.10	24.97	2.47	1
74.20	24.97	0.448	1
95.14	25.09	0.0172	1
24.93	44.98	15.0	1
50.10	44.98	4.62	1
74.20	44.97	0.756	1
95.14	45.19	0.0294	1
24.93	65.11	24.5	1
50.10	65.10	8.03	1
74.20	65.07	1.23	1
95.14	85.00	0.0488	1

Table 71 (continued)
SOLUBILITIES OF THE AMINO ACIDS IN GRAMS PER 100 GRAMS OF WATER-ETHANOL MIXTURES

L-Isoleucine

Per cent ethanol by volume	Temp. °C	Grams amino acid per 100 grams solvent	Ref. No.
80	20	0.46	4
80	78—80	1.16	4

L-allo-Isoleucine

Per cent ethanol by volume	Temp. °C	Grams amino acid per 100 grams solvent	Ref. No.
80	20	0.81	4
80	78—80	1.97	4

DL-Leucine

Per cent ethanol by volume	Temp. °C	Grams amino acid per 100 grams solvent	Ref. No.
24.93	0.00	0.251	1
50.10	0.00	0.118	1
74.50	0.00	0.0693	1
95.14	0.00	0.0116	1
10	25	0.771	2
24.93	24.97	0.493	1
50.10	24.97	0.318	1
74.20	24.97	0.175	1
95.14	25.09	0.0258	1
24.93	45.15	0.853	1
50.10	45.24	0.633	1
74.20	45.04	0.323	1
95.14	45.18	0.0471	1
24.93	45.18	1.45	1
50.10	65.16	1.16	1
74.20	65.20	0.584	1
95.09	65.07	0.0844	1

L-Leucine

Per cent ethanol by volume	Temp. °C	Grams amino acid per 100 grams solvent	Ref. No.
20	25	1.33	2
60	25	0.641	2
90	25	0.123	2

L-Proline

Per cent ethanol by volume	Temp. °C	Grams amino acid per 100 grams solvent	Ref. No.
100	19	1.5	5

DL-Serine

Per cent ethanol by volume	Temp. °C	Grams amino acid per 100 grams solvent	Ref. No.
24.93	0.00	0.1530	1
50.10	0.00	0.146	1
74.50	0.00	0.0304	1
95.14	0.00	0.0008	1
24.93	25.14	1.54	1
50.10	25.14	0.461	1
74.50	25.10	0.0840	1
95.14	25.09	0.0028	1
24.93	45.15	3.14	1
50.10	45.04	0.985	1
74.20	45.04	0.185	1
95.14	45.18	0.0058	1
24.93	65.26	5.99	1
50.10	65.25	1.88	1
74.50	65.24	0.318	1
95.14	65.01	0.0152	1

DL-Threonine

Per cent ethanol by volume	Temp. °C	Grams amino acid per 100 grams solvent	Ref. No.
95	25	0.07*	6

DL-allo-Threonine

Per cent ethanol by volume	Temp. °C	Grams amino acid per 100 grams solvent	Ref. No.
95	25	0.03*	6

L-Tyrosine

Per cent ethanol by volume	Temp. °C	Grams amino acid per 100 grams solvent	Ref. No.
95	17	0.10	7

DL-Tyrosine

Per cent ethanol by volume	Temp. °C	Grams amino acid per 100 grams solvent	Ref. No.
95.09	0.00	0.0031	8
25.28	24.85	0.0285	8
50.99	24.75	0.0226	8
74.63	24.75	0.0117	8
95.09	25.24	0.0032	8
25.28	45.15	0.0630	8
50.99	45.16	0.0513	8
74.63	44.98	0.0230	8
95.09	44.98	0.0035	8
95.09	65.06	0.0067	8

DL-Valine

Per cent ethanol by volume	Temp. °C	Grams amino acid per 100 grams solvent	Ref. No.
24.93	0.02	2.10	1
50.10	0.02	0.769	1
74.20	0.02	0.269	1
95.14	0.01	0.0277	1
10	25	5.50	2
25.28	24.85	3.30	1
50.99	24.93	1.53	1
74.35	25.04	0.570	1
95.14	44.91	0.0569	1
24.55	44.92	5.10	1
50.25	44.92	2.74	1
74.35	45.21	0.999	1
95.14	44.92	0.0979	1
24.55	65.07	7.44	1
50.10	64.94	4.49	1
74.20	64.34	1.62	1
95.09	65.15	0.167	1

L-Valine

Per cent ethanol by volume	Temp. °C	Grams amino acid per 100 grams solvent	Ref. No.
20	25	5.11	2
40	25	2.93	2
60	25	1.61	2
80	25	0.52	2

* Grams per 100 ml. of solution.

References

1. Dunn, M. S., and Ross, F. J., J. Biol. Chem., 125, 309 (1938).
2. Cohn, E. J., McMeekin, T. L., Edsall, J. T., and Weare, J. H., J. Am. Chem. Soc., 56, 2270 (1934).
3. McMeekin, T. L., Cohn, E. J., and Weare, J. H., J. Am. Chem. Soc., 57, 626 (1935).
4. Abderhalden, E., and Zeisset, W., Z. physiol. Chem., 196, 121 (1931).
5. Kapfhammer, J., and Eck, R., Z. physiol. Chem., 170, 294 (1927).
6. West, H. D., and Carter, H. E., J. Biol. Chem., 119, 109 (1937).
7. Stutzer, A., Z. anal. Chem., 31, 501 (1892).
8. Dunn, M. S. and Ross, F. J., unpublished data.

From Weast, R. C., Ed., CRC Handbook of Chemistry and Physics, 61st ed., CRC Press, Boca Raton, Fla., 1980, C-714. With permission.

Table 72

SOLUBILITIES OF THE AMINO ACIDS IN GRAMS PER 100 GRAMS OF ORGANIC SOLVENT

Solvent	Grams amino acid per 100 grams solvent	Temp. °C	Ref. No.
DL-Alanine			
Ethanol	0.0087	25	1
L-Aspartic acid			
Ethanol	0.000196	25	2
L-Glutamic acid			
Ethanol	0.000347	25	2
Ethanol	0.0056	44.93	3
Glycine			
Acetone	0.000291	25	4
Butanol	0.000892	25	4
Ethanol	0.0037	25	1
Formamide	0.558	25	4
Methanol	0.0407	25	4
L-Isoleucine			
Ethanol	0.09	20	5
Ethanol	0.13	78—80	5
L-allo-Isoleucine			
Ethanol	0.13	20	5
Ethanol	0.19	78—80	5
L-Leucine			
Ethanol	0.0217	25	1
L-Proline			
Ethanol	1.5	19	6
DL-Valine			
Ethanol	0.0136	0.03	3
Ethanol	0.019	25	1

References

1. Cohn, E. J., McMeekin, T. L., Edsall, J. T., and Weare, J. H., J. Am. Chem. Soc., **56**, 2270 (1934).
2. McMeekin, T. L., Cohn, E. J., and Weare, J. H., J. Am. Chem. Soc., **57**, 626 (1935).
3. Dunn, M. S., and Ross, F. J., J. Biol. Chem., **125**, 309 (1938).
4. McMeekin, T. L., Cohn, E. J., and Weare, J. H., J. Am. Chem. Soc., **58**, 2173 (1936).
5. Abderhalden, E., and Zeisset, W., Z. physiol. Chem., **196**, 121 (1931).
6. Kapfhammer, J., and Eck, R., Z. physiol. Chem., **170**, 294 (1927).

From Weast, R. C., Ed., *CRC Handbook of Chemistry and Physics*, 61st ed., CRC Press, Boca Raton, Fla., 1970, C-715. With permission.

Table 73

IONIZATION CONSTANTS AND pH VALUES AT THE ISOELECTRIC POINTS OF THE AMINO ACIDS IN WATER AT 25°C

The majority of the recorded values are true thermodynamic constants calculated from electrometric force measurements of cells without liquid junctions. The values for the constants given in the table were derived from the classical, the zwitterionic (Bjerrum), and the acidic (Brönsted) formulations of ionization and the corresponding mass law expressions. pH values at the isoelectric points were calculated from the expression, $pI = \frac{1}{2}(pk_{a1} + pk_w - pk_{b1})$. The error is approximately 0.5 per cent when this expression is used to calculate pI values for cystine, tyrosine, and diiodotyrosine.

Amino acid	Classical				Zwitterionic				Acidic				pI	Ref. no.
	pk_{a1}	pk_{a2}	pk_{b1}	pk_{b2}	pK_{A1}	pK_{A2}	pK_{B1}	pP_{B2}	pK_1	pK_2	pK_3	pK_4		
DL-Alanine	9.866		11.649	11.99	2.348		4.131		2.348	9.866			6.107	1
L-Arginine	12.48		4.96		2.01		1.52	4.96	2.01	9.04	12.48		10.76	2
L-Aspartic acid	3.86	9.82	11.93		2.10	3.86	4.18		2.10	3.86	9.82		2.98	3
L-Cystine	8.00	10.25	11.95	12.96	1.04	2.05	6.18	6.00	1.04	2.05	8.00	10.25	5.02	4
Diiodo-L-tyrosine	6.48	7.82	11.88		2.12	6.48	4.53		2.12	6.48	7.82		4.29	5,6
L-Glutamic acid	4.07	9.47	11.90		2.10	4.07	4.219		2.10	4.07	9.47		3.08	7
Glycine	9.778		11.647		2.350		4.82		2.350	9.778			6.064	8
L-Histidine	9.18		7.90		1.77		4.27	7.90	1.77	6.10	9.18		7.64	8
Hydroxy-L-proline	9.73		12.08	12.23	1.92				1.92	9.73			5.82	1
DL-Isoleucine	9.758		11.679		2.318		4.239		2.318	9.758			6.038	2
DL-Leucine	9.744		11.669		2.328		4.253		2.328	9.744			6.036	1
L-Lysine	10.53		5.05	11.82	2.18		3.47	5.05	2.18	8.95	10.53		9.47	2
DL-Methionine	9.21		11.72		2.28		4.79		2.28	9.21			5.74	10
DL-Phenylalanine	9.24		11.42		2.58		4.76		2.58	9.24			5.91	11
L-Proline	10.60		12.0		2.00		3.40		2.00	10.60			6.3	12
DL-Serine	9.15		11.79		2.21		4.85		2.21	9.15			5.68	9
L-Tryptophan	9.39		11.62		2.38		4.61		2.38	9.39			5.88	13
L-Tyrosine	9.11	10.07	11.80		2.20	9.11	3.93		2.20	9.11	10.07		5.63	6
DL-Valine	9.719		11.711		2.286		4.278		2.286	9.719			6.002	1

References

1. Smith, P. K., Taylor, A. C., and Smith, E. R. B., J. Biol. Chem., 122, 109 (1937–38).
2. Schmidt, C. L. A., Kirk, P. L., and Appleman, W. K., J. Biol. Chem., 88, 285 (1930).
3. Greenstein, J. P., J. Biol. Chem., 93, 479 (1931).
4. Borsook, H., Ellis, E. L., and Huffman, H. M., J. Biol. Chem., 117, 281 (1937).
5. Dalton, J. B., Kirk, P. L., and Schmidt, C. L. A., J. Biol. Chem., 88, 589 (1930).
6. Winnek, P. S., and Schmidt, C. L. A., J. Gen. Physiol., 18, 889 (1935).
7. Simms, H. S., J. Gen. Physiol., 11, 629 (1928); 12, 231 (1928).
8. Owen, B. B., J. Am. Chem. Soc., 56, 24 (1934).
9. Kirk, P. L., and Schmidt, C. L. A., J. Biol. Chem., 81, 237 (1929).
10. Emerson, O. H., Kirk, P. L., and Schmidt, C. L. A., J. Biol. Chem., 92, 449 (1931).
11. Miyamoto, S., and Schmidt, C. L. A., J. Biol. Chem., 90, 165 (1931).
12. McCay, C. M., and Schmidt, C. L. A., J. Gen. Physiol., 9, 333 (1926).
13. Schmidt, C. L. A., Appleman, W. K., and Kirk, P. L., J. Biol. Chem., 85, 137 (1929–30).

From Weast, R. C., Ed., CRC Handbook of Chemistry and Physics, 61st ed., CRC Press, Boca Raton, Fla., 1980, C-712. With permission.

Table 74
IONIZATION CONSTANTS OF THE AMINO ACIDS IN AQUEOUS FORMALDEHYDE SOLUTION

Amino acid	Mole per cent formaldehyde				
	0.99	3.95	5.60	10.0	17.9
DL-Alanine	8.36	7.42	6.96[b]	6.56	6.10
L-Arginine	....	3.45[c]	3.40[d]	....	....
L-Aspartic acid	....	......	7.21[d]	≲3.8[e] 6.85[f]	....
L-Glutamic acid	....	......	6.91[d]	≲4.2[e] 6.8[f]	....
Glycine	7.16	6.08	5.92[b]	5.34	5.04
L-Histidine	....	7.90[e]	7.90[d]	....	....
Hydroxy-L-Proline	....	......	7.19[d]	....	....
L-Leucine	8.44	7.50	6.92[d]	6.62	6.20
DL-Leucine	8.44	7.48	......	6.60	6.20
L-Lysine	....	7.35[c]	7.15[d]	....	....
L-Phenylalanine	....	......	6.62[d]	5.9[e]	....
DL-Phenylalanine	8.09	7.16	6.80[b]	6.35	6.15
L-Proline	....	......	7.78[d]	....	....
DL-Serine	6.66	5.74	5.63[b]	....	4.94
L-Tryptophan	....	......	6.88[d]	....	....
L-Tyrosine	....	......	7.50[d]	6.2[e] >9[f]	....
DL-Valine	8.52	7.65	7.47[b]	....	6.52

[a] Dunn and Weiner (1), pK_2 at 22°.
[b] Dunn and Loshakoff (2), pK_2 at 22°.
[c] Levy (3) pK_2 at 30° for arginine and pK_3 at 30° for histidine and lysine.
[d] Levy and Silberman (4), pK_2 at 30°, pK_3 at 30° for histidine and lysine.
[e] Harris (5), pK_2 at 25° for aspartic acid, glutamic acid, phenylalanine and tyrosine.
[f] Harris (5), pK_3 at 30° for aspartic acid, glutamic acid, and tyrosine.

References

1. Dunn, M. S., and Weiner, J. G., J. Biol. Chem., **117**, 381 (1937).
2. Dunn, M. S., and Loshakoff, A., J. Biol. Chem., **113**, 691 (1936).
3. Levy, M., J. Biol. Chem., **109**, 365 (1935).
4. Levy M., and Silberman, D. E., J. Biol. Chem., **118**, 723 (1937).
5. Harris, L. J., Proc. Roy. Soc. London, Series B, **95**, 440 (1923—24).

From Weast, R. C., Ed., *CRC Handbook of Chemistry and Physics,* 61st ed., CRC Press, Boca Raton, Fla., 1980, C-712. With permission.

Table 75

PHYSICAL PROPERTIES OF SELECTED ORGANIC SOLVENTS

Substance	Temp., °C	Density, g·mL⁻³	Refractive Index (n_D)	Viscosity, mN·s·m⁻²	Dielectric Constant (ε)	Dipole Moment, D	Surface Tension, dyn·cm⁻¹ a	Surface Tension, dyn·cm⁻¹ b
Acetaldehyde	20	0.7780	1.3311	0.244	21.1	2.69	23.90	0.1360
Acetamide	110	0.9681	1.4158	1.46[12]	59[83]	3.44	47.66	0.1021
Acetic acid	20	1.0492	1.3719	1.314[15]	6.15	1.74	29.58	0.0994
Acetic anhydride	20	1.0811	1.3904	0.971[15]	20.7	2.8	35.52	0.1436
Acetone	20	0.7908	1.3588	0.337[15]	20.70[25]	2.88	26.26	0.112
Acetonitrile	20	0.7822	1.3441	0.375	37.5	3.92	29.58	0.1178
Acetophenone	25	1.0238	1.5322	1.642	17.39	3.02	41.92	0.1154
Acetyl bromide	16	1.663	1.4537		16.2[20]	2.45		
Acetyl chloride	20	1.105	1.3898		15.8	2.72		
Acrolein	20	0.8389	1.4017			2.90[25]		
Acrylic acid	20	1.0511	1.4224				(28.1[30])	0.1178
Acrylonitrile	20	0.8060	1.3911	0.35	33.0	3.87	29.58	0.0973
Adiponitrile	20	0.950	1.4597				47.88	0.1186
Allyl acetate	20	0.9256	1.4040	0.207[30]			28.73	0.1287
Allylamine	20	0.7629	1.4205	0.375[25]		1.31	27.49	0.1117
2-Aminoethanol	25	1.0116	1.4521	19.35	37.72	2.27	51.11	0.1092
1-Amino-2-methylpropane	25	0.7297	1.3945	21.7	4.43[21]	1.27	24.48	0.1085
Aniline	20	1.0217	1.5855	4.400	6.89	1.53	44.83	0.1090
Benzaldehyde	20	1.0447	1.5455	1.321[25]	17.8	2.77	40.72	(27.56[30])
Benzene	25	0.8737	1.4979	0.6028	2.275	0	(28.88[20])	0.1202
Benzenethiol	20	1.0766	1.5897	1.239	4.38[25]	1.23	41.41	0.1159
Benzonitrile	20	1.0051	1.5282	1.447[15]	25.20[25]	4.18	41.69	0.1084
Benzoyl chloride	20	1.211	1.5525		23	3.2	41.34	0.1381
Benzyl acetate	20	1.055	1.5200	1.399[45]	5.1	1.80	38.25	0.1065
Benzyl alcohol	20	1.045	1.5403	7.760[15]	13.1	1.66	48.07	0.1227
Benzyl benzoate	25	1.1121	1.5685[21]	8.292	4.9[20]	1.90	39.92	(29.97[40])
Benzyl chloride	20	1.0993	1.5390	1.400	7.0[13]	1.85		
Benzyl ethyl ether	20	0.9478	1.4958		3.9		(32.82[20])	0.0951
Bicyclohexane	20	0.8862	1.4800	3.75		<0.4	34.64	
3-Bromoaniline	20	1.579	1.6260		13.0	2.66		0.1160
Bromobenzene	25	1.4882	1.5571	0.985[30]	5.40	1.55	38.14	

1-Bromobutane	20	1.2758	1.4394	0.633		2.08	28.71	0.1126
2-Bromobutane	20	1.255	1.4369		7.7[0.1]	2.04	27.48	0.1107
Bromoethane	15	1.4708	1.4276	0.418	9.39[20]	2.03	26.52	0.1159
Bromoethene	20	1.517	1.4380		4.78[25]	1.42		
1-Bromohexane	20	1.176	1.4475		5.82[25]	2.00	29.81	0.09669
1-Bromonaphthalene	20	1.4834	1.6582	5.99[15]	5.12	1.29[liq]	46.44	0.1018
1-Bromopropane	15	1.3597	1.4370	0.539	8.09[25]	2.18	28.30	0.1218
2-Bromopropane	15	1.3222	1.4285	0.536	9.46[25]	2.21	26.21	0.1183
o-Bromotoluene	20	1.422	1.5608		4.28[8]	1.45	36.62	0.09979
1-Butanal	20	0.8016	1.3791	0.455	13.4[26]	2.72	26.67	0.0925
2-Butanal	20	0.7891	1.3727			2.58		
1-Butanamine	20	0.7392	1.4014	0.681	4.88	1.37	26.24	0.1122
2-Butanamine	20	0.7246	1.3932			1.28	23.75	0.1057
1,3-Butanediol	20	1.0053	1.441	130.3		2.5	(37.8[25])	
Butanenitrile	15	0.7954	1.3860	0.624	20.3[21]	4.07	29.51	0.1037
1-Butanethiol	20	0.8416	1.4430	0.501	5.07[25]	1.53	28.07	0.1142
Butanoic acid	20	0.9582	1.3980	1.814[15]	2.97	1.65	28.35	0.0920
1-Butanol	20	0.8097	1.3993	3.379[15]	17.5[25]	1.66	27.18	0.08983
2-Butanol	20	0.8069	1.3972	4.210	16.56	1.66		
2-Butanone	20	0.8047	1.3788	0.423[15]	18.51	2.75	26.77	0.1122
cis-2-Butene-1,4-diol	20	1.0740	1.4793			2.48		
trans-2-Butene-1,4-diol	20	1.0685	1.4779			2.45		
2-Butoxyethanol	25	0.8964	1.4177	3.15	9.30	2.08	28.18	0.0816
Butyl acetate	20	0.8813	1.3941	0.734	5.01	1.84	27.55	0.1068
Butylbenzene	20	0.8601	1.4898	1.035	2.36	0.36[liq]	31.28	0.1025
sec-Butylbenzene	20	0.8621	1.4902	28.53	2.36	0.37[liq]	30.48	0.0979
tert-Butylbenzene	20	0.8665	1.4927	28.13	2.25	0	30.10	0.0985
Butyl ethyl ether	20	0.7495	1.3818	0.421		1.24	22.75	0.1049
Butyl formate	20	0.8917	1.3890	0.704	2.43[80]		27.08	0.1026
Butyl oleate	20	0.864	1.4522		4.0			
Butyl stearate	25	0.8540	1.4422	8.26	3.11[30]	1.88	(33.0[25])	(32.7[30])
Butyric anhydride	20	0.9668	1.4124	1.615	12.9		(28.93[20])	(28.44[25])
γ-Butyrolactone	25	1.1254	1.4348	1.7	39[20]	4.12		
D-Camphor	20	0.9920		9[78]	11.35	3.10		
ε-Caprolactam		1.027[77]	1.4935[40]			3.88		
Carbon disulfide	20	1.2628	1.6280	0.363	2.641	0.06	35.29	0.1484
Carbon tetrachloride	20	1.5842	1.4603	0.965	2.238	0	29.49	0.1224
Chloroacetic acid	65	1.370	1.4297	3.15[50]	12.3[60]	2.31	43.27	0.1117
o-Chloroaniline	25	1.2077	1.5859	0.925	13.4	1.81	42.46	0.08667
Chlorobenzene	20	1.1063	1.5248	0.799	5.62[25]	1.69	35.97	0.1191
1-Chlorobutane	20	0.8864	1.4021	0.469[15]	7.39	2.05	25.97	0.1117

Table 75 (continued)
PHYSICAL PROPERTIES OF SELECTED ORGANIC SOLVENTS

Substance	Temp., °C	Density, g·mL⁻³	Refractive Index (n_p)	Viscosity, mN·s·m⁻²	Dielectric Constant (ϵ)	Dipole Moment, D	Surface Tension, dyn·cm⁻¹ a	Surface Tension, dyn·cm⁻¹ b
2-Chlorobutane	20	0.8732	1.3971	0.439^{15}	7.09^{30}	2.04	24.40	0.1118
1-Chloro-2,3-epoxypropane	25	1.1746	1.4358	1.03	22.6^{22}	1.8	39.76	0.1360
Chloroethane		0.90928^{15}	1.37381^{10}	0.2791^{0}	9.45^{20}	2.05	(21.18^{5})	(20.58^{10})
2-Chloroethanol	15	1.2072	1.4438	3.913	25.8^{25}	1.88	(38.9^{20})	
bis(2-Chloroethyl)ether	25	1.2130	1.4553	2.14	21.2^{20}	2.58	40.57	0.1306
Chloroform	15	1.4985	1.4486	0.596	4.806^{20}	1.01	29.91	0.1295
1-Chloro-2-methylpropane	15	0.8829	1.4010	0.471	6.49	2.00	24.40	0.1099
2-Chloro-2-methylpropane	20	0.8414	1.3856	0.543^{15}	9.96	2.13	(20.06^{15})	(18.35^{30})
1-Chloronaphthalene	25	1.1930	1.6332^{20}	2.940	5.04	1.59	44.12	0.1035
1-Chloropentane	20	0.8840	1.4118	0.580	6.6^{11}	2.16	27.09	0.1076
o-Chlorophenol		1.2410^{18}	1.54734^{0}	2.250^{45}	6.31^{25}	2.19	42.5	0.1122
p-Chlorophenol		1.2651^{40}	1.55794^{0}	6.018^{45}	9.47^{55}	2.11	46.0	0.1049
1-Chloropropane	20	0.8923	1.3880	0.372^{15}	7.7	2.05	24.41	0.1246
2-Chloropropane	20	0.8617	1.3777	0.335^{15}	9.82	2.17	21.37	0.0883
3-Chloro-1-propene	20	0.9376	1.4151	0.347^{15}	8.2	1.94	25.50	0.0946
o-Chlorotoluene	20	1.0817	1.5238		4.45	1.56		
p-Chlorotoluene	20	1.0697	1.5199^{19}		6.08	2.21	34.93	0.1082
1,8-Cineole	25	0.9192	1.4555		4.57			
Cinnamaldehyde	20	1.0497	1.6195		16.9^{24}			
o-Cresol	46	1.0230	1.5336	3.506	11.5^{25}	1.41	39.43	0.1011
m-Cresol	15	1.0380	1.5415^{20}	24.67	11.8^{25}	1.54	38.00	0.09237
p-Cresol	46	1.0140	1.5287	5.607	9.91^{58}	1.54	38.58	0.0962
Crotonaldehyde	20	0.8516	1.4373			3.50		
Crotonic acid	77	0.9604	1.4249			2.13		
Cyclohexanamine	20	0.8671	1.4593	1.662	4.73	1.32	34.19	0.1188
Cyclohexane	20	0.7786	1.4262	0.980	2.02	0	27.62	0.1188
Cyclohexanol	30	0.9416	1.4629	41.07	15.0^{25}	1.86	35.33	0.0966
Cyclohexanone	20	0.9462	1.4510	2.453^{15}	18.3	3.01	37.67	0.1242
Cyclohexene	20	0.8110	1.4465	0.650	2.20	0.55	29.23	0.1223
Cyclohexylbenzene	20	0.9427	1.5263	3.681^{0}				
Cyclopentane	20	0.7454	1.4065	0.439	1.965	0	25.53	0.1462
p-Cymene	20	0.8573	1.4909	3.402	2.253	0	(29.44^{20})	

Compound	t (°C)	Density	n_D					
cis-Decahydronaphthalene	20	0.8967	1.4810	3.381	2.197	0	(32.18^{20})	(31.01^{30})
trans-Decahydronaphthalene	20	0.8697	1.4693	2.128	2.172	0	(29.89^{20})	(28.87^{30})
Decane	20	0.7301	1.4119	0.928	1.991	0	25.67	0.09197
1-Decanol	20	0.8297	1.4371			1.71	30.34	0.07324
1-Decene	20	0.7408	1.4215	0.805		0.42	25.84	0.09190
Dibenzylamine	20	1.0278	1.57432^{22}			0.97^{liq}	43.27	0.1086
Dibenzyl ether	25	0.9974	1.5385	3.71^{35}	3.6	1.39	(38.23^{35})	
1,2-Dibromoethane	25	2.1687	1.5360		4.78	1.23	35.43	0.1428
cis-1,2-Dibromoethene	20	2.2464	1.5428	1.490^{30}	7.08	1.35		
trans-1,2-Dibromoethene	20	2.2308	1.55051^{18}		2.88	0		
Dibromomethane	20	2.4921	1.5419		7.77^{10}	1.43	42.77	0.1488
1,2-Dibromotetrafluoroethane	25	2.163	1.367	0.72	2.34	0	(18.92^{20})	(18.1^{25})
Dibutylamine	25	0.7619	1.4177	0.95	2.978	1.04	26.50	0.0952
Dibutyl ether	25	0.7646	1.3969	0.602^{30}	3.06	1.22	24.78	0.0934
Dibutyl maleate	20	0.9950	1.4454	5.63			32.46	0.0865
Dibutyl o-phthalate	25	1.0426	1.4901	16.47	6.436^{30}	2.4	(33.40^{20})	(35.55^{30})
Dibutyl sebacate	25	0.9324	1.4397	7.96	4.540^{30}	2.48	(26.84^{20})	0.1147
o-Dichlorobenzene	25	1.3003	1.5491	1.324	9.93	2.50	38.30	0.0879
m-Dichlorobenzene	25	1.2828	1.5434	1.04	5.04	1.72	34.66	0.1186
p-Dichlorobenzene	60	1.2417	1.5285	0.720^{70}	2.41^{50}	0.0	27.03	0.1428
1,1-Dichloroethane	15	1.1835	1.4198	0.505^{25}	10.0^{18}	2.06	35.43	0.1284
1,2-Dichloroethane	15	1.2600	1.4476	0.887	10.36^{25}	1.20		0.1240
cis-1,2-Dichloroethene	25	1.2736	1.4490^{15}	0.444	9.20	1.90	(25^{20})	0.1233
trans-1,2-Dichloroethene	20	1.2546	1.4462	0.404	2.14^{25}	0	30.41	
Dichloromethane	15	1.3348	1.4246^{20}	0.449	9.08^{20}	1.60	31.42	
1,2-Dichloropropane	20	1.1558	1.4390	0.769^{15}		1.46⁷²	36.40	
1,3-Dichloropropane	20	1.1859	1.4469		11.37	2.08	(23.62^{20})	(22.53^{30})
2,2-Dichloropropane	20	1.0912	1.4093			2.27		
Diethanolamine	30	1.0899	1.4747	380	2.81^{25}			
1,1-Diethoxyethane	20	0.8254	1.3805		3.80^{25}	1.38	(21.26^{22})	0.1143
Diethylamine	20	0.7056	1.3864	0.388^{10}	3.6^{22}	0.92	22.71	0.1100
Diethyl carbonate	15	0.9804	1.3854	0.868	2.82^{20}	1.10	28.62	0.0908
Diethyl ether	15	0.7193	1.3556	0.247	4.335^{20}	1.15^{20}	18.92	0.1039
Diethyl maleate	25	1.0637	1.4383	3.14	8.58	2.54	34.67	0.1042
Diethyl malonate	20	1.0550	1.4136	2.15	7.87^{25}	2.54	33.91	0.1119
Diethyl oxalate	15	1.0843	1.4124	2.311	8.1^{21}	2.49^{25}	34.32	0.0976
Diethyl sulfate	20	1.1774	1.4004		29.2		35.47	
Diethyl sulfide	20	0.8367	1.4425	0.446	5.72^{25}	1.54	(25.28^{20})	(24.16^{30})
Diiodomethane	25	3.3078	1.7380	2.392^{30}	5.316	1.08	70.21	0.1613
Diisoamyl ether	20	0.7777	1.4085	1.40^{11}		1.23	24.76	0.0871
Diisopropylamine	20	0.7153	1.3924	0.40^{25}	2.82		21.83	0.1077

Table 75 (continued)
PHYSICAL PROPERTIES OF SELECTED ORGANIC SOLVENTS

Substance	Temp., °C	Density, $g \cdot mL^{-3}$	Refractive Index (n_D)	Viscosity $mN \cdot s \cdot m^{-2}$	Dielectric Constant (ϵ)	Dipole Moment, D	Surface Tension, $dyn \cdot cm^{-1}$ a	b
Diisopropyl ether	25	0.7325	1.3681	0.379	3.88	1.22	19.89	0.1048
1,2-Dimethoxybenzene	25	1.0819	1.5323	3.281	4.09	1.32	34.4	0.0642
1,2-Dimethoxyethane	25	0.8621	1.3781	0.455	7.20	1.71		
Dimethoxymethane	15	0.8665	1.3563	0.340	2.7^{20}	0.74	23.59	0.1199
N,N-Dimethylacetamide	25	0.9366	1.4356	0.838^{30}	37.78	3.72	(32.43^{30})	(29.50^{50})
Dimethylamine	15	1.6616	1.3501^{17}	0.207	5.26^{25}	1.0	29.50	0.1265
N,N-Dimethylaniline	20	0.9559	1.5584	1.285^{25}	36.7^{25}	1.68	38.14	0.1049
2,2-Dimethylbutane	25	0.6445	1.3660	0.351	1.873		18.29	0.0990
2,3-Dimethylbutane	25	0.6570	1.3723	0.361	1.890		19.38	0.09998
2,2-Dimethyl-1-butanol	20	0.8286	1.4208					
2,3-Dimethyl-1-butanol	20	0.8300	1.4205				26.22	0.0992
3,3-Dimethyl-2-butanol	20	0.8179	1.4151					
N,N-Dimethylformamide	25	0.9445	1.4269	0.802	36.71	3.86	(36.76^{20})	(34.40^{40})
Dimethyl maleate	20	1.1513	1.4422	3.54		2.48	40.73	0.1220
2,3-Dimethylpentane	20	0.6951	1.3920	0.406	1.939	0	19.94	0.09565
2,4-Dimethylpentane	20	0.6727	1.3815	0.361	1.914	0	20.09	0.09715
Dimethyl o-phthalate	21	1.1905	1.5155	9.18^{35}	8.5^{24}		(12.05^{20})	(10.98^{30})
2,2-Dimethylpropane	20	0.5910	1.342	0.303^{5}	1.80	~0		
Dimethyl sulfate	20	1.3322	1.3874		42.6		41.26	0.1163
Dimethyl sulfoxide	25	1.0958	1.4773	1.996	46.6	3.9	(43.54^{20})	(42.41^{30})
1,4-Dioxane	25	1.0280	1.4203	1.439^{15}	2.209	0	36.23	0.1391
Dipentyl ether	25	0.7790	1.4099	0.922^{30}	2.77	1.20	26.66	0.0925
Diphenyl ether	30	1.0661	1.5763	1.158	2.68^{20}(s)	1.16	35.17	0.1104
Diphenylmethane	20	1.0060	1.5768		2.57^{25}			
Dipropylamine	20	0.7375	1.4043	0.534	3.07	1.03	24.86	0.1022
Dipropyl ether	15	0.7518	1.3830	0.448	3.39^{26}	1.21	22.60	0.1047
Dodecane	20	0.7487	1.4216	1.508	2.002^{30}	0	27.12	0.08843
1-Dodecanol	20	0.8343	1.4428		5.15		31.25	0.0748
1,2-Epoxybutane	20	0.8297	1.3840	0.41		2.01		
1,2-Ethanediamine	20	0.8977	1.4568	1.54^{25}	14.2	1.99	44.77	0.1398
1,2-Ethanediol	15	1.1171	1.4331	26.09	37.7^{25}	2.28	50.21	0.0890
1,2-Ethanediol diacetate	20	1.1043	1.4150	3.13	10^{-54}(s)	2.34		

Compound								
Ethanol	25	0.7851	1.3594	1.078	24.55	1.69	24.05	0.0832
Ethoxybenzene	20	0.9651	1.5074	1.364^{15}	4.22	1.36	35.17	0.1104
2-Ethoxyethanol	20	0.9295	1.4077	2.05	29.6^{24}	2.08	30.59	0.0897
2-(2-Ethoxyethoxy)ethanol	25	0.9841	1.4254	3.71			(31.8^{25})	(27.2^{25})
2-(2-Ethoxyethoxy)ethyl acetate	20	1.0096	1.4213	2.8				
2-Ethoxyethyl acetate	25	0.9730^{20}	1.4023	1.025	7.567^{30}	2.25	(31.8^{25})	0.1161
Ethyl acetate	25	0.8946	1.3698	0.426	6.02	1.81	26.29	0.1015
Ethyl acetoacetate	20	1.025	1.4198	1.508^{25}	15.7	3.22^{keto} 2.06^{enol}	34.42	
Ethyl acrylate	20	0.9234	1.4068	0.678				
Ethylbenzene	20	0.8670	1.4959		2.41	0.59	31.48	0.1094
Ethyl benzoate	20	1.0465	1.5052	2.407^{15}	6.02	1.99	37.16	0.1059
2-Ethyl-1-butanol	20	0.8330	1.4224	5.892^{25}	6.19^{90}		(25.06^{15})	(24.32^{25})
Ethyl butyrate	20	0.8794	1.3922	0.672	5.10^{18}	1.74	26.55	0.1045
Ethyl cinnamate	20	1.0494	1.5598	8.7	6.1^{18}	2.14	39.99	0.1045
Ethyl cyanoacetate	20	1.0648	1.4175	2.50^{25}	26.9	2.17	38.80	0.1092
Ethylcyclohexane	20	0.7879	1.4330	0.843	2.054	0	27.78	0.1054
Ethylene carbonate	40	1.3208^{25}	1.4199		89.6	4.87	47.33	0.0880
2,2'-(Ethylenedioxy)diethanol	20	1.1235	1.4561	49.0	23.69	5.58^{liq}		
Ethylenimine	25	0.832	1.4123	0.418	18.3	1.77		
Ethyl formate	20	0.9160	1.3599	0.419^{15}	7.16^{25}	1.94	26.47	0.1315
2-Ethyl-1-hexanol	20	0.8332	1.4231	9.8	4.41^{90}	1.74		
2-Ethylhexyl acetate	20	0.8718	1.4204	1.5				
bis(2-Ethylhexyl) phthalate	20	0.9843	1.4859	81.4	5.3	2.84	30.72	0.0983
Ethyl lactate	25	1.0299	1.4121	2.44	13.1^{20}	2.4	25.79	0.1006
Ethyl 3-methylbutanoate	20	0.8657	1.3962		4.7^{18}			
Ethyl propanoate	15	0.8957	1.3864	0.564	5.65^{19}	1.74	26.72	0.1168
Ethyl salicylate	20	1.1362	1.5251	1.772^{45}	7.99^{30}		41.00	0.1091
Fluorobenzene	20	1.0240	1.4657	0.620^{15}	5.42^{25}	1.61	29.67	0.1204
o-Fluorotoluene	17	1.0014	1.4716	0.680^{20}	4.22^{30}	1.37	32.31	0.1257
m-Fluorotoluene	20	0.9974	1.4691	0.608	5.42^{30}	1.86	30.44	0.1109
p-Fluorotoluene	20	0.9975	1.4688	0.622	5.86^{30}	2.00		
Formamide	20	1.1334	1.4475	3.764	111.0	3.73	59.13	0.0842
Formic acid	25	1.2141	1.3694	1.966	58.5^{16}	1.41	39.87	0.1098
2-Furaldehyde	20	1.1616	1.5262	1.495^{25}	41.9	3.61	46.41	0.1327
Furan	20	0.9378	1.4214	0.380	2.94	0.66	(24.10^{20})	(23.38^{25})
Furfuryl alcohol	20	1.1285	1.4868	4.62^{25}		1.92	$(ca\ 38^{20})$	
Glycerol	25	1.2582	1.4730	945	42.5	2.56	(63.14^{17})	(62.5^{25})
Heptane	25	0.6795	1.3851	0.397	1.92^{20}	0	22.10	0.0980
1-Heptanol	25	0.8223	1.4242			1.73		
2-Heptanol	25	0.8139	1.4190	5.06	9.21^{22}	1.71		

Table 75 (continued)
PHYSICAL PROPERTIES OF SELECTED ORGANIC SOLVENTS

Substance	Temp. °C	Density, g·mL⁻³	Refractive Index (n_D)	Viscosity, mN·s·m⁻²	Dielectric Constant (ε)	Dipole Moment D	Surface Tension, dyn·cm⁻¹ a	b
1-Heptene	20	0.6970	1.3999	0.35	2.07	0.34[liq]	22.28	0.09908
1-Hexadecanol	60	1.4355	0.8116			1.7		
Hexafluorobenzene	20	1.6182	1.3781			0	(22.6[20])	
Hexamethylphosphoric triamide	20	1.027	1.4588	3.47	30	4.31[liq]	(33.8[20])	
Hexane	20	0.6594	1.3749	0.313	1.89[25]	0.08	20.44	0.1022
Hexanenitrile	20	0.8052	1.4069	1.041[25]	17.26[25]		29.64	0.0907
Hexanoic acid	25	0.9230	1.4149	2.814	2.63[71]	1.13[liq]	(28.05[20])	(27.55[25])
1-Hexanol	25	0.8162	1.4161	4.592	13.3	1.55	27.81	0.0801
2-Hexanol	20	0.8144	1.4147				26.44	0.0869
3-Hexanol	20	0.8185	1.4160				26.27	0.0880
1-Hexene	20	0.6732	1.3879	0.26	2.051	0.34	20.47	0.10271
4-Hydroxy-4-methyl-2-pentanone	25	0.9341	1.4213	2.9[20]	18.2	3.24	(31.0[20])	0.1123
Iodobenzene	20	1.8307	1.5714	1.774[17]	4.63	1.71	41.52	0.1123
Iodoethane	20	1.9358	1.5137	0.617[15]	7.82	1.91	31.67	0.1286
Iodomethane	20	2.2790	1.5315	0.518[15]	7.00	1.62	33.42	0.1234
1-Iodopropane	20	1.7489	1.5041	0.837[15]	7.00	2.04	31.64	0.1136
2-Iodopropane	20	1.7025	1.4992	0.732[15]	8.19	1.95	29.35	0.1107
Isobutylamine	20	0.7346	1.3972	0.553[25]	4.43[21]	1.27	24.48	0.1092
Isobutyronitrile	25	0.7656	1.3712	0.456[30]	20.4[24]	3.61	(24.93[20])	(23.84[30])
Isopropyl acetate	20	0.8718	1.3773	0.569		1.86	24.44	0.1072
Isopropylamine	20	0.6875	1.3742	0.36[25]	5.45		19.91	0.09719
Isopropylbenzene	20	0.8618	1.4915	0.791	2.38	0.79	30.32	0.1054
Isoquinoline	25	1.0986[20]	1.6223		10.7	2.73		
Lactic acid	25	1.2060	1.4392[20]	40.33	22[17]			
Methacrylic acid	20	1.0153	1.4314			1.65		
Methacrylonitrile	20	0.8001	1.4007	0.392		3.69	(24.4[20])	
Methanol	25	0.7866	1.3265	0.544	32.70	1.70	24.00	0.0773
Methoxybenzene	25	0.9893	1.5143	0.789[30]	4.33	1.38	38.11	0.1204
2-Methoxyethanol	20	0.9646	1.4021	1.72	16.9	2.04	33.30	0.0984
2-(2-Methoxyethoxy)ethanol	25	1.0167	1.4245	3.48			(34.8[25])	(29.9[75])

Compound	t (°C)	d	n					
2-Methoxyethyl acetate	20	1.0049	1.4022	0.981	8.25	2.13	32.47	0.1164
bis(2-Methoxyethyl) ether	25	0.9440	1.4043			1.97	(33.67³⁰)	(30.62⁵⁰)
N-Methylacetamide	35	0.9460	1.4253	3.23	191.33³²	4.39	27.95	0.1289
Methyl acetate	25	0.9273	1.3614²⁰	0.362	6.68	1.72	34.98	0.0944
Methyl acetoacetate	20	1.0747	1.4186	1.704			40.10	0.1171
Methyl acrylate	20	0.9535	1.4117¹⁸	1.398			17.20	0.1103
Methyl benzoate	15	1.0933	1.5205	2.298	6.59²⁰	1.86	28.89	0.0917
2-Methylbutane	20	0.6197	1.3537	0.225	1.84	0.13		0.0886
4-Methylbutanenitrile	20	0.8035	1.4061	0.980	15.5	3.53		0.0820
2-Methylbutanoic acetate	20	0.8719	1.4054	0.872	4.63³⁰	1.8	(24.62²¹)	0.0748
3-Methylbutanoic acid	15	0.9308	1.4064	2.731	2.64²⁰	0.63	27.28	
2-Methyl-1-butanol	20	0.8190	1.4107	5.50	14.7²⁵	1.82	25.76	0.0989
3-Methyl-1-butanol	20	0.8103	1.4072	4.81¹⁵	14.7²⁵	1.7	24.18	0.1145
2-Methyl-2-butanol	20	0.8090	1.4050	5.48¹⁵	5.82²⁵	1.82	(23.02⁵)	0.1074
3-Methyl-2-butanol	20	0.8179	1.4096	3.51²⁵	4.63³⁰	1.72	26.75	0.1130
3-Methylbutyl acetate	25	0.8664	1.3981	0.790	5.6		27.48	0.0770
Methyl butyrate	25	0.8984²⁰	1.3870	0.543			41.32	0.0629
Methyl cyanoacetate	25	1.1225	1.4166	2.793²⁰	29.30²⁰		26.11	0.0690
Methylcyclohexane	20	0.7694	1.4231	0.734	2.02	0	32.45	0.1163
cis-2-Methylcyclohexanol	20	0.9254	1.4609	18.08²⁵	13.3*	1.95*	29.08	
trans-2-Methylcyclohexanol	20	0.9247	1.4616	37.13²⁵				
cis-3-Methylcyclohexanol	20	0.9168	1.4576	19.7²⁵	16.47	1.91	29.07	
trans-3-Methylcyclohexanol	25	0.9214	1.4580	25.1²⁵	8.05	1.75		
cis-4-Methylcyclohexanol	25	0.9122	1.4565	0.247²⁵	13.3*	1.9*		
trans-4-Methylcyclohexanol	20	0.9080	1.4544	0.385				
Methylcyclopentane	20	0.7486	1.4097	0.507	1.985	0	24.63	
N-Methylformamide	25	0.9988	1.4300	1.65	182.4	3.86	(37.96³⁰)	(35.02⁵⁰)
Methyl formate	20	0.9742	1.3433	0.328²⁵	8.5	1.77	28.29	0.1572
2-Methylhexane	20	0.6786	1.3849	0.378	1.92	0	21.22	0.09635
3-Methylhexane	20	0.6871	1.3886	0.372	1.93	0	21.73	0.09699
Methyl methacrylate	20	0.9433	1.4146	0.632	2.9	1.68		
Methyl oleate	20	0.8702²⁵	1.4521	4.88³⁰	3.211		(31.32⁵)	(25.4¹⁰⁰)
2-Methylpentane	20	0.6532	1.3715	0.310	1.88		19.37	0.09967
3-Methylpentane	20	0.6643	1.3765	0.307²⁵	1.895		20.26	0.1060
2-Methyl-1-pentanol	20	0.8242	1.4190				26.98	0.0819
3-Methyl-1-pentanol	20	0.8237	1.4193				26.92	0.07894

*Mixed isomers.

Table 75 (continued)
PHYSICAL PROPERTIES OF SELECTED ORGANIC SOLVENTS

Substance	Temp., °C	Density, $g \cdot mL^{-3}$	Refractive Index (n_D)	Viscosity, $mN \cdot s \cdot m^{-2}$	Dielectric Constant (ε)	Dipole Moment, D	Surface Tension, $dyn \cdot cm^{-1}$ a	b
4-Methyl-1-pentanol	20	0.8130	1.4154				25.93	0.07434
2-Methyl-2-pentanol	20	0.8136	1.4113				25.07	0.08606
3-Methyl-2-pentanol	20	0.8291	1.4197				27.14	0.0919
4-Methyl-2-pentanol	20	0.8076	1.4112	4.074^{25}			24.67	0.0821
2-Methyl-3-pentanol	20	0.8239	1.4168				26.43	0.0914
3-Methyl-3-pentanol	20	0.8281	1.4186				25.48	0.0888
4-Methyl-2-pentanone	20	0.8006	1.3958	0.542^{25}	13.11		(23.64^{20})	(19.62^{60})
2-Methylpropanamine	20	0.7346	1.3970		4.4	1.28	24.48	0.1092
2-Methylpropanoic acid	20	0.9682	1.3930	1.213^{25}	2.73^{40}	1.08	(25.55^{20})	(25.13^{25})
2-Methyl-1-propanol	25	0.7978	1.3938	3.91	17.93	1.79	24.53	0.0795
2-Methyl-2-propanol	25	0.7812	1.3852	3.316^{30}	12.47	1.66	(20.02^{15})	(19.10^{30})
N-Methylpropionamide	25	0.9305	1.4345	5.215	172.2	3.59	(31.20^{30})	(29.12^{50})
Methyl propionate	15	0.9221	1.3793	0.477	6.21^{20}	1.7	27.58	0.1258
1-Methylpropyl acetate	20	0.8720	1.3894				25.72	0.1054
2-Methylpropyl acetate	20	0.8745	1.3902	0.697	5.29	1.85	25.59	0.1013
2-Methylpropyl formate	20	0.8854	1.3855	0.680	6.41	1.88	26.14	0.1122
2-Methylpyridine	20	0.9444	1.5010	0.805	9.80	1.92	36.11	0.1243
3-Methylpyridine	20	0.9566	1.5068		9.80	2.40	37.35	0.1153
4-Methylpyridine	20	0.9548	1.5058			2.60	37.71	0.1141
1-Methyl-2-pyrrolidinone	25	1.0279	1.4680	1.666	32.0	4.09		
Methyl salicylate	20	1.1831	1.5365		9.41^{30}	2.53	42.15	0.1174
Morpholine	15	1.0050	1.4573	38.27	7.42^{25}	1.50	(37.63^{20})	(36.24^{30})
Naphthalene	85	0.9752	1.5898	0.780^{100}	2.54	0.0	42.84	0.1107
o-Nitroanisole	25	1.2408	1.5597			4.83	48.62	0.1185
Nitrobenzene	20	1.2033	1.5526	1.634	34.82^{25}	4.22	46.34	0.1157
Nitroethane	25	1.0382	1.3902	0.661	28.06^{30}	3.65	35.27	0.1255
Nitromethane	25	1.1312	1.3795	0.595^{30}	35.87^{30}	3.46	40.72	0.1678
1-Nitro-2-methoxybenzene	20	1.2527	1.5619		23.24^{30}	4.81	48.62	0.1185
1-Nitropropane	25	0.9955	1.3994	0.798		3.66	32.62	0.1009
2-Nitropropane	25	0.9821	1.3921	0.750	25.52	3.73	32.18	0.1158
Nonane	20	0.7176	1.4054	0.7160	1.972	0	24.72	0.09347

1-Nonanol	20	0.8280	1.4338			1.72	29.79	0.07589
1-Nonene	20	0.7922	1.4157	0.620		0.59	24.90	0.09379
1-Octadecanol	20	0.8123	1.4388			1.74		0.09509
Octane	20	0.70252	1.3974	0.546	1.95	0	23.52	0.0802
Octanenitrile	30	0.8905	1.4163	1.356	13.90[25]		29.61	(28.7[25])
Octanoic acid	20	0.9106	1.4279	5.828	2.45[25]	1.15[liq]	(29.2[20])	0.0795
1-Octanol	20	0.8258	1.4296	6.125[30]	10.34	1.68	29.09	0.08197
2-Octanol	20	0.8207	1.4261			1.65	27.96	
3-Octanol	20	0.8216	1.423					
4-Octanol	20	0.8192	1.425					
1-Octene	20	0.7149	1.4087	0.470	2.084	0.34[liq]	23.68	0.09581
Oleic acid	20	0.8906	1.4599	38.80	2.46	1.18	(32.80[20])	(27.94[90])
2,2'-Oxybis(chloroethane)	20	1.2192	1.4575	2.41	21.2	2.58	46.97	0.0880
2,2-Oxydiethanol	20	1.1167	1.4475	35.7	31.69	2.31	37.09	0.1178
Pentachloroethane	15	1.6881	1.5054	2.751	3.73[20]	0.92		
cis-1,3-Pentadiene	25	0.6859	1.4329		2.32			
trans-1,3-Pentadiene	25	0.6710	1.4267					
2,3-Pentadiene	25	0.6900	1.4251					
Pentane	25	0.6214	1.3547	0.225	1.84	0	18.25	0.11021
2,4-Pentanedione	25	0.9721[25]	1.4518[19]	0.779	25.7	3.05	33.28	0.1144
Pentanenitrile	15	0.8035	1.3991	2.359[15]	17.4[21]	3.57	29.28	0.0937
1-Pentanoic acid	20	0.9392	1.4080	3.347	2.66		28.90	0.0887
1-Pentanol	25	0.8112	1.4080	2.780[30]	13.9	1.7	27.54	0.0874
2-Pentanol	25	0.8053	1.4044	3.306[30]	13.82[22]	1.66	25.96	0.1004
3-Pentanol	25	0.8160	1.4079	0.478	13.02[22]	1.64	(24.60[20])	(23.76[30])
3-Pentanone	20	0.8144	1.3924		17.00	2.70	27.36	0.1047
2-Pentanone	20	0.8095	1.3903			2.72	24.89	0.06547
1-Pentene	20	0.6405	1.3715	0.24[0]	2.017	0.34[liq]	18.20	0.1099
cis-2-Pentene	20	0.6556	1.3830				19.73	0.1172
trans-2-Pentene	20	0.6482	1.3793				18.90	0.09972
Pentyl acetate	20	0.8753	1.4028	0.924	4.75	1.91	27.66	0.09943
Phenol	46	1.0533	1.5396	4.076	9.78[60]	1.45	43.54	0.1068
Phenylacetonitrile	25	1.0125	1.5209	1.93	18.7[27]	3.5	44.57	0.1155
D-Pinene	20	0.8600	1.4658	1.61[25]	2.64[25]	0.80	28.35	0.09444
L-Pinene	20	0.8590	1.4666	1.41[25]	2.76		28.26	0.09343
Piperidine	20	0.8613	1.4525	1.362[25]	5.8	1.19	31.79	0.1153
1-Propanal	20	0.7970	1.3619	0.317[27]	18.5[17]	2.52		
1,2-Propanediol	20	1.0364	1.4329	56.0	32.0	2.25	(72.0[25])	0.0903
1,3-Propanediol	20	1.0538	1.4396	46.6	35.0	2.50	47.43	0.1037
Propanenitrile	20	0.7911	1.3838	0.624[15]	20.3[21]	3.57	29.51	0.0777
1-Propanol	25	0.7995	1.3837	2.004	20.33	1.68	25.26	0.0777

Table 75 (continued)
PHYSICAL PROPERTIES OF SELECTED ORGANIC SOLVENTS

Substance	Temp., °C	Density, g·ml⁻³	Refractive Index (n_D)	Viscosity, mN·s·m⁻²	Dielectric Constant (ε)	Dipole Moment, D	Surface Tension, dyn·cm⁻¹ a	b
2-Propanol	25	0.7813	1.3752	1.765[30]	19.92	1.66	22.90	0.0789
2-Propen-1-ol	15	0.8551	1.4135[20]	1.486	21.6	1.63	27.53	0.0902
Propionic acid	20	0.9934	1.3865	1.175[15]	3.44[40]	1.75	28.68	0.0993
Propionic anhydride	20	1.0110	1.4046	1.144	18.3[16]		(30.30[20])	(29.70[25])
Propionitrile	20	0.7818	1.3681[15]	0.454[15]	27.2	3.56	29.63	0.1153
Propyl acetate	20	0.8883	1.3844	0.585	6.00	1.86	26.60	0.1120
Propylamine	20	0.7173	1.3882	0.353[25]	5.31	1.26	24.86	0.1243
Propyl benzoate	20	1.0232	1.5003				36.55	0.1069
Propylene oxide*	20	0.8287	1.3660	0.327		2.01		
Propyl formate	20	0.9006	1.3769	0.574	7.72	1.89	26.77	0.1119
2-Propyn-1-ol	20	0.9478	1.4320	1.68	24.5	1.78	38.59	0.1270
1-Propynyl acetate	20	0.9982	1.4187				(32.81[20])	(30.20[40])
Pyridine	20	0.9832	1.5102	0.952	12.3	2.20	39.82	0.1306
Pyrrole	20	0.9699	1.5102	1.352	7.48[18]	1.80	39.81	0.1100
2-Pyrrolidinone	25	1.107	1.486	13.3		3.55		
Quinoline	15	1.0977	1.6293	4.354	9.00[25]	2.18	42.25	0.1063
Salicylaldehyde	20	1.1574	1.5718	2.90	13.9	2.86	45.38	0.1242
Succinonitrile	60	0.9867	1.4173	2.591	56.5[57]	3.68	53.26	0.1079
Sulfolane	30	1.2614	1.4820	10.286	43.3	4.81	(35.53[30])	
Styrene	20	0.9060	1.5468	0.751	2.426	0.13[liq]	(32.32[20])	(30.98[30])
1,1,2,2-Tetrabromoethane	20	2.9640	1.6353	9.79	7.0[22]	1.29	52.37	0.1463
1,1,2,2-Tetrachloro-difluoroethane	25	1.6447	1.4130	1.21	2.52	1.32	(22.73[30])	(21.56[40])
1,1,2,2-Tetrachloroethane	15	1.6026	1.4968	1.844	8.20[20]		38.75	0.1268
1,1,2,2-Tetrachloroethene	15	1.6311	1.5076	1.932	2.30[25]	0	(32.86[15])	(31.27[30])
1-Tetradecanol	50	0.8151	1.4358	0.55		1.7		
Tetrahydrofuran	20	0.8889	1.4072	0.55	7.58[25]	1.75	(26.5[25])	
Tetrahydrofurfuryl alcohol	20	1.0524	1.4520	6.24	13.61[23]	2.12[liq]	39.96	0.1008
1,2,3,4-Tetrahydro-naphthalene	20	0.9702	1.5414	2.202	2.77	0.60[liq]	35.55	0.0954
Tetrahydropyran	25	0.8772	1.4195	0.764	5.61	1.55		

Compound	t (°C)	d	n			μ		
Tetrahydrothiophene	25	0.9938	1.5257	0.971		1.90	38.44	0.1342
1,1,2,2-Tetramethylurea	25	0.9654	1.4493	0.373	23.06	3.47		
Tetranitromethane		1.6372^{21}	1.4399^{17}	0.638	2.32^{20}			
2-Thiabutane	20	0.8422	1.4403		5.72^{25}		(24.9^{20})	(23.4^{30})
Thiacyclobutane	20	1.0200	1.5102	1.042	6.2	1.78	(36.3^{20})	(35.0^{30})
Thiacyclohexane	20	0.9861	1.5070				(36.06^{20})	(33.74^{40})
Thiacyclopentane	20	0.9987	1.5048	0.440		1.61	(35.8^{20})	(34.6^{30})
2-Thiapentane	20	0.8424	1.4442				(25.2^{20})	(23.9^{30})
3-Thiapentane	20	0.8363	1.4430				(26.50^{11})	(23.33^{23})
2-Thiapropane	20	0.8483	1.4353	0.289		1.45	34.00	0.1328
Thiophene	20	1.0649	1.5289	0.654	2.76^{16}	0.55	30.90	0.1189
Toluene	25	0.8623	1.4941	0.552	2.38	0.36	42.87	0.1094
o-Toluidine	15	1.0028	1.5685	5.195	6.34	1.60	40.33	0.0979
m-Toluidine	15	0.9930	1.5638	4.418	5.95	1.45	39.58	0.0957
p-Toluidine	60	0.9538	1.5532	1.557	4.98	1.52	48.14	0.1308
Tribromomethane	15	2.9035	1.6005	2.152	4.39^{20}	0.99	(26.2^{20})	(25.8^{25})
Tri-n-butyl borate	20	0.8580	1.4092	1.776		0.78	28.71	0.0666
Tri-n-butyl phosphate	25	0.9760	1.4226	3.39	7.95^{30}	3.07	28.28	0.1242
Trichloroacetonitrile		1.4403^{25}	1.4409^{20}	0.903^{15}	7.85^{19}	1.79	37.40	0.1351
1,1,1-Trichloroethane	20	1.3492	1.4377	0.119	7.6		(29.5^{20})	(28.8^{25})
1,1,2-Trichloroethane	20	1.4424	1.4706	0.566	8.78^{23}		(37.8^{20})	(37.05^{25})
1,1,2-Trichloroethene	20	1.4679	1.4775	18.834	3.42^{16}	0.9	27.73	0.08719
1,2,3-Trichloropropane	20	1.3880	1.4834				28.01	0.08839
Tridecane	20	0.7563	1.4256			0	22.70	0.0992
1-Tridecene	25	0.7653	1.4334				15.64	0.08444
Triethanolamine	20	1.1196	1.4835	613.6	29.36	3.57	30.91	0.1040
Triethylamine	20	0.7281	1.4010	0.394^{15}	2.42^{25}	0.77	31.76	0.1025
Trifluoroacetic acid	20	1.4890	1.2850	0.926	39.5	2.28^{100}	29.79	0.08966
1,2,3-Trimethylbenzene	20	0.8944	1.5139	0.895^{15}	2.64	0.56	20.70	0.09726
1,2,4-Trimethylbenzene	20	0.8758	1.5048	1.154	2.38	0.30		
1,3,5-Trimethylbenzene	20	0.8652	1.4994	0.579	2.28	0		
2,2,3-Trimethylbutane	20	0.6901	1.3894	0.632	1.93	0		
cis-1,3,5-Trimethylcyclohexane	20	0.7705	1.4245	0.714		0		
trans-1,3,5-Trimethyl-cyclohexane	20	0.7789	1.4286	0.598				
2,2,3-Trimethylpentane	20	0.7160	1.4030	0.504	1.962	0	22.46	0.08950
2,2,4-Trimethylpentane	20	0.6919	1.3915	11.855	1.940	0	20.55	0.08876
Undecane	20	0.7402	1.4173				26.46	0.09010

* Mixed isomers.

Table 75 (continued)
PHYSICAL PROPERTIES OF SELECTED ORGANIC SOLVENTS

Substance	Temp., °C	Density, g·mL^{-3}	Refractive Index (n_p)	Viscosity mN·s·m^{-2}	Dielectric Constant (ϵ)	Dipole Moment, D	Surface Tension, dyn·cm^{-1}	
							a	b
1-Undecanol	20	0.8324	1.4402			1.74		
Vinyl acetate	20	0.9312	1.3959	23.95		1.79	(23.95[20])	(22.54[30])
o-Xylene	20	0.8802	1.5054	0.809	2.57	0.62	32.51	0.1101
m-Xylene	20	0.8642	1.4972	0.617	2.37	0.37	31.23	0.1104
p-Xylene	20	0.8611	1.4958	0.644	2.27	0	30.69	0.1074

From Dean, J. A., Ed., *Lange's Handbook of Chemistry*, 12th ed., McGraw-Hill, New York, 1979, 10-103. With permission.

Table 76
PROPERTIES OF COMBUSTIBLE MIXTURES

Fuel	Flame temperature, °C (stoichiometric mixture)		Lowest ignition temperature, °C		Flammability limit, % by volume fuel (25°C, 760 mm)				Maximum burning velocity, cm/sec	
					Air		Oxygen			
	Air	Oxygen	Air	Oxygen	Lower	Upper	Lower	Upper	Air	Oxygen
Acetone	2105		700	568	2.5	13.0			46	
Acetylene	2400	3140	335	350	2.4	52.3	3.5	89.4	266	2480
Acrylonitrile	2444				4.6				43	
Ammonia					15.5	27.0	13.5	79.0		
Amyl alcohol			409	390	1.2					
Aniline			770	530						
Benzene	2290		740	662	1.2	9.1			40	
n-Butane	1895	2900	490	460	1.9	8.4			83	
1-Butene	1930				1.8	12.0			43	
Carbon disulfide					1.2	73.0			49	
Carbon monoxide	1960	2600	610	590	15.7	70.9	16.7	93.5	60	250
Cyanogen		4500			6.6	42.6				270
Cyclopropane	2310				2.4	10.4	2.5	63.1	50	
Decane	2265		463	202	0.6	5.4			37	
Diethyl ether	2235		343	178	1.8	36.5	2.1	82.0	40	
Ethane	1895		530	500	4.2	9.5	4.1	50.5	86	
Ethene	1975		540	485	4.0	28.6	2.9	79.9	67	
Ethyl alcohol			558	425	3.3	18.9				
Hexane	2220		487	268	1.3	8.6			40	
Hydrogen	2045	2660	530	450	9.5	65.2	9.4	91.6	440	1140
Hydrogen sulfide					4.3	45.5				
Isopropyl alcohol			590	512	1.7				34	
Methane	1875	2680	645	645	6.3	11.9	6.5	59.2	70	
Methyl alcohol				555	6.7	50			49	
Natural gas	1700–1910	2740	560	450	9.8	24.8	10.0	73.6	55	
Pentane	2230				1.3	4.9			40	
Propane	1925	2850	510	490	2.4	9.5			43	390
Propene	1935				2.2	12.1	2.1	52.8	43	
Propyl alcohol			505	445	2.2	13.5				
Toluene	2325		810	552	1.0	7.3	1.3	6.8	37	

	Fluorine									
Hydrogen	4030									

	Nitrous oxide									
Acetylene	2800								160	
Hydrogen	2690								390	

From Dean, J. A., Ed., *Lange's Handbook of Chemistry,* 12th ed., McGraw-Hill, New York, 1979, 11-10. With permission.

Table 77

ORGANIC SOLVENTS ARRANGED BY BOILING POINTS

Name	B. P. 760 mm °C	Name	B.P. 760 mm °C
Ethyl chloride	13	Allyl alcohol	96.6
Ethylene oxide	14	1,2-Dichloropropane	96.8
Furan	31—2	n-Propyl alcohol	97.8
Methyl formate	32	n-Heptane	98.4
Diethyl ether	34.6	Ethyl propionate	99.1
Propylene oxide	35	sec-Butyl alcohol	99.5
n-Pentane	36.1	Isoamyl chloride	99.7
Ethyl bromide	38.4	Ligroin (high boiling)	100—150
Methylene chloride	40	Formic acid	100.8
Methylal	42.3	Methyl-cyclohexane	100.9
Carbon disulfide	46.3	Dioxane (1,4)	101.1
Ethyl formate	54	Nitromethane	101.5
Acetone	56.5	n-Propyl acetate	101.6
Methyl acetate	57.1	Diethyl ketone	101.7
Ethylidene dichloride	57.3	tert-Amyl alcohol	102
Acetylene dichloride*	59—61	Acetal	102.2
Ligroin (low boiling)	60—120	n-Butyl formate	106.9
Chloroform	61.2	Isobutyl alcohol	107—8
Methyl alcohol	64.7	Acetylene dibromide*	110
Tetrahydrofuran	65—6	Toluene	110.6
Di-isopropyl ether	68.5—69.0	sec-Butyl acetate	112—3
n-Hexane	68.7	Trichloroethane (1,1,2)	113.5
Isobutyl chloride	68.9	Nitroethane	114.8
Trichloroethane (1,1,1)	74.1	Pyridine	115—6
Dioxolane	75—6	Pentan-3-ol	115.6
Carbon tetrachloride	76.8	Epichlorohydrin	117
Ethyl acetate	77.1	n-Butyl alcohol	117
n-Butyl chloride	77.9	Isobutyl acetate	118
Ethyl alcohol	78.4	Methyl isobutyl ketone	118
Methyl-ethyl ketone	79.6	Acetic acid	118.1
2-Methyl-tetrahydrofuran	80.0	Propylene glycol monomethyl ether	119
Benzene	80.1		
Cyclohexane	80.7	Ethyl n-butyrate	120—1
n-Propyl formate	81.3	2-Nitropropane	120.3
Acetonitrile	82	Isoamyl bromide	120.4
		Tetrachloroethylene	120.8
Isopropyl alcohol	82.5	Di-isopropyl ketone	123.7
tert-Butyl alcohol	82.9	Ethylene glycol diethyl ether (Diethyl Cellosolve)	124
Cyclohexene	83.3		
Ethylene chloride	83.7	Ethylene glycol monomethyl ether (Methyl Cellosolve)	124—5
Thiophene	84		
		n-Octane	125.7
Trichloroethylene	87.2	Diethyl carbonate	126
Isopropyl acetate	88.4		
Isobutyl bromide	91.5	n-Butyl acetate	126
2,5-Dimethyl-furan	93—4	sec-Propylene chlorohydrin	127
Ethyl chloroformate	94—5	Ethylene chlorohydrin	128.8
		Mesityl oxide	130.1
		Ethylene bromide	131.5

Table 77 (continued)
ORGANIC SOLVENTS ARRANGED BY BOILING POINTS

Name	B. P. 760 mm °C	Name	B. P. 760 mm °C
1-Nitropropane	131.6	Ethylene glycol mono-*n*-butyl-ether	171.2
2-Methyl-pentan-4-ol	131.8		
Isobutyl carbinol	132.0	Phenetole	172
Chlorobenzene	132.1	Di-isoamyl ether	173.4
Xylene*	*ca* 133	*n*-Decane	174.0
		Glycol diformate	174
Cyclohexylamine	134		
Ethylene glycol monoethyl ether (Cellosolve)	135.1	Cyclohexyl acetate	174—5
		2,6-Dimethylheptan-4-ol	174—5
Ethylbenzene	136.2	α,α'-Dichlorohydrin	174.3
n-Amyl alcohol (*prim*.)	138	Furfuryl acetate	175—7
Acetic anhydride	139.6	Methyl *p*-tolyl ether	176
Di-isopropyl carbinol	140	Eucalyptole	176—7
Acetylacetone	140.5	Methyl *m*-tolyl ether	177
Isoamyl acetate	142	*p*-Cymene	177.1
Di-*n*-butyl ether	142.4	Dichloroethyl ether (*sym*)	178.5
Ethylene glycol mono-isopropyl ether	144	*iso*-Amyl *n*-butyrate	178.6
Monomethylglycol acetate (Methyl-cellosolve Acetate)	144.5	*o*-Dichlorobenzene	179
		Octan-2-ol	179—80
Acetylene tetrachloride	146.3	Ethyl acetoacetate	180
2-Methylpentan-1-ol	148	Ethylene glycol mono-isoamyl ether	181
3-Methylol-pentane	148.9		
n-Amyl acetate	149	Phenol	181.4
Ethyl-*n*-butyl ketone	149—50	2-Ethylhexan-2-ol	184
Bromoform	150.5	Aniline	184.4
n-Nonane	150.8	Diethyl oxalate	186
Methyl-*n*-amyl ketone	151.5	Diethylene glycol diethyl ether	188
Isopropylbenzene	152.4	α-Propylene glycol	188—9
Anisole	154—5	Ethyl benzyl ether	190
Ethyl lactate	155	Glycol diacetate	190.5
Heptan-4-ol	156	Benzonitrile	190.7
Cyclohexanone	155—6	Decalin*	191.7
Heptan-4-ol	156	Dimethylaniline	193
Bromobenzene	156.2	Isoamyl isovalerate	194
Monoethyl-glycol acetate	156.5	Octan-1-ol	194—5
Hexan-1-ol	157.2	Acetonyl acetone	194.1
Trichloropropane (1,2,3)	158	Diethylene glycol monomethyl ether	194.2
Ethylene glycol mono-isobutyl ether	158.8		
		Diethylene glycol monoethyl ether	195
Cyclohexanol	160—1		
Isoamyl propionate	160.2	Glycol	197.4
Heptan-2-ol	160.4	Methyl benzoate	198—9
Furfural	161.7	Diethyl malonate	198.9
Pentachloroethane	162	*o*-Toluidine	199.7
		"Carbitol"	200.1
Diacetone alcohol	167.9		
Di-isobutyl ketone	168.1	*p*-Toluidine	200.3
Methyl acetoacetate†	169—70	Acetophenone	203
Furfuryl alcohol	170	Ethylene glycol dibutyl ether	203
Methyl *o*-tolyl ether	171—2	Methyl-phenyl carbinol	203.9
		Benzyl alcohol	204.7

Table 77 (continued)
ORGANIC SOLVENTS ARRANGED BY BOILING POINTS

Name	B. P. 760 mm °C	Name	B. P. 760 mm °C
Tetralin	206—7	Diethylene glycol monobutyl ether acetate	246.4
Butan-1,3-diol	206.5		
γ-Valerolactone	207—8	Triethylene glycol monoethyl ether	248
Ethylene glycol monomethyl ether acetal	207.2		
Camphor	209.1	n-Butyl benzoate	250
		Diethylene glycol mono-n-hexyl ether	252
Diethylene glycol monomethyl ether acetate	209.1		
o-Chloroaniline	210.5	Diethylene glycol di-n-butyl ether (Dibutyl Carbitol)	255
Nitrobenzene	210.9	Triacetin	258—9
Ethyl benzoate	211—2	α-Chloronaphthalene	259.3
Isophorone	215—6		
		Isoamyl benzoate	262
Diethylene glycol monoethyl ether acetate	217.7	α-Monobutyrin	269—71
		Ethyl cinnamate	271
Naphthalene	217.9	o-Nitroanisole	272—3
Acetamide	222	Tetraethylene glycol dimethyl ether	275.8
Methyl salicylate	222.2		
Diethyl maleate	225		
n-Propyl benzoate	231	Isoamyl salicylate	277—8
Tributyl borate	231	α-Bromonaphthalene	281.1
Decan-1-ol	232.9	Dimethyl phthalate	282
Benzyl cyanide	233—4	Glycerol	290
Quinoline	238	Triethylene glycol	290
Ethylene glycol monophenyl ether (Phenyl Cellosolve)	244.7	Diethyl phthalate	298—9
		Benzyl benzoate	323—4
Diethylene glycol	244.8	Tetraethylene glycol dibutyl ether	330
n-Dibutyl oxalate	245.5		
		n-Dibutyl phthalate	340

Note: For methods of testing the purity, the purification, and the drying of many of the solvents listed, the reader is referred to Riddick and Bunger, Organic solvents, in *Techniques of Chemistry*, 3rd ed., Weissberger, A., ed., Vol. II, Wiley-Interscience, New York, 1970.

From Dean, J. A., Ed., *Lange's Handbook of Chemistry*, 12th ed., McGraw-Hill, New York, 1979, 10-60. With permission.

Table 78
SOME STANDARD STOCK SOLUTIONS

Element	Procedure
Aluminum	Dissolve 1.000 g Al wire in minimum amount of 2 M HCl; dilute to volume.
Antimony	Dissolve 1.000 g Sb in (1) 10 ml HNO₃ plus 5 m𝓁 HCl, and dilute to volume when dissolution is complete; or (2) 18 ml HBr plus 2 m𝓁 liquid Br₂; when dissolution is complete add 10 m𝓁 HClO₄, heat in a well-ventilated hood while swirling until white fumes appear and continue for several minutes to expel all HBr, then cool and dilute to volume.
Arsenic	Dissolve 1.3203 g of As₂O₃ in 3 ml 8 M HCl and dilute to volume; or treat the oxide with 2 g NaOH and 20 m𝓁 water; after dissolution dilute to 200 m𝓁 neutralize with HCl (pH meter), and dilute to volume.

Table 78 (continued)
SOME STANDARD STOCK SOLUTIONS

Element	Procedure
Barium	(1) Dissolve 1.7787 g $BaCl_2 \cdot 2H_2O$ (fresh crystals) in water and dilute to volume. (2) Dissolve 1.516 g $BaCl_2$ (dried at 250°C for 2 hr) in water and dilute to volume. (3) Treat 1.4367 g $BaCO_3$ with 300 ml water, slowly add 10 ml of HCl and, after the CO_2 is released by swirling, dilute to volume.
Beryllium	(1) Dissolve 19.655 g $BeSO_4 \cdot 4H_2O$ in water, add 5 ml HCl (or HNO_3), and dilute to volume. (2) Dissolve 1.000 g Be in 25 ml 2 MHCl, then dilute to volume.
Bismuth	Dissolve 1.000 g Bi in 8 ml of 10 M HNO_3, boil gently to expel brown fumes, and dilute to volume.
Boron	Dissolve 5.720 g fresh crystals of H_3BO_3 and dilute to volume.
Bromine	Dissolve 1.489 g KBr (or 1.288 g NaBr) in water and dilute to volume.
Cadmium	(1) Dissolve 1.000 g Cd in 10 ml of 2 M HCl; dilute to volume. (2) Dissolve 2.282 g $3CdSO_4 \cdot 8H_2O$ in water; dilute to volume.
Calcium	Place 2.4973 g $CaCO_3$ in volumetric flask with 300 ml water, carefully add 10 ml HCl; after CO_2 is released by swirling, dilute to volume.
Cerium	(1) Dissolve 4.515 g $(NH_4)_4Ce(SO_4)_4 \cdot 2H_2O$ in 500 ml water to which 30 ml H_2SO_4 had been added, cool, and dilute to volume. Advisable to standardize against As_2O_3. (2) Dissolve 3.913 g $(NH_4)_2Ce(NO_3)_6$ in 10 ml H_2SO_4, stir 2 min, cautiously introduce 15 ml water and again stir 2 min. Repeat addition of water and stirring until all the salt has dissolved, then dilute to volume.
Cesium	Dissolve 1.267 g CsCl and dilute to volume. Standardize: Pipette 25 ml of final solution to Pt dish, add 1 drop H_2SO_4, evaporate to dryness, and heat to constant weight at ⟩800°C. Cs (in μg/ml) = (40)(0.734)(wt of residue)
Chlorine	Dissolve 1.648 g NaCl in water and dilute to volume.
Chromium	(1) Dissolve 2.829 g $K_2Cr_2O_7$ in water and dilute to volume. (2) Dissolve 1.000 g Cr in 10 ml HCl, and dilute to volume.
Cobalt	Dissolve 1.000 g Co in 10 ml of 2 MHCl, and dilute to volume.
Copper	(1) Dissolve 3.929 g fresh crystals of $CuSO_4 \cdot 5H_2O$, and dilute to volume. (2) Dissolve 1.000 g Cu in 10 ml HCl plus 5 ml water to which HNO_3(or 30%H_2O_2) is added dropwise until dissolution is complete. Boil to expel oxides of nitrogen and chlorine, then dilute to volume.
Dysprosium	Dissolve 1.1477 g Dy_2O_3 in 50 ml of 2 MHCl; dilute to volume.
Erbium	Dissolve 1.1436 g Er_2O_3 in 50 ml of 2 MHCl; dilute to volume.
Europium	Dissolve 1.1579 g Eu_2O_3 in 50 ml of 2 MHCl; dilute to volume.
Fluorine	Dissolve 2.210 g NaF in water and dilute to volume.
Gadolinium	Dissolve 1.152 g Gd_2O_3 in 50 ml of 2 MHCl; dilute to volume.
Gallium	Dissolve 1.000 g Ga in 50 ml of 2 MHCl; dilute to volume.
Germanium	Dissolve 1.4408 g GeO_2 with 50 g oxalic acid in 100 ml of water; dilute to volume.
Gold	Dissolve 1.000 g Au in 10 ml of hot HNO_3 by dropwise addition of HCl, boil to expel oxides of nitrogen and chlorine, and dilute to volume. Store in amber container away from light.
Hafnium	Transfer 1,000 g Hf to Pt dish, add 10 ml of 9 $M$$H_2SO_4$, and then slowly add HF dropwise until dissolution is complete. Dilute to volume with 10% H_2SO_4.
Holmium	Dissolve 1.1455 g Ho_2O_3 in 50 ml of 2 MHCl; dilute to volume.
Indium	Dissolve 1.000 gIn in 50 ml of 2 MHCl; dilute to volume.
Iodine	Dissolve 1.308 g KI in water and dilute to volume.
Iridium	(1) Dissolve 2.465 g Na_3IrCl_6 in water and dilute to volume. (2) Transfer 1.000 g Ir sponge to a glass tube, add 20 ml of HCl and 1 ml of $HClO_4$. Seal the tube and place in an oven at 300°C for 24 hr. Cool, break open the tube, transfer the solution to a volumetric flask, and dilute to volume. Observe all safety precautions in opening the glass tube.
Iron	Dissolve 1.000 g Fe wire in 20 ml of 5 MHCl; dilute to volume.
Lanthanum	Dissolve 1.1717 g La_2O_3 (dried at 110°C) in 50 ml of 5 MHCl, and dilute to volume.
Lead	(1) Dissolve 1.5985 g $Pb(NO_3)_2$ in water plus 10 ml HNO_3, and dilute to volume. (2) Dissolve 1.000 g Pb in 10 ml HNO_3, and dilute to volume.
Lithium	Dissolve a slurry of 5.3228 g Li_2CO_3 in 300 ml of water by addition of 15 ml HCl; after release of CO_2 by swirling, dilute to volume.
Lutetium	Dissolve 1.6079 g $LuCl_3$ in water and dilute to volume.

Table 78 (continued)
SOME STANDARD STOCK SOLUTIONS

Element	Procedure
Magnesium	Dissolve 1.000 g Mg in 50 m*l* of 1 *M* HCl and dilute to volume.
Manganese	(1) Dissolve 1.000 g Mn in 10 m*l* HCl plus 1 m*l* HNO$_3$, and dilute to volume. (2) Dissolve 3.0764 g MnSO$_4$·H$_2$O (dried at 105°C for 4 hr) in water and dilute to volume. (3) Dissolve 1.5824 g MnO$_2$ in 10 HCl in a good hood, evaporate to gentle dryness, dissolve residue in water and dilute to volume.
Mercury	Dissolve 1.000 g Hg in 10 m*l* of 5 *M* HNO$_3$ and dilute to volume.
Molybdenum	(1) Dissolve 2.0425 g (NH$_4$)$_2$MoO$_4$ in water and dilute to volume. (2) Dissolve 1.5003 g MoO$_3$, in 100 m*l* of 2 *M* ammonia, and dilute to volume.
Neodymium	Dissolve 1.7373 g NdCl$_3$ in 100 m*l* 1 *M* HCl and dilute to volume.
Nickel	Dissolve 1.000 g Ni in 10 m*l* hot HNO$_3$, cool, and dilute to volume.
Niobium	Transfer 1.000 g Nb (or 1.4305 g Nb$_2$O$_5$) to Pt dish, add 20 ml HF, and heat gently to complete dissolution. Cool, add 40 m*l* H$_2$SO$_4$, and evaporate to fumes of SO$_3$. Cool and dilute to volume with 8 *M* H$_2$SO$_4$.
Osmium	Dissolve 1.3360 g OsO$_4$ in water and dilute to 100 m*l*. Prepare only as needed as solution loses strength on standing unless Os is reduced by SO$_2$ and water is replaced by 100 m*l* 0.1 *M* HCl.
Palladium	Dissolve 1.000 g Pd in 10 m*l* of HNO$_3$ by dropwise addition of HCl to hot solution; dilute to volume.
Phosphorus	Dissolve 4.260 g (NH$_4$)$_2$HPO$_4$ in water and dilute to volume.
Platinum	Dissolve 1.000 g Pt in 40 m*l* of hot aqua regia, evaporate to incipient dryness, add 10 ml HCl and again evaporate to moist residue. Add 10 m*l* HCl and dilute to volume.
Potassium	Dissolve 1.9067 g KCl(or 2.8415 g KNO$_3$) in water and dilute to volume.
Praseodymium	Dissolve 1.1703 g Pr$_2$O$_3$ in 50 m*l* of 2 *M* HCl; dilute to volume.
Rhenium	Dissolve 1.000 g Re in 10 m*l* of 8 *M* HNO$_3$ in an ice bath until initial reaction subsides, then dilute to volume.
Rhodium	Dissolve 1.000 g Rh by the sealed-tube method described under iridium.
Rubidium	Dissolve 1.4148 g RbCl in water. Standardize as described under cesium. Rb (in μg/m*l* = (40)(0.320)(wt of residue).
Ruthenium	Dissolve 1.317 g RuO$_2$ in 15 m*l* of HCl; dilute to volume.
Samarium	Dissolve 1.1596 g Sm$_2$O$_3$ in 50 m*l* of 2 *M* HCl; dilute to volume.
Scandium	Dissolve 1.5338 g Sc$_2$O$_3$ in 50 m*l* of 2 *M* HCl; dilute to volume.
Selenium	Dissolve 1.4050 g SeO$_2$ in water and dilute to volume or dissolve 1.000 g Se in 5 m*l* of HNO$_3$, then dilute to volume.
Silicon	Fuse 2.1393 g SiO$_2$ with 4.60 g Na$_2$CO$_3$, maintaining melt for 15 min in Pt crucible. Cool, dissolve in warm water, and dilute to volume. Solution contains also 2000 μg/m*l* sodium.
Silver	(1) Dissolve 1.5748 g AgNO$_3$ in water and dilute to volume. (2) Dissolve 1.000 g Ag in 10 m*l* of HNO$_3$; dilute to volume. Store in amber glass container away from light.
Sodium	Dissolve 2.5421 g NaCl in water and dilute to volume.
Strontium	Dissolve a slurry of 1.6849 g SrCO$_3$ in 300 m*l* of water by careful addition of 10 m*l* of HCl; after release of CO$_2$ by swirling, dilute to volume.
Sulfur	Dissolve 4.122 g (NH$_4$)$_2$SO$_4$ in water and dilute to volume.
Tantalum	Transfer 1.000 g Ta (or 1.2210 g Ta$_2$O$_5$) to Pt dish, add 20 m*l* of HF, and heat gently to complete the dissolution. Cool, add 40 m*l* of H$_2$SO$_4$ and evaporate to heavy fumes of SO$_3$. Cool and dilute to volume with 50% H$_2$SO$_4$.
Tellurium	(1) Dissolve 1.2508 g TeO$_2$ in 10 m*l* of HCl; dilute to volume. (2) Dissolve 1.000 g Te in 10 m*l* of warm HCl with dropwise addition of HNO$_3$, then dilute to volume.
Terbium	Dissolve 1.6692 g of TbCl$_3$ in water, add 1 m*l* of HCl, and dilute to volume.
Thallium	Dissolve 1.3034 g TlNO$_3$ in water and dilute to volume.
Thorium	Dissolve 2.3794 g Th(NO$_3$)$_4$·4H$_2$O in water, add 5 m*l* HNO$_3$, and dilute to volume.
Thulium	Dissolve 1.142 g Tm$_2$O$_3$ in 50 m*l* of 2 *M* HCl; dilute to volume.
Tin	Dissolve 1.000 g Sn in 15 m*l* of warm HCl; dilute to volume.
Titanium	Dissolve 1.000 g Ti in 10 m*l* of H$_2$SO$_4$ with dropwise addition of HNO$_3$; dilute to volume with 5% H$_2$SO$_4$.
Tungsten	Dissolve 1.7941 g of Na$_2$WO$_4$·2H$_2$O in water and dilute to volume.
Uranium	Dissolve 2.1095 g UO$_2$(NO$_3$)$_2$·6H$_2$O (or 1.7734 g uranyl acetate dihydrate) in water and dilute to volume.

Table 78 (continued)
SOME STANDARD STOCK SOLUTIONS

Element	Procedure
Vanadium	Dissolve 2.2963 g NH_4VO_3 in 100 ml of water plus 10 ml of HNO_3; dilute to volume.
Ytterbium	Dissolve 1.6147 g $YbCl_3$ in water and dilute to volume.
Yttrium	Dissolve 1.2692 g Y_2O_3 in 50 ml of 2 MHCl and dilute to volume.
Zinc	Dissolve 1.000 g Zn in 10 ml of HCl; dilute to volume.
Zirconium	Dissolve 3.533 g $ZrOCl_2 \cdot 8H_2O$ in 50 ml of 2 MHCl, and dilute to volume. Solution should be standardized.

* 1000 μg/ml as the element in final volume of 1 liter unless stated otherwise.

From Dean, J. A. and Rains, T. C., Standard solutions for flame spectrometry, in *Flame Emission and Atomic Absorption Spectrometry*, Vol. II, Dean, J. A. and Rains, T. C., Eds., Marcel Dekker, New York, 1971. With permission.

Table 79
CARBOHYDRATES: PART I. NATURAL MONOSACCHARIDES: ALDOSES AND KETOSES

	Substance (Synonym)	Chemical Formula	Melting Point °C	Specific Rotation $[\alpha]_D$
	(A)	(B)	(C)	(D)
		Aldoses		
1	D-Glyceraldehyde	$C_3H_6O_3$	....	+13.5 ± 0.5 (syrup)
2	D-Glyceraldehyde, 3-deoxy-3,3-C-bis-(hydroxymethyl)- (Cordycepose)	$C_5H_{10}O_4$		−26 (c 0.6, C_2H_5OH
3	D-Glyceraldehyde, 3,3-bis(C-hydroxymethyl)- (Apiose)	$C_5H_{10}O_5$	....	+5.6 (c 10) [15°] syrup
4	β-D-Arabinose	$C_5H_{10}O_5$	155	−175 → −103
5	D-Arabinose, 2-O-methyl-	$C_6H_{12}O_5$	Syrup	−102
6	α-L-Arabinose	$C_5H_{10}O_5$	158 amorphous	+55.4 → +105
7	β-L-Arabinose	$C_5H_{10}O_5$	160	+190.6 → +104.5
8	DL-Arabinose	$C_5H_{10}O_5$	163.5—164.5	None
9	α-L-Lyxose	$C_5H_{10}O_5$	105	+5.8 → +13.5
10	L-Lyxose, 5-deoxy-3-C-formyl- (Streptose)	$C_6H_{10}O_5$	....	....
11	L-Lyxose, 3-C-formyl- (Hydroxystreptose)	$C_6H_{10}O_6$	....	....
12	Pentose, 4,5-anhydro-5-deoxy-D-*erythro*-	$C_5H_8O_3$		
13	Pentose, 2-deoxy-D-*erythro*-	$C_5H_{10}O_4$	96—98	−91 → −58
14	D-Ribose	$C_5H_{10}O_5$	87	−23.1 → −23.7
15	D-Ribose, 2-C-hydroxymethyl- (Hamamelose)	$C_6H_{12}O_6$	....	−7.1 [λ578]
16	α-D-Xylose	$C_5H_{10}O_5$	145	+93.6 → +18.8
17	D-Xylose, 5-deoxy-	$C_5H_{10}O_4$		+16
18	β-D-Xylose, 2-O-methyl-	$C_6H_{12}O_5$	137—138	−21 → +34
19	α-D-Xylose, 3-O-methyl-	$C_6H_{12}O_5$	95	+45 → +19
20	D-Allose, 6-deoxy-	$C_6H_{12}O_5$	140—143 / 146—148	+1.6 [18°] (c 0.6) / −4.7 → 0
21	D-Allose, 6-deoxy-2,3-di-O-methyl- (Mycinose)	$C_8H_{16}O_5$	102—106	−46 → −29
22	Amicetose (a trideoxy hexose)	$C_6H_{12}O_3$	Oil, b.p. 65—70	+28.6 ($CHCl_3$)
23	Antiarose	$C_6H_{12}O_5$	....	Levo

Table 79 (continued)
CARBOHYDRATES: PART I. NATURAL MONOSACCHARIDES: ALDOSES AND KETOSES

	Substance (Synonym)	Chemical Formula	Melting Point °C	Specific Rotation $[\alpha]_D$
	(A)	(B)	(C)	(D)
	Aldoses (Con't)			
24	α-D-Galactose	$C_6H_{12}O_6$	167	$+150.7 \to +80.2$
25	β-D-Galactose	$C_6H_{12}O_6$	143–145	$+52.8 \to +80.2$
26	D-Galactose, 3,6-anhydro-	$C_6H_{10}O_5$	....	$+21.3 \,[10°]$
27	α-D-Galactose, 6-deoxy- (D-Fucose; Rhodeose)	$C_6H_{12}O_5$	140–145	$+127 \to +76.3 \,(c\ 10)$
28	D-Galactose, 6-deoxy-3-O-methyl- (Digitalose)	$C_7H_{14}O_5$	106[1], 119[2]	$+106$
29	D-Galactose, 6-deoxy-4-O-methyl-	$C_7H_{14}O_5$	131–132	$+82$
30	D-Galactose, 6-deoxy-2,3-di-O-methyl-	$C_8H_{16}O_5$	....	$+73$
31	α-D-Galactose, 3-O-methyl-	$C_7H_{14}O_6$	144–147	$+150.6 \to +108.6$
32	α-D-Galactose, 6-O-methyl-	$C_7H_{14}O_6$	122–123	$+117 \to +77.3$
33	L-Galactose	$C_6H_{12}O_6$	....	*See* D-Galactose
34	α-L-Galactose, 3,6-anhydro-	$C_6H_{10}O_5$	....	$-39.4 \to -25.2$
35	α-L-Galactose, 6-deoxy- (L-Fucose)	$C_6H_{12}O_5$	145	$-124.1 \to -76.4$
36	L-Galactose, 6-deoxy-2-O-methyl-	$C_7H_{14}O_5$	149–150	$-75 \pm 4 \,(c\ 0.5)$
37	L-Galactose, 6-sulfate	$C_6H_{12}O_9S$	....	$-47 \,(c\ 0.2)$ (Na salt)
38	DL-Galactose	$C_6H_{12}O_6$	143–144, 163	None (racemic)
39	α-D-Glucose	$C_6H_{12}O_6$	146, 83 (H_2O)	$+112 \to +52.7$
40	β-D-Glucose	$C_6H_{12}O_6$	148–150	$+18.7 \to +52.7$
41	D-Glucose, 6-acetate	$C_7H_{14}O_7$	135	$+48$
42	D-Glucose, 2,3-di-O-methyl-	$C_8H_{16}O_6$	85–86, 121	$+50$
43	D-Glucose, 6-O-benzoyl- (Vaccinin)	$C_{13}H_{16}O_7$	Amorphous	$+48 \,(C_2H_5OH)$
44	α-D-Glucose, 6-deoxy- (Chinovose; Epirhamnose; Glucomethylose; Isorhamnose; Isorhodeose; Quinovose)	$C_6H_{12}O_5$	139–140	$+73.3 \to +29.7 \,(c\ 8)$
45	α-D-Glucose, 6-deoxy-3-O-methyl- (D-Thevetose)	$C_7H_{14}O_5$	116	$+84 \to +33$
46	D-Glucose, 6-sulfonic acid, 6-deoxy- (6-Sulfoquinovose)	$C_6H_{12}O_8S$	173–174	$+87[3]$
47	D-Glucose, 3-O-methyl-	$C_7H_{14}O_6$	162–167	$+98 \to +59.5$
48	α-L-Glucose	$C_6H_{12}O_6$	141–143	$-95.5 \to -51.4$
49	L-Glucose, 6-deoxy-3-O-methyl- (L-Thevetose)	$C_7H_{14}O_5$	126–129	-36.9 ± 2
50	D-Gulose, 6-deoxy-	$C_6H_{12}O_5$	....	....
51	Hexose, 2-deoxy-D-*arabino*-[4]	$C_6H_{12}O_5$	148	$+46.6 \,[18°]$
52	Hexose, 2,6-dideoxy-3-O-methyl-D-*arabino*- (D-Oleandrose)	$C_7H_{14}O_4$	....	-11
53	Hexose, 3,6-dideoxy-D-*arabino*- (Tyvelose)	$C_6H_{12}O_4$	....	$+24 \pm 2$
54	Hexose, 2,6-dideoxy-3-O-methyl-L-*arabino*- (L-Oleandrose)	$C_7H_{14}O_4$	62–63	$+11.9 \pm 2.5$
55	Hexose, 3,6-dideoxy-L-*arabino*- (Ascarylose)	$C_6H_{12}O_4$	....	-24 ± 2
56	Hexose, 2,6-dideoxy-3-O-methyl-D-*lyxo*- (Diginose)	$C_7H_{14}O_4$	90–92	$+56 \pm 4$
57	Hexose, 2,6-dideoxy-L-*lyxo*- (L-Fucose, 2-deoxy-)	$C_6H_{12}O_4$	103–106	-61.6
58	Hexose, 2,6-dideoxy-3-O-methyl-L-*lyxo*-	$C_7H_{14}O_4$	78–85	-65
59	Hexose, 2,6-dideoxy-D-*ribo*- (Digitoxose; D-Altrose, 2,6-dideoxy-)	$C_6H_{12}O_4$	110	$+46.4$

Table 79 (continued)
CARBOHYDRATES: PART I. NATURAL MONOSACCHARIDES: ALDOSES AND KETOSES

Substance (Synonym)	Chemical Formula	Melting Point °C	Specific Rotation [α]$_D$
(A)	(B)	(C)	(D)
Aldoses (Con't)			
60 Hexose, 2,6-dideoxy-3-*O*-methyl-D-*ribo*- (Cymarose)	$C_7H_{14}O_4$	93	+52
61 Hexose, 3,6-dideoxy-D-*ribo*- (Paratose)	$C_6H_{12}O_4$	....	+10 ± 2 (*c* 0.9)
62 Hexose, 4,6-dideoxy-3-*O*-methyl-D-*ribo*- (D-Gulose, 4,6-dideoxy-3-*O*-methyl-; Chalcose)	$C_7H_{14}O_4$	96–99	+120 → +76
63 Hexose, 2,6-dideoxy-D-*xylo*- (Boivinose)	$C_6H_{12}O_4$	96–98	−3.9 → +3.9
64 Hexose, 2,6-dideoxy-3-*O*-methyl-D-*xylo*- (Sarmentose)	$C_7H_{14}O_4$	78–79	+12 → +15.8
65 Hexose, 3,6-dideoxy-D-*xylo*- (Abequose)	$C_6H_{12}O_4$	....	−3.2 ± 0.6
66 Hexose, 2,6-dideoxy-3-*C*-methyl-L-*xylo*- (Mycarose)	$C_7H_{14}O_4$	129–129	−31.1
67 Hexose, 2,6-dideoxy-3-*C*-methyl-3-*O*-methyl-L-*xylo*-(Cladinose)	$C_8H_{16}O_4$	oil, b.p. 120–132 (0.25 mm)	−23.1
68 Hexose, 3,6-dideoxy-L-*xylo*- (Colitose)	$C_6H_{12}O_4$	....	+4 (H_2O); −51 ± 2 (CH_3OH)
69 D-Idose[5]	$C_6H_{12}O_6$	....	....
70 L-Idose, 1,6-anhydro-	$C_6H_{10}O_5$	....	....
71 α-D-Mannose	$C_6H_{12}O_6$	133	+29.3 → +14.5
72 β-D-Mannose	$C_6H_{12}O_6$	132	−16.3 → +14.5
73 D-Mannose, 6-deoxy- (D-Rhamnose)	$C_6H_{12}O_5$	86–90	−7.0
74 α-L-Mannose, 6-deoxy-monohydrate (L-Rhamnose)	$C_6H_{14}O_6$	93–94	−8.6 → +8.2
75 β-L-Mannose, 6-deoxy-	$C_6H_{12}O_5$	123–125	+38.4 → +8.9
76 L-Mannose, 6-deoxy-2-*O*-methyl-	$C_7H_{14}O_5$	....	....
77 L-Mannose, 6-deoxy-3-*O*-methyl- (L-Acofriose)	$C_7H_{14}O_5$	114–115	+30 [18°]
78 L-Mannose, 6-deoxy-2,4-di-*O*-methyl-	$C_8H_{16}O_5$	82	−19 [16°]
79 L-Mannose, 6-deoxy-5-*C*-methyl-4-*O*-methyl-(Noviose)	$C_8H_{16}O_5$	128–130	+19.9 (50% C_2H_5OH)
80 Rhodinose (a 2,3,6-trideoxyhexose)	$C_6H_{12}O_3$	....	−11 ± 1.6
81 D-Talose	$C_6H_{12}O_6$	128–132	+16.9
82 D-Talose, 6-deoxy- (D-Talomethylose)	$C_6H_{12}O_5$	129–131	+20.6
83 L-Talose, 6-deoxy- (L-Talomethylose)	$C_6H_{12}O_5$	116–118	−19.5 ± 2 [18°]
84 L-Talose, 6-deoxy-2-*O*-methyl- (L-Acovenose)	$C_7H_{14}O_5$	....	−19.4
85 Heptose, D-*glycero*-D-*galacto*-	$C_7H_{14}O_7$	139–140	+47 → +64 (*c* 0.5)
86 Heptose, D-*glycero*-D-*manno*-	$C_7H_{14}O_7$	....	....
87 Heptose, D-*glycero*-L-*manno*-	$C_7H_{14}O_7$	....	....
Ketoses			
88 Dihydroxyacetone	$C_3H_6O_3$	80 (dimer)	None
89 Tetrulose, L-*glycero*-[8] (L-Erythrulose; Ketoerythritol; L-Threulose)	$C_4H_8O_4$	Syrup	+12
90 Pentulose, D-*erythro*- (Adonose; D-Ribulose)	$C_5H_{10}O_5$	Syrup	+16.6 [27°]
91 Pentulose, L-*erythro*- (L-Ribulose)	$C_5H_{10}O_5$	....	−16.6
92 Pentulose, D-*threo*- (D-Xylulose)	$C_5H_{10}O_5$	....	−33
93 Pentulose, 5-deoxy-D-*threo*-	$C_5H_{10}O_4$	....	−5 ± 1 (CH_3OH)

Table 79 (continued)
CARBOHYDRATES: PART I. NATURAL MONOSACCHARIDES: ALDOSES AND KETOSES

Substance (Synonym)	Chemical Formula	Melting Point °C	Specific Rotation $[\alpha]_D$
(A)	(B)	(C)	(D)
Ketoses (Con't)			
94 Pentulose, L-*threo*- (L-Xylulose; L-Lyxulose; Xyloketose)	$C_5H_{10}O_5$	Syrup	+33.1
95 Hexulose, β-D-*arabino*-(β-D-Fructose; Levulose)	$C_6H_{12}O_6$	102–104[7]	−133.5 → −92
96 Hexulose, 6-deoxy-D-*arabino*- (D-Rhamnulose)	$C_6H_{12}O_5$	....	−13 ± 2
97 Hexulose, D-*lyxo*- (D-Tagatose)	$C_6H_{12}O_6$	131–132	+2.7 → −4, −5
98 5-Hexulose, D-*lyxo*-	$C_6H_{12}O_6$	158	−86.6
99 Hexulose, 6-deoxy-L-*lyxo*- (L-Fuculose)	$C_6H_{12}O_5$	....	
100 Hexulose, D-*ribo*- (D-Psicose)	$C_6H_{12}O_6$	Amorphous	+4.7
101 Hexulose, L-*xylo*- (L-Sorbose)	$C_6H_{12}O_6$	159–161	−43.1
102 Hexulose, 6-deoxy-L-*xylo*-	$C_6H_{12}O_5$	88	−25 ± 2 (c 0.7)
103 Heptulose, D-*altro*- (Sedoheptulose; Sedoheptose)	$C_7H_{14}O_7$	Amorphous	+2.5 (c 10)
104 Heptulose·hemihydrate, L-*galacto*- (Perseulose)	$C_7H_{14}O_7 \cdot \frac{1}{2}H_2O$	110–115	−90 → −80
105 Heptulose, L-*gulo*-	$C_7H_{14}O_7$	....	−28
106 Heptulose, D-*ido*-	$C_7H_{14}O_7$	172	−34 ± 8 (c 0.3)
107 Heptulose, D-*manno*- (Mannoketoheptose; D-Mannotagatoheptose)	$C_7H_{14}O_7$	152	+29.4
108 Heptulose, D-*talo*-	$C_7H_{14}O_7$	....	....
109 Octulose, D-*glycero*-L-*galacto*-	$C_8H_{16}O_8$	....	−57, −43.4 → −13.4
110 Octulose, D-*glycero*-D-*manno*-	$C_8H_{16}O_8$	....	+20 (CH_3OH)

[1] Original melting point. [2] Melting point after four-months' storage. [3] As a methyl glycoside cyclohexylamine salt. [4] Included because of speculations concerning it in biological processes. [5] Either D-idose or L-altrose is in the polysaccharide varianose. [6] Early literature refers to this as D-erythrose. [7] The $\cdot\frac{1}{2}H_2O$ and $\cdot 2H_2O$ forms also exist.

Note: All data are for crystalline substances, unless otherwise specified. Selection of substances was restricted to natural carbohydrates found free (or in chemical combination and released on hydrolysis) and to biological oxidation products of the natural carbohydrates. The nomenclature conforms with that of the British-American report as published in the *Journal of Organic Chemistry,* 28:281 (1963). Substances have been arranged alphabetically under the name of the parent sugar within groups formulated according to increasing carbon content (excluding carbon in substituents), with synonymous common names in parentheses. **Melting Point:** b.p. = boiling point; d. = decomposes; s. = sinters. **Specific Rotation** was determined in water at concentrations of 1 to 5 g per 100 m*l* of solution and at 20° to 25°C, unless otherwise specified; other temperatures or wavelengths are shown in brackets; c = grams solute per 100 m*l* of solution.

From Weast, R. C., Ed., *CRC Handbook of Chemistry and Physics,* 61st ed., CRC Press, Boca Raton, Fla., 1980, C-716. With permission.

Table 80
CARBOHYDRATES: PART II. NATURAL MONOSACCHARIDES: AMINO SUGARS

	Substance (Synonym)	Chemical Formula	Melting Point °C	Specific Rotation $[\alpha]_D$
	(A)	(B)	(C)	(D)
	Aldosamines			
1	D-Ribose, 3-amino-3-deoxy-	$C_5H_{11}NO_4$	158–158.5 d.	−24.6 (hydrochloride)
2	D-Galactose, 2-amino-2-deoxy- (Galactosamine; Chondrosamine)	$C_6H_{13}NO_5$	185	+121 → +80 (hydrochloride)
3	α-L-Galactose, 2-amino-2,6-dideoxy- (L-Fucosamine)	$C_6H_{13}NO_4$	192–193 d.	−119 → −92 [27°] (hydrochloride)
4	α-D-Glucose, 2-amino-2-deoxy- (Glucosamine; Chitosamine)	$C_6H_{13}NO_5$	88	+100 → +47.5
5	β-D-Glucose, 2-amino-2-deoxy-	$C_6H_{13}NO_5$	110–111	+28 → +47.5
6	D-Glucose, 3-amino-3-deoxy- (Kanosamine)	$C_6H_{13}NO_5$	128 d.	+19 [14°]
7	D-Glucose, 6-amino-6-deoxy-	$C_6H_{13}NO_5$	161–162 d.	+23 → +50.1 (hydrochloride)
8	D-Glucose, 2,6-diamino-2,6-dideoxy- (Neosamine C)	$C_6H_{14}N_2O_4$	>230	+61.5 (dihydrochloride)
9	D-Glucose, 3,6-dideoxy-3-dimethylamino- (Mycaminose)	$C_8H_{17}NO_4$	115–116	+31 (hydrochloride)
10	D-Glucose, 4,6-dideoxy-4-dimethylamino-	$C_8H_{17}NO_4$	192–193	+45.5 (hydrochloride)
11	L-Glucose, 2-deoxy-2-methylamino-	$C_7H_{15}NO_5$	130–132	−64
12	D-Gulose, 2-amino-1,6-anhydro-2-deoxy-	$C_6H_{11}NO_4$	250–260 d.	+41 ± 2 (hydrochloride)
13	D-Gulose, 2-amino-2-deoxy-	$C_6H_{13}NO_5$	152–162 d.	+5.6 → −18.7 (hydrochloride)
14	Hexose, 3,4,6-trideoxy-3-dimethylamino-D-*xylo*- (Desosamine; Picrocine)	$C_8H_{17}NO_3$	189–191 d.	+49.5 (*c* 10) (hydrochloride)
15	Hexose, a 4-acetamido-2-amino-2,4,6-trideoxy-	$C_8H_{16}N_2O_4$	216–219	+115 → +94 [26°] (*c* 0.05)
16	Hexose, an amino-deoxy-3-*O*-carboxyethyl-	$C_9H_{17}NO_7$	. . . .	. . . .
17	Hexose, a 2,6-diamino-2,6-dideoxy- (Neosamine B; Paramose)	$C_6H_{14}N_2O_4$	135–150 d.	+17.5 (*c* 0.9 (hydrochloride)
18	Hexose, a 3-dimethylamino-2,3,6-trideoxy- (Rhodosamine)	$C_8H_{17}NO_3$	. . . .	. . . .
19	D-Mannose, 2-amino-2-deoxy- (Mannosamine)	$C_6H_{13}NO_5$	142 d.	−4.3 (*c* 9) (hydrochloride)
20	D-Mannose, 3-amino-3,6-dideoxy- (Mycosamine)	$C_6H_{13}NO_4$	162	−11.5 (hydrochloride)
21	D-Talose, 2-amino-2-deoxy- (Talosamine)	$C_6H_{13}NO_5$	151–153	+3.4 → −5.7 (*c* 0.9) (hydrochloride)
22	L-Talose, 2-amino-2,6-dideoxy- (Pneumosamine)	$C_6H_{13}NO_4$	162–163	+6.9 → +10.4 (hydrochloride)
	Ketosamines			
23	Pentulose, 1-(*o*-carboxyanilino)-1-deoxy-D-*erythro*-	$C_{12}H_{14}NO_6$	. . . .	. . . .
24	Hexulose, 1-(*o*-carboxyanilino)-1-deoxy-D-*arabino*-	$C_{13}H_{16}NO_7$	. . . .	. . . .
25	Hexulose, 5-amino-5-deoxy-L-*xylo*-	$C_6H_{13}NO_5$	174–176	−62
26	Hexulose, 6-deoxy-6-(*N*-methylacetamido)-L-*xylo*-	$C_9H_{17}NO_6$	. . . .	. . . .

From Weast, R. C., Ed., *CRC Handbook of Chemistry and Physics*, 61st ed., CRC Press, Boca Raton, Fla., 1980, C-719. With permission.

Table 81
CARBOHYDRATES: PART III. NATURAL ALDITOLS AND INOSITOLS
(WITH INOSOSES AND INOSAMINES)

Substance (Synonym)	Chemical Formula	Melting Point °C	Specific Rotation $[\alpha]_D$
(A)	(B)	(C)	(D)
Alditols			
1 Glycerol	$C_3H_8O_3$	20	None
2 Glycerol, 1-deoxy- (1,2-Propane-diol)[1]	$C_3H_8O_2$	Oil, b.p. 188–189	None (racemic)
3 Erythritol	$C_4H_{10}O_4$	118–120	None (meso)
4 Erythritol, 1,4-dideoxy- (2,3-Butylene-glycol)	$C_4H_{10}O_2$	25, 34	None (meso)
5 D-Threitol, 1,4-dideoxy-	$C_4H_{10}O_2$	19	−13.0
6 L-Threitol, 1,4-dideoxy-	$C_4H_{10}O_2$	....	+10.2
7 DL-Threitol, 1,4-dideoxy-	$C_4H_{10}O_2$	7.6	None (racemic)
8 D-Arabinitol	$C_5H_{12}O_5$	103	+7.82 (c 8, borax solution)
9 L-Arabinitol	$C_5H_{12}O_5$	101–102	−32 (c 0.4, 5% molybdate)
10 Ribitol (Adomitol)	$C_5H_{12}O_5$	102	None (meso)
11 Galactitol (Dulcitol)	$C_6H_{14}O_6$	186–188	None (meso)
12 D-Glucitol (Sorbitol)	$C_6H_{14}O_6$	112	−1.8 [15°]
13 D-Glucitol, 1,5-anhydro- (Polygalitol)	$C_6H_{12}O_5$	140–141	+42.4
14 L-Iditol	$C_6H_{14}O_6$	73.5	−3.5 (c 10)
15 D-Mannitol	$C_6H_{14}O_6$	166	−0.21
16 D-Mannitol, 1,5-anhydro- (Styracitol)	$C_6H_{12}O_5$	157	−49.9
17 Heptitol, D-*glycero*-D-*galacto*- (Heptitol, L-*glycero*-D-*manno*-; Perseitol)	$C_7H_{16}O_7$	183–185, 188	−1.1
18 Heptitol, D-*glycero*-D-*gluco*- (Heptitol, L-*glycero*-D-*talo*-; β-Sedoheptitol)	$C_7H_{16}O_7$	131–132	+46 (5% NH₄ molybdate)
19 Heptitol, D-*glycero*-D-*manno*- (Heptitol, D-*glycero*-D-*talo*-; Volemitol)	$C_7H_{16}O_7$	153	+2.65
20 Octitol, D-*erythro*-D-*galacto*-	$C_8H_{18}O_8 \cdot H_2O$	169–170	−11 (5% NH₄ molybdate)
Inositols			
21 Betitol (a dideoxy inositol)	$C_6H_{12}O_4$	224	....
22 Bioinosose (*scyllo*-Inosose; *myo*-Inosose-2; a deoxy keto inositol)	$C_6H_{10}O_6$	198–200	None (meso)
23 *h*-Bornesitol (a *myo*-inositol mono-methyl ether)	$C_7H_{14}O_6$	200	+31.6
24 *l*-Bornesitol (a *myo*-inositol mono-methyl ether)	$C_7H_{14}O_6$	205–206	−32.1
25 Conduritol (a 2,3-dehydro-2,3-di-deoxyinositol)	$C_6H_{10}O_4$	142–143	None (meso)
26 Cordycepic acid (a tetrahydroxycyclo-hexanecarboxylic acid)[2]	$C_7H_{12}O_6$	....	....
27 Dambonitol (a *myo*-inositol dimethyl ether)	$C_8H_{16}O_6$	206	None (meso)
28 DL-Inositol	$C_6H_{12}O_6$	253	None (racemic)
29 *d*-Inositol	$C_6H_{12}O_6$	....	+60
30 *l*-Inositol	$C_6H_{12}O_6$	240	−65
31 Laminitol (a C-methyl *myo*-inositol)	$C_7H_{14}O_6$	266–269	−3
32 Liriodendritol (a *myo*-inositol dimethyl ether)	$C_8H_{16}O_6$	224	−25
33 *muco*-Inositol monomethyl ether	$C_7H_{14}O_6$	322–325	....
34 *myo*-Inositol (*meso*-Inositol)	$C_6H_{12}O_6$	217–218	None (meso)
35 *d-myo*-Inosose-1 (a deoxy keto inositol)	$C_6H_{10}O_6$	138–139	+19.6
36 Mytilitol (a C-methyl *scyllo*-inositol)	$C_7H_{14}O_6$	259	None (meso)

Table 81 (continued)
CARBOHYDRATES: PART III. NATURAL ALDITOLS AND INOSITOLS
(WITH INOSOSES AND INOSAMINES)

Substance (Synonym)	Chemical Formula	Melting Point °C	Specific Rotation $[\alpha]_D$
(A)	(B)	(C)	(D)
Inositols (Con't)			
37 neo-Inosamine-2 (a deoxy amino inositol)	$C_6H_{13}O_5N$	239–241 d.	None (meso)
38 d-Ononitol (a myo-inositol mono-methyl ether)	$C_7H_{14}O_6$	172	+6.6
39 h-Pinitol (a dextro-inositol monomethyl ether)	$C_7H_{14}O_6$	186	+65.5
40 l-Pinitol (a levo-inositol monomethyl ether)	$C_7H_{14}O_6$	186	−65
41 l-Quebrachitol (a levo-inositol mono-methyl ether)	$C_7H_{14}O_6$	190–191	−80.2 [28°]
42 d-Quercitol (a deoxy dextro-inositol)	$C_6H_{12}O_5$	235	+24.2
43 d-Quinic acid (a trideoxy carboxy dextro-inositol)	$C_7H_{12}O_6$	164	+44 (c 10)
44 l-Quinic acid (a trideoxy carboxy levo-inositol)	$C_7H_{12}O_6$	162	−42.1
45 Quinic acid, 5-dehydro-	$C_7H_{10}O_6$	140–142 (138 s.)	−82.4 [28°]
46 Scyllitol (scyllo-Inositol; Cocositol)	$C_6H_{12}O_6$	352–353	None (meso)
47 Sequoyitol (a myo-inositol monomethyl ether)	$C_7H_{14}O_6$	234–235	None (meso)
48 Shikimic acid (a 3,4-anhydro-quinic acid)	$C_7H_{10}O_5$	183–184	−200 [16°]
49 Shikimic acid, 5-dehydro-	$C_7H_8O_5$	150–152	−57.5 [28°] (EtOH)
50 Streptamine (2,4-diaminodideoxy-scyllitol)	$C_6H_{14}O_4N_2$	88, 210–250 d.	None (meso)
51 Streptamine, 2-deoxy-	$C_6H_{14}O_3N_2$	....	None (meso)
52 Streptadine (1,3-Dideoxy-1,3-diguani-dino-scyllitol)	$C_8H_{18}N_6O_4$	....	None (meso)
53 Viburnitol (a deoxy levo-inositol)[3]	$C_6H_{12}O_5$	174	−73.9

[1] The 1-phosphate ester of this diol is said to occur in brain tissue and sea-urchin eggs. [2] Strong evidence that cordycepic acid is really D-mannitol. [3] Not an enantiomorph of d-quercitol; other isomeric relationship is involved.

From Weast, R. C., Ed., *CRC Handbook of Chemistry and Physics,* 61st ed., CRC Press, Boca Raton, Fla., 1980, C-720. With permission.

Table 82

CARBOHYDRATES: PART IV. NATURAL ALDONIC, URONIC, AND ALDARIC ACIDS

Substance (Synonym)	Chemical Formula	Melting Point °C	Specific Rotation $[\alpha]_D$
(A)	(B)	(C)	(D)
Aldonic Acids			
1 D-Glyceric acid	$C_3H_6O_4$	Gum	Dextro
2 L-Glyceric acid	$C_3H_6O_4$	Gum	Levo
3 D-Arabinonic acid	$C_5H_{10}O_6$	114–116	+10.5 (c 6)
4 L-Arabinonic acid	$C_5H_{10}O_6$	118–119	$-9.6 \rightarrow -41.7^1$
5 L-Arabinonic-1,4-lactone	$C_5H_8O_5$	97–99	−72
6 D-Ribonic acid	$C_5H_{10}O_6$	112–113	−17.0
7 D-Xylonic acid	$C_5H_{10}O_6$	....	$-2.9 \rightarrow +20.1^1$
8 L-Xylonic acid	$C_5H_{10}O_6$	....	-91.8^1
9 D-Altronic acid	$C_6H_{12}O_7$	....	$+11.5 \rightarrow +24.8^1$ (Ca salt, N HCl)
10 D-Galactonic acid	$C_6H_{12}O_7$	122	$-11.2 \rightarrow +57.6^1$
11 D-Gluconic acid	$C_6H_{12}O_7$	130–132 (110–112 s.)	$-6.7 \rightarrow +11.9^1$
12 L-Gulonic acid	$C_6H_{12}O_7$	Exists only in soln.	[ca. 0°]
13 Hexsonic acid, 2-deoxy-D-*arabino*-	$C_6H_{12}O_6$	93–95	+68 (lactone)
14 2-Hexulosonic acid, D-*arabino*-	$C_6H_{10}O_7$	....	−81.7 (Na salt)
15 2-Hexulosonic acid, 3-deoxy-D-*erythro*-	$C_6H_{10}O_6$	....	−29.2 (c 6, Ca salt)
16 2-Hexulosonic acid, D-*lyxo*-	$C_6H_{10}O_7$	169	−5
17 5-Hexulosonic acid, D-*arabino*-	$C_6H_{10}O_7$	108–109	....
18 5-Hexulosonic acid, D-*xylo*-	$C_6H_{10}O_7$	....	−14.5
19 D-Mannonic acid	$C_6H_{12}O_7$	....	−15.6
20 D-Gluconic acid, O-β-D-galactopyranosyl- (1 → 4)- (Lactobionic acid)	$C_{12}H_{22}O_{12}$	....	+25.1 (Ca salt)
Uronic Acids			
21 L-Lyxuronic acid	$C_5H_8O_6$	....	....
22 β-D-Galacturonic acid	$C_6H_{10}O_7$	160	$+27 \rightarrow +55.6$
23 α-D-Galacturonic acid·monohydrate	$C_6H_{12}O_8$	159–160 (110–115 s.)	$+97.9 \rightarrow +50.9$
24 D-Galacturonic acid, 2-amino-2-deoxy-	$C_6H_{11}O_6N$	160 d.	+84.5 (pH 2 HCl)
25 β-D-Glucuronic acid	$C_6H_{10}O_7$	156	$+11.7 \rightarrow +36.3$
26 D-Glucuronic acid, 2-amino-2-deoxy-	$C_6H_{11}O_6N$	120–172 d.	+55
27 D-Glucuronic acid, 3-O-methyl-	$C_7H_{12}O_7$	Syrup	+6
28 L-Guluronic acid	$C_6H_{10}O_7$	....	....
29 L-Iduronic acid	$C_6H_{10}O_7$	....	+30
30 β-D-Mannuronic acid	$C_6H_{10}O_7$	165–167	$-47.9 \rightarrow -23.9$
31 α-D-Mannuronic acid·monohydrate	$C_6H_{12}O_8$	110 s., 120–130 d.	$+16 \rightarrow -6.1$ (c 6.8)
Aldaric Acids			
32 D-Tartaric acid	$C_4H_6O_6$	170	−15
33 L-Tartaric acid	$C_4H_6O_6$	170	+15 [15°]
34 L-Malic acid	$C_4H_6O_5$	100	−2.3 (c 8.4)

[1] Equilibrates with the lactone.

From Weast, R. C., Ed., *CRC Handbook of Chemistry and Physics,* 61st ed., CRC Press, Boca Raton, Fla., 1980, C-722. With permission.

Table 83

PHYSICAL AND CHEMICAL PROPERTIES OF AMINO ACIDS

Name	Abbreviation*	Formula and mol.wt.	Structure	Elementary composition (%) C	H	N	Solubility (grams per 100 g water at 25°C)	Specific rotation Temperature (°C)	Concentration**	Solvent	$[\alpha]_D$	Special properties / Organism for microbiological assay	Special occurrence and biological function (for references see page 358)
α-Alanine (α-amino-propionic acid)	Ala.	$C_3H_7NO_2$ 89.10	$CH_3 \cdot CH(NH_2) \cdot COOH$	40.44	7.92	15.72	16.72	25 25 20	2.06 10.00 1.78	6-N HCl Water 3-N NaOH	+ 13.70 + 2.41 + 3.0	*Leuconostoc (Lactobacillus) citrovorum* 8081	
Arginine (α-amino-δ-guanido-n-valeric acid)	Arg.	$C_6H_{14}N_4O_2$ 174.21		41.37	8.10	32.16	15	23.3 20 20	1.65 3.48 0.87	6-N HCl Water 0.5-N NaOH	+ 27.58 + 12.5 + 11.8	Basic. Gives SAKAGUCHI color reaction with α-naphthol and sodium hypohalite. *Streptococcus faecalis* 9790; *Leuconostoc citrovorum*	Intermediate in ornithine cycle of urea synthesis (see page 445) and in creatine synthesis (see page 440)
Asparagine (aspartic acid β-monoamide; α-amino-β-carbamyl-propionic acid)	Asp-NH₂	$C_4H_8N_2O_3$ 132.12	$NH_2CO \cdot CH_2 \cdot CH(NH_2) \cdot COOH$	36.37	6.11	21.21	2.46	20 20 20	2.24 1.41 11.23	3.4-N HCl Water 2.5-N NaOH	+ 34.26 − 5.30 − 6.35	Hydrolyzed by hot acid or specific enzymes to give NH_3 + aspartic acid	Found in free state in many plant tissues, especially etiolated seedlings
Aspartic acid (aminosuccinic acid)	Asp.	$C_4H_7NO_4$ 133.11	$HOOC \cdot CH_2 \cdot CH(NH_2) \cdot COOH$	36.09	5.30	10.52	0.50	24 18 18	2.02 1.33 1.33	6-N HCl Water 3-N NaOH	+ 24.6 + 4.7 − 1.7	Acidic. Yields 2 moles of CO_2 and 1 mole of NH_3 in ninhydrin reaction. *Leuconostoc mesenteroides* P-60, 8042	Involved in transformation of citrulline to arginine (see page 445), and in biosynthesis of purines and pyrimidines (see pages 438 and 442)
Cysteine (α-amino-β-thiolpropionic acid)	CySH	$C_3H_7NO_2S$ 121.16	$HS \cdot CH_2 \cdot CH(NH_2) \cdot COOH$ Sulfur 26.46	29.73	5.82	11.56	very soluble	26	12.1	1-N HCl	+ 7.6	Readily autoxidized in neutral or basic solution to cystine	Interconvertible with cystine by oxido-reduction (see page 409). Component of glutathione (see page 441). Some aromatic compounds are excreted in urine as derivatives of N-acetylcysteine (mercapturic acids) (see pages 445 and 448)
Cystine (di-(α-amino-propionic)-β-disulfide)	Cys.	$C_6H_{12}N_2O_4S_2$ 240.31	$SCH_2 \cdot CH(NH_2) \cdot COOH$ $SCH_2 \cdot CH(NH_2) \cdot COOH$ Sulfur 26.68	29.99	5.03	11.66	0.011	24 18.5	1.0 0.4	1-N HCl 0.2-N NaOH	−214.4 − 70.0	Readily reduced to cysteine. *Leuconostoc mesenteroides* P-60, 8042; *Lactobacillus arabinosus*	Occurs abundantly in hair, keratin and insulin. The disulfide bond links together different polypeptide chains or different parts of the same polypeptide chain within the protein molecule
3,5-Diiodo-tyrosine***		$C_9H_9NO_3I_2$ 433.00	$HO-\text{(benzene ring, I, I)}-CH_2 \cdot CH(NH_2) \cdot COOH$ Iodine 58.62	24.97	2.10	3.23	0.062	20 20	5.08 4.41	1.1-N HCl 3.4-N NH₄OH	+ 2.89 + 2.27		Occurrence confined to protein of thyroid gland[9] (thyroid hormone; see pages 443 and 479)

Table 83 (continued)

PHYSICAL AND CHEMICAL PROPERTIES OF AMINO ACIDS

Name	Abbreviation*	Formula and mol.wt.	Structure	Elementary composition (%)			Solubility (grams per 100 g water at 25°C)	Specific rotation				Special properties / Organism for microbiological assay	Special occurrence and biological function (for references see page 358)
				C	H	N		Temperature (°C)	Concentration**	Solvent	$[\alpha]_D$		
Glutamic acid (α-aminoglutaric acid)	Glu.	$C_5H_9NO_4$ 147.14	$HOOC(CH_2)_2CH(NH_2)COOH$	40.81	6.17	9.52	0.843	22.4 / 18 / 18	1.00 / 1.47 / 1.47	6-N HCl / Water / 1-N NaOH	+31.2 / +11.5 / +10.96	Acidic. On boiling in solution over a wide pH range (4–10) is cyclized to pyrrolidone-carboxylic acid. Leuconostoc mesenteroides P-60, 8042; Lactobacillus arabinosus	Component of glutathione (see page 441) and of the folic acid vitamins (see page 461). Present in high concentration in tissues. More readily dehydrogenated in animal tissue than any other amino acid, and also more reactive in enzymic transamination reactions
Glutamine (glutamic acid β-monoamide; α-amino-γ-carbamyl-butyric acid)	Glu-NH₂	$C_5H_{10}N_2O_3$ 146.15	$NH_2CO(CH_2)_2CH(NH_2)COOH$	41.09	6.90	19.17	3.6 (at 18°C)	22	3.6	Water	+5.0	On heating in solution to ca. 100°C at near neutrality cyclizes to ammonium salt of pyrrolidonecarboxylic acid. Amide group reacts with nitrous acid in presence of acetic acid to release N_2. Hydrolyzed by specific enzyme to ammonium glutamate	Occurs in the free state in animal tissues and many plants, e.g. sugar beet. Phenylactic acid is excreted by man as phenacetylglutamine (see pages 445 and 448)
Glycine (aminoacetic acid)	Gly.	$C_2H_5NO_2$ 75.07	NH_2CH_2COOH	32.00	6.72	18.66	24.99					Optically inactive. Gives green color with o-phthalaldehyde. Leuconostoc mesenteroides P-60, 8042	In many animals benzoic acid is excreted as benzoylglycine (hippuric acid) (see pages 445 and 448). Component of glutathione (see page 441). Metabolite in synthesis of creatine, porphyrins, purines, serine (see page 437)
Histidine (β-[4-imidazole]-propionic acid)	His.	$C_6H_9N_3O_2$ 155.16	$HC{=}C{-}CH_2CH(NH_2)COOH$, $HN\ \ C\ H$ (imidazole)	46.44	5.85	27.09	4.29	25 / 25 / 20	1.00–4.05 / 0.75–3.77 / 0.77	6.1-N HCl / Water / 0.5-N NaOH	+13.34 / -38.95 / -10.9	Basic. Gives biuret test; couples with diazotized sulfanilic acid to give intense red color (PAULY reaction). Leuconostoc mesenteroides P-60, 8042	Decarboxylated to histamine. Constituent of carnosine (β-alanylhistidine) found in muscle
δ-Hydroxylysine (α,ε-diamino-δ-hydroxy-n-caproic acid)		$C_6H_{14}N_2O_3$ 162.19	$NH_2CH_2CH(OH)(CH_2)_2CH(NH_2)COOH$	44.43	8.70	17.28		25	2.0	6-N HCl	+14.5	Reacts with periodate to yield formaldehyde and ammonia	Has been found as protein constituent only in collagen and gelatin. Phosphate ester occurs naturally[4]
Hydroxyproline (γ-hydroxy-pyrrolidine-α-carboxylic acid)	Hypro.	$C_5H_9NO_3$ 131.14	(pyrrolidine ring structure)	45.80	6.92	10.68	36.11	20 / 22.5 / 20	1.31 / 1.00 / 0.65	1-N HCl / Water / 0.5-N NaOH	-47.3 / -75.2 / -70.6	Has no primary α-amino group and therefore differs in many respects from primary amino acids, e.g. with nitrous acid does not release nitrogen. Oxidized by hypochlorite to give hydroxypyrroline. Decarboxylated by ninhydrin to give reddish color below pH 4.4, yellowish color at higher pH. Forms bright blue condensation product with isatin	Present only in gelatin and collagen

Name	Abbr.	Formula / M.W.	Structural formula	C	H	N	Solubility	Temp. °C	—	Solvent	[α]	Occurrence	Remarks
Leucine (α-amino-isocaproic acid)	Leu.	C₆H₁₃NO₂ 131.18	$(CH_3)_2CHCH_2CH(NH_2)COOH$	54.94	9.99	10.68	2.19	25 / 25 / 20	2.00 / 2.00 / 1.31	6-N HCl / Water / 3-N NaOH	+15.20 / −10.57 / +7.6	*Lactobacillus arabinosus* 17-5, 8014; *Lactobacillus helveticus*; *Streptococcus faecalis*; *Leuconostoc mesenteroides* P-60, 8042	
*iso*Leucine (α-amino-β-methyl-n-valeric acid)	Ileu.	C₆H₁₃NO₂ 131.18	$CH_3CH_2CH(CH_3)CH(NH_2)COOH$	54.94	9.99	10.68	2.93 (at 20°C)	20 / 20 / 20	5.09 / 3.10 / 3.34	6.1-N HCl / Water / 0.33-N NaOH	+40.61 / +11.29 / +11.09	*Lactobacillus arabinosus* 17-5, 8014; *Lactobacillus helveticus*; *Streptococcus faecalis*; *Leuconostoc mesenteroides* P-60, 8042	
Lysine (α,ε-diamino-n-caproic acid)	Lys.	C₆H₁₄N₂O₂ 146.19	$NH_2(CH_2)_4CH(NH_2)COOH$	49.29	9.65	19.16	very soluble	23 / 20	2.00 / 6.50	6-N HCl / Water	+25.9 / +14.6	*Streptococcus faecalis* 9790; *Leuconostoc mesenteroides* P-60, 8042	Basic. Can be precipitated with phosphotungstic acid. Dry heating of proteins containing lysine causes an apparent loss of lysine
Methionine (α-amino-γ-methylthiol-n-butyric acid)	Met.	C₅H₁₁NO₂S 149.22	$CH_3SCH_2CH_2CH(NH_2)COOH$	40.24	7.43	9.39 Sulfur 21.49	3.35 (for DL-acid)	20 / 25	5.00 / 0.80	3-N HCl / Water	+23.40 / −8.11	*Leuconostoc mesenteroides* P-60, 8042; *Lactobacillus fermenti* 36 (for DL-acid)	Provides the sulfur atom for cysteine biosynthesis (see Cystathionine, Table 8, page 360). Carrier of "active" methyl groups
3-Monoiodo-tyrosine		C₉H₁₀NO₃I 307.10	(structure) $—CH_2CH(NH_2)COOH$ with I, HO	35.15	3.27	4.57 Iodine 41.2		20	5.0	1-N HCl	−4.4		Occurrence confined to protein of thyroid gland? (thyroid hormone; see pages 443 and 447?)
Norleucine (α-aminocaproic acid)		C₆H₁₃NO₂ 131.18	$CH_3(CH_2)_3CH(NH_2)COOH$	54.94	9.99	10.68	1.15 (at 18°C)	20	4.3	6-N HCl	+21.3		Has not been proved to be a protein constituent[4]
Phenylalanine (α-amino-β-phenyl-propionic acid)	Phe.	C₉H₁₁NO₂ 165.20	$—CH_2CH(NH_2)COOH$	65.44	6.71	8.48	2.965	20	1.93	Water	−35.14	*Leuconostoc mesenteroides* P-60, 8042	Can be converted to tyrosine in the human body (see page 410)
Proline (pyrrolidine-α-carboxylic acid)	Pro.	C₅H₉NO₂ 115.14	(structure)	52.16	7.88	12.17	162.3	20 / 23.4 / 20	0.57 / 1.00 / 2.42	0.5-N HCl / Water / 0.6-N KOH	−52.6 / −85.0 / −93.0	*Leuconostoc mesenteroides* P-60, 8042; *Lactobacillus brevis*	Neutral. Soluble in alcohol. Chemically very similar to hydroxyproline
Serine (α-amino-β-hydroxy-propionic acid)	Ser.	C₃H₇NO₃ 105.10	$HOCH_2CH(NH_2)COOH$	34.28	6.71	13.33	5.023 (for DL-acid)	25 / 20	9.34 / 10.41	1-N HCl / Water	+14.95 / −6.83	*Leuconostoc mesenteroides* P-60, 8042; *Lactobacillus delbrueckii* LD 5; *Lactobacillus helveticus*	Split by periodate or lead tetraacetate to formaldehyde and glyoxylic acid. Gives biuret test. Acid hydrolysis of protein leads to partial destruction. Most of the serine in phosphoproteins (vitellin, casein) occurs in form of phosphoserine.[6] Phosphatidylserine is a component of some phospholipids

Table 83 (continued)
PHYSICAL AND CHEMICAL PROPERTIES OF AMINO ACIDS

Name	Abbreviation*	Formula and mol.wt.	Structure	Elementary composition (%) C	H	N	Solubility (grams per 100 g water at 25°C)	Specific rotation Temperature (°C)	Concentration**	Solvent	$[\alpha]_D$	Special properties Organism for microbiological assay	Special occurrence and biological function (for references see page 358)
Threonine (β-methylserine; α-amino-β-hydroxy-butyric acid)	Thr.	$C_4H_9NO_3$ 119.12	$CH_3CH(OH)CH(NH_2)COOH$	40.33	7.62	11.76		26	1.63 (g solute per 100 g solution)	Water	− 9.1	Some properties similar to those of serine. Oxidized by periodate to acetaldehyde and glyoxylic acid. *Streptococcus faecalis* 9790; *Leuconostoc mesenteroides* P-60, 8042	Phosphothreonine has been found in casein hydrolysates[7]
Thyroxine (3,5,3',5'-tetraiodo-thyronine)		$C_{15}H_{11}NO_4I_4$ 776.90		23.17	1.43	1.80 Iodine 65.35	0.001		3 (g solute per 100 g solution)	0.13-N NaOH in 70% ethanol	− 4.4		Occurrence confined to protein of thyroid gland[3] (thyroid hormone; see pages 443 and 479)
3,5,3'-Triiodo-thyronine		$C_{15}H_{12}NO_4I_3$ 651.00		27.6	1.8	2.2 Iodine 58.5		29.5	4.75 (as hydrochloride)	1-N HCl-ethanol (1:2)	+ 21.5		Occurrence confined to protein of thyroid gland[3] (thyroid hormone; see pages 443 and 479)
Tryptophan (α-amino-β-[3-indolyl]-propionic acid)	Try.	$C_{11}H_{12}N_2O_2$ 204.23		64.69	5.93	13.72	1.14	20 22.7 20	1.02 1.00 2.42	0.5-N HCl Water 0.5-N NaOH	+ 2.4 − 31.5 + 6.17	Destroyed on prolonged heating in hot acid. Gives MILLON reaction (see Tyrosine, below) and FOLIN test. Gives color reaction with p-dimethyl-aminobenzaldehyde and nitrous acid (EHRLICH test). *Lactobacillus arabinosus* 17-5, 8014; *Streptococcus faecalis*	
Tyrosine (α-amino-β-[p-hydroxyphenyl]-propionic acid)	Tyr.	$C_9H_{11}NO_3$ 181.20		59.66	6.12	7.73	0.045	20 18	4.40 0.90	6.3-N HCl 3.0-N NaOH	− 8.64 − 13.2	Reacts with Hg salts and nitrous acid to give red color (MILLON reaction). Other less specific color tests are: reaction with α-nitroso-β-naphthol; reaction with FOLIN phenol reagent. *Leuconostoc mesenteroides* P-60, 8042	Precursor of thyroxine[3] (see page 443), adrenaline (see page 442), melanin (see page 443)
Valine (α-amino-isovaleric acid)	Val.	$C_5H_{11}NO_2$ 117.15	$(CH_3)_2CHCH(NH_2)COOH$	51.26	9.46	11.96	8.85	20 20	3.4 3.58	6-N HCl Water	+ 28.8 + 6.42	*Lactobacillus arabinosus* 17-5, 8014; *Lactobacillus helveticus*; *Streptococcus faecalis*	

Name	Formula and mol. wt.	Structure	Elementary composition (%)			Solubility (grams per 100 g water at 25°C)	Specific rotation				Special properties	Occurrence and biological function (*for references see page 360*)
			C	H	N		Temperature (°C)	Concentration (g per 100 ml solution)	Solvent	$[\alpha]_D$		
β-Alanine (β-amino-propionic acid)	$C_3H_7NO_2$ 89.10	$NH_2 \cdot CH_2 \cdot CH_2 \cdot COOH$	40.44	7.92	15.72	very soluble						Breakdown product of pyrimidines (see page 413). Occurs as constituent of pantothenic acid, coenzyme A, carnosine and anserine
α-Aminoadipic acid	$C_6H_{11}NO_4$ 161.16	$HOOC \cdot (CH_2)_3 \cdot CH(NH_2) \cdot COOH$	44.71	6.88	8.69	0.22 (at 20°C)					Decomposes on heating to α-piperidone-α'-carboxylic acid	Intermediate in breakdown of lysine (see page 410)
α-Amino-n-butyric acid	$C_4H_9NO_2$ 103.12	$CH_3 \cdot CH_2 \cdot CH(NH_2) \cdot COOH$	46.60	8.80	13.59	28 (for DL-acid)	20	5.46	Water	+ 7.86		Found in brain preparations[2]. Occurs as constituent of tripeptide ophthalmic acid in lens tissue[2]
γ-Amino-n-butyric acid	$C_4H_9NO_2$ 103.12	$NH_2 \cdot (CH_2)_3 \cdot COOH$	46.60	8.80	13.59							Found in brain[3], lung and heart[3] preparations
β-Aminoisobutyric acid	$C_4H_9NO_2$ 103.12	$NH_2 \cdot CH_2 \cdot CH(CH_3) \cdot COOH$	46.60	8.80	13.59							Breakdown product of thymine (see page 413)
δ-Aminolevulic acid (γ-keto-δ-amino-n-valeric acid)	$C_5H_9NO_3$ 131.14	$NH_2 \cdot CH_2 \cdot CO \cdot CH_2 \cdot CH_2 \cdot COOH$	45.82	6.90	10.71						Reduces BENEDICT's solution in the cold. Split by periodate to formaldehyde and succinic acid	Intermediate in porphyrin biosynthesis (see page 440)
Carbamylaspartic acid (ureidosuccinic acid)	$C_5H_8N_2O_5$ 176.14	$HOOC \cdot CH_2 \cdot CH(COOH) \cdot NH \cdot CO \cdot NH_2$	34.10	4.58	15.91	0.4 (at 20°C)	25		Water (Ba salt)	+ 24.1		Intermediate in biosynthesis of pyrimidines from aspartic acid in mammals and bacteria (see page 442)
Citrulline (α-amino-δ-ureido-n-valeric acid)	$C_6H_{13}N_3O_3$ 175.19	$NH_2 \cdot CO \cdot NH \cdot (CH_2)_3 \cdot CH(NH_2) \cdot COOH$	41.13	7.47	23.98		21.23 / 21.23	5.00 / 5.00	0.3-N HCl / Water	+ 17.9 / + 3.5	Converted to ornithine by alkaline hydrolysis	Intermediate in ornithine cycle of urea synthesis (see page 445 and 446)
Creatine (methyl-glycocyamine)	$C_4H_9N_3O_2$ 131.14	$\mathrm{HN{=}C(NH_2){-}N(CH_3) \cdot CH_2 \cdot COOH}$	36.64	6.92	32.05	1.35 (at 18°C)					Weakly basic. Gives creatinine on heating with dilute acids	Cell constituent. Creatine phosphate acts as store of "high energy" phosphate in vertebrate muscle (see page 440)
Creatinine (1-methylglyco-cyamidine)	$C_4H_7N_3O$ 113.12	$\mathrm{HN{=}C{-}N(CH_3) \cdot CH_2 \cdot CO}$ (ring, N–H)	42.48	6.24	37.15	8.7 (at 16°C)					Strongly basic	Present in urine

Table 83 (continued)
PHYSICAL AND CHEMICAL PROPERTIES OF AMINO ACIDS

Name	Formula and mol. wt.	Structure	Elementary composition (%)			Solubility (grams per 100 g water at 25°C)	Specific rotation				Special properties	Occurrence and biological function (for references see page 360)
			C	H	N		Temperature (°C)	Concentration (g per 100 ml solution)	Solvent	$[\alpha]_D$		
Cystathionine	$C_7H_{14}N_2O_4S$ 222.27	HOOC CH(NH₂) CH₂ S CH₂ CH₂ CH(NH₂) COOH	37.82	6.35	12.60		22	1.0	1-N HCl	+ 23.7		Intermediate in transsulfuration of methionine with serine (see page 409)
			Sulfur 14.45									
Cysteic acid	$C_3H_7NO_5S$ 169.16	HO₃S CH₂ CH(NH₂) COOH	21.27	4.16	8.26		28	6.0	Water	+ 7.8		Intermediate in formation of taurine, a bile constituent, from cysteine
			Sulfur 18.90									
Ergothioneine (betaine of thiolhistidine)	$C_9H_{15}N_3O_2S$ 229.31	(structure with N(CH₃)₃ imidazole and SH)	47.14	6.54	18.35		21	5.0	Water	+116.0	Basic. Stable to alkalis; thiol group is readily oxidized to sulfate under acid conditions. Gives reddish purple color with diazotized sulfanilic acid and alkali	Occurs in erythrocytes, liver, kidney, other tissues, and in urine and semen[4]. Constituent of ergot
			Sulfur 13.95									
Glycocyamine (guanidinoacetic acid)	$C_3H_7N_3O_2$ 117.11	HN=C(NH₂) NH CH₂ COOH	30.77	6.02	35.88	slightly soluble						In urine. Formed in kidney from arginine and glycine. Precursor of creatine and creatinine (see page 440)
Homoserine (α-amino-γ-hydroxy-n-butyric acid)	$C_4H_9NO_3$ 119.12	HO CH₂ CH₂ CH(NH₂) COOH	40.33	7.62	11.76		24–26	0.25	5-N HCl	+ 18.3		Intermediate in methionine metabolism (see page 409)
Ornithine (2,5-diamino-n-valeric acid)	$C_5H_{12}N_2O_2$ 132.17	NH₂ (CH₂)₃ CH(NH₂) COOH	45.42	9.15	21.19	very soluble	20	0.84	0.45-N HCl	+ 14.1	Formed from arginine by alkaline hydrolysis	Intermediate in ornithine cycle of urea synthesis (see page 445). Benzoic acid is excreted by the fowl as N,N'-dibenzoylornithine
Taurine (2-aminoethanesulfonic acid)	$C_2H_7NO_3S$ 125.15	NH₂ CH₂ CH₂ SO₃H	19.19	5.64	11.19	8.78 (at 20°C)						In muscle tissue of invertebrates. Formed in mammalian liver from cysteine (see page 441). Component of taurocholic acid (bile acid)
			Sulfur 25.62									

* Used in the description of amino acid sequences in polypeptide and protein molecules³.

** As grams per 100 ml solution, unless otherwise stated.

*** For chromatographic separation from other iodinated amino acids and estimation see BLOCK and WEISS².

REFERENCES

1. Walker, D. M., *Biochem. J.*, 52, 679 (1952).
2. Waley, S. G., *Biochem. J.*, 64, 715 (1956).
3. Udenfriend, S., *J. biol. Chem.*, 187, 65 (1950).
4. For review of recent work see Bell, D. J., *Ann. Rep. Progr. Chem.*, 52, 285 (1955).

REFERENCES

1. Brand and Edsall, *Ann. Rev. Biochem.*, 16, 223 (1947).
2. Block and Weiss, *Amino Acid Handbook*, Springfield, Ill., 1956, page 37; Roche et al., in Glick, D. (Ed.), *Methods of Biochemical Analysis*, vol. I, New York, 1954, page 243.
3. Roche and Michel, *Ann. Rev. Biochem.*, 23, 481 (1954).
4. Astrup et al., *Acta physiol scand*, 24, 202 (1952).
5. Consden et al., *Biochem. J.*, 39, 251 (1945).
6. Agren et al., *Acta chem. scand*, 5, 324 (1951).
7. de Verdier, C.-H., *Nature*, 170, 804 (1952).

† Page numbers in the last column refer to the original reference.

From Diem, K., Ed., *Documenta Geigy Scientific Tables*, 6th ed., Geigy Pharmaceuticals, Ardsley, N.Y., 1962, 355. With permission.

Table 84
CHEMICAL AND PHYSICAL PROPERTIES OF LIPIDS

Name of acid	Formula and mol. wt.	Structure	Physical properties	Remarks
		Saturated straight-chain monocarboxylic acids		
Formic acid (methanoic acid)	CH_2O_2 46.03	H COOH	m.p. 8.6° C b.p. 100.8° C d^{20} 1.220 n_D^{20} 1.3714	Occurs in human urine and many plant materials
Acetic acid (ethanoic acid)	$C_2H_4O_2$ 60.05	CH_3 COOH	m.p. 16.5° C b.p. 118.1° C d^{20} 1.0492 n_D^{25} 1.36976	Present in most biological materials. Formed from ethanol by many species of aerobic bacteria and from pentoses by some anaerobic species
Propionic acid (propanoic acid)	$C_3H_6O_2$ 74.08	CH_3 CH_2 COOH	m.p. − 22° C b.p. 140.9° C d^{20} 0.992 $n_D^{19.9}$ 1.38736	Formed by bacterial decomposition of carbohydrates
n-Butyric acid (butanoic acid)	$C_4H_8O_2$ 88.11	CH_3 $(CH_2)_2$ COOH	m.p. − 7.9° C b.p. 163° C d^{20} 0.9587 n_D^{20} 1.39906	Occurs in traces in many fats
n-Valeric acid (pentanoic acid)	$C_5H_{10}O_2$ 102.14	CH_3 $(CH_2)_3$ COOH	m.p. − 34.5° C b.p. 186.4° C d^{20} 0.9387 n_D^{20} 1.4086	
Caproic acid (hexoic acid, hexanoic acid)	$C_6H_{12}O_2$ 116.16	CH_3 $(CH_2)_4$ COOH	m.p. − 4° C b.p. 205° C · d^{20} 0.929 n_D^{20} 1.41635	Occurs in traces in many fats
Enanthic acid (heptanoic acid)	$C_7H_{14}O_2$ 130.19	CH_3 $(CH_2)_5$ COOH	m.p. − 7.46° C b.p. 223° C d^{14} 0.9216 $n_D^{19.8}$ 1.42162	
Caprylic acid (octanoic acid)	$C_8H_{16}O_2$ 144.22	CH_3 $(CH_2)_6$ COOH	m.p. 16° C b.p. 239° C d^{20} 0.9088 n_D^{20} 1.4285	Constituent of many fats
Pelargonic acid (nonanoic acid)	$C_9H_{18}O_2$ 158.24	CH_3 $(CH_2)_7$ COOH	m.p. 12.3° C b.p. 254° C d^{20} 0.9055 n_D^{20} 1.4330	Occurs in oil of rue, Japan wax, fusel oils, leaves of *Pelargonium roseum*
Capric acid (decanoic acid)	$C_{10}H_{20}O_2$ 172.27	CH_3 $(CH_2)_8$ COOH	m.p. 31.3° C b.p. 269° C d^{40} 0.8858 n_D^{40} 1.42855	Component of many animal and vegetable fats
Undecylic acid (hendecanoic acid)	$C_{11}H_{22}O_2$ 186.30	CH_3 $(CH_2)_9$ COOH	m.p. 28.5° C b.p. 284° C d^{20} 0.8905 $n_D^{45.3}$ 1.4294	Found in *Pseudomonas*
Lauric acid (dodecanoic acid)	$C_{12}H_{24}O_2$ 200.32	CH_3 $(CH_2)_{10}$ COOH	m.p. 43.5° C b.p. 225° C/100 d^{20} 0.883 $n_D^{62.1}$ 1.4183	Major component of vegetable fats (esp. laurel). In smaller quantities in depot fat of animals, milk fat, fishliver oils
Tridecylic acid (tridecanoic acid)	$C_{13}H_{26}O_2$ 214.35	CH_3 $(CH_2)_{11}$ COOH	m.p. 51° C b.p. 312.4° C n_D^{20} 1.4249	Occurs in animal fats in very small traces
Myristic acid (tetradecanoic acid)	$C_{14}H_{28}O_2$ 228.38	CH_3 $(CH_2)_{12}$ COOH	m.p. 54.4° C b.p. 250.5° C/100 d^{54} 0.8622 n_D^{60} 1.4308	Component of almost all animal fats (1–5%) and vegetable fats, esp. milk fat, fish oils, palm oil, nutmegs

Table 84 (continued)
CHEMICAL AND PHYSICAL PROPERTIES OF LIPIDS

Name of acid	Formula and mol. wt.	Structure	Physical properties	Remarks
Pentadecylic acid (pentadecanoic acid)	$C_{15}H_{30}O_2$ 242.41	$CH_3(CH_2)_{13}COOH$	m.p. 52.1° C b.p. 257° C/100 d^{80} 0.8423 n_D^{70} 1.4270	Occurs in traces in animal fats, esp. liver fats
Palmitic acid (hexadecanoic acid)	$C_{16}H_{32}O_2$ 256.43	$CH_3(CH_2)_{14}COOH$	m.p. 62.85° C b.p. 268.5° C/100 d^{70} 0.8487 $n_D^{70.8}$ 1.4273	Widely distributed in nature. Present in almost all fats
Margaric acid (heptadecanoic acid)	$C_{17}H_{34}O_2$ 270.46	$CH_3(CH_2)_{15}COOH$	m.p. 62° C b.p. 227° C/100 d^{80} 0.8579 n_D^{70} 1.4319	Occurs in traces in mutton fat
Stearic acid (octadecanoic acid)	$C_{18}H_{36}O_2$ 284.49	$CH_3(CH_2)_{16}COOH$	m.p. 69.6° C b.p. 298° C/100 d^{20} 0.9408 $n_D^{80.3}$ 1.4299	Found abundantly in important edible fats. Also occurs in vegetable fats
Nondecylic acid (nonadecanoic acid)	$C_{19}H_{38}O_2$ 298.51	$CH_3(CH_2)_{17}COOH$	m.p. 68–69° C b.p. 298° C/100	
Arachidic acid (eicosanoic acid)	$C_{20}H_{40}O_2$ 312.54	$CH_3(CH_2)_{18}COOH$	m.p. 75.4° C b.p. 328° C d^{100} 0.824 n_D^{100} 1.4250	Occurs in traces in many seed and animal fats
Heneicosanoic acid	$C_{21}H_{42}O_2$ 326.57	$CH_3(CH_2)_{19}COOH$	m.p. 75.1° C	
Behenic acid (docosanoic acid)	$C_{22}H_{44}O_2$ 340.59	$CH_3(CH_2)_{20}COOH$	m.p. 80° C b.p. 306° C/60 d^{100} 0.8221 n_D^{100} 1.4270	Present in traces in animal fats and seed fats. Constitutes 50% of the spleen cerebrosides in GAUCHER's disease (see page 388)
Tricosanoic acid	$C_{23}H_{46}O_2$ 354.62	$CH_3(CH_2)_{21}COOH$	m.p. 79.1° C	
Lignoceric acid (tetracosanoic acid)	$C_{24}H_{48}O_2$ 368.65	$CH_3(CH_2)_{22}COOH$	m.p. 84.2° C d^{20} 0.8207 n_D^{100} 1.4287	Component of sphingomyelins and of kerasin (spleen cerebroside in GAUCHER's disease; see page 388). Also found in some vegetable fats and bacterial and insect waxes
Pentacosanoic acid	$C_{25}H_{50}O_2$ 382.68	$CH_3(CH_2)_{23}COOH$	m.p. 83° C	
Cerotic acid (hexacosanoic acid)	$C_{26}H_{52}O_2$ 396.70	$CH_3(CH_2)_{24}COOH$	m.p. 87.7° C d^{100} 0.8198 n_D^{100} 1.4301	Occurs free and combined. In Chinese wax (cetyl ester), beeswax, wool fat
Heptacosanoic acid	$C_{27}H_{54}O_2$ 410.73	$CH_3(CH_2)_{25}COOH$	m.p. 87.6° C	
Montanic acid (octacosanoic acid)	$C_{28}H_{56}O_2$ 424.76	$CH_3(CH_2)_{26}COOH$	m.p. 90.9° C d^{100} 0.8191 n_D^{100} 1.4313	Component of montan wax, beeswax, Chinese wax
Nonacosanoic acid	$C_{29}H_{58}O_2$ 438.78	$CH_3(CH_2)_{27}COOH$	m.p. 90.3° C	
Melissic acid (triacontanoic acid)	$C_{30}H_{60}O_2$ 452.81	$CH_3(CH_2)_{28}COOH$	m.p. 93.6° C n_D^{100} 1.4323	Occurs in beeswax

Table 84 (continued)
CHEMICAL AND PHYSICAL PROPERTIES OF LIPIDS

Name of acid	Formula and mol. wt.	Structure	Physical properties	Remarks
Lacceroic acid (dotriacontanoic acid)	$C_{32}H_{64}O_2$ 480.86	$CH_3 \cdot (CH_2)_{30} \cdot COOH$	m.p. 96.2° C	Occurs in stick-lac wax (from *Tachardia lacca*) and other natural waxes
		Unsaturated (mono-olefinic) straight-chain monocarboxylic acids		
Acrylic acid (propenoic acid)	$C_3H_4O_2$ 72.07	$CH_2 = CH \cdot COOH$	m.p. 13° C b.p. 141° C d^{16} 1.062 n_D^{20} 1.4224	
trans-(α-)Crotonic acid (*trans*-butenoic acid)	$C_4H_6O_2$ 86.09	$CH_3 \cdot CH$ $HC \cdot COOH$	m.p. 72° C b.p. 189° C d^{72} 0.973 $n_D^{79.7}$ 1.4228	Constituent of croton oil (from *Croton tiglium* seeds)
iso-(β-)Crotonic acid (*cis*-butenoic acid)	$C_4H_6O_2$ 86.09	$HC \cdot CH_3$ $HC \cdot COOH$	m.p. 15.5° C b.p. 169° C d^{15} 1.0312 n_D^{20} 1.4457	Readily isomerizes to the *trans*-acid
Δ^2-Hexenoic acid	$C_6H_{10}O_2$ 114.15	$CH_3 \cdot (CH_2)_2 \cdot CH = CH \cdot COOH$	m.p. 32° C b.p. 217° C d^{40} 0.9627 n_D^{40} 1.4601	Occurs in Japanese peppermint oil
Δ^4-Decenoic acid (obtusilic acid)	$C_{10}H_{18}O_2$ 170.25	$CH_3 \cdot (CH_2)_4 \cdot CH = CH \cdot (CH_2)_2 \cdot COOH$	b.p. 148—150° C/13 d^{20} 0.9197 n_D^{20} 1.4497	Occurs in seed fat of *Lindera obtusiloba*
Δ^9-Decenoic acid	$C_{10}H_{18}O_2$ 170.25	$CH_2 = CH \cdot (CH_2)_7 \cdot COOH$	b.p. 143—148° C/15 d^{15} 0.9238 n_D^{20} 1.4488	Occurs in butter and milk fats and in sperm head oil
Δ^4-Dodecenoic acid (linderic acid)	$C_{12}H_{22}O_2$ 198.31	$CH_3 \cdot (CH_2)_6 \cdot CH = CH \cdot (CH_2)_2 \cdot COOH$	m.p. 1—1.3° C b.p. 170—172° C/13 d^{20} 0.9081 n_D^{20} 1.4529	Occurs in various seed oils, e.g. *Lindera obtusiloba*
Δ^5-Dodecenoic acid (lauroleic acid)	$C_{12}H_{22}O_2$ 198.31	$CH_3 \cdot (CH_2)_5 \cdot CH = CH \cdot (CH_2)_3 \cdot COOH$	d^{15} 0.9130 n_D^{15} 1.4535	Occurs in sperm blubber and head oil
Δ^9-Dodecenoic acid	$C_{12}H_{22}O_2$ 198.31	$CH_3 \cdot CH_2 \cdot CH = CH \cdot (CH_2)_7 \cdot COOH$		Occurs in fat of cow's milk
Δ^4-Tetradecenoic acid (tsuzuic acid)	$C_{14}H_{26}O_2$ 226.36	$CH_3 \cdot (CH_2)_8 \cdot CH = CH \cdot (CH_2)_2 \cdot COOH$	m.p. 18—18.5° C b.p. 185—188° C/13 d^{20} 0.9024 n_D^{20} 1.4557	Occurs in various tropical plant oils, esp. tsuzu oil
Δ^5-Tetradecenoic acid (physeteric acid)	$C_{14}H_{26}O_2$ 226.36	$CH_3 \cdot (CH_2)_7 \cdot CH = CH \cdot (CH_2)_3 \cdot COOH$	d^{20} 0.9046 n_D^{20} 1.4552	Occurs in whale blubber and sardine oil
Δ^9-Tetradecenoic acid (myristoleic acid)	$C_{14}H_{26}O_2$ 226.36	$CH_3 \cdot (CH_2)_3 \cdot CH = CH \cdot (CH_2)_7 \cdot COOH$	d^{20} 0.9018 n_D^{20} 1.4549	Occurs in milk fat and depot and liver fat of many animals
Δ^9-Hexadecenoic acid (palmitoleic acid)	$C_{16}H_{30}O_2$ 254.42	$CH_3 \cdot (CH_2)_4 \cdot CH = CH \cdot (CH_2)_7 \cdot COOH$	m.p. 1° C b.p. 218—220° C d^{15} 0.9003	Widely distributed. In marine oils (15—20% of total fatty acids), in depot and milk fat of animals, vegetable oils and fats

Table 84 (continued)
CHEMICAL AND PHYSICAL PROPERTIES OF LIPIDS

Name of acid	Formula and mol. wt.	Structure	Physical properties	Remarks* (for references see page 395)
cis-Δ^6-Octadecenoic acid (petroselinic acid)	$C_{18}H_{34}O_2$ 282.47	$CH_3(CH_2)_{10}CH=CH(CH_2)_4COOH$	m.p. 32–33° C b.p. 208–210° C/10 d^{25} 0.8824 n_D^{47} 1.4535	Occurs in seeds of aromatic plants (parsley, celery, etc.) and in some umbellate fats
Oleic acid (*cis*-Δ^9-octadecenoic acid)	$C_{18}H_{34}O_2$ 282.47	$CH(CH_2)_7COOH$ $\parallel$ $CH(CH_2)_7CH_3$	m.p. 13° C b.p. 286° C/100 d^{20} 0.895 n_D^{20} 1.45823	Most abundant of the unsaturated fatty acids. Present in nearly all natural fats (one-third of fatty acids of cow's milk; phosphatides). Occurs in traces in human urine
Elaidic acid (*trans*-Δ^9-octadecenoic acid)	$C_{18}H_{34}O_2$ 282.47	$CH_3(CH_2)_7CH$ $\parallel$ $CH(CH_2)_7COOH$	m.p. 44–45° C b.p. 288° C/100 d^{79} 0.851 n_D^{70} 1.4405	Formed by isomerization of oleic acid
trans-Vaccenic acid (*trans*-Δ^{11}-octadecenoic acid)	$C_{18}H_{34}O_2$ 282.47	$CH_3(CH_2)_5CH$ $\parallel$ $CH(CH_2)_9COOH$	m.p. 42.5° C d^{70} 0.8560 n_D^{60} 1.4439	Occurs in many animal fats and vegetable oils
cis-Vaccenic acid (*cis*-Δ^{11}-octadecenoic acid)	$C_{18}H_{34}O_2$ 282.47	$CH(CH_2)_9COOH$ $\parallel$ $CH(CH_2)_5CH_3$	m.p.12.4–13° C	Has been shown to be the hemolytic acid occurring in plasma and various animal tissues[1]. Also present in *Lactobacillus* species[2]
Δ^{12}-Octadecenoic acid	$C_{18}H_{34}O_2$ 282.47	$CH_3(CH_2)_4CH=CH(CH_2)_{10}COOH$		Occurs in partially hydrogenated peanut oil
Gadoleic acid (Δ^9-eicosenoic acid)	$C_{20}H_{38}O_2$ 310.52	$CH_3(CH_2)_9CH=CH(CH_2)_7COOH$	m.p. 24.5° C	*Cis*- and *trans*-forms. In many fish and marine animal oils, in vegetable oils, in brain phosphatides
Δ^{11}-Eicosenoic acid	$C_{20}H_{38}O_2$ 310.52	$CH_3(CH_2)_7CH=CH(CH_2)_9COOH$	m.p. *cis* 22° C *trans* 52–53° C	Principal acid of jojoba nuts ("goat nuts"), also in seed oil of *Conringia orientalis,* rape and mustard seed oils, fish oils
Cetoleic acid (Δ^{11}-docosenoic acid)	$C_{22}H_{42}O_2$ 338.58	$CH_3(CH_2)_9CH=CH(CH_2)_9COOH$	m.p. 32–33° C	Occurs in various marine oils
Erucic acid (*cis*-Δ^{13}-docosenoic acid)	$C_{22}H_{42}O_2$ 338.58	$CH(CH_2)_{11}COOH$ $\parallel$ $CH(CH_2)_7CH_3$	m.p. 33.5° C b.p. 281° C/30 d^{55} 0.860 n_D^{54} 1.4480	Occurs in seed oils, esp. rapeseed oil
Brassidic acid (*trans*-Δ^{13}-docosenoic acid)	$C_{22}H_{42}O_2$ 338.58	$CH_3(CH_2)_7CH$ $\parallel$ $CH(CH_2)_{11}COOH$	m.p. 61.5° C b.p. 282° C/30 d^{57} 0.8585 n_D^{100} 1.4347	Formed by isomerization of erucic acid
Selacholeic acid (nervonic acid, *cis*-Δ^{15}-tetracosenoic acid)	$C_{24}H_{46}O_2$ 366.63	$CH(CH_2)_{13}COOH$ $\parallel$ $CH(CH_2)_7CH_3$	m.p. 40.5–41° C n_D^{46} 1.4535	Occurs in shark and ray liver oils, in brain cerebrosides (nervone) and sphingomyelins[3]
Ximenic acid (Δ^{17}-hexacosenoic acid)	$C_{26}H_{50}O_2$ 394.69	$CH_3(CH_2)_7CH=CH(CH_2)_{15}COOH$	m.p. 45° C	Occurs in *Ximenia americana* (tallow-wood). A hexacosenoic acid is found with nervonic acid in brain cerebrosides

Table 84 (continued)
CHEMICAL AND PHYSICAL PROPERTIES OF LIPIDS

Name of acid	Formula and mol. wt.	Structure	Physical properties	Remarks
Unsaturated (polyolefinic) straight-chain monocarboxylic acids				
Sorbic acid ($\Delta^{2,4}$-hexadienoic acid)	$C_6H_8O_2$ 112.13	$CH_3 \cdot CH=CH \cdot CH=CH \cdot COOH$	m.p. 134.5° C b.p. 228° C (decomp.)	Occurs as lactone in oil of unripe mountain ash berries
Linoleic acid (*cis-cis*-$\Delta^{9,12}$-octa-decadienoic acid)	$C_{18}H_{32}O_2$ 280.45	$CH_3 \cdot (CH_2)_4 \cdot CH$ $\|$ $CH \cdot CH_2 \cdot CH$ $\|$ $CH \cdot (CH_2)_7 \cdot COOH$	m.p. $-11(-5)°$ C b.p. 230° C/16 d^{20} 0.9025 n_D^{20} 1.4699	Widely distributed in plants, esp. in linseed, hemp and cottonseed oils. Also in depot fat of animals (component of phosphatides)
Hiragonic acid ($\Delta^{6,10,14}$-hexa-decatrienoic acid)	$C_{16}H_{26}O_2$ 250.38	$CH \cdot (CH_2)_2 \cdot CH=CH \cdot CH_3$ $\|$ $CH \cdot (CH_2)_2 \cdot CH=CH \cdot (CH_2)_4 \cdot COOH$	d^{20} 0.9288 n_D^{20} 1.4855	Occurs in sardine oil
α-Eleostearic acid (*cis*-$\Delta^{9,11,13}$-octa-decatrienoic acid)	$C_{18}H_{30}O_2$ 278.44	$CH_3 \cdot (CH_2)_3 \cdot (CH=CH)_3 \cdot (CH_2)_7 \cdot COOH$	m.p. 48° C b.p. 235° C/12 d^{46} 0.8980 n_D^{66} 1.5080	Occurs in vegetable oils, esp. tung oil
β-Eleostearic acid (*trans*-$\Delta^{9,11,13}$-octa-decatrienoic acid)	$C_{18}H_{30}O_2$ 278.44		m.p. 71° C d^{80} 0.8839 n_D^{74} 1.5000	Formed from α-eleostearic acid by action of light, heat and chemical reagents
Linolenic acid ($\Delta^{9,12,15}$-octa-decatrienoic acid)	$C_{18}H_{30}O_2$ 278.44	$CH \cdot CH_2 \cdot CH=CH \cdot CH_2 \cdot CH_3$ $\|$ $CH \cdot CH_2 \cdot CH=CH \cdot (CH_2)_7 \cdot COOH$	m.p. -11.2 to $-11°$ C b.p. 230–232° C/17 d^{20} 0.9046 n_D^{20} 1.4780	Occurs in many vegetable oils, esp. drying oils such as linseed oil. Also in traces in animal fats (phosphatides)
Stearidonic acid (moroctic acid, $\Delta^{4,8,12,15}$-octa-decatetraenoic acid)	$C_{18}H_{28}O_2$ 276.42	$CH_3 \cdot CH_2 \cdot CH=CH \cdot CH_2 \cdot CH$ $CH \cdot (CH_2)_2 \cdot CH=CH \cdot (CH_2)_2 \cdot CH$ $\|$ $CH \cdot (CH_2)_2 \cdot COOH$	d^{20} 0.9297 n_D^{20} 1.4911	Occurs in fish oils. The position of the double bonds is not confirmed
Timnodonic acid ($\Delta^{4,8,12,16,18}$-eicosapentaenoic acid)	$C_{20}H_{30}O_2$ 302.46	$CH_3 \cdot CH=CH \cdot CH_2 \cdot CH=CH$ $\|$ $CH \cdot (CH_2)_2 \cdot CH=CH \cdot CH_2$ $\|$ $CH \cdot (CH_2)_2 \cdot CH=CH \cdot (CH_2)_2 \cdot COOH$		Occurs in sardine oil, cod-liver oil, pilot whale oil and oil from *Squalus sucklei* (spiny dog fish)
Arachidonic acid ($\Delta^{5,8,11,14}$-eicosa-tetraenoic acid)	$C_{20}H_{32}O_2$ 304.48	$CH_3 \cdot (CH_2)_4 \cdot CH=CH \cdot CH_2 \cdot CH$ $\|$ $CH \cdot CH_2 \cdot CH=CH \cdot CH_2 \cdot CH$ $\|$ $CH \cdot (CH_2)_3 \cdot COOH$	m.p. $-49.5°$ C n_D^{20} 1.8482	Occurs in animal fats (liver, phosphatides) and in fish oils
Clupanodonic acid ($\Delta^{4,8,12,16,19}$-doco-sapentaenoic acid)	$C_{22}H_{34}O_2$ 330.51	$CH_3 \cdot CH_2 \cdot CH=CH \cdot (CH_2)_2 \cdot CH$ $\|$ $CH \cdot (CH_2)_2 \cdot CH=CH \cdot CH_2 \cdot CH$ $\|$ $CH \cdot (CH_2)_2 \cdot CH=CH \cdot (CH_2)_2 \cdot COOH$	m.p. $< -78°$ C b.p. 236° C/5 d^{20} 0.9290 n_D^{20} 1.4868	Occurs in fish oils
Nisinic acid ($\Delta^{4,8,12,15,18,21}$-tetracosahexaenoic acid)	$C_{24}H_{36}O_2$ 356.55	$CH_3 \cdot CH_2 \cdot CH=CH \cdot CH_2 \cdot CH=CH \cdot CH_2$ $CH \cdot (CH_2)_2 \cdot CH=CH \cdot CH_2 \cdot CH=CH$ $\|$ $CH \cdot (CH_2)_2 \cdot CH=CH \cdot (CH_2)_2 \cdot COOH$		Occurs in tunny oil
Thynnic acid (Δ^2-hexacosahexa-enoic acid)	$C_{26}H_{40}O_2$ 384.61		d^{20} 0.9433 n_D^{20} 1.5022	Occurs in tunny oil

Table 84 (continued)
CHEMICAL AND PHYSICAL PROPERTIES OF LIPIDS

Name of acid	Formula and mol. wt.	Structure	Physical properties	Remarks
		Unsaturated (acetylenic) straight-chain monocarboxylic acids		
Tariric acid (6-stearolic acid, 6-octadecynoic acid)	$C_{18}H_{32}O_2$ 280.45	$CH_3 \cdot (CH_2)_{10} \cdot C \equiv C \cdot (CH_2)_4 \cdot COOH$	m.p. 50.5° C	Occurs in fat of *Picramnia* spp. (tariri) (bitterbush oil)
Stearolic acid (9-octadecynoic acid)	$C_{18}H_{32}O_2$ 280.45	$CH_3 \cdot (CH_2)_7 \cdot C \equiv C \cdot (CH_2)_7 \cdot COOH$	m.p. 48.5° C b.p. 260° C	Formed by oxidation of oleic or elaidic acid
Behenolic acid (13-docosynoic acid)	$C_{22}H_{40}O_2$ 336.56	$CH_3 \cdot (CH_2)_7 \cdot C \equiv C \cdot (CH_2)_{11} \cdot COOH$	m.p. 57.5° C	Formed by oxidation of erucic or brassidic acid
		Branched-chain monocarboxylic acids		
*iso*Butyric acid (2-methyl-propanoic acid)	$C_4H_8O_2$ 88.11	$\begin{array}{c} CH_3 \\ \diagdown CH \cdot COOH \\ CH_3 \diagup \end{array}$	m.p. − 47° C b.p. 154.4° C d^{20} 0.949 n_D^{20} 1.393	Occurs free in carob beans *(Ceratonia siliqua)*, as ethyl ester in croton oil; also in feces and as product of enzymic breakdown of proteins. Intermediate in metabolism of valine (see page 408)
*iso*Valeric acid (3-methylbutanoic acid)	$C_5H_{10}O_2$ 102.14	$\begin{array}{c} CH_3 \\ \diagdown CH \cdot CH_2 \cdot COOH \\ CH_3 \diagup \end{array}$	m.p. − 51° C b.p. 176.7° C d^{16} 0.937 $n_D^{22.4}$ 1.40178	Occurs in root of valerian, tobacco leaves, volatile oils, depot fat of dolphins and porpoises, as glyceride in human feces. Formed from leucine in bacterial degradation of proteins. Intermediate in metabolism of leucine (see page 408)
Tiglic acid (*cis*-2-methyl-Δ^2-butenoic acid)	$C_5H_8O_2$ 100.12	$\begin{array}{c} CH_3 \cdot CH = C \cdot COOH \\ \quad\quad\quad CH_3 \end{array}$	m.p. 64.5° C b.p. 198.5° C d^{78} 0.964 n_D^{81} 1.4342	Occurs in croton oil (glyceride); in Roman cumin oil (esters), in geranium oils. Intermediate in metabolism of isoleucine (see page 408)
Tuberculostearic acid (D-(−)10-methyl-octadecanoic acid)	$C_{19}H_{38}O_2$ 298.51	$\begin{array}{c} CH_3 \cdot (CH_2)_7 \cdot CH \cdot (CH_2)_8 \cdot COOH \\ \quad\quad\quad CH_3 \end{array}$	m.p. 12.5–12.9 (23.5–25.8)° C b.p. 180° C/0.1 d^{24} 0.8771 n_D^{25} 1.4512 $[\alpha]_D^{19}$ − 0.08°	Occurs free in lipids of tubercle bacilli and *Mycobacterium leprae*[4]
Mycolipenic acid ([+]-2,4,6-trimethyltetracos-2-enoic acid)	$C_{27}H_{52}O_2$ 408.71	$\begin{array}{c} CH_3 \cdot (CH_2)_{17} \cdot CH \cdot CH_2 \cdot CH \cdot CH = C \cdot COOH \\ \quad\quad CH_3 \quad\quad CH_3 \quad CH_3 \end{array}$		The so-called "phthioic acid" of tubercle bacilli has been shown to be a mixture of these acids and a third component[5]
Mycoceranic acid	$C_{32}H_{64}O_2$ 480.87	$\begin{array}{c} CH_3 \cdot (CH_2)_{22} \cdot CH \cdot CH_2 \cdot CH \cdot CH_2 \cdot CH \cdot COOH \\ \quad\quad CH_3 \quad\quad CH_3 \quad CH_3 \end{array}$		
		Hydroxy acids		
Ricinoleic acid (*cis*-12-hydroxy-Δ^9-octadecenoic acid)	$C_{18}H_{34}O_3$ 298.47	$\begin{array}{l} CH_3 \cdot CH_2 \cdot CH(OH) \cdot (CH_2)_5 \cdot CH_3 \\ \quad || \\ CH \cdot (CH_2)_7 \cdot COOH \end{array}$	m.p. 5; 7.7; 16° C (3 forms) b.p. 250° C/15 n_D^{20} 1.4711 $[\alpha]_D^{20}$ + 7.8°	As glyceride, chief constituent of castor oil
Cerebronic acid (phrenosinic acid, 2-hydroxytetra-cosanoic acid)	$C_{24}H_{48}O_3$ 384.65	$CH_3 \cdot (CH_2)_{21} \cdot CH(OH) \cdot COOH$	m.p. 90–93 (102)° C $[\alpha]_D^{22}$ + 3.33°	Component of cerebroside phrenosin (cerebron). The natural product contains 15% of the corresponding hexacosanoic acid[6]
2-Hydroxynervonic acid (2-hydroxy-Δ^{15}-tetracosenoic acid)	$C_{24}H_{46}O_3$ 382.63	$CH_3 \cdot (CH_2)_7 \cdot CH = CH \cdot (CH_2)_{12} \cdot CH(OH) \cdot COOH$	m.p. 65° C $[\alpha]_D^{20}$ + 2.87°	Component of cerebroside hydroxynervone, of which the isomeric Δ^{17}-acid is also a component

References

1) LASER, H., *J. Physiol. (Lond.)*, **110**, 338 (1949); MORTON and TODD, *Biochem. J.*, **47**, 327 (1950). 2) BOUNDS et al., *J. chem. Soc.*, **1954**, 448. 3) HOFMANN and SAX, *J. biol. Chem.*, **205**, 55 (1953). 4) SCHMIDT and SHIRLEY, *J. Amer. chem. Soc.*, **71**, 3804 (1949); LINSTEAD et al., *J. chem.* *Soc.*, **1951**, 1130. 5) POLGÁR, N., *J. chem. Soc.*, **1954**, 1008, 1011; ASSELINEAU et al., *Acta chem. scand.*, **11**, 196 (1957); LEDERER, E., *Angew. Chem.*, **72**, 372 (1960). 6) CHIBNALL et al., *Biochem. J.*, **55**, 707 (1953); MISLOW and BLEICHER, *J. Amer. chem. Soc.*, **76**, 2825 (1954).

From Diem, K., Ed., *Documenta Geigy Scientific Tables,* 6th ed., Geigy Pharmaceuticals, Ardsley, N.Y., 1962, 390. With permission.

Table 85

PHYSICAL AND CHEMICAL PROPERTIES OF STEROID HORMONES

	Name		Biology		Chemistry					
No.	Trivial	Chemical	Main source	Action	No.	Formula	Mol. wt.	Melting point, °C	Identification (O.R.D., etc.)	Abs. max.
1	Hydrocortisone (cortisol)	11β,17α,21-trihydroxy-pregn-4-ene-3,20-dione	Adrenal cortex	Glucocorticoid. Anti-inflammatory. Rx as ester	1	$C_{21}H_{30}O_5$	362.47	217—220	$[\alpha]^{25}D + 150$—$156°$ (dioxane) green fluorescence in H_2SO_4	242 mμ (methanol)
2	Cortisone 11-dehydrohydro-cortisone	17α,21-dihydroxy-pregn-4-ene-3,11,20-trione	Metabolism of (1) synthetic	Glucocorticoid. Anti-inflammatory. Rx as ester. 0.7 times potency of hydrocortisone in rats and humans	2	$C_{21}H_{28}O_5$	360.44	220—224	$[\alpha]^{25}D + 209°$ (ethanol) green fluorescence in H_2SO_4	237 mμ
3	Corticosterone	11β,21-dihydroxy-pregn-4-ene-3,20-dione	Adrenal cortex	Replaces (1) in some species (rats, mice, birds)	3	$C_{21}H_{30}O_4$	346.45	180—182	$[\alpha]^{55}D + 223°$ (ethanol) green fluorescence in H_2SO_4	240 mμ
4	Aldosterone	11β,21-dihydroxy-18-formyl-pregn-4-ene-3, 20-dione	Adrenal cortex (synthetic)	Mineralocorticoid. Potency: about 25 times deoxycorticosterone re NaCl retention	4	$C_{21}H_{28}O_5$	360.45	108—112 (hydrate)	$[\alpha]^{25}D + 161°$ (chloroform)	240 mμ
5	Deoxycorticosterone (DOC)	21-hydroxy-pregn-4-ene-3,20-dione	Adrenal cortex (synthetic)	Mineralocorticoid. Rx as esters	5	$C_{21}H_{30}O_3$	330.45	141—142	$[\alpha]^{22}D + 178°$ (ethanol) no H_2SO_4 fluorescence	240 mμ

6	9α fluorohydrocortisone (fluorinef)	9α-fluoro-11β,17α,21-trihydroxy-pregn-4-ene-3,20-dione	Synthetic only	Mineralocorticoid (and glucocorticoid). NaCl retaining potency equal to aldosterone	$C_{21}H_{29}FO_5$	380.46	260—262 [a]	$[\alpha]^{28}D + 139°$ (ethanol)	239 mμ
7	Prednisolone (Δ-1 hydrocortisone)	11β,17α,21-trihydroxy-pregn-1,4-diene-3,20-dione	Synthetic only	Glucocorticoid. Anti-inflammatory potency: 4 times hydrocortisone in man; 3.5 times in rats.	$C_{21}H_{28}O_5$	360.44	240—241	$[\alpha]^{25}D + 102°$ (dioxane)	242 mμ
8	Methyl prednisolone (6α methyl Δ-1 hydrocortisone)	11β,17α,21-trihydroxy-6α methyl-pregn-1,4-diene-3,20-dione	Synthetic only	Same. Rx as alcohol or ester. Anti-inflammatory potency: 5 times hydrocortisone in man and rats	$C_{22}H_{30}O_5$	374.46	228—237	$[\alpha]^{20}D + 83°$ (dioxane)	243 mμ
9	Dexamethasone (16α methyl-9α-fluoro Δ-1 hydrocortisone)	9α fluoro-11β,17α,21-trihydroxy-16α-methyl-pregn-1,4-diene-3,20-diene	Synthetic only	Same. Anti-inflammatory potency: 30 times hydrocortisone in man; 150 times in rats	$C_{22}H_{29}FO_5$	392.45	262—264	$[\alpha]^{25}D + 73°$ (chloroform) for 21-acetate	239 mμ
10	Triamcinolone (16α-hydroxy-9α fluoro Δ-1 hydrocortisone)	9α fluoro-11β, 16α-17α 21-tetrahydroxy-pregn-1,4-diene-3,20-dione	Synthetic only	Same. Anti-inflammatory potency: 4 times hydrocortisone in man; 3 times in rats	$C_{25}H_{31}FO_8$	478.52	186—188	$[\alpha]^{25}D + 22°$ (chloroform)	239 mμ
11	Estradiol 17β	3β,17β dihydroxy-estr-1,3,5-triene	Ovary	Estrogenic: the principal biological estrogen. Rx as esters	$C_{18}H_{24}O_2$	272.37	173—179	$[\alpha]^{25}D + 76$—8 3° (dioxane)	225, 280 mμ

Table 85 (continued)
PHYSICAL AND CHEMICAL PROPERTIES OF STEROID HORMONES

	Name		Biology				Chemistry			
No.	Trivial	Chemical	Main source	Action	No.	Formula	Mol. wt.	Melting point, °C	Identification (O.R.D., etc.)	Abs. max.
12	Estrone (theelin)	3β hydroxy-estr-1,3,5-triene-17-one	Ovary, placenta	Estrogenic (weak)	12	$C_{18}H_{22}O_2$	270.36	258—262[a]	$[α]^{25}$D+158—168° (dioxane); precipitated by digitonin	283, 285 mμ
13	Estriol	3β,16α,17β,tri-hydroxy-estr-1,3,5-triene	Ovary	Estrogenic (weak)	13	$C_{18}H_{24}O_3$	288.37	282	$[α]^{25}$D+58° (dioxane); precipitated by digitonin	280 mμ
14	Progesterone	Pregn-4-ene-3,20-dione	Ovary, placenta	Progestogen. Maintains secretory endometrium	14	$C_{21}H_{30}O_2$	314.45	121	$[α]^{20}$D+172—182° (dioxane)	240 mμ
15	Pregnanediol	3α,20α,-dihydroxy-pregnane	Metabolism of (14)	Metabolite of progesterone; presence in urine used in pregnancy test	15	$C_{21}H_{36}O_2$	320.50	238	$[α]^{20}$D+27.4° (ethanol) water soluble not precipitated by digitonin	
16	Testosterone	17β-hydroxy-4-androsten-3-one	Testis, synthetic	Androgen; the principal biologic androgen. Rx as esters	16	$C_{19}H_{28}O_2$	288.41	155	$[α]^{24}$D+109° (ethanol)	238 mμ

[a]   Decomposition point.

From Dean, J. A., Ed., *Lange's Handbook of Chemistry*, 12th ed., McGraw-Hill, New York, 1979, 7-420. With permission.

Table 86
MOLECULAR WEIGHTS OF SOME MEDICALLY IMPORTANT COMPOUNDS

Compound	Formula	Mol wt
Acetanilide	$C_6H_5NHCOCH_3$	135.17
Acetophenetidine	$C_6H_4(NHCOCH_3)(OC_2H_5)\text{-}1,4$	179.22
Acetylcholine chloride	$[CH_3COOCH_2CH_2N(CH_3)_3]Cl$	181.67
Acetylsalicylic acid	$C_6H_4(COOH)(OCOCH_3)\text{-}1,2$	180.16
Aconitine	$C_{34}H_{49}O_{11}N$	647.77
Adrenaline	See Epinephrine	
Allyl isothiocyanate	$CH_2{:}CHCH_2NCS$	99.16
Amidopyrine	See Aminopyrine	
Aminophylline	$\{[N(CH_3)CON(CH_3)C{:}CCO\ N{:}CHNH]_2\cdot\ NH_4CH_2NH_2 + 2\,H_2O\}$	456.48
Aminopyrine (Dimethyl-aminoantipyrine)	$N(C_6H_5)N(CH_3)C(CH_3){:}C[N(CH_3)_2]CO$	231.30
Amphetamine	$C_6H_5CH_2CH(NH_2)CH_3$	135.21
Amyl nitrite	$(CH_3)_2CHCH_2CH_2ONO$	117.15
Antimony potassium tartrate (Tartar emetic)	$[KOOC(CHOH)_2\ COOSbO] + \frac12\,H_2O$	333.94
Antipyrine	$N(C_6H_5)N(CH_3)C(CH_3){:}CHCO$	188.23
Apomorphine hydrochloride	$[C_{17}H_{17}O_2N{\cdot}HCl] + \frac12\,H_2O$	312.80
Atropine	$C_{17}H_{23}O_3N$	289.38
Barbital(Barbitone)	$NHCONHCOC(C_2H_5)_2CO$	184.20
Barbital sodium	$NHC(ONa){:}NCOC(C_2H_5)_2CO$	206.18
Bromoform	$CHBr_3$	252.77
Butacaine sulfate	$[H_2NC_6H_4COO(CH_2)_3N(C_4H_9)_2\text{-}1,4]_2{\cdot}H_2SO_4$	710.99
Caffeine	$[N(CH_3)CON(CH_3)C{:}CCO\ N{:}CHN(CH_3)] + H_2O$	212.22
Calcium gluconate	$[CH_2OH(CHOH)_4COO]_2Ca] + H_2O$	448.40
– glycerophosphate	$[C_3H_5(OH)_2OPO_3Ca] + H_2O$	228.16
– lactate	$[(CH_3CH(OH)COO)_2Ca] + 5\,H_2O$	308.31
– mandelate	$[C_6H_5CH(OH)COO]_2Ca$	342.37
Camphor	$C(CH_2)_2CH_2CH_2CHC(CH_3)_2\ CH_2CO$	152.24
N-Carbamyl-arsanilic acid	$C_6H_4[AsO(OH)_2][NHCONH_2]\text{-}1,4$	260.08
Carbromal	$(C_2H_5)_2CBrCONHCONH_2$	237.11
Chloral hydrate	$CCl_3CH(OH)_2$	165.42
Chloramphenicol	$CH(OH)(C_6H_4\text{-}p\text{-}NO_2){\cdot}CH(NHCOCHCl_2){\cdot}CH_2OH$	323.15
Chloroform	$CHCl_3$	119.39
Chloroquine	$C_{18}H_{26}N[Cl][NHCH(CH_3){\cdot}(CH_2)_3{\cdot}N(C_2H_5)_2]\text{-}1,7,4$	319.89
Cholesterol	$C_{27}H_{45}OH$	386.67
Cinchophen	$CH{:}CH{\cdot}CH{:}CH{\cdot}CH{:}C{\cdot}C{\cdot}C(COOH){:}CH{\cdot}C(C_4H_5){:}N$	249.27
Ether	$(C_2H_5)_2O$	74.12
Ethylmorphine hydrochloride	$[C_{19}H_{23}O_3N]{\cdot}HCl + 2\,H_2O$	385.90
N-Ethylpiperidine	$CH_2CH_2CH_2CH_2N{\cdot}C_2H_5$	113.21
Eucalyptol	$(CH_3)_2CCHCH_2CH_2C(O)(CH_3)_2CH_2CH_2$	154.25
Eucatropine	$\{COOCHCH_2C(CH_3)_2N(CH_3)CH(CH_2)CH_2\ /\ C_6H_5CH(OH)\}$	291.40
Formaldehyde	$HCHO$	30.03
Girard's reagent	$[(CH_3)_3NCH_2CONHNH_2]Cl$	167.65
Glycerol	$HOCH_2CH(OH)CH_2OH$	92.10
Histamine	$N{:}CHNHCH{:}CCH_2CH_2NH_2$	111.15
Homatropine	$C_{16}H_{21}O_3N$	275.35
Hyoscine	See Scopolamine	
Hyoscyamine	$C_{17}H_{23}O_3N$	289.38
Iodoform	CHI_3	393.75
Lactic acid	$CH_3CH(OH)COOH$	90.08
Lanatoside C	$C_{49}H_{76}O_{20}$	985.15
Mandelic acid	$C_6H_5CH(OH)COOH$	152.15
Menthol	$(CH_3)_2CHCHCH(OH)CH_2CH(CH_2CH_2)CH_3$	156.27
Mepacrine	See Quinacrine	
Meperidine	$C_6H_5{\cdot}C(COOC_2H_5){\cdot}CH_2CH_2N(CH_2)CH_2CH_3$	247.34
Methanol (Methyl alcohol)	CH_3OH	32.04
Methylene blue	$[C_{16}H_{18}ClN_3S]+3\,H_2O$	373.92
Morphine	$[C_{17}H_{19}O_3N]+H_2O$	303.36
Neostigmine bromide	$C_6H_4[N(CH_3)_3Br][OCON(CH_3)_2]\text{-}1,3$	303.22
Nikethamide	$CH{:}CHCH{:}NCH{:}CCON(C_2H_5)_2$	178.24
Nitrogen mustard	See Allyl isothiocyanate	
Ouabain (Strophanthin-G)	$[C_{29}H_{44}O_{12}]+8\,H_2O$	728.80
Papaverine	$[C_{20}H_{21}O_4N]$	339.40
Penicillin-G	$C_{16}H_{17}N_2NaO_4S$	356.39
Pentobarbital sodium	$NHCONHCOC[CH(CH_3)C_3H_7][C_2H_5]CO$	226.28
Phenacetin	See Acetophenetidine	
Phenazone	See Antipyrine	
Phenobarbital	$NHCONHCOC(C_2H_5)(C_6H_5)CO$	232.24
– sodium	$NHC(ONa){:}NCOC(C_2H_5)(C_6H_5)CO$	254.23
Phenytoin	$(C_6H_5)_2CNHCONHCO$	252.28
Physostigmine	$C_{15}H_{21}O_2N_3$	275.36
Picrotoxin	$C_{30}H_{34}O_{13}$	602.60
Pilocarpine	$C_{11}H_{16}O_2N_2$	208.27
Piperocaine	$C_6H_5COO(CH_2)_3N{\cdot}(CH_2)_4{\cdot}CH(CH_3)$	261.37

Table 86 (continued)
MOLECULAR WEIGHTS OF SOME MEDICALLY IMPORTANT COMPOUNDS

Compound	Formula	MW
Cocaine	$C_{17}H_{21}O_4N$	303.36
Codeine	$[C_{18}H_{21}O_3N]+H_2O$	317.39
Colchicine	$C_{22}H_{25}O_6N$	399.45
Coumarin	$CH:CHCH:CHC:C \cdot O \cdot CO \cdot CH:CH$	146.15
Creatinine	$CH_3NC(:NH)NHCOCH_2$	113.12
Cresol	$CH_3C_6H_4OH$	108.14
Cyclopropane	$CH_2CH_2CH_2$	42.08
Dextrose (D-Glucose)	$[CH_2(OH) \cdot CH(OH) \cdot CH(OH) \cdot (HO)CH \cdot CH(OH) \cdot CHO]+H_2O$	198.18
Diethylbarbituric acid	See Barbital	
Diethylstilbestrol	See Stilbestrol	
Digitonin	$C_{56}H_{90}O_{29}$	1215.33
Digitoxin	$C_{41}H_{64}O_{13}$	764.96
Dihydromorphinone	$C_{17}H_{19}O_3N$	285.35
Dihydrostreptomycin hydrochloride – sulfate	$(C_{21}H_{41}O_{12}N_7)_2 \cdot 3H_2SO_4$	1461.48
Diiododydroxy-quinoline	$CI:CHCI:C(OH)C:CCH \cdot CHCH:N$	396.97
Dimercaprol(BAL)	$CH_2(SH)CH(SH)CH_2OH$	124.23
m-Dinitrobenzene	$C_6H_4(NO_2)_2$-1,3	168.11
Dioxan	$OCH_2CH_2OCH_2CH_2$	88.11
Diphenylhydantoin	See Phenytoin	
Emetine hydrochloride	$C_{29}H_{40}O_4N_2.2HCl$	553.59
Ephedrine hydrochloride	$C_6H_5CH(OH)CH(NHCH_3)CH_3.HCl$	201.70
Epinephrine hydrochloride	$C_6H_3[CH(OH)CH_2NHCH_3][OH]_2.HCl$-1,3,4	219.68
Ergonovine (Ergometrine)	$C_{19}H_{23}O_2N_3$	325.42
Ergotamine – tartrate	$C_{33}H_{35}N_5O_5$	581.68
	$(C_{33}H_{35}N_5O_5)_2 \cdot H_2C_4H_4O_6$	1313.46
Ergotoxine	$C_{35}H_{41}O_6N_5$	627.75
Ethanol (Ethyl alcohol)	C_2H_5OH	46.07
Procaine hydrochloride	$C_6H_4[COOCH_2CH_2N(C_2H_5)_2][NH_2].HCl$-1,4	272.78
Quinacrine	$C_{23}H_{30}ClON_3$	399.97
Quinidine	$C_{20}H_{24}O_2N_2$	324.43
Quinine	$[C_{20}H_{24}O_2N_2]+3H_2O$	378.48
Resorcinol	$C_6H_4(OH)_2$-1,3	110.11
Rochelle salt (Potassium sodium tartrate)	$[CH(OH)(COOK) \cdot CH(OH)(COONa)]+4H_2O$	282.23
Rutin	$[C_{27}H_{30}O_{16}]+3H_2O$	664.59
Salicylic acid	$C_6H_4(COOH)(OH)$-1,2	138.13
Santonin	$CH:CH \cdot CO \cdot C(CH_3):C \cdot C(CH_3) \cdot CH_2 \cdot CH_2 \cdot CH \cdot CH \cdot O \cdot CO \cdot CO \cdot CH \cdot CH_3$	246.31
Scopolamine (Hyoscine)	$C_{17}H_{21}O_4N$	303.36
Sodium lactate	$CH_3CH(OH)COONa$	112.06
Stilbestrol	$p\text{-}HOC_6H_4(C_2H_5)C:C(C_2H_5)C_6H_4OH\text{-}p'$	268.36
Streptomycin hydrochloride – sulfate	$C_{21}H_{39}O_{12}N_7.3HCl$	690.99
	$(C_{21}H_{39}O_{12}N_7)_2 \cdot 3H_2SO_4$	1457.44
Strophanthin-G	See Ouabain	
Strychnine	$C_{21}H_{22}O_2N_2$	334.42
Sulfadiazine	$p\text{-}NH_2 \cdot C_6H_4 \cdot SO_2NHC:N \cdot CH:CH:CH:N$	250.29
Sulfamerazine	$p\text{-}NH_2 \cdot C_6H_4 \cdot SO_2NHC:N \cdot CH:CH \cdot C(CH_3):N$	264.32
Sulfanilamide	$p\text{-}NH_2 \cdot C_6H_4 \cdot SO_2NH_2$	172.21
Sulfapyridine	$p\text{-}NH_2 \cdot C_6H_4 \cdot SO_2NHC:N \cdot CH:CH:CH:CH$	249.30
Sulfathiazole	$p\text{-}NH_2 \cdot C_6H_4 \cdot SO_2NHC:N \cdot CH:CH \cdot S$	255.33
Theobromine	$NH \cdot CO \cdot N(CH_3) \cdot C:C \cdot CO \ N:CH \cdot NCH_3$	180.17
Theophylline	$N(CH_3) \cdot CO \cdot N(CH_3) \cdot C:C \cdot CO \ N:CH \cdot NH \cdot H_2O$	198.19
Tryparsamide	$[p\text{-}NH_2COCH_2NH \cdot C_6H_4AsO(OH)ONa]+\frac{1}{2}H_2O$	305.09
Tubocurarine chloride	$[C_{38}H_{44}O_6N_2]Cl_2+5H_2O$	785.78
Undecylenic acid	$CH_2:CH \cdot (CH_2)_8 \cdot COOH$	184.28
Urea (Carbamide)	$CO(NH_2)_2$	60.06
Urethane	$H_2NCOOC_2H_5$	89.10
Uric acid	$NH \cdot CO \cdot NH \cdot C:C \cdot CO \ NH \cdot CO \cdot NH$	168.12

From Diem, K., Ed., *Documenta Geigy Scientific Tables*, 6th ed., Geigy Pharmaceuticals, Ardsley, N.Y., 1962, 311. With permission.

Table 87 gives absorbance values to four significant figures corresponding to percent transmittance values which are given to three significant figures. The values of transmittance are given in the left-hand column and in the top row. For example, 8.4 percent transmittance corresponds to an absorbance of 1.076.

Interpolation is facilitated and accuracy is maximized if the percent transmittance is between 1 and 10, by multiplying its value by 10, finding the absorbance corresponding to the result, and adding 1. For example, to find the absorbance corresponding to 8.45% transmittance, note that 84.5% transmittance corresponds to an absorbance of 0.0731, so that 8.45% transmittance corresponds to an absorbance of 1.0731. For percent transmittance values between 0.1 and 1, multiply by 100, find the absorbance corresponding to the result, and add 2.

Conversely, to find the percent transmittance corresponding to an absorbance between 1 and 2, subtract 1 from the absorbance, find the percent transmittance corresponding to the result, and divide by 10. For example, an absorbance of 1.219 can best be converted to percent transmittance by noting that an absorbance of 0.219 would correspond to 60.4% transmittance; dividing this by 10 gives the desired value, 6.04% transmittance. For absorbance values between 2 and 3, subtract 2 from the absorbance, find the percent transmittance corresponding to the result, and divide by 100.

Table 87
TRANSMITTANCE-ABSORBANCE CONVERSION TABLE

% Transmittance	0.0	0.1	0.2	0.3	0.4	0.5	0.6	0.7	0.8	0.9
0	......	3.000	2.699	2.523	2.398	2.301	2.222	2.155	2.097	2.046
1	2.000	1.959	1.921	1.886	1.854	1.824	1.796	1.770	1.745	1.721
2	1.699	1.678	1.658	1.638	1.620	1.602	1.585	1.569	1.553	1.538
3	1.523	1.509	1.495	1.481	1.469	1.456	1.444	1.432	1.420	1.409
4	1.398	1.387	1.377	1.367	1.357	1.347	1.337	1.328	1.319	1.310
5	1.301	1.292	1.284	1.276	1.268	1.260	1.252	1.244	1.237	1.229
6	1.222	1.215	1.208	1.201	1.194	1.187	1.180	1.174	1.167	1.161
7	1.155	1.149	1.143	1.137	1.131	1.125	1.119	1.114	1.108	1.102
8	1.097	1.092	1.086	1.081	1.076	1.071	1.066	1.060	1.056	1.051
9	1.046	1.041	1.036	1.032	1.027	1.022	1.018	1.013	1.009	1.004
10	1.000	0.9957	0.9914	0.9872	0.9830	0.9788	0.9747	0.9706	0.9666	0.9626
11	0.9586	0.9547	0.9508	0.9469	0.9431	0.9393	0.9355	0.9318	0.9281	0.9245
12	0.9208	0.9172	0.9136	0.9101	0.9066	0.9031	0.8996	0.8962	0.8928	0.8894
13	0.8861	0.8827	0.8794	0.8761	0.8729	0.8697	0.8665	0.8633	0.8601	0.8570
14	0.8539	0.8508	0.8477	0.8447	0.8416	0.8386	0.8356	0.8327	0.8297	0.8268
15	0.8239	0.8210	0.8182	0.8153	0.8125	0.8097	0.8069	0.8041	0.8013	0.7986
16	0.7959	0.7932	0.7905	0.7878	0.7852	0.7825	0.7799	0.7773	0.7747	0.7721
17	0.7696	0.7670	0.7645	0.7620	0.7595	0.7570	0.7545	0.7520	0.7496	0.7471
18	0.7447	0.7423	0.7399	0.7375	0.7352	0.7328	0.7305	0.7282	0.7258	0.7235
19	0.7212	0.7190	0.7167	0.7144	0.7122	0.7100	0.7077	0.7055	0.7033	0.7011
20	0.6990	0.6968	0.6946	0.6925	0.6904	0.6882	0.6861	0.6840	0.6819	0.6799
21	0.6778	0.6757	0.6737	0.6716	0.6696	0.6676	0.6655	0.6635	0.6615	0.6596
22	0.6576	0.6556	0.6536	0.6517	0.6498	0.6478	0.6459	0.6440	0.6421	0.6402
23	0.6383	0.6364	0.6345	0.6326	0.6308	0.6289	0.6271	0.6253	0.6234	0.6216
24	0.6198	0.6180	0.6162	0.6144	0.6126	0.6108	0.6091	0.6073	0.6055	0.6038
25	0.6021	0.6003	0.5986	0.5969	0.5952	0.5935	0.5918	0.5901	0.5884	0.5867

Table 87 (continued)
TRANSMITTANCE-ABSORBANCE CONVERSION TABLE

% Trans-mittance	0.0	0.1	0.2	0.3	0.4	0.5	0.6	0.7	0.8	0.9
26	0.5850	0.5834	0.5817	0.5800	0.5784	0.5766	0.5751	0.5735	0.5719	0.5702
27	0.5686	0.5670	0.5654	0.5638	0.5622	0.5607	0.5591	0.5575	0.5560	0.5544
28	0.5528	0.5513	0.5498	0.5482	0.5467	0.5452	0.5436	0.5421	0.5406	0.5391
29	0.5376	0.5361	0.5346	0.5331	0.5317	0.5302	0.5287	0.5272	0.5258	0.5243
30	0.5229	0.5214	0.5200	0.5186	0.5171	0.5157	0.5143	0.5129	0.5114	0.5100
31	0.5086	0.5072	0.5058	0.5045	0.5031	0.5017	0.5003	0.4989	0.4976	0.4962
32	0.4949	0.4935	0.4921	0.4908	0.4895	0.4881	0.4868	0.4855	0.4841	0.4828
33	0.4815	0.4802	0.4789	0.4776	0.4763	0.4750	0.4737	0.4724	0.4711	0.4698
34	0.4685	0.4672	0.4660	0.4647	0.4634	0.4622	0.4609	0.4597	0.4584	0.4572
35	0.4559	0.4547	0.4535	0.4522	0.4510	0.4498	0.4486	0.4473	0.4461	0.4449
36	0.4437	0.4425	0.4413	0.4401	0.4389	0.4377	0.4365	0.4353	0.4342	0.4330
37	0.4318	0.4306	0.4295	0.4283	0.4271	0.4260	0.4248	0.4237	0.4225	0.4214
38	0.4202	0.4191	0.4179	0.4168	0.4157	0.4145	0.4134	0.4123	0.4112	0.4101
39	0.4089	0.4078	0.4067	0.4056	0.4045	0.4034	0.4023	0.4012	0.4001	0.3989
40	0.3979	0.3969	0.3958	0.3947	0.3936	0.3925	0.3915	0.3904	0.3893	0.3883
41	0.3872	0.3862	0.3851	0.3840	0.3830	0.3820	0.3809	0.3799	0.3788	0.3778
42	0.3768	0.3757	0.3747	0.3737	0.3726	0.3716	0.3706	0.3696	0.3686	0.3675
43	0.3665	0.3655	0.3645	0.3635	0.3625	0.3615	0.3605	0.3595	0.3585	0.3575
44	0.3565	0.3556	0.3546	0.3536	0.3526	0.3516	0.3507	0.3497	0.3487	0.3478
45	0.3468	0.3458	0.3449	0.3439	0.3429	0.3420	0.3410	0.3401	0.3391	0.3382
46	0.3372	0.3363	0.3354	0.3344	0.3335	0.3325	0.3316	0.3307	0.3298	0.3288
47	0.3279	0.3270	0.3261	0.3251	0.3242	0.3233	0.3224	0.3215	0.3206	0.3197
48	0.3188	0.3179	0.3170	0.3161	0.3152	0.3143	0.3134	0.3125	0.3116	0.3107
49	0.3098	0.3089	0.3080	0.3072	0.3063	0.3054	0.3045	0.3036	0.3028	0.3019
50	0.3010	0.3002	0.2993	0.2984	0.2976	0.2967	0.2958	0.2950	0.2941	0.2933
51	0.2924	0.2916	0.2907	0.2899	0.2890	0.2882	0.2874	0.2865	0.2857	0.2848
52	0.2840	0.2832	0.2823	0.2815	0.2807	0.2798	0.2790	0.2782	0.2774	0.2765
53	0.2757	0.2749	0.2741	0.2733	0.2725	0.2716	0.2708	0.2700	0.2692	0.2684
54	0.2676	0.2668	0.2660	0.2652	0.2644	0.2636	0.2628	0.2620	0.2612	0.2604
55	0.2596	0.2588	0.2581	0.2573	0.2565	0.2557	0.2549	0.2541	0.2534	0.2526
56	0.2518	0.2510	0.2503	0.2495	0.2487	0.2480	0.2472	0.2464	0.2457	0.2449
57	0.2441	0.2434	0.2426	0.2418	0.2411	0.2403	0.2396	0.2388	0.2381	0.2373
58	0.2366	0.2358	0.2351	0.2343	0.2336	0.2328	0.2321	0.2314	0.2306	0.2299
59	0.2291	0.2284	0.2277	0.2269	0.2262	0.2255	0.2248	0.2240	0.2233	0.2226
60	0.2218	0.2211	0.2204	0.2197	0.2190	0.2182	0.2175	0.2168	0.2161	0.2154
61	0.2147	0.2140	0.2132	0.2125	0.2118	0.2111	0.2104	0.2097	0.2090	0.2083
62	0.2076	0.2069	0.2062	0.2055	0.2048	0.2041	0.2034	0.2027	0.2020	0.2013
63	0.2007	0.2000	0.1993	0.1986	0.1979	0.1972	0.1965	0.1959	0.1952	0.1945
64	0.1938	0.1931	0.1925	0.1918	0.1911	0.1904	0.1898	0.1891	0.1884	0.1878
65	0.1871	0.1864	0.1858	0.1851	0.1844	0.1838	0.1831	0.1824	0.1818	0.1811
66	0.1805	0.1798	0.1791	0.1785	0.1778	0.1772	0.1765	0.1759	0.1752	0.1746
67	0.1739	0.1733	0.1726	0.1720	0.1713	0.1707	0.1701	0.1694	0.1688	0.1681
68	0.1675	0.1669	0.1662	0.1656	0.1649	0.1643	0.1637	0.1630	0.1624	0.1618
69	0.1612	0.1605	0.1599	0.1593	0.1586	0.1580	0.1574	0.1568	0.1561	0.1555
70	0.1549	0.1543	0.1537	0.1530	0.1524	0.1518	0.1512	0.1506	0.1500	0.1494
71	0.1487	0.1481	0.1475	0.1469	0.1463	0.1457	0.1451	0.1445	0.1439	0.1433
72	0.1427	0.1421	0.1415	0.1409	0.1403	0.1397	0.1391	0.1385	0.1379	0.1373
73	0.1367	0.1361	0.1355	0.1349	0.1343	0.1337	0.1331	0.1325	0.1319	0.1314
74	0.1308	0.1302	0.1296	0.1290	0.1284	0.1278	0.1273	0.1267	0.1261	0.1255
75	0.1249	0.1244	0.1238	0.1232	0.1226	0.1221	0.1215	0.1209	0.1203	0.1198

Table 87 (continued)
TRANSMITTANCE-ABSORBANCE CONVERSION TABLE

% Trans-mittance	0.0	0.1	0.2	0.3	0.4	0.5	0.6	0.7	0.8	0.9
76	0.1192	0.1186	0.1180	0.1175	0.1169	0.1163	0.1158	0.1152	0.1146	0.1141
77	0.1135	0.1129	0.1124	0.1118	0.1113	0.1107	0.1101	0.1096	0.1090	0.1085
78	0.1079	0.1073	0.1068	0.1062	0.1057	0.1051	0.1046	0.1040	0.1035	0.1029
79	0.1024	0.1018	0.1013	0.1007	0.1002	0.0996	0.0991	0.0985	0.0980	0.0975
80	0.0969	0.0964	0.0958	0.0953	0.0947	0.0942	0.0937	0.0931	0.0926	0.0921
81	0.0915	0.0910	0.0904	0.0899	0.0894	0.0888	0.0883	0.0878	0.0872	0.0867
82	0.0862	0.0857	0.0851	0.0846	0.0841	0.0835	0.0830	0.0825	0.0820	0.0814
83	0.0809	0.0804	0.0799	0.0794	0.0788	0.0783	0.0778	0.0773	0.0768	0.0762
84	0.0757	0.0752	0.0747	0.0742	0.0737	0.0731	0.0726	0.0721	0.0716	0.0711
85	0.0706	0.0701	0.0696	0.0691	0.0685	0.0680	0.0675	0.0670	0.0665	0.0660
86	0.0655	0.0650	0.0645	0.0640	0.0635	0.0630	0.0625	0.0620	0.0615	0.0610
87	0.0605	0.0600	0.0595	0.0590	0.0585	0.0580	0.0575	0.0570	0.0565	0.0560
88	0.0555	0.0550	0.0545	0.0540	0.0535	0.0531	0.0526	0.0521	0.0516	0.0511
89	0.0506	0.0501	0.0496	0.0491	0.0487	0.0482	0.0477	0.0472	0.0467	0.0462
90	0.0458	0.0453	0.0448	0.0443	0.0438	0.0434	0.0429	0.0424	0.0419	0.0414
91	0.0410	0.0405	0.0400	0.0395	0.0391	0.0386	0.0381	0.0376	0.0372	0.0367
92	0.0362	0.0357	0.0353	0.0348	0.0343	0.0339	0.0334	0.0329	0.0325	0.0320
93	0.0315	0.0311	0.0306	0.0301	0.0297	0.0292	0.0287	0.0283	0.0278	0.0273
94	0.0269	0.0264	0.0259	0.0255	0.0250	0.0246	0.0241	0.0237	0.0232	0.0227
95	0.0223	0.0218	0.0214	0.0209	0.0205	0.0200	0.0195	0.0191	0.0186	0.0182
96	0.0177	0.0173	0.0168	0.0164	0.0159	0.0155	0.0150	0.0146	0.0141	0.0137
97	0.0132	0.0128	0.0123	0.0119	0.0114	0.0110	0.0106	0.0101	0.0097	0.0092
98	0.0088	0.0083	0.0079	0.0074	0.0070	0.0066	0.0061	0.0057	0.0052	0.0048
99	0.0044	0.0039	0.0035	0.0031	0.0026	0.0022	0.0017	0.0013	0.0009	0.0004

From Dean, J. A., Ed., *Lange's Handbook of Chemistry,* 12th ed., McGraw-Hill, New York, 1979, 2-21. With permission.

Table 88
TRANSMISSION OF CORNING COLORED FILTERS

Supplied by R. G. Saxton

If I_o is the intensity of radiation entering a layer of some medium and I the intensity reaching the opposite surface, the ratio I/I_o is called the transmittance. In practice the ratio of intensity of radiation passing through a glass sample to that incident on its surface is often measured and plotted as transmission. The transmission is the result of two factors, the transmittance of the glass and the losses by reflection. These losses amount to about 4 % for each glass-air surface; the transmission of a sample is about 92 % of its transmittance. Since the reflection losses differ slightly with different samples, the correction is often determined and applied when the transmission is measured. Values in this table have been corrected for reflection losses.

The identifying glass number, CS number, color and properties, and nominal thickness for the Corning glasses in this table are:

Glass No.	CS	Color and properties	Nominal thickness
0160	0-54	Clear; Ultraviolet transmitting	2.0
2030	2-64	Red; Sharp cut	3.0
2403	2-58	Red; Sharp cut	3.0
2404	2-59	Red; Sharp cut	3.0
2408	2-60	Red; Sharp cut	3.0
2412	2-61	Red; Sharp cut	3.0
2418	2-62	Red; Sharp cut	3.0
2424	2-63	Red; Sharp cut	3.0
2434	2-73	Red; Sharp cut	3.0
2540	7-56	Black; IR transmitting; Visible absorbing	2.5
2550	7-57	Black; IR transmitting; Visible absorbing	2.0
2600	7-69	Black; IR transmitting; Visible absorbing	3.0
3060	3-75	Straw	2.0
3304	3-76	Dark amber	3.0
3307	3-77	Dark amber	3.0
3384	3-70	Yellow	3.0
3385	3-71	Yellow	3.0
3387	3-72	Straw	3.0
3389	3-73	Straw	3.0
3391	3-74	Straw	3.0
3480	3-66	Yellow; Sharp cut	3.0
3482	3-67	Yellow; Sharp cut	3.0
3484	3-68	Yellow; Sharp cut	3.0
3486	3-69	Yellow; Sharp cut	3.0
3718	3-94	Yellow	3.0
3750	3-79	Yellow; Yellow green fluorescing	5.0
3780	3-80	Yellow	2.0
3850	0-51	Clear; UV transmitting	4.0
3961	1-56	Bluish; IR absorbing; Visible transmitting	2.5
3962	1-57	Bluish; IR absorbing; Visible transmitting	2.5
3965	1-58	Bluish; IR absorbing; Visible transmitting	2.5
3966	1-59	Bluish; IR absorbing; Visible transmitting	2.5
4010	4-64	Green	4.0
4015	4-65	Yellow green	3.0
4060	4-67	Green	2.0
4034	4-68	Green	4.5
4303	4-72	Blue green	4.0
4305	4-71	Blue green	4.0
4308	4-70	Blue green	4.0
4309	4-69	Blue green	4.0
4445	4-74	Green	2.5
4602	1-75	Bluish; IR absorbing; Visible transmitting	3.0
4784	4-94	Blue green	5.0
5030	5-57	Blue	5.0
5031	5-56	Blue	4.5
5070	7-62	Amethyst	3.9
5071	7-63	Amethyst	3.9
5073	7-64	Amethyst	3.9
5113	5-58	Blue	4.0
5120	1-60	Smoky violet; Absorbs yellow	5.2
5300	4-106	Green	3.9
5330	1-64	Blue	4.5
5433	5-59	Blue	5.0
5543	5-60	Blue	5.0
5562	5-61	Blue	5.0
5572	1-61	Blue	5.0
5840	7-60	Black; UV transmitting; Visible absorbing	4.5
5850	7-59	Purple; UV transmitting; Visible absorbing	4.0
5860	7-37	Black; UV transmitting; Visible absorbing	5.0
5874	7-39	Black; UV transmitting; Visible absorbing	5.0
5900	1.62	Blue	5.5
5970	7-51	Black; UV transmitting; Visible absorbing	5.0
7380	0-52	Clear; UV transmitting	2.0
7740	0-53	Clear; UV transmitting	2.0
7905	9-30	Clear; UV transmitting; Long Range IR transmitting	2.0
7910	9-54	Clear; UV transmitting	2.0
8364	7-98	Gray	2.0
9780	4-76	Blue green	5.0
9782	4-96	Blue green	5.0
9788	4-97	Blue green	5.0
9830	4-77	Green	3.4
9863	7-54	Black; UV transmitting; Visible absorbing	3.0

Table 88 (continued)
TRANSMISSION OF CORNING COLORED FILTERS

						Transmittance						
						Corning Glass Number						
λ(μ)	0160	2030	2403	2404	2408	2412	2418	2424	2434	2540	2550	2600
.22	.000	.000	.000	.000	.000	.000	.000	.000	.000	.000	.000	.000
.24	.000	.000	.000	.000	.000	.000	.000	.000	.000	.000	.000	.000
.26	.000	.000	.000	.000	.000	.000	.000	.000	.000	.000	.000	.000
.28	.000	.000	.000	.000	.000	.000	.000	.000	.000	.000	.000	.000
.30	.005	.000	.000	.000	.000	.000	.000	.000	.000	.000	.000	.000
.32	.642	.000	.000	.000	.000	.000	.000	.000	.000	.000	.000	.000
.34	.850	.000	.000	.000	.000	.000	.000	.000	.000	.000	.000	.000
.36	.882	.000	.000	.000	.000	.000	.000	.000	.000	.000	.000	.000
.38	.890	.000	.000	.000	.000	.000	.000	.000	.000	.000	.000	.000
.40	.892	.000	.000	.000	.000	.000	.000	.000	.000	.000	.000	.000
.41	.893	.000	.000	.000	.000	.000	.000	.000	.000	.000	.000	.000
.42	.896	.000	.000	.000	.000	.000	.000	.000	.000	.000	.000	.000
.43	.896	.000	.000	.000	.000	.000	.000	.000	.000	.000	.000	.000
.44	.898	.000	.000	.000	.000	.000	.000	.000	.000	.000	.000	.000
.45	.899	.000	.000	.000	.000	.000	.000	.000	.000	.000	.000	.000
.46	.900	.000	.000	.000	.000	.000	.000	.000	.000	.000	.000	.000
.47	.900	.000	.000	.000	.000	.000	.000	.000	.000	.000	.000	.000
.48	.900	.000	.000	.000	.000	.000	.000	.000	.000	.000	.000	.000
.49	.900	.000	.000	.000	.000	.000	.000	.000	.000	.000	.000	.000
.50	.900	.000	.000	.000	.000	.000	.000	.000	.000	.000	.000	.000
.51	.900	.000	.000	.000	.000	.000	.000	.000	.000	.000	.000	.000
.52	.900	.000	.000	.000	.000	.000	.000	.000	.000	.000	.000	.000
.53	.900	.000	.000	.000	.000	.000	.000	.000	.000	.000	.000	.000
.54	.900	.000	.000	.000	.000	.000	.000	.000	.000	.000	.000	.000
.55	.900	.000	.000	.000	.000	.000	.000	.000	.000	.000	.000	.000
.56	.901	.000	.000	.000	.000	.000	.000	.000	.000	.000	.000	.000
.57	.904	.000	.000	.000	.000	.000	.000	.000	.005	.000	.000	.000
.58	.904	.000	.000	.000	.000	.000	.000	.005	.200	.000	.000	.000
.59	.908	.000	.000	.000	.000	.000	.008	.170	.615	.000	.000	.000
.60	.910	.000	.000	.000	.000	.006	.250	.575	.808	.000	.000	.000
.61	.910	.000	.000	.000	.018	.190	.660	.790	.856	.000	.000	.000
.62	.910	.000	.000	.015	.265	.625	.822	.848	.872	.000	.060	.000
.63	.910	.000	.018	.295	.670	.828	.862	.870	.881	.000	.000	.000
.64	.910	.006	.260	.660	.828	.868	.874	.880	.887	.000	.001	.000
.65	.910	.028	.675	.796	.866	.881	.881	.887	.892	.000	.003	.000
.66	.910	.110	.838	.828	.877	.885	.885	.893	.895	.000	.005	.000
.67	.910	.305	.871	.842	.883	.887	.887	.897	.897	.000	.006	.000
.68	.910	.550	.880	.847	.886	.889	.889	.900	.899	.000	.009	.000
.69	.910	.735	.885	.851	.888	.900	.900	.901	.900	.000	.012	.000
.70	.910	.820	.886	.852	.888	.900	.900	.903	.900	.000	.017	.000
.71	.910	.853	.888	.854	.888	.889	.889	.903	.900	.000	.023	.000
.72	.910	.864	.889	.853	.888	.888	.888	.903	.899	.000	.031	.040
.73	.910	.867	.900	.851	.887	.887	.887	.903	.897	.000	.041	.175
.74	.910	.867	.900	.850	.885	.886	.886	.903	.896	.000	.055	.372
.75	.910	.866	.900	.849	.884	.885	.885	.902	.895	.000	.069	.547
.80	.910	.839	.875	.827	.870	.858	.857	.881	.866	.005	.225	.770
1.00	.912	.801	.840	.772	.840	.828	.822	.857	.842	.562	.780	.350
1.20	.908	.799	.845	.786	.845	.837	.827	.859	.849	.790	.870	.000
1.40	.909	.811	.854	.809	.854	.848	.840	.862	.857	.850	.895	.000
1.60	.913	.839	.873	.837	.873	.869	.858	.880	.879	.872	.904	.000
1.80	.909	.844	.870	.829	.870	.864	.854	.877	.872	.880	.900	.000
2.00	.904	.841	.868	.827	.868	.861	.851	.874	.871	.880	.897	.000
2.20	.888	.833	.820	.773	.825	.820	.810	.837	.835	.860	.875	.005
2.40	.875	.832	.803	.757	.809	.803	.792	.818	.812	.868	.870	.049
2.60	.868	.822	.750	.695	.754	.750	.723	.772	.767	.858	.850	.058
2.80	.690	.600	.100	.100	.100	.100	.050	.050	.260	.450	.480	.030
3.00	.630	.470	.070	.020	.070	.070	.072	.122	.400	.465	.383	.022
3.20	.500	.340	.140	.074	.150	.140	.142	.180	.470	.390	.330	.020
3.40	.379	.260	.140	.078	.140	.140	.120	.150	.350	.310	.245	.018
3.60	.320	.247	.000	.000	.000	.000	.000	.006	.015	.280	.220	.021
3.80	.310	.257	.000	.000	.000	.000	.000	.000	.020	.285	.250	.035
4.00	.311	.274	.000	.000	.000	.000	.000	.000	.015	.285	.275	.068
4.20	.251	.200	.000	.000	.000	.000	.000	.000	.017	.190	.190	.065
4.40	.110	.060	.000	.000	.000	.000	.000	.000	.002	.050	.100	.020
4.60	.012	.000	.000	.000	.000	.000	.000	.000	.000	.000	.000	.000
4.80	.004	.000	.000	.000	.000	.000	.000	.000	.000	.000	.000	.000
5.00	.000	.000	.000	.000	.000	.000	.000	.000	.000	.000	.000	.000

Table 88 (continued)
TRANSMISSION OF CORNING COLORED FILTERS

	Transmittance											
						Corning Glass Number						
λ(μ)	3060	3304	3307	3384	3385	3387	3389	3391	3480	3482	3484	3486
.22	.000	.000	.000	.000	.000	.000	.000	.000	.000	.000	.000	.000
.24	.000	.000	.000	.000	.000	.000	.000	.000	.000	.000	.000	.000
.26	.000	.000	.000	.000	.000	.000	.000	.000	.000	.000	.000	.000
.28	.000	.000	.000	.000	.000	.000	.000	.000	.000	.000	.000	.000
.30	.000	.000	.000	.000	.000	.000	.000	.000	.000	.000	.000	.000
.32	.000	.000	.000	.000	.000	.000	.005	.000	.000	.000	.000	.000
.34	.000	.000	.038	.000	.000	.000	.010	.000	.000	.000	.000	.000
.36	.000	.000	.050	.000	.005	.000	.015	.000	.000	.000	.000	.000
.38	.060	.000	.027	.005	.010	.010	.020	.000	.000	.000	.000	.000
.40	.410	.000	.016	.011	.016	.020	.026	.075	.000	.000	.000	.005
.41	.517	.000	.014	.011	.016	.020	.025	.425	.000	.000	.000	.005
.42	.604	.000	.014	.010	.015	.019	.105	.655	.000	.000	.000	.005
.43	.665	.000	.016	.009	.013	.017	.437	.747	.000	.000	.000	.005
.44	.710	.000	.022	.005	.011	.050	.620	.801	.000	.000	.000	.005
.45	.748	.000	.033	.003	.010	.325	.714	.838	.000	.000	.000	.005
.46	.778	.000	.049	.002	.008	.565	.780	.860	.000	.000	.000	.004
.47	.800	.000	.070	.001	.060	.690	.820	.874	.000	.000	.000	.003
.48	.819	.000	.101	.005	.410	.763	.848	.884	.000	.000	.000	.003
.49	.836	.003	.143	.088	.640	.803	.866	.890	.000	.000	.000	.002
.50	.850	.009	.193	.350	.727	.834	.878	.895	.000	.000	.000	.001
.51	.860	.019	.250	.595	.780	.854	.886	.898	.000	.000	.000	.045
.52	.870	.037	.315	.725	.817	.868	.890	.900	.000	.000	.003	.425
.53	.875	.063	.379	.789	.840	.876	.892	.901	.000	.000	.175	.710
.54	.881	.102	.447	.825	.856	.883	.894	.902	.000	.015	.600	.792
.55	.884	.146	.504	.846	.866	.887	.895	.903	.000	.230	.774	.823
.56	.886	.200	.560	.860	.873	.889	.894	.902	.020	.675	.818	.844
.57	.885	.255	.607	.869	.876	.890	.893	.901	.325	.850	.839	.859
.58	.883	.310	.648	.873	.878	.889	.892	.900	.710	.885	.854	.868
.59	.882	.360	.680	.876	.877	.887	.890	.898	.829	.894	.865	.876
.60	.882	.404	.705	.877	.877	.884	.886	.896	.858	.900	.873	.882
.61	.882	.438	.722	.877	.877	.881	.884	.893	.869	.903	.880	.886
.62	.882	.466	.735	.875	.876	.875	.880	.890	.876	.905	.885	.888
.63	.882	.488	.744	.871	.874	.871	.876	.886	.881	.906	.888	.889
.64	.883	.505	.748	.865	.872	.866	.872	.884	.884	.907	.890	.890
.65	.885	.519	.750	.860	.867	.860	.868	.881	.885	.908	.892	.890
.66	.886	.531	.750	.856	.863	.856	.865	.876	.886	.908	.893	.890
.67	.888	.543	.749	.850	.858	.851	.860	.873	.886	.908	.894	.890
.68	.890	.552	.745	.844	.853	.846	.856	.869	.886	.908	.893	.889
.69	.891	.561	.740	.837	.847	.839	.852	.865	.885	.907	.892	.887
.70	.892	.569	.734	.831	.842	.834	.847	.860	.884	.907	.891	.885
.71	.893	.574	.727	.825	.837	.827	.842	.856	.882	.906	.890	.883
.72	.893	.575	.720	.819	.831	.822	.837	.852	.880	.905	.888	.880
.73	.892	.576	.712	.813	.825	.816	.831	.848	.877	.905	.886	.877
.74	.891	.574	.702	.807	.820	.810	.826	.844	.874	.904	.885	.875
.75	.890	.570	.694	.800	.814	.805	.820	.840	.870	.903	.882	.873
.80	.871	.526	.642	.770	.865	.780	.775	.815	.837	.878	.846	.829
1.00	.830	.435	.516	.715	.830	.725	.716	.772	.801	.857	.811	.781
1.20	.860	.429	.500	.718	.858	.735	.730	.782	.807	.859	.819	.793
1.40	.901	.475	.540	.750	.900	.768	.768	.810	.828	.870	.837	.817
1.60	.917	.580	.635	.795	.918	.812	.812	.843	.852	.884	.856	.841
1.80	.916	.627	.675	.808	.915	.818	.820	.849	.847	.882	.852	.834
2.00	.908	.620	.668	.800	.909	.822	.817	.846	.847	.884	.852	.835
2.20	.900	.630	.675	.802	.900	.823	.810	.840	.805	.865	.829	.798
2.40	.885	.651	.690	.800	.885	.825	.811	.842	.787	.853	.817	.777
2.60	.860	.650	.690	.785	.858	.815	.800	.840	.725	.818	.757	.718
2.80	.550	.345	.390	.325	.670	.360	.340	.440	.050	.060	.110	.060
3.00	.379	.320	.360	.318	.348	.348	.322	.423	.080	.088	.190	.088
3.20	.315	.240	.290	.268	.332	.324	.290	.395	.150	.158	.270	.145
3.40	.250	.151	.190	.218	.289	.288	.255	.353	.120	.132	.140	.090
3.60	.231	.130	.150	.217	.266	.280	.249	.351	.000	.000	.000	.000
3.80	.258	.140	.160	.228	.270	.298	.267	.376	.000	.000	.000	.000
4.00	.283	.140	.175	.220	.280	.290	.260	.365	.000	.000	.000	.000
4.20	.200	.090	.125	.143	.210	.210	.178	.288	.000	.000	.000	.000
4.40	.100	.020	.030	.025	.080	.070	.040	.115	.000	.000	.000	.000
4.60	.008	.008	.010	.007	.002	.003	.000	.009	.000	.000	.000	.000
4.80	.000	.005	.009	.000	.000	.000	.000	.000	.000	.000	.000	.000
5.00	.000	.001	.008	.000	.000	.000	.000	.000	.000	.000	.000	.000

Table 88 (continued)
TRANSMISSION OF CORNING COLORED FILTERS

					Transmittance							
						Corning Glass Number						
$\lambda(\mu)$	3718	3750	3780	3850	3961	3962	3965	3966	4010	4015	4060	4084
.22	.000	.000	.000	.000	.000	.000	.000	.000	.000	.000	.000	.000
.24	.000	.000	.000	.000	.000	.000	.000	.000	.000	.000	.000	.000
.26	.000	.000	.000	.000	.000	.000	.000	.000	.000	.000	.000	.000
.28	.000	.000	.000	.000	.000	.000	.000	.000	.000	.000	.000	.000
.30	.000	.000	.000	.000	.000	.000	.000	.000	.000	.000	.000	.000
.32	.004	.000	.000	.000	.000	.000	.018	.055	.000	.000	.000	.000
.34	.030	.000	.000	.000	.000	.018	.192	.375	.000	.000	.001	.018
.36	.550	.215	.000	.005	.020	.125	.430	.630	.000	.000	.021	.128
.38	.665	.327	.000	.350	.085	.270	.558	.710	.000	.000	.080	.248
.40	.480	.113	.000	.675	.185	.395	.636	.781	.000	.000	.178	.216
.41	.443	.088	.000	.749	.218	.426	.651	.788	.000	.000	.228	.180
.42	.465	.088	.000	.788	.248	.453	.666	.795	.000	.000	.281	.151
.43	.560	.135	.000	.812	.269	.474	.678	.800	.000	.000	.335	.136
.44	.675	.255	.006	.828	.290	.494	.693	.806	.000	.000	.388	.140
.45	.748	.410	.028	.841	.313	.519	.709	.816	.000	.005	.434	.163
.46	.780	.472	.058	.850	.331	.538	.724	.824	.006	.025	.473	.200
.47	.803	.570	.092	.858	.346	.556	.737	.831	.021	.073	.506	.247
.48	.800	.555	.088	.865	.361	.570	.748	.836	.050	.145	.527	.303
.49	.802	.550	.095	.870	.370	.582	.756	.840	.100	.245	.535	.370
.50	.824	.597	.152	.874	.376	.590	.762	.842	.160	.350	.528	.430
.51	.862	.720	.325	.878	.377	.594	.765	.843	.220	.455	.503	.465
.52	.894	.825	.595	.881	.373	.593	.765	.841	.252	.537	.460	.467
.53	.904	.853	.717	.883	.364	.588	.761	.838	.247	.582	.400	.438
.54	.905	.860	.763	.884	.354	.579	.764	.833	.207	.594	.325	.376
.55	.906	.864	.783	.884	.342	.569	.746	.827	.153	.572	.252	.303
.56	.907	.867	.795	.883	.331	.559	.736	.821	.096	.525	.183	.225
.57	.907	.869	.799	.882	.317	.547	.724	.813	.054	.457	.125	.157
.58	.907	.870	.804	.880	.298	.529	.706	.802	.026	.385	.083	.107
.59	.907	.870	.811	.877	.276	.508	.685	.790	.011	.311	.053	.073
.60	.908	.876	.826	.876	.251	.481	.662	.775	.004	.245	.033	.048
.61	.908	.880	.835	.875	.225	.452	.636	.757	.000	.190	.020	.034
.62	.909	.881	.842	.874	.299	.423	.610	.739	.000	.145	.012	.024
.63	.909	.884	.848	.875	.217	.392	.577	.719	.000	.115	.007	.018
.64	.910	.885	.854	.876	.147	.359	.546	.698	.000	.098	.004	.015
.65	.910	.887	.859	.877	.125	.326	.515	.675	.000	.084	.001	.012
.66	.911	.891	.864	.880	.104	.297	.482	.652	.000	.075	.000	.010
.67	.913	.896	.869	.883	.086	.265	.450	.630	.000	.075	.000	.009
.68	.914	.900	.873	.885	.063	.235	.418	.603	.000	.071	.000	.008
.69	.915	.901	.877	.887	.055	.206	.385	.576	.000	.065	.000	.007
.70	.915	.904	.880	.888	.042	.279	.352	.550	.000	.067	.000	.007
.71	.915	.905	.883	.888	.032	.155	.322	.524	.000	.070	.000	.007
.72	.915	.906	.885	.889	.025	.133	.294	.496	.000	.075	.000	.007
.73	.915	.907	.885	.888	.018	.114	.266	.472	.000	.080	.000	.007
.74	.915	.907	.882	.887	.014	.097	.242	.448	.000	.084	.000	.008
.75	.914	.906	.882	.886	.010	.084	.220	.424	.000	.086	.000	.008
.80	.902	.875	.855	.863	.000	.033	:120	.310	.000	.109	.000	.013
1.00	.899	.860	.882	.820	.000	.002	.038	.158	.000	.215	.018	.100
1.20	.898	.882	.898	.850	.000	.002	.040	.161	.007	.393	.158	.303
1.40	.880	.810	.855	.894	.002	.018	.100	.270	.058	.549	.404	.548
1.60	.882	.805	.855	.905	.007	.050	.190	.390	.162	.663	.612	.710
1.80	.900	.844	.907	.904	.008	.057	.201	.408	.299	.740	.740	.791
2.00	.897	.819	.909	.895	.011	.070	.228	.435	.422	.783	.817	.830
2.20	.888	.720	.900	.870	.021	.105	.277	.475	.518	.803	.840	.792
2.40	.865	.668	.890	.850	.037	.140	.320	.512	.597	.817	.862	.808
2.60	.840	.570	.860	.800	.050	.005	.339	.515	.634	.813	.870	.740
2.80	.460	.075	.620	.200	.022	.092	.135	.085	.270	.460	.520	.010
3.00	.282	.028	.465	.165	.033	.133	.200	.230	.260	.418	.620	.033
3.20	.252	.016	.420	.120	.054	.150	.247	.294	.204	.345	.642	.128
3.40	.175	.007	.370	.070	.048	.112	.165	.200	.131	.235	.631	.130
3.60	.150	.003	.351	.045	.003	.008	.016	.018	.121	.192	.634	.006
3.80	.168	.000	.368	.067	.007	.015	.035	.048	.125	.200	.600	.016
4.00	.182	.000	.370	.080	.006	.010	.020	.022	.123	.200	.557	.006
4.20	.120	.000	.300	.040	.005	.013	.022	.021	.070	.130	.422	.001
4.40	.020	.000	.140	.005	.001	.001	.000	.001	.002	.015	.135	.000
4.60	.002	.000	.010	.000	.000	.000	.000	.000	.000	.000	.012	.000
4.80	.000	.000	.000	.000	.000	.000	.000	.000	.000	.000	.002	.000
5.00	.000	.000	.000	.000	.000	.000	.000	.000	.000	.000	.000	.000

Table 88 (continued)
TRANSMISSION OF CORNING COLORED FILTERS

	Transmittance											
						Corning Glass Number						
λ(μ)	4303	4305	4308	4309	4445	4602	4784	5030	5031	5070	5071	5073
.22	.000	.000	.000	.000	.000	.000	.000	.000	.000	.000	.000	.000
.24	.000	.000	.000	.000	.000	.000	.000	.000	.000	.000	.000	.000
.26	.000	.000	.000	.000	.000	.000	.000	.000	.000	.000	.000	.000
.28	.000	.000	.000	.000	.000	.000	.000	.000	.000	.000	.000	.000
.30	.000	.000	.000	.000	.000	.001	.000	.000	.016	.000	.000	.000
.32	.000	.000	.000	.022	.001	.106	.000	.000	.145	.012	.045	.090
.34	.011	.060	.190	.394	.018	.505	.009	.038	.420	.310	.330	.540
.36	.188	.380	.580	.740	.114	.755	.200	.285	.685	.628	.340	.757
.38	.390	.590	.740	.831	.248	.827	.450	.595	.820	.745	.420	.810
.40	.545	.723	.826	.884	.400	.835	.596	.770	.884	.712	.665	.830
.41	.588	.750	.840	.887	.454	.845	.627	.799	.894	.600	.712	.786
.42	.624	.770	.850	.890	.505	.846	.648	.808	.895	.430	.716	.705
.43	.654	.786	.857	.892	.550	.851	.666	.797	.890	.290	.694	.620
.44	.680	.798	.862	.894	.593	.856	.680	.767	.872	.170	.655	.525
.45	.698	.809	.867	.897	.631	.857	.697	.738	.865	.097	.612	.436
.46	.712	.815	.869	.898	.659	.856	.717	.702	.864	.055	.568	.365
.47	.715	.814	.866	.897	.678	.861	.735	.628	.845	.035	.531	.313
.48	.705	.802	.856	.894	.689	.866	.750	.522	.805	.023	.501	.275
.49	.678	.780	.838	.885	.687	.869	.763	.406	.750	.017	.482	.252
.50	.636	.740	.810	.872	.673	.870	.768	.288	.684	.015	.469	.237
.51	.570	.685	.770	.850	.641	.869	.767	.186	.601	.013	.463	.231
.52	.480	.610	.714	.817	.586	.866	.753	.105	.495	.013	.461	.230
.53	.387	.525	.650	.781	.520	.863	.725	.053	.388	.014	.464	.235
.54	.288	.430	.576	.736	.437	.865	.676	.022	.295	.016	.473	.245
.55	.205	.340	.502	.683	.355	.869	.615	.007	.198	.018	.486	.260
.56	.132	.255	.422	.627	.275	.868	.525	.000	.113	.023	.502	.278
.57	.082	.184	.345	.565	.202	.863	.427	.000	.057	.030	.522	.300
.58	.047	.127	.277	.505	.144	.856	.328	.000	.025	.038	.540	.324
.59	.026	.087	.218	.447	.102	.848	.235	.000	.008	.048	.555	.348
.60	.013	.057	.170	.393	.068	.838	.157	.000	.000	.058	.571	.373
.61	.006	.036	.131	.341	.046	.824	.102	.000	.000	.070	.587	.395
.62	.001	.022	.100	.296	.031	.806	.058	.000	.000	.083	.600	.415
.63	.000	.013	.074	.256	.020	.787	.032	.000	.000	.096	.612	.435
.64	.000	.007	.056	.221	.013	.767	.017	.000	.000	.108	.622	.450
.65	.000	.004	.042	.191	.008	.745	.007	.000	.000	.120	.633	.466
.66	.000	.002	.033	.167	.006	.722	.002	.000	.000	.135	.644	.482
.67	.000	.001	.025	.146	.003	.695	.000	.000	.000	.148	.658	.498
.68	.000	.000	.020	.128	.001	.665	.000	.000	.000	.165	.674	.515
.69	.000	.000	.016	.116	.000	.634	.000	.000	.000	.182	.686	.531
.70	.000	.000	.013	.104	.000	.600	.000	.000	.000	.200	.700	.548
.71	.000	.000	.010	.096	.000	.565	.000	.000	.000	.220	.712	.566
.72	.000	.000	.009	.088	.000	.531	.000	.004	.024	.245	.725	.586
.73	.000	.000	.007	.083	.000	.496	.000	.047	.119	.268	.736	.606
.74	.000	.000	.006	.079	.000	.463	.000	.190	.330	.295	.749	.675
.75	.000	.000	.005	.075	.000	.430	.000	.440	.580	.323	.759	.642
.80	.000	.000	.008	.080	.003	.258	.000	.890	.917	.505	.815	.750
1.00	.000	.005	.045	.188	.056	.019	.000	.753	.868	.860	.885	.872
1.20	.013	.060	.166	.375	.269	.009	.000	.455	.720	.890	.897	.890
1.40	.080	.180	.342	.542	.527	.016	.003	.100	.285	.892	.902	.892
1.60	.210	.342	.500	.658	.701	.035	.038	.056	.162	.890	.902	.892
1.80	.350	.482	.617	.732	.792	.059	.142	.052	.140	.878	.890	.877
2.00	.475	.590	.690	.772	.839	.048	.275	.075	.175	.860	.865	.860
2.20	.560	.653	.720	.778	.850	.038	.345	.172	.295	.840	.830	.840
2.40	.635	.704	.752	.791	.870	.045	.404	.330	.483	.812	.795	.805
2.60	.663	.710	.748	.770	.861	.066	.340	.382	.530	.804	.752	.775
2.80	.260	.370	.370	.250	.280	.018	.045	.030	.030	.550	.500	.500
3.00	.249	.250	.203	.212	.395	.000	.000	.010	.001	.390	.308	.340
3.20	.202	.212	.145	.168	.451	.000	.000	.065	.026	.222	.145	.180
3.40	.135	.136	.084	.107	.449	.000	.000	.002	.006	.120	.070	.090
3.60	.124	.120	.078	.077	.470	.000	.000	.000	.000	.078	.032	.063
3.80	.132	.125	.082	.083	.470	.000	.000	.000	.000	.078	.029	.060
4.00	.132	.131	.094	.090	.448	.000	.000	.000	.000	.093	.037	.075
4.20	.078	.073	.050	.042	.320	.000	.000	.000	.000	.079	.020	.048
4.40	.008	.009	.008	.002	.100	.000	.000	.000	.000	.020	.004	.014
4.60	.000	.000	.000	.000	.007	.000	.000	.000	.000	.002	.000	.000
4.80	.000	.000	.000	.000	.000	.000	.000	.000	.000	.000	.000	.000
5.00	.000	.000	.000	.000	.000	.000	.000	.000	.000	.000	.000	.000

Table 88 (continued)
TRANSMISSION OF CORNING COLORED FILTERS

						Transmittance						
						Corning Glass Number						
λ(μ)	5113	5120	5300	5330	5433	5543	5562	5572	5840	5850	5860	5874
.22	.000	.000	.000	.000	.000	.000	.000	.000	.000	.000	.000	.000
.24	.000	.000	.000	.000	.000	.000	.000	.000	.000	.000	.000	.000
.26	.000	.000	.000	.000	.000	.000	.000	.000	.000	.000	.000	.000
.28	.000	.000	.000	.000	.000	.000	.000	.000	.000	.000	.000	.000
.30	.000	.000	.000	.002	.000	.000	.000	.000	.001	.039	.000	.000
.32	.000	.000	.000	.250	.000	.000	.000	.045	.242	.490	.008	.031
.34	.000	.000	.000	.622	.000	.000	.012	.325	.600	.790	.179	.228
.36	.035	.018	.000	.796	.100	.120	.205	.660	.682	.858	.340	.447
.38	.200	.540	.000	.835	.350	.380	.495	.805	.392	.850	.085	.378
.40	.371	.670	.000	.865	.585	.600	.717	.874	.000	.788	.000	.032
.41	.371	.790	.000	.850	.636	.635	.748	.873	.000	.720	.000	.004
.42	.337	.805	.000	.823	.665	.646	.761	.865	.000	.630	.000	.000
.43	.272	.560	.000	.783	.674	.635	.759	.857	.000	.522	.000	.000
.44	.198	.386	.000	.725	.665	.602	.742	.845	.000	.410	.000	.000
.45	.118	.485	.000	.650	.635	.550	.713	.832	.000	.290	.000	.000
.46	.055	.475	.000	.555	.577	.465	.662	.808	.000	.175	.000	.000
.47	.013	.370	.008	.455	.467	.335	.565	.765	.000	.125	.000	.000
.48	.000	.385	.026	.355	.327	.190	.435	.693	.000	.022	.000	.000
.49	.000	.685	.085	.270	.205	.090	.300	.605	.000	.005	.000	.000
.50	.000	.660	.125	.197	.120	.040	.197	.523	.000	.000	.000	.000
.51	.000	.390	.106	.145	.060	.013	.110	.430	.000	.000	.000	.000
.52	.000	.305	.094	.110	.024	.002	.051	.328	.000	.000	.000	.000
.53	.000	.175	.064	.085	.008	.000	.022	.247	.000	.000	.000	.000
.54	.000	.610	.149	.068	.005	.000	.012	.216	.000	.000	.000	.000
.55	.000	.817	.147	.055	.006	.000	.015	.246	.000	.000	.000	.000
.56	.000	.230	.087	.043	.007	.000	.018	.285	.000	.000	.000	.000
.57	.000	.125	.013	.032	.000	.000	.011	.258	.000	.000	.000	.000
.58	.000	.000	.000	.022	.000	.000	.002	.175	.000	.000	.000	.000
.59	.000	.006	.000	.016	.000	.000	.000	.116	.000	.000	.000	.000
.60	.000	.180	.000	.012	.000	.000	.000	.113	.000	.000	.000	.000
.61	.000	.545	.000	.009	.000	.000	.000	.123	.000	.000	.000	.000
.62	.000	.825	.000	.010	.000	.000	.000	.126	.000	.000	.000	.000
.63	.000	.838	.000	.013	.000	.000	.000	.120	.000	.000	.000	.000
.64	.000	.878	.000	.015	.000	.000	.000	.111	.000	.000	.000	.000
.65	.000	.893	.000	.015	.000	.000	.000	.120	.000	.000	.000	.000
.66	.000	.883	.000	.014	.000	.000	.000	.165	.000	.000	.000	.000
.67	.000	.820	.000	.013	.000	.000	.000	.265	.000	.000	.000	.000
.68	.000	.705	.000	.015	.000	.000	.000	.425	.000	.000	.000	.000
.69	.000	.743	.000	.022	.000	.000	.001	.615	.000	.029	.000	.000
.70	.000	.860	.000	.041	.000	.000	.004	.756	.007	.160	.000	.017
.71	.000	.876	.000	.085	.000	.001	.004	.837	.020	.385	.000	.075
.72	.000	.815	.000	.165	.000	.002	.004	.874	.037	.615	.000	.168
.73	.000	.435	.000	.285	.000	.002	.002	.889	.060	.760	.000	.257
.74	.000	.045	.000	.460	.000	.002	.002	.895	.086	.843	.000	.320
.75	.000	.055	.000	.645	.000	.001	.001	.898	.080	.878	.000	.335
.80	.000	.030	.000	.900	.005	.002	.003	.895	.009	.890	.000	.218
1.00	.000	.860	.003	.800	.010	.015	.020	.880	.000	.716	.000	.050
1.20	.000	.770	.026	.570	.030	.020	.047	.690	.000	.169	.000	.011
1.40	.000	.550	.072	.410	.060	.040	.095	.640	.004	.042	.004	.012
1.60	.000	.620	.200	.405	.116	.060	.154	.625	.002	.036	.002	.010
1.80	.000	.580	.265	.425	.172	.090	.223	.635	.000	.048	.000	.010
2.00	.010	.580	.374	.476	.343	.247	.410	.750	.000	.168	.000	.018
2.20	.071	.747	.533	.535	.475	.400	.528	.780	.000	.338	.000	.036
2.40	.190	.400	.370	.552	.575	.520	.600	.780	.000	.492	.000	.074
2.60	.203	.530	.523	.540	.580	.532	.603	.745	.000	.510	.000	.088
2.80	.100	.250	.265	.007	.162	.132	.360	.470	.000	.170	.000	.003
3.00	.080	.080	.202	.012	.195	.151	.280	.315	.000	.128	.000	.020
3.20	.068	.072	.180	.082	.131	.105	.172	.200	.000	.084	.000	.053
3.40	.030	.029	.131	.003	.079	.065	.100	.100	.000	.065	.000	.050
3.60	.021	.017	.103	.000	.061	.030	.053	.049	.002	.070	.000	.000
3.80	.023	.021	.093	.000	.068	.032	.040	.037	.001	.082	.000	.000
4.00	.040	.025	.112	.000	.071	.039	.050	.042	.005	.090	.000	.000
4.20	.019	.010	.069	.000	.020	.015	.017	.020	.002	.055	.000	.000
4.40	.001	.001	.007	.000	.000	.000	.000	.002	.000	.002	.000	.000
4.60	.000	.000	.000	.000	.000	.000	.000	.000	.000	.000	.000	.000
4.80	.000	.000	.000	.000	.000	.000	.000	.000	.000	.000	.000	.000
5.00	.000	.000	.000	.000	.000	.000	.000	.000	.000	.000	.000	.000

Table 88 (continued)
TRANSMISSION OF CORNING COLORED FILTERS

					Transmittance							
						Corning Glass Number						
λ (μ)	5900	5970	7380	7740	7905	7910	8364	9780	9782	9788	9830	9863
.22	.000	.000	.000	.000	.000	.012	.000	.000	.000	.000	.000	.000
.24	.000	.000	.000	.000	.360	.505	.000	.000	.000	.000	.000	.054
.26	.000	.000	.000	.000	.495	.780	.000	.000	.000	.000	.000	.482
.28	.000	.000	.000	.004	.590	.855	.000	.000	.000	.000	.000	.731
.30	.000	.000	.000	.321	.720	.877	.000	.000	.000	.000	.000	.831
.32	.000	.138	.000	.722	.825	.900	.000	.000	.000	.000	.000	.862
.34	.008	.600	.000	.851	.880	.903	.002	.015	.000	.060	.001	.854
.36	.150	.799	.440	.889	.910	.905	.083	.290	.060	.470	.059	.816
.38	.445	.742	.795	.900	.915	.906	.136	.590	.445	.770	.160	.620
.40	.678	.190	.892	.916	.920	.920	.296	.725	.747	.885	.130	.090
.41	.688	.029	.904	.915	.920	.920	.273	.705	.790	.895	.044	.018
.42	.635	.000	.910	.915	.920	.921	.232	.770	.818	.902	.004	.003
.43	.586	.000	.913	.914	.920	.923	.191	.778	.836	.905	.000	.000
.44	.522	.000	.915	.913	.920	.924	.157	.801	.847	.906	.000	.000
.45	.458	.000	.916	.913	.922	.925	.144	.814	.855	.906	.000	.000
.46	.400	.000	.917	.914	.922	.925	.140	.823	.860	.906	.000	.000
.47	.350	.000	.917	.915	.922	.925	.141	.832	.863	.906	.000	.000
.48	.306	.000	.917	.915	.923	.925	.146	.839	.863	.905	.000	.000
.49	.275	.000	.918	.915	.930	.926	.152	.843	.859	.904	.000	.000
.50	.246	.000	.918	.915	.925	.926	.166	.843	.848	.900	.014	.000
.51	.223	.000	.919	.915	.925	.927	.178	.838	.825	.893	.180	.000
.52	.196	.000	.919	.916	.925	.928	.190	.824	.784	.880	.175	.000
.53	.172	.000	.919	.916	.923	.928	.196	.798	.720	.862	.018	.000
.54	.154	.000	.919	.916	.926	.929	.198	.756	.627	.831	.000	.000
.55	.148	.000	.920	.917	.923	.929	.197	.697	.515	.787	.050	.000
.56	.151	.000	.920	.917	.925	.930	.199	.615	.380	.728	.265	.000
.57	.146	.000	.919	.918	.925	.930	.206	.518	.255	.655	.165	.000
.58	.125	.000	.918	.919	.925	.930	.217	.414	.150	.570	.035	.000
.59	.102	.000	.918	.920	.925	.930	.222	.302	.075	.475	.004	.000
.60	.093	.000	.920	.920	.925	.930	.215	.215	.032	.380	.000	.000
.61	.087	.000	.920	.920	.925	.930	.196	.135	.010	.290	.000	.000
.62	.081	.000	.920	.920	.925	.930	.175	.080	.002	.210	.000	.000
.63	.070	.000	.920	.919	.926	.930	.161	.042	.000	.145	.000	.000
.64	.061	.000	.920	.919	.927	.931	.156	.021	.000	.094	.000	.000
.65	.055	.000	.920	.918	.927	.931	.162	.008	.000	.059	.000	.000
.66	.055	.000	.920	.917	.927	.932	.176	.003	.000	.035	.000	.000
.67	.059	.000	.920	.916	.928	.932	.200	.000	.000	.020	.000	.000
.68	.065	.000	.920	.916	.927	.932	.228	.000	.000	.010	.075	.022
.69	.068	.007	.921	.915	.927	.932	.248	.000	.000	.005	.380	.106
.70	.068	.036	.921	.915	.928	.932	.251	.000	.000	.001	.642	.234
.71	.066	.085	.922	.914	.926	.933	.237	.000	.000	.000	.694	.332
.72	.064	.145	.922	.912	.926	.933	.223	.000	.000	.000	.666	.383
.73	.060	.222	.922	.910	.926	.933	.210	.000	.000	.000	.607	.384
.74	.057	.323	.921	.909	.927	.934	.197	.000	.000	.000	.531	.358
.75	.055	.385	.921	.907	.928	.934	.180	.000	.000	.000	.445	.322
.80	.050	.287	.918	.890	.930	.932	.110	.000	.000	.000	.045	.175
1.00	.085	.032	.910	.860	.930	.928	.032	.000	.000	.000	.000	.119
1.20	.180	.021	.910	.860	.925	.928	.032	.000	.000	.005	.007	.016
1.40	.295	.109	.906	.870	.925	.930	.062	.018	.000	.080	.000	.005
1.60	.405	.088	.910	.892	.931	.930	.137	.131	.011	.266	.000	.007
1.80	.495	.040	.903	.896	.931	.930	.182	.317	.085	.430	.157	.011
2.00	.590	.008	.900	.897	.934	.929	.171	.440	.216	.512	.041	.029
2.20	.628	.002	.898	.875	.934	.835	.184	.440	.278	.455	.000	.048
2.40	.640	.009	.890	.850	.930	.890	.225	.440	.325	.433	.000	.060
2.60	.630	.018	.860	.820	.920	.780	.258	.280	.212	.252	.057	.051
2.80	.470	.030	.375	.140	.908	.180	.145	.060	.040	.060	.020	.000
3.00	.248	.030	.425	.360	.880	.695	.130	.000	.000	.000	.000	.000
3.20	.121	.030	.380	.490	.861	.760	.125	.000	.000	.000	.000	.000
3.40	.032	.020	.310	.270	.670	.620	.099	.000	.000	.000	.000	.000
3.60	.010	.015	.270	.010	.111	.080	.115	.000	.000	.000	.000	.000
3.80	.010	.020	.275	.040	.270	.240	.142	.000	.000	.000	.000	.000
4.00	.010	.030	.260	.013	.170	.150	.172	.000	.000	.000	.000	.000
4.20	.006	.017	.180	.026	.250	.230	.158	.000	.000	.000	.000	.000
4.40	.001	.002	.040	.004	.085	.080	.078	.000	.000	.000	.000	.000
4.60	.000	.000	.000	.000	.050	.020	.010	.000	.000	.000	.000	.000
4.80	.000	.000	.000	.000	.000	.000	.003	.000	.000	.000	.000	.000
5.00	.000	.000	.000	.000	.000	.000	.000	.000	.000	.000	.000	.000

From Weast, R. C., Ed., *CRC Handbook of Chemistry and Physics*, 61st ed., CRC Press, Boca Raton, Fla., 1980, E-402. With permission.

TRANSMISSION OF WRATTEN FILTERS

The following pages give (1) percentage luminous transmittance at wave lengths from 400 to 700 μ at intervals of 10 μ for the standard illuminant "C" adopted by the International Commission of Illumination, (2) dominant wave length in millimicrons, and (3) percentage of excitation purity. Values of wave length followed by "c" indicate the complementary wave lengths of purple filters which do not have a dominant wave length.

All colorimetric specifications are based on the 1931 standard ICI colorimetric and luminosity data.

The transmittance data are given as representing standard samples of the filters. They are intended only for the information of users in choosing filters which will meet their requirements. Values taken from the tables of data should not be used by research workers as representing precisely the absorption characteristics of a particular filter. If such precise data are needed, they should be determined for the particular filter being used.

Where the spectra extend into the ultraviolet this fact is indicated by an asterisk(*) in the transmission tables immediately beneath the filter number, and quantitative data are not given. The manufacturer should be consulted for this information. Transmission in the ultraviolet of wave lengths less than 330 μ will be eliminated in the case of cemented filters, as glass absorbs ultraviolet radiation of wave lengths shorter than about 330 μ.

Stability ratings are given as three letter combinations following the filter description in the table below. In establishing the stability classifications each filter is exposed to a selected light souce for a specific time interval. The following grading system is used to describe the result:

Class A-stable
Class B-relatively stable
Class C-somewhat unstable
Class D-unstable

The classification letters, for example, AAA, describe the stability to the following three exposure tests in this order:

1. Two weeks' exposure to daylight in a south window
2. Twenty-four hours' exposure to a "Fade-Ometer"
3. Two weeks' exposure at two feet from a 1000-watt tungsten lamp

Filters are supplied in two forms; as lacquered gelatin film, or as a gelatin film cemented between pieces of optical glass. Filters in glass are cemented between sheets of plane-parallel glass, which is surfaced in quantities and is of sufficient accuracy for general photographic work, and for most scientific purposes.

Most Wratten Gelatin Filters are stocked in 2- or 3-inch squares. Stocks of 2- or 3-inch square filters cemented in glass are maintained only in filters usually used for general photographic work.

The booklet "Kodak Filters and Lens Attachments" gives more valuable information on this subject.

Table 89
TRANSMISSION OF WRATTEN FILTERS

No.	Description, use, and stability
	Colorless
0	For compensating thickness of other gelatin filters in optical systems, AAA.
1	Absorbs ultraviolet below 360 mμ, DDD.
1A	Kodak Skylight Filter—Reduces excess bluishness in outdoor color photographs in open shade under a clear, blue sky, ACA.
	Yellows
2B	Absorbs ultraviolet below 410 mμ, ACA.
3	Light yellow, CCD.
3N5	No. 3 plus 0.5 neutral density, AAA.
4	Light yellow—Approximate correction on panchromatic materials for outdoor scenes, including sky, CCC.
6	K1—Light yellow—Partial correction outdoors, BBA.
8	K2—Yellow—Full correction outdoors on Type B panchromatic materials. Widely used for proper sky, cloud, and foliage rendering. Green separation for Fluorescence Process, AAA.
8N5	No. 8 plus 0.5 neutral density, AAA.
9	K3—Deep yellow. Moderate contrast in outdoor photography (with black-and-white films), AAA.
11	X1—Greenish yellow. Correction for tungsten light on Type B panchromatic materials; also for daylight correction with Type C panchromatic materials in making outdoor portraits, darkening skies, or lightening foliage, AAA.
12	Minus blue. Haze cutting in aerial photography, AAA.
13	X2—Yellow green. Correction for Type C panchromatic materials in tungsten light, ABA.
15	G—Deep yellow. Overcorrection in landscape photography. Contrast control in copying and in aerial infrared photography, AAA.
	Oranges and Reds
16	Blue absorption, AAB.
18A	Transmits ultraviolet and infrared only (glass), AAA.
21	Blue and blue-green absorption, CBB.
22	Yellow-orange. For increasing contrast in blue preparations in microscopy. Mercury yellow, BAC.
23A	Light red. Two-color projection—contrast effects, BAB.
24	Red for two-color photography (daylight or tungsten). White-flame-arc tricolor projection, AAB.
25	A—Tricolor red for direct color separation. Contrast effects in commercial photography and in outdoor scenes. Two-color general viewing. Aerial infrared photography and haze cutting, AAA.
26	Stereo red, AAA.
29	Red color separation from transparencies and for the Kodak Fluorescence Process. Strong contrast effects. Copying blue-prints. Tungsten tricolor projection, AAA.
	Magentas and Violets
30	Green absorption, BBC.
31	Green absorption, CCA.
32	Minus green, CCD.
33	Strong green absorption, CCB.
34	Violet, CDD.
34A	Blue separation—Kodak Fluorescence Process, DCC.
35	Contrast in microscopy, CDD.
36	Dark violet, CCC.
	Blues and Blue-greens
38	Red absorption, BCA.
38A	Red absorption. Increasing contrast in visual microscopy, BBB.
39	Contrast control in printing motion-picture duplicates (glass), AAA.
40	Green for two-color photography (tungsten), CBC.
44	Minus red—Two-color general viewing, DDD.
44A	Minus red, DDD.
45	Contrast in microscopy, DDD.
45A	Highest resolving power in visual microscopy, CDC.
46	Blue projection (experimental), DDD.
47	Tricolor blue for direct color separation and from Kodak Ektacolor Film for Dye Transfer. Contrast effects in commercial photography. Tungsten and white-flame-arc tricolor projection, BBC.
47B	Tricolor blue for color separation from transparencies and from Kodak Ektacolor Film for Graphic Arts, BBB.
48	Green and red absorption, CBC.
48A	Green and red absorption, AAB.
49	Dark blue, BCB.
49B	Very dark blue, BBB.
50	Very dark blue. Mercury violet, CCC.
	Greens
52	Light green, AAB.
53	Medium green, CCB.
54	Very dark green, AAA.
55	Stereo green, BBC.
56	Very light green, CBC.

No.	Description, use, and stability
	Greens (Continued)
57	Green for two-color photography (daylight), CBC.
57A	Light green, BBC.
58	Tricolor green for direct color separation. Contrast effects in commercial photography and microscopy, BBC.
59	Green for tricolor projection (white-flame-arc), BBB.
59A	Very light green, BBB.
60	Green for two-color photography (tungsten), BDC.
61	Green color separation from transparencies and Kodak Ekta-color Film. Tricolor projection (tungsten), ABC.
64	Red absorption (light), CDB.
65	Red absorption, ADB.
65A	Red absorption, CCD.
66	Contrast effects in microscopy and medical photography, DDC.
67A	Red absorption (light). Two-color projection, CDC.
	Narrow-band
70	Dark red. Infrared photography. Color separation for Kodak Ektacolor Film (with tungsten), ABC.
72B	Dark orange-yellow, CCC.
73	Dark yellow-green, ABB.
74	Dark green. Mercury green, BBC.
75	Dark blue-green, ACC.
76	Dark violet (compound filter), DDD.
77	Transmits 546 mµ mercury line (glass plus gelatin), AAA.
77A	Transmits 546 mµ mercury line (glass plus gelatin), AAA.
	Photometrics
78	Bluish. Photometric filter (visual), BAB.
78AA	Bluish. Photometric filter (visual), BAA.
78A	Bluish. Photometric filter (visual), AAA.
78B	Bluish. Photometric filter (visual), AAA.
78C	Bluish. Photometric filter (visual), BAA.
86	Yellowish. Photometric filter (visual), BBA.
86A	Yellowish. Photometric filter (visual), AAA.
86B	Yellowish. Photometric filter (visual), BCA.
86C	Yellowish. Photometric filter (visual), AAA.

No.	Description, use, and stability
	Light Balancing
80A	For Kodachrome Film, Daylight Type, and photographic flood lamps, ABA.
81	Yellowish. For warmer color rendering.
81A	Yellowish. For Kodak Ektachrome Film, Type B, with photographic flood lamps.
81B	Yellowish. For warmer color rendering.
81C	Yellowish. For Kodachrome Film, Type A, with flash lamps.
81D	Yellowish. For Kodachrome Film, Type A, with flash lamps.
81EF	Yellowish. For Kodak Ektachrome Film, Type B, with flash lamps.
82	Bluish. For cooler color rendering.
82A	Bluish. For Kodachrome Film, Type A, with 3200 K lamps.
82B	Bluish. For cooler color rendering.
82C	Bluish. For cooler color rendering.
83	Yellowish. For 16 mm Commercial Kodachrome Film and daylight exposure, BBB.
85	Orange. For Type A Kodak color films and daylight exposure, BAA.
85B	Orange. For Kodak Ektachrome Film, Type B, and daylight exposure, BAB.
	Miscellaneous
79	Photographic sensitometry. Corrects 2360 K to 5500 K, AAA.
87	For infrared photography. Absorbs visual.
87C	Absorbs visual, transmits infrared.
88A	For infrared photography. Absorbs visual.
89B	For infrared photography, AAA.
90	Narrow-band viewing filter for judging brightness scale of scenes, CCD.
96	Neutral filters for controlling luminance, AAB.
97	Dichroic absorption, AAA.
102	Correction filter for Barrier-layer photocell, ABA.
106	Correction filter for S-4 type photocell, AAA.

Table 89 (continued)
TRANSMISSION OF WRATTEN FILTERS

Wave length	No. 0 *	No. 1 *	No. 1A *	No. 2B	No. 3	No. 3N5 *	No. 4 *	No. 6 *	No. 8	No. 8N5	No. 9	No. 11 *	No. 12 *
400	88.0	85.0	59.0	19.0				7.40					
10	88.5	85.5	76.0	48.0				8.32					
20	88.9	86.0	82.0	67.0				10.4					
30	89.3	86.5	84.6	75.3	0.36			13.5					
40	89.6	87.4	86.0	80.0	1.78			18.9	0.25			0.16	
50	89.8	87.8	86.8	83.0	11.5	1.59		27.6	5.50	0.16		0.29	
60	89.9	88.2	87.2	85.2	38.0	9.40	6.9	39.0	19.0	2.0	1.78	0.56	
70	90.1	88.5	87.5	86.7	68.0	18.5	42.0	52.3	41.0	6.3	8.31	1.32	
80	90.3	88.7	87.3	88.1	80.8	23.5	74.0	65.8	63.5	13.2	20.7	4.00	
90	90.4	88.8	86.8	88.8	85.2	25.5	84.7	76.5	78.0	20.3	34.5	12.0	1.50
500	90.5	89.1	86.3	89.5	86.9	26.3	87.5	83.5	84.1	24.3	48.0	26.0	17.3
10	90.6	89.3	85.5	89.9	87.8	26.7	88.5	87.0	86.5	26.7	62.0	43.7	55.0
20	90.7	89.5	84.8	90.3	88.4	27.0	89.1	88.4	87.7	28.0	76.0	55.0	77.0
30	90.7	89.7	84.3	90.5	89.0	27.2	89.4	89.0	88.4	28.6	83.5	60.0	86.0
40	90.8	89.9	84.0	90.6	89.5	27.5	89.6	89.4	88.8	29.0	87.0	60.2	88.4
50	90.8	90.1	83.9	90.7	89.8	27.8	89.8	89.7	89.2	29.3	88.0	57.8	89.4
60	90.9	90.2	84.1	90.8	90.1	27.0	90.0	89.9	89.5	29.5	88.3	54.2	89.7
70	90.9	90.3	84.8	90.9	90.4	28.0	90.2	90.1	89.8	29.6	88.8	50.0	90.1
80	91.0	90.4	86.0	90.9	90.5	28.4	90.4	90.3	90.1	29.8	89.1	44.8	90.3
90	91.0	90.4	87.4	91.1	90.7	29.0	90.6	90.5	90.3	29.9	89.3	38.5	90.4
600	91.0	90.5	88.5	91.1	90.8	29.5	90.8	90.6	90.5	29.4	89.5	33.1	90.5
10	91.0	90.5	89.5	91.2	90.9	29.5	90.9	90.8	90.7	29.1	89.7	27.6	90.7
20	91.0	90.6	90.2	91.3	91.0	29.3	91.0	90.9	90.9	28.9	89.8	22.7	90.8
30	91.1	90.7	90.6	91.3	91.1	29.0	91.1	91.0	91.0	28.9	89.9	19.0	90.9
40	91.1	90.7	90.8	91.4	91.2	29.4	91.2	91.2	91.1	29.2	90.0	14.9	91.0
50	91.1	90.8	91.0	91.4	91.3	29.6	91.3	91.3	91.1	29.4	90.1	11.4	91.0
60	91.1	90.8	91.1	91.5	91.4	29.8	91.4	91.3	91.2	29.5	90.1	9.10	91.2
70	91.1	90.9	91.1	91.5	91.5	30.0	91.5	91.4	91.3	29.7	90.2	8.05	91.2
80	91.1	90.9	91.1	91.7	91.6	30.2	91.5	91.5	91.4	30.2	90.2	7.50	91.2
90	91.1	91.0	91.1	91.8	91.7	31.0	91.6	91.5	91.5		90.3	7.05	91.3
700	91.1										90.3	6.50	
Luminous transmit.	90.8	89.9	85.9	90.5	88.3	27.4	87.8	87.5	82.7	27.0	76.6	40.2	73.8
Dominant wave lgth.	571.0	575.0	498.0	570.0	569.5	570.5	569.5	570.3	571.8	572.0	574.4	550.3	576.1
Excitation purity.	0.8	1.5	1.2	5.7	50.0	56.3	64.0	44.7	85.2	84.0	91.4	60.7	97.8

Percent transmittance

Percent transmittance

Wave length	No. 13*	No. 15*	No. 16	No. 18A*	No. 21	No. 22	No. 23A	No. 24	No. 25	No. 26	No. 29	No. 30	No. 31*
400												48.6	13.8
10												47.4	14.5
20	0.18											48.5	16.4
30	0.50											50.1	25.5
40	1.35											49.4	42.7
50	4.08											43.0	50.2
60	11.0											26.5	40.4
70	23.5											13.8	22.6
80	39.0											5.00	8.20
90	50.8											0.63	1.85
500	55.2	1.00	3.00										0.12
10	56.5	16.0	22.0										
20	55.0	52.1	48.0										
30	51.0	70.7	69.5										
40	46.0	84.3	79.5		2.50	0.25							
50	39.2	87.5	84.0		29.0	19.0							
60	32.0	88.7	86.3		65.0	60.0	11.0						
70	25.1	89.3	87.8		80.6	81.0	47.0	4.55				0.10	
80	18.2	89.7	89.0		85.4	87.0	69.6	37.3	12.6	2.90		10.0	0.63
90	13.5	90.0	89.6		87.3	88.5	82.7	72.3	50.0	30.0		45.0	26.0
600	9.60	90.1	90.0		88.1	89.0	85.8	82.9	75.0	63.2	10.0	76.0	67.2
10	6.40	90.2	90.2		88.7	89.5	87.2	86.4	82.6	78.9	45.3	87.4	84.0
20	3.66	90.3	90.3		89.0	89.8	87.9	87.8	85.5	84.0	71.4	89.5	88.1
30	2.20	90.4	90.4		89.5	90.0	88.5	88.5	86.7	86.1	82.7	90.2	89.8
40	1.58	90.5	90.5		89.9	90.1	89.0	89.0	87.6	87.2	86.6	90.7	90.2
50	1.74	90.6	90.6		90.2	90.2	89.4	89.3	88.2	88.1	88.4	90.8	90.4
60	2.62	90.6	90.7		90.4	90.3	89.6	89.7	88.5	88.5	89.4	90.9	90.5
70	3.55	90.7	90.8		90.5	90.4	89.8	89.9	89.0	88.9	90.0	91.0	90.7
80	4.48	90.7	90.8		90.6	90.5	90.0	90.2	89.3	89.2	90.3	91.1	90.8
90	5.25	90.8		0.25	90.6	90.6	90.2	90.3	89.5	89.5	90.4	91.1	91.0
700				1.20									
Luminous transmit.	34.5	66.2	57.7	0.0014	45.6	35.8	25.0	17.8	14.0	11.7	6.3	26.6	12.9
Dominant wave lgth.	542.0	579.3	582.7	700.0	588.9	595.1	602.7	610.6	615.1	619.0	631.6	498.6c	513.1c
Excitation purity.	57.5	99.0	99.3	100.0	99.9	99.9	100.0	100.0	100.0	100.0	100.0	62.4	81.9

*Some transmission below 400 mμ.   Consult the manufacturer.

Table 89 (continued)
TRANSMISSION OF WRATTEN FILTERS

Percent transmittance

Wave length	No. 32 *	No. 33 *	No. 34 *	No. 34A *	No. 35 *	No. 36 *	No. 38 *	No. 38A *	No. 39 *	No. 40 *	No. 44 *	No. 44A *	No. 45 *
400	38.0	0.85	64.0		48.0	36.5	60.5	33.4	85.2		0.44	2.52	
10	37.9	0.71	70.1	0.1	57.0	45.5	66.5	41.2	78.2		0.36	3.39	
20	40.0	1.17	72.0	40.0	57.6	45.5	72.5	53.0	70.5		0.63	6.30	
30	43.0	1.69	68.4	69.7	47.5	32.7	75.3	58.0	63.0		3.63	17.4	5.00
40	55.5	5.36	58.2	68.7	29.5	15.2	76.2	58.8	53.6	3.16	13.1	32.7	19.0
50	66.0	14.3	42.3	56.2	12.3	3.7	75.0	57.6	42.5	21.6	25.4	41.8	29.5
60	66.0	12.4	25.2	40.5	3.5	0.35	74.8	55.2	28.5	44.7	36.5	48.1	34.4
70	57.0	5.00	12.1	23.8	0.25		73.4	51.9	17.3	61.4	46.5	51.7	35.7
80	40.0	0.50	2.7	9.2			71.6	48.5	10.2	70.2	53.6	52.9	34.5
90	21.0		0.2	2.3			69.5	44.6	4.00	72.4	56.8	52.2	29.7
500	9.56			0.33			66.7	40.2	1.33	70.5	55.8	49.8	21.5
10	2.51						63.9	35.8	0.35	64.8	50.9	44.8	11.5
20	0.13						60.8	31.7		55.5	42.1	36.8	3.80
30							57.0	27.2		44.2	30.5	26.8	0.85
40							52.6	22.3		32.5	18.6	16.8	
50			0.4	0.13	0.1	0.21	48.0	17.6		20.3	8.99	8.20	
60			4.0	1.0	3.0	7.5	42.8	12.9		9.56	3.59	2.95	
70			20.7	6.3	19.0	29.0	37.0	8.78		3.20	0.80	0.91	
80			45.2	22.0	43.5	55.0	30.6	5.65		1.10		0.10	
90			66.5	45.0	66.0	71.3	25.5	3.48		0.32			
600	6.04	0.80	78.8	65.0	77.7		20.9	2.09					
10	41.0	24.9	85.0	77.3			16.8	1.15					
20	75.0	60.8		85.0			12.9	0.59					
30	86.1	78.0		88.2			10.0	0.28					
40	89.0	85.0		89.8			7.79	0.13					
50	90.6	87.5					6.68						
60	90.7	88.7					6.20						
70	90.8	89.4					5.91		0.50	0.80	0.18		
80	90.9	89.8					5.41		4.06	6.99	1.60		
90	91.0	90.0					4.90		17.8	23.5			1.00
700							5.00						
Luminous transmit.	12.5	5.2	1.3	2.9	0.45	0.25	42.5	17.3	1.2	33.6	15.6	14.4	5.2
Dominant wave lgth.	551.7c	498.0c	424.0	564.8c	566.8c	566.4c	483.5	478.9	450.6	516.2	589.1	483.4	481.5
Excitation purity	79.6	88.3	94.4	91.4	96.3	97.8	41.8	69.8	98.9	48.5	72.9	77.2	88.4

Percent transmittance

Wave length	No. 45A	No. 46*	No. 47*	No. 47B*	No. 48*	No. 48A*	No. 49*	No. 49B*	No. 50	No. 52*	No. 53*	No. 54	No. 55
400		1.20	7.80	16.0	0.96	5.65	3.30	1.70	0.45	2.18			
10		0.60	17.4	29.5	3.16	10.0	4.28	2.00	0.39	1.51			
20	1.00	0.80	34.0	43.6	8.25	16.0	6.93	3.55	0.59	0.80			
30	8.81	5.98	47.0	50.0	15.0	21.0	11.2	7.00	2.63	0.44			
40	17.4	19.0	50.3	47.2	22.6	25.0	18.9	13.0	8.90	0.41			
50	20.9	30.1	48.3	36.0	30.3	26.2	25.6	17.4	14.0	0.69			
60	21.6	33.8	43.4	25.0	33.2	22.9	24.0	14.8	12.3	1.45	0.10		
70	20.5	32.1	36.2	13.2	29.6	16.5	15.7	7.60	5.36	2.70	0.71		0.20
80	18.0	27.0	28.5	4.5	22.4	9.55	6.93	2.76	1.55	4.90	2.14		2.90
90	14.4	20.2	19.6	1.3	14.1	4.27	2.14	0.40	0.10	8.50	4.47		13.1
500	10.1	11.1	11.3	0.17	7.30	1.58	0.46			13.3	7.24	0.10	34.2
10	5.60	4.39	5.64		2.64	0.48				18.2	10.7	0.31	53.4
20	2.52	1.66	1.91		0.50					23.7	14.0	0.64	67.0
30	0.04	0.35	0.36							28.5	16.6	0.89	69.3
40	0.10									32.1	17.3	0.93	65.1
50										33.1	15.4	0.62	56.7
60										31.0	11.4	0.21	45.0
70										25.6	6.90		33.1
80										19.1	3.60		20.7
90										12.6	1.41		9.00
600										7.78	0.40		2.70
10										4.17	0.15		0.40
20										2.34			
30										1.38			
40										0.80			
50										0.54			
60										0.36			
70										0.27			
80	0.20	0.25								0.23			0.66
90	0.24	0.85								0.19			6.90
700										0.17			27.8
Luminous transmit.	2.8	2.4	2.8	0.78	1.86	0.88	0.69	0.36	0.26	20.1	9.0	0.032	31.4
Dominant wave lgth.	477.6	470.4	463.7	479.8	466.5	458.0	457.9	455.5	455.9	553.3	551.1	546.1	530.2
Excitation purity.	89.7	94.9	95.8	69.1	96.1	98.3	98.9	99.3	99.4	77.3	89.7	97.0	68.4

* Some transmission below 400 mμ. Consult the manufacturer.

Table 89 (continued)
TRANSMISSION OF WRATTEN FILTERS

Percent transmittance

Wave length	No. 56	No. 57	No. 57A	No. 58	No. 59*	No. 59A*	No. 60	No. 61	No. 64*	No. 65	No. 65A*	No. 66*	No. 67A*
400									9.00			12.3	1.10
10									9.20			13.0	0.93
20			0.19						8.75			15.0	1.28
30			0.87			0.16			9.20	0.23		18.4	3.16
40			2.56			0.37	0.19		11.3	0.61	0.16	23.2	6.40
50			7.80		0.40	1.26	1.38		15.5	1.58	1.32	31.2	10.5
60	0.16	0.44	21.6	0.23	1.90	4.57	5.38		23.3	4.10	5.50	42.2	17.7
70	3.12	3.10	41.7	1.38	7.70	13.2	15.0		34.4	9.00	13.0	55.5	28.5
80	13.0	13.1	58.8	4.90	21.5	30.0	32.0	0.33	46.8	16.8	24.9	68.4	41.4
90	34.5	31.9	67.9	17.7	41.5	50.8	48.4	4.00	56.6	24.9	36.6	77.6	52.1
500	59.0	50.5	70.1	38.8	59.0	66.0	57.2	16.6	62.1	31.3	45.1	82.7	57.9
10	73.0	60.6	67.6	52.2	67.9	73.0	59.5	32.3	62.9	33.7	39.7	84.0	58.8
20	79.0	63.3	61.8	53.6	69.8	75.1	55.5	40.0	59.1	32.4	29.7	84.0	55.4
30	79.9	61.0	53.5	47.6	67.2	73.2	47.5	39.6	51.6	27.5	17.9	82.6	47.5
40	77.5	55.0	43.3	38.4	61.5	68.5	36.8	34.5	41.3	20.7	7.90	79.1	36.0
50	72.6	47.1	31.6	27.8	54.0	62.0	25.2	26.3	28.0	13.7	2.40	73.7	25.0
60	66.1	37.3	19.4	17.4	45.0	54.4	14.4	17.3	16.2	6.50	0.32	67.1	14.2
70	58.0	26.5	9.70	9.0	35.0	44.5	6.3	9.70	7.95	1.66		58.8	5.50
80	46.1	16.6	4.50	3.50	24.0	33.0	1.82	4.40	3.10	0.40		47.2	1.40
90	33.8	8.69	2.00	1.50	14.0	22.0	0.48	1.66	0.80			34.4	0.28
600	24.0	3.70	0.87	0.41	7.95	14.6	0.10	0.38				24.4	
10	18.7	1.60	0.22		4.90	10.5						18.7	
20	13.2	0.49			2.70	6.92						13.7	
30	7.22				1.00	3.16						7.70	
40	3.02				0.17	1.07						3.00	
50	1.48					0.50						1.46	
60	1.91					0.91						1.91	
70	7.95				0.63	3.00						6.17	
80	23.0		0.16		4.00	10.0						19.9	
90	44.1		1.15		12.0	20.0	2.10		0.10		0.20	42.6	
700	64.8		3.17	0.53	22.6	30.0	8.70		4.50		2.18	63.1	0.40
Luminous transmit.	52.8	32.5	37.2	23.7	38.7	45.8	26.1	16.8	25.0	9.6	9.8	58.3	22.4
Dominant wave lgth.	552.3	536.4	534.0	540.2	538.3	541.4	525.7	536.8	497.3	496.6	492.7	512.3	499.8
Excitation purity.	78.2	69.2	62.1	88.1	66.0	59.3	62.2	85.4	55.0	67.8	77.4	21.5	55.8

Percent transmittance

Wave length	No. 70	No. 72B	No. 73	No. 74	No. 75	No. 76 *	No. 77	No. 77A	No. 78	No. 78AA *	No. 78A *	No. 78B *
400						0.22			37.2	43.0	56.0	64.1
10						0.18			41.7	46.0	58.6	66.5
20						0.29			44.2	48.7	61.0	68.4
30						1.38			44.6	49.8	61.8	69.5
40						3.50			44.2	49.7	61.0	70.0
50						3.50			41.7	48.0	61.0	69.4
60					1.97	1.92			38.0	44.9	58.7	67.5
70					10.0	0.51			33.8	40.3	55.0	65.4
80					17.4				27.5	35.6	51.0	62.9
90					18.0				23.5	30.9	47.1	59.8
500					13.0				19.5	26.5	43.5	57.0
10		1.26		0.96	7.35		0.30	0.10	15.8	23.4	40.0	54.2
20		5.89	2.24	7.95	3.20		9.10	5.35	13.8	20.3	36.9	51.4
30		5.25	5.97	14.6	0.83		13.5	1.90	11.8	17.8	34.4	49.3
40		2.88	4.56	12.9	0.14		46.0	35.0	10.5	16.6	32.7	48.1
50		1.26	2.00	7.60			78.0	71.8	9.56	14.9	31.2	46.7
60		0.48	0.56	3.06			75.8	63.1	8.53	13.2	29.4	45.0
70		0.14	0.10	0.83			8.00		7.77	12.1	28.5	43.6
80				0.12			1.00		7.41	11.6	27.0	43.1
90							0.32		6.93	11.1	27.0	42.9
600							16.2	1.60	6.45	10.40	26.0	41.8
10							52.1	32.1	5.50	9.20	24.1	40.0
20							83.0	78.0	4.80	7.70	21.8	37.6
30							84.9	79.5	3.94	6.50	19.7	35.5
40							88.1	86.5	3.46	5.60	18.6	34.2
50	0.63						89.8	89.2	3.24	5.50	18.4	33.6
60	10.5						89.9	89.0	3.16	5.60	18.5	34.1
70	35.0						85.5	89.5	3.39	5.80	18.7	34.1
80	55.2						76.0	62.5	3.45	6.10	19.0	34.5
90	70.0					0.13	75.0	62.4	3.51	6.10	19.3	34.8
700	79.0				0.14	1.24	86.5	83.0	3.90	6.50	20.2	36.0
Luminous transmit.	0.31	0.74	1.3	4.0	1.9	0.046	32.3	25.5	10.7	15.8	31.6	46.7
Dominant wave lgth.	675.6	604.9	574.9	538.6	487.7	449.2	579.9	581.5	471.1	473.4	475.7	477.2
Excitation purity.	100.0	100.0	100.0	96.7	90.4	99.7	99.0	99.1	63.0	54.5	33.7	20.7

* Some transmission below 400 mμ.   Consult the manufacturer.

Table 89 (continued)
TRANSMISSION OF WRATTEN FILTERS

Percent transmittance

Wave length	No.78C *	No.79 *	No.80A *	No.81 *	No.81A *	No.81B *	No.81C *	No.81D *	No.81EF *	No.82 *	No.82A *	No.82B *
400	74.9	24.0	67.6	77.7	65.1	55.1	46.1	38.2	30.7	83.0	80.1	76.7
10	76.6	26.0	73.1	78.1	65.9	55.8	46.6	38.4	31.5	83.7	80.8	78.0
20	77.9	29.0	76.8	79.0	67.6	57.7	49.0	41.0	34.3	84.6	81.6	79.2
30	78.9	31.0	77.7	80.5	70.2	61.0	52.5	45.0	38.6	85.1	82.2	79.7
40	79.4	32.2	76.5	81.9	72.8	64.5	57.2	50.0	43.2	85.4	82.4	79.2
50	79.5	32.7	73.0	83.0	74.8	67.2	60.5	53.9	47.1	85.6	82.4	79.2
60	79.3	31.4	69.0	83.7	76.0	69.1	63.0	56.5	50.2	85.0	81.7	78.0
70	78.6	28.8	63.6	84.6	77.1	70.6	64.2	58.1	52.0	84.6	80.7	76.3
80	77.8	25.6	57.3	84.8	77.8	71.3	65.0	59.0	53.0	84.0	79.3	74.4
90	76.7	22.2	51.3	85.3	78.3	71.8	65.7	60.8	54.0	83.3	78.0	72.1
500	75.5	19.3	45.2	85.4	78.6	72.9	66.5	60.8	55.4	82.6	76.6	70.2
10	74.2	16.8	39.4	85.5	79.0	73.2	67.0	61.1	56.2	81.4	75.3	68.3
20	73.0	14.2	34.2	86.0	79.5	74.5	68.8	61.6	57.0	81.0	74.0	66.5
30	72.1	12.7	30.0	86.5	80.4	76.0	71.0	62.5	59.5	84.0	73.1	65.5
40	71.5	11.0	27.1	86.8	81.5	77.0	72.0	66.1	62.7	80.8	72.4	65.0
50	70.7	9.76	24.8	87.0	82.3	77.6	72.5	67.3	64.5	80.6	72.7	64.5
60	69.8	8.81	23.5	87.1	82.6	77.8	72.7	68.0	65.3	80.4	71.8	63.8
70	69.0	8.50	22.6	87.4	82.8	78.0	73.0	68.3	66.0	80.2	71.5	63.2
80	68.8	8.29	22.6	87.4	83.1	78.2	74.0	68.5	66.5	80.3	71.5	63.4
90	68.6	7.56	23.2	87.6	84.0	79.1	75.6	69.5	68.1	80.2	71.5	63.0
600	68.0	6.45	23.7	88.1	85.0	81.0	78.5	72.0	71.6	79.3	70.3	61.5
10	66.7	5.13	23.2	88.8	86.1	83.1	80.8	75.0	74.7	78.4	68.5	59.0
20	65.0	4.17	21.0	89.2	87.0	84.2	82.1	78.0	77.0	77.5	66.9	56.9
30	63.8	3.47	18.2	89.4	87.4	85.1	83.2	79.8	78.4	76.8	65.5	55.5
40	63.0	3.16	15.8	89.5	87.7	85.6	83.5	80.8	79.2	76.5	64.6	54.1
50	62.7	3.09	14.5	89.8	88.0	86.5	84.8	81.5	80.1	76.2	64.6	53.7
60	63.3	3.16	13.8	89.8	88.2	87.0	85.1	83.7	80.9	76.1	64.5	53.7
70	63.4	3.16	13.4	90.0	88.5	87.5	86.1	83.7	81.8	76.1	64.4	53.5
80	63.4	3.16	12.7	90.1	89.0	87.5	86.1	84.6	82.9	76.2	64.2	53.4
90	63.6	3.16	11.7	90.3	89.0	87.5	86.1	84.6	82.9	76.2	64.2	53.4
700	65.0	3.31	11.5	90.5	89.2	88.0	86.8	85.5	84.0	77.1	64.6	54.1
Luminous transmit.	70.4	11.3	28.4	86.8	82.0	76.9	72.0	67.4	64.0	80.7	72.5	64.6
Dominant wave lgth.	479.8	474.8	471.7	576.7	577.5	577.8	577.4	579.5	579.0	477.5	476.6	475.6
Excitation purity.	6.8	52.8	45.9	2.9	6.0	8.7	10.7	14.7	19.0	3.0	6.3	10.2

Wave length	Percent transmittance												
	No. 82C*	No. 83*	No. 85*	No. 85B	No. 86	No. 86A	No. 86B	No. 86C	No. 89B	No. 90	No. 96	No. 97	No. 102*
400	73.4	13.5	6.0	1.59	0.50	8.00	20.0	44.0	....	....	4.28	....	1.12
10	75.0	13.1	18.0	9.32	0.81	12.2	26.1	55.0	....	....	4.91	....	0.96
20	76.4	13.5	28.4	15.5	1.55	16.7	31.6	62.0	....	....	5.50	....	0.89
30	77.2	14.1	33.4	19.0	2.88	21.5	37.5	66.6	....	....	6.17	....	0.96
40	77.2	15.6	36.2	20.8	5.50	27.8	44.0	70.8	....	....	6.92	....	1.86
50	76.6	17.8	38.1	22.1	9.10	34.2	50.1	74.3	....	....	7.50	....	3.23
60	75.2	21.0	40.4	24.3	13.5	40.4	55.4	76.8	....	....	7.81	....	6.45
70	73.2	25.0	43.0	27.5	17.8	45.0	59.5	78.7	....	....	8.15	0.22	14.0
80	70.7	30.2	45.3	30.9	21.3	48.7	62.5	80.2	....	....	8.47	0.43	21.6
90	68.1	35.8	47.2	34.3	24.5	51.2	64.6	81.2	....	....	8.60	0.39	30.7
500	65.7	43.5	48.2	38.3	26.8	52.8	66.4	81.9	....	....	8.73	0.15	41.4
10	63.5	46.3	49.2	40.7	27.9	53.4	66.6	82.0	....	....	8.85	....	51.3
20	61.5	47.2	48.3	40.6	28.6	53.7	67.0	82.4	....	....	8.96	....	59.4
30	59.9	48.3	49.2	40.7	30.4	55.0	69.2	83.0	....	....	9.01	....	64.2
40	59.1	49.6	51.0	41.6	32.5	56.5	70.2	83.5	....	....	9.07	....	66.7
50	58.3	51.8	55.8	43.2	35.0	58.5	73.0	84.6	....	9.00	9.20	....	66.3
60	57.2	56.5	64.5	47.1	41.2	63.0	78.1	86.8	....	30.5	9.30	....	63.0
70	56.2	65.0	75.0	56.0	53.0	70.9	84.0	88.9	....	34.3	9.20	....	58.0
80	56.1	75.5	83.0	68.1	67.5	79.0	87.5	89.9	....	25.2	9.19	....	51.9
90	56.0	83.0	87.2	78.1	76.5	85.2	89.3	90.6	....	16.1	9.54	....	45.2
600	55.0	87.3	88.9	85.0	85.0	88.1	90.3	91.1	....	11.3	9.64	....	37.8
10	53.0	89.3	90.0	88.0	88.1	89.8	90.7	91.2	....	7.40	9.73	....	30.5
20	50.2	90.4	90.5	89.6	89.6	90.5	90.9	91.3	....	2.94	9.55	....	25.0
30	47.4	90.8	90.7	90.3	90.4	90.8	91.1	91.4	....	0.76	9.27	....	20.6
40	45.2	91.0	90.9	90.7	90.7	91.1	91.2	91.5	....	0.29	9.10	....	17.5
50	44.1	91.1	91.0	90.9	91.0	91.2	91.3	91.6	....	0.41	9.07	....	15.2
60	43.6	91.5	91.0	91.2	91.2	91.3	91.4	91.6	....	2.30	9.00	....	13.7
70	43.5	91.5	91.0	91.2	91.3	91.4	91.5	91.6	....	9.52	9.13	....	13.2
80	43.1	91.5	91.0	91.3	91.3	91.4	91.6	91.6	0.10	28.5	9.08	0.44	12.8
90	42.8	91.5	91.0	91.3	91.3	91.5	91.6	91.6	1.58	51.9	9.21	5.02	12.1
700	43.5	91.5	91.0	91.3	91.3	91.5	91.6	91.6	11.2		9.52	18.7	12.0
Luminous transmit.	58.1	61.4	62.5	55.5	49.7	67.1	75.5	85.4	0.017	9.8	9.1	0.041	50.8
Dominant wave lgth.	477.2	581.5	587.7	585.7	585.7	581.7	579.6	577.6	700	583.1	572.4	555.0e	564.9
Excitation purity.	14.5	55.4	30.3	48.0	69.7	37.1	24.1	9.0	100	100.0	12.1	48.0	80.0

*Some transmission below 400 mμ.   Consult the manufacturer.

Table 89 (continued)
TRANSMISSION OF WRATTEN FILTERS

Percent transmittance

Wave length	No. 106	CC-05R	CC-10R	CC-20R	CC-30R	CC-40R	CC-50R	CC-05B	CC-10B	CC-20B	CC-30B	CC-40B	CC-50B
400	....	81.0	73.0	61.5	51.6	42.5	36.4	87.0	85.5	82.2	80.2	77.0	74.1
10	....	81.0	72.4	60.0	50.0	40.0	33.9	87.5	86.4	84.0	82.5	80.3	78.4
20	0.10	81.1	72.0	58.6	48.2	38.2	31.9	87.7	87.5	85.0	84.0	82.2	80.7
30	0.20	81.2	71.6	57.7	47.0	36.8	31.5	88.0	87.5	85.3	84.3	82.5	81.1
40	0.35	81.4	71.5	57.2	46.4	36.0	30.5	88.1	87.2	85.0	83.5	81.3	79.8
50	0.58	81.7	71.6	57.2	46.4	36.1	29.7	88.1	87.0	83.9	81.9	78.7	76.6
60	0.98	81.7	72.4	57.1	47.5	37.5	29.6	87.9	86.4	82.5	79.5	75.9	72.9
70	1.5	82.3	73.7	58.5	49.9	40.0	31.0	87.5	85.3	80.3	76.2	72.0	67.0
80	2.3	82.8	74.9	60.6	52.0	42.5	33.6	87.0	84.0	77.8	72.5	67.5	62.7
90	3.5	83.3	75.8	62.0	53.9	44.8	35.5	86.2	82.4	74.2	68.3	62.3	56.6
500	5.2	83.0	76.0	63.5	55.5	46.1	37.0	85.2	80.5	71.2	63.8	56.7	50.1
10	7.7	83.3	76.1	64.0	54.5	46.0	39.4	84.4	78.6	67.7	58.7	51.0	44.5
20	10.7	82.4	74.9	64.0	51.6	42.5	38.5	83.5	77.0	64.0	54.4	46.0	38.6
30	15.1	81.6	73.5	61.5	48.5	38.5	35.0	82.6	75.2	61.5	50.7	41.6	34.1
40	20.7	81.2	72.5	59.4	46.6	36.6	31.5	82.1	73.9	59.5	48.3	39.0	31.3
50	25.7	81.1	72.4	57.1	45.6	36.4	29.4	81.5	73.0	58.0	46.6	36.9	29.5
60	31.0	81.4	74.5	57.1	46.2	39.2	28.7	81.4	72.7	57.5	45.9	35.5	28.6
70	35.6	82.5	74.3	58.0	49.2	45.5	29.7	81.4	73.0	57.9	46.3	36.1	28.7
80	43.2	83.9	77.3	60.0	54.6	53.5	32.7	82.7	73.9	59.3	47.9	37.8	30.5
90	53.8	85.7	80.7	65.0	61.8	61.8	38.6	83.4	75.1	61.6	50.3	40.8	33.4
600	65.0	87.0	84.0	71.0	70.5	64.0	47.0	83.6	76.3	63.5	53.0	43.5	36.0
10	77.0	89.0	86.5	77.0	77.8	73.0	58.7	83.5	76.7	64.3	53.6	44.7	37.5
20	82.8	90.6	88.9	82.0	83.5	80.8	69.8	83.2	76.5	63.1	54.4	44.3	37.5
30	86.6	90.1	89.9	86.2	87.2	85.1	78.3	82.8	76.0	61.6	53.2	42.5	35.6
40	87.6	91.6	90.5	88.1	89.2	88.0	84.2	82.5	74.5	60.6	51.5	40.2	33.7
50	88.7	91.2	90.8	89.8	90.3	89.5	87.5	82.5	74.0	60.1	50.3	39.0	32.4
60	89.5	91.3	91.1	90.5	90.6	90.4	89.2	82.3	73.8	59.6	49.5	38.4	31.8
70	90.5	91.4	91.3	90.8	90.9	90.8	90.1	82.0	73.3	58.6	49.0	37.7	31.0
80	90.8	91.5	91.5	91.0	91.1	91.3	90.7	81.9	72.8	58.1	48.2	36.5	30.0
90	91.0	91.7	91.7	91.4	91.4	91.4	91.1	82.2	72.5	58.1	47.2	35.4	29.0
700	91.0	91.9	91.9	91.5	91.5	91.4	91.2	82.2	73.0	58.5	47.5	35.6	29.0
Luminous transmit.	34.6	83.7	77.0	65.3	55.9	47.3	41.3	82.8	75.5	62.3	52.0	42.8	35.7
Dominant wave lgth.	589.4	605.0	597.8	604.2	605.8	605.5	608.5	459.0	462.0	460.0	461.0	463.2	462.5
Excitation purity.	95.2	2.0	4.7	8.5	12.3	17.3	21.4	2.8	6.3	13.2	20.2	27.7	34.2

Percent transmittance

Wave length	CC-05G	CC-10G	CC-20G	CC-30G	CC-40G	CC-50G	CC-05Y *	CC-10Y *	CC-20Y *	CC-30Y *	CC-40Y *	CC-50Y *
400	80.0	73.1	58.8	48.0	39.7	32.0	81.0	74.5	61.3	50.5	43.0	34.5
10	80.7	72.9	57.8	46.5	38.1	30.3	80.6	73.2	59.0	47.4	39.5	30.5
20	81.0	72.8	57.3	45.8	37.3	29.5	80.4	72.6	57.8	46.0	37.5	29.0
30	81.6	72.7	57.0	45.5	36.5	29.0	80.4	72.5	57.5	45.6	36.5	28.7
40	81.6	73.0	57.3	45.8	36.6	29.1	80.6	72.8	57.8	46.5	36.8	29.5
50	82.1	73.0	58.4	46.9	38.1	30.6	81.2	74.0	59.5	48.5	38.5	31.5
60	83.0	75.5	61.4	50.3	41.5	34.3	82.5	76.0	63.0	52.5	42.5	36.5
70	84.4	78.0	65.0	55.8	47.0	40.5	83.9	78.5	67.5	58.2	48.8	43.5
80	85.6	80.4	70.0	61.8	53.5	47.0	85.0	81.2	72.3	64.9	56.2	54.0
90	86.8	83.0	73.2	68.9	61.3	57.0	87.0	84.4	78.0	72.4	66.0	64.0
500	87.9	85.9	80.3	76.4	70.7	68.0	88.4	87.2	84.0	81.0	77.0	75.5
10	88.7	87.5	83.8	80.6	77.8	73.3	89.5	89.0	88.0	86.6	85.5	84.2
20	89.0	88.1	84.6	82.3	79.5	77.0	90.0	90.0	89.6	89.1	89.0	88.5
30	89.0	88.6	83.7	81.7	79.4	74.8	90.4	90.4	90.6	89.7	89.9	89.6
40	88.6	87.1	82.4	80.5	77.8	72.2	90.7	90.7	90.8	90.4	90.2	90.0
50	88.1	86.3	80.9	78.6	75.8	68.8	91.0	90.9	90.9	90.8	90.4	90.3
60	87.5	85.3	79.0	76.2	72.9	64.3	91.3	91.0	91.1	90.9	90.7	90.6
70	87.0	84.1	77.0	73.5	69.3	60.3	91.4	91.4	91.2	91.0	90.8	90.7
80	86.4	82.8	74.5	70.4	65.3	55.9	91.4	91.4	91.3	91.2	90.8	90.7
90	85.7	81.5	72.0	67.2	61.9	51.7	91.4	91.4	91.4	91.3	90.9	90.8
600	85.0	80.5	69.5	64.1	57.7	47.3	91.4	91.4	91.4	91.3	90.9	90.9
10	85.0	78.5	66.5	60.7	53.7	42.5	91.4	91.4	91.4	91.3	90.9	90.9
20	84.0	76.9	63.8	57.2	49.8	38.0	91.5	91.5	91.4	91.3	91.0	91.0
30	83.2	75.6	61.5	53.7	45.5	34.6	91.5	91.5	91.4	91.3	91.0	91.1
40	82.2	74.9	60.1	50.8	42.0	32.5	91.5	91.5	91.4	91.3	91.1	91.1
50	81.9	74.4	59.4	49.1	39.5	31.5	91.5	91.5	91.4	91.3	91.1	91.2
60	81.5	74.0	58.8	48.0	38.5	31.0	91.5	91.5	91.4	91.4	91.2	91.2
70	81.4	73.5	58.1	47.5	37.6	29.9	91.5	91.5	91.4	91.4	91.2	91.3
80	81.1	73.2	57.6	46.0	36.5	28.9	91.5	91.5	91.4	91.4	91.3	91.3
90	81.5	73.5	58.0	46.4	36.5	28.7	91.5	91.5	91.4	91.4	91.3	91.3
700	81.5	73.5	58.0	46.4	36.5	28.7	91.5	91.5	91.4	91.4	91.3	91.3
Luminous transmit.	87.2	84.5	77.8	72.2	67.7	63.3	90.4	90.1	89.1	88.2	87.4	86.0
Dominant wave lgth.	553.0	555.5	555.0	554.0	554.3	553.4	572.0	571.3	571.4	571.3	571.3	571.2
Excitation purity.	2.3	5.2	10.9	15.8	21.1	25.9	5.3	9.6	18.8	28.3	35.7	42.0

* Some transmission below 400 mμ. Consult the manufacturer.

Table 89 (continued)
TRANSMISSION OF WRATTEN FILTERS

Percent transmittance

Wave length	CC-05M	CC-10M	CC-20M	CC-30M	CC-40M	CC-50M	CC-05C	CC-10C	CC-20C	CC-30C	CC-40C	CC-50C
400	87.6	86.6	85.6	84.2	82.3	80.9	87.3	86.0	83.9	82.3	80.4	78.8
10	88.2	87.7	86.6	85.7	84.6	83.6	88.2	87.5	85.2	84.5	83.4	82.7
20	88.6	88.0	87.0	85.9	85.2	84.4	88.7	88.1	86.5	86.0	85.3	84.8
30	88.7	88.0	86.0	85.6	84.4	83.6	89.0	88.6	87.5	87.0	86.3	85.9
40	88.7	87.9	86.0	84.7	82.5	81.4	89.3	89.0	87.7	87.5	86.6	86.0
50	88.4	87.5	84.0	82.8	80.0	78.1	89.3	89.1	87.8	87.3	86.6	86.1
60	88.4	86.5	83.1	80.0	76.1	73.7	89.6	89.1	87.7	87.0	86.4	85.7
70	87.8	85.2	80.8	76.1	71.3	68.0	89.7	89.0	87.5	87.0	85.8	85.2
80	87.0	83.6	77.4	72.1	65.8	61.7	89.7	89.0	87.2	86.5	85.3	84.4
90	86.0	81.8	74.4	67.0	60.0	55.0	89.0	89.0	87.0	86.0	84.4	83.4
500	85.0	79.7	70.5	61.7	53.7	48.1	89.6	89.7	86.5	85.2	83.5	82.3
10	83.8	77.5	66.7	56.5	47.7	41.6	89.6	88.5	86.0	84.4	82.4	80.8
20	82.7	75.3	63.4	52.0	42.8	36.3	89.5	88.5	85.2	83.5	81.1	79.2
30	81.8	73.7	60.5	48.6	39.0	31.9	89.4	88.0	84.3	82.4	79.6	77.3
40	81.3	72.5	58.6	46.6	36.7	29.8	89.2	87.0	83.4	81.0	77.7	75.0
50	81.2	72.2	58.0	46.0	36.0	29.1	88.9	86.1	82.3	79.0	75.3	72.2
60	81.5	72.8	58.3	46.5	36.7	29.7	88.5	85.0	80.5	76.7	73.1	69.0
70	82.5	74.6	60.5	49.8	40.2	32.3	88.0	83.8	78.5	74.0	69.3	65.0
80	84.0	77.3	64.9	55.6	46.2	39.0	87.5	82.5	76.1	70.9	65.6	60.5
90	85.8	80.8	70.6	63.3	54.9	48.7	87.0	81.0	73.9	67.5	61.6	55.8
600	88.0	84.5	77.1	71.6	64.9	59.9	86.4	79.5	71.2	64.1	57.6	51.3
10	89.3	87.0	82.2	79.2	74.4	70.7	85.5	77.9	68.5	60.4	53.4	46.2
20	90.2	88.9	86.1	84.1	81.4	79.2	84.5	76.3	65.5	56.7	49.2	42.3
30	90.6	90.0	88.7	87.4	86.0	84.5	83.8	75.1	62.7	53.1	45.0	38.0
40	90.8	90.5	90.0	89.3	88.7	87.6	83.3	74.4	60.8	50.4	42.0	34.9
50	91.0	90.7	90.5	90.2	90.0	89.7	82.8	74.0	59.5	48.8	40.2	32.9
60	91.1	91.0	90.8	90.8	90.4	90.4	82.5	73.6	57.9	48.0	39.4	32.0
70	91.2	91.2	91.2	91.0	90.7	90.7	82.4	73.0	57.9	47.2	38.6	31.0
80	91.3	91.3	91.3	91.1	91.3	91.3	82.0	72.8	57.4	46.0	37.5	29.9
90	91.4	91.4	91.4	91.3	91.3	91.3	82.0	72.8	57.4	45.5	36.7	29.1
700	91.5	91.5	91.5	91.5	91.5	91.5	82.5	74.0	58.5	46.4	37.3	29.8
Luminous transmit.	84.2	77.9	67.1	58.1	50.0	44.0	88.0	85.1	78.9	74.8	70.5	66.7
Dominant wave lgth.	541.0.	547.5.	551.2.	550.0.	550.3.	551.2.	489.2	487.5	486.5	486.2	486.1	485.5
Excitation purity	3.5	7.4	14.4	21.5	28.3	34.0	1.6	4.1	8.9	12.8	17.5	20.2

Wave length	Percent transmittance			
	No. 87	No. 87C	No. 88A	No. 89B
700	....	....	....	11.2
10	....	....	....	32.4
20	....	....	....	57.6
30	0.10	....	7.4	69.1
40	2.19	....	32.8	77.6
50	7.95	....	56.8	83.1
60	17.4	....	69.2	85.0
70	31.6	....	74.2	86.1
80	43.7	....	77.6	87.0
90			79.7	87.7
800	53.8	0.32	81.4	88.1
10	61.7	3.20	82.6	88.4
20	69.2	8.90	83.7	88.6
30	74.1	17.8	84.7	88.8
40	77.7	28.2	85.5	89.0
50	81.4	41.0	86.1	89.2
60	84.0	53.8	86.6	89.4
70	85.4	61.6	87.2	89.6
80	86.8	69.2	87.5	89.8
90	87.8	74.1	87.8	89.9
900	88.4	78.5	88.0	90.0
10	88.8	81.5	88.2	90.1
20	89.1	83.6	88.4	90.2
30	89.1	85.1	88.6	90.3
40	89.1	86.0	88.8	90.4
50	89.1	87.0	89.0	90.5

* Some transmission below 400 mμ.   Consult the manufacturer.

Note: Data in Table 89 were condensed from *Kodak Wratten Filters for Scientific and Technical Use*, published by Eastman Kodak Company. The data were compiled by Allen C. Peed for the Eastman Kodak Company.

From Weast, R. C., Ed., *CRC Handbook of Chemistry and Physics*, 61st ed., CRC Press, Boca Raton, Fla., 1980, E-409. With permission.

Table 90
FLAME EMISSION SPECTRA

Note: This table lists the lines and band heads, in order of wave length in ångströms, recorded in air/acetylene and oxygen/acetylene flames, together with the atom, ion, or molecule responsible. Wave lengths have been rounded off to the first decimal place. The various symbols have the following meanings:

Column 1: "R" or "V" following the wave length indicates that it is a band head, with the band fading in the direction of longer or shorter wave length, reespectively.

Column 2: "F " indicates that the band is due to the flame itself, and is followed by the ion responsible.

Column 3: "A" indicates that the line or band appears in the air/acetylene flame, "O" that it appears in the oxygen/acetylene flame. "K" indicates that the line appears only in the inner cone of the flame; "Res" indicates a resonance line.

Wave length Å	Due to	Remarks (see above)	Wave length Å	Due to	Remarks (see above)	Wave length Å	Due to	Remarks (see above)	Wave length Å	Due to	Remarks (see above)
2428.0	Au	A, Res	3039.4	In	A	3391.1	Ni	A	3510.3	Ni	A
47.9	Pd	O	44.0	Co	A	93.0	Ni	A	12.6	Co	A
76.4	Pd	O	47.6	Fe	A	95.4	Co	A	12.9 R	SnO	A
78.6	C	O, K	50.8	Ni	O	96.9	Rh	A, Res	13.5	Co	A
83.3	Fe	O	57.6	Ni	O	3404.6	Pd	A, Res	15.1	Ni	A, Res
88.1	Fe	O	59.1	Fe	A	05.1	Co	A	16.9	Pd	A
90.6	Fe	O	64 R	(F, OH)	A, O	06.9 R	SnO	A	19.2	Tl	A
2522.8	Fe	O	64.6	Ni	O	09.2	Co	A	20.1	Co	A
27.4	Fe	O	64.7	Pt	A, Res	12.3	Co	A	21.3	Fe	A
36.5	Hg	A, Res	75.9	Zn	A, K, Res	12.6	Co	A	21.6	Co	O
2609 R	(F, OH)	O, K	3122 R	(F, OH)	A, O	14.8	Ni	A, Res	23.4	Co	O
13.7	Pb	O	32.6	Mo	A, K	15.8 R	SnO	A	24.5	Ni	A, Res
14.2	Pb	O	34.1	Ni	O	17.2	Co	A	26.8	Co	O
59.4	Pt	O	44	(F, CH)	A, O	21.2	Pd	A, Res	28.0	Rh	A
61.2	Sn	A, K	58.2	Mo	A, K	23.7	Ni	A	29.0	Co	O
76.0	Au	A, Res	70.3	Mo	A, K	28 R	(F, OH)	O	29.4	Tl	A
77 R	(F, OH)	O, K	75.0	Sn	A, K, Res	28.3	Ru	A	29.8	Co	A
2706.5	Sn	A, K, Res	85 R	(F, OH)	A, O	31.6	Co	A	33.4	Co	A
53 R	(F, OH)	O, K	94.0	Mo	A, K	33.0	Co	A	38.1	Rh	O
61.8	Sn	A, K	3205.8 R	SnO	A	33.4	Pd	O	42.4 R	SnO	A
63.1	Pd	O	08.8	Mo	A, K	33.6	Ni	A	43.9	Rh	O
67.9	Tl	A	32.6	Li	A	34.9	Rh	A	49.5	Rh	O
76.7	Mg	A, K	33.0	Ni	O	36.7	Ru	A	50.6	Co	A
78.3	Mg	A, K	42.7	Pd	A	37.3	Ni	A	53.1	Pd	A
79.8	Mg	A, K	43.1	Ni	O	40.6	Fe	A	65.4	Fe	A
79.8	Sn	A, K	47.5	Cu	A, Res	41.0	Fe	A	66.2 V	LaO	A
81.4	Mg	A, K	51.4 R	SnO	A	41.4	Pd	O	66.4	Ni	A
83.0	Mg	A, K	54 R	(F, OH)	O	43.6	Co	A	69.4	Co	A
85.0	Sn	A, K	56.1	In	A	44.6 R	SnO	A	70.1	Fe	A
94.8	Mn	A	58.6	In	A	46.3	Ni	A	70.2	Rh	A
95.5	Mg II	A	61.1	Cd	A, Res	46.4	K	A	71.2	Pd	O
98.3	Mn	A	62.3	Sn	A, K	47.4	K	A	71.9	Ni	A
2801.1	Mn	A	62.4	SnO	A	49.2	Co	A	75.0	Co	A
02.7	Mg II	A	74.0	Cu	A, Res	49.4	Co	A	75.4	Co	A
11 R	(F, OH)	A, K; O	80.7	Ag	A, Res	52.9	Ni	A	78.7	Cr	A
13.6	Sn	A, K	82.3	Zn	A, K	53.5	Co	A	81.2	Fe	A
33.1	Pb	A, Res	91.8 R	SnO	A	55.2	Co	A	83.1	Rh	A
40.0	Sn	A, K, Res	3302.1	Pd	A	55.2	Rh	O	83.9	(F, CN)	A, K
50.6	Sn	A, K	02.3	Na	A	58.5	Ni	A	85.4 R	SnO	A
52.1	Mg	A, Res	03.0	Na	A	60.8	Pd	A	85.9	(F, CN)	A, K
63.3	Sn	A, K, Res	15.7	Ni	O	61.7	Ni	A	87.1	Rb	A
74.2	Ga	O	23.1	Rh	O	62.0	Rh	O	87.2	Co	A
75 R	(F, OH)	O	23.4 R	SnO	A	62.8	Co	A	89.2	Ru	A
2943.6	Ga	O	29.9	Mg	A, K	65.8	Co	A	90.4	(F, CN)	A, K
44.2	Ga	O	30.6	Sn	A, K	65.9	Fe	A	91.6	Rb	A
45 R	(F, OH)	O	32.2	Mg	A, K	72.5	Ni	A	93.0	Ru	A
66.9	Fe	A	34.1	Co	A	74.0	Co	A, Res	93.5	Cr	A
73.1	Fe	A	36.7	Mg	A, K	75.4	Fe	A	94.9	Co	A
73.2	Fe	A	45.0	Zn	A, K	81.2	Pd	A	96.2	Rh	O
83.6	Fe	A	45.6	Zn	A, K	83.4	Co	A	96.2	Ru	O
94.4	Fe	A	45.9	Zn	A, K	83.8	Ni	A	97.1	Rh	O
			54.4	Co	A	84.5 R	SnO	A	97.7	Ni	A
3000.9	Fe	A	60	(F, NH)	A, K; O	85 R	(F, OH)	O	3602.1	Co	A
02.5	Ni	O	67.1	Co	A	90.6	Fe	A	04.6 V	LaO	A
03.6	Ni	O	69.6	Ni	A	93.0	Ni	A, Res	05.3	Cr	A
08.1	Fe	A	73.0	Pd	A	95.7	Co	A	08.1 V	LaO	A
09.1	Sn	A, K, Res	80.6	Ni	A	98.7	Rh	O	08.9	Fe	A
12.0	Ni	O	80.9	Ni	O	98.9	Ru	A	09.5	Pd	A
20.5	Fe	A	81.7 R	SnO	A	3502.3	Co	A	10.5	Ni	A
20.6	Fe	A	82.9	Ag	A, Res	02.5	Rh	A	11.5 V	LaO	A
34.1	Sn	A, K, Res	85.2	Co	A	06.3	Co	A	12.5	Rh	O
37.4	Fe	A	88.2	Co	A	07.3	Rh	O	12.7	Ni	A
37.9	Ni	O	88.3 R	SnO	A	09.8	Co	A	14.7 R	SnO	A

Table 90 (continued)
FLAME EMISSION SPECTRA

Wave length Å	Due to	Remarks (see opposite page)	Wave length Å	Due to	Remarks (see opposite page)	Wave length Å	Due to	Remarks (see opposite page)	Wave length Å	Due to	Remarks (see opposite page)
3614.9 V	LaO	A	3805.0	MgO	A	4071.7	Fe	O	4648.2 R	AlO	A
18.8	Fe	A	06.8	Rh	O	77.7	Sr II	A	63.5 R	BaO	A
19.4	Ni	A	07.1	Ni	A	80.6	Ru	A	72.0 R	AlO	A
26.6	Rh	O	11.3	MgO	A	82.8	Rh	O	78.6 V	(F, C_2)	A, K
27.8	Co	A	18.2	Rh	O	4101.8	In	A, Res	80.3 R	BaO	A
31.5	Fe	A	20.4	Fe	A	08.9 R	SnO	A	84.8 V	(F, C_2)	A, K
34.7	Pd	A, Res	22.3	Rh	O	12.7	Ru	O	94.6 R	AlO	A
34.9	Ru	O	23.3	Mg	A, K	18.8	Co	A	97.6 V	(F, C_2)	A, K
39.6	Pb	A, Res	25.9	Fe	O	21.3	Co	A	4715.2 V	(F, C_2)	A, K
42.7	MgO	A	27.8	Fe	O	28.9	Rh	O	15.5 R	AlO	A
47.7	Co	A	28.5	Rh	O	35.3	Rh	O	22.2	Zn	A, K
47.8	Fe	O	29.4	Mg	A, K	44.4 R	SnO	A	22.7 R	BaO	A
47.9	Fe	A	32.3	Mg	A, K	72.1	Ga	A, Res	35.8 R	AlO	A
48.6	MgO	A	32.3	Pd	O	96.5	Rh	O	37.1 V	(F, C_2)	A, K
52.5	Co	A	33.2 R	SnO	A	98.9	Ru	O	41.0 R	BaO	A
58.0	Rh	A	33.9	Rh	O	4201.9	Rb	O	70.1 R	(F, C_2)	O, K
61.4	Ru	A	34.2	Fe	A	06.0	Ru	O	84.1 R	BaO	A
66.2	Rh	O	38.3	Mg	A, K	11.1	Rh	O	4810.5	Zn	A, K
74.8	MgO	A	40.4	Fe	A	12.1	Ru	O	30.0 R	BaO	A
79.9	Fe	A	42.1	Co	A	13.0	Pd	O	36.1 R	(F, C_2)	O, K
81.6	MgO	A	48.5	Fe	A	15.5	Sr II	A	42.3 R	AlO	A
83.0	Fe	A	49.8	MgO	A	15.6	Rb	A	50.6 R	BaO	A
83.5	Pb	A, Res	54.7	(F, CN)	A, K	26.7	Ca	A, Res	66.4 R	AlO	A
90.3	Pd	O	56.4	Fe	A, Res	40.3 R	SnO	A	73.8 R	BaO	A
90.7	Rh	A	56.5	Rh	O	54.3	Cr	A, Res	96.5 R	BaO	A
91.4 R	SnO	A	58.3	Ni	A	71.8	Fe	O	4911.0 R	(F, C_2)	O, K
92.4	Rh	A	59.9	Fe	A, Res	74.8	Cr	A, Res	34.1	Ba II	A
3700.9	Rh	A	61.9	(F, CN)	A, K	89.7	Cr	A, Res	41.7 R	BaO	A
05.6	Fe	A	64.1	Mo	A, K, Res	4307.9	Fe	O	65.4 R	BaO	A
09.6 V	LaO	A	71.4	(F, CN)	A, K	12.5 V	(F, CH)	A, K	96.7 R	(F, C_2)	O, K
13.0	Rh	O	72 R	(F, CH)	A, K	24	(F, CH)	A, K	5012.4 R	BaO	A
18.9	Pd	O	73.1	Co	A	25.8	Fe	O	86.7 R	BaO	A
20.0	Fe	A	74.0	Co	A	52.6 R	AlO	A	97.7 V	(F, C_2)	A, K; O, K
20.7	MgO	A	76.4	Cs	A	65.2 V	(F, C_2)	A, K	5129.3 V	(F, C_2)	A, K; O, K
21.0	MgO	A	78.6	Fe	A, Res	71.4 V	(F, C_2)	A, K	37	Ba	
21.2 R	SnO	A	81.9	Co	A	72.0 R	LaO	A	58	MnCl₂	
22.6	Fe	A	84.3	(F, CH)	A, K	73.7 R	AlO	A	65.2 V	(F, C_2)	A, K; O, K
24.4	MgO	A	86.3	Fe	A, Res	74.8	Rh	O	93	B	
24.8	MgO	A	88.7	Cs	A	75.8 R	LaO	A	93	MnCl₂	
25.9	MgO	A	94.1	Co	A	79.7 R	LaO	A	5204.5	Cr	A
26.9	Ru	A, Res	94.2	Pd	O	82.5 V	(F, C_2)	A, K	06.0	Cr	A
28.0	Ru	A, Res	95.7	Fe	A, Res	83.5	Fe	O	08.4	Cr	A
30.4	Ru	O	99.3 R	SnO	A	83.5 R	LaO	A	14.7 R	BaO	A
33.3	Fe	O	99.7	Fe	A, Res	87.6 R	LaO	A	30	MnCl₂	
34.9	Fe	A	3902.9	Fe	A	91.6 R	LaO	A	81.2 R	CaO	A
37.1	Fe	A	03.0	Mo	A, K, Res	93.8 R	AlO	A	5346.0 R	CaO	A
42.3	Ru	A	05.1	MgO	A	95.7 R	LaO	A	47	Ba	
45.6	Fe	A	10.9	MgO	A	4404.7	Fe	O	49.7 R	BaO	A
48.2	Rh	O	20.3	Fe	A, Res	18.2 R	LaO	A	50.5	Tl	A, Res
48.3	Fe	O	22.9	Fe	A, Res	23.2 R	LaO	A	56.4 R	CrO	A
49.5	Fe	A	25.9	Ru	A	28.1 R	LaO	A	59.4 R	MnO	A
52.3 R	SnO	A	27.9	Fe	A, Res	33.0 R	LaO	A	60	MnCl₂	
55.9	Ru	O	30.3	Fe	A, Res	38.0 R	LaO	A	66.7 R	BaO	A
58.2	Fe	A	31.8	Ru	A	43.0 R	LaO	A	82.5 R	LaO	A
63.8	Fe	A	33.7	Ca II	A	48.0 R	LaO	A	89.4 R	MnO	A
65.1	Rh	O	34.2	Rh	O	70.5 R	AlO	A	92	MnCl₂	
66.5	MgO	A	38.4	MgO	A	94.0 R	AlO	A	94.7	Mn	A
67.2	Fe	A	43.7	MgO	A	4511.3	In	A, Res	5402.7 R	BaO	A
72.2	MgO	A	44.0	Al	A, Res	16.4 R	AlO	A	07.7 R	LaO	A
75.7	Tl	A, Res.	51.0 R	SnO	A	24.0 R	BaO	A	16.5 R	CrO	A
78.2	MgO	A	58.6	Pd	O	24.7	Sn	A, K	19.8 R	BaO	A
86.1	Ru	O	58.9	Rh	O	28.7	Rh	O	23.8 R	MnO	A
87.9	Fe	A	61.5	Al	A, Res	37.6 R	AlO	A	24	MnCl₂	
88.5	Rh	O	68.5	Ca II	A	37.8 R	BaO	A	32.5	Mn	A
90.5	Ru	A	83.9 R	SnO	A	54.0	Ba II	A	33.1 R	LaO	A
93.2	Rh	O	95.3	Co	A	54.5	Ru	O	40	B	
98.1	Ru	A				55.4	Cs	A	54.6 R	BaO	A
98.3	Mo	A, K, Res	4030.8	Mn	A, Res	57.6 R	AlO	A	58.7 R	LaO	A
98.9	Ru	A	33.0	Ga	A, Res	76.3 R	AlO	A	70.3 V	(F, C_2)	A, K; O, K
99.1	MgO	A	33.1	Mn	A, Res	79.4 R	BaO	A	75.0 R	CaO	A
99.2	Pd	O	34.5	Mn	A, Res	84.4	Ru	A	79.2 R	CrO	A
99.3	Ru	A	44.2	K	A	93.2	Cs	A	81	B	
99.3	Rh	O	45.8	Fe	O	4607.3	Sr	A	92.7 R	BaO	A
3801.0	Sn	A, K	47.2	K	A	21.1 R	BaO	A	92.7 R	BaO	A
02.7 R	SnO	A	57.8	Pb	A, Res	36.8 R	BaO	A	5501.9 V	(F, C_2)	A, K; O, K

Table 90 (continued)
FLAME EMISSION SPECTRA

Wave length Å	Due to	Remarks (see page 268)	Wave length Å	Due to	Remarks (see page 268)	Wave length Å	Due to	Remarks (see page 268)	Wave length Å	Due to	Remarks (see page 268)
5506.5 R	CaO	A	6002.2 V	CaO	A	6643.6 R	CrO	A	7610.1 R	BaO	A
09.7 R	BaO	A	03 V	NdO	A	50 R	NdO	A	18.9	Rb	A
27.9 R	FeO	A	04.9 V	(F, C_2)	A,K;O,K	77.3 V	(F, C_2)	O, K	25.0 R	LaO	A
35.6	Ba	A, Res	39.6 R	BaO	A	6700 R	FeO	A	57.8 R	LaO	A
40.7 V	(F, C_3)	A,K;O,K	42.8 R	CrO	A	04.5 R	BaO	A	64.9	K	A, Res
43.2 R	FeO	A	51.6 R	CrO	A	07.9	Li	A, Res	87.8 R	CaO	A
43.7 R	CrO	A	59.7 V	(F, C_2)	A,K;O,K	15.1 R	CrO	A	90 R	FeO	A
44	Ca		60	Sr		23.3	Cs		91.1 R	LaO	A
47.8 R	BaO	A	85.1 V	SrO	A	47	Sr		99.0	K	A, Res
64.1 R	CrO	A	90.2 V	SrO	A	50 V	CaO	A	7712.2 R	CaO	A
82.5 R	FeO	A	94.8 R	FeO	A	63.2 V	(F, C_2)	O, K	13.9 R	CrO	A
85.5 V	(F, C_3)	A,K;O,K	96.5 V	SrO	A	71.8 R	CrO	A	15.5 R	CaO	A
86.4 R	MnO	A	6101.3 V	SrO	A	82.8 R	BaO	A	21.1 R	CaO	A
92	MnCl₂		02.3 R	BaO	A	6829.2 R	CrO	A	25.0 R	LaO	A
5602.4 R	LaO	A	03.6	Li		57.2 R	BaO	A	57.7	Rb	
02.4 R	BaO	A	07.5 V	SrO	A	58.8 V	(F, C_2)	O, K	75 R	FeO	A
09.5 R	MnO	A	08.0 R	CrO	A	61.4 V	SrO	A	78.1 R	CrO	A
13.9 R	FeO	A	09.3 R	FeO	A	67.9 V	SrO	A	7800.3	Rb	A, Res
23.3 R	CrO	A	09.9 V	SrO	A	75.6 V	SrO	A	15.2 R	BaO	A
28.6 R	LaO	A	11.0 R	BaO	A	84.5 V	SrO	A	42.8 R	CrO	A
35.5 V	(F, C_3)	A,K;O,K	11.9 V	SrO	A	91.5 R	CrO	A	52.8 R	SrO	A
38.9 R	MnO	A	16.2 V	SrO	A	6931.4 R	BaO	A	82.3 R	SrO	A
44.1 R	BaO	A	16.2 R	BaO	A	39.0	K	A	7902.0 R	SrO	A
54.8 R	LaO	A	22.1 V	(F, C_2)	A,K;O,K	57.2 R	CrO	A	05.1 R	BaO	A
59.0 R	BaO	A	54.9 R	MnO	A	64.7	K	A	08.0 R	CrO	A
72.9 R	BaO	A	65.1 R	BaO	A	73.3	Cs		10.5 R	LaO	A
81.1 R	LaO	A	67.3 R	CrO	A	7007.1 R	BaO	A	44.9 R	LaO	A
85.1 R	CrO	A	69.6 R	BaO	A	11.2 R	LaO	A	47.6	Rb	A, Res
99.0	Ru	A	75.9 R	MnO	A	22 R	FeO	A	79.7 R	LaO	A
5710.0 R	BaO	A	80.7 R	FeO	A	27.5 R	CrO	A	8014.8 R	LaO	A
13.7 R	BaO	A	82	Ca		40.8 R	LaO	A	15.7	Cs	A
19.3 R	CaO	A	91.2 V	(F, C_2)	A,K;O,K	70.8 R	LaO	A	50.2 R	LaO	A
48.7 R	CrO	A	6203	Ca		97.4 R	BaO	A	78.9	Cs	A
58.4 R	BaO	A	03.2 R	MnO	A	7101.0 R	LaO	A	79.0	Cs	A
70.1 R	BaO	A	13	Cs		31.6 R	LaO	A	86.1 R	LaO	A
89.6 R	FeO	A	18.3 R	FeO	A	62.6 R	LaO	A	8112 R	FeO	A
94.4 R	CrO	A	24.7 R	BaO	A	76.6 R	BaO	A	22.2 R	LaO	A
5805.1 R	BaO	A	29.1 R	CrO	A	87.1 R	CrO	A	53.1 R	CaO	A
05.6 R	FeO	A	69.0 V	CaO	A	93.7 R	LaO	A	59.0 R	LaO	A
12.2 R	CaO	A	91.0 R	BaO	A	7208.0 V	SrO	A	64.7 R	CaO	A
17.6 R	BaO	A	94.1 R	CrO	A	49.1 R	CrO	A	67.3 R	CaO	A
18.4 R	FeO	A	6358.0 R	BaO	A	54.3 R	BaO	A	83.3	Na	A
29.2 R	BaO	A	63.7 R	CrO	A	64.5 V	SrO	A	94.8	Na	A
52.1 R	CrO	A	94.3 R	CrO	A	86.0 V	SrO	A	96.1 R	LaO	A
59.6 R	MnO	A	6411 R	NdO	A	7308.3 R	CaO	A	8230 R	FeO	A
64.5 R	BaO	A	23.1 R	BaO	A	13.8 R	CrO	A	33.1 R	LaO	A
67.6 R	FeO	A	25 R	NdO	A	18.5 R	CaO	A	57.8 R	SrO	A
69.5 R	LaO	A	40 R	NdO	A	26 R	CaO	A	70.7 R	LaO	A
75.3 R	BaO	A	42.3 V	(F, C_2)	O, K	36.9 R	BaO	A	72.2 R	SrO	A
80.3 R	MnO	A	51.5 R	CrO	A	75.3 R	CrO	A	8302 R	FeO	A
86.9 R	BaO	A	80.5 V	(F, C_2)	O, K	7403.5 R	LaO	A	60.3 R	CrO	A
90.0	Na	A, Res	93.1 R	BaO	A	28 R	FeO	A	8453.5 R	LaO	A
95.9	Na	A, Res	6500 V	CaO	A	34.3 R	LaO	A	90.0 R	LaO	A
96.7 R	LaO	A	11 R	SmO	A	36 V	SrO	A	8521.1	Cs	A, Res
5902.6 R	FeO	A	11.8 R	CrO	A	39.5 R	CrO	A	26.6 R	LaO	A
09.1 R	MnO	A	33 R	SmO	A	40.4 R	BaO	A	63.5 R	LaO	A
13.1 R	CrO	A	33.7 V	(F, C_2)	O, K	65.2 R	LaO	A	78 R	FeO	A
23.4 V	(F, C_3)	A,K;O,K	57 R	SmO	A	89.7 V	SrO	A	8600.8 R	LaO	A
24.0 R	LaO	A	63.2 R	BaO	A	96.5 R	LaO	A	38.5 R	LaO	A
25.1 R	BaO	A	70 R	SmO	A	7500.6 R	SrO	A	52.2 R	CaO	A
51.3 R	LaO	A	75.1 R	CrO	A	22.8 R	SrO	A	76.6 R	LaO	A
58.7 V	(F, C_3)	A,K;O,K	80 R	NdO	A	23.5 R	BaO	A	8700.0 R	SrO	A
75 R	NdO	A	99.2 V	(F, C_2)	O, K	27 R	FeO	A	22.5 R	SrO	A
75.9 R	CrO	A	6600 R	NdO	A	28.2 R	LaO	A	61.4	Cs	A
76.3 R	BaO	A	20 R	NdO	A	41.6 R	SrO	A	90 R	FeO	A
78.8 R	LaO	A	28	Sr		60.0 R	LaO	A	8943.5	Cs	A, Res
84.9 R	BaO	A	34.5 R	BaO	A	92.3 R	LaO	A			

From Diem, K., Ed., *Documenta Geigy Scientific Tables*, 6th ed., Geigy Pharmaceuticals, Ardsley, N.Y., 1962, 268. With permission.

INFRARED SPECTROSCOPY: ABSORPTION FREQUENCIES OF SINGLE BONDS TO HYDROGEN

Tables 91 through 110 are from Williams and Fleming, *Spectroscopic Methods in Organic Chemistry*, 3rd ed., McGraw-Hill, 1980. They were reproduced with permission.

Table 91
SATURATED C−H AND C−C

Group	Band	Remarks
>CH₂ ⎱ −CH₃ ⎰	2960—2850(s)	Two or three bands usually;
		→C−H stretching
→CH	2890—2880(w)	
>CH₂	1470—1430(m)	→C−H deformations
−CH₃		
−CH₃	1390—1370(m)	−CH₃ symmetrical deformation
>CH₂	∼720(w)	>CH₂ rocking

Table 92
MISCELLANEOUS C−H

Group	Band	Remarks
Cyclopropane C−H ⎱ Epoxide C−H −CH₂-halogen ⎰	∼3050(w)	C−H stretching; cf. alkenes
−CO−CH₃	3100—2900(w)	Often very weak
−CHO	2900—2700(w)	Usually two bands, one near 2720 cm⁻¹
−O−CH₃	2850—2810(m)	
−O−CH₂−O−	2790—2770(m)	
N−CH₃ and N−CH₂−	2820—2780(m)	
−C(CH₃)₃	1395—1385(m) 1365(s)	
>C(CH₃)₂	∼1380(m)	A roughly symmetrical doublet
−O−CO−CH₃	1385—1365(s)	The high intensity of these
−CO−CH₃	1360—1355(s)	bands often dominates this region of the spectrum

Table 93
ALKENE AND AROMATIC C—H

Group	Band	Remarks
—C≡C— H	~3300(s)	
>C=C⟨ (H, H)	3095—3075(m)	C—H stretching; sometimes obscured by the much stronger bands of saturated C—H groups which occur below 3000 cm⁻¹
>C=C⟨ (H)	3040—3010(m)	
Aryl— H	3040—3010(w)	Often obscured
R>C=C<R (H, H)	970—960(s)	C—H out-of-plane deformation. When the double bond is conjugated with, for example, a C=O group this band is shifted towards 990 cm⁻¹
RCH=CH₂	995—985(s) and 940—900(s)	
R₂C=CH₂	895—885(s)	
R₂C=C⟨ (H, R)	840—790(m)	
R>C=C<R (H, H)	730—675(m)	

Note: See also Tables 103 and 104 for the corresponding double bond absorptions, and Table 105 for the aromatic C—H out-of-plane bending vibrations.

Much is known about the precise position of the various CH, CH_2 and CH_3, symmetrical and unsymmetrical vibration frequencies. C—H bonds do not take part in hydrogen bonding and, therefore, their position is little affected by the state of measurement or their chemical environment. C—C vibrations, which absorb in the fingerprint region, are generally weak and not practically useful. Since most organic molecules possess alkane residues, the groups of saturated C—H absorption bands given above are of little diagnostic value; their general appearance may be seen in the Nujol spectrum. The absence of saturated C—H absorption in a spectrum is, of course, diagnostic evidence for the absence of such a part structure in the corresponding compound. Unsaturated and aromatic C—H stretching frequencies can be distinguished from the saturated C—H absorption since the latter occurs below 3000 cm⁻¹ (see, however, Table 92) while the former gives rise to much less intense absorption above 3000 cm⁻¹. Alkene and aromatic C—H absorption is covered by Table 93.

A few special structural features in saturated C—H groupings give rise to characteristic absorption bands. These are summarized in Table 92.

Table 94
ALCOHOL AND PHENOL −O−H

Group	Band	Remarks
Water in solution	3710	
Free−OH	3650—3590(v)	Sharp; O−H stretching
H-bonded −OH (solid, liquid and dilute solution)	3600—3200(s)	Often broad but may be sharp for some intramolecular single bridge H bonds; the lower the frequency the stronger the H bond
Intramolecular H-bonded −OH in chelate form	3200—2500(v)	Broad; the lower the frequency the stronger the H bond; sometimes so broad as to be overlooked
Water of crystallization (solid state spectra)	3600—3100(w)	Usually a weak band at 1640—1615 cm⁻¹ also; water in trace amounts in KBr discs shows a broad band at 3450 cm⁻¹
−O−H	1410—1260(s)	O−H bending
→C−OH	1150—1040(s)	C−O stretching

Note: The value of the O—H stretching frequency has been used for many years as a test for and measure of the strength of hydrogen.

Table 95
AMINE, IMINE, AMMONIUM, AND AMIDE N−H N−H STRETCHING

Group	Band	Remarks
Amine and imine >N−H =N−H	3500—3300(m)	Primary amines show two bands in this range: the un-symmetrical and symmetrical stretching. Secondary amines absorb weakly. The pyrrole and indole N−H band is sharp
−NH⁺₃;Amino acids AminoSalts	3130—3030(m) ∼3000(m)	Values for solid state; broad; bands also (but not always) near 2500 and 2000 cm⁻¹
>NH⁺₂ →NH⁺ =⁺NH	2700—2250(m)	Values for solid state; broad, due to the presence of over-tone bands, etc.
Primary amide −CONH₂	∼3500(m) ∼3400(m)	Lowered ∼150 cm⁻¹ in the solid state and on H bonding; often several bands 3200—3050 cm⁻¹
Secondary amide −CONH−	3460—3400(m)	Two bands: lowered on H bonding and in the solid state. Only one band with lactams
	3100—3070(w)	A weak extra band with bonded and solid state samples

Note: Much is known of amide N-H absorptions, the appearance of two bands being ascribed to forms 1 and 2. The carbonyl region of many amides (Table 100) also shows two bands.

Hydrogen bonding lowers and broadens N—H stretching frequencies to a lesser extent than was the case with O—H groups. The intensity of N—H absorption is usually less than that of O—H absorption.

Table 96
N—H BENDING

Group	Band	Remarks
—NH$_2$	1650—1560(m)	
>NH	1580—1490(w)	Often too weak to be noticed
—NH$_3^+$	1600(s)	Secondary amine salts have the
	1500(s)	1600 cm^{-1} band

Note: See also Tables 2-10 for amide absorptions in this region.

Table 97
MISCELLANEOUS R—H

Group	Band	Remarks
—S—H	2600—2550(w)	Weaker than O—H and less affected by H-bonding
P—H	2440—2350(m)	Sharp
$\overset{O}{\underset{OH}{\diagup P}}$	2700—2560(m)	Associated OH
R—D	1/1 37 times the corresponding R—H frequency	Useful when assigning R—H bands, deuteration leading to a known shift to lower frequency

INFRARED SPECTROSCOPY: ABSORPTION FREQUENCIES OF TRIPLE BONDS AND CUMULATED DOUBLE BONDS

Table 98
TRIPLE BONDS

Group	Band	Remarks
$-C\equiv C-H$	3300(m)	C—H stretching
	2140—2100(w)	C≡C stretching
$-C\equiv C-$	2260—2150(v)	See note
$-C\equiv N$	2260—2200(v)	C≡N stretching; stronger and to the lower end of the range when conjugated; occasionally very weak or absent; for example, some cyanohydrins show no C≡N absorption
Diazonium salts		
$R-^+N\equiv N$	~2260	
Thiocyanates		
$R-S-C\equiv N$	2175—2140(s)	Aryl thiocyanates at upper end of the range, alkyl at the lower end

Note: Conjugation with olefinic or acetylenic groups lowers the frequency and raises the intensity. Conjugation with carbonyl groups usually has little effect on the position of absorption.

Symmetrical and nearly symmetrical substitution makes the C≡C stretching frequency inactive in the infrared. It is, however, seen clearly in the Raman spectrum.

When more than one acetylenic linkage is present, and sometimes when there is only one, there are frequently more absorption bands in this region than there are triple bonds to account for them.

The ranges quoted in Table 99 are tentative, since relatively few compounds in some of these classes have been examined.

The unusually high double bond frequencies encountered in the X=Y=Z systems are believed to arise from strong coupling of the two separate stretching vibrations, the asymmetrical and symmetrical stretching frequencies becoming widely separated. This type of coupling occurs only when two groups with similar high frequency vibrations and the same symmetry are situated near one another. Other examples in which such coupling is found are the amide group and the carboxylate ion (Table 100).

Table 99
CUMULATED DOUBLE BONDS

Group	Band	Remarks
Carbon dioxide		
O=C=O	2349(s)	Appears in many spectra due to inequalities in path length
Isocyanates		
$-N=C=O$	2275—2250(s)	Very high intensity; position unaffected by conjugation

Table 99 (continued)
CUMULATED DOUBLE BONDS

Group	Band	Remarks
Azides		
$-N_3$	2160—2120(s)	
Carbodiimides		
$-N=C=N-$	2155—2130(s)	Very high intensity; split into an unsymmetrical doublet by conjugation with aryl groups
Ketenes		
$>C=C=O$	~2150(s)	
Isothiocyanates		
$-N=C=S$	2140—1990(s)	Broad and very intense
Diazoalkanes		
$R_2C= \overset{+}{N}= \overset{-}{N}$	~2100(s)	
Ketenimines		
$C=C=N-$	~2000(s)	
Allenes		
$C=C=C$	~1950(m)	Two bands when terminal allene or when bonded to electron attracting groups, e.g., $-CO_2H$

INFRARED SPECTROSCOPY: ABSORPTION FREQUENCIES OF DOUBLE BOND REGION

Table 100
CARBONYL ABSORPTION $>C=O$ ALL BANDS QUOTED ARE STRONG

Groups	Band	Remarks
Acid anhydrides		
$-CO-O-CO-$		
Saturated	1850—1800 1790—1740	Two bands usually separated by about 60 cm^{-1}. The higher frequency band is more intense in acyclio anhydrides and the lower frequency band is more intense in cyclic anhydrides
Aryl and $\alpha\beta$-unsaturated	1830—1780 1770—1710	
Saturated five-ring	1870—1820 1800—1750	
All classes	1300—1050	One or two strong bands due to C—O stretching
Acid chlorides $-COCl$		
Saturated	1815—1790	Acid fluorides higher, bromides and iodides lower
Aryl and $\alpha\beta$-unsaturated	1790—1750	
Acid peroxides		
$-CO-O-O-CO-$		
Saturated	1820—1810 1800—1780	

Table 100 (Continued)
CARBONYL ABSORPTION>C=O ALL BANDS QUOTED ARE STRONG

Groups	Band	Remarks
Aryl and $\alpha\beta$-unsaturated	1805—1780	
	1785—1755	
Esters and lactones –CO–O–		
Saturated	1750—1735	
Aryl and $\alpha\beta$-unsaturated	1730—1715	
Aryl and vinyl esters C=C–O–CO–Alkyl	1800—1750	The C=C stretching band also shifts to higher frequency
Esters with electronegative α-substituents; e.g., >CCl–CO–O–	1770—1745	
α-keto esters	1755—1740	
Six-ring and larger lactones	Similar values to the corresponding open chain esters	
Five-ring lactone	1780—1760	
$\alpha\beta$-unsaturated five-ring lactone	1770—1740	When α-C–H present there are two bands, the relative intensity depending on the solvent
$\beta\gamma$-unsaturated five-ring lactone; i.e., vinyl ester type	~1800	
Four-ring lactone	~1820	
β-keto ester in H-bonding enol form	~1650	Keto form normal; chelate-type H bond causes shift to lower frequency than the normal ester. The C=C is usually near 1630(s) cm^{-1}
All classes	1300—1050	Usually two strong bands due to C–O stretching
Aldehydes –CHO All values given below are lowered in liquid film or solid state spectra by about 10—20 cm^{-1}. Vapour phase spectra have values raised about 20 cm^{-1}		
Saturated	1740—1720	
Aryl	1715—1695	*Ortho* hydroxy or amino groups shift this value to 1655—1625 cm^{-1} due to intramolecular H-bonding
$\alpha\beta$-unsaturated	1705—1680	
$\alpha\beta,\gamma\delta$-unsaturated	1680—1660	
β-ketoaldehyde in enol form	1670—1645	Lowering caused by chelate-type H-bonding
Ketones>C=O All values given below are lowered in liquid film or solid state spectra by about 10—20 cm^{-1}. Vapour phase spectra have values raised about 20 cm^{-1}		

Table 100 (continued)
CARBONYL ABSORPTION >C=O ALL BANDS QUOTED ARE STRONG

Groups	Band	Remarks
Saturated	1725—1705	
Aryl	1700—1680	
$\alpha\beta$-unsaturated	1685—1665	
$\alpha\beta,\alpha'\beta'$-unsaturated and diaryl	1670—1660	
Cyclopropyl	1705—1685	
Six-ring ketones and larger	Similar values to the corresponding open chain ketones	
Five-ring ketones	1750—1740	$\alpha\beta$-Unsaturation, etc., has a similar effect on these values
Four-ring ketones	~1780	as on those of open chain ketones. Affected by conforma-
α-Halo ketones	1745—1725	tion; highest values are obtained when both halogens are
α,α'-Dihalo ketones	1765—1745	in the same plane as the C=O
1,2-Diketones s-*trans:* (i.e., open chains)	1730—1710	Antisymmetrical stretching frequency of both C=O's. The symmetrical stretching is inactive in the infrared but active in the Raman
1,2-Diketones s-*cis*, six-ring	1760 and 1730	
1,2-Diketones s-*cis*, five-ring	1775 and 1760	
o-Amino- or *o*-hydroxy-aryl ketones	1655—1635	Low due to intramolecular H-bonding. Other substituents and steric hindrance, etc., affect the position of the band
Quinones	1690—1660	C=C usually near 1600(s) cm^{-1}
Extended quinones	1655—1635	
Tropone	1650	Near 1600 cm^{-1} when lowered by H-bonding as in tropolones
Carboxylic acids $-CO_2H$		
All types	3000—2500	O—H stretching; a characteristic group of small bands due to combination bands, etc.
Saturated	1725—1700	The monomer is near 1760 cm^{-1} but is rarely observed. Occasionally both bands, the free monomer and the H-bonded dimer can be seen in solution spectra. Ether solvents give one band near 1730 cm^{-1}
$\alpha\beta$-unsaturated	1715—1690	
Aryl	1700—1680	
α-Halo-	1740—1720	
Carboxylate ions $-CO_2^-$		
For amino acids, see text below		
Most types	1610—1550	Antisymmetrical and symmetrical stretching, respectively
	1420—1300	
Amides $-CO-N<$		
Primary $-CONH_2$		
In solution	~1690	Amide I; C=O stretching
Solid state	~1650	

Table 100 (continued)
CARBONYL ABSORPTION >C=O ALL BANDS QUOTED ARE STRONG

Groups	Band	Remarks
In solution	~1600	Amide II; mostly N–H bending
Solid state	~1640	
		Amide I is generally more intense than amide II. (In the solid state amide I and II may overlap.)
Secondary –CONH–		
In solution	1700—1670	Amide I
Solid state	1680—1630	
In solution	1550—1510	Amide II: found in open chain amides only
Solid state	1570—1515	
		Amide I is generally more intense than amide II
Tertiary	1670—1630	Since H-bonding is absent solid and solution spectra are much the same
Lactams		
Six- and larger rings	~1670	
Five-ring	~1700	Shifted to higher frequency when the N atom is in a bridged system
Four-ring	~1745	
R–CO–N–C=C		Shifted + 15 cm^{-1} by the additional double bond
C=C–CO–N		Shifted by up to + 15 cm^{-1} by the additional double bond. This is an unusual effect for $\alpha\beta$-unsaturation. It is said to be due to the inductive effect of the C=C on the well-conjugated CO–N system, the usual conjugation effect being less important in such a system
Imides –CO–N–CO–		
Cyclic six-ring	~1710 and ~1700	Shift of + 15 cm^{-1} with $\alpha\beta$-unsaturation
Cyclic five-ring	~1770 and ~1700	
Ureas N–CO–N		
RNHCONHR	~1660	
Six-ring	~1640	
Five-ring	~1720	
Urethanes		
R–O–CO–N	1740—1690	Also shows amide II band when non- or mono-substituted on N
Thioesters and Acids		
RCO–S–R′		
RCOSH	~1720	$\alpha\beta$-unsaturated or aryl acid or ester shifted ~ –25 cm^{-1}
RCOS–alkyl	~1690	
RCOS–aryl	~1710	

Intensities of carbonyl bands. Acids generally absorb more strongly than esters, and esters more strongly than ketones or aldehydes. Amide absorption is usually similar in intensity to that of ketones but is subject to much greater variations.

Position of carbonyl absorption. The general trends of structural variation on the position of C=O stretching frequencies may be summarized as follows:

1. The more electronegative the group X in the system R−CO−X−, the higher is the frequency.
2. $\alpha\beta$-unsaturation causes a lowering of frequency of 15 to 40 cm⁻¹, except in amides, where little shift is observed and that usually to higher frequency.
3. Further conjugation has relatively little effect.
4. Ring strain in cyclic compounds causes a relatively large shift to higher frequency. This phenomenon provides a remarkably reliable test of ring size, distinguishing clearly between four, five, and larger membered ring ketones, lactones, and lactams. Six-ring and larger ketones, etc., show the normal frequency found for the open chain compounds.
5. Hydrogen bonding to a carbonyl group causes a shift to lower frequency of 40 to 60 cm⁻¹. Acids, amides, enolized β-keto carbonyl systems and o-hydroxy- and o-aminophenyl carbonyl compounds show this effect. All carbonyl compounds tend to give slightly lower values for the carbonyl stretching frequency in the solid state compared with the value for dilute solutions.
6. Where more than one of the structural influences on a particular carbonyl group is operating, the net effect is usually close to additive.

Table 101
IMINES, OXIMES, ETC. >C=N−

Group	Band	Remarks
>C=N−H	3400—3300(m)	N−H stretching; lowered on H-bonding
>C=N−	1690—1640(v)	Difficult to identify due to large variations
$\alpha\beta$-Unsaturated	1660—1630(v)	in intensity and the closeness to C=C
Conjugated cyclic systems	1660—(1480v)	stretching region. Oximes usually give very weak bands.

Table 102
AZO COMPOUNDS −N=N−

Group	Band	Remarks
−N=N−	∼1575(v)	Very weak or inactive in infrared. Sometimes seen in Raman
− N=N− (with O⁻)	∼1570	

Table 103
ALKENES$>$C$=$C$<$

Group	Band	Remarks
Non-conjugated $>$C$=$C$<$	1680—1620(v)	May be very weak if more or less symmetrically substituted
Conjugated with aromatic ring	∼1625(m)	More intense than with unconjugated double bonds
Dienes, trienes, etc.	1650(s) and 1600(s)	Lower frequency band usually more intense and may hide or overlap the higher frequency band
$\alpha\beta$-Unsaturated carbonyl compounds	1640—1590(s)	Usually much weaker than the C$=$O band
Enol esters, enol ethers and enamines	1690—1650(s)	

Note: See also Table 93 for the $=$C$-$H absorptions of alkenes.

The most substituted double bonds tend to absorb at the high frequency end of the range, the least substituted at the low frequency end. The absorption may be very weak when the double bond is more or less symmetrically substituted, but the vibration frequency can then be detected and measured in the Raman spectrum. For the same reason *trans*-double bonds tend to absorb less strongly than *cis*-double bonds. Table 93 should be consulted for the $=$C$-$H vibration frequencies, which may give additional structural information.

A general trend which has been observed is the effect caused by strain on the stretching frequency of double bonds. A double bond exocyclic to a ring shows the same pattern as cyclic ketones: that is, the frequency rises as the ring size decreases. A double bond within a ring shows the opposite trend: that is, the frequency falls as the ring size decreases. The C$-$H stretching frequency rises slightly as ring strain increases.

The two or three bands in the 1600-1500 cm^{-1} region are shown by most six-membered aromatic ring systems such as benzenes, polycyclic systems, and pyridines. They constitute a valuable identification of such a system. Further bands are shown by aromatic rings in the fingerprint region between 1225 and 950 cm^{-1} which are of little diagnostic value. The weak overtone and combination bands in the 2000 to 1660 cm^{-1} region have been already mentioned. A fourth group of bands below 900 cm^{-1} is produced by the out-of-plane C$-$H bending vibrations (see Table 105.)

Table 104
AROMATIC COMPOUNDS

Group	Band	Remarks
Aromatic rings	∼1600(m)	
	∼1580(m)	Stronger when the ring is further conjugated
	∼1500(m)	This is usually the strongest of the two or three bands

Note: See also Table 93 and Table 105 for aryl$-$H vibration frequencies.

Table 105
SUBSTITUTION PATTERNS OF THE BENZENE RING

Group	Band	Remarks
Five adjacent H	770—730(s) and 720—680(s)	Monosubstituted
Four adjacent H	770—735(s)	*Ortho*-disubstituted
Three adjacent H	810—750(s)	*Meta*-disubstituted, etc., and 1, 2, 3-trisubstituted
Two adjacent H	860—800(s)	*Para*-disubstituted, etc.
Isolated H	900—800(w)	*Meta*-disubstituted, etc.; usually not strong enough to be useful

The frequency of the C—H out-of-plane vibration is determined by the number of adjacent hydrogen atoms on the ring and hence the frequency is a means of determining the substitution pattern. This does not work as well in practice as one might hope. These strong bands are not always the only — or even the strongest — bands in the region (for example, *C*-halogen frequencies interfere particularly) so that assignments based on this evidence alone should be treated with caution. For example, the spectrum of tryptophan and the spectrum of *p*-acetamidobenzaldehyde show only the characteristic absorption of *ortho-* and *para*-disubstituted benzene rings respectively; but the spectrum of p-nitrophenylpropiolic acid shows not only a band at 860 cm⁻¹, consistent with its being a *p*-disubstituted benzene, but also bands at 750 and 685 cm⁻¹, consistent with its being monosubstituted. It is obvious that a positive assignment of substitution pattern is not possible in the latter case.

The values in Table 105 hold reasonably well for condensed ring systems and for pyridines. Powerful electron withdrawing substituents tend to shift the values to higher frequency.

Table 106
NITRO, NITROSO, ETC. N=O

Group	Band	Remarks
C—NO$_2$	~1560(s) ~1350(s)	Lowered ~30 cm⁻¹ when conjugated. The two bands are due to asymmetrical and symmetrical stretching of the NO bonds
Nitrates		
O—NO$_2$	1650—1600(s) 1270—1250(s)	
Nitramines		
N—NO$_2$	1630—1550(s) 1300—1250(s)	
C—N=O	1600—1500(s)	
O—N=O	1680—1610(s)	Two bands
N—N=O	1500—1430(s)	
$\overset{+}{N}$—$\overset{-}{O}$		
Aromatic	1300—1200(s)	Very strong bands
Aliphatic	970—950(s)	
NO$^-_3$	1410—1340 860—800	

Table 107
SULFUR COMPOUNDS

Group	Band	Remarks
−S−H	2600—2550(w)	S−H stretching; weaker than O−H and less affected by H-bonding. This absorption is strong in the Raman
>C=S	1200—1050(s)	
>C−N< (‖S)	~3400	N−H stretching; lowered to ~3150 cm^{-1} in the solid state
	1550—1460(s)	Amide II
	1300—1100(s)	Amide I
>S=O	1060—1040(s)	
>SO₂	1350—1310(s)	
	1160—1120(s)	
−SO₂−N<	1370—1330(s)	
	1180—1160(s)	
−SO₂−O−	1420—1330(s)	
	1200—1145(s)	

Table 108
PHOSPHORUS COMPOUNDS

Group	Band	Remarks
P−H	2440—2350(s)	Sharp
P−Ph	1440(s)	Sharp
P−O-alkyl	1050—1030(s)	
P−O-aryl	1240—1190(s)	
P=O	1300—1250(s)	
P−O−P	970—910	Broad
P(=O)−OH	2700—2560	H-bonded O−H
	1240—1180(s)	P=O stretching

Table 109
ETHERS

Group	Band	Remarks
C−O−C	1150—1070(s)	C−O stretching
=C−O−C	1275—1200(s)	
	1075—1020(s)	
C−O−CH₃	2850—2810(m)	C−H stretching; aryl ethers at higher end of the range
>C—C< (\O/)	~1250	
	~900	
	~800	

Table 110
HALOGEN COMPOUNDS

Group	Band	Remarks
C−F	1400—1000(s)	
C−Cl	800—600(s)	
C−Br	750—500(s)	
C−I	~500(s)	

Table 111
INORGANIC IONS

Group	Band	Remarks
Ammonium	3300—3030	All bands strong
Cyanide, Thiocyanate, Cyanate	2200—2000	
Carbonate	1450—1410	
Sulphate	1130—1080	
Nitrate	1380—1350	
Nitrite	1250—1230	
Phosphates	1100—1000	

From Dean, J. A., Ed., *Lange's Handbook of Chemistry,* 12th ed., McGraw-Hill, New York, 1979. With permission.

RADIOISOTOPES IN MEDICINE AND BIOLOGY

The following text and Tables 112 through 117 are from Diem, K. and Lentner, C., Eds., *Documenta Geigy Scientific Tables,* 7th ed., S.A. Basle, Switzerland, 1970. This material was reproduced with permission.

Basic Concepts

The term *nuclide* indicates a species of atom having specified numbers of protons and neutrons in its nucleus. Nuclides of one and the same chemical element, i.e., nuclides with the same number of protons and differing only in the number of neutrons, are known as *isotopes* of the element concerned. In some nuclides various energy states of the nucleus with finite lifetimes are possible. These states are called *isomers* of the nuclide. Isomeric nuclides have the same numbers of protons and neutrons and differ only in their energy content and thus their lifetime.

The nature of a nuclide is indicated unambiguously by the chemical symbol of the element and the number of nucleons (sum of the protons and neutrons = mass number) shown as an upper index to the left of the element symbol (e.g., ^{12}C, ^{32}P). Additionally, the number of protons (atomic number) can be given as a lower index on the left. Isomers in an excited, metastable state are indicated by a right upper index 'm' (e.g., $^{99}Tc^m$).

Radioactivity and Law of Disintegration

Radioactivity is the property of certain nuclides of spontaneously emitting either particles or gamma rays from the nucleus (nuclear radiation) or X-rays from the shell after capture of an electron from the shell by the nucleus (characteristic X radiation). Except for isomeric transitions, this process always results in a change in the nature of the nuclide (radioactive transformation or radioactive disintegration). Nuclides possessing this property are known as radionuclides.

It is impossible to predict the time when an individual atom will disintegrate; the occurrence of disintegrations is statistically distributed, i.e., it is a stochastic process. For a large number of atoms of the same radionuclide disintegration is governed by the empirical law stating that the number dN of atoms disintegrating in the time dt is at all times proportional to the number N of atoms not yet disintegrated. The proportionality factor is known as the *decay constant* λ; this is a characteristic constant of the nuclide concerned:

$$- dN = \lambda N dt$$

If at time zero N_o atoms of an isolated radionuclide are present the number N_t of atoms not yet disintegrated at any time t is given by

$$N_t = N_0 \, e^{-\lambda t}$$

In equal time intervals the number of radioactive atoms decreases by the same proportion; the time interval during which the number decreases by half is known as the *half-life* ($T_{1/2}$):

$$T_{1/2} = \frac{\ln 2}{\lambda} = \frac{0.693}{\lambda}$$

The reciprocal of the decay constant λ has the dimension of time and is known as the *mean lifetime* τ. τ is the time during which the number of atoms of a radionuclide falls to the fraction $1/e$ ($\approx 37\%$) of its original value.

Activity

The quantity $-(dN/dt) = \lambda N$, i.e., the number of radioactive transformations taking place in a sample during the time dt divided by this time interval, is called the *activity A*. It is a measure of the 'strength' of the radioactive sample.

The unit of activity in the International System of Units is the reciprocal second $(s^{-1})^*$; the commonly used unit is the curie (Ci):

$$1 \text{ Ci} = 3.7 \times 10^{10} s^{-1}**$$

Decimal multiples and submultiples of the curie are

1 megacurie(MCi) $= 10^6$ Ci $= 3.7 \times 10^{16} s^{-1}$
1 kilocurie (kCi) $= 10^3$ Ci $= 3.7 \times 10^{13} s^{-1}$
1 millicurie (mCi) $= 10^{-3}$ Ci $= 3.7 \times 10^7 s^{-1}$
1 microcurie (μCi) $= 10^{-6}$ Ci $= 3.7 \times 10^4 s^{-1}$
1 nanocurie (nCi) $= 10^{-9}$ Ci $= 3.7 \times 10^1 s^{-1}$
1 picocurie (pCi) $= 10^{-12}$ Ci $= 3.7 \times 10^{-2} s^{-1}$

Specific Activity

The *specific activity a of a radioactive material* (for instance a radioactive solution) is the activity A of the radionuclide contained in it divided by the mass m of the material:

$$a = \frac{A}{m}$$

The commonly used unit of specific activity is the curie per gramm (Ci g^{-1}) or a decimal multiple of it.

The *specific activity of a radionuclide* is obtained by dividing the activity $A = \lambda N$ of the radionuclide by the mass of N of its atoms. This quantity is a characteristic constant of the radionuclide expressing the maximum specific activity attainable (i.e., in the carrier-free state):

$$a = \frac{\lambda N_A}{M}$$

$$= 1.63 \times 10^{13} \frac{\lambda}{A_r} \text{ Ci g}^{-1}$$

where N_A is the AVOGADRO constant, M the molar mass in gramme per mole of the radionuclide, A_r its relative atomic mass and λ the value of its decay constant in s^{-1}.

The *activity concentration* of a radioactive material (liquid or gaseous, at a given temperature and pressure) is the ratio of the activity of the contained radionuclide to

---

* In English-speaking countries other units in common use are dps (disintegrations per second) and dpm (disintegrations per minute): 1 dpm = 0.0167 dps = 0.45 pCi; 1 pCi = 0.037 dps = 2.22 dpm.
** The 12th General Conference of Weights and Measures (1964) decided that the curie so defined should be retained as a special unit of activity.

Table 112
RECIPROCALS OF THE SPECIFIC ACTIVITIES OF SOME RADIONUCLIDES[2]

Nuclide	$T\frac{1}{2}$	1 α in g Ci^{-1}
^{24}Na	14.8 h	0.000000113
^{131}I	8.06 d	0.0000081
^{32}P	14d	0.00000352
^{45}Ca	164d	0.0000566
^{14}C	5570a	0.187

the volume of the material. The commonly used unit is the curie per litre (Cil^{-1}) or a decimal multiple of it. A special unit of activity concentration used in balneology for the activity concentration of water containing ^{222}Rn is the eman:

$$1 \text{ eman} = 10^{-10} \text{ Ci } l^{-1}$$

The MACHE unit formerly in common use is equal to 3.64 eman.

Radiation Dosimetry

A long discussion on radiological quantities and units ended provisionally in 1962 with the general acceptance of the definitions recommended by the International Commission on Radiological Units and Measurements (ICRU)[3]. These are formulated in accordance with the customary physical principles, and the units have been assimilated into the International System of Units of the Metric Convention. For comprehensive surveys of the physical concepts and quantities in the dosimetry of ionizing radiations as well as of the quantities characterizing radiation sources and radiation fields see the literature.[3,5]

In accordance with the fundamental Grotthus-Draper law for radiation of any kind, when matter is traversed by energy-rich radiation only that part of the energy that is absorbed can have an action on the matter. With ionizing radiations this absorption of energy occurs in several stages[6] before it becomes evident biologically. It has been agreed internationally that 'energy imparted to matter' shall be understood to mean only that energy manifested as excitation, ionization, or change in the chemical bond energy of the atoms or molecules. This dosimetrically important quantity is defined as follows:

The *energy E_D imparted* by ionizing radiation to the matter in a volume is the difference between the sum E_{in} of the energies (exclusive of rest energies) of all the directly and indirectly ionizing particles which have entered the volume and the sum E_{ex} of the energies (exclusive of rest energies) of all those which have left it, minus the energy equivalent Q of any increase in rest mass that took place in nuclear or elementary particle reactions within the volume:

$$E_D = \Sigma E_{in} - \Sigma E_{ex} + \Sigma Q$$

Whereas there can be no confusion concerning the energy totals E_{in} and E_{ex} it is necessary in the case of Q to be quite clear as to whether the nuclear or elementary particle reaction is exothermic or endothermic, i.e., whether Q is positive or negative. For example, the absorption of a photon in the volume concerned may produce an electron pair (electron + positron), an endothermic process; for this reaction therefore $Q = + 2m_e c^2$ (m_e is the rest mass of the electron, c the velocity of light) for each interaction.

The *absorbed dose* is the amount of energy E_D imparted to the matter divided by the mass m of the matter (see below). The most important task of dosimetry is to determine this absorbed dose, which is now regarded as the most meaningful quantity to which the observable chemical and biological effects can be related. The absorbed dose is the result of certain physical reactions between radiation and matter, reactions that are in turn dependent on the nature, intensity, and spectral energy distribution of the radiation and the atomic composition of the material.

Radiation fields in the body are usually nonuniform in space as well as in time. Thus there may be nonuniform distribution of the absorbed dose at the boundary surfaces of soft tissues or bones, while the pulsed electrons from particle accelerators constitute radiation nonuniform in time. The quantities concerned must therefore be determined for regions of space or intervals of time so small that any further reduction would not appreciably change the values of the quotients measured. This requirement necessitates the use of some limiting procedure, and in the ICRU definitions the quantities are presented as quotients of small differences. As the ICRU Reports point out, the region of space considered also has a lower limit of size, for it must still be large enough to contain many interactions and be traversed by many particles. If it is impossible to find a mass fulfilling both these conditions the dose has to be deduced from multiple measurements involving extrapolation or averaging procedures. The symbol Δ is placed before symbols for quantities concerned in such averaging procedures.

Radiation Field Quantities

A radiation field is a region in vacuum or matter that is traversed by radiation.

$$\text{Particle fluence } \Phi = \frac{\Delta N}{\Delta a}$$

where ΔN is the number of particles* entering a sphere of cross-sectional area Δa.

$$\text{Particle flux density or particle fluence rate } \varphi = \frac{\Delta \Phi}{\Delta t}$$

where $\Delta \Phi$ is the particle fluence in time Δf.

$$\text{Energy fluence } \Psi = \frac{\Delta E\psi}{\Delta a}$$

where $\Delta E\psi$ is the sum of the energies, exclusive of rest energies, of all the particles entering a sphere of cross-sectional area Δa.

$$\text{Energy flux density or energy fluence rate } \psi = \frac{\Delta \Psi}{\Delta t}$$

where $\Delta \psi$ is the energy fluence in the time Δt.

---

* In this section the expression 'particle' is understood to include not only corpuscles like electrons, protons, neutrons, etc. but also photons.

Interactions

Since the great majority of radiations used in medicine are X rays, gamma rays or electrons discussion will be limited here to the interactions of photons and electrons with matter. Neutron/matter interactions and neutron dosimetry fall outside the scope of the present article.

When photons collide with atoms or molecules, electrons are liberated (as a result of the photoelectric effect, Compton effect and pair production) and absorb some of the energy of the photons.

$$\text{Mass energy transfer coefficient} \quad \frac{\mu_K}{\rho} = \frac{1}{E\rho} \cdot \frac{\Delta E_K}{\Delta l}$$

where ΔE_K is the sum of the kinetic energies of the secondary electrons liberated in a layer of thickness Δl and density ϱ, and E is the sum of the energies (excluded rest energies) of the photons incident normally upon the layer.

When charged particles collide with atoms or molecules, part of their kinetic energy is lost in collisions with atoms or molecules due to ionization, electronic excitation and production of bremsstrahlung.

$$\text{Mass stopping power} \quad \frac{S}{\rho} = \frac{1}{\rho} \cdot \frac{\Delta E}{\Delta l}$$

where ΔE is the average amount of energy lost by a charged particle of energy E when traversing a path of length Δl in a layer of density ϱ.

$$S_e/\varrho = \text{electron mass stopping power}$$

Quantities and Units of Dose

Absorbed Dose and Absorbed Dose Rate

The absorbed dose* D produced by ionizing radiation in matter is the quotient of ΔE_D by Δm where ΔE_D is the energy imparted by the radiation to the matter in a volume element and $\Delta m = \varrho \times \Delta V$ is the mass of the matter in that volume element:

$$D = \frac{\Delta E_D}{\Delta m} = \frac{1}{\rho} \cdot \frac{\Delta E_D}{\Delta V}$$

The expression 'integral absorbed dose' still in common use thus simply means the amount of energy imparted to matter (see under 1. above):

$$E_D = \sum_i (Di \cdot \Delta mi)$$

In medical radiology 'matter' could for instance be a single organ or the whole body. The term 'energy imparted to matter' is much to be preferred from the point of view of clarity.

---

* The designation 'absorbed dose' has been criticized on the grounds that an 'absorbed dose' can be produced only in matter and not in vacuum. For this reason this quantity is known in the German literature as 'energy dose'.

The special unit of absorbed dose is the rad (rd):

$$1 \text{ rd} = 0.01 \text{ J kg}^{-1} = 100 \text{ erg g}^{-1} = 2.388 \times 10^{-6} \text{ cal}_{IT} \text{ g}^{-1}$$

$$= 6.242 \times 10^{13} \text{ eV g}^{-1}$$

The absorbed dose rate $\dot{D}$ is the quotient of ΔD by Δt, where ΔD is the increment in absorbed dose in the time Δt:

$$\dot{D} = \frac{\Delta D}{\Delta t}$$

When the conditions are such that there is no variability in time $\dot{D} = D/t$.

Special units of absorbed dose rate are rad per second (rd s^{-1}), rad per minute (rd min^{-1}), rad per hour (rd h^{-1}), etc.:

$$1 \text{ rd s}^{-1} = 0.01 \text{ W kg}^{-1}$$

The direct measurement of absorbed dose or absorbed dose rate is possible only by means of calorimetry in phantoms and is very time-consuming. In practical dosimetry indirect methods are used, particularly those based on ionization measurements in air, in which the absorbed dose is obtained by simple calculation.

Exposure and Exposure Rate

The exposure (X) is the quotient of ΔQ by Δm, where ΔQ is the sum of the electrical charges on all the ions of one sign produced in air when all the electrons (negatrons and positrons), liberated by photons in a volume element of air whose mass is Δm, are completely stopped in air:

$$X = \frac{\Delta Q}{\Delta m}$$

The special unit of exposure is the roentgen* (R), defined as

$$1 \text{ R} = 2.58 \times 10^{-4} \text{ C kg}^{-1} \text{ (exactly)}$$

From the definition of the roentgen and the elementary charge $\approx 1.602 \times 10^{-19}$ C it follows that an exposure of 1 R produces 1.610×10^{12} ion pairs per gramme (2.082×10^{9} ion pairs per cubic centimetre) of air at its normal density of 1.293 mg cm^{-3}.

In German-speaking countries a quantity equivalent to exposure, the equilibrium ion dose (J_s) is used in medical radiology; this is defined as the ion dose produced by photons at the point of interest when there is secondary electron equilibrium at this point. Equilibrium ion dose is measured by the same methods of exposure and has the same numerical value when expressed in roentgen.

Secondary electron equilibrium exists at a point in matter when the sum of the kinetic energies of the photon-produced secondary electrons entering a volume containing this point is equal to the sum of the kinetic energies of the secondary electrons

---

* This unit is numerically identical with the old roentgen (r), defined as 1 electrostatic unit of charge per 1.293 mg of air.

Table 113
CONVERSION OF COMMON UNITS OF EXPOSURE RATE AND ION DOSE RATE

	mR h^{-1}	μRs^{-1}	Rh^{-1}	R min^{-1}	Rs^{-1}
1mRH^{-1} = 1		2.8×10^{-1}	10^{-3}	1.7×10^{-3}	2.8×10^{-7}
1μRs^{-1} = 3.6		1	3.6×10^{-3}	6×10^{-5}	10^{-6}
1Rh^{-1} = 10^3		2.8×10^2	1	1.7×10^{-2}	2.8×10^{-4}
1Rmin^{-1} = 6×10^4		1.7×10^4	60	1	1.7×10^{-2}
1Rs^{-1} = 3.6×10^6		10^6	3.6×10^3	60	1

leaving this volume. This equilibrium can be established in an ionization chamber by enclosing the volume of air by a wall equivalent to air, for instance graphite, of a thickness at least equal to the range of the secondary electrons in this wall. A further condition is that the mean range of the photon-produced secondary electrons is small compared to $1/\mu$ (μ being the linear attenuation coefficient for the photons). Since this second condition is approximately fulfilled only for photons of energies up to about 3 MeV the equilibrium ion dose can be measured only when the photon energy is below this level. The introduction of the ionization chamber must not noticeably disturb the radiation field of the photons.

The exposure rate ($\dot{X}$) is the quotient of $\triangle X$ by $\triangle t$, where $\triangle X$ is the increment in exposure in time $\triangle t$:

$$\dot{X} = \frac{\triangle X}{\triangle t}$$

When the conditions are such that there is no variability in time $\dot{X} = X/t$.

Special units of exposure rate are roentgen per second (R s^{-1}), roentgen per minute (R min^{-1}), roentgen per hour (R h^{-1}), etc.

$$1 \text{ R/s} = 2.58 \times 10^{-4} \text{ A kg}^{-1}$$

Not included in the ICRU Reports but appearing in the appropriate German DIN Standard is the quantity 'ion dose', applicable to all kinds of radiation except neutrons.

The ion dose J produced by ionizing radiation in matter is the quotient of $\triangle Q$ by $\triangle m_A$, where $\triangle Q$ is the electric charge of the ions of one sign formed directly or indirectly by the radiation in air in a volume element $\triangle V$, and $\triangle m_A$ is the mass of the air of density ϱ_A in that volume element:

$$J = \frac{\triangle Q}{\triangle m_A} = \frac{1}{\rho_A} \cdot \frac{\triangle Q}{\triangle V}$$

The special unit of ion dose is likewise the roentgen (see above). From the definition of the roentgen and the elementary charge $e \approx 1.602 \times 10^{-19}$ C it follows that an ion dose of 1 R produces 1.610×10^{12} ion pairs per gramme (2.082×10^9 ion pairs per cubic centimetre) of air at its normal density of 1.293 mg cm^{-3}.

The ion dose rate $\dot{J}$ is the quotient of $\triangle J$ by $\triangle t$, where $\triangle J$ is the increment in ion dose in time $\triangle t$:

$$\dot{j} = \frac{\triangle J}{\triangle t}$$

When the conditions are such that there is no variability in time $J = J/t$.

The special units of ion dose rate are roentgen per second (R s^{-1}), roentgen per minute (R min^{-1}), roentgen per hour (R h^{-1}), etc.

The cavity ion dose J_c is the ion dose produced by photon or electron irradiation in an air-filled cavity surrounded by matter of any kind when the Bragg-Gray conditions are fulfilled.

If a cavity within a material A is filled with a material B (for instance air) the Bragg-Gray conditions are fulfilled when

1. The flux density of the first generation of electrons and their energy distribution remain unchanged by the cavity filled with material B
2. The energy of the secondary electrons produced by the photons in material B is negligible in comparison with the energy imparted to material B
3. The flux density of the electrons of all generations within the material B is uniform throughout

These conditions can be approximately met if the cavity contains air and its linear dimensions are small compared with $1/\mu$ (μ being the linear attenuation coefficient for the photons) and compared with the mean range of the secondary electrons. The walls of such a cavity ionization chamber must either be very thin or have values for mass energy transfer coefficient $\mu K/\varrho$ and electron mass stopping power S_e/ϱ deviating only slightly from those of the surrounding material A; in other words, the ionization of the air molecules in the cavity by the photons must be due predominantly to the secondary electrons produced in the surrounding material A. In order to reduce boundary layer effects between the material of the wall and the air in the cavity resulting from low-energy delta-electrons the inner side of the wall must be covered with a graphite layer about 1 μm thick. If this is not done the mean cavity ion dose will be dependent on the volume in which the dose is being measured.

Conversion of Dose Quantities

The absorbed dose D_A for air is obtained from the ion dose J (measured as J_s* or J_e):

$$D_A = U_{1A} \cdot J$$

where $U_{1A} = 0.869$ rd/R is the ionization constant of air, obtained from the average energy E_1 ($= 33.7$ eV) required for the formation of an ion pair in air, from the elementary charge e and from the relationship $1\ V = 1\ J/1\ C = 2.58 \times 10^{-4}$ rd R^{-1}. Above about 10 keV, U_{1A} remains practically constant over a wide energy range.

For photon irradiation the absorbed dose D_Z at the point of interest in material Z is obtained from the absorbed dose D_A at the same point with secondary electron equilibrium in air in accordance with the relationship

$$D_Z = D_A \cdot (\mu_K/\rho)z/(\mu_K/\rho)_A$$

where $(\mu K/\varrho)_Z$ are the mass energy transfer coefficients of the material Z (for instance body tissues) and air, respectively, for photons of energy E. For a photon spectrum the values $(\mu K/\varrho)_Z$ and $(\bar{\mu} K/\varrho)_A$ averaged over the spectrum must be used instead.

---

* Or exposure X.

Table 114
CONVERSION FACTOR $f = D/X$

		f in rd R^{-1} for		
E in MeV	Air	Water[a]	Soft tissues[b]	Bone (compact)[b]
0.010	0.869	0.912	0.925	3.54
0.015	0.869	0.890	0.916	3.97
0.020	0.869	0.877	0.916	4.23
0.030	0.869	0.870	0.910	4.39
0.04	0.869	0.873	0.919	4.14
0.05	0.869	0.893	0.926	3.58
0.06	0.869	0.915	0.929	2.91
0.08	0.869	0.937	0.939	1.91
0.10	0.869	0.942	0.948	1.45
0.15	0.869	0.964	0.956	1.05
0.20	0.869	0.971	0.963	0.979
0.30	0.869	0.964	0.957	0.938
0.4	0.869	0.967	0.954	0.928
0.5	0.869	0.964	0.957	0.925
0.6	0.869	0.964	0.957	0.925
0.8	0.869	0.967	0.956	0.920
1.0	0.869	0.967	0.956	0.922
1.5	0.869	0.966	0.958	0.920
2.0	0.869	0.966	0.954	0.921
3.0	0.869	0.964	0.954	0.928

[a] From National Bureau of Standards, Report 8681, U.S. Government Printing Office, Washington, D.C., 1965.
[b] From National Bureau of Standards, Physical Aspects of Irradiation, ICRU Report 10b, 1962, Handbook 85, U.S. Government Printing Office, Washington, D.C., 1964.

For the exposure X^* the conversion equations are as follows: (a) for photons of uniform energy E:

$$D_Z = f \cdot X, \text{ with } f = U_{1A} \cdot (\mu_K/\rho)z/(\mu_K/\rho)_A$$

(b) for a photon spectrum:

$$D_Z = \bar{f} \cdot X, \text{ with } \bar{f} = U_{1A} \cdot (\bar{\mu}_K/\rho)z/(\mu_K/\rho)_A$$

Values of the conversion factors f and $\bar{f}$ for air, water, soft tissues and bone are given in Tables 114 and 115.

For photon and electron irradiation the absorbed dose D_Z at the point of interest in the material Z is obtained from the absorbed dose D_A for air measured at the same point under Bragg-Gray conditions in accordance with the relationship

$$D_Z = D_A \cdot (\bar{S}e/\rho)z/(\bar{S}e/\rho)_A$$

* Or equilibrium ion dose J_s, in which case J_s must replace X in the formula.

Table 115
CONVERSION FACTOR $f = D/X$

Tube potential in kV	Radiation				f in rd R^{-1} for			
	Filter		Half-value layer				Soft tissues	Bone (compact)
	mm Al	mm Cu	mm Al	mm Cu	Air	Water		
50	1.4	—	1.2	0.03	0.87	0.88	0.93	4.2
100	—	0.2	4.2	0.18	0.87	0.89	0.92	3.6
150	—	0.5	—	0.75	0.87	0.92	0.94	2.3
200	—	1.0	—	1.45	0.87	0.94	0.95	1.6
250	—	1.5	—	2.35	0.87	0.95	0.95	1.4
300	—	3.0	—	3.5	0.87	0.96	0.95	1.2
400	—	3.0	—	4.2	0.87	0.96	0.96	1.1

Note: When calculating the absorbed dose for soft tissues embedded in bone from the measured exposure a value of f should be chosen lying between those for soft tissues and bone and depending on the distance of the bone from the point of measurement.

From National Bureau of Standards, Physical Aspects of Irradiation, ICRU Report 106, 1962, Handbook 85, U.S. Government Printing Office, Washington, D.C., 1964.

where $(\bar{S}e/\varrho)_Z$ and $(\bar{S}e/\varrho)_A$ are the electron mass stopping powers of the material Z and of air averaged over the electron spectrum. Using the cavity ion dose J_c the conversion equation is as follows:

$$D_Z = \bar{g} \cdot Jc, \text{ with } \bar{g} = U_{1A} \cdot (\bar{S}e/\rho)z/(\bar{S}e/\rho)_A$$

Values of the conversion factor $\bar{g}$ for air, water and soft tissues are given in Table 116.

Relation of Absorbed Dose to Radiation Field

For photons of uniform energy the energy flux density ψ_{ph} of the photons at the point of interest is related to the absorbed dose rate $\dot{D}$ at the same point when there is secondary electron equilibrium as follows:

$$\dot{D} = (\mu_K/\rho) \cdot \psi_{ph}$$

Similarly, for the absorbed dose D and the energy fluence ψ_{ph} of the photons,

$$D = (\mu_K/\rho) \cdot \Psi_{ph}$$

where μ_K/ϱ is the mass energy transfer coefficient of the material for photons of this energy.

For electrons of uniform energy the particle flux density ψ_e of the electrons at the point of interest is related to the absorbed dose rate D at the same point under Bragg-Gray conditions as follows:

$$\dot{D} = (S_e/\rho) \cdot \varphi_e$$

Similarly, for the absorbed dose D and the particle fluence ϕ_e of the electrons,

Table 116
CONVERSION FACTOR $\bar{g}$ = D/J_c

Radiation		$\bar{g}$ in rd R^{-1} for		
Quantum energy or electron energy	Half-value layer or radionuclide	Air	Water	Soft tissues
(a) Bremsstrahlung at 400 kV tube potential	4.2 mm Cu	0.87	1.01	1.00
0.66 MeV	^{137}Cs	0.87	1.00	1.00
1.25 MeV	^{60}Co	0.87	0.99	0.99
Bremsstrahlung 15 MeV	—	0.87	0.98	0.97
Bremsstrahlung 30 MeV	—	0.87	0.95	0.94
Bremsstrahlung 45 MeV	—	0.87	0.94	0.93
(b) Electrons 5 MeV	—	0.87	0.92	0.91
10 MeV	—	0.87	0.88	0.87
20 MeV	—	0.87	0.84	0.83
30 MeV	—	0.87	0.82	0.81
40 MeV	—	0.87	0.81	0.80
50 MeV	—	0.87	0.80	0.79

Note: When calculating the absorbed dose for soft tissues embedded in bone, the factor for the latter should be used since the effect of bone has already been allowed for in measurement of the cavity ion dose.

Values (a) from National Bureau of Standards, Physical Aspects of Irradiation, ICRU Report 10b, 1962, Handbook 85, U.S. Government Printing Office, Washington, D.C.,1964; (b) calculated from Berger and Seltzer, Tables of Energy Losses and Ranges of Electrons and Positrons, NASA SP-3012, National Aeronautics and Space Administration, Washington, D.C., 1964, and Additional Stopping Power and Range Tables for Protons, Mesons and Electrons, NASA SP-3036, National Aeronautics and Space Administration, Washington, D.C., 1966.

$$D = (S_e/\rho) \cdot \Phi_e$$

where S_e/ϱ is the electron mass stopping power of the material for electrons of this energy.

Since the coefficients $\mu x/\varrho$ and S_e/ϱ are themselves functions of the photon and electron energy respectively the absorbed dose rates for two different particle energies generally differ even when the energy flux density or particle flux density, as the case may be, is the same. In practice, radiation is usually not of uniform energy but has an energy spectrum, in which case mean values of the coefficients over the spectral range must be used.

Relative Biological Effectiveness (RBE)

Radiobiological studies have shown that different kinds of ionizing radiation can produce biological effects of different intensity even when the absorbed dose in the biological material being irradiated, and all other conditions, are the same. The absorbed dose is therefore still not an adequate physical quantity from which all the biological effects can be deduced. This has led to the introduction of the concept of relative biological effectiveness (RBE), with the dimensionless RBE factor ξ defined as

$$\xi = D_0 / D$$

where D is the absorbed dose of the radiation under consideration that produces a particular biological effect, and D_o the absorbed dose of a standard radiation (at present hard filtered 200 kV X rays) that produces the same effect under otherwise identical conditions. The RBE factor is not a constant for a particular kind of radiation since different values are obtained depending on the nature of the irradiation reaction being observed, on the kind of biological system under study, on the stage of development of the object being irradiated, and on the distribution of the absorbed dose in space and time.[7] Since the RBE factor as such is unsuitable for use in the field of radiation protection the ICRU[3] has recommended that the term RBE should be employed in radiobiology only.

Dose Equivalent and Quality Factor

In radiation protection the place of the RBE factor ξ is taken by the quality factor* q, and that of the absorbed dose of the standard radiation — which is not used in radiation protection — by the dose equivalent* D_q, defined as follows:

$$D_q = q \cdot D$$

The concept of dose equivalent is intended for use in radiation protection only. The quality factor q is a dimensionless number whose magnitude depends mainly on the nature of the radiation, the particle energy and the conditions under which the irradiation takes place. In practice, agreed conventional values of q are used based on the relative biological effectiveness ξ. The dose equivalent is equal to the absorbed dose produced by a standard radiation with a quality factor $q = 1$ (at present 200 kV X rays); this absorbed dose is considered from the point of view of risk to be the same as the absorbed dose produced by the actual radiation with a quality factor $q \neq 1$.

If a number of different radiations are present simultaneously the total dose equivalent is the sum of the dose equivalents of the individual radiations:

$$D_q = \sum_i D_{qt} = \sum_i (D_i q_i)$$

For dose equivalents the unit rad is given the special name rem (symbol rem):

$$1 \text{ rem} = 1 \text{ rd}$$

The term rem is reserved exclusively for expressing dose equivalents, so that data given in this unit are immediately recognizable as such.

---

* The symbol QF used in the ICRU Reports, like the symbol DE for dose equivalent, is inconvenient for use in formula.

Table 117
SPECIFIC GAMMA RAY CONSTANTS OF SOME RADIONUCLIDES

Γ in $Rh^{-1}m^2Ci^{-1}$

^{22}Na	^{24}Na	^{42}K	^{54}Fe	^{58}Co	^{60}Co	^{64}Cu	^{130}I	^{131}I	^{137}Cs $+ ^{137}$Ba^m	^{192}Ir	^{198}Au
1.19	1.84	0.14	0.63	0.55	1.31	0.12	1.22	0.22	0.31	0.50	0.23

Specific Gamma Ray Constant

The specific gamma ray constant Γ of a gamma-emitting radionuclide is the quotient of $P \times \Delta\dot{X}$ by the activity A of the nuclide, where $\Delta\dot{X}$ is the exposure rate at a distance l from a point source of the nuclide and the gamma rays are assumed to undergo no absorption either in the sample or over the distance l:

$$\Gamma = \frac{\dot{X} \cdot l^2}{A}$$

$\dot{X}$ is normally only the exposure rate resulting from gamma radiation and from the annihilation radiation of positron-emitting nuclides. If the X rays due to internal conversion or electron capture are not included this must be clearly stated when giving the specific gamma ray constant.

In the case of radionuclides with short-lived decay products Γ is given for the state of radioactive equilibrium; here $\dot{X}$ is the exposure rate resulting from the gamma rays emitted by all members of the series, A the activity of the parent nuclide. For radium the specific gamma ray constant is related to the mass m_{Ra} of the radium nuclide ^{226}Ra in equilibrium with its decay products and enclosed in a platinum envelope of 0.5 mm thickness:

$$\Gamma_{Ra} = \frac{\dot{X} \cdot l^2}{m}; \text{ numerically } \Gamma_{Ra} = 0.825 \text{ R } h^{-1} \text{ m}^2 \text{ g}^{-1}$$

The special unit of specific gamma ray constant is

$$\frac{\text{roentgen} \times \text{square metre}}{\text{hour} \times \text{curie}} \quad (R \text{ } h^{-1} \text{ m}^2 \text{ Ci}^{-1})$$

For ^{226}Ra the unit is

$$\frac{\text{roentgen} \times \text{square metre}}{\text{hour} \times \text{gramme}} \quad (R \text{ } h^{-1} \text{ m}^2 \text{ g}^{-1})$$

If the activity A or the mass m_{Ra} of the radium is known the exposure rate at the distance l can therefore be calculated provided absorption of the gamma rays in the source and intervening air is neglected:

$$\dot{X} = \Gamma \frac{A}{l^2} \text{ or } \dot{X} = \Gamma_{Ra} \frac{m_{Ra}}{l^2}$$

REFERENCES

1. Conférence Générale des Poids et Mesures, *Comptes rendus des séances de la 12ᵉ Conférence générale des Poids et Mesures,* Paris 1964, Gauthier-Villars, Paris, 1964, 94.
2. **Quimby et al.** *Radioactive Isotopes in Clinical Practice,* Lea & Febiger, Philadelphia, 1958.
3. **National Bureau of Standards,** *Radiation Quantities and Units,* ICRU Report 10a, 1962, Handbook 84, U.S. Government Printing Office, Washington, D.C., 1962; International Commission on Radiological Units, *Radiation Quantities and Units* (1968), ICRU Report No. 11, ICRU Publications, Washington, D.C., 1968.
4. Deutscher Normenausschuβ, DIN 6809, October 1963; Entwurf DIN 6814, Blatt 3, November 1968.
5. **Attix and Roesch, Eds.,** *Radiation Dosimetry,* Vol. 1, 2nd ed., Academic Press, New York, 1968; National Bureau of Standards, *Physical Aspects of Irradiation,* ICRU Report 10b, 1962, U.S. Government Printing Office, Washington, D.C., 1964.
6. United Nations Scientific Committee on the Effects of Atomic Radiation, *Report to the General Assembly,* Seventeenth Session, Suppl. No. 16, United Nations, New York, 1962; Deeley and Wood, Eds., *Modern Trends in Radiotherapy,* Vol. 1, Butterworth, London, 1967.
7. Relative Biological Effectiveness Committee, *Health Phys.,* 9, 357, 1963.
8. **Berger et al.,** *Strahlentherapie,* 131, 143, 1966.

Table 118
WAVE LENGTHS OF VARIOUS RADIATIONS

	Ångstroms
Cosmic rays	0.0005
Gamma rays	0.005–1.40
X-rays	0.1–100
Ultra violet, below	4000
Limit of sun's U.V. at earth's surface	2920
Visible spectrum	4000–7000
Violet, representative, 4100, limits	4000–4240
Blue, representative, 4700, limits	4240–4912
Green, representative, 5200, limits	4912–5750
Maximum visibility	5560
Yellow, representative, 5800, limits	5750–5850
Orange, representative, 6000, limits	5850–6470
Red, representative, 6500, limits	6470–7000
Infra red, greater than	7000
Hertzian waves, beyond	2.20×10^6

From Weast, R. C., Ed., *CRC Handbook of Chemistry and Physics,* 61st ed., CRC Press, Boca Raton, Fla., 1980, E-216. With permission.

Table 119
NUCLEAR DATA OF THE MOST IMPORTANT ARTIFICIAL RADIO-ISOTOPES IN MEDICINE AND BIOLOGY

Isotope	Half-life[1]	Mode of decay	Maximum energy of β-radiation[1] MeV	Mean energy of β-radiation MeV	Maximum range of β-radiation in H_2O mm	Energy of γ-radiation[1] MeV	Fraction disintegrating per day (cf. Table 2, col. 3)
^{3}H	12.26 y	β^-	0.018	0.006	0.003		2×10^{-4}
^{11}C	20 m	β^+	0.98	0.38	4		1.0
^{14}C	5760 y[a]	β^-	0.155	0.05	0.3		4×10^{-1}
^{13}N	10 m	β^+	1.25	0.48	5.6		1.0
^{18}F	1.87 h	β^+	0.65	0.25	2.5		1.0
^{22}Na	2.6 y	β^+; γ	0.54	0.2	2	1.28	7×10^{-4}
^{24}Na	15.0 h	$\beta-$; γ	1.39	0.54	6	1.37;2.76	0.68
^{31}Si	2.62 h	β^-;γ	1.47	0.6	7	1.26	0.998
^{32}P	14.2 d	β^-	1.71	0.68	8		0.05
^{35}S	87 d	β^-	0.167	0.05	0.3		0.008
^{36}Cl	3.1×10^8y	β^-	0.714	0.24	2.5		5×10^{-9}
^{38}Cl	37.3 m	β^-; γ	4.8; 2.8; 1.1	1.39	27	1.6; 2.15	1.0
^{48}K	12.5 h	β^-; γ	3.6; 2.0	1.4	19	1.53	0.74
^{42}Ca	153 d	β^-	0.25	0.09	0.6		0.004
^{61}Cr	27.8d	γ; K		0.01		0.323	0.025
^{83}Mn	5.7d	β^+; γ; K	0.58	0.2	2.2	0.7; 0.9; 1.5	0.10
^{86}Mn	291 d	γ; K		0.005		0.84	0.002
^{83}Fe	2.94 y	K		0.006			6×10^{-4}
^{60}Fe	45d	β^-; γ	0.46; 0.27; 1.56	0.12	1.6	1.1; 1.3; 0.19	0.015
^{86}Co	5.25 y	β^-; γ	0.306	0.1	0.8	1.17; 1.33	4×10^{-4}
^{82}Zn	38 m	β^+; γ; K	2.36; 1.4	0.96	11	1.0; 1.9; 2.6	1.0
^{86}Zn	245 d	β^+; γ; K	0.325	0.1	0.8	1.11	0.003
^{83}Br	36 h	β^-; γ	0.44	0.15	1.4	0.55—1.48	0.40
^{85}Kr	10.6 y	β^-; γ	0.67	0.2	2.5	0.52	2×10^{-6}
^{89}Sr	51 d	β^-; γ	1.46	0.55	7	0.91	0.013
^{131}I	8 d	β^-; γ	0.61; 0.81	0.2	2	0.36; 0.6—0.72	0.08
^{190}Au	2.69 d	β^-; γ	0.96; 0.29; 1.37	0.34	3.8	0.41; 0.67; 1.1	0.26

1. The data are from the Table of Isotopes of Strominopr et al., *Rev. Mod. Phrs.,* 30, 585 (1958), and the *International Directory of Radioisotopes,* vol. 1, International Atomic Energy Agency, Vienna, 1959.
2. Redetermined value of National Bureau of Standards, *Nat. Bur. Stand. Techn. News Bull.,* 45, 21 (1961).

From Diem, K., Ed., *Documenta Geigy Scientific Tables,* 6th ed., Geigy Pharmaceuticals, Ardsley, N.Y., 1962, 271. With permission.

Table 120
CONVERSION OF ROENTGEN DOSE RATES

A	B				
	mr/hr	μr/sec	r/hr	r/min	r/sec
mr/hr	1×1	2.7×10^{-1}	1×10^{-3}	1.6×10^{-5}	2.7×10^{-7}
μr/sec	3.6×1	1×1	3.6×10^{-3}	6×10^{-5}	1×10^{-6}
r/hr	1×10^3	2.7×10^2	1×1	1.6×10^{-2}	$2.7' \times 10^{-4}$
r/min	6×10^4	1.6×10^4	6×10	1×1	1.6×10^{-2}
r/sec	3.6×10^6	1×10^8	3.6×10^3	6×10	1×1

Note: To convert from a unit under A into any unit under B, multiply by the factor given in the appropriate column under B.

From Diem, K., Ed., *Documenta Geigy Scientific Tables*, 6th ed., Geigy Pharmaceuticals, Ardsley, N.Y., 1962, 237. With permission.

Table 121
FACTORS IN ROENTGENS PER MILLICURIE AND HOUR
AT A DISTANCE OF 1 CM

Nuclide	Half-life T	Energy of γ-radiation MeV	kr/mch at 1 cm	Ref.
^{22}Na	2.6 y	1.28	11.6	1,2
^{24}Na	15 h	1.37;2.76	18.7	3
^{42}K	12.5 h	1.53	1.5[a]	5
^{51}Cr	27.8 d	0.323	1.8	6
^{52}Mn	5.7 d	0.7;0.9;1.5	19.5	6
^{54}Mn	291 d	0.84	4.7	1,2
^{56}Mn	2.58 h	2.06;1.77;0.822	9.4	4
^{59}Fe	45 d	1.3;1.1;0.19	6.2	1,2
^{60}Co	5.25 y	1.33;1.17	12.8±0.1	3,7
			13.0	1,2
			13.5±0.3	7
^{65}Zn	245 d	1.11	2.8	1,2
^{74}As	17.5 d	0.635;0.596;0.511	5.1	6
^{76}As	26.8 h	1.70;1.2;0.55	2.3	1,2
^{82}Br	36 h	0.55—1.48	14.6	1,2
^{124}Sb	60 d	0.21—2.04	8.9	1
^{131}I	8.04 d	0.08(0.3%);0.284(3%) 0.364(79%);0.368 (12.5%);0.72(5.2%)	2.2	1,2
^{132}I	2.4h	0.69;1.41;2.0	11.9	1
^{137}Cs	30 y	0.661	3.1	1,2
^{170}Tm	129 d	0.084	0.4[a]	1,2
^{182}Ta	112 d	0.066—1.223(eff. 1.13)	6.1	4
^{192}Ir	74.4 d	0.136—0.613(eff. 0.6)	5.0	3
^{198}Au	2.69 d	0.41;0.67;1.1	2.35	6
^{203}Hg	47.9 d	0.279	1.6	6
^{226}Ra	1600y	13 lines from 0.184 to 2.45	8.25±0.08[b]	8
			7.60±0.05[c]	9

[a]   In practice probably much higher since bremsstrahlung (from the β-radiation) has been ignored.

[b]   With 0.5 mm platinum filtering.

[c]   With 1.0 mm platinum filtering.

Table 121 (continued)
FACTORS IN ROENTGENS PER MILLICURIE AND HOUR AT A DISTANCE OF 1 CM

References: 1) Calculated from the decay scheme and quoted by Faires and Parks, *Radioisotope Laboratory Techniques*, London, 1958, 40. 2) Calculated from the decay scheme and quoted by Price et al., *Radiation Shielding*, London, 1957, 305. 3) Experimental value of the National Physical Laboratory Standards Committee (London). Cf. Perry, W. E., *Acta radiol. (Stockh.)*, Suppl. 117. 105 (1954). 4) Rajewsky, B., *Strahlendosis und Strablenwirkung*, 2nd ed., Stuttgart, 1956, N13-N21. 5) Calculated from the decay scheme and quoted by Sinclair, W. K., in Hine and Brownell (Eds.), *Radiation Dosimetry*, New York, 1956, 518. 6) Calculated from the decay scheme and quoted by Hine, G. J., in Hine and Brownell, loc. cit., 899. 7) Robinson, B. W., *Acta radiol. (Stockh.)*, Suppl. 117, 71, (1954). 8) Experimental value of Garrett, C., *Canad. J. Phys.*, 36, 149, 1958. 9) Lüthy, H., personal communication.

From Diem, K., Ed., *Documenta Geigy Scientific Tables*, 6th ed., Geigy Pharmaceuticals, Ardsley, N.Y., 1962, 237. With permission.

Table 122
VALUES OF THE FACTOR f(ABSORBED DOSE IN RAD PER ROENTGEN) FOR MONOCHROMATIC PHOTON BEAMS[a]

Photon energy MeV	Factor f for				Photon energy MeV	Factor f for			
	Air[b]	Water	Muscle	Bone		Air[b]	Water	Muscle	Bone
0.010	0.87_7	0.92_0	0.93_3	3.58	0.20	0.87_7	0.98_2	0.97_2	0.98_8
0.015	0.87_7	0.89_7	0.92_5	4.00	0.30	0.87_7	0.97_7	0.96_5	0.94_7
0.020	0.87_7	0.88_7	0.92_5	4.27	0.40	0.87_7	0.97_5	0.96_3	0.93_6
0.030	0.87_7	0.87_7	0.91_9	4.43	0.50	0.87_7	0.97_4	0.96_6	0.93_3
0.040	0.87_7	0.88_7	0.92_8	4.18	0.60	0.87_7	0.97_5	0.96_6	0.93_3
0.050	0.87_7	0.90_0	0.93_4	3.61	0.80	0.87_7	0.97_4	0.96_5	0.92_9
0.060	0.87_7	0.91_3	0.93_7	2.94	1.0	0.87_7	0.97_4	0.96_5	0.92_7
0.080	0.87_7	0.94_0	0.94_8	1.93	1.5	0.87_7	0.97_3	0.96_6	0.92_9
0.10	0.87_7	0.95_7	0.95_7	1.47	2.0	0.87_7	0.97_4	0.96_3	0.92_9
0.15	0.87_7	0.97_1	0.96_4	1.06	3.0	0.87_7	0.97_1	0.96_3	0.93_7

[a] From ICRU Report 1956.

[b] The factor 0.87_7 is calculated from the roentgen as follows: 1 ion pair carries an electric charge of 4.80273×10^{-10} esu. 1 r therefore produces 2.08215×10^9 ion pairs in 0.001293 g air, or 1.61032×10^{12} ion pairs in 1 g air. The energy required for the production of one ion pair in air by electrons for radiations above 20 keV is 34 eV = 5.4469×10^{-11} erg. The production of 1.61032×10^{12} ion pairs in 1 g air therefore requires 87.712, or roughly 88 ergs. This energy is absorbed from the radiation by the air and is the so-called energy equivalent of the roentgen. From the definition of the rad it is equal to 0.87_7 rad (air).

From Diem, K., Ed., *Documenta Geigy Scientific Tables*, 6th ed., Geigy Pharmaceuticals, Ardsley, N.Y., 1962, 238. With permission.

Table 123

VALUES OF THE FACTOR *f*(AVERAGE ABSORBED DOSE IN RAD PER ROENTGEN) FOR VARIOUS PRIMARY X-RAY SPECTRA

Radiation				Factor *f* for			
Tube potential kV	Filter mm	Half-value layer mm	Spectrum	Air[b]	Water	Muscle	Bone
100	0.18 Cu	0.25 Cu or 5.5 Al	measured	0.87_7	0.91	0.94	3.10
100	as above	as above	calculated[a]	0.87_7	0.91	0.94	3.13
150	0.075 Cu	0.2 Cu	calculated[a]	0.87_7	0.92	0.94	2.69
200	0.20 Cu	0.5 Cu	calculated[a]	0.87_7	0.94	0.95	2.05
250	0.17 Cu + 3.0 Al	1.0 Cu	calculated[a]	0.87_7	0.95	0.95	1.76
250	0.9 Cu + 3.0 Al	2.0 Cu	calculated[a]	0.87_7	0.96	0.96	1.42
280	—	1.7 Cu	measured	0.87_7	0.96	0.96	1.44
280	—	2.5 Cu	measured	0.87_7	0.97	0.96	1.22
280	—	3.1 Cu	measured	0.87_7	0.97	0.96_5	1.13
400	—	4.16 Cu	measured	0.87_7	0.97	0.97	1.11

[a]   Kramers, H. A., *Phil. Mag.*, 46, 836 (1923).
[b]   See Footnote to Table 122.

From Diem, K., Ed., *Documenta Geigy Scientific Tables,* 6th ed., Geigy Pharmaceuticals, Ardsley, N.Y., 1962, 238. With permission.

Table 124

VALUES OF THE FACTOR *f*(ABSORBED DOSE IN RAD PER 1 ESU)

Radiation		Factor *f'* for		
Tube potential kV	Half-value layer mm Cu	Air[a]	Water	Soft tissues[b]
150	0.2	0.87_7	1.03	1.02
200	0.5	0.87_7	1.03	1.02
250	1.0	0.87_7	1.03	1.01
280	1.7	0.87_7	1.03	1.01
400	4.2	0.87_7	1.02	1.01
15000	—	0.87_7	0.98	0.96
Isotope	Quantum energy MeV			
[137]Cs	0.67	0.87_7	1.02	1.00
[60]Co	1.25	0.87_7	1.01	1.00

[a]   See the footnote to Table 122.
[b]   Values of *f'* for soft tissues are also valid for bone embedded in soft tissue since the effect of the bone substance has already been taken into account in measuring the charge Q with the thin-walled chamber.

(From Röntgen und gammastahlen in der Medizin und Biologie: Regeln für die Dosimetrie, DIN 6809 (Vornorm), Berlin 1958)

From Diem, K., Ed., *Documenta Geigy Scientific Tables,* 6th ed., Geigy Pharmaceuticals, Ardsley, N.Y., 1962, 238. With permission.

Table 125
DOSIMETRY OF SOME ARTIFICIAL RADIOISOTOPES

(1) Isotope	(2) Half-life T	(3) Fraction disintegrating per day f_d	(4) Mean energy of β-radiation $\bar{E}_\beta$ (MeV)	(5) Dose factor K_β (rad)	(6) Maximum permissible concentration S_β (μc/kg)	(7) Critical organ[1]	(8) Weight of critical organ[1] (g)	(9) Maximum permissible concentration in the total body[1] (μc)
³H	12.26 y	2 × 10⁻⁴	0.006	2000	125	total body	7 × 10⁴	10³
¹¹C	20 m	1.0	0.38	0.4	125	–	–	–
¹⁴C	5760 y	4 × 10⁻⁷	0.05	8 × 10⁶	16	fat; total body	10⁴; 7 × 10⁴	300; 400
¹³N	10 m	1.0	0.48	0.25	200	–	–	–
¹⁸F	1.87 h	1.0	0.25	1.5	33	–	–	–
²²Na	2.6 y	7 × 10⁻⁴	0.2	1.5 × 10⁴	5	total body	7 × 10⁴	10
²⁴Na	15.0 h	0.68	0.54	24	3	total body	7 × 10⁴	7
³¹Si	2.62 h	0.998	0.6	5	10	total body	7 × 10⁴	30
³²P	14.2 d	0.05	0.68	720	1	bone	7 × 10³	6
³⁵S	87 d	0.008	0.05	320	20	skin; testis	2 × 10³; 40	100
³⁶Cl	3.1 × 10⁵ y	5 × 10⁻⁹	0.24	3.10⁹	3	total body	7 × 10⁴	80
³⁸Cl	37.3 m	1.0	1.39	2.7	19	–	–	–
⁴²K	12.5 h	0.74	1.4	60	1	muscle	3 × 10⁴	20
⁴⁵Ca	153 d	0.004	0.09	1000	12	bone	7 × 10³	30
⁵¹Cr	27.8 d	0.025	0.01	17	300	total body	7 × 10⁴	800
⁵²Mn	5.7 d	0.10	0.2	8	60	total body	7 × 10⁴	9
⁵⁴Mn	291 d	0.002	0.005	115	220	liver	1.7 × 10³	20
⁵⁵Fe	2.94 y	6 × 10⁻⁴	0.006	480	170	spleen; blood	150; 5 × 10³	1000
⁵⁹Fe	45 d	0.015	0.12	400	8	spleen; blood	150; 5 × 10³	20
⁶⁰Co	5.25 y	4 × 10⁻³	0.1	1.4 × 10⁴	9	total body	7 × 10⁴	10
⁶³Zn	38 m	1.0	0.96	2	25	–	–	–
⁶⁵Zn	245 d	0.003	0.1	185	90	total body	7 × 10⁴	60
⁸²Br	36 h	0.40	0.15	17	7	total body	7 × 10⁴	10
⁸⁵Kr	10.6 y	2 × 10⁻⁴	0.2	5 × 10⁴	5	–	–	–
⁸⁹Sr	51 d	0.013	0.55	2200	2	bone	7 × 10³	4
¹³¹I	8 d	0.08	0.2	120	5	thyroid	20	0.7
¹⁹⁸Au	2.69 d	0.26	0.34	67	3	kidney	3 × 10³	20

Column 3: Fraction of isotope disintegrating in 24 hours: $f_d = (1 - e^{-0.693/T})$, where $T =$ half-life in days (see also pages 277–291).

4: In the case of ⁵¹Cr, ⁵⁴Mn and ⁵⁵Fe, which have no β-radiation, the data are for the γ-radiation following K-electron capture.

5: $K_\beta = 74 \cdot \bar{E}_\beta \cdot T =$ dose in rad for one microcurie completely disintegrated per gram of tissue. The values are based on the physical data given in the table.

6: The maximum permissible concentration in microcuries per gram of tissue which delivers the maximum permissible dose of 0.05 rad/day is $S_\beta = \dfrac{0.05 \times 1000}{K_\beta \times f_d}$. In calculating the concentrations for simultaneous β- and γ-emitters, only the β-component has been taken into consideration*. The values are based on the physical data given in the table.

* Dose rates for a number of γ-emitting isotopes are given in Table 3.

1) The data in columns 7–9 are from National Bureau of Standards Handbook No. 52, *Maximum Permissible Amounts of Radioisotopes in the Human Body and Maximum Permissible Concentrations in Air and Water*, Washington, 1953, pages 15–16, and from *Recommendations of the International Commission on Radiological Protection*, Pergamon Press, London and New York, 1959, as amended by *ICRP Publication 2* (Report of Committee II on permissible dose from internal radiation, 1959), Pergamon Press, London and New York, 1960.

Table 126
DOSE RATES FROM A 1 mc POINT
SOURCE OF SOME γ-EMITTING
ISOTOPES

Isotope	Half-life (approx.)	γ-energy in MeV (approx.)	Dose rate** in mr/h from 1 mc at 1 m	Dose rate** in r/h from 1 mc at 1 cm
^{22}Na	2.6 y	1.3	1.16	11.6
^{24}Na	15.0 h	1.4; 2.8	1.87	18.7
^{38}Cl	37 m	1.6; 2.2	0.76	7.6
^{42}K	12.5 h	1.5	0.15	1.5
^{51}Cr	27.8 d	0.3	0.18	1.8
^{52}Mn	5.7 d	0.7; 0.9; 1.5	1.95	19.5
^{56}Mn	2.6 h	0.8; 1.8; 2.1	0.94	9.4
^{59}Fe	45 d	1.1; 1.3; 0.2	0.62	6.2
^{60}Co	5.3 y	1.2; 1.3	1.31	13.1
^{63}Zn	38 m	1.0; 1.9; 2.6	0.69	6.9
^{65}Zn	245 d	1.1	0.28	2.8
^{82}Br	36 h	0.5–1.5	1.46	14.6
^{131}I	8 d	0.4; 0.6–0.7	0.22	2.2
^{137}Cs	30 y	0.7	0.31	3.1
^{182}Ta	112 d	effective 1.1	0.61	6.1
^{192}Ir	74.4 d	effective 0.6	0.50	5.0
^{198}Au	2.7 d	0.4; 0.7; 1.1	0.24	2.35

* For further information see under Radiological Dose Units, page 237.
** For the sources of these data see the table of dose rates (*k*-factors) on page 237.

From Diem, K., Ed., *Documenta Geigy Scientific Tables*, 6th ed., Geigy Pharmaceuticals, Ardsley, N.Y., 1962, 273. With permission.

Table 127

NALGENE LABWARE CHEMICAL RESISTANCE CHART

This Chemical Resistance Chart is intended to be used as a general guide only. Since each pair of ratings listed is for ideal conditions, all factors affecting chemical resistance (see facing page) must be considered. First letter of each pair applies to conditions at 20°C; the second to those at 50°C.

CAUTION

Do not store strong oxidizing agents in plastic labware except that made of Teflon FEP. Prolonged exposure causes embrittlement and failure.

Do not place plastic labware in a direct flame or on a hot plate.

1st letter: at 20°C → EG ← 2nd letter: at 50°C

CHEMICAL	RESINS						
	CPE	LPE	PP/PA	PMP	FEP/ETFE/TFE	PC	PVC
Acetaldehyde	GN	GF	GN	GN	EE	FN	GN
Acetamide, Sat.	EE	EE	EE	EE	EE	NN	NN
Acetic Acid, 5%	EE	EE	EE	EE	EE	EG	EE
Acetic Acid, 50%	EE	EE	EE	EE	EE	EG	EG
Acetone	EG	EE	EE	EE	EE	NN	NN
Adipic Acid	EE	EE	EE	EE	EE	EE	EG
Alanine	EE	EE	EE	EE	EE	EE	NN
Allyl Alcohol	EG	EE	EG	EG	EE	GF	GF
Aluminum Hydroxide	EE	EE	EG	FN	EE	FN	EG
Aluminum Salts	EE	EE	EE	EG	EE	EG	EE
Amino Acids	EE	EE	EE	EE	EE	GF	EE
Ammonia	EE	EE	EE	EE	EE	NN	EG
Ammonium Acetate, Sat.	EG	EE	EG	EG	EE	GF	EG
Ammonium Glycolate	EE	EE	EG	EG	EE	FN	EE
Ammonium Hydroxide, 5%	EE	EE	EG	EG	EE	NN	EE
Ammonium Hydroxide, 30%	EG	EE	EG	EG	EE	NN	EG
Ammonium Oxalate	EG	EE	EE	EE	EE	EG	EG
Ammonium Salts	GF	EG	GF	GF	EE	EG	EG
n-Amyl Acetate	NN	FN	NN	NN	EE	NN	NN
Amyl Chloride	EG	EE	EG	GF	EE	FN	NN
Aniline	EG	EE	EG	EG	EE	FN	NN
Benzaldehyde	FN	GG	GF	EG	EE	NN	NN
Benzene	EE	EE	EG	GF	EE	EG	EG
Benzoic Acid, Sat.	EG	EE	EG	EG	EE	EG	EG
Benzyl Acetate	EG	EE	EE	NN	EE	FN	NN
Benzyl Alcohol	NN	FN	NN	NN	EE	GF	GF
Bromine	NN	FN	NN	NN	EE	FN	GN
Bromobenzene	NN	NN	NN	NN	EE	NN	NN
Bromoform	NN	FN	NN	NN	EE	NN	NN
Butadiene	GF	EG	GF	GF	EE	NN	FN
n-Butyl Acetate	GF	EE	GF	EG	EE	NN	GF
n-Butyl Alcohol	EE	EE	EE	EE	EE	GF	GF

CHEMICAL	RESINS						
	CPE	LPE	PP/PA	PMP	FEP/ETFE/TFE	PC	PVC
Formic Acid, 3%	EG	EE	EG	EG	EE	EG	GF
Formic Acid, 50%	EG	EE	EG	EG	EE	EG	FN
Formic Acid, 98-100%	EG	EE	EG	EF	EE	EF	FN
Fuel Oil	FN	GF	GF	GF	EE	FF	GN
Gasoline	FN	GG	EG	GF	EE	NN	EG
Glacial Acetic Acid	EG	EE	EG	EG	EE	EE	EG
Glycerine	EE	EE	EE	EE	EE	EE	EG
Hexane	NN	GF	FF	FN	EE	FN	GF
Hydrochloric Acid, 1-5%	EE	EE	EE	EE	EE	EG	EE
Hydrochloric Acid, 20%	EE	EE	EE	EE	EE	GF	EG
Hydrochloric Acid, 35%	EG	EE	EG	EG	EE	NN	EG
Hydrofluoric Acid, 4%	EE	EE	EE	EE	EE	GF	EE
Hydrofluoric Acid, 48%	EG	EE	EG	EG	EE	NN	EG
Hydrogen Peroxide, 3%	EG	EE	EG	EG	EE	GF	EG
Hydrogen Peroxide, 30%	EG	EE	EG	EG	EE	EG	EG
Hydrogen Peroxide, 90%	EG	EE	EG	EG	EE	EG	EG
Isobutyl Alcohol	EE	EE	EE	EE	EE	GF	EE
Isopropyl Acetate	GF	EG	GF	GF	EE	NN	NN
Isopropyl Alcohol	EE	EE	EE	EE	EE	GF	EG
Isopropyl Benzene	FN	GF	EE	NN	EE	NN	EG
Kerosene	FN	GG	EG	GF	EE	GF	EE
Lactic Acid, 3%	EE	EE	EE	EE	EE	EG	GF
Lactic Acid, 85%	EE	EE	EE	EE	EE	EG	GF
Methoxyethyl Oleate	EE	EE	EG	EF	EE	GF	EF
Methyl Alcohol	EG	EE	EE	EE	EE	GF	NN
Methyl Ethyl Ketone	EG	EE	EG	EG	EE	NN	NN
Methyl Isobutyl Ketone	GF	EG	GF	FF	EE	NN	NN
Methyl Propyl Ketone	GF	EG	GF	FF	EE	NN	EG
Methylene Chloride	FN	NN	FN	FN	EE	NN	NN
Mineral Oil	GN	EE	EE	NN	EE	EG	EG
Nitric Acid, 1-10%	EE	EE	EE	EE	EE	EG	EG

Table 127 (continued)
NALGENE LABWARE CHEMICAL RESISTANCE CHART

1st letter: at 20°C → EG ← 2nd letter: at 50°C

RESINS

CHEMICAL	CPE	LPE	PP/PA	PMP	FEP/ETFE/TFE	PC	PVC
sec-Butyl Alcohol	EG	EE	EG	EG	EE	GF	GG
tert-Butyl Alcohol	EG	EE	EG	EG	EE	GF	EG
Butyric Acid	NN	NN	NN	NN	EE	FN	GN
Calcium Hydroxide, Conc.	EE	EE	EE	EE	EE	NN	EE
Calcium Hypochlorite, Sat.	EE	EE	EG	EG	EE	FN	GF
Carbazole	NN	NN	NN	NN	EE	NN	NN
Carbon Disulfide	FN	FN	GF	NN	EE	GF	NN
Carbon Tetrachloride	EG	GF	GF	NN	EE	NN	GF
Cedarwood Oil	EG	EE	EG	EG	EE	GF	FN
Cellosolve Acetate	GN	EF	GN	GN	EE	FN	FN
Chlorine, 10% in Air	NN	NN	GN	GN	EE	EG	EE
Chlorine, 10% (Moist)	EE	EE	EG	EG	EE	GF	EG
Chloroacetic Acid	EE	EE	EG	EE	EE	FN	EG
p-Chloroacetophenone	FN	GF	FN	FN	EE	GF	NN
Chloroform	NN	NN	FN	NN	EE	NN	NN
Chromic Acid, 10%	EE	EE	EE	EE	EE	GF	EG
Chromic Acid, 50%	EE	FF	GF	GF	EE	FN	EF
Cinnamon Oil	NN	FN	FN	NN	EE	GF	NN
Citric Acid, 10%	EE	EE	EE	EE	EE	EG	GG
Cresol	NN	NN	GF	NN	EE	NN	NN
Cyclohexane	FN	FN	GF	EE	EE	EG	GF
Decalin	GF	GG	FN	FN	EE	NN	EG
o-Dichlorobenzene	FN	FF	FN	NN	EE	NN	NN
p-Dichlorobenzene	NN	FN	GF	NN	EE	FN	NN
Diethyl Benzene	NN	FN	NN	NN	EE	FN	NN
Diethyl Ether	FN	FN	GF	GF	EE	NN	FN
Diethyl Ketone	GF	GG	GG	EG	EE	NN	FN
Diethyl Malonate	EE	EE	EE	EE	EE	FN	GN
Diethylene Glycol	EE	EE	EE	EG	EE	GF	FN
Diethylene Glycol Ethyl Ether	EE	EE	EE	EE	EE	GF	FN
Dimethyl Formamide	EE	EE	GF	NN	EE	FN	FN
Dimethylsulfoxide	GF	GG	GF	GF	EE	GF	GF
1,4-Dioxane	EE	EE	EE	EE	EE	FN	FN
Ether	NN	NN	NN	NN	EE	GF	GF
Ethyl Acetate	EG	FN	EG	EG	EE	GF	FN
Ethyl Alcohol	EG	EE	EG	EG	EE	EG	EG
Ethyl Alcohol, 40%	EG	GF	EG	EG	EE	GF	EG
Ethyl Benzene	FN	GF	FN	FN	EE	NN	NN
Ethyl Benzoate	GN	GG	GF	GF	EE	NN	NN
Ethyl Butyrate	FN	FN	FN	GF	EE	FN	FN
Ethyl Chloride, Liquid	NN	FN	EE	NN	EE	FN	FN
Ethyl Cyanoacetate	EE	EE	EE	EE	EE	GF	FN
Ethyl Lactate	EE	EE	EE	EE	EE	NN	NN
Ethylene Chloride	GN	GF	FN	NN	EE	FN	NN
Ethylene Glycol	EE	EE	EE	EE	EE	GF	EE

RESINS

CHEMICAL	CPE	LPE	PP/PA	PMP	FEP/ETFE/TFE	PC	PVC
Nitric Acid, 50%	GG	GN	FN	GN	EE	GF	GF
Nitric Acid, 70%	FN	FN	FN	GF	EE	NN	FN
Nitrobenzene	NN	NN	NN	NN	EE	NN	NN
n-Octane	FN	EE	EE	EE	EE	GF	FN
Orange Oil	EG	EE	GF	FF	EG	FF	FN
Ozone	EG	EE	EG	GG	EE	GF	EG
Perchloric Acid	NN	NN	GN	NN	EE	NN	GN
Perchloroethylene	GN	GF	GN	NN	GF	EN	NN
Phenol, Crystals	EE	GF	FG	FG	EE	EN	FN
Phosphoric Acid, 1-5%	EE	EE	EE	EE	EE	EE	EG
Phosphoric Acid, 85%	EE	EE	EG	EG	EE	EG	EG
Pine Oil	GN	GF	GN	NN	EE	GF	EG
Potassium Hydroxide, 1%	EE	EE	EE	EE	EE	FN	FN
Potassium Hydroxide, Conc.	NN	NN	NN	NN	EE	NN	FN
Propane Gas	EG	EE	EG	EG	EE	GF	GF
Propylene Glycol	EE	EE	EG	EG	EE	EG	EE
Propylene Oxide	EG	EE	GF	GF	EE	GF	EE
Resorcinol, Sat.	EE	EE	EG	EG	EE	EG	GG
Resorcinol, 5%	EG	EE	EG	EE	EE	EE	EG
Salicylaldehyde	EG	EG	EG	EG	EE	EG	GF
Salicylic Acid, Powder	EE	EE	EE	EE	EE	EG	GF
Salicylic Acid, Sat.	EE	EE	EG	EG	EE	EG	EE
Salt Solutions, Metallic	EE	EE	EG	EG	EE	EG	EE
Silver Acetate	EE	EE	EG	EG	EE	EG	GG
Silver Nitrate	EG	EE	EG	EE	EE	EG	EG
Sodium Acetate, Sat.	EE	EE	EE	EE	EE	EG	GF
Sodium Hydroxide, 1%	EE	EE	EE	EE	EE	FN	EE
Sodium Hydroxide, 50% to Sat.	EE	EE	EG	EG	EE	GF	EG
Sodium Hypochlorite, 15%	EE	EE	EG	EG	EE	GF	EG
Stearic Acid, Crystals	EE	EE	EG	EG	EE	EG	EG
Sulfuric Acid, 1-6%	EE	EE	EE	EE	EE	GF	EG
Sulfuric Acid, 20%	EE	EE	EG	EG	EE	GF	EG
Sulfuric Acid, 60%	EE	EE	EG	EG	EE	GF	EG
Sulfuric Acid, 98%	GG	FN	GG	GG	EE	NN	GN
Sulfur Dioxide, Liq., 46 psi	NN	GF	NN	NN	EE	EG	FN
Sulfur Dioxide, Wet or Dry	FN	EE	FN	FN	EE	GN	EG
Sulfur Salts	EE	EE	EE	EE	EE	EG	EG
Tartaric Acid	FN	GF	GF	FF	EE	FN	EG
Tetrahydrofuran	NN	NN	NN	NN	EE	NN	NN
Thionyl Chloride	NN	NN	NN	NN	EE	NN	NN
Toluene	GF	EG	GF	GF	EE	FN	FN
Tributyl Citrate	NN	FN	GF	NN	EE	NN	NN
Trichloroethane	NN	NN	NN	NN	EE	NN	NN
Trichloroethylene	EE	EE	EE	EE	EE	EG	EG
Triethylene Glycol	EE	EE	EE	EE	EE	EG	EG
Tripropylene Glycol	EE	EE	EE	EE	EE	EG	GF

Ethylene Glycol Methyl Ether	EE	EE	EE	EE	EE	FN	FN
Ethylene Oxide	FF	GF	FF	FN	FN	FN	FN
Fluorides	EE	EE	FN	FN	EE	EE	EE
Fluorine	FN	GN	FN	FN	EG	GF	GF
Formaldehyde, 10%	EE	EE	EG	EG	EE	EG	GF
Formaldehyde, 40%	EG	EE	EG	EG	EE	EG	GF

Turpentine	FN	GG	GF	FF	EE	FN	GF
Undecyl Alcohol	EF	EG	EG	EG	EE	GF	EF
Urea	EE	EE	EE	EG	EE	NN	GN
Vinylidene Chloride	NN	FN	NN	FN	EE	NN	NN
Xylene	GN	GF	FN	FN	EE	NN	NN
Zinc Stearate	EE	EE	EE	EE	EE	EE	EG

Table 128

CHEMICAL RESISTANCE AND PHYSICAL PROPERTIES OF THE RESINS USED IN NALGENE LABWARE

Interpretation of Chemical Resistance

The Chemical Resistance Chart on the facing page and this Chemical Resistance Summary Chart are general guides only. Because so many factors can affect the chemical resistance of a given product, you should test under your own conditions. If any doubt exists about specific applications of Nalgene products, please contact Technical Service, Nalgene Labware Department, Nalge Company, Box 365, Rochester, New York 14602, or call (716) 586-8800.

Effects of Chemicals on Plastics

Chemicals can affect the strength, flexibility, surface appearance, color, dimensions or weight of plastics. The two basic modes of interaction which cause these changes are: (1) chemical attack on the polymer chain, including oxidation; reaction of functional groups in or on the chain; or depolymerization, with resultant reduction in physical properties; and (2) physical change: absorption of solvents, resulting in softening and swelling, or permeation of solvent through the plastic; dissolving in a solvent; cracking from interaction of a "stress-cracking agent" with molded-in stresses.

The combination of compounds of two or more classes may cause a synergistic or undesirable chemical effect. Other factors affecting chemical resistance include temperature, pressure and other stresses (e.g., centrifugation), length of exposure and concentration of the chemical. **As temperature increases, resistance to attack decreases.**

Caution

Do not store strong oxidizing agents in plastic labware except that made of Teflon FEP. Prolonged exposure causes embrittlement and failure. While prolonged storage may not be intended at time of filling, a forgotten container will fail in time and result in leakage of contents. Do not place plastic labware in a direct flame or on a hot plate.

Resin Codes

CPE: Conventional (Low Density) Polyethylene
LPE: Linear (High Density) Polyethylene
PP: Polypropylene
PA: Polyallomer
PMP: Polymethylpentene ("TPX")
FEP: Teflon FEP (fluorinated ethylene propylene)

TFE: Teflon TFE (tetrafluoroethylene)
ETFE: Tetzel ETFE (ethylene-tetrafluoroethylene)
PC: Polycarbonate
PVC: Polyvinyl Chloride
PSF: Polysulfone
PS: Polystyrene

Chemical Resistance Classification:

E—30 days of constant exposure cause no damage. Plastic may even tolerate for years.
G—Little or no damage after 30 days of constant exposure to the reagent.
F— Some effect after 7 days of constant exposure to the reagent. Depending on the plastic, the effect may be crazing, cracking, loss of strength or discoloration. Solvents may cause softening, swelling and permeation losses with CPE, LPE, PP and PMP. The solvent

Nalgene Chemical Resistance Summary

Classes of Substances 20°C	CPE	LPE	PP/PA	PMP	FEP/ETFE/TFE	PC	PSF	PVCbottles**	PS
Acids, dilute or weak	E	E	E	E	E	E	E	E	E
Acids*, strong and concentrated	E	E	E	E	E	N	G	E	F
Alcohols	G	G	E	E	E	G	G	E	E
Aldehydes	G	G	G	E	E	F	F	N	N
Bases	E	E	E	E	E	N	E	N	E
Esters	G	G	G	G	E	G	N	N	E
Hydrocarbons, Aliphatic	F	G	F	F	E	F	E	E	N

effects on these four resins are normally reversible; the part will usually return to its normal condition after solvent loss.

N—Not recommended for continuous use. Immediate damage may occur. Depending on the plastic, the effect may be a more severe crazing, cracking, loss of strength, discoloration or dissolution. Solvents may cause softening, swelling and permeation losses with CPE, LPE, PP and PMP. The solvent effects on these four resins are more severe than the effects of "F" substances. However, they are normally reversible; the part will usually return to its normal condition after solvent loss.

Physical Properties

	Max. Use Temp. (°C)	Trans-parency	Sterilization*: Auto-clavable	Gas	Dry Heat	Chemical	Specific Gravity	Flexi-bility	Brittle-ness Temp. (°C)	Permeability (approx.) { cc–mm / sec–cm²–cm Hg } × 10⁻¹⁰ Units: N₂	O₂	CO₂	Tensile Strength (psi)	Water Absorp-tion (%)	Hydrocarbons, Aromatic	Hydrocarbons, Halogenated	Ketones	Oxidizing Agents, Strong
CPE	80	Transluc	No	Yes	No	Yes	0.92	excel	−100	20	60	280	2000	<0.01	F	N	G	F
LPE	120	Transluc	With caution°°	Yes	No	Yes	0.95	rigid	−100	3	10	45	4000	<0.01	G	F	G	F
PP	135	Transluc	Yes	Yes	No	Yes	0.90	rigid	0	4	25	90	5000	<0.02	F	N	G	F
PMP	175	Clear	Yes	Yes	No	Yes	0.83	rigid	—	65	270	—	4000	<0.01	F	G	G	F
FEP	205	Transluc	Yes	Yes	Yes	Yes	2.15	excel	−270	20	60	135	3000	<0.01	E	G	F	E
ETFE	150	Transluc	Yes	Yes	Yes	Yes	1.70	mod	−100	—	—	—	6500	0.1	N	N	F	E
PC	135	Clear	Yes†	Yes	No	Yes	1.20	rigid	−135	3	20	85	8000	0.35	N	F	F	N
PVC	70††	Clear	No††	Yes	No	Yes	1.34	rigid	−30	0.5–2	1–6	10–35	6500	0.06	N	F	F	E
PA	130	Transluc	Yes	Yes	No	Yes	0.90	mod	−40	6	30	100	4000	<0.02	F	N	N	N
PSF	165	Clear	Yes	Yes	Yes	Yes	1.24	rigid	−100	3	15	60	10,000	0.30	N	N	G	G

*Except for oxidizing acids; for oxidizing acids, see "Oxidizing Agents, Strong."

°°For PVC tubing, see pages 34 through 36.

*Sterilization:
- Autoclaving—Clean and rinse item with distilled water before autoclaving. Certain chemicals which have no appreciable effect on resins at room temperature may cause deterioration at autoclaving temperatures unless removed with distilled water beforehand.
- Gas—Ethylene oxide.
- Dry heat—at 160°C.
- Chemical—Benzalkonium chloride, formalin, ethanol, etc.

°° Can be autoclaved at 121°C for 20 minutes (containers must be empty and uncovered).

† Autoclaving reduces mechanical strength. Do not use PC vessels for vacuum applications if they have been autoclaved. (See also comments for Sterilization)

†† Except for the PVC in tubing, which will withstand temperatures to 121°C and can be autoclaved.

Table 129
CENTRIFUGAL FORCE CHART

RADII IN INCHES

IF R.P.M. IS IN THIS
COLUMN DIVIDE FORCE BY 100.

IF R.P.M. IS IN THIS
COLUMN MULTIPLY FORCE BY 100

MULTIPLY FORCE BY 39370
IF RADIUS IS GIVEN IN CENTIMETERS

MULTIPLY FORCE BY 2 IF RADIUS IS IN THIS COLUMN →

MULTIPLY FORCE BY 5 IF RADIUS IS IN THIS COLUMN →

Table 130*
TEMPERATURE CORRECTION FOR GLASS VOLUMETRIC APPARATUS

Temperature in degrees C.	2,000 ml	1,000 ml	500 ml	400 ml	300 ml	250 ml	Temperature in degrees C.	2,000 ml	1,000 ml	500 ml	400 ml	300 ml	250 ml
15	+0.25	+0.12	+0.06	+0.05	+0.04	+0.031	23	− .15	− .08	− .04	− .03	− .02	− .019
16	+ .20	+ .10	+ .05	+0.04	+ .03	+ .025	24	− .20	− .10	− .05	− .04	− .03	− .025
17	+ .15	+ .08	+ .04	+ .03	+ .02	+ .019	25	− .25	− .12	− .06	− .05	− .04	− .031
18	+ .10	+ .05	+ .02	+ .02	+ .02	+ .012	26	− .30	− .15	− .08	− .06	− .04	− .038
19	+ .05	+ .02	+ .01	+ .01	+ .01	+ .006	27	− .35	− .18	− .09	− .07	− .05	− .044
							28	− .40	− .20	− .10	− .08	− .06	− .050
21	− .05	− .02	− .01	− .01	− .01	− .006	29	− .45	− .22	− .11	− .09	− .07	− .056
22	− .10	− .05	− .02	− .02	− .02	− .012	30	− .50	− .25	− .12	− .10	− .08	− .062

Note: This table gives the correction to be added to actual capacity (determined at certain temperatures) to give the capacity at the standard temperature 20°C. Conversely, by subtracting the corrections from the indicated capacity of an instrument standard at 20°C the corresponding capacity at other temperatures is obtained. The table assumes for the cubical degree centigrade. The coefficients of expansion of glasses used for volumetric instruments vary from 0.000023 to 0.000028.

TEMPERATURE CORRECTION FOR VOLUMETRIC SOLUTIONS

Table 131*
CORRECTION FACTORS FOR WATER

Temperature of measurement, °C.	Capacity of apparatus in milliliters at 20°C.							Temperature of measurement, °C.	Capacity of apparatus in milliliters at 20°C.						
	2,000	1,000	500	400	300	250	150		2,000	1,000	500	400	300	250	150
	Correction in milliliters to give volume of water at 20°C.								Correction in milliliters to give volume of water at 20°C.						
15	+1.54	+0.77	+0.38	+0.31	+0.23	+0.19	+0.12	23	−1.18	− .59	− .30	− .24	− .18	− .15	− .09
16	+1.28	+ .64	+ .32	+ .26	+ .19	+ .16	+ .10	24	−1.61	− .81	− .40	− .32	− .24	− .20	− .12
17	+ .99	+ .50	+ .25	+ .20	+ .15	+ .12	+ .07	25	−2.07	−1.03	− .52	− .41	− .31	− .26	− .15
18	+ .68	+ .34	+ .17	+ .14	+ .10	+ .08	+ .05	26	−2.54	−1.27	− .64	− .51	− .38	− .32	− .19
19	+ .35	+ .18	+ .09	+ .07	+ .05	+ .04	+ .03	27	−3.03	−1.52	− .76	− .61	− .46	− .38	− .23
								28	−3.55	−1.77	− .89	− .71	− .53	− .44	− .27
21	− .37	− .18	− .09	− .07	− .06	− .05	− .03	29	−4.08	−2.04	−1.02	− .82	− .61	− .51	− .31
22	− .77	− .38	− .19	− .15	− .12	− .10	− .06	30	−4.62	−2.31	−1.16	− .92	− .69	− .58	− .35

Note: This table gives the correction to various observed volumes of water measured at the designated temperatures to give the volume at the standard temperature, 20°C. Conversely, by subtracting the corrections from the volume desired at 20°C, the volume that must be measured out at the designated temperatures in order to give the desired volume at 20°C, will be obtained. It is assumed that the volumes are measured in glass apparatus having a coefficient of cubical expansion of 0.000025 per degree centigrade. The table is applicable to dilute aqueous solutions having the same coefficient of expansion as water.

Table 132*
CORRECTION FACTORS FOR SOME CHEMICAL SOLUTIONS

Solution	Normality		
	N	N/2	N/10
HNO_3	50	25	6
H_2SO_4	45	25	5
NaOH	40	25	5
KOH	40	20	4

Note: In using Table 131 to correct the volume of certain standard solutions to 20°C more accurate results will be obtained if the numerical values of the corrections are increased by the percentages given above.

* All tables from Weast, R. C. Ed., CRC *Handbook of Chemistry and Physics,* 61st ed., CRC Press, Boca Raton, Fla., 1980, F-2 and F-3. With permission.

Table 133
CALIBRATION OF VOLUMETRIC GLASSWARE FROM THE WEIGHT OF THE CONTAINED WATER OR MERCURY WHEN WEIGHED IN AIR

$t°C$	W_t	W_{18}	W_{25}	M_t	M_{18}	M_{25}
0	1.001 220	1.001 396	1.001 466	0.0735 519	0.0735 648	0.0735 698
1	1.001 161	1.001 327	1.001 395	0.0735 653	0.0735 775	0.0735 825
2	1.001 120	1.001 276	1.001 345	0.0735 787	0.0735 902	0.0735 952
3	1.001 096	1.001 242	1.001 311	0.0735 920	0.0736 028	0.0736 078
4	1.001 088	1.001 225	1.001 293	0.0736 054	0.0736 154	0.0736 205
5	1.001 096	1.001 223	1.001 291	0.0736 188	0.0736 281	0.0736 332
6	1.001 120	1.001 237	1.001 306	0.0736 322	0.0736 408	0.0736 458
7	1.001 158	1.001 265	1.001 334	0.0736 456	0.0736 535	0.0736 585
8	1.001 211	1.001 309	1.001 377	0.0736 590	0.0736 662	0.0736 712
9	1.001 279	1.001 367	1.001 435	0.0736 724	0.0736 789	0.0736 839
10	1.001 360	1.001 438	1.001 506	0.0736 858	0.0736 915	0.0736 966
11	1.001 455	1.001 523	1.001 592	0.0736 992	0.0737 042	0.0737 093
12	1.001 563	1.001 622	1.001 690	0.0737 125	0.0737 168	0.0737 218
13	1.001 684	1.001 733	1.001 801	0.0737 259	0.0737 295	0.0737 345
14	1.001 816	1.001 855	1.001 923	0.0737 393	0.0737 422	0.0737 472
15	1.001 961	1.001 990	1.002 059	0.0737 526	0.0737 548	0.0737 598
16	1.002 118	1.002 138	1.002 206	0.0737 660	0.0737 674	0.0737 725
17	1.002 286	1.002 296	1.002 364	0.0737 794	0.0737 801	0.0737 852
18	1.002 466	1.002 466	1.002 534	0.0737 928	0.0737 928	0.0737 978
19	1.002 658	1.002 648	1.002 717	0.0738 062	0.0738 055	0.0738 105
20	1.002 859	1.002 839	1.002 908	0.0738 196	0.0738 182	0.0738 232
21	1.003 072	1.003 043	1.003 111	0.0738 330	0.0738 308	0.0738 359
22	1.003 294	1.003 255	1.003 323	0.0738 463	0.0738 434	0.0738 485
23	1.003 528	1.003 479	1.003 548	0.0738 597	0.0738 561	0.0738 611
24	1.003 771	1.003 712	1.003 781	0.0738 731	0.0738 688	0.0738 738
25	1.004 024	1.003 955	1.004 024	0.0738 864	0.0738 814	0.0738 864
26	1.004 287	1.004 209	1.004 277	0.0738 998	0.0738 940	0.0738 991
27	1.004 560	1.004 472	1.004 540	0.0739 132	0.0739 067	0.0739 118
28	1.004 842	1.004 744	1.004 813	0.0739 266	0.0739 194	0.0739 244
29	1.005 133	1.005 025	1.005 094	0.0739 400	0.0739 321	0.0739 371
30	1.005 434	1.005 316	1.005 385	0.0739 534	0.0739 447	0.0739 498
31	1.005 743	1.005 615	1.005 684	0.0739 669	0.0739 575	0.0739 626
32	1.006 060	1.005 923	1.005 991	0.0739 801	0.0739 700	0.0739 750
33	1.006 388	1.006 241	1.006 310	0.0739 934	0.0739 826	0.0739 876
34	1.006 723	1.006 566	1.006 635	0.0740 068	0.0739 953	0.0740 003
35	1.007 066	1.006 899	1.006 968	0.0740 202	0.0740 079	0.0740 130
36	1.007 418	1.007 242	1.007 311	0.0740 335	0.0740 205	0.0740 256
37	1.007 780	1.007 593	1.007 669	0.0740 469	0.0740 332	0.0740 382
38	1.008 149	1.007 952	1.008 021	0.0740 603	0.0740 459	0.0740 509
39	1.008 525	1.008 318	1.008 387	0.0740 737	0.0740 585	0.0740 636
40	1.008 910	1.008 694	1.008 762	0.0740 871	0.0740 712	0.0740 763
41	1.009 303	1.009 077	1.009 146	0.0741 007	0.0740 841	0.0740 891
42	1.009 703	1.009 467	1.009 536	0.0741 139	0.0740 966	0.0741 016
43	1.010 112	1.009 866	1.009 935	0.0741 273	0.0741 092	0.0741 143
44	1.010 528	1.010 272	1.010 341	0.0741 407	0.0741 219	0.0741 270
45	1.010 951	1.010 685	1.010 754	0.0741 541	0.0741 346	0.0741 396
46	1.011 382	1.011 106	1.011 175	0.0741 675	0.0741 473	0.0741 523
47	1.011 820	1.011 534	1.011 603	0.0741 810	0.0741 600	0.0741 651
48	1.012 266	1.011 970	1.012 039	0.0741 944	0.0741 727	0.0741 778
49	1.012 719	1.012 413	1.012 482	0.0742 078	0.0741 854	0.0741 904
50	1.013 180	1.012 864	1.012 933	0.0742 213	0.0741 981	0.0742 031

A borosilicate glass vessel containing g_t grams of water at a temperature of $t°C$ has, at the same temperature, a volume $V_t = W_t \times g_t$ cubic centimeters. Similarly when filled with G_t grams of mercury at a temperature of $t°C$ the volume at the same temperature is given by $V_t = M_t \times G_t$ cubic centimeters.

When filled with g_t grams of water at a temperature of $t°C$ the volume of the vessel at 18°C is given by $V_{18} = W_{18} \times g_t$ cubic centimeters and the true volume at 25°C is given by $V_{25} = W_{25} \times g_t$. The volumes at 18°C and 25°C are given similarly when using mercury by using the values under M_{18} and M_{25}, respectively.

The data on water are adapted from the data of G. S. Kell, *Journal of Chemical and Engineering Data*, 12, 67–68 (1967) (Table on p. F5, 52nd Edition, this handbook) and the data on mercury are adapted from *Smithsonian Tables*, Ninth Revised Edition, Volume 120, Publication No. 4169. The coefficient of linear expansion for borosilicate glass used here is 32.5×10^{-7} deg^{-1} and the volume coefficient of expansion is 97.5×10^{-7} deg^{-1}.

From Swinehart, D. F., in Weast, R. C., Ed., *CRC Handbook of Chemistry and Physics,* 61st ed., CRC Press, Boca Raton, Fla., 1980, D-144. With permission.

Table 134
TOLERANCES FOR ANALYTICAL WEIGHTS

Denomination	Class M		Class S		Class S-1, individual tolerance, mg	Class P, individual tolerance, mg
	Individual tolerance, mg	Group tolerance, mg	Individual tolerance, mg	Group tolerance, mg		
100 g	0.50	None	0.25	None	1.0	2.0
50 g	0.25	specified	0.12	specified	0.60	1.2
30 g	0.15		0.074		0.45	0.90
20 g	0.10		0.074	0.154	0.35	0.70
10 g	0.050		0.074		0.25	0.50
5 g	0.034		0.054		0.18	0.36
3 g	0.034	0.065	0.054	0.105	0.15	0.14
2 g	0.034		0.054		0.13	0.26
1 g	0.034		0.054		0.10	0.20
500 mg	0.0054		0.025		0.080	0.16
300 mg	0.0054	0.0105	0.025	0.055	0.070	0.14
200 mg	0.0054		0.025		0.060	0.12
100 mg	0.0054		0.025		0.050	0.10
50 mg	0.0054		0.014		0.042	0.085
30 mg	0.0054	0.0105	0.014	0.034	0.038	0.076
20 mg	0.0054		0.014		0.035	0.070
10 mg	0.0054		0.014		0.030	0.060
5 mg	0.0054		0.014		0.028	0.055
3 mg	0.0054		0.014		0.026	0.052
2 mg	0.0054	0.0105	0.014	0.034	0.025	0.050
1 mg	0.0054		0.014		0.025	0.050
½ mg	0.0054		0.014		0.025	.....

From Dean, J. A., Ed., *Lange's Handbook of Chemistry*, 12th ed., McGraw-Hill, New York, 1979, 2-79. With permission.

This table gives the individual and group tolerances established by the National Bureau of Standards (Washington, D.C.) for classes M, S, S-1, and P weights. Individual tolerances are "acceptance tolerances" for new weights. Group tolerances are defined by the National Bureau of Standards as follows: "The corrections of individual weights shall be such that no combination of weights that is intended to be used in a weighing shall differ from the sum of the nominal values by more than the amount listed under the group tolerances."

For class S-1 weights, two-thirds of the weights in a set must be within one-half of the individual tolerances given below. No group tolerances have been specified for class P weights. See Natl. Bur. Standards Circ. 547, sec. 1 (1954).

Table 135
TOLERANCES FOR VOLUMETRIC BURETS AND PIPETS

Capacity, up to and including, ml	Volumetric flasks, calibrated		Transfer pipets and burets	Measuring pipets
	To contain	To deliver		
1	±0.01	......	....	......
2	......	......	±0.006*	±0.01
3	0.015	......	....	......
5	0.02	......	0.01	0.02
10	0.02	±0.04	0.02	0.03
25	0.03	0.05	....	......
30	......	......	0.03	0.05
50	0.05	0.10	0.05	0.08
100	0.08	0.15	0.08†	0.15
200	0.10	0.20	0.10*	......
300	0.12	0.25	....	......
500	0.15	0.30	....	......
1000	0.30	0.50	....	......
2000	0.50	1.00	....	......
3000	0.75	1.50	....	......
4000	1.00	2.0	....	......
5000	1.2	2.4	....	......

Deviations, ml, permitted for

* Applies to pipets only.
† ±0.10 ml. for burets.

From Dean, J. A., Ed., *Lange's Handbook of Chemistry,* 12th ed., McGraw-Hill, New York, 1979, 2-78. With permission.

BAROMETRY AND BAROMETRIC CORRECTIONS *

In principle, the mercurial barometer balances a column of pure mercury against the weight of the atmosphere. The height of the column above the level of the mercury in the reservoir can be measured and serves as a direct index of atmospheric pressure. The space above the mercury in a barometer tube should be a Torricellian vacuum, perfect except for the practically negligible vapor pressure of mercury. The perfection of the vacuum is indicated by the sharpness of the click noted when the barometer tube is inclined. A barometer should be in a vertical position, suspended rather than fastened to a wall, and in a good light but not exposed to direct sunlight or too near a source of heat. The standard conditions for barometric measurements are 0°C., and gravity as at 45° latitude and sea level. There are numerous sources of error, but corrections for most of these are readily applied. Some of the corrections are very small, and their application may be questionable in view of the probably larger errors. The degree of consistency to be expected in careful measurements is about 0.13 mm with a 6.4-mm tube, increasing to 0.04 mm with a tube 12.7 mm in diameter.

In reading a barometer of the Fortin type (the usual laboratory instrument for precision measurements), the procedure should be as follows: (1) Observe and record the temperature as indicated by the thermometer attached to the barometer. The temper-

---

* From Dean, J. A., Ed., *Lange's Handbook of Chemistry,* 12th ed., McGraw-Hill, New York, 1979, 2-66. With permission.

ature correction is very important and may be affected by heat from the observer's body. (2) Set the mercury in the reservoir at zero level, so that the point of the pin above the mercury just touches the surface, making a barely noticeable dimple therein. Tap the tube at the top and verify the zero setting. (3) Bring the vernier down until the view at the light background is cut off at the highest point of the meniscus. Record the reading.

The corrections to be made on the reading are as follows: (1) Temperature, to correct for the difference in thermal expansion of the mercury and the brass (or glass) to which the scale is attached. This correction converts the reading into the value of 0°C. The brass scale table is applicable to the Fortin barometer (see Table 136). (2a) Latitude-gravity correction, and (2b) altitude-gravity correction, to compensate for differences in gravity, which would affect the height of the mercury column by variation in mass (Table 137). If local gravity is unknown, an approximate correction may be made from the tables. Local values of gravity are often subject to irregularities which lead to errors even when the corrections here provided are made. It is, therefore, advisable to determine the local value of gravity, from which the correction can be effected in the following manner:

$$Bt = Br + \left(\frac{g_1 - g_0}{g_0}\right) \times Br$$

in which Bt and Br are the true and the observed heights of the barometer, respectively, g_0 is standard gravity (980 665 cm · sec^{-2}), and g_1 is the local gravity. It may be noted that for most localities, g_1 is smaller than g_0, which makes the correction negative. These corrections compensate the reading to gravity at 45° latitude and sea level. (3) Correction for capillary depression of the level of the meniscus. This varies with the tube diameter and actual height of the meniscus in a particular case. Some barometers are calibrated to allow for an average value of the latter and approximating the correction (see Table 138). (4) Correction for vapor pressure of mercury. This correction is usually negligible, being only 0.001 mm at 20°C. and 0.006 mm at 40°C. This correction is added. See table of vapor pressure of mercury.*

The corrections above do not apply to aneroid barometers. These instruments should be calibrated at regular intervals by checking them against a corrected mercurial barometer.

For records on weather maps, meteorologists customarily correct barometer readings to sea level, and some barometers may be calibrated accordingly. Such instruments are not suitable for laboratory use where true pressure under standard conditions is required. Scale corrections should be specified in the maker's instructions with the instrument, and are also indicated by the lack of correspondence between a gauge mark usually placed exactly 76.2 cm from the zero point and the 76.2-cm scale graduation.

---

* Can be found in Lange's Handbook of Chemistry, 12th ed., Dean, J. A., Ed., McGraw-Hill, New York, 1979, 10-22.

Table 136

MERCURY BAROMETER — TEMPERATURE CORRECTION (1 TO 40°C, 600 TO 700 mm Hg)

Brass Scale

Barometer temperature t = °C	600	610	620	630	640	650	660	670	680	690	700	710	720	730	740	750	760	770	780	790	Correction factor f_t
	Barometer reading β_t in millimeters — Amount $\Delta\beta_t$ in millimeters to be subtracted																				
1	0.10	0.10	0.10	0.10	0.10	0.11	0.11	0.11	0.11	0.11	0.11	0.12	0.12	0.12	0.12	0.12	0.12	0.13	0.13	0.13	0.999 837
2	0.20	0.20	0.20	0.21	0.21	0.21	0.22	0.22	0.22	0.23	0.23	0.23	0.24	0.24	0.24	0.25	0.25	0.25	0.26	0.26	999 673
3	0.29	0.29	0.30	0.30	0.31	0.31	0.32	0.32	0.33	0.33	0.34	0.34	0.35	0.35	0.36	0.36	0.36	0.37	0.37	0.38	999 520
4	0.39	0.40	0.41	0.41	0.42	0.43	0.43	0.44	0.44	0.45	0.46	0.46	0.47	0.48	0.48	0.49	0.50	0.50	0.51	0.52	999 346
5	0.49	0.50	0.51	0.51	0.52	0.53	0.54	0.55	0.55	0.56	0.57	0.58	0.59	0.60	0.60	0.61	0.62	0.63	0.64	0.64	0.999 184
6	0.59	0.60	0.61	0.62	0.63	0.64	0.65	0.66	0.67	0.68	0.69	0.70	0.70	0.71	0.72	0.73	0.74	0.75	0.76	0.77	999 021
7	0.69	0.70	0.71	0.72	0.73	0.74	0.75	0.77	0.78	0.79	0.80	0.81	0.82	0.83	0.85	0.86	0.87	0.88	0.89	0.90	998 858
8	0.78	0.80	0.81	0.82	0.84	0.85	0.86	0.87	0.89	0.90	0.91	0.93	0.94	0.95	0.97	0.98	0.99	1.00	1.02	1.03	998 695
9	0.88	0.90	0.91	0.92	0.94	0.95	0.97	0.98	1.00	1.01	1.03	1.04	1.06	1.07	1.09	1.10	1.12	1.13	1.15	1.16	998 532
10	0.98	0.99	1.01	1.03	1.04	1.06	1.08	1.09	1.11	1.13	1.14	1.16	1.17	1.19	1.21	1.22	1.24	1.26	1.27	1.29	0.998 369
11	1.08	1.09	1.11	1.13	1.15	1.17	1.18	1.20	1.22	1.24	1.26	1.27	1.29	1.31	1.33	1.35	1.36	1.38	1.40	1.42	998 206
12	1.17	1.19	1.21	1.23	1.25	1.27	1.29	1.31	1.33	1.35	1.37	1.39	1.41	1.43	1.45	1.47	1.49	1.51	1.53	1.55	998 044
13	1.27	1.29	1.31	1.33	1.36	1.38	1.40	1.42	1.44	1.46	1.48	1.50	1.53	1.55	1.57	1.59	1.61	1.63	1.65	1.67	997 881
14	1.37	1.39	1.41	1.44	1.46	1.48	1.51	1.53	1.55	1.57	1.60	1.62	1.64	1.67	1.69	1.71	1.73	1.76	1.78	1.80	997 718
15	1.47	1.49	1.52	1.54	1.56	1.59	1.61	1.64	1.66	1.69	1.71	1.74	1.76	1.78	1.81	1.83	1.86	1.88	1.91	1.93	0.997 556
16	1.56	1.59	1.62	1.64	1.67	1.69	1.72	1.75	1.77	1.80	1.82	1.85	1.88	1.90	1.93	1.96	1.98	2.01	2.03	2.06	997 393
17	1.66	1.69	1.72	1.74	1.77	1.80	1.83	1.86	1.88	1.91	1.94	1.97	1.99	2.02	2.05	2.08	2.10	2.13	2.16	2.19	997 231
18	1.76	1.79	1.82	1.85	1.88	1.91	1.94	1.96	1.99	2.02	2.05	2.08	2.11	2.14	2.17	2.20	2.23	2.26	2.29	2.32	997 068
19	1.86	1.89	1.92	1.95	1.98	2.01	2.04	2.07	2.10	2.13	2.17	2.20	2.23	2.26	2.29	2.32	2.35	2.38	2.41	2.44	996 906
20	1.95	1.99	2.02	2.05	2.08	2.12	2.15	2.18	2.21	2.25	2.28	2.31	2.34	2.38	2.41	2.44	2.47	2.51	2.54	2.57	0.996 744
21	2.05	2.08	2.12	2.15	2.19	2.22	2.26	2.29	2.32	2.36	2.39	2.43	2.46	2.50	2.53	2.56	2.60	2.63	2.67	2.70	996 582
22	2.15	2.18	2.22	2.26	2.29	2.33	2.36	2.40	2.43	2.47	2.51	2.54	2.58	2.61	2.65	2.69	2.72	2.76	2.79	2.83	996 420
23	2.25	2.28	2.32	2.36	2.39	2.43	2.47	2.51	2.54	2.58	2.62	2.66	2.69	2.73	2.77	2.81	2.84	2.88	2.92	2.96	996 258
24	2.34	2.38	2.42	2.46	2.50	2.54	2.58	2.62	2.66	2.69	2.73	2.77	2.81	2.85	2.89	2.93	2.97	3.01	3.05	3.08	996 095
25	2.44	2.48	2.52	2.56	2.60	2.64	2.68	2.72	2.76	2.81	2.85	2.89	2.93	2.97	3.01	3.05	3.09	3.13	3.17	3.21	0.995 934
26	2.54	2.58	2.62	2.66	2.71	2.75	2.79	2.83	2.88	2.92	2.96	3.00	3.04	3.09	3.13	3.17	3.21	3.26	3.30	3.34	995 772
27	2.63	2.68	2.72	2.77	2.81	2.85	2.90	2.94	2.99	3.03	3.07	3.12	3.16	3.20	3.25	3.29	3.34	3.38	3.42	3.47	995 610
28	2.73	2.78	2.82	2.87	2.91	2.96	3.00	3.05	3.10	3.14	3.19	3.23	3.28	3.32	3.37	3.41	3.46	3.51	3.55	3.60	995 448
29	2.83	2.88	2.92	2.97	3.02	3.06	3.11	3.16	3.21	3.25	3.30	3.35	3.39	3.44	3.49	3.54	3.58	3.63	3.68	3.72	995 286
30	2.93	2.97	3.02	3.07	3.12	3.17	3.22	3.27	3.32	3.36	3.41	3.46	3.51	3.56	3.61	3.66	3.71	3.75	3.80	3.85	0.995 125
31	3.02	3.07	3.12	3.17	3.22	3.27	3.32	3.37	3.43	3.48	3.53	3.58	3.63	3.68	3.73	3.78	3.83	3.88	3.93	3.98	994 963
32	3.12	3.17	3.22	3.28	3.33	3.38	3.43	3.48	3.54	3.59	3.64	3.69	3.74	3.80	3.85	3.90	3.95	4.00	4.06	4.11	994 801
33	3.22	3.27	3.32	3.38	3.43	3.48	3.54	3.59	3.64	3.70	3.75	3.81	3.86	3.91	3.97	4.02	4.07	4.13	4.18	4.23	994 640
34	3.31	3.37	3.42	3.48	3.53	3.59	3.64	3.70	3.75	3.81	3.86	3.92	3.98	4.03	4.09	4.14	4.20	4.25	4.31	4.36	994 479
35	3.41	3.47	3.52	3.58	3.64	3.69	3.75	3.81	3.86	3.92	3.98	4.03	4.09	4.15	4.21	4.26	4.32	4.38	4.43	4.49	0.994 317
36	3.51	3.56	3.62	3.68	3.74	3.80	3.86	3.92	3.97	4.03	4.09	4.15	4.21	4.27	4.32	4.38	4.44	4.50	4.56	4.62	994 156
37	3.60	3.66	3.72	3.78	3.84	3.90	3.96	4.02	4.08	4.14	4.20	4.26	4.32	4.38	4.44	4.50	4.56	4.62	4.68	4.74	993 995
38	3.70	3.76	3.82	3.89	3.95	4.01	4.07	4.13	4.19	4.26	4.32	4.38	4.44	4.50	4.56	4.63	4.69	4.75	4.81	4.87	993 833
39	3.80	3.86	3.92	3.99	4.05	4.11	4.18	4.24	4.30	4.37	4.43	4.49	4.56	4.62	4.68	4.75	4.81	4.87	4.94	5.00	993 672
40	3.89	3.96	4.02	4.09	4.15	4.22	4.28	4.35	4.41	4.48	4.54	4.61	4.67	4.74	4.80	4.87	4.93	5.00	5.06	5.13	0.993 511

Glass Scale

Barometer temperature t = °C	600	610	620	630	640	650	660	670	680	690	700	710	720	730	740	750	760	770	780	790	Correction factor f_t
1	0.10	0.11	0.11	0.11	0.11	0.11	0.11	0.12	0.12	0.12	0.12	0.12	0.12	0.13	0.13	0.13	0.13	0.13	0.13	0.14	0.999 827
2	0.21	0.21	0.21	0.22	0.22	0.22	0.23	0.23	0.24	0.24	0.24	0.25	0.25	0.25	0.26	0.26	0.26	0.27	0.27	0.27	999 654
3	0.31	0.32	0.32	0.33	0.33	0.34	0.34	0.35	0.35	0.36	0.36	0.37	0.37	0.38	0.38	0.39	0.40	0.40	0.41	0.41	999 480
4	0.42	0.42	0.43	0.44	0.44	0.45	0.46	0.46	0.47	0.48	0.49	0.49	0.50	0.51	0.51	0.52	0.53	0.53	0.54	0.55	999 307
5	0.52	0.53	0.54	0.55	0.55	0.56	0.57	0.58	0.59	0.60	0.61	0.61	0.62	0.63	0.64	0.65	0.66	0.67	0.68	0.68	0.999 134
6	0.62	0.63	0.64	0.65	0.66	0.68	0.69	0.70	0.71	0.72	0.73	0.74	0.75	0.76	0.77	0.78	0.79	0.80	0.81	0.82	998 961
7	0.73	0.74	0.75	0.76	0.78	0.79	0.80	0.81	0.82	0.84	0.85	0.86	0.87	0.88	0.90	0.91	0.92	0.93	0.95	0.96	998 788
8	0.83	0.84	0.86	0.87	0.89	0.90	0.91	0.93	0.94	0.95	0.97	0.98	1.00	1.01	1.02	1.04	1.05	1.07	1.08	1.09	998 616
9	0.93	0.95	0.97	0.98	1.00	1.01	1.03	1.04	1.06	1.07	1.09	1.11	1.12	1.14	1.15	1.17	1.18	1.20	1.21	1.23	998 443
10	1.04	1.06	1.07	1.09	1.11	1.12	1.14	1.16	1.18	1.19	1.21	1.23	1.25	1.26	1.28	1.30	1.31	1.33	1.35	1.37	0.998 270
11	1.14	1.16	1.18	1.20	1.22	1.24	1.26	1.27	1.29	1.31	1.33	1.35	1.37	1.39	1.41	1.43	1.45	1.46	1.48	1.50	998 098
12	1.25	1.27	1.29	1.31	1.33	1.35	1.37	1.39	1.41	1.43	1.45	1.47	1.49	1.51	1.54	1.56	1.58	1.60	1.62	1.64	997 925
13	1.35	1.37	1.39	1.42	1.44	1.46	1.48	1.51	1.53	1.55	1.57	1.60	1.62	1.64	1.66	1.69	1.71	1.73	1.75	1.78	997 752
14	1.45	1.48	1.50	1.52	1.55	1.57	1.60	1.62	1.65	1.67	1.69	1.72	1.74	1.77	1.79	1.82	1.84	1.86	1.89	1.91	997 580
15	1.56	1.58	1.61	1.63	1.66	1.68	1.71	1.74	1.76	1.79	1.81	1.84	1.87	1.89	1.92	1.94	1.97	2.00	2.02	2.05	0.997 408
16	1.66	1.69	1.71	1.74	1.77	1.80	1.82	1.85	1.88	1.91	1.94	1.96	1.99	2.02	2.05	2.07	2.10	2.13	2.16	2.18	997 235
17	1.76	1.79	1.82	1.85	1.88	1.91	1.94	1.97	2.00	2.03	2.06	2.09	2.11	2.14	2.17	2.20	2.23	2.26	2.29	2.32	997 063
18	1.87	1.90	1.93	1.96	1.99	2.02	2.05	2.08	2.11	2.15	2.18	2.21	2.24	2.27	2.30	2.33	2.36	2.39	2.43	2.46	996 891
19	1.97	2.00	2.03	2.07	2.10	2.13	2.17	2.20	2.23	2.26	2.30	2.33	2.36	2.40	2.43	2.46	2.49	2.53	2.56	2.59	996 719
20	2.07	2.11	2.14	2.18	2.21	2.24	2.28	2.31	2.35	2.38	2.42	2.45	2.49	2.52	2.56	2.59	2.62	2.66	2.69	2.73	0.996 547
21	2.18	2.21	2.25	2.28	2.32	2.36	2.39	2.43	2.47	2.50	2.54	2.57	2.61	2.65	2.68	2.72	2.76	2.79	2.83	2.86	996 375
22	2.28	2.32	2.35	2.39	2.43	2.47	2.51	2.54	2.58	2.62	2.66	2.70	2.73	2.77	2.81	2.85	2.89	2.92	2.96	3.00	996 203
23	2.38	2.42	2.46	2.50	2.54	2.58	2.62	2.66	2.70	2.74	2.78	2.82	2.86	2.90	2.94	2.98	3.02	3.06	3.10	3.14	996 031
24	2.48	2.53	2.57	2.61	2.65	2.69	2.73	2.77	2.82	2.86	2.90	2.94	2.98	3.02	3.06	3.11	3.15	3.19	3.23	3.27	995 859
25	2.59	2.63	2.67	2.72	2.76	2.80	2.85	2.89	2.93	2.98	3.02	3.06	3.11	3.15	3.19	3.23	3.28	3.32	3.36	3.41	0.995 687
26	2.69	2.74	2.78	2.83	2.87	2.92	2.96	3.00	3.05	3.09	3.14	3.18	3.23	3.27	3.32	3.36	3.41	3.45	3.50	3.54	995 515
27	2.79	2.84	2.89	2.93	2.98	3.03	3.07	3.12	3.17	3.21	3.26	3.31	3.35	3.40	3.45	3.49	3.54	3.59	3.63	3.68	995 344
28	2.90	2.95	2.99	3.04	3.09	3.14	3.19	3.24	3.28	3.33	3.38	3.43	3.48	3.53	3.57	3.62	3.67	3.72	3.77	3.82	995 170
29	3.00	3.05	3.10	3.15	3.20	3.25	3.30	3.35	3.40	3.45	3.50	3.55	3.60	3.65	3.70	3.75	3.80	3.85	3.90	3.95	995 001
30	3.10	3.15	3.21	3.26	3.31	3.36	3.41	3.46	3.52	3.57	3.62	3.67	3.72	3.77	3.83	3.88	3.93	3.98	4.03	4.09	0.994 829
31	3.21	3.26	3.31	3.37	3.42	3.47	3.53	3.58	3.63	3.69	3.74	3.79	3.85	3.90	3.95	4.01	4.06	4.11	4.17	4.22	994 658
32	3.31	3.36	3.42	3.47	3.53	3.58	3.64	3.69	3.75	3.80	3.86	3.91	3.97	4.02	4.08	4.13	4.19	4.25	4.30	4.36	994 487
33	3.41	3.47	3.52	3.58	3.64	3.70	3.75	3.81	3.87	3.92	3.98	4.04	4.09	4.15	4.21	4.26	4.32	4.38	4.43	4.49	994 315
34	3.51	3.57	3.63	3.69	3.75	3.81	3.86	3.92	3.98	4.04	4.10	4.16	4.22	4.27	4.33	4.39	4.45	4.51	4.57	4.63	994 144
35	3.62	3.68	3.74	3.80	3.86	3.92	3.98	4.04	4.10	4.16	4.22	4.28	4.34	4.40	4.46	4.52	4.58	4.64	4.70	4.76	0.993 974
36	3.72	3.78	3.84	3.90	3.97	4.03	4.09	4.15	4.21	4.28	4.34	4.40	4.46	4.52	4.59	4.65	4.71	4.77	4.83	4.90	993 802
37	3.82	3.89	3.95	4.01	4.08	4.14	4.20	4.27	4.33	4.39	4.46	4.52	4.59	4.65	4.71	4.78	4.84	4.90	4.97	5.03	993 631
38	3.92	3.99	4.05	4.12	4.19	4.25	4.32	4.38	4.45	4.51	4.58	4.64	4.71	4.77	4.84	4.91	4.97	5.04	5.10	5.17	993 460
39	4.03	4.09	4.16	4.23	4.30	4.36	4.43	4.50	4.56	4.63	4.70	4.76	4.83	4.90	4.97	5.03	5.10	5.17	5.23	5.30	993 289
40	4.13	4.20	4.27	4.34	4.40	4.47	4.54	4.61	4.68	4.75	4.82	4.89	4.96	5.02	5.09	5.16	5.23	5.30	5.37	5.44	0.993 118

Note: Subtract from the barometer reading the amount $\Delta\beta_t$ corresponding to the actual temperature t, or multiply the barometer reading by the correction factor f_t. The vapor pressure correction for air fully saturated with water vapor is already included in the conversion tables for gas volumes. The correction factor f_t is calculated from the formula $f_t = 1 - (\beta - \alpha)t/1 + \beta_t$, where $\beta = 1.818 \times 10^{-3}$°$C^{-1} \approx$ volume expansion coefficient of mercury, $\alpha = 1.84 \times 10^{-5}$°$C^{-1} \approx$ linear expansion coefficient of brass, or $\alpha = 8.5 \times 10^{-6}$°$C^{-1} \approx$ linear expansion coefficient of glass.

From Diem, K., Ed., *Documenta Geigy Scientific Tables,* 6th ed., Geigy Pharmaceuticals, Ardsley, N.Y., 1962, 295. With permission.

Table 137
BAROMETRIC LATITUDE-GRAVITY TABLE — METRIC UNITS (SMITHSONIAN TABLES)

Deg. Lat.	Barometer readings, millimeters					
	680	700	720	740	760	780
	mm.	mm.	mm.	mm.	mm.	mm.
0	1.82	1.87	1.93	1.98	2.04	2.09
5	1.79	1.85	1.90	1.95	2.00	2.06
10	1.71	1.76	1.81	1.86	1.92	1.97
15	1.58	1.63	1.67	1.72	1.77	1.81
20	1.40	1.44	1.49	1.53	1.57	1.61
21	1.36	1.40	1.44	1.48	1.52	1.56
22	1.32	1.36	1.40	1.44	1.48	1.51
23	1.28	1.31	1.35	1.39	1.43	1.46
24	1.23	1.27	1.30	1.34	1.37	1.41
25	1.18	1.22	1.25	1.29	1.32	1.36
26	1.13	1.17	1.20	1.23	1.27	1.30
27	1.08	1.12	1.15	1.18	1.21	1.24
28	1.03	1.06	1.09	1.12	1.15	1.18
29	0.98	1.01	1.04	1.07	1.10	1.12
30	0.93	0.95	0.98	1.01	1.04	1.06
31	0.87	0.90	0.92	0.95	0.98	1.00
32	0.82	0.84	0.86	0.89	0.91	0.94
33	0.76	0.78	0.80	0.83	0.85	0.87
34	0.70	0.72	0.74	0.76	0.79	0.81
35	0.64	0.66	0.68	0.70	0.72	0.74
36	0.58	0.60	0.62	0.64	0.65	0.67
37	0.52	0.54	0.56	0.57	0.59	0.60
38	0.46	0.48	0.49	0.51	0.52	0.53
39	0.40	0.42	0.43	0.44	0.45	0.46
40	0.34	0.35	0.36	0.37	0.38	0.39
41	0.28	0.29	0.30	0.30	0.31	0.32
42	0.22	0.22	0.23	0.24	0.24	0.25
43	0.16	0.16	0.16	0.17	0.17	0.18
44	0.09	0.10	0.10	0.10	0.10	0.11
45	0.03	0.03	0.03	0.03	0.03	0.04
46	0.03	0.03	0.03	0.03	0.04	0.04
47	0.09	0.10	0.10	0.10	0.10	0.11
48	0.16	0.16	0.17	0.17	0.18	0.18
49	0.22	0.23	0.23	0.24	0.25	0.25
50	0.28	0.29	0.30	0.31	0.31	0.32
51	0.34	0.35	0.36	0.37	0.38	0.39
52	0.40	0.42	0.43	0.44	0.45	0.46
53	0.46	0.48	0.49	0.51	0.52	0.53
54	0.52	0.54	0.56	0.57	0.59	0.60
55	0.58	0.60	0.62	0.64	0.65	0.67
56	0.64	0.66	0.68	0.70	0.72	0.74
57	0.70	0.72	0.74	0.76	0.78	0.80
58	0.76	0.78	0.80	0.82	0.85	0.87
59	0.81	0.84	0.86	0.89	0.91	0.93
60	0.87	0.89	0.92	0.94	0.97	1.00
61	0.92	0.95	0.98	1.00	1.03	1.06
62	0.97	1.00	1.02	1.05	1.08	1.11
63	1.03	1.06	1.09	1.12	1.15	1.18
64	1.08	1.11	1.14	1.17	1.20	1.23
65	1.13	1.16	1.19	1.22	1.26	1.29
66	1.17	1.21	1.24	1.28	1.31	1.35
67	1.22	1.25	1.29	1.33	1.36	1.40
68	1.26	1.30	1.34	1.37	1.41	1.45
69	1.31	1.34	1.38	1.42	1.46	1.50
70	1.35	1.39	1.43	1.47	1.51	1.55
72	1.42	1.47	1.51	1.55	1.59	1.63
75	1.53	1.57	1.62	1.66	1.71	1.75
80	1.66	1.71	1.76	1.81	1.86	1.90
85	1.74	1.79	1.84	1.90	1.95	2.00
90	1.77	1.82	1.87	1.93	1.98	2.03

Note: The values above are to be subtracted from the barometric reading for latitudes from 0 to 45° inclusive, and are to be added from 46 to 90°.

Table 137 (continued)
BAROMETRIC LATITUDE-GRAVITY TABLE — METRIC UNITS (SMITHSONIAN TABLES)

Height above sea-level meters	Observed barometer height in millimeters								
	400 mm.	450 mm.	500 mm.	550 mm.	600 mm.	650 mm.	700 mm.	750 mm.	800 mm.
100	....	....	....	....	....	....	0.02	0.02	0.02
200	....	....	....	....	....	....	0.04	0.05	0.05
300	....	....	....	....	....	....	0.07	0.07	0.07
400	....	....	....	....	....	....	0.09	0.10	0.10
500	....	....	....	....	....	....	0.11	0.12	0.13
600	....	....	....	....	....	0.12	0.13	0.14	....
700	....	....	....	....	....	0.14	0.15	0.16	....
800	....	....	....	....	....	0.16	0.18	0.19	....
900	....	....	....	....	....	0.18	0.20	0.22	....
1000	....	....	....	0.18	0.19	0.20	0.22	0.24	....
1100	....	....	....	0.19	0.21	0.22	0.24	....	....
1200	....	....	....	0.21	0.23	0.24	0.26	....	....
1300	....	....	....	0.22	0.24	0.26	0.29	....	....
1400	....	....	....	0.24	0.26	0.28	0.31	....	....
1500	....	....	0.24	0.26	0.28	0.30	0.33	....	....
1600	....	....	0.25	0.28	0.30	0.32	....	....	....
1700	....	....	0.27	0.30	0.32	0.34	....	....	....
1800	....	....	0.28	0.31	0.34	0.36	....	....	....
1900	....	....	0.30	0.33	0.36	0.39	....	....	....
2000	....	0.28	0.31	0.34	0.38	0.41	....	....	....
2100	....	0.30	0.33	0.36	0.40	....	....	....	....
2200	....	0.31	0.35	0.38	0.41	....	....	....	....
2300	....	0.32	0.36	0.40	0.43	....	....	....	....
2400	....	0.34	0.38	0.42	0.45	....	....	....	....
2500	0.31	0.35	0.39	0.43	0.47	....	....	....	....
2600	0.33	0.37	0.41	....	....	....	....	....	....
2800	0.35	0.40	0.44	....	....	....	....	....	....
3000	0.38	0.42	0.47	....	....	....	....	....	....
3200	0.40	0.46	....	....	....	....	....	....	....
3400	0.43	0.48	....	....	....	....	....	....	....

Note: The values in the table above are to be subtracted from the readings taken on a mercurial barometer to correct for the decrease in gravity with increase in altitude.

From Dean, J. A., Ed., *Lange's Handbook of Chemistry,* 12th ed., McGraw-Hill, New York, 1979, 2-70. With permission.

Table 138
CORRECTION OF A BAROMETER FOR CAPILLARITY (SMITHSONIAN TABLES)

Diameter of tube, millimeters	Height of meniscus in millimeters							
	0.4	0.6	0.8	1.0	1.2	1.4	1.6	1.8
	Correction to be added in millimeters							
4	0.83	1.22	1.54	1.98	2.37	....	....	....
5	0.47	0.65	0.86	1.19	1.45	1.80	....	....
6	0.27	0.41	0.56	0.78	0.98	1.21	1.43	....
7	0.18	0.28	0.40	0.53	0.67	0.82	0.97	1.13
8	....	0.20	0.29	0.38	0.46	0.56	0.65	0.77
9	....	0.15	0.21	0.28	0.33	0.40	0.46	0.52
10	....	....	0.15	0.20	0.25	0.29	0.33	0.37
11	....	....	0.10	0.14	0.18	0.21	0.24	0.27
12	....	....	0.07	0.10	0.13	0.15	0.18	0.19
13	....	....	0.04	0.07	0.10	0.12	0.13	0.14

From Dean, J. A., Ed., *Lange's Handbook of Chemistry,* 12th ed., McGraw-Hill, New York, 1979, 2-69. With permission.

Table 139
CORRECTED PARTIAL PRESSURES OF GAS MIXTURES AT 37°C

Atmospheric Pressure (mm Hg)	Corrected Partial Pressures (mm Hg)						
	5%	7%	10%	12%	15%	20%	21%
720.0	33.65	47.11	67.29	80.75	100.94	134.59	141.32
721.0	33.70	47.18	67.39	80.87	101.09	134.79	141.53
722.0	33.75	47.25	67.49	80.99	101.24	134.99	141.74
723.0	33.80	47.32	67.59	81.11	101.39	135.19	141.95
724.0	33.85	47.39	67.69	81.23	101.54	135.39	142.16
725.0	33.90	47.46	67.79	81.35	101.69	135.59	142.37
726.0	33.95	47.53	67.89	81.47	101.84	135.79	142.58
727.0	34.00	47.60	67.99	81.59	101.99	135.99	142.79
728.0	34.05	47.67	68.09	81.71	102.14	136.19	143.00
729.0	34.10	47.74	68.19	81.83	102.29	136.39	143.21
730.0	34.15	47.81	68.29	81.95	102.44	136.59	143.42
731.0	34.20	47.88	68.39	82.07	102.59	136.79	143.63
732.0	34.25	47.95	68.49	82.19	102.74	136.99	143.84
733.0	34.30	48.02	68.59	82.31	102.89	137.19	144.05
734.0	34.35	48.09	68.69	82.43	103.04	137.39	144.26
735.0	34.40	48.16	68.79	82.55	103.19	137.59	144.47
736.0	34.45	48.23	68.89	82.67	103.34	137.79	144.68
737.0	34.50	48.30	68.99	82.79	103.49	137.99	144.89
738.0	34.55	48.37	69.09	82.91	103.64	138.19	145.10
739.0	34.60	48.44	69.19	83.03	103.79	138.39	145.31
740.0	34.65	48.51	69.29	83.15	103.94	138.59	145.52
741.0	34.70	48.58	69.39	83.27	104.09	138.79	145.73
742.0	34.75	48.65	69.49	83.39	104.24	138.99	145.94
743.0	34.80	48.72	69.59	83.51	104.39	139.19	146.15
744.0	34.85	48.79	69.69	83.63	104.54	139.39	146.36
745.0	34.90	48.86	69.79	83.75	104.69	139.59	146.57
746.0	34.95	48.93	69.89	83.87	104.84	139.79	146.78
747.0	35.00	49.00	69.99	83.99	104.99	139.99	146.99
748.0	35.05	49.07	70.09	84.11	105.14	140.19	147.20
749.0	35.10	49.14	70.19	84.23	105.29	140.39	147.41
750.0	35.15	49.21	70.29	84.35	105.44	140.59	147.62
751.0	35.20	49.28	70.39	84.47	105.59	140.79	147.83
752.0	35.25	49.35	70.49	84.59	105.74	140.99	148.04
753.0	35.30	49.42	70.59	84.71	105.89	141.19	148.25
754.0	35.35	49.49	70.69	84.83	106.04	141.39	148.46
755.0	35.40	49.56	70.79	84.95	106.19	141.59	148.67
756.0	35.45	49.63	70.89	85.07	106.34	141.79	148.88
757.0	35.50	49.70	70.99	85.19	106.49	141.99	149.09
758.0	35.55	49.77	71.09	85.31	106.64	142.19	149.30
759.0	35.60	49.84	71.19	85.43	106.79	142.39	149.51
760.0	35.65	49.91	71.29	85.55	106.94	142.59	149.72
761.0	35.70	49.98	71.39	85.67	107.09	142.79	149.93
762.0	35.75	50.05	71.49	85.79	107.24	142.99	150.14
763.0	35.80	50.12	71.59	85.91	107.39	143.19	150.35
764.0	35.85	50.19	71.69	86.03	107.54	143.39	150.56
765.0	35.90	50.26	71.79	86.15	107.69	143.59	150.77
766.0	35.95	50.33	71.89	86.27	107.84	143.79	150.98
767.0	36.00	50.40	71.99	86.39	107.99	143.99	151.19
768.0	36.05	50.47	72.09	86.51	108.14	144.19	151.40
769.0	36.10	50.54	72.19	86.63	108.29	144.39	151.61

Table 139 (continued)
CORRECTED PARTIAL PRESSURES OF GAS MIXTURES AT 37°C

Atmospheric Pressure (mm Hg)	Corrected Partial Pressures (mm Hg)						
	5%	7%	10%	12%	15%	20%	21%
770.0	36.15	50.61	72.29	86.75	108.44	144.59	151.82
771.0	36.20	50.68	72.39	86.87	108.59	144.79	152.03
772.0	36.25	50.75	72.49	86.99	108.74	144.99	152.24
773.0	36.30	50.82	72.59	87.11	108.89	145.19	152.45
774.0	36.35	50.89	72.69	87.23	109.04	145.39	152.66
775.0	36.40	50.96	72.79	87.35	109.19	145.59	152.87
776.0	36.45	51.03	72.89	87.47	109.34	145.79	153.08
777.0	36.50	51.10	72.99	87.59	109.49	145.99	153.29
778.0	36.55	51.17	73.09	87.71	109.64	146.19	153.50
779.0	36.60	51.24	73.19	87.83	109.79	146.39	153.71

Note: Partial pressures (mm Hg) of indicated % V:V gases, saturated with water vapor are given for atmospheric barometric pressures from 720 to 779 mm Hg. Partial pressures are corrected for the vapor pressure of water at 37°C (47.067 mm Hg).

Courtesy of Robert Rej.

THERMOMETER CALIBRATION*

Tests of a liquid-in-glass thermometer consist of comparisons of readings at an adequate number of points on its scale, usually at intervals from 40 to 100 divisions, with those on a standard instrument. Such a standard instrument may be either a standard platinum resistance thermometer or a suitable mercury-in-glass thermometer which has been standardized on the International Temperature Scale. Tests may also be made at the ice and steam points and a number of secondary fixed points if the equipment for attaining the temperatures of the fixed points is available.

Thermometers are subject to changes with time and use primarily as a result of changes in the volume of the bulb. For well-annealed thermometers, not subject to excessively high temperatures, bulb changes are small. After a thermometer has been heated, the bulb does not at once return to its original volume but remains somewhat larger thus temporarily lowering the readings; and if a thermometer is kept at a high temperature for long periods the ice point and the boiling point may be permanently lowered. All such changes may be determined by making a test at some reference point and changing all readings by the amount of the observed change.

Use of a thermometer at pressures differing appreciably from those prevailing during test will yield different readings. A greater pressure on the bulb will result in higher readings and a lesser pressure in lower readings. Such changes for cylindrical bulbs having a diameter of from 5 to 7 mm are of the order of 0.1°C (0.2°F) per atmosphere of pressure. See also section on Correction for Emergent Stem.

When a thermometer which has been standardized for total immersion is used with a part of the liquid column at a temperature below that of the bulb, the reading is low and a correction must be applied. For this correction the following formula is employed:

$$T_c = T_o + f \times l \times (T_o - T_m)$$

---

* From Dean, J. A., Ed., *Lange's Handbook of Chemistry*, 12th ed., McGraw-Hill, New York, 1979, 11. With permission.

Table 140
CORRECTION FOR EMERGENT STEM OF LIQUID-IN-GLASS THERMOMETERS

	Values of f for various glasses				
Tm °C	Corning 0041	Corning 8800	Corning 8810	Jena 16 III	Jena 59 III
50	0.000157	0.000166	0.000156	0.000158	0.000164
150	0.000159	0.000167	0.000157	0.000158	0.000165
250	0.000163	0.000168	0.000161	0.000161	0.000170
350	0.000168	0.000173	0.000166	—	0.000177
450	—	0.000180	0.000174	—	0.000187
500	—	—	—	—	0.000195

From Dean, J. A., Ed., *Lange's Handbook of Chemistry*, 12th ed., McGraw-Hill, New York, 1979, 11. With permission.

where T_c = corrected temperature

T_o = observed temperature

l = length of column in degrees above the surface of the liquid the temperature of which is being taken

T_m = mean temperature of mercury (or other thermometer liquid) column; i.e., the temperature of the middle point of the mercury (or other liquid) column as read from another thermometer

f = correction factor as given in the table below. In calculating the emergent stem correction for thermometers containing organic liquids (alcohol, pentane, toluene) it is sufficient to use the approximate value, $f = 0.001$. In such thermometers the value of f is practically independent of the kind of glass

CORRECTIONS FOR OSMOTIC PRESSURE MEASUREMENTS

Measures of Concentration

The concentration of a dilute solution is usually expressed as its molality (mole per 1000 g solvent) or molarity (mole per 1000 ml solution, analogous with the normality).

The use of 1000 ml solution as reference unit has great advantages in volumetric analysis but also the disadvantage of being temperature-dependent. Giving concentrations in equivalents instead of moles (normality instead of molarity) is always necessary when valency or valency change is involved, particularly in acid-base reactions, oxidations and reductions, but it should be borne in mind that the normality of a solution can differ in different types of reaction.

Freezing-point depression data are always given for concentrations expressed as molality.

In extremely dilute aqueous solutions the molality and molarity can be assumed to be equal; with rising concentration they deviate more and more to an extent depending on the specific volume of the solute(s). Thus the molarity of serum differs from its molality owing to the high specific volume of proteins. In order to calculate the molality of any particular serum component, its concentration, e.g., in milligrams per liter serum, must be converted into its concentration in the serum water.

This conversion can be made by means of either the specific gravity or the protein content of serum. The former method is the more accurate. On the basis of protein content the conversion is made by means of the following formula[1]:

$$\begin{array}{l} \text{Water content} \\ \text{of serum in g/}\ell \\ \quad \text{serum} \end{array} \quad = 984.0 - (0.718 \times \text{protein content} \\ \qquad\qquad\qquad\qquad \text{in g/}\ell \text{ serum})$$

(The value of 984 instead of 1000 represents a correction for the volume occupied by inorganic and other constituents.)

Example. The freezing-point depression of serum is 0.56°C, corresponding to a molality of 300 mmol (300 mmol/1000 g water). The molarity of a normal serum is accordingly $300 \times 0.940 = 282$ mmol/ℓ serum.

In the following only the molality will be used. It should be noted that molality and molarity are sometimes confused even by reputable authors, so that it is advisable always to examine carefully what is meant by the expression 'molarity' in any particular case.

In order to avoid this confusion the molarity and molality should always be related to the undissociated solute; otherwise they should be clearly specified, for instance 'the molality of all osmotically active particles'.

Osmolarity, Osmolality

These terms indicate respectively the molarity and molality an ideal solution of a nondissociating substance must possess in order to exert the same osmotic pressure as the solution under consideration. Osmolarity and osmolality are not used in the physicochemical field but find considerable application in the sphere of biology and medicine.[2] As is clear from the definition, the (real) osmolality is a quantity capable of experimental determination. It can also be calculated from the molality of the solution provided (1) the number of molecular fragments (for weak electrolytes the degree of dissociation α and (2) the correction factor (osmotic coefficient g from the ideal to the real state are known.

If weak electrolytes are excluded, the ideal osmolality can be obtained by multiplying the molality by the number of molecular fragments. Multiplication of this by the osmotic coefficient g gives the (real) osmolality as defined above:

$$\text{ideal osmolality} = m_2 \nu \tag{1}$$
$$\text{(real) osmolality} = \text{ideal osmolality} \times g = m_2 \nu g = \Delta T/1.86$$

For mixed solutions $m_2\nu$ is replaced by the sum $\sum\limits_{i=2}^{n} m_i\nu_i = m_2\nu_2 + m_3\nu_3 + \ldots$ for each of the component solutes. For the sake of simplicity it is assumed that there is no change in the osmotic coefficient when passing from a simple to a mixed solution.

In analogy with the mole, the unit of osmolarity and osmolality is the osmole (osm).

Applications

To obtain freezing-point depression and osmotic pressure from osmolality see Table 141.

Osmolality of blood serum from freezing-point depression (0.56°C) (Table 141)— Columns 5 and 6 of the table show that the (real) osmolality of serum is 302.1 mmol.

Sodium chloride and glucose solutions (Table 142) — (a) The weights of NaCl and glucose (or fructose) corresponding to given ideal osmolalities are obtained from columns 1/2 and 1/6. Column 7 gives the corresponding calorific values for glucose and fructose.

Table 141
AQUEOUS SOLUTIONS — CALCULATION OF FREEZING-POINT DEPRESSION AND OSMOTIC PRESSURE

Real osmolality (mmol/1000 g water)	Freezing-point depression ($\Delta T °C$)	Osmotic pressure at 0 °C (atm)	Osmotic pressure at 38 °C* (atm)	Freezing-point depression ($\Delta T °C$)	Real osmolality (mmol/1000 g water)	Osmotic pressure at 0 °C (atm)	Osmotic pressure at 38 °C* (atm)
1	2	3	4	5	6	7	8
10	0.019	0.22	0.26	0.01	5.4	0.12	0.14
20	0.037	0.45	0.51	02	10.7	0.24	0.28
30	0.056	0.67	0.77	03	16.1	0.36	0.41
40	0.074	0.90	1.01	04	21.5	0.48	0.55
50	0.093	1.12	1.27	0.05	26.9	0.60	0.69
60	0.112	1.35	1.52	06	32.3	0.72	0.82
70	0.130	1.57	1.78	07	37.6	0.84	0.95
80	0.149	1.79	2.03	08	43.0	0.97	1.09
90	0.167	2.02	2.28	09	48.4	1.09	1.23
100	0.186	2.24	2.54	0.10	53.8	1.21	1.37
10	0.205	2.47	2.79	11	59.2	1.33	1.50
20	0.223	2.69	3.05	12	64.6	1.45	1.64
30	0.242	2.91	3.30	13	70.0	1.57	1.78
40	0.260	3.14	3.55	14	75.3	1.69	1.92
50	0.279	3.36	3.80	0.15	80.7	1.81	2.05
60	0.297	3.59	4.06	16	86.1	1.93	2.18
70	0.316	3.81	4.31	17	91.5	2.05	2.32
80	0.334	4.03	4.57	18	96.9	2.17	2.46
90	0.353	4.26	4.83	19	102.3	2.29	2.59
200	0.371	4.48	5.07	0.20	107.7	2.41	2.73
10	0.390	4.71	5.33	21	113.0	2.53	2.87
20	0.408	4.93	5.58	22	118.4	2.65	3.01
30	0.427	5.16	5.84	23	123.8	2.77	3.14
40	0.445	5.38	6.09	24	129.2	2.89	3.28
50	0.464	5.60	6.34	0.25	134.6	3.02	3.42
60	0.482	5.83	6.59	26	140.0	3.14	3.56
70	0.501	6.05	6.85	27	145.4	3.26	3.68
80	0.519	6.28	7.10	28	150.8	3.38	3.82
90	0.537	6.50	7.36	29	156.2	3.50	3.96
300	0.556	6.72	7.62	0.30	161.6	3.62	4.10
10	0.574	6.95	7.87	31	167.0	3.74	4.23
20	0.593	7.17	8.12	32	172.4	3.86	4.37
30	0.611	7.40	8.37	33	177.8	3.98	4.51
40	0.630	7.62	8.63	34	183.2	4.10	4.65
50	0.648	7.84	8.88	0.35	188.6	4.23	4.78
60	0.667	8.07	9.14	36	194.0	4.35	4.92
70	0.685	8.29	9.39	37	199.4	4.47	5.06
80	0.704	8.52	9.64	38	204.8	4.59	5.20
90	0.722	8.74	9.90	39	210.2	4.71	5.33
400	0.741	8.97	10.15	0.40	215.5	4.84	5.47
10	0.759	9.19	10.40	41	220.9	4.96	5.61
20	0.778	9.41	10.66	42	226.3	5.08	5.75
30	0.796	9.64	10.92	43	231.8	5.20	5.88
40	0.815	9.86	11.16	44	237.2	5.32	6.02
50	0.833	10.09	11.42	0.45	242.6	5.44	6.16
60	0.851	10.31	11.67	46	248.0	5.56	6.30
70	0.870	10.53	11.93	47	253.4	5.68	6.43
80	0.887	10.76	12.17	48	258.8	5.80	6.57
90	0.906	10.98	12.43	49	264.2	5.92	6.71
500	0.925	11.21	12.69	0.50	269.6	6.04	6.85
10	0.943	11.43	12.94	51	275.0	6.16	6.98
20	0.962	11.66	13.20	52	280.4	6.28	7.11
30	0.980	11.88	13.45	53	285.8	6.40	7.25
40	0.998	12.10	13.70	54	291.2	6.52	7.39
50	1.017	12.33	13.95	0.55	296.7	6.64	7.53
60	1.035	12.55	14.21	56	302.1	6.77	7.66
70	1.054	12.78	14.47	57	307.5	6.89	7.80
80	1.072	13.00	14.72	58	312.9	7.02	7.94
90	1.090	13.22	14.97	59	318.3	7.14	8.07
600	1.109	13.45	15.22	0.60	323.7	7.26	8.21
10	1.127	13.67	15.48	61	329.2	7.38	8.35
20	1.146	13.90	15.73	62	334.6	7.49	8.49
30	1.164	14.12	15.99	63	340.0	7.62	8.62
40	1.182	14.34	16.24	64	345.4	7.74	8.76
50	1.201	14.57	16.49	0.65	350.8	7.86	8.90
60	1.219	14.79	16.75	66	356.2	7.98	9.04
70	1.238	15.02	17.00	67	361.6	8.10	9.17
80	1.256	15.24	17.26	68	367.0	8.22	9.31
90	1.274	15.47	17.51	69	372.5	8.34	9.45
700	1.292	15.69	17.77	0.70	377.9	8.47	9.59
10	1.311	15.91	18.01	71	383.3	8.59	9.72
20	1.329	16.14	18.27	72	388.7	8.71	9.86
30	1.347	16.36	18.52	73	394.2	8.84	10.00
40	1.365	16.59	18.78	74	399.6	8.96	10.14

* Normal blood temperature = ca. 38 °C = 311.15 K.

From Diem, K., Ed., *Documenta Geigy Scientific Tables,* 7th ed., Geigy Pharmaceuticals, Ardsley, N.Y., 1970, 272. With permission.

Table 142
AQUEOUS SOLUTIONS — CALCULATION OF SALT AND GLUCOSE CONCENTRATIONS

Values in columns 3–5 and 8–10 have been calculated for the osmolalities in column 1 read as real osmotic concentrations (millimoles or grammes per 1000 g water). The osmotic coefficients g have been obtained by interpolation from the data of Scatchard and Prentiss, *J. Amer. chem. Soc.*, 55, 4355 (1933), for NaCl and Roth, W. A., *Z. phys. Chem.*, 43, 539 (1903), for glucose

Osmolality (ideal)	Common salt (NaCl, mol. wt. 58.443)				D-Glucose* ($C_6H_{12}O_6$, mol. wt. 180.16)					Common salt (NaCl)		D-Glucose* ($C_6H_{12}O_6$)	
	Corresponds to a weight of	Weight necessary when the mmol in column 1 are required to be added to bring the total osmolality to 300 mmol	Osmotic coefficient corresponding to the osmolality in column 1 read as real osmolality		Corresponds to a weight of	Corresponds to a calorific value of	Weight necessary when the mmol in column 1 are required to be added to bring the total osmolality to 300 mmol	Osmotic coefficient corresponding to the osmolality in column 1 read as real osmolality		Corresponds to an ideal osmolality of	Corresponds to an ideal osmolality of	Corresponds to a calorific value of	
mmol	g	g	g	1/g	g	cal**	g	g	1/g	g	mmol	mmol	cal**
1	2	3	4	5	6	7	8	9	10	11	12	13	14
10	0.292	0.315	0.9778	1.0227	1.802	7.53	1.777	1.0005	0.9995	1	34.22	5.55	4.18
20	0.584	0.630	0.9703	1.0306	3.603	15.07	3.554	1.0009	0.9991	2	68.44	11.10	8.36
30	0.877	0.947	0.9653	1.0359	5.405	22.60	5.331	1.0014	0.9986	3	102.66	16.65	12.55
40	1.169	1.262	0.9612	1.0404	7.206	30.14	7.108	1.0018	0.9982	4	136.89	22.20	16.73
50	1.461	1.577	0.9579	1.0440	9.008	37.67	8.885	1.0023	0.9977	5	171.11	27.75	20.91
60	1.753	1.892	0.9550	1.0471	10.810	45.21	10.662	1.0028	0.9972	6	205.33	33.30	25.09
70	2.046	2.208	0.9525	1.0499	12.611	52.74	12.439	1.0032	0.9968	7	239.55	38.85	29.27
80	2.338	2.524	0.9503	1.0523	14.413	60.27	14.217	1.0037	0.9963	8	273.77	44.41	33.46
90	2.630	2.839	0.9482	1.0546	16.214	67.81	15.994	1.0041	0.9959	9	307.99	49.96	37.64
100	2.922	3.154	0.9463	1.0567	18.016	75.34	17.771	1.0046	0.9954	10	342.22	55.51	41.82
10	3.214	3.469	0.9448	1.0584	19.818	82.88	19.548	1.0051	0.9949	11	376.44	61.06	46.00
20	3.507	3.785	0.9432	1.0602	21.619	90.41	21.325	1.0055	0.9945	12	410.66	66.61	50.18
30	3.799	4.101	0.9418	1.0618	23.421	97.95	23.102	1.0060	0.9940	13	444.88	72.16	54.37
40	4.091	4.416	0.9405	1.0633	25.222	105.48	24.879	1.0064	0.9936	14	479.10	77.71	58.55
50	4.383	4.731	0.9392	1.0647	27.024	113.01	26.656	1.0069	0.9931	15	513.32	83.26	62.73
60	4.675	5.047	0.9380	1.0661	28.826	120.55	28.433	1.0074	0.9927	16	547.54	88.81	66.91
70	4.968	5.362	0.9368	1.0675	30.627	128.08	30.210	1.0078	0.9923	17	581.77	94.36	71.09
80	5.260	5.678	0.9357	1.0687	32.429	135.62	31.987	1.0083	0.9918	18	615.99	99.91	75.28
90	5.552	5.993	0.9347	1.0699	34.230	143.15	33.764	1.0087	0.9914	19	650.21	105.46	79.46
200	5.844	6.308	0.9337	1.0710	36.032	150.68	35.541	1.0092	0.9909	20	684.43	111.01	83.64
10	6.137	6.624	0.9328	1.0720	37.834	158.22	37.318	1.0097	0.9904	21	718.65	116.56	87.82
20	6.429	6.939	0.9319	1.0731	39.635	165.75	39.096	1.0101	0.9900	22	752.87	122.11	92.00
30	6.721	7.255	0.9311	1.0740	41.437	173.29	40.873	1.0106	0.9895	23	787.09	127.66	96.19
40	7.013	7.570	0.9304	1.0748	43.238	180.82	42.650	1.0110	0.9891	24	821.32	133.22	100.37
50	7.305	7.885	0.9297	1.0756	45.040	188.36	44.427	1.0115	0.9886	25	855.54	138.77	104.55
60	7.598	8.201	0.9290	1.0764	46.842	195.89	46.204	1.0120	0.9881	26	889.76	144.32	108.73
70	7.890	8.516	0.9283	1.0772	48.643	203.43	47.981	1.0124	0.9878	27	923.98	149.87	112.91
80	8.182	8.832	0.9276	1.0780	50.445	210.96	49.758	1.0129	0.9873	28	958.20	155.42	117.10
90	8.474	9.147	0.9270	1.0787	52.246	218.49	51.535	1.0133	0.9869	29	992.42	160.97	121.28
300	8.766	9.463	0.9264	1.0794	54.048	226.03	53.312	1.0138	0.9864	30	1026.65	166.52	125.46
10	9.059		0.9258	1.0801	55.850	233.56		1.0143	0.9859	31	1060.87	172.06	129.64
20	9.351		0.9252	1.0808	57.651	241.10		1.0147	0.9855	32	1095.09	177.62	133.82
30	9.643		0.9246	1.0815	59.453	248.63		1.0152	0.9850	33	1129.31	183.17	138.01
40	9.935		0.9241	1.0821	61.254	256.17		1.0156	0.9846	34	1163.53	188.72	142.19
50	10.228		0.9236	1.0827	63.056	263.70		1.0161	0.9842	35	1197.75	194.27	146.37
60	10.520		0.9232	1.0832	64.858	271.23		1.0166	0.9837	36	1231.97	199.82	150.55
70	10.812		0.9227	1.0838	66.659	278.77		1.0170	0.9833	37	1266.20	205.37	154.73
80	11.104		0.9223	1.0842	68.461	286.30		1.0175	0.9828	38	1300.42	210.92	158.92
90	11.396		0.9219	1.0847	70.262	293.84		1.0179	0.9824	39	1334.64	216.47	163.10
400	11.689		0.9215	1.0852	72.064	301.37		1.0183	0.9820	40	1368.86	222.03	167.28
10	11.981		0.9211	1.0857	73.866	308.91		1.0187	0.9816	41	1403.08	227.58	171.46
20	12.273		0.9207	1.0861	75.667	316.44		1.0192	0.9812	42	1437.30	233.13	175.64
30	12.565		0.9204	1.0864	77.469	323.98		1.0196	0.9808	43	1471.52	238.68	179.82
40	12.857		0.9200	1.0868	79.270	331.51		1.0201	0.9803	44	1505.75	244.23	184.00
50	13.150		0.9196	1.0874	81.072	339.04		1.0205	0.9799	45	1539.97	249.78	188.19
60	13.442		0.9192	1.0878	82.874	346.58		1.0209	0.9795	46	1574.19	255.33	192.37
70	13.734		0.9189	1.0882	84.675	354.11		1.0214	0.9790	47	1608.41	260.88	196.55
80	14.026		0.9185	1.0887	86.477	361.64		1.0218	0.9787	48	1642.63	266.43	200.74
90	14.319		0.9182	1.0891	88.278	369.18		1.0222	0.9783	49	1676.85	271.98	204.92
500	14.611		0.9180	1.0893	90.080	376.72		1.0226	0.9779	50	1711.08	277.53	209.10
10	14.903		0.9177	1.0897	91.882	384.25		1.0230	0.9775	51	1745.30	283.08	213.28
20	15.195		0.9174	1.0900	93.683	391.78		1.0234	0.9771	52	1779.52	288.63	217.46
30	15.487		0.9172	1.0903	95.485	399.32		1.0238	0.9767	53	1813.74	294.18	221.65
40	15.780		0.9170	1.0905	97.286	406.85		1.0242	0.9764	54	1847.96	299.73	225.82
50	16.072		0.9167	1.0908	99.088	414.39		1.0245	0.9761	55	1882.18	305.28	230.01
60	16.364		0.9165	1.0911	100.890	421.92		1.0249	0.9757	56	1916.40	310.84	234.19
70	16.656		0.9163	1.0913	102.691	429.45		1.0253	0.9753	57	1950.63	316.39	238.37
80	16.948		0.9161	1.0916	104.493	436.98		1.0256	0.9750	58	1984.85	321.94	242.56
90	17.241		0.9159	1.0918	106.294	444.52		1.0260	0.9747	59	2019.07	327.49	246.73
600	17.533		0.9157	1.0921	108.096	452.06		1.0263	0.9744	60	2053.29	333.04	250.92
10	17.825		0.9155	1.0923	109.898	459.59		1.0267	0.9740	61	2087.51	338.59	255.10
20	18.117		0.9153	1.0925	111.699	467.13		1.0270	0.9737	62	2121.73	344.14	259.28
30	18.410		0.9152	1.0927	113.501	474.66		1.0273	0.9734	63	2155.95	349.69	263.47
40	18.702		0.9150	1.0929	115.302	482.19		1.0276	0.9731	64	2190.18	355.24	267.65
50	18.994		0.9148	1.0931	117.104	489.73		1.0279	0.9729	65	2224.40	360.79	271.83
60	19.286		0.9146	1.0934	118.906	497.26		1.0282	0.9726	66	2258.62	366.34	276.01
70	19.578		0.9145	1.0935	120.707	504.80		1.0285	0.9723	67	2292.84	371.89	280.19
80	19.871		0.9144	1.0936	122.509	512.33		1.0288	0.9720	68	2327.06	377.44	284.38
90	20.163		0.9142	1.0938	124.310	519.87		1.0291	0.9717	69	2361.28	382.99	288.56
700	20.455		0.9140	1.0941	126.112	527.40		1.0293	0.9715	70	2395.51	388.55	292.74
10	20.747		0.9139	1.0942	127.914	534.93		1.0296	0.9713	71	2429.73	394.09	296.92
20	21.039		0.9137	1.0945	129.715	542.47		1.0298	0.9711	72	2463.95	399.65	301.10
30	21.332		0.9135	1.0947	131.517	550.00		1.0300	0.9709	73	2498.17	405.20	305.29
40	21.624		0.9134	1.0948	133.318	557.54		1.0302	0.9707	74	2532.39	410.75	309.47

Note (column 3): These values are obtained by dividing the values in column 2 by 0.9264 or multiplying them by 1.079 4 (g or 1/g for NaCl at an osmolality of 300 mmol or mosm).

Note (column 8): These values are obtained by dividing the values in column 6 by 1.0138 or multiplying them by 0.986 4 (g or 1/g for glucose at an osmolality of 300 mmol or mosm).

* Since the elementary composition of fructose is the same as that of glucose, columns 6, 7, 13 and 14 can also be used for fructose. Note that the osmotic coefficient g for fructose is not the same as that for glucose.

** Loewy's value for the calorific equivalent of carbohydrates (4.182 calories per gramme) has been used.

From Diem, K., Ed., *Documenta Geigy Scientific Tables*, 7th ed., Geigy Pharmaceuticals, Ardsley, N.Y., 1970, 273. With permission.

(b)The ideal osmolalities corresponding to given weights of NaCl and glucose (or fructose) are obtained from columns 11/12 and 11/13. The corresponding calorific values for glucose and fructose are given in column 14.

(c)The osmotic coefficient g (or 1/g) is obtained for NaCl from columns 1/4 (or 1/5) and for glucose from columns 1/9 (or 1/10), the values in column 1 in this case being read as real osmolalities.

Example 1. Required is the weight of NaCl necessary to yield a solution with a real osmolality of 500 mmol.

From Equation 1 the ideal osmolality = real osmolality/g = 500/0.9180 = 500 × 1.0893 = 544.65 mmol. The corresponding weight of NaCl from (a) above lies between 15.780 and 16.072 at ca.15.9 g. This is therefore the quantity of NaCl which must be dissolved in 1000 g water in order to yield an osmolality of 500 mmol.

The inconvenience of first calculating the ideal osmolality can be avoided by calculating the required weight direct from column 2 (for NaCl) or column 6 (for glucose). To do this, the weight given in column 2 obtained by entering the real osmolality in column 1 is multiplied by the corresponding value for 1/g. The above example, with a real osmolality of 500 mmol, thus gives the result: 14.611 × 1.0893 = 15.916 g NaCl.

Example 2. It is required to increase to 500 mmol the osmolality of a solution of 400 mmol by addition of NaCl. Here the calculation is simplified by assuming that the factor g does not change when the solution becomes a mixed solution. 1/g for NaCl for 500 mmol/1000 g water is 1.0893. Since 100 mmol are to be added by means of NaCl, the required weight (see example 1) is 2.922 × 1.0893 = 3.183 g NaCl.

(d)Isotonic solutions. The concentrations required to yield these can be calculated as in examples 1 and 2 or simply read off from the table in columns 1/3 (NaCl) and 1/8 (glucose).

Example 1. In order to obtain an isotonic NaCl or glucose solution (osmolality 300 mmol), 9.463 g NaCl or 53.312 g glucose must be dissolved in 1000 g water.

Example 2. It is required to render a solution of osmolality 200 mmol isotonic with serum by addition of NaCl, i.e., to increase the osmolality to 300 mmol. Since the additional osmolality is 100 mmol (column 1) the necessary weight of NaCl is 3.154 g (column 3). The figure for glucose can be calculated in a similar manner.

REFERENCES

1. Welt, L. G., in *Diseases of Metabolism,* 5th ed., Duncan, G. G., Ed., Saunders, Philadelphia, 1964, 449.
2. Netter, H., *Theoretische Biochemie,* Springer, Berlin, 1959, 108.

Table 143
WAVE LENGTHS FOR SPECTROSCOPE CALIBRATION

Source	Wave length	Source	Wave length
Potassium flame........	0.7699μ	E, solar..............	0.5270μ
Potassium flame........	0.7665	b₁, solar or magnesium	
Mercury I arc..........	0.6907	flame..............	0.5184
B, solar................	0.6869	b₂, solar or magnesium	
Lithium flame..........	0.6708	flame..............	0.5173
C, solar or hydrogen tube.	0.6563	Mercury I arc........	0.4960
Mercury I arc..........	0.6234	Mercury I arc.........	0.4916
D₁, solar or sodium flame.	0.5896	F, solar or hydrogen tube	0.4861
D₂, solar or sodium flame.	0.5890	Strontium flame.......	0.4608
Mercury I arc..........	0.5791	Mercury I arc.........	0.4358
Mercury I arc..........	0.5770	G, solar or hydrogen tube	0.4340
Mercury I arc..........	0.5461	Mercury I arc.........	0.4047
Thallium flame.........	0.5351	H₁, solar.............	0.3969
		K, solar..............	0.3934

From Weast, R. C., Ed., *CRC Handbook of Chemistry and Physics,* 61st ed., CRC Press, Boca Raton, Fla., 1980, E-216. With permission.

Table 144
CONVERSION OF SPECIFIC GRAVITY/ SERUM WATER/SOLUTES

Specific gravity	Water per kilogramme serum (g)	Water per litre serum (g)	Factor
1	2	3	4
1.015 ...	.... 952	.... 966	... 1.035
1.016 ...	.... 948	.... 964	... 1.038
1.017 ...	.... 945	.... 961	... 1.040
1.018 ...	.... 942	.... 959	... 1.043
1.019 ...	.... 939	.... 957	... 1.045
1.020 ...	.... 936	.... 954	... 1.048
1.021 ...	.... 932	.... 952	... 1.051
1.022 ...	.... 929	.... 949	... 1.053
1.023 ...	.... 926	.... 947	... 1.056
1.024 ...	.... 923	.... 945	... 1.059
1.025 ...	.... 919	.... 942	... 1.061
1.026 ...	.... 916	.... 940	... 1.064
1.027 ...	.... 913	.... 938	... 1.067
1.028 ...	.... 910	.... 935	... 1.069
1.029 ...	.... 906	.... 933	... 1.072
1.030 ...	.... 903	.... 930	... 1.075
1.031 ...	.... 900	.... 928	... 1.078
1.032 ...	.... 897	.... 925	... 1.081
1.033 ...	.... 894	.... 923	... 1.083
1.034 ...	.... 890	.... 921	... 1.086

Note: The table can be used to convert

(a) The concentration of a substance per litre serum into its concentration per kilogramme serum water: Multiply by the factor in column 4,

(b) The concentration of a substance per kilogramme serum water into its concentration per litre serum: Multiply by the figure in column 3 and divide by 1000.

From Diem, K. and Lentner, C., Eds., *Documenta Geigy Scientific Tables,* 7th ed., S. A. Basle, Switzerland, 1970, 557. With permission

TABLE OF CROSS-REFERENCES

This table is a cross-referenced index of the physical and chemical data, of interest to clinical chemists and other workers in the clinical laboratory, contained in the five commonly encountered handbooks listed at the end of the table. Arranged alphabetically by subject the table lists page numbers (in the specific editions of these handbooks given), where information on the desired subject can be obtained. Only first pages are tabulated, however, and in many instances an individual reference may lead to a lengthy and comprehensive listing.

Subject	A	B	C	D	E
Abbreviations and symbols, physics, and chemistry	F259	2-5	6	199	xxi
		2-49			
	F279	4-10	104		
Absorption data of biochemical substances					81
Absorption data, infrared	F224	8-48			
Activity coefficients of acids, bases, and salts	D132				
Alphabet table, Greek	F278	2-53			
Air contaminants, limits for human exposure	D87				
Atomic weights of the elements, international	B1	3-2		250	
				740	
Azeotropes-binary and ternary systems	D1	10-63			
Barometer, correction to latitude	E44	2-70			
Barometer, reduction to sea level	E42	2-73			
Barometric readings, conversion table	E38				
Baths, low temperature liquid	D191				
Boiling point					
Correction to standard pressure	D155	10-55			
Index of organic compounds	C652				
Molecular elevation	D154	10-73			
Of azeotropes-binary and ternary systems	D1	10-63			
Of the elements	B1	3-2		252	
	B4				
	D153				
Of inorganic compounds	B63	4-14			
Of organic compounds	C75	7-56			
Of organometallic compounds	C680				
Of water	D149	10-54		297	
Bond lengths and angles of elements and compounds	F200				
Buffer solutions					
Common	D113	5-77	358	314	22
			362		
			529		
Definition of pH	D112	5-69	353		19
		5-82			
Standard aqueous, properties at 25°C	D112				
Standard, values of pH at 0—95°C	D114	5-74	354	315	
			360		
Calibration of volumetric glassware with water or mercury	D109	2-80			
Composition of body fluids, foods, and tissues				500	640
					1080
					1110
Concentrative properties of aqueous solutions, conversion tables	D194				
Conversion factors table	F223	2-11		203	
	F282				
Conversion formulas for solutions expressed in various ways	D119				

TABLE OF CROSS-REFERENCES (continued)

Subject	A	B	C	D	E
Corning color filters, transmission	E230				80
Critical data, sources	F323				
Deci-normal solutions of oxidation and reduction reagents	D104		352	313	
Deci-normal solutions of salts and other reagents	D102		352	313	
Definitions and formulas	F81	2-2	93	199	
		2-7			
Density					
Gradients		10-97	426	320	
Of air	F9	10-92		245	
Of elements	B235	3-2		252	
	B240				
Of inorganic compounds	B63	4-14			
Of mercury	F6	10-91		259	
Of organic compounds	C75	7-56			
Of organometallic compounds	C680				
Of various liquids, solutions, and biological materials	F3	10-78	404	245	918
	F4	10-91	411	317	1070
	F7		429		
	F8		520		
			540		
Of water	F4	10-91			
Specific gravity of aqueous sucrose solutions	D242	10-97	414		
Units and hydrometers, hydrometer conversion tables	F3	10-86		319	
Dielectric constants	E55	5-45	522		
Dissociation constants					
Of aqueous ammonia from 0—50°C	D130				
Of inorganic acids in aqueous solutions	D130	5-14	307		
Of inorganic bases in aqueous solutions	D128				
Of organic acids in aqueous solutions	D129	5-17	309		43
			315		45
			347		
Of organic bases in aqueous solutions	D126		322		43
Drying agents, efficiency	E41	10-85			
Dyestuff intermediates, trade names	C754				
Electrochemical series	D120	6-2	123		85
Electron affinities	E67	3-10			
Electronic configuration of the elements	B2	3-2		267	
The elements, a description	B4				
Equivalent conductance of some electrolytes in aqueous solutions	D132	6-38			
Equivalent conductance of the separate ions	D132				
Fats and oils	D192	7-428			
Flame and bead tests	D93				
Formula index of organic compounds	C609	7-27			
Formulas and definitions	F81	2-2	93	199	
		2-7			
Formulas for calculating titration date	D119				
Formulas of organic compounds, structural	C543		153		
Freezing point depression, molecular	D154	10-80		325	
Gamma energies and intensities of radionuclides	B333	3-16		271	4
Half wave potentials		6-23			
Hydrometers and density units, hydrometer conversion tables	F3	10-86		319	
Illuminants, efficacies of	E203				
Illumination conversion factors	207	2-34			
Index of refraction					
Of aqueous solutions	E223		387		

TABLE OF CROSS-REFERENCES (continued)

Subject	A	B	C	D	E
Of elements	E218	10-94			
Of inorganic compounds	E218	4-14			
Of organic compounds	E220	7-432	386		
		10-103	388		
			395		
			404		
			408		
			540		
Of water	E222	10-99	386		
Indicators					
Acid base	D115	5-86	357	316	53
			378		
Fluorescent	D117	5-91	383		
Oxidation-Reduction		6-20	383		
Inflammability of gases and vapors, limits	D85	11-16			
Infrared absorption data	F224	8-48			
Inorganic acids and bases, composition	D131	11-25	384		
Ionic exchange resins, cation and anion	C775				
Ionic radii of elements	F198	3-120			
Ionization constants					
Of acids in water at various temperatures	D131	5-42			
Of amino acids	C741		318		
Of water (K$_w$)	D131	5-7			
Ionization potentials of the elements	E68	3-7			
Ionization potentials of molecules	E74	3-9			
Isotopes, table	B248	3-16		277	4
Laboratory reagents and solutions, preparation	D94	11-22			11
Luminance of various light sources, approximate	E204				
Magnetic rotatory power	E248				
Melting point and boiling points of the elements	B1	3-2		252	
	B4				
	D153				
Freezing point depression, molecular	D154	10-80		325	
Index of organic compounds	C639	7-48			
Of inorganic compounds	B63	4-14			
Of organic compounds	C75	7-56			
Of organometallic compounds	C680				
Miscibility of organic solvent pairs	C720				
Nomenclature					
Of biochemistry			8		
Of carbohydrates				336	180
Of inorganic compounds	B39	4-2			
Of organic compounds	C1	7-2	20		172
Prefix names of organic radicals	C46	7-18			175
Nuclear magnetic resonance data	E69	8-41			
	F243				
Optical density to transparency conversion table	E250	2-91			
Organic ring compounds	C50	7-9			
Osmotic parameters and electrical conductivities of aqueous solutions table	D237				
Oxidation-reduction potentials of reactions					93
Parameters, general physical and chemical					
Of alkaloids		7-394	326		1018
Of amino acids	C741		111	354	
Of carbohydrates	C745	7-410	201	336	190
			317		

TABLE OF CROSS-REFERENCES (continued)

Subject	A	B	C	D	E
Of enzymes				396	205
Of hormones	C723	7-418	348	473	
Of inorganic compounds	B63	4-14			
Of lipids	C723	7-418	348	385	
		7-428		486	
Of nucleotides			330	361	199
Of organic compounds	C75	7-56			
Of organometallic compounds	C680				
Of porphyrins			131	376	183
			139		616
			348		
Of vitamins		7-446	63	449	
			348		
Periodic table of the elements	B3	inside cover		249	
Persistent lines of the elements	E211	8-18		268	10
pH values of acids, bases, or biological materials, approximate	D114		305	540	
				571	
				594	
Photometric quantities	E207	2-8			
Units and standards and radiometric quantities, standard units, symbols, and defining equations	E203 E206				
Physical constants, fundamental	F222	2-3		245	
Plastics, properties	C765	7-453			
Plastics, trade names, composition and manufacturers	C757				
Publications, national standard reference data system	F335				
Radionuclides, gamma energies and intensities	B333	3-16		271	4
Radionuclides, permissible intakes	B406			272	8
Reagents, organic analytical	D105				
Solubility					
Of alkaloids		7-394			
Of amino acids	C743		107	355	
			115		
Of gases		10-3			
Of hormones		7-418		355	
Of inorganic compounds	B63 D110	4-14			
Of organic compounds	C75	7-54			
		7-56			
		10-7			
Of resins		7-436			
Miscibility of organic solvent pairs	C720				
Solubility products	B232	5-7			
Specific gravity of aqueous sucrose solutions	D242	10-97	414		
Specific rotations	C742	7-394	385	355	
	E247	7-421	393		
Spectra index of organic compounds	C659				
Standard wavelengths	E208				
Structural formulas of organic compounds	C543		153		
Temperature conversion table	F122	2-54		211	
Temperature correction					
Brass scale	E39	2-68		295	
Emergent stem correction for liquid-in-glass thermometers	D158	11-15			
For glass volumetric apparatus	F3				
For volumetric solutions	F2				
Glass scale, metric units	E41	2-67		295	

TABLE OF CROSS-REFERENCES (continued)

Subject	A	B	C	D	E
Transmission of corning color filters	E230				80
Transmission of wratten filters	E237				
Transparency to optical density conversion table	E250	2-91			
Vapor pressure					
Of carbon dioxide	D161	10-30			
Of the elements and inorganic compounds	D162	10-22			
	D190	10-29			
Lowering by salts in aqueous solutions	E1				
Of organic compounds	D170	10-37			
Of water at various temperatures	D158	10-24		296	
Viscosity					
Conversion tables	F47	2-76			
Of liquids, solutions, and biological materials	F50	10-97	405	548	
		10-100	415		
		10-103	425		
			521		
Of water at various temperatures	F47	10-99			
	F49				
Wavelengths of various radiations	E205				
Wavenumber-wavelength conversion		2-94			
Waxes	C753	7-428			
Weighings, reductions in air to vacuo	D111	2-77			
Wratten filters, transmission	E237				
X-ray wavelengths	E143	8-3			

REFERENCES

A. **Weast, R. C., Ed.,** *CRC Handbook of Chemistry and Physics,* 55th ed., CRC Press, Boca Raton, Fla., 1974.

B. **Dean, J. A., Ed.,** *Lange's Handbook of Chemistry,* 12th ed., McGraw-Hill, New York, N.Y., 1979.

C. **Diem, K., Ed.,** *Documenta Geigy Scientific Tables,* 6th ed., Geigy Pharmaceuticals, Ardsley, N.Y., 1962.

D. **Fasman, G. D., Ed.,** *CRC Handbook of Biochemistry and Molecular Biology,* 3rd ed., Vol. 1, CRC Press, Boca Raton, Fla., 1975.

E. **Long, C., Ed.,** *Biochemists' Handbook,* D. Van Nostrand Co., Princeton, N.J., 1961.

Reagents, Specimen Collection, Calibrators,
Reference Material and Standards

INTRODUCTION

Pierre W. Keitges and Robert J. Mohrbacher (Editors)
with the assistance of the following contributors:
Kaiser J. Aziz and Robert S. Melville

This area brings fundamental components of clinical chemistry together. As they change, replacement of old technology with new or improved must be ensured to maintain quality at the state of the art. Accordingly, some classical procedures are treated more briefly than newer materials and methods, which now allow comprehensive process control, as well as the formulation and the attainment of well-defined performance standards.

Advances of generally used materials are illustrated in many ways. Disposable glass and plastic containers avoid the contamination problems encountered with washed containers. Mechanical dispensers and dilutors have largely replaced manual pipettes and ensure precision and microcapability. Separation aids have been developed which remove most particulate matter from water, and simple reliable deionized water systems have eliminated the need to purify water through distillation. Similar separation aids can remove particulate matter from reagent solutions. Other developments in reagent manufacture have made commercial chemicals available, which meet or exceed the standards of the American Chemical Society, and obviate the need to prepare and purify chemicals in the clinical laboratory. Similarly, assay mixtures prepared in-house are being replaced by commercial products. This is especially true for enzyme assays where some constituents tend to be unstable and have been difficult to prepare in pure form. Typically, the enzyme kits offer the convenience of "single vial reagents", but the use of such products may be confined to a single manufacturer's instrument. Finally, blood collection devices have been developed which assist in the separation of plasma or serum from blood cells. These containers also eliminate the need for "pour-over" tubes and, so, enhance positive sample identification.

Other advances have had a more focused effect. In most chemical reactions, pH is important, but some variation in hydrogen ion concentration can usually be tolerated. Only enzyme analysis crucially requires the capability to verify pH value as precisely as has become possible. Water removal from reagents generally is not required in the clinical laboratory, since the stability of commercial reagents is ensured by lyophilization or by the addition of preservatives. On the other hand, removal of water from the carrier gases used in column chromatography is a necessity. Solvent extraction is avoided by most of today's methodology, but continues to be utilized in the analysis of urine and in the assay of drugs in blood.

Indissolubly linked with technical progress in the clinical laboratory is the emergence of formal standards and performance requirements. To comply with the latter, the typical clinical chemistry laboratory today performs about a third of its tests as quality control or standard assays. Since improved technology inherently provides more reliable methods, the extent and the cost of this effort merits continuous and critical evaluation. Such an analysis can be approached from two major points of view. Some believe that governmental regulation calling for ideal performance is desirable; others maintain that "consensus laboratory standards" which take practical limitations into account are more likely to enhance quality.

REFERENCES

1. Gordon, A. J. and Ford, R. A., *The Chemist's Companion: A Handbook of Practical Data, Techniques and References,* John Wiley & Sons, New York, 1972.
2. Sober, H. A., *Handbook of Biochemistry,* 2nd ed., CRC Press, Cleveland, 1970.
3. Weast, R. D., *Handbook of Chemistry and Physics,* 61st Ed., CRC Press, Boca Raton, Fla., 1980.
4. Werner, M., *Microtechniques for the Clinical Laboratory: Concepts and Applications,* John Wiley & Sons, New York, 1976.

REAGENTS, SPECIMEN COLLECTION, CALIBRATORS, REFERENCE MATERIALS AND STANDARDS

Pierre W. Keitges and Robert J. Mohrbacher

GLASSWARE

Types of Glassware
There are five common types of glassware which may be found in the clinical laboratory.[1]

1. *High thermal resistant glass* is a borosilicate glass free of magnesia-lime-zinc group elements and heavy metals. It resists heat, corrosion, and thermal shock. Pyrex®, Kimax®, Corex®, and Vycor®-brand glass are the most common borosilicate glassware found in the laboratory. Vycor® is the most rugged.
2. *High silica glass* is similar to fused quartz in its physical and chemical properties.
3. *Boron-free glassware* was developed for use with strong alkaline solutions. It has a low thermal resistance.
4. *Low actinic glassware* is used in the handling of materials sensitive to light (bilirubin, carotene, or Vitamin A).
5. *Standard flint glass* is a soda-lime glass. This type of glass should be used with caution in the clinical laboratory since it may interfere with the assay.

Pipettes
There are two types of pipettes routinely used in clinical chemistry:

1. **Volumetric or transfer pipettes** — These must meet National Bureau of Standards (NBS) Class A specifications and are designed to deliver a fixed volume of liquid without blowing out what remains in the tip. They are used for measurements of low viscosity liquids and have limits of error from 0.8 to 1.2% (Table 1), depending on the pipette volume. To calibrate, the weighing of deionized water is acceptable. A code of colored rings for easy identification of different sizes has been developed.

2. **Graduated or serological pipettes** — This is a graduated piece of glass tubing. It must be blown out to deliver the entire volume. These pipettes are used where limits of error from 0.6 to 5% (Table 2) are acceptable. To calibrate, the weighing of deionized water is acceptable.

Volumetric Flasks
Volumetric flasks are found in the following sizes: 1, 2, 5, 10, 25, 20, 100, 200, 250, 500, 1000, 2000, and 4000 ml. They are primarily used in preparing solutions of known concentration and must meet or exceed NBS Class A specifications. Limits of error range from 0.2% for a 10-ml flask to 0.025% for a 2l-flask.

Glassware Cleaning Solutions
Clean glassware is essential in the clinical laboratory. The National Bureau of Standards prefers a chromic-sulfuric acid mixture. These mixtures are commercially available as Dichlean Acid Dichromate, Curtin-Matheson Scientific; Dichrol, Scientific Products; Chromic Sulfuric Acid, Fisher Scientific Company, and others. Manufacturers' instructions must be strictly adhered to, to ensure proper cleaning.

Table 1
VOLUMETRIC PIPETTES

Capacity (ml)	Accuracy (ml)	Minimum flow time (sec)	Color code rings
0.5	± 0.006	10	Two black
1	± 0.006	10	Blue
2	± 0.006	10	Orange
3	± 0.01	15	Black
4	± 0.01	15	Two red
5	± 0.01	15	White
10	± 0.02	15	Red
15	± 0.03	25	Green
20	± 0.03	25	Yellow
25	± 0.03	25	Blue
50	± 0.05	30	Red
100	± 0.08	40	Yellow

From **Anon.**, Sales Catalog, Fisher Scientific, Pittsburgh, 1979, 938. With permisson.

Table 2
SEROLOGICAL PIPETTES

Capacity (ml)	Accuracy (ml)	Minimum flow time (sec)	Color code rings
1/10 in 1/100	± 0.0025	2	White
2/10 in 1/1000	± 0.004	—	Blue
2/10 in 1/100	± 0.004	2	Black
2/10 in 1/100	± 0.005	2	Two yellow
2/10 in 1/20	± 0.005	2	Two black
1 in 1/100	± 0.01	3	Yellow
1 in 1/10	± 0.01	3	Red
2 in 1/10	± 0.015	3	Green
2 in 1/100	± 0.015	3	Two white
5 in 1/10	± 0.02	5	Blue
10 in 1/10	± 0.03	8	Orange
25 in 1/10	± 0.10	15	White

From Anon., Sales Catalog, Fisher Scientific, Pittsburgh, 1979, 939. With permission.

PLASTICWARE

The use of plasticware in the laboratory must be carefully assessed, since "plasticizers" are known to cause interference in gas-liquid chromatography and in some enzyme reactions. Also, various plastics are readily soluble in many organic solvents. Basically, the use of plasticware should be discouraged, and if used, it must be verified that it will not create problems. Refer to the appropriate tables in the area on Physical and Chemical Data.

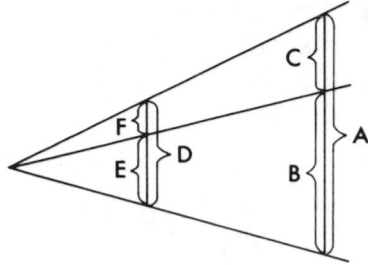

$$E_d \ (\%) \ = \ \frac{C}{A} \times 100$$

$$E_a \ (\%) \ = \ \frac{F}{B + D} \times 100$$

$$= \ \frac{D}{B + D} \times 100 \ \left(\frac{C}{A} \times 100 \right)$$

FIGURE 1.   Influence of pipetting errors E_d on the analytical error E_a. Line A represents the correct sample volume. If a delivery error C occurs, volume B is aliquoted. Line D represents the reagent volume required by the method. However, with a sample volume B, correct proportions of sample and reagent would require an amount E of reagent instead. The same analysis is applicable when there is an excess of sample and, therefore, an insufficient volume of reagent. For a discussion of different ratios of sample and reagent volumes A/D, see text.

MICRODISPENSERS AND DILUTORS[4]

Strategy

In assaying concentrations, only the ratio of sample volume to reagent volume in the final reaction mixture must be correct; absolute volumes are immaterial. Figure 1 shows three rays intersecting two parallel vertical lines representing two liquid volumes which are to be mixed, the sample volume A and the reagent volume D. If a sampling error C occurs, such that only volume B instead of A is delivered, the relative delivery error E_d is

$$E_d \ (\%) \ = \frac{C}{A} \times 100 \tag{1}$$

The geometric rule of similarity shows that, if the reagent volume D were correctly dispensed, a relative excess of reagent F would be added to the reaction mixture, causing a relative analytical error E_a:

$$E_a \ (\%) \ = \frac{F}{B + D} \times 100 \tag{2}$$

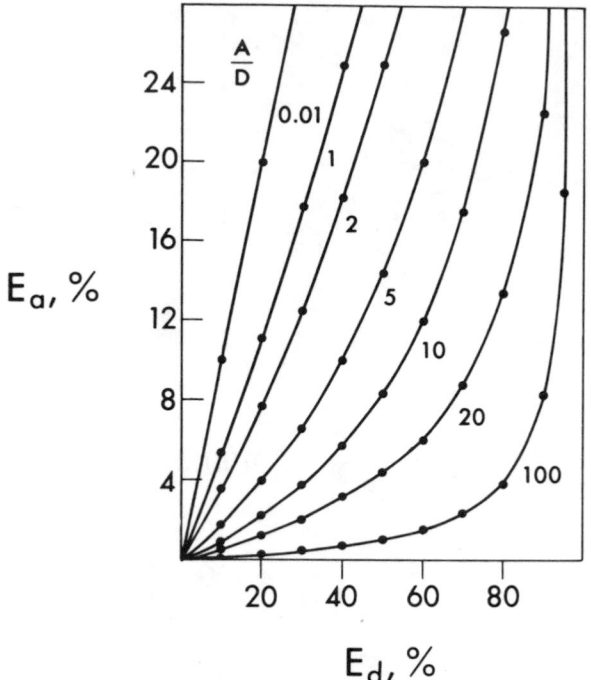

FIGURE 2.   Influence of pipetting errors E_d on the analytical error E_a. Both errors are given in percent of the correct amount. Curves represent different ratios of sample volume A and reagent volume D.

but since

$$F = \frac{DC}{A} \tag{3}$$

by substituting (4) into (3) we obtain:

$$E_a\ (\%) = \frac{D}{B + D}\left(\frac{C}{A} \times 100\right) \tag{4}$$

As the quantity in brackets in (4) is the delivery error E_d shown in (1), the analytical error E_a is equal to the delivery error multiplied by the reagent volume D divided by the reaction volume B + D. Thus, the delivery error appears only as a multiplicative quantity in the analytical error and, if the reagent volume D is small with respect to the delivered sample volume B, the multiplier D/(B + D) is a small fractional number, such that only a small fraction of the delivery error appears as analytical error. However, if the sample volume is smaller than the reagent volume, the analytical error fully reflects any delivery error.

Figure 2 plots delivery errors, in percent of the correct amount, against the resulting analytical error, in percent of the correct result. Different curves represent different ratios A/D of sample volume and reagent volume. When this ratio increases, the effect of delivery errors on the analytical error decreases.

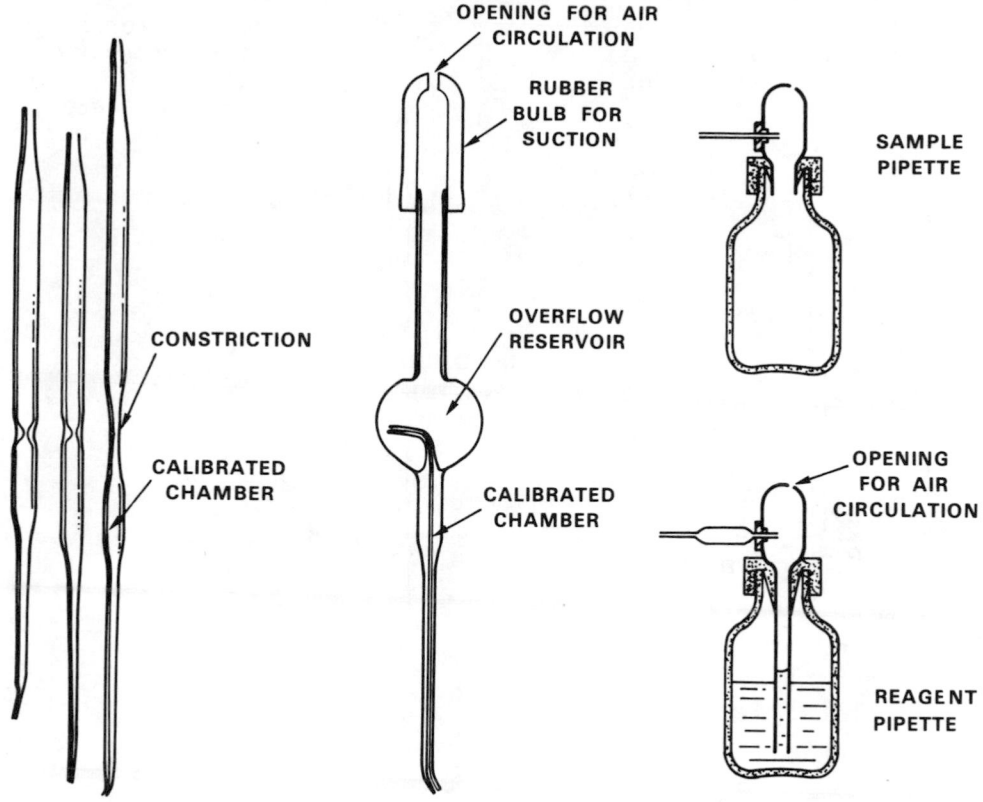

FIGURE 3. Self adjusting constriction pipettes and overflow pipettes. Left — Lang-Levy constriction pipettes. Center — Grunbaum overflow pipette. Right — Sanz overflow pipettes, a sample pipette is shown at the top and a reagent pipette at the bottom.

Capillary Containers

Self-adjusting constriction pipettes of the Lang-Levy type have probably been used longer than any other microdispensers (Figure 3). Constriction pipettes of glass or of plastic may be constructed for set volumes as small as a few nanoliters. However, commercial constriction pipettes available from many sources are more typically in the range of 1 to 250 μl. Depending on their size, the pipettes are filled by capillary action or suction with a mouthpiece, or by a rubber bulb or a syringe assembly. Surface tension, which holds the liquid level at the constriction, ensures reproducible filling. The sample is expelled from the pipette by positive pressure. Cleaning between deliveries of different samples is necessary, as in traditional pipettes. Because liquid adheres to the walls of the pipette after delivery of the sample, most glass micropipettes are calibrated "to contain" rather than "to deliver". The difference between these two values is exaggerated on the microscale, because the surface/volume ratio is high. Further, the liquid film remaining in the pipette after delivery may vary with different sample viscosities, particularly with serum. Therefore, these pipettes should be used with washout techniques.

Overflow pipettes apply the same basic principle as self-adjusting constriction pipettes (Figure 3): surface tension holds the liquid sample in a calibrated space. However, a mechanism to create positive or negative pressure for filling and emptying, and a container for the overflow of excess sample or for the storage of reagent liquid, are

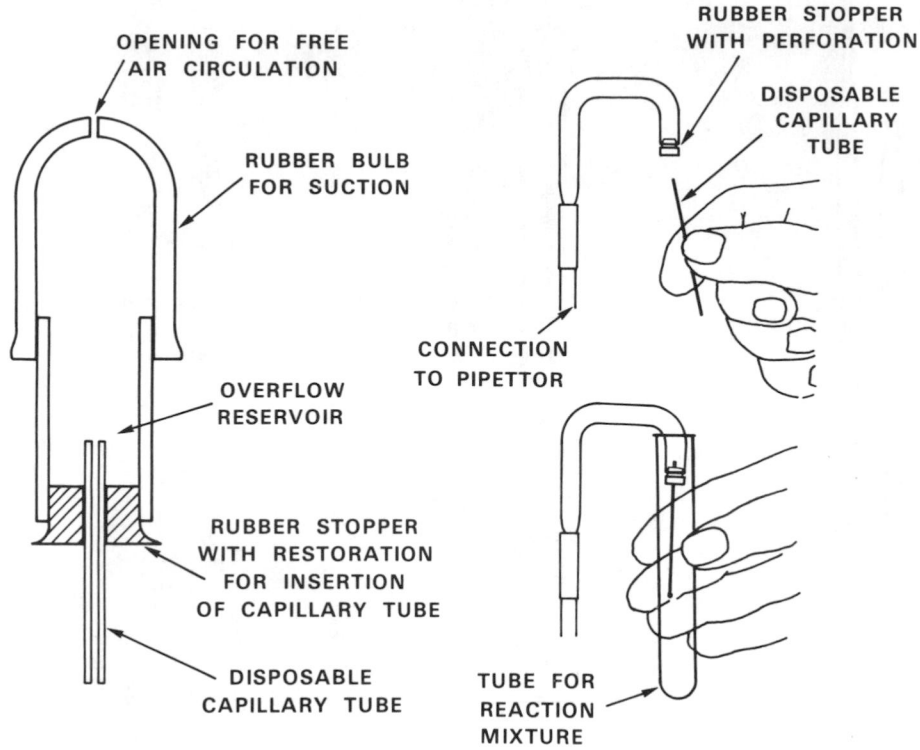

FIGURE 4.   Use of calibrated glass capillary tubes as aliquoting devices. Left — disposable capillary and rubber bulb assembly for filling, emptying, and rinsing tube. Right — disposable capillary used with a mechanical pipettor. (From Martens, E. H., *Am. J. Clin. Pathol.,* 55, 779, 1971. With permission.)

integral parts of the design. Overflow pipettes can be self-rinsing when an excess of sample is drawn through them, or can be used as constant-volume reagent dispensers.

The Grunbaum® pipette is a glass dispenser which can be used for aliquoting samples or to deliver a reagent placed in the glass bulb that is part of its construction. The smaller pipette sizes (under 25 $\mu\ell$) are self-filling; for the larger sizes, an attached rubber bulb provides both suction and pressure. The Sanz pipette consists of a plastic bottle and a screw cap which comprises the calibrated pipetting device. This is either a nonwettable plastic capillary tube for aliquoting samples or a larger plastic nozzle to deliver reagents. The seal between bottle and cap is airtight. A vent hole in the cap establishes equal pressure inside and outside the device when open and can be closed with a fingertip. It is possible to generate pressure or suction in the pipetting device by squeezing or releasing the flexible bottle.

Calibrated capillary tubes in the range 0.5 to over 100 $\mu\ell$ capacity, and color-coded according to size, are commercially available as aliquoting devices. For the sampling of hematological specimens, capillary tubes treated with dipotassium EDTA and heparin have been developed. Not only the calibration of the tubes, but also their uniformity, depends on the manufacturing process. Today, volumetric tolerance as precise as 0.25% is available for capillary tubes in the range of 1 to 10 $\mu\ell$.

The tubes may be filled by capillary action and emptied by gravity flow. Alternatively, the sample can be aspirated into and expelled from the tube by creating positive or negative pressure with a small rubber bulb assembly (Figure 4) or a calibrated syringe (Figure 5).

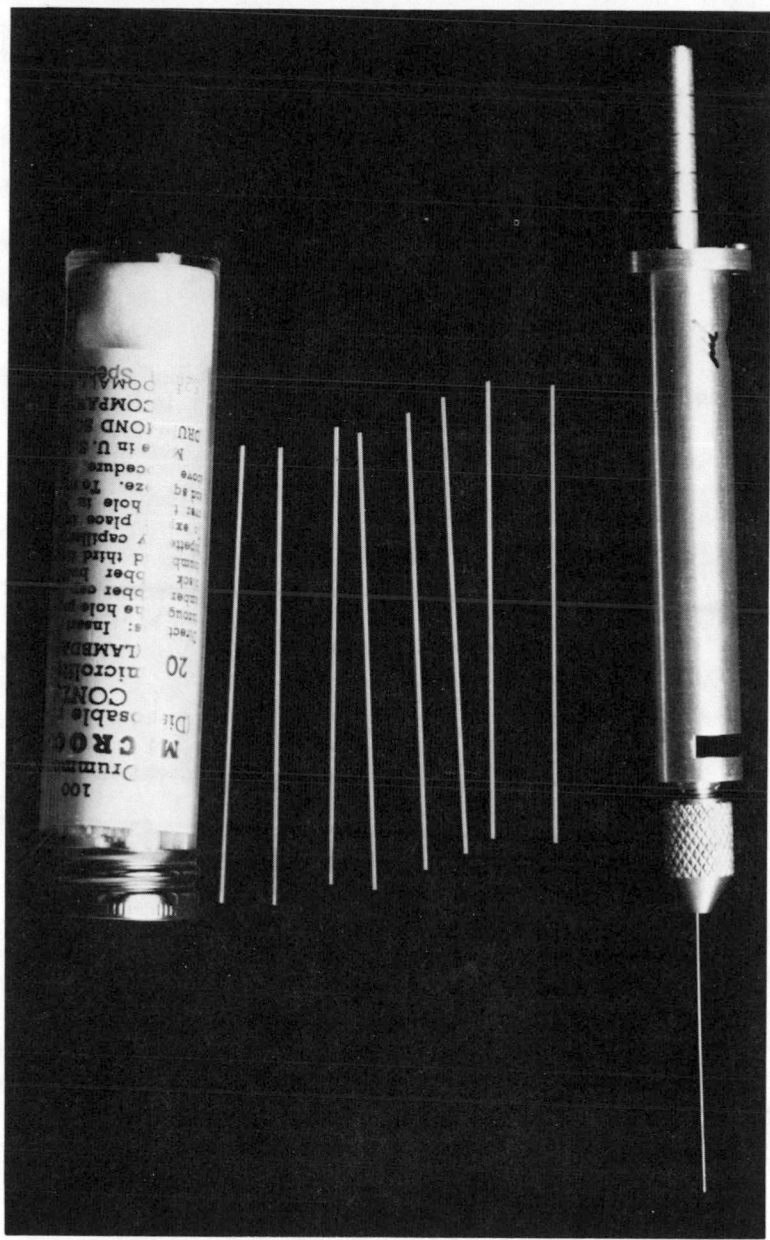

FIGURE 5.   Capillary tubes used with a calibrated syringe.

Air-Displacement Piston Pipettes

In the range of 1 to 1000 μl capacity, these pipettes are of recent development.
Usually, pipettes with a fixed delivery volume are used, but pipettes that are continuously adjustable by a digital micrometer are also available. While such adjustment may simplify dilutions and calculations, it may also introduce a variable difficult to control in a busy laboratory. The typical tolerance of piston pipettes is $\pm$ 1% for aqueous solutions.

Aliquoting is controlled by a push button which moves a piston in the pipette barrel, and the sample is picked up in a disposable, nonwettable, polypropylene plastic tip so that the specimen never comes in contact with the pipette mechanism (Figure 6). This eliminates cross-contamination between specimens. Other problems related to reproducibility may be introduced if the plastic tips do not consistently fit properly and air leaks occur, or if the orifice of the tip is not patent. With serum samples, the liquid film remaining on the inside wall of the pipette tip can be as high as 3% of the dispensed volume, if the pipette is filled and emptied only once. By filling the tip twice in succession with the same specimen and using the second fill as the sample for analysis, a dispensing accuracy in excess of 99% may be achieved.

Microsyringes Equipped with Aliquoting Adaptors

These provide adjustable "stops" in any number of positions (Figure 7). The stops are used to bring the syringe piston to a calibrated position. In this way, one or more reagents and the sample are aspirated in succession and then expelled together into the reactor vessel. As the sample is flushed out with reagent, this dilutor is self rinsing, but the need for adequate flushing confines the volume ratios of sample and reagents to certain limits. On the one hand, a reagent volume at least four times as large as the sample volume is required to provide adequate flushing and avoid carryover from one sample to the next. On the other hand, it is generally not practicable to achieve more than a 20-fold dilution of the sample with reagent, since the same syringe is used to aspirate both. If larger dilutions are desired, two consecutive dilution steps must be performed. It is also possible to use microsyringes as dispensers, rather than as dilutors, by fitting the metal tip with a disposable Teflon® delivery tip (Teflon® standard-wall spaghetti tubing size 28 SW, minimal I.D. 0.013 in., maximal I.D. 0.017 in.).

Two-Barrel Dilutors

These are extremely flexible, since they contain separate syringes for sample and reagent. In commercial dilutors, this type of design has largely become the standard. Compared to single-barrel devices, the two-barrel design permits the mixing of microliter samples with milliliter volumes of diluent or reagent. Compared to calibrated pipette tip burettes, the two-barrel design permits continuous, independent adjustment of both sample and reagent volumes. However, because of the increased number of movable parts, ruggedness in operation always comes at a price in two-barrel dilutors. Moreover, as in all other dilutors, cross-contamination between successive samples occurs unless an adequate amount of reagent is used to flush out the sample.

Manually operated two-barrel microdilutors and a large number of mechanized instruments are commercially available. Mechanized dilutors typically have a volumetric accuracy of 0.5 to 2.0%, while the precision of delivery may vary as much as 0.1 to 1.0% among different makes. As a rule, repeatability is a weightier consideration than absolute accuracy in selecting a dilutor. The required precision is determined by the type of assay. Given the wide biological variability of serum enzymes or lipids even in health, most commercial dilutors appear to be sufficiently reliable for their assay. However, only the best currently available dilutors provide satisfactory reliability for serum sodium or chloride assays.

Other considerations in selecting a microdilutor are self-priming properties, self-rinsing of the entire delivery system and absence of "dead corners", cycle speed or speed of delivery rapidity of changeover from one method to another, overall dimensions of the instrument, and the availability of foot-pedal control. However, the overriding consideration remains ruggedness with prolonged use.

Mechanized Microdispensers without Pump Bodies

One mechanized dispenser that offers push-button selection of delivered volumes

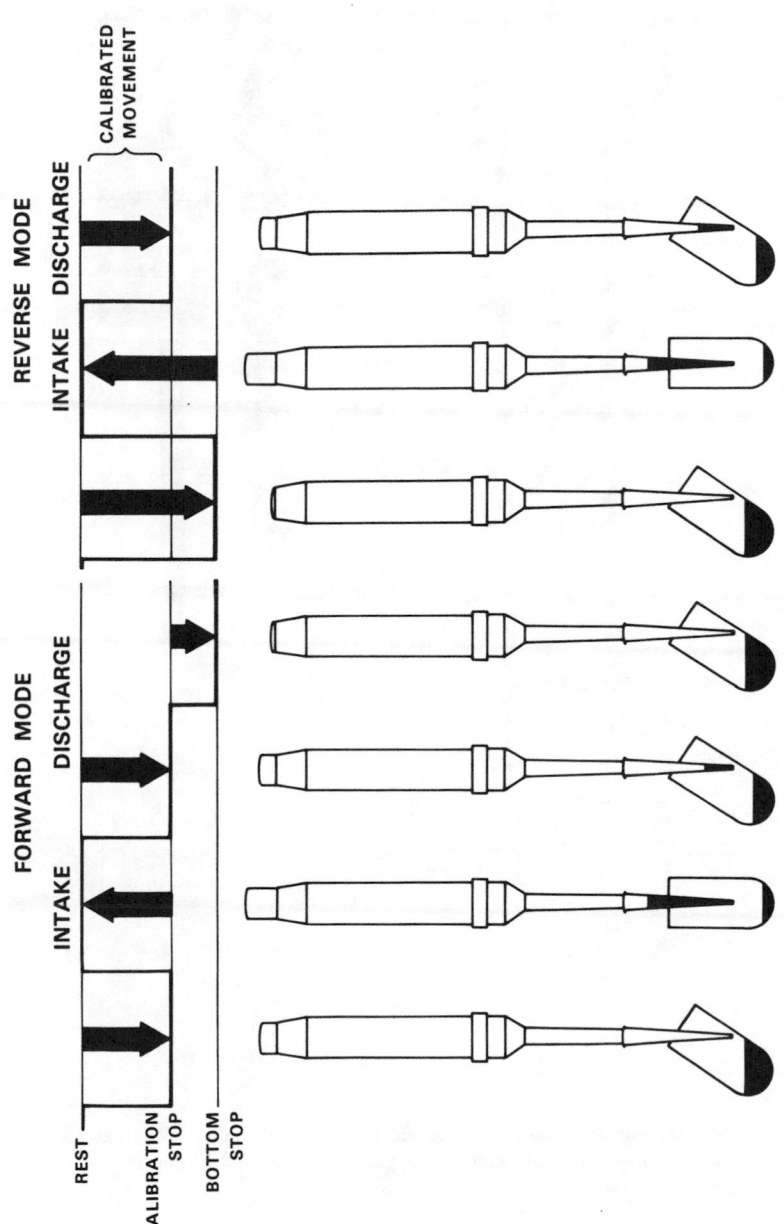

FIGURE 6. Forward mode and reverse mode of operation for air-displacement piston pipettes. At rest, the plunger is in the high position. When the plunger is depressed, an upper positive stop, the calibration stop, and a bottom stop are encountered. In the forward mode, the tip is filled by moving the plunger between the calibration stop and the rest position. The sample is discharged in two successive motions. First, the plunger is depressed to the calibration stop, discharging almost the entire sample. Second, with the tip still inside the recipient vessel, the plunger is pressed down to the bottom stop. In the reverse mode, the tip is filled by moving the plunger to the calibration stop. This discharges the calibrated volume, while leaving excess liquid in the tip.

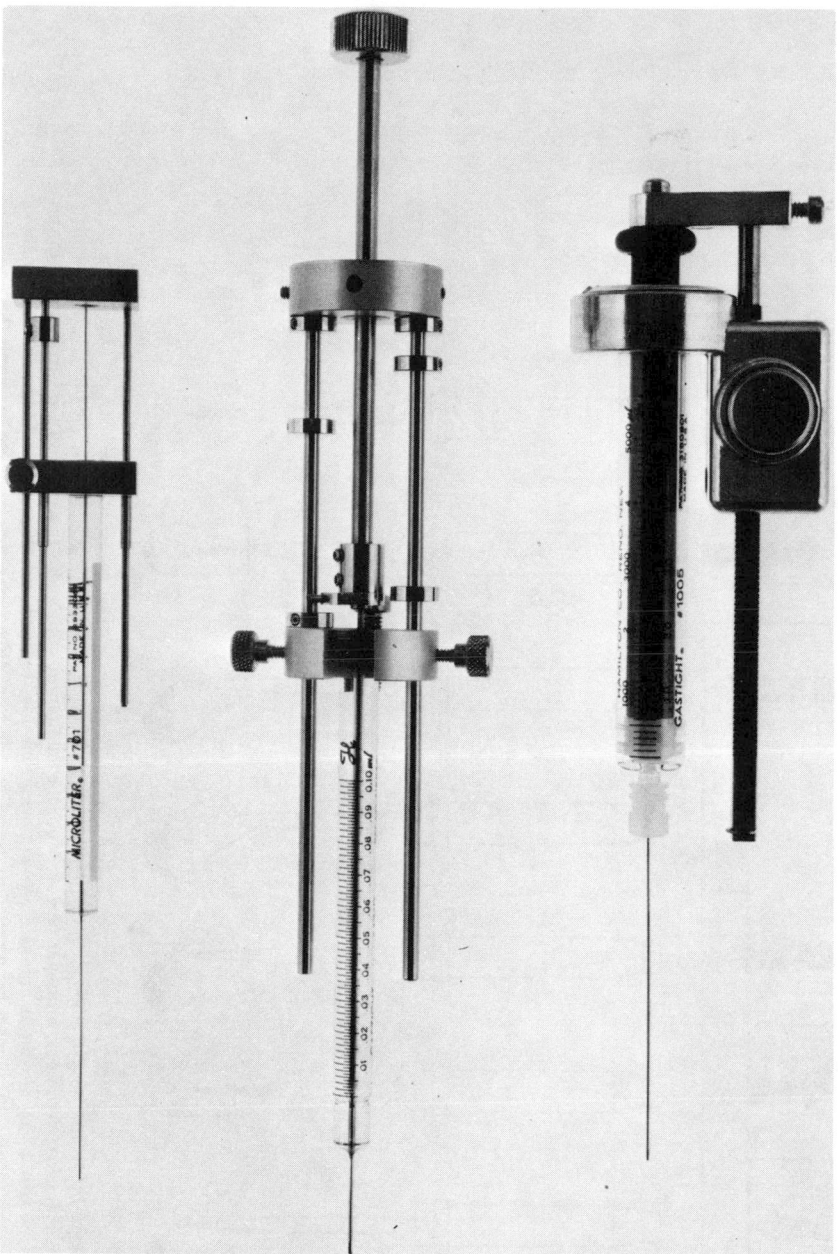

FIGURE 7.   Microsyringes with aliquoting adaptors. Left — Chaney adaptor with single stop. Middle — Siggard-Anderson adaptor with multiple stops. Right — Push-button repeating dispenser. (Courtesy of the Hamilton Company, Reno, Nev.)

between 5 and 100 mℓ contains no pump barrels or seals (Compu-Pet 100). Another mechanized dispenser, capable of delivering quantities between 1 nℓ and 1 mℓ with 1% accuracy, operates by generating identical drops which are then counted electronically. There are models for different sample volumes, with maximum aliquot (which can be delivered in less than 1 sec) 1000 times larger than the minimum aliquot in all models (Princeton Fluidics Corporation).

CHEMICALS

Laboratory chemicals are supplied in several grades. The most common designations follow.[5]

Reagent grade or analytical reagent grade (AR) — The specifications of this reagent group have been prepared by the Committee on Analytical Reagents of the American Chemical Society (ACS). Manufacturers of ACS chemicals check each lot in a control laboratory and place ACS labels on those lots that meet the Society's published specifications. This grade is suitable for the clinical laboratory.

Chemically pure grade (CP) — "Highest purity" and "chemically pure" are equivalent terms. This designation fails to reveal impurities tolerated and manufacturers' use is not uniform. Chemicals in this category should be analyzed before use.

USP and NF grade — These chemicals meet specifications set down in the U.S. Pharmacopeia (USP) or the National Formulary (NF). Chemical purity is not specified and these chemicals should be restricted to pharmaceutical use.

Purified, practical, or pure grade — These chemicals can be used for chemical synthesis. These chemicals should not be used in clinical analysis.

Technical or commercial grade — These chemicals are used in the manufacturing of purer forms.

Purification of Organic Reagents

The purity of organic reagents is usually inferior to that of inorganic reagents.[5] The presence of impurities may be a source of analytical error. Impure reagents may result in high background absorbance masking the actual analyte. Isomers may be a source of difficulty in that a formed complex may have a different absorbance resulting in a depressed signal. The organic reagent must be purified if any of these problems exist. The reader is referred to Reference 6.

DISTILLED AND DEIONIZED WATER

Distilled or deionized water is necessary for the preparation of all reagents and solutions in the laboratory.[7] Deionized water has been purified by a process of ion exchange. The quality of deionized water depends on the water supply and the type of ion-exchange resins utilized. A conductivity light is installed between two of the ion-exchange tanks. The light goes out when the resistance of the effluent drops below 200,000 Ω/cm-resistance. A second tank "polishes" the water to give a resistance of 15 to 18 MΩ/cm. To remove organics, an activated carbon tank is used. This system can be expected to give water that has a total solid content of less than 1 ppm and a pH near 7. It is not pure, however, since it is contaminated by dissolved gases and by material from the container in which it is stored. Storage in Pyrex® or Vycor® glass eliminates the latter contamination. When water of the highest purity is required, deionized water is fed into a still resulting in ultrapure water. The simplest check on the quality of distilled water is the specific conductance. This is monitored by a conductivity warning light installed in the effluent water line. Resistance of the water should be over 1 million Ω/cm (Table 3).

In a publication of the College of American Pathologists, the specifications are given for a higher grade of reagent water that may be required for atomic absorption spectrophotometry, enzyme measurements, pH determinations, or flame photometry. This water should have a resistance of over 10 million Ω/cm.

Table 3
GENERAL ANALYTICAL REQUIREMENTS OF DISTILLED OR DEIONIZED WATER

Property	Requirement
Residue after evaporation	Not more than 1 mg/l after drying for 1 hr at 105°C
Chloride content	Not more than 0.1 mg/l
Ammonia content	Not more than 0.1 mg/l
Heavy metals (e.g., lead)	Not more than 0.01 mg/l
Consumption of permanganate	When 0.03 ml of 0.1 molar potassium permanganate is added to 500 ml of water containing 1 ml of concentrated sulfuric acid, the color should not completely disappear on standing for 1 hr at room temperature
Silicate	Not more than 0.01 mg/l
Sodium	Not more than 0.1 mg/l
Carbon dioxide	Not more than 3 mg/l

From Tietz, N. W., *Fundamentals of Clinical Chemistry*, 2nd ed., W.B. Saunders, Philadelphia, 1976, 27. With permission.

SEPARATION AIDS

Separation techniques may be necessary in the preparation of a reagent or in the analysis of a substance. Solvent extraction, distillation, diffusion, sublimation, gel filtration, centrifugation, electrophoresis and exclusion-, thin layer, ion exchange, gas-liquid, pressure, and affinity chromatography are presented in Dorner's chapter (Volume II). Presented here are: filter papers, glass fiber filters, microporous polymeric and cellulose membranes, "skinned" membranes, and powders.

Paper

The clinical chemistry laboratory should have several grades of filter paper available to meet unanticipated demands for the removal of precipitates from reagents (Tables 4 and 5).

A water-repellant, silicon-treated, phase-separator filter paper is also available from Whatman®. This filter paper, #IPS, can be used when solvent extractions are made from aqueous solutions or from organic solutions. It provides complete and rapid separation, with complete retention of the aqueous phase.

Glass Fiber Filters

Glass fiber filters are available for clinical techniques (Table 6). These filters are produced entirely from borosilicate fibers and when used in a Gooch, Büchner, or similar filtering apparatus, they give a combination of fine retention with extremely rapid filtering speed not usually found in any cellulose grade. Glass fiber filters are well suited for filtration of heavy viscous solutions or gels, since they do not clog as quickly as open cellulose papers and yet can retain very fine particles.

Microporous Polymeric and Cellulose Membranes

Millipore® membrane filters (Millipore Corporation, Bedford, Mass.) are polymeric microporous or cellulose membranes (Table 7). These filters enable the analyst to concentrate efficiently many particles by pore size (molecular size). By selecting the appropriate pore size, separation can be achieved by exclusion of the unwanted product from the eluate or washing of the wanted product from the filter.

Table 4
WHATMAN® FILTER PAPER GUIDE

Grade	Wt (g/m²)	Thickness (mm)	Ash (%)	Retention (liquid) (μm)	Initial filtration speed[a]	Loading[b] capacity	Wet[c] strength
			Qualitative				
1	87	0.18	0.06	11	40	N	L
2	97	0.19	0.06	8	55	N	L
2V[d]	97	0.20	0.06	8	55	N	L
3	185	0.40	0.06	6	90	H	M
4	92	0.21	0.06	20—25	12	N	L
5	100	0.20	0.06	2.5	250	N	L
6	100	0.18	0.1—0.2	3	175	N	L
			Qualitative Wet Strengthened[d]				
111	75	0.19	0.06	12	20	N	H
111V[d]	75	0.19	0.06	12	20	N	H
113	125	0.44	0.06	30	8	VH	M
113V[d]	125	0.44	0.06	30	8	VH	H
114	75	0.22	0.06	20—25	8	N	H
114V[d]	75	0.22	0.06	20—25	8	N	H
			Ashless Quantitative				
40	96	0.20	0.010	8	75	N	L
41	85	0.21	0.010	20—25	12	N	L
42	100	0.20	0.010	2.5	240	N	L
43	96	0.21	0.010	16	40	N	L
44	80	0.17	0.010	3	175	N	L
			Hardened				
50	97	0.12	0.025	2.7	250	N	H
52	98	0.17	0.025	7	55	N	H
54	93	0.18	0.025	20—25	10	N	H
			Ashless Hardened				
540	90	0.15	0.008	8	55	N	H
541	82	0.16	0.008	20—25	12	N	H
542	97	0.15	0.008	2.7	250	N	H

[a] Filtration speed — Time taken for 100 m*l* of clean water to pass through 15 cm quadrant folded circle, according to a modified ASTM method D981-56.

[b] Loading capacity — N: normal loading capacity before severe blockage occurs; H: high loading capacity; and VH: very high loading capacity.

[c] Wet strength — L: suitable for gravity or low suction filtration. Care must be taken to prevent rupture of the paper; M: will withstand all normal use on Buchner funnels or other devices under suction or moderate pressure; and H: high wet strength and high chemical resistance.

[d] V Grades are pre-pleated.

From **Anon.**, Sales Catalog, Fisher Scientific, Pittsburgh, 1979, 939. With permission.

"Skinned" Membranes

Anisotropic "skinned" membranes formed from a noncellulose polymer have macrosolute retentions ranging from 500 to 300,000 dalton mol wt as calibrated with

Table 5

SCHLEICHER AND SCHUELL® FILTER PAPER GUIDE

Grade	Weight (g/m²)	Thickness (mm)	Gurley[a]	Average filtration speed[b]	Ash[c] (%)	Wet strength	Surface	Description and applications
Quantitative								
589 Black ribbon	78	0.185	15.0	12.0	0.005	Low	Rough	Acid-washed; used for rapid filtration of coarse and gelatinous precipitates [e.g., $Al(OH)_3$, $Fe(OH)_3$, CuS, CoS, etc.]; also used for air pollution analysis
589 White ribbon	86	0.187	53.0	52.0	0.008	Low	Rough	Acid-washed; medium speed and retention; used for filtration of medium fine precipitates (e.g., hot-precipitated CaC_2O_4, $BaSO_4$, CdS, and PbS, etc.); used most frequently for quantitative gravimetric analysis
589 Blue ribbon	90	0.203	167.0	109.0	0.005	Low	Rough	Acid-washed; slow; dense texture; used for filtration of fine precipitates (e.g., cold-precipitated $RaSO_4$, $PbSO_4$, H_2SNO_3, CUO, and NiS
589 Red ribbon	89	0.174	129.0	88.0	0.009	High	Smooth	Acid-washed; slow; smooth; thin; extra dense version of #589 blue ribbon; for finest precipitates not retained by other quantitative papers (e.g., cold-precipitated CaC_2O_4)
Qualitative								
597	88	0.190	49.0	31.0	0.029	Low	Smooth	Moderately rapid; medium thickness and retention; most suitable grade for general qualitative work
General purpose								
SharkSkin[d]	44	0.181	42.0	22.0	—	High	Rough	Medium speed, creped, good wet strength; resistant to acids and alkalis; recommended for student work and high volume process filtration
595	72	0.138	33.0	24.0	—	Very high	Smooth	Medium speed; good wet strength; pure white filter fairly retentive for filtering solutions which entail no particular difficulties; for student work and hospital laboratories
Special analysis								
9 Grid	63	0.180	16.0	8.4	—	High	Smooth	White filter paper with 5 mm equidistant green grid markings; rulings are resistant to aqueous and nonaqueous solutions

a   Gurley (air permeability): The time in seconds for 100 mℓ of air, pushed with a 5 oz cylinder, to pass through a 0.1 in.^{2} filter area. ASTM D726-58 (TAPPI T460M-49).

b   Average filtration speed (rapidity): The time in seconds for 10 mℓ of distilled water to pass through an 11 cm folded circle. ASTM D981-56 (TAPPI T471M-47).

c   Percentage of ash is determined by measuring the remaining residue after complete combustion of paper at 950° ± 25°C. ASTM D586-63 (TAPPI T41305-59).

From Anon., Sales Catalog, Scientific Products, McGaw Park, Ill., 1975, 394. With permission.

Table 6

WHATMAN® GLASS MICROFIBER FILTERS

Type	Thickness (mm)	Effective retention (μm)	Flow rate (ml/sec)	Loading capacity	Water Absorption (ml/m²)	Wt (g/m²)	Use
GF/A	0.25	16	13	High	275	52	Biochemistry: retains proteins, cells, cell debris; radioimmunoassay: retains weak beta emitters for scintillation counting; air pollution monitoring: retains oil mists, polycylic hydrocarbons
GF/B	0.73	1.0	5.5	Very high	275	150	Biochemistry: polymers (proteins, nucleic acids, precipitated by denaturation), collection of gases absorbed into solid particles, biological fluids
GF/C	0.26	1.2	10.5	Good	250	55	Water analysis: pollutants such as Hg; marine chemistry: particulate components in sea water; biochemistry: harvesting lymphocytes during incorporation trials; RIA procedures
GF/D	0.67	2.7	16.5	Very high	900	100	High volume, high efficiency for repetitive routine laboratory filtering; ideal as prefilter to extend life of membrane filters
GF/F	0.44	0.7	6.0	Extremely high	400	75	Use in place of costly membranes in some applications; biochemistry/biology: (use with GF/D as prefilter) clarification of dilute aqueous solutions containing strongly oxidizing, acidic or alkaline components prior to Raman lasar spectroscopy; extremely fine proteins such as IgG

From Anon., Sales Catalog, Fisher Scientific, Pittsburgh, 1979, 332. With permission.

Table 7

MEMBRANE FILTER SPECIFICATIONS

Filter type	Pore size	Flow Rate[a] Water[b]	Air[c]	Porosity (%)	Auto-clavable	Refractive index	Bubble point[d] (kg/cm²)	(psi)	Water extractables (W/W)[e]
MF-Millipore (mixed cellulose acetate and nitrate)										
SC	8.0 μm	630	65	84	Yes	1.515	0.28	4	6.0%
SM	5.0 μm	400	32	84	Yes	1.495	0.42	6	6.0%
SS	3.0 μm	296	30	83	Yes	1.495	0.70	10	6.0%
RA	1.2 μm	222	20	82	Yes	1.512	0.84	12	5.0%
AA	0.80μm	157	16	82	Yes	1.510	1.12	16	4.0%
(black)		157	16	82	No	N.A.	1.12	16	N.A.
DA	0.65 μm	111	9	81	Yes	1.510	1.34	19	3.0%
HA	0.45 μm	38.5	4	79	Yes	1.510	2.32	33	2.5%
(black)		38.5	4	79	No	N.A.	2.32	33	N.A.
PH	0.30 μm	29.6	3	77	Yes	1.510	2.81	40	2.0%
GS	0.22 μm	15.6	2	75	Yes	1.510	3.87	55	2.0%
VC	0.10 μm	1.5	0.6	74	Yes	1.500	17.6	250	1.5%
VM	0.05 μm	0.74	0.5	72	Yes	1.500	26.4	375	1.5%
VS	0.025 μm	0.15	0.2	70	Yes	1.500	35.2	500	1.5%
Duralon (nylon)										
NC	14.0 μm	756	75	68	No	See	0.18	2.5	1.5%
NS	7.0 μm	452	50	65	No	note[f]	0.28	4.0	2.5%
NR	1.0 μm	118	29	63	No		0.56	8.0	2.5%
Fluoropore (PTFE)										
FA	1.0 μm	90	16	85	Yes	See	0.21	3	—
FH	0.5 μm	40	8	85	Yes	note[g]	0.49	7	—
FG	0.2 μm	15	3	70	Yes		0.91	13	—
Celotate (cellulose acetate)										
EA	1.0 μm	178	7	74	No	1.47	0.98	14	4.0%
EH	0.5 μm	49.6	4.5	73	No	1.47	1.97	28	2.5%
EG	0.2 μm	15.6	2.2	71	No	1.47	3.87	55	2.0%
Mitex (Teflon®)										
LC	10.0 μm	126	14	68	Yes	N.A.	0.04	0.5	—
LS	5.0 μm	51.9	9	60	Yes	N.A.	0.06	0.9	—
Polyvic (polyvinyl chloride)										
BS	2.0 μm	231	19	79	No	1.528	0.28	4	3.2%
BD	0.6 μm	33.3	3	73	No	1.528	0.70	10	3.2%

[a] Flow rates listed are based on measurement with clean water and air, and represent *initial* flow rates for a liquid of 1 cp viscosity at the start of filtration, before filter plugging is detectable. Actual initial flow rates may vary from the average values given here. Variability depends on filter type and is roughly proportional to pore size.

[b] Water flow rates are milliliters per minute per cm² of filtration area, at 25°C with a differential pressure of 52 cm Hg (10 psi). Flow rates for Fluoropore and Polyvic filters are based on methanol instead of water.

Table 7 (continued)
MEMBRANE FILTER SPECIFICATIONS

* Air flow rates are liters per minute per cm² of filtration area, at 20°C with a differential pressure of 52 cm Hg (10 psi) and exit pressure of 1 atmosphere (14.7 psia).
* Bubble point pressure is the differential pressure required to force air through the pores of a water-wet filter (except methanol-wet for Fluoropore, Mitex, and Polyvic filters).
* See also description of Low Water-Extractable (TF) Filters under Special Disc Filters.
* Duralon filters cannot be made completely transparent with immersion oil.
* Crystalline and amorphous regions of Fluoropore filters have differing refractive indexes, and it is, therefore, not possible to obtain uniform clearing.

From Catalog #MC 179/V, Millipore Corporation, Bedford, Mass., 1979, 13. With permission.

globular macrosolutes (Amicon Corporation, Lexington, Mass.) and minimize the eventual membrane plugging experienced with microporous filtration. Concentration polarization can be prevented by using stirred cells, thin-channel systems, or hollow fiber systems.

Several nonagitated systems are available offering macrosolute retentions from 500 to 125,000 mol wt. These include pressurized cells, passive filtration, and "cones" which are centrifuged. Their uses are listed in Table 8.

Powders

Cellulose powders for column chromatography meet the demand for large-scale separations as listed in Table 9.

Cellex particle lengths vary from 20 to 300 μm with an average of 80, specific gravity varies from 1.45 to 1.52, moisture content is 3 to 18% by weight, capacity ranges from 0.2 to 0.9 meq/g, the void volume is approximately 40% and the flow rate is dependent on the bed. The pK, settled volume, and use of the various beds are presented in Table 9.

SOLUTIONS

Dilute, Saturated, Concentrated, and Standard[14]

A dilute solution contains a relatively small proportion of solute, and a concentrated solution contains a relatively large proportion of solute. Concentrated solutions are possible only when the solute is very soluble. A saturated solution exists when the molecules of the solute in solution are in equilibrium with the excess undissolved molecules. Since temperature affects solubility, the exact temperature of the solution should be specified.

A supersaturated solution exists when there is more solute in solution than is present in a saturated solution of the same substance at the same temperature and pressure. Supersaturated solutions are unstable and cannot exist in equilibrium with the solid phase.

Standard solutions are an integral part of every quantitative analysis in the clinical chemistry laboratory and are used during the assay of an unknown sample. This is a solution whose precise concentration is known and which is used as a reference to assign concentrations to the assayed unknown samples.

Expressing Concentrations of Solutions in Physical Units

Refer to the section on Nomenclature, Quantities, and Units.

Weight of solute per unit volume of solution (mass concentration) — This is the

Table 8

ULTRA FILTRATION USING ANISOTROPIC "SKINNED" MEMBRANES

System	MW cutoffs	Conc. of solution (%)	Volumes	Maximum rate	Concentrates	Retains (%)	Sample types
Stirred cells	500—300,000	<5.0	60 µl—2000 ml	10l/day	100×	Up to 99	Proteins, spinal fluids, urine, and many other fluids
Thinchannel	500—300,000	<40.0	2—600 ml	20l/day	100×	Up to 99	Recommended for preparation and/or recovery of biologicals, recovery of metabolic products from fermentation broths, removal of bacteria, virus, pryogens, etc., processing of suspended cellular material, clarification of sera, syrups, extracts
Hollow fiber	5,000—100,000	<30.0	200—2000 ml	24l/day	20×	Up to 90	Recommended for rapid procesing of solutions containing proteins, enzymes, virus, toxoids, blood fractions, extracts, colloidal products
Nonagiitated	500—300,000	<1.0	0.5—5.0 ml	Volume dependent	100×	Up to 99	Ideal for concentration of urine and CSF proteins, serum isoenzymes, antibodies, antigens, etc., prior to electrophoresis or other analytical procedure
Filter cones	25,000 or 50,000	<1.0	Up to 7 ml	Centrifuge dependent	Volume dependent	Up to 99	Preparation of protein-free filtrates, partitioning free from protein-bound species, or concentrating dilute biologicals nearly to dryness; especially valuable when recovery of ultrafiltrate is vital

From Anon., Catalog #447D, Amicon Corporation, Lexington, Mass., 1979. With permission.

Table 9
CELLULOSE POWDERS

Bio-Rad product	pK	Settled bed volume (m*l*/dry g)	Use
Cellex D (low capacity)	9.5	8	
Cellex D (standard capacity)	9.5	6.25	Separation and purification of many enzymes, hormones, serum components acidic, and basic proteins, etc
Cellex D (high capacity)	9.5	7.8	
Cellex E	7.5	3.5	Separation of viruses, nucleic acids, and nucleoproteins
Cellex T	9.5	6	Chromatography of certain acidic proteins and folic acid derivatives
Cellex QAE	Strong	8	Separation of weakly acidic materials and partition chromatography in polar solvents
Cellex BD (supplied hydrated)	9	16	Separations of RNA fractions
Cellex CM	4	7	Bind and separate polycations
Cellex P	1.5,6	4.5	Bind and can be used under pH 4
Cellex N-1	—	3	Stationary phase in electrophoresis, partition chromatography, and affinity chromatography
Cellex 410	—	3	

From Anon., Catalog Price List F, Bio-Rad Laboratories, Richmond, Calif., 1980, 33. With permission.

mass of the component divided by the volume of the mixture. The use of kilogram per liter (kg/*l*) and fractions of kilograms is preferred. Thus, kg/*l*, g/*l*, mg/*l*, μg/*l*, and ng/*l* are preferred units.

Amount (mass) of solute per mass of solution — The use of kilograms (and fractions of kilograms) per kilogram is preferred (kg/kg). Thus, kg/kg, g/kg, mg/kg, μg/kg, and ng/kg, are preferred units.

Volume of solute per volume of solution — When the solution of solvent are both liquids, as in alcohol solutions, the concentration of such a solution is frequently expressed in terms of volume per volume (v/v). By adding 70 m*l* of alcohol to a flask and filling it up to 100 m*l* with water, one would achieve a solution whose concentration could be expressed as 700 m*l*/*l*.

Expressing Concentration in Chemical Units

A *molar solution* contains 1 mol of the solute in 1*l* of solution; i.e., a 1-*M* solution of H_2SO_4 contains 98.08 g of H_2SO_4/*l* of solution, since the molecular weight of H_2SO_4 is 98.08. A 0.50 molar solution contains 0.50 × 98.08 g = 40.04 g H_2SO_4/*l* of solution.

A *molal solution* contains 1 mol of solute in 1 kg of solvent.

A *normal solution* contains 1 g-equivalent weight of the solute in 1*l* of solution.

Commercial concentrations of acids and bases are listed in Table 10 and some ionization constants are listed in Table 11.

Accurately prepared solutions of hydrochloric and sulfuric acids, potassium hydrogen phthalate, and sodium hydroxide are necessary in a clinical chemistry laboratory. These solutions are used to establish the molarity of all acids and bases in the laboratory.

Table 10
CONCENTRATION OF ACIDS AND BASES COMMON COMMERCIAL STRENGTHS

	(Mol wt)	(Mol/l)	(g/l)	Wt (%)	Specific gravity
Acetic acid glacial	60.06	17.4	1.045	99.5	1.05
Acetic acid	60.05	6.27	376	36	1.045
Ammonium hydroxide	35.05	14.8	251	28	0.89
Butyric acid	88.1	10.3	912	95	0.96
Formic acid	46.02	23.4	1.080	90	1.20
	—	5.75	264	25	1.06
Hydroidic acid	127.9	7.57	969	57	1.70
	—	5.51	705	47	1.50
	—	0.86	110	10	1.1
Hydrobromic acid	80.92	8.89	720	48	1.50
	—	6.82	552	40	1.38
Hydrochloric acid	36.5	11.6	424	36	1.18
	—	2.9	105	10	1.05
Hydrocyanic acid	27.03	25	676	97	0.697
	—	0.74	19.9	2	0.996
Hydrofluoric acid	20.01	32.1	642	55	1.167
	—	28.8	578	50	1.155
Hydrofluosilicic acid	144.1	2.65	382	30	1.27
Hypophosphorous acid	66.0	9.47	625	50	1.25
	—	5.14	339	30	1.13
	—	1.57	104	10	1.04
Lactic acid	90.1	11.3	1.020	85	1.2
Nitric acid	63.02	15.99	1.008	71	1.42
	—	14.9	938	67	1.40
	—	1.3	837	61	1.37
Perchloric acid	100.5	11.65	1.172	70	1.67
	—	9.2	923	60	1.54
Phosphoric acid	80	18.1	1.445	85	1.70
Potassium hydroxide	56.1	13.5	757	50	1.52
	—	1.94	109	10	1.09
Sodium carbonate	106.0	1.04	110	10	1.10
Sodium hydroxide	40.0	19.1	763	50	1.53
	—	2.75	111	10	1.11
Sulfuric acid	98.1	18.0	1.766	96	1.84
Sulfurous acid	82.1	0.74	61.2	6	1.02

From Sober, H. A., *Handbook of Biochemistry*, 2nd ed., CRC Press, Boca Raton, Fla., 1970, J-245. With permission.

Potassium hydrogen phthalate may be obtained from the National Bureau of Standards in very pure form.

Deci-normal solutions of salts and selected reagents are listed in Table 12.

Deci-normal solutions of selected oxidation and reduction reagents are listed in Table 13.

Buffer Solutions

Refer to the appropriate tables in the area on Physical and Chemical Data.

Table 11

IONIZATION CONSTANTS FOR ROUTINELY USED
ACIDS AND BASES (25°C)

Acid or base	K	pKa	Molarity	Specific gravity
Acetic acid	1.76×10^{-5}	4.75	17.5	1.05
Ammonium hydroxide	1.79×10^{-5}	4.75	14.8	0.90
Hydrochloric acid	—	—	12.0	1.18
Lactic acid	1.37×10^{4}	3.79	—	—
Nitric acid	$4.36 \times 10^{+1}$	−1.64	15.9	1.42
Phosphoric acid	7.52×10^{3}	2.12	14.7	1.7
Sodium hydroxide	—	—	19.5	1.54
Sulfuric acid	>1	—	18.0	1.84

Note: For an expanded list refer to the section on Physical and Chemical Data.

Media and Reagents

Tissue Culture Media Compositions such as Conaught 198 and Eagle basal medium are described in the *Biochemist's Handbook.*[18]

Protein fractionation reagents such as ammonium sulfate are described in *Data for Biochemical Research.*[19] Qualitative or quantitative reagents such as Benedict's Reagent and Fehling's Solution are also described in Ref. 19, pp. 617-622, as are electron donors, carriers, and acceptors, pp. 436-442; peptidase substrates, pp. 444-446; glycosidase substrates, pp. 448-451; phosphatase substrates, pp. 452-454; sulphatase substrates, pp. 456-457; esterase and lipase substrates, pp. 458-459; enzyme activators, pp. 460-461; and protein modifiers, pp. 462-465.

These lists are not inclusive, as recent advances include n-acetyl-cysteine, an enzyme activator, di-adenosine pentaphosphate, an adenylate kinase inhibitor, and the immunological inactivation of enzymes.

MEASUREMENT OF pH

Refer to the appropriate tables in the area on Physical and Chemical Data.

DRYING AGENTS

Water must be removed from most organic solvents, since their use frequently is in systems that do not tolerate its presence. Dehydrating reagents should combine quickly and irreversibly with water. Predrying may be necessary, since the best dehydrating reagents will react violently when a large amount of water is present (Table 14).

Many times it is necessary to remove the analyte from an aqueous matrix. This can be done by the proper choice of solvent. Table 15 lists the boiling points, water solubility, and hazardous characteristics of a number of solvents.

Solvents useful in crystallization can be found in the *Chemists Companion.*[20] A description of their boiling points, flammability, toxicity, and use are given.

BLOOD COLLECTION

A good description of proper blood drawing can be found in *The Fundamentals of Clinical Chemistry.*[21] Also, the publications TSH-3 and TSH-4 of the National Com-

Table 12

DECI-NORMAL SOLUTIONS OF SALTS AND OTHER REAGENTS

Name	Formula	Atomic or molecular weight	Hydrogen equivalent	0.1 Hydrogen equivalent in g
Acetic acid	$HC_2H_3O_2$	60.0530	$HC_2H_3O_2$	6.0053
Ammonia	NH_3	17.0306	NH_3	1.7031
Ammonium ion	NH_4^+	18.0386	NH_4^+	1.8039
Ammonium chloride	NH_4Cl	53.4916	NH_4Cl	5.3492
Ammonium sulfate	$(NH_4)_2SO_4$	132.1388	$\frac{1}{2}(NH_4)_2SO_4$	6.6069
Ammonium thiocyanate	NH_4CNS	76.1204	NH_4CNS	7.6120
Barium	Ba	137.34	$\frac{1}{2}Ba$	6.867
Barium carbonate	$BaCO_3$	197.3494	$\frac{1}{2}BaCO_3$	9.8675
Barium chloride hydrate	$BaCl_2 \cdot 2H_2O$	244.2767	$\frac{1}{2}BaCl_2 \cdot 2H_2O$	12.2138
Barium hydroxide	$Ba(OH)_2$	171.3547	$\frac{1}{2}Ba(OH)_2$	8.5677
Barium oxide	BaO	153.3394	$\frac{1}{2}BaO$	7.6670
Bromine	Br	79.909	Br	7.9909
Calcium	Ca	40.08	$\frac{1}{2}Ca$	2.004
Calcium carbonate	$CaCO_3$	100.0894	$\frac{1}{2}CaCO_3$	5.0045
Calcium chloride	$CaCl_2$	110.9860	$\frac{1}{2}CaCl_2$	5.5493
Calcium chloride hydrate	$CaCl_2 \cdot 6H_2O$	219.0150	$\frac{1}{2}CaCl_2 \cdot 6H_2O$	10.9508
Calcium hydroxide	$Ca(OH)_2$	74.0947	$\frac{1}{2}Ca(OH)_2$	3.7047
Calcium oxide	CaO	56.0794	$\frac{1}{2}CaO$	2.8040
Chlorine	Cl	35.453	Cl	3.5453
Citric acid	$C_6H_8O_7 \cdot H_2O$	210.1418	$\frac{1}{3}C_6H_8O_7 \cdot H_2O$	7.0047
Cobalt	Co	58.9332	$\frac{1}{2}Co$	2.9466
Copper	Cu	63.54	$\frac{1}{2}Cu$	3.177
Copper oxide (cupric)	CuO	79.5394	$\frac{1}{2}CuO$	3.9770
Copper sulfate hydrate	$CuSO_4 \cdot 5H_2O$	249.6783	$\frac{1}{2}CuSO_4 \cdot 5H_2O$	12.4839
Cyanogen	$(CN)_2$	26.0179	CN	2.6018
Hydrochloric acid	HCl	36.4610	HCl	3.6461
Hydrocyanic acid	HCN	27.0258	HCN	2.7026
Iodine	I	126.9044	I	12.6904
Lactic acid	$C_3H_6O_3$	90.0795	$C_3H_6O_3$	9.0080
Malic acid	$C_4H_6O_5$	134.0894	$\frac{1}{2}C_4H_6O_5$	6.7045
Magnesium	Mg	24.312	$\frac{1}{2}Mg$	1.2156
Magnesium carbonate	$MgCO_3$	84.3214	$\frac{1}{2}MgCO_3$	4.2161
Magnesium chloride	$MgCl_2$	95.2180	$\frac{1}{2}MgCl_2$	4.7609
Magnesium chloride hydrate	$MgCl_2 \cdot 6H_2O$	203.2370	$\frac{1}{2}MgCl_2 \cdot 6H_2O$	10.1623
Magnesium oxide	MgO	40.3114	$\frac{1}{2}MgO$	2.0156
Manganese	Mn	54.938	$\frac{1}{2}Mn$	2.7469
Manganese sulfate	$MnSO_4$	150.9996	$\frac{1}{2}MnSO_4$	7.5500
Mercuric chloride	$HgCl_2$	271.4960	$\frac{1}{2}HgCl_2$	13.5748
Nickel	Ni	58.71	$\frac{1}{2}Ni$	2.9356
Nitric acid	HNO_3	63.0129	HNO_3	6.3013
Nitrogen	N	14.3067	N	1.4007
Nitrogen pentoxide	N_2O_5	108.0104	$\frac{1}{2}N_2O_5$	5.4005
Oxalic acid	$H_2C_2O_4$	90.0358	$\frac{1}{2}H_2C_2O_4$	4.5018
Oxalic acid hydrate	$H_2C_2O_4 \cdot 2H_2O$	126.0665	$\frac{1}{2}H_2C_2O_4 \cdot 2H_2O$	6.3033
Oxalic acid anhydride	C_2O_3	72.0205	$\frac{1}{2}C_2O_3$	3.6010
Phosphoric acid	H_3PO_4	97.9953	$\frac{1}{3}H_3PO_4$	3.2665
Potassium	K	39.102	K	3.9102
Potassium bicarbonate	$KHCO_3$	100.1193	$KHCO_3$	10.0119
Potassium carbonate	K_2CO_3	138.2134	$\frac{1}{2}K_2CO_3$	6.9106
Potassium chloride	KCl	74.5550	KCl	7.4555
Potassium cyanide	KCN	65.1199	KCN	6.5120
Potassium hydroxide	KOH	56.1094	KOH	5.6109
Potassium oxide	K_2O	94.2034	$\frac{1}{2}K_2O$	4.7102
Potassium permanganate for Co estimation	$KMnO_4$	158.0376	$\frac{1}{6}KMnO_4$	2.6339
Potassium permanganate for Mn estimation	$KMnO_4$	158.0376	$\frac{1}{3}KMnO_4$	5.2678
Potassium tartrate	$K_2H_4C_4O_6$	226.2769	$\frac{1}{2}K_2H_4C_4O_6$	11.3139
Silver	Ag	107.87	Ag	10.787
Silver nitrate	$AgNO_3$	169.8749	$AgNO_3$	16.9875
Sodium	Na	22.9898	Na	2.2990
Sodium bicarbonate	$NaHCO_3$	84.0071	$NaHCO_3$	8.4007
Sodium carbonate	Na_2CO_3	105.9890	$\frac{1}{2}Na_2CO_3$	5.2995
Sodium chloride	$NaCl$	58.4428	$NaCl$	5.8443
Sodium hydroxide	$NaOH$	39.9972	$NaOH$	3.9997
Sodium oxide	Na_2O	61.9790	$\frac{1}{2}Na_2O$	3.0990
Sodium sulfide	Na_2S	78.0436	$\frac{1}{2}Na_2S$	3.9022
Succinic acid	$H_2C_4H_4O_4$	118.0900	$\frac{1}{2}H_2C_4H_4O_4$	5.9045
Sulfuric acid	H_2SO_4	98.0775	$\frac{1}{2}H_2SO_4$	4.9039
Sulfur trioxide	SO_3	80.0622	$\frac{1}{2}SO_3$	4.0031
Tartaric acid	$C_4H_6O_6$	150.0888	$\frac{1}{2}C_4H_6O_6$	7.5044
Zinc	Zn	65.37	$\frac{1}{2}Zn$	3.269
Zinc sulfate	$ZnSO_4 \cdot 7H_2O$	287.5390	$\frac{1}{2}ZnSO_4 \cdot 7H_2O$	14.3769

Note: Atomic and molecular weights in the above table are based upon the 1965 atomic weight scale and the isotope C-12. The weight in grams of the compound in 1 cc of the above deci-normal solutions is found by dividing the H equivalent in the last column by 1000.

From Weast, R. C., Ed., *Handbook of Chemistry and Physics*. 61st ed., CRC Press, Boca Raton, Fla., 1980, D-116. With permission.

Table 13
DECI-NORMAL SOLUTIONS OF OXIDATION AND REDUCTION REAGENTS

Name	Formula	Atomic or molecular weight	Hydrogen equivalent	0.1 Hydrogen equivalent in g
Antimony	Sb	121.75	$\frac{1}{2}Sb$	6.0875
Arsenic	As	74.9216	$\frac{1}{2}As$	3.7461
Arsenic trisulfide	As_2S_3	246.0352	$\frac{1}{4}As_2S_3$	6.1509
Arsenous oxide	As_2O_3	197.8414	$\frac{1}{4}As_2O_3$	4.9460
Barium peroxide	BaO_2	169.3388	$\frac{1}{2}BaO_2$	8.4669
Barium peroxide hydrate	$BaO_2 \cdot 8H_2O$	313.4615	$\frac{1}{2}BaO_2 \cdot 8H_2O$	15.6730
Calcium	Ca	40.08	$\frac{1}{2}Ca$	2.004
Calcium carbonate	$CaCO_3$	100.0894	$\frac{1}{2}CaCO_3$	5.0045
Calcium hypochlorite	$Ca(OCl)_2$	142.9848	$\frac{1}{4}Ca(OCl)_2$	3.5746
Calcium oxide	CaO	56.0794	$\frac{1}{2}CaO$	2.8040
Chlorine	Cl	35.453	Cl	3.5453
Chromium trioxide	CrO_3	99.9942	$\frac{1}{3}CrO_3$	3.3331
Ferrous ammonium sulfate	$FeSO_4(NH_4)SO_4 \cdot 6H_2O$	392.0764	$FeSO_4(NH_4)_2SO_4 \cdot 6H_2O$	39.2076
Hydroferrocyanic acid	$H_4Fe(CN)_6$	215.9860	$H_4Fe(CN)_6$	21.5986
Hydrogen peroxide	H_2O_2	34.0147	$\frac{1}{2}H_2O_2$	1.7007
Hydrogen sulfide	H_2S	34.0799	$\frac{1}{2}H_2S$	1.7040
Iodine	I	126.9044	I	12.6904
Iron	Fe	55.847	Fe	5.5847
Iron oxide (ferrous)	FeO	71.8464	FeO	7.1846
Iron oxide (ferric)	Fe_2O_3	159.6922	$\frac{1}{2}Fe_2O_3$	7.9846
Lead peroxide	PbO_2	239.1888	$\frac{1}{2}PbO_2$	11.9594
Manganese dioxide	MnO_2	86.9368	$\frac{1}{2}MnO_2$	4.3468
Nitric acid	HNO_3	63.0129	$\frac{1}{3}HNO_3$	2.1004
Nitrogen trioxide	N_2O_3	76.0116	$\frac{1}{4}N_2O_3$	1.9002
Nitrogen pentoxide	N_2O_5	108.0104	$\frac{1}{6}N_2O_5$	1.8001
Oxalic acid	$C_2H_2O_4$	90.0358	$\frac{1}{2}C_2H_2O_4$	4.5018
Oxalic acid hydrate	$C_2H_2O_4 \cdot 2H_2O$	126.0665	$\frac{1}{2}C_2H_2O_4 \cdot 2H_2O$	6.3033
Oxygen	O	15.9994	$\frac{1}{2}O$	0.8000
Potassium dichromate	$K_2Cr_2O_7$	294.1918	$\frac{1}{6}K_2Cr_2O_7$	4.9032
Potassium chlorate	$KClO_3$	122.5532	$\frac{1}{6}KClO_3$	2.0425
Potassium chromate	K_2CrO_4	194.1076	$\frac{1}{3}K_2CrO_4$	6.4733
Potassium ferrocyanide	$K_4Fe(CN)_6$	368.3621	$K_4Fe(CN)_6$	36.8362
Potassium ferrocyanide	$K_4Fe(CN)_6 \cdot 3H_2O$	422.4081	$K_4Fe(CN)_6 \cdot 3H_2O$	42.2408
Potassium iodide	KI	166.0064	KI	16.6006
Potassium nitrate	KNO_3	101.1069	$\frac{1}{3}KNO_3$	3.3702
Potassium perchlorate	$KClO_4$	138.5526	$\frac{1}{8}KClO_4$	1.7319
Potassium permanganate	$KMnO_4$	158.0376	$\frac{1}{5}KMnO_4$	3.1608
Sodium chlorate	$NaClO_3$	106.4410	$\frac{1}{6}NaClO_3$	1.7740
Sodium nitrate	$NaNO_3$	84.9947	$\frac{1}{3}NaNO_3$	2.8332
Sodium thiosulfate	$Na_2S_2O_3 \cdot 5H_2O$	248.1825	$Na_2S_2O_3 \cdot 5H_2O$	24.8183
Stannous chloride	$SnCl_2$	189.5960	$\frac{1}{2}SnCl_2$	9.4798
Stannous oxide	SnO	134.6894	$\frac{1}{2}SnO$	6.7345
Sulfur dioxide	SO_2	64.0628	$\frac{1}{2}SO_2$	3.2031
Tin	Sn	118.69	$\frac{1}{2}Sn$	5.935

Note: Atomic anc molecular weights in the above tab le are based upon the 1965 atomic weight scale and the isotope C-12. The weight in grams of the compound in 1 cc of the above deci-normal solutions is found by dividing the H equivalent in the last column by 1000.

From Weast, R. C., Ed., *Handbook of Chemistry and Physics,* 61st ed., CRC Press, Boca Raton, Fla., 1980, D-117. With permission.

Table 14

SELECTED DEHYDRATING REAGENTS:
EFFICIENCY AND CAUTIONS

Reagent	Residual water (mg/l of dry air)	Cautions
P_2O_5	<1 mg in 40,000 l	Very reactive; predrying may be necessary; do not use with reactive compounds
Mg $(ClO_4)_2$	<1 mg in 210 l	Same as for P_2O_5
BaO	<0.0007	Slow, becomes strongly basic
CaSO4 (Drierite)	<0.006	Fast, changes from blue to pink
CaCl₂	<1.6	Fast, but inefficient

Table 15

SOLVENTS FOR EXTRACTION OF AQUEOUS SOLUTIONS

Solvent	Toxicity	BP (°C)	Flammability	Dielectric constant	Precautions
2-Butanol	High	99.5	Low	15.8	Toxic
Chloroform	Very high	61.7	—	4.7	Toxic
Cyclohexane	Low	80.7	High	2.0	
Ether, diethyl	Moderate	34.5	Very high	4.1	Explosive
Ethyl acetate	Low	77.1	High	6.0	
Hexane	Low	69	Very high	1.9	Flammable
Heptane	Low	98.4	High	1.9	
Isoamyl alcohol	High	137	High	13.9	
Methylene chloride	Moderate	40	—	8.9	
Pentane	Low	36.1	Very high	1.8	Flammable
Toluene	Moderate	110.6	High	2.4	

mittee for Clinical Laboratory Standards (see Appendix) propose tentative standard procedures for blood collection.

Evacuated tubes for blood collection are available in a wide variety of sizes and formulations. Sterile tubes are coming increasingly into use as the remote, but definite hazard, of backflow from a contaminated tube is appreciated.

The tubes are made of glass which has to be free of contamination and has to withstand sterilization by cobalt irradiation without becoming cloudy. This allows inspection of the contents. The tubes may be uncoated, silicone coated, or lined with silica particles dissolved in a water-soluble coating which speeds coagulation and reduces latent fibrin formation.

Tubes are available with no anticoagulant, with thrombin as a coagulant accelerator, with lithium iodoacetate as an antiglycolytic agent, or with a variety of other anticoagulants suited to chemistry, serology, hematology, coagulation studies, microbiology, and blood banking. In addition, serum separation tubes are available with gels, either silicone or an inert radioresistant polymer, which have a density between that of plasma or serum and red cells. After centrifugation, such tubes develop a barrier at the serum red cell interface which protects serum from contamination with diffusible red cell constituents.

The stopper or closure carries a color-coding indicating the nature of the tube filling. The stopper has to be inert and cell repellent so that red cells do not collect at the rim

Table 16
VACUTAINER® TUBES[a]

Color	Additive	Use
Lavender	EDTA	Plasma or whole blood
Red	—	Serum
Orange/black	Serum separator (SST)®	Serum
Yellow	—	Sterile interior-serum
Gray	Oxalate/fluoride	Plasma or whole blood (glucose)
Light blue	Lithium iodoacetate	Plasma or whole blood (glucose)
Green	Heparin	Plasma or whole blood
Blue	Citrate	Plasma or whole blood
Royal blue	—	Serum (trace elements)
White (600 μl)	Serum separator	Serum (pediatrics)

[a]   Manufactured by the Becton Dickinson Company, East Rutherford, N.J.

of the tube during centrifugation. The stopper is left in place in order to eliminate aerosol formation and evaporation. Stoppers do require lubrication, and either silicone or glycerol is used for this purpose. Zinc is a natural constituent of ordinary rubbers. Plasticizers in the rubber lead to spurious peaks when samples for toxicology are subjected to gas chromatography, and some stoppers contain a heparin inhibitor. In order to eliminate specific interferences, special stoppered tubes are available.

The approved standard (publication ASH-1) of the National Committee for Clinical Laboratory Standards (NCCLS) on the evacuated tubes for blood specimen collection is an in-depth assessment of what single-use evacuated blood collection tubes should conform to. This includes material, size, design, construction, limits on interference, additives, stability of additives, loss of vacuum limits, and other pertinent data.

Table 16 provides a limited compilation of stopper colors, additives, and principal uses of Vacutainer® tubes (Beckton Dickinson Company, Stanley Street, East Rutherford, N.J.).

There are other stopper colors with variations on additive content and Vacutainers® can be manufactured to the customer's specifications. "Corvac" blood collection tubes are available from Monoject, Division of Sherwood Medical, A Brunswick Company, St. Louis, Mo. The Corvac® system contains a serum-separator gel similar to that of the SST® system.

Prior to the availability of the Corvac® and SST® tubes, serum separation was facilitated by the use of barrier beads, filter separators, and Sure-Sep® (General Diagnostics, Morris Plains. N.J.), a silicone-based gel that is introduced into the blood tube on centrifugation.

The handling and transportation of blood specimens is available from NCCLS as a tentative standard, publication TSH-5.

CALIBRATORS AND REFERENCE MATERIALS

Calibrators are serum materials which have been assigned values determined by reference to primary standards where possible. Ideally, a calibrator is a primary standard, but since it lacks a serum matrix which usually influences quantitation, a primary standard is not appropriate.

Control materials are serum based as is a calibrator and is, in fact, identical. The difference between the two is that the assigned value of the control is determined by assay against the calibrator and not a primary standard.

NCCLS publications TSC-5 and ASC-2 outline proposed standards for calibration and control materials.

Standard reference materials, primary standards, are available from the National Bureau of Standards.

Calibrators and controls should never originate from the same serum pool. Ideally, the use of different manufacturers ensures this. Manufacturers' values on calibrators should be validated by assaying against the previous calibrator and discussing any discrepancies with the manufacturer.

STANDARD MATERIALS

The International Standards Organization (ISO) has tentatively defined a reference material (RM) to be "a material or substance, one or more physical or chemical properties of which are sufficiently well-established to be used for the calibration of an apparatus or for the verification of a measurement method. Generally, any reasonably small part of a RM sample should exhibit the property value(s) established by the RM as a whole within the stated uncertainty limits."[22] A certified reference material is defined to be "a reference material accompanied by or traceable to a certificate stating the property value(s) concerned and issued by an organization which is generally accepted as technically competent."

The International Federation of Clinical Chemistry defines a primary standard material to be "a substance of known chemical composition and sufficient purity to be used in preparing a primary standard solution."[23]

The NCCLS has approved a written standard for reference materials which are used for either calibration or control purposes.[24] A calibration reference material is one whose constituents are established with sufficient accuracy and precision to be useful for calibration purposes. The NCCLS standard provides that calibration reference materials have values of constituents determined by comparison to primary standards (primary reference materials) whenever possible. The NCCLS describes both calibration materials and control materials as substances used to simulate analytical samples which are usually complex mixtures and are commonly derived, at least partially, from biological sources.[25]

The above definitions, as well as others, are not entirely consistent with each other and some confusion results. Acknowledging this, the National Bureau of Standards, in a prepared position paper for the national conference dedicated to a national understanding for the development of reference materials and methods for clinical chemistry (held in Atlanta, Ga., in November of 1977), presented a table on reference material hierarchy (see Table 17).

The acknowledged Pure Primary Reference Materials are the NBS Standard Reference Materials (SRM). The first clinical SRM (Cholesterol 911) was issued in 1967, and was followed by Urea (SRM 912), Uric Acid (SRM 913), and Creatinine (SRM 914) in 1968. The 29 currently available SRMs are listed (see Table 18).

In addition, the National Bureau of Standards makes available Spectrophotometric Standards. The spectrophotometric SRMs are intended primarily for use in verifying the accuracy of the transmittance scale of spectrophotometers. The luminescence SRM provides relative admission spectra to determine spectral responsivity and to verify the accuracy of spectrofluorometers. The spectrophotometric standards are listed in Table 19. The constituents and makeup of these standards are taken from the NBS Standard Reference Materials for Clinical Laboratory Measurements brochure.

In addition, the NBS offers a Standards Information Service. Through this service, the NBS maintains a reference collection of standards which includes over 240,000

Table 17
REFERENCE MATERIAL HIERARCHY

Material name	Synonym/Organization	Characterization by	Property value name	Major use
1. Primary reference material				
a. Pure primary reference material	Standard reference material (SRM)/NBS Reference material/ISO Primary standard material/IFCC	Definitive method(s) and/or a number of independent methods of high accuracy	Certified value	Assure measurement compatibility through transfer of accuracy to secondary reference materials and/or reference methods
b. Matrix primary reference material	Standard reference material (SRM)/NBS Reference material/ISO	Same as 1.a. Homogeneity, stability, sterility, etc., considerations take on added importance	Certified value	Same as 1.a.
2. Secondary reference material	Reference material/ISO Reference material (calibration)/NCCLS	Preferably reference method and/or comparison primary reference material	Measured value by a specified method (preferably reference method)	Calibration standard; transfer accuracy to laboratory methods
3. Control reference	Reference material (control)/NCCLS Control material/IFCC	Laboratory method (if measured)	May or may not have measured values	Internal quality control

Table 18
NATIONAL BUREAU OF STANDARDS PURE PRIMARY REFERENCE MATERIALS

SRM[a]	Type	Purity(%)	Wt/unit
900	Antiepilepsy drug level assay	4 drugs/3 levels	Set of 4 vials
911a	Cholesterol	99.8	2 g
912	Urea	99.7	25 g
913	Uric acid	99.7	10 g
914	Creatinine	99.8	10 g
915	Calcium carbonate	99.9	20 g
916	Bilirubin	99.0	100 mg
917	D-Glucose	99.9	25 g
918	Potassium chloride	99.9	30 g
919	Sodium chloride	99.9	30 g
920	D-Mannitol	99.8	50 g
921	Cortisol	98.9	1 g
922	Tris (hydroxymethyl) aminomethane	99.9	25 g
923	Tris (hydroxymethyl) aminomethane HCl	99.7	35 g
924	Lithium carbonate	100.0	30 g
925	VMA (4-hydroxy-3-methoxymandelic acid)	99.4	1 g
926	Bovine serum albumin (powder)	Not certified	5 g
927	Bovine serum albumin (7% solution)	Not certified	10 vials, 2.15 ml each
928	Lead nitrate	100.00	30 g
929	Magnesium gluconate	100.00	5 g
933	Clinical laboratory thermometers	—	Set of 3
934	Clinical laboratory thermometer	—	1 each
937	Iron metal	99.90	50 g
1968	Callium melting point	—	1 each

[a] Standard reference material.

Table 19
NATIONAL BUREAU OF STANDARDS SPECTROPHOTOMETRIC STANDARDS

SRM[a]	Type	Unit
930D	Glass filters for spectrophotometry	Set: 3 filters, 4 holders
931b	Liquid filters for spectrophotometry	Set: 12 vials
932	Quartz cuvette for spectrophotometry	1 each
935	Crystalline potassium dichromate for use as an ultraviolet absorbance standard	15 g
936	Quinine sulfate dihydrate	1 g
2031	Metal-on-quartz filters for spectrophotometry	In preparation
2032	Potassium iodide for use as a stray light standard	In preparation

[a] Standard reference material.

standards, specifications, test methods, codes, and recommended practices issued by U.S. technical societies, professional organizations and trade associations, state purchasing offices, U.S. government agencies and foreign, national, and international standardization, 150 periodicals and newsletters, and visual search microfilm files. Included within this system are standards pertaining to clinical laboratories.

Requests for free lists of standards compiled by this system, together with names of

Table 20
STANDARD SOLUTIONS OF THE COLLEGE OF AMERICAN
PATHOLOGISTS

	Concentration	Constituents
Calcium	10 mg/dℓ	Calcium carbonate
		Sodium chloride
		Potassium chloride
		Preservative
Chloride	100 mmol/ℓ	Sodium chloride, preservative
Cholesterol	400 mg/dℓ	Cholesterol, ethyl and methyl, alcohol
Creatinine	100 mg/dℓ	Creatinine hydrochloric acid
Sodium and potassium	140 mmol/ℓ Na +	Sodium chloride, potassium chloride, preservative
	4.0 mmol/ℓ K +	
Sodium and potassium	130 mmol/ℓ Na +	Sodium chloride, potassium chloride, preservative
	2.0 mmol/ℓ K +	
Sodium and potassium	150 mmol/ℓ Na +	Sodium chloride, potassium chloride, preservative
Glucose	200 mg/dℓ	Dextrose, preservative
Nitrogen	20 mg/dℓ	Ammonium sulfate, preservative
Phosphate	8.0 mg/dℓ	Potassium acid, phosphate, sulfuric acid
Urea nitrogen	50 mg/dℓ	Urea, phenyl mercuric acetate, sulfuric acid
Uric acid	100 mg/dℓ	Uric acid, preservative
Ethyl alcohol	150 mg/dℓ	Ethyl alcohol, preservative
pH Control	7.00 @ 37°C	Buffer, preservative
Protein	Approx. 6.25 g/dℓ	Bovine albumin
Triglyceride	200 mg/dℓ ± 5%	Triolein, isoprophyl alcohol, preservative

organizations where copies of the standards can be obtained, should be as specific as possible and include all laboratory terms necessary to locate the standards. One can take advantage of this service by writing the Standards Information Service, Room B-162, Building 225, National Bureau of Standards, Washington, D.C. 20234. Other reference materials have been developed to become an integral part of the clinical laboratory quality control system. They are used to improve individual laboratory results and interlaboratory comparability. These are also available at a lower cost than certified standard reference materials, since laboratories cannot practically use SRMs on a routine basis. An example of such a program is the College of American Pathologists (CAP) Standard Solutions. CAP Clinical Standard Solutions are assayed directly against Certified Reference Materials (CRM) of the National Bureau of Standards. The clinical standard solutions are assayed by at least three laboratories according to a detailed protocol reported in the *American Journal of Clinical Pathology*.[26,27] Table 20 is a list of the CAP Standard Solutions currently available, including their concentration and constituents.

The International Committee for Standardization in Hematology (ICSH) distinguished itself by making available an international cyanmethemoglobin standard for the determination of hemoglobin by the hemoglobincyanide method. In 1964, the Subcommittee on Hemoglobinometry of ICSH proposed that hemoglobinometry be standardized by international agreement. This was done in 1967.

The final recommendations of the Standard Materials Working Group of the Conference on "A National Understanding for the Development of Reference Materials and Methods for Clinical Chemistry" included the following recommendations.[28]

Recommendation 1: Nomenclature

Three classes of reference materials should be designated for use in clinical chemistry. They are: (1) Certified Reference Material (CRM); (2) Calibration and Test Material (CTM); and (3) Control Material (no acronym). Only CRMs and CTMs should be considered as "standards".

Recommendation 2: Characterization and Use of Reference Materials

CRMs should be characterized by definitive methods (when available) and/or exhaustive characterization by independent reliable techniques. Major use should be to assure the long-term adequacy and integrity of measurement systems, for the validation of CTMs and to aid in establishing the performance characteristics of methods.

CTMs must be traceable to a CRM (if available) by a reference method or well-accepted methods. Major use should be to help maintain calibration of individual laboratory measurement systems. Where CRMs and acceptable measurement techniques do not exist, then the term "Provisional CTM" would be appropriate.

Recommendation 3: General Technical Criteria for Reference Materials

The following general criteria should apply to CRMs: (1) specific composition; (2) hemogeneity; (3) physical characteristics (e.g., color, particle size — importance of these will depend on specific reference material); (4) other chemical characteristics (e.g., reactivity); (5) biological characteristics (e.g., biological hazards, sterility); (6) stability; and (7) preservatives, if any.

Recommendation 4: National Committee for Clinical Laboratory Standards Establishes Technical Criteria

NCCLS is the appropriate body to develop general technical criteria for the characterization and certification of CRMs by a consensus process and to establish specific criteria for individual materials.

Recommendation 5: Certification of CRMs

CRMs should be certified by NBS or other qualified organizations recognized by NCCLS.

STANDARD METHODS

Standards for reference methods constitute even more complex semantic problems than materials do. At the national conference on an understanding for the development of reference materials and methods for clinical chemistry,[28] it is noted that the term "definitive method" first appeared in a glossary for clinical chemistry among the provisional recommendations of the International Federation for Clinical Chemistry Committee on Terminology.[29] That definition: "a method which after exhaustive investigation is found to have no known source of inaccuracy or ambiguity", was revised by the National Bureau of Standards in its position paper to include the degree of accuracy needed for the intended use of the definitive method. The NBS definition was "a definitive method is an analytical method that is capable of providing the highest accuracy among all methods for determining that analyte, and its accuracy must be adequate for its stated end-purposes." NBS further stated that important end-purposes for the definitive method for an analyte are used in evaluating accuracy of the reference method and used in determining definitive method values for matrix primary reference materials. It is because definitive methods are too expensive and inefficient to use for standardization end-purposes that the more practical reference methods are essential. At the national conference, the NBS proposed the following definitions for a reference method: "a reference method is an analytical method whose inaccuracy and imprecision are small enough as demonstrated by direct comparison with the definitive method and whose low incidence of susceptibility to known interferences is thoroughly documented that the stated end-purposes of the reference method may be achieved.'' In a definitive method, systematic error is either absent or accurate corrections can be applied. In a reference method, not all of the systematic error may be known and corrected or correctable and, hence, it may exhibit some bias. Both methods must necessarily be capable of being performed with high precision.

At the same conference, the CAP stated that the definitive methods offer unique opportunity to measure the magnitude of inaccuracy. However, CAP felt that the application of definitive values to routine laboratory purposes is limited because of the cost and complexity of definitive value technology and the time constraints required for routine application. With regard to reference methods, many laboratorians feel that the role of the reference methods should be more clearly defined with regard to their use in the clinical laboratory. This has become especially important due to the fact that reference methods are becoming the focus about which many regulatory pressures in the clinical laboratory revolve.

It is noted above that it has been suggested that reference methods be documented by comparison to definitive methods, and hierarchies are proposed to fix the relationship of each component in the measuring system. The pathologist organizations feel that this would substantially lessen the chances for any further evolution in the designation of a reference method and that the establishment of reference methods can become a costly undertaking. For this reason, many professionals feel that flexibility and plurality of approaches should be preserved. The CAP suggests that a reference method is valuable when inaccuracy has been demonstrated to be a significant problem in routine clinical practice and when the inaccuracy is detrimental to patient care. The National Bureau of Standards agrees that the system should be flexible to accommodate new findings and to make appropriate modifications when necessary. The Center for Disease Control states that when a reference method is being evaluated for acceptance, it should be independently validated to assure: (1) its conformity with general guidelines and content in format, and (2) its accuracy, precision, specificity, and workability as a procedure. It further feels that transferability testing should take place with all proposed and validated reference methods. The debate will continue. However, in this observer's opinion, the course established by the NCCLS is most reasonable and will be most effective. The NCCLS recognizes a profound limitation in the resources of people, time, and money in developing definitive methods and standardized reference methods. NCCLS proposes that before any standards development project proceeds, the following questions be asked and answered affirmatively:

1. Is the project or issue medically useful and clinically relevant?
2. Is consensus involved?
3. Are the funds available to support the project?
4. Is sufficient time available to complete the project reasonably?

If standards development processes are geared to analytical goals such as those defined by the Aspen Conference on Analytical Goals in Clinical Chemistry sponsored by the College of American Pathologists, then the primary motivating mechanism for the standards development process will be medical usefulness and clinical relevance and, thereby, improved patient care.

REGULATORY STANDARDS

For several reasons, some documented by scientific data and others not, laboratory performance and function was subjected to regulatory control. This provided further impetus to standards development. The two most significant pieces of legislation in the past two decades were the Clinical Laboratory Improvement Act of 1967. This Act addressed the safety and efficacy of medical devices including in-vitro diagnostic devices. The Food and Drug Administration (FDA) presently requires that all medical devices be classified in one of three classes depending upon the degree of risks to health. Thus, a Class I medical device would be required to meet general manufacturing principles as established by the FDA. A Class II medical device would be required to meet existing or proposed performance standards judged necessary by the FDA, to control their safety and effectiveness. Since there exist certain devices for which standards do not exist and cannot be written, a different method of control must be followed, which is designated as a Class III category. Every device in this class (III) requires approval by the FDA prior to being made available by industry for in-vitro diagnostic use.

During the process of classification, considerations included the following factors: (a) whether the in-vitro diagnostic product is potentially hazardous to life, health, or well-being when put to its intended use; and (b) whether the results obtained by use of the product directly affect treatment. As a result of this classification process, many in-vitro diagnostic products or devices were placed in Class II since they represent potential hazards to a person's well-being (over-treatment/under-treatment).

The following sections discuss the necessay performance characteristics of laboratory results, performance standards development (FDA and voluntary), manufacturing practices, product labeling, reporting of product problems, and the future course of performance standards—all of which are relevant to the director of a clinical chemistry laboratory.

Performance Characteristics of Laboratory Results

Laboratory results are expected to be accurate, dependable and reproducible, Therefore, the fundamental question concerning test results is "How accurate and dependable must the test be?" Limits must be established for the accuracy, precision, sensitivity and specificity of each procedure. This is frequently established by comparative studies of the device methodology and the use of a standard reference material. Accuracy and precision of the testing procedure must be reviewed with respect to the range that is covered. Sensitivity and specificity must be reviewed with respect to the diagnostic use of the product. These points must be addressed in the development of performance standards.

Clinical Chemistry Device Performance Standards Development

Developing a performance standard for a diagnostic device by the FDA requires a great deal of help from many sources. Any interested individual or professional group may recommend a performance standard to the agency. Presently, there are three predominant mechanisms available to the agency for the development of performance standards: (1) an existing standard may be utilized, such as one developed by the College of American Pathologists (CAP) or National Committee for Clinical Laboratory Standards (NCCLS); (2) interested individuals outside the FDA may develop a proposed performance standard; or (3) the FDA may develop the standard.

In an effort to review the adequacy of existing standards and as a guide for the development of new standards, the agency annually prepares the "Medical Device and Diagnostic Products Standards Survey". The survey is published in two parts: the International Edition in January, and the National Edition in July. The information serves as a basis for the continuing review of the availability and adequacy of existing standards and as a guideline for the development of future standards.

The FDA has a memorandum-of-understanding with the Center for Disease Control (CDC) for scientific and technical advice; it also works closely with the National Bureau of Standards (NBS). Additionally, the Bureau of Medical Devices (BMD) has access to the regional laboratory system to various referencce methodologies and diagnostic devices. Frequently, these are the subject of complaints which must be investigated. It is planned that, at a later date, these facilities may validate the testing and conformance of performance standards. Many members of the College of American Pathologists (CAP) and the American Association of Clinical Chemists (AACC) serve actively on BMD's in-vitro device advisory panels as members or consultants. It is hoped that in the future, professional organizations such as CAP and AACC will be interested in contributing valuable data from Clinical Chemistry and Therapeutic Drug Monitoring (TDM) Surveys for FDA standards development activities.

Voluntary Standards Development as Related to FDA Standards Program

Voluntary standards organizations or other interested parties are invited to send copies of standards and pertinent information to the FDA. They must include the following;

1.   A rationale for the establishment of the standard
2.   Statements regarding reasonable assurance of safety and effectiveness
3.   Development process which provides for reasonable technical and scientific information input
4.   Development process which allows participation of all concerned interests
5.   A standard which does not contribute any bias
6.   A standard which is subject to periodic review
7.   A standard which emphasizes performance-oriented require requirements
8.   A standard which manufacturers are likely to comply with
9.   A standard which identifies testing criteria to determine device conformance where applicable

A review will be made by the FDA's Division of Standards. The initial review will identify any deficiencies that are to be corrected by the standards developer before proceeding to detailed scientific review. A detailed scientific review will then be performed for each voluntary standard. For voluntary standards found adequate by the FDA, a recommended action will be taken in conformance with FDA's future voluntary standards policy.

There are several compelling reasons why great reliance has to be placed on future use of voluntary standards rather than a great number of mandatory standards. This is consistent with the present objectives of reducing the cost of regulation, allowing greater flexibility, allowing technological innovation in the private sector, and the elimination of redundant and obsolete federal standard setting. In that regard, FDA's policy directly follows guidelines issued in the Office of Management and Budget Circular A-119, January 17, 1980. This circular recognizes the value of voluntary standards and provides guidance for federal agency interaction with voluntary standards organizations. The present environment will require mandatory standards only when there are no feasible voluntary standards available or under development or when there is an absolute necessity for regulatory approach in the interest of safety and effectiveness.

Good Manufacturing Practices for Medical Devices

"The Act" authorizes the FDA to prescribe good manufacturing practice regulations (GMPs) concerning the manufacturing process, packaging, storage, and installation. It further requires a complete record of specific manufacturing processes as applicable to a particular quality assurance system. The law requires the FDA to publish the Clinical Chemistry Panel's recommendations and a proposed regulation classifying each device in the Federal Register allowing sufficient opportunity for interested persons to comment on the proposed regulations.

In-Vitro Diagnostic Product Instructions (Labeling)

A product package insert is a mini-brochure. Its chief purpose is to provide more information about a product than can be placed on a container label. The label information is usually limited to the product's name, the manufacturer, and an expiration date when appropriate.

The labeling (package insert) for in-vitro products includes intended use and type of procedure (qualitative or quantitative); summary and explanation of test (including literature references, special merits, limitations, deviations from original methods and its effects on results); test principle (chemical and physiological reactions and techniques).

For reagents: (concentration, activity, reactive ingredients, buffers, preservatives, stabilizers); for instruments (principles of operation, instructions for use, calibration, installation service and maintenance, performance specifications); speciment collection and preparation (collection and handling precautions, necessary preservations and precaution from interfering substances); procedural outline (with details of calibration, reference sample, details of quality control procedures with performance limits); results (detailed explanation of calculations for determination of value of unknown analyte or reactive substance, adequate description of expected results); limitations of procedure (interfering substances and indication of further required testing); expected values (studies of various populations, how range was established, identification of population on which range was established; specific performance characteristics (accuracy, precision, specificity, and sensitivity based upon generally accepted method using specimens from normal and abnormal populations); bibliography.

Medical Device and Laboratory Product Problem Reporting Program

Another program of the Agency is the Laboratory Product Problem Reporting (LPPR) Program. This program is aimed at identifying problems and improving in-vitro diagnostic products performance through user's participation. A contract was awarded to the US Pharmacopeia (USP) to administer the program. All Laboratory Product Problem Report forms are sent to the USP, and replacement forms are supplied to users by the USP. When the LPPR is received by the USP, a copy is sent to the manufacturer and a copy is sent to the FDA. The Food and Drug Administration may also send a copy to CDC for evaluation. The LPPR form is a simple one-sheet form requesting the following information: product name, lot number, date purchased, source, user's name, lab, name, phone number, and the nature of the problem.

This program is a simple and effective channel for the user to report pfoduct problems and plays an important part in their correction. The exchange of meaningful and reasonable information among health care professionals, industry and government is an integral part of improving the safety, efficacy and quality of diagnostic products.

Future Course of Performance Standards

The law requires the FDA to publish all Advisory Panel recommendations and proposed regulations classifying each device in the federal register and to provide an opportunity for interested persons to comment on the proposed regulations.

Because of the importance of final classification of products in clinical chemistry, the agency is particularly interested in receiving comments from consumers, health professionals, and manufacturers. To be most influential, comments should be specific and alternative suggestions should be offered. A public hearing may be scheduled to provide interested parties with an additional opportunity for input on the pertinent issues. It is therefore obvious that performance standards can be influenced by all concerned parties.

STANDARDS FOR LABORATORY INSPECTION AND ACCREDITATION

Proficiency testing and inspection and accreditation of laboratories have been viable tools in clinical laboratory quality assurance programs since the 1960's. However, with the development of increased regulatory impact upon laboratories, these modalities have attracted increased attention and usage. As a result, standards have developed concerning laboratory inspection and accreditation. The College of American Pathologists Commission on Laboratory Inspection and Accreditation has developed a set of standards for accreditation of medical laboratories, and they are generically listed as follows:

Standard I.   The pathology and clinical laboratories shall have sufficient space, equipment and facilities for the performance of the required volume of work with optimum accuracy, precision, efficiency, and safety.

Standard II.   A hospital department of pathology and clinical laboratory shall be directed by a physician who is qualified to assume professional, organizational, and administrative responsibility for the department.

Standard III.   The director of an independent laboratory must be a physician with training and experience in clinical laboratory work to meet the requirements of a laboratory director under the Laboratory Improvement Act of 1967. Sufficient personnel with training and experience adequate to supervise and conduct the work of the clinical laboratories shall be provided.

Standard IV.   Channels of communication within the laboratories as well as with all other closely affiliated sections shall be appropriate for the size and complexity of the organization.

Standard V.   The quality control system of the laboratory shall be designed to assure the medical reliability of laboratory data.

All specimens removed at operation must be clearly identified and sent to the pathologists accompanied by the pertinent clinical information. The extent of the examination is to be determined by the pathologist.

Standards are also available from the Center for Disease Control dealing with the inspection and accreditation of laboratories. These were published in the *Federal Register* under Part 74, Title 42 of the Public Health Service Code, in Volume 33, pages 15 through 303.

In addition to standards dealing with material and methods, other standards have been developed which do not have specific application to a material or a method. These have received a variety of names, such as performance standards, "how to" standards, standards of practice, protocol standards, etc. A classical example of such a standard would be the NCCLS ASM-2, dealing with antimicrobial disc susceptibility testing in microbiology.

CURRENT CLINICAL LABORATORY STANDARDS

Table 21 shows an abbreviated list of standards basically developed over the last decade by certain United States, international, and national professional organizations dealing with methods, materials, protocols, and systems potentially pertinent to clinical chemistry. Readers are referred to the *Guide to Biomedical Standards*[30] for a more complete listing. Personnel standards are not addressed in this chapter.

Most of the listed standards are relevant to safety, not to the solution of analytical problems. NCCLS is one organization which is addressing the latter. Areas covered include: water works, electricity, compressed gases, heating, communication, lighting, fire protection, sanitation, nuclear material, and others.

Standards for Blood Banks and Transfusion Services can be obtained from the Committee on Standards, American Association of Blood Banks, 1828 L Street, N.W., Washington, D.C. 20036.

Table 21

SELECTED BIOMEDICAL STANDARDS OF THE U.S. INTERNATIONAL AND NATIONAL PROFESSIONAL ORGANIZATIONS PERTINENT TO CLINICAL CHEMISTRY

1. UNITED STATES STANDARDS

American National Standards Institute (ANSI)
ANSI, Inc., 1430 Broadway, New York, N.Y.10018

The Institute represents America in the International Organization for Standardization (ISO) and in the Pan American Standards Commission (COPANT). The Medical Devices Standards Management Board (MDSMB) of the Institute sets medical product standards.

ANSI/CGA C4	Portable Compressed Gas Containers to Identify the Material Contained, Method of marking, 1954, 1978.
ANSI/iCGA V-1 B57.1	Compressed Gas Cylinder Valve Outlet and Inlet Connections, 1977.
ANSI/NCCLS ASI-1	Preparation of Manuals for Installation Operation, and Repair of Laboratory Instruments, 1972.
ANSI/NFPA 10	Portable Fire Extinguishers, Installation Maintenance, and Use of, 1975.
ANSI/NFA 70	National Electrical Code, 1978.
ANSI/NFPA 56C	Laboratories in Health-Related Institutions, 1975.

American Society for Quality Control
806 15th Street, N.W., Washington, D.C. 20036

ASQC STD A--01	Definitions for Control Chart (Revision) Z1.5
ASQC STD A-2	Definitions for Acceptance Sampling by Attributes (Revision Z1.6)
ASQC STD A-3	Glossary-General Terms used in Q.C. (Revision Z1.7)
ASQC STD A-4	Standard on Generic Terms (New)
ANSI STD Z-1,9	(MIL STD 414) Sampling Inspection by Variables
ASQC	Sampling Plan by Attributes (New)
Z 1.15	Generic Guidelines for Quality Systems (a Z-1 Committee project)
A-5	Standard on Design of Experiments Terminology
ASQC STD C-1	Specification of General Requirements for a Quality Program (Reaffirmation)

Reliability System Standard
Z1.4 (MIL STD 105D) Revision
Small Lots (LTPD/AOQL) Sampling Plan
A-6 Reliability Terminology
Quality System Standard for Test Laboratories
Metrology Quality Standard I
   (Setup and Operation) (From Z-1)
Metrology Quality Standard II (Maintenance and Calibration)
   (From Z-1)
Nuclear Inspection Planning Standard
Quality System Standard for Non-Nuclear Power Generation
Sampling Plan for Sterile Processes

American Society for Testing and Materials (ASTM)
ASTM, 1916 Race Street, Philadelphia, Pa. 19103

ASTM Committee on Medical and Surgical Materials and Devices is F-4. Subcommittees are F-4.10 Administrative; F4.11 Ophthalmology; F4.20 Resources; F4.30 Orthopaedics; F4.40 Cardiovascular; F4.50 Neurosurgical; F4.60 Plastic and Reconstructive Surgery; F4.70 Otolaryngology; F4.80 Medical/Surgical Instruments; F4.90 Executive; and F4.92 Planning.

Association for the Advancement of Medical Instrumentation (AAMI) AAMI, 1901 N. Ft. Myer Drive, Suite 602, Arlington, Va. 22209.

College of American Pathologists (CAP)
College of American Pathologists, 7400 North Skokie Boulevard,
Skokie, Ill. 60077.

Standards for Laboratory Inspection and Accreditation, June 1978.
Product Evaluation Program Protocol, March 1978.
Workload Recording Method for Clinical Laboratories, 1979.
Standard for Cyanmethemoglobin Determinations, 1968.
Standard for Bilirubin (Crystalline) Determinations.
Criteria for Certification of a Standard Cholesterol for Use in Serum Cholesterol Measurements by Medical Laboratories, 1966.

Food and Drug Administration (FDA)

To keep current with FDA regulations, subscribe to the *Federal Register*. Order from: Superintendent of Documents, U.S. Government Printing Office, Washington, D.C. 20402.

Title 21, Code of Federal Regulations

Title 21, CFR, consists of 7 volumes containing all regulations. These volumes are updated each year. The current edition may be purchased as follows:

Vol. 1	Parts 1	through 9	-	General regulations
2	100	199	-	Food for Human Consumption
3	200	299	-	Drugs, General
4	300	499	-	Drugs for human use.
5	500	599	-	Animal drugs, feed, and related products.
6	600	1299	-	Biologics; Cosmetics; Medical Devices; Radiological Health and Regulations Under Certain Other Acts Administered by FDA.
7	1300	End	-	Regulations issued by Drug Enforcement Administration, Department of Justice; Special Action office for Drug Abuse Prevention.

Chapter 1 of Title 21 of the CFR is now organized as Subchapter H, Parts 800-999 of Title 21.

Federal Register

All FDA regulations, proposed regulations, and other notices are published in the *Federal Register*. Subscription available from U.S. Government Printing Office, Washington, D.C. 20402.

Bureau of Medical Devices (BMD) FDA's Bureau of Medical Devices, 8757 Georgia Avenue, Silver Spring, Md. 20910.

The Bureau's authority derives from the Medical Device Amendments of 1976, Public Law 94-295. Order copies from: Bureau of Medical Devices.

FR* 37:819	In Vitro Diagnostic Products for Human Use, January 1972.
FR 39:2089	In Vitro Diagnostic Products for Human Use, January 1974.
FR 39:8610-8611	In Vitro Diagnostic Products for Human Use, March 1974.
FR 39:9217-9219	In Vitro Diagnostic Products for Human Use, March 1974.
FR 37:16613-16617	In Vitro Diagnostic Products for Human Use, August 1972.

*   FR-refers to *Federal Register* publication.

FR 37:20040	In Vitro Diagnostic Products for Human Use, September 1972.
FR 38:7096-7012	In Vitro Diagnostic Products for Human Use:
	Labeling Requirement for Development of Standards for In Vitro Diagnostic Products for Human Use, March 1973.
FR 38:13558	In Vitro Diagnostic Products for Human Use, May 1973.
FR 38:30290-30292	In Vitro Diagnostic Products for Human Use, Oct. 19
FR 38:8650	Labeling Requirements and Procedures for
	Development of Standards for In Vitro
	Diagnostic Prodcts for Human Use: Correction, April 1973.
FR 38:9665	In Vitro Diagnostic Products for Human Use, April 1973.
21 CFR S 809	"In Vitro Glucoses for Human Use"
	(Contains General Provisions: Labeling; Requirements for Manufacturers and Procedures).

Bureau of Radiological Health (BRH)
FDA's Bureau of Radiological Health, 5600 Fishers Lane,
Rockville, Md. 20911.

The Bureau derives its authority from the "Radiation Control for Health and Safety Act of 1968".

FR 34:8952	Regulations for the Administration and Enforcement of the Radiation Control for Health and Safety Act of 1968.
	FR-2073-4
	FR 35:980-4
	7699
	8383-5
	15642
	16795

Other Publications of FDA
Order from:Bureau of Medical Devices, 8757 Georgia Avenue,
Silver Spring, Md. 20910

Medical Device Standards Survey — Current national and international standards.

Includes: voluntary and regulatory standards, tentative and recommended practices, purchasing specifications, policy statements and glossaries of technical terms.

Medical Device Classification Panel Reports — Outline the classification process. Request reports from the Hearing Clerk, Food and Drug Administration, Room 4-65, 5600 Fishers Lane, Rockville, Md. 20852.

Medical Device Experience Monitoring Network (MDEMN) — Reporting Forms — MDEMN processes experience reports. Forms are available for reporting product experience data

Federal Food, Drug, and Cosmetic Act — Contains the text of the Federal Food, Drug, and Cosmetic Act.

Freedom of Information Regulations — Regulations issued under Title 21 of the Code of Federal Regulations for Freedom of Information.

Imports and FDA — Describes the laws enforced by FDA and products covered. Shows the steps an import must go through.

Regulations for the Administration and Enforcement of the Radiation Control for Health and Safety Act of 1968 — Regulations of the Bureau of Radiological Health, for electronic and radiological products.

Health, Education, and Welfare, Department of (HEW) Free single copies of the publications (up to 15 titles) may be obtained from: Technical Publications Section,

National Health Planning Information Center, Division of Planning Methods and Technology, Center Building, Room 5-22, 3700 East-West Highway, Hyattsville, Md. 20782.

Joint Commission on Accreditation of Hospitals (JCAH) JCAH, 875 North Michigan Avenue, Chicago, Ill. 60611

H-103	Accreditation Manual for Hospitals, 1980 edition.
H-104	Hospital Survey Profile, 1979.
H-110	Guidelines for the Application of Hospital Accreditation Standards in University Hospitals, 1977.
LTC-200A	Accreditation Manual for Long Term Care Facilities, 1975.
AHC-501	Accreditation Manual for Ambulatory Health Care, 1978.
DEP-110	Program on Hospital Accreditation Standards (PHAS) Manual, 6th Edition, 1979.

PHAS Manual Supplements:

DEP-115	Pathology and Medical Laboratory Services, 6th Edition.

National Committee for Clinical Laboratory Standards (NCCLS) NCCLS, 881 East Lancaster Avenue, Villanova, Pa. 19085.

The status of the Standards is indicated by the first letter if the Code: "A" is an Approved Standard, "T" is Tentative, and "P" is Proposed.

Clinical Chemistry

ASC-1	Standardized Protein Solution (Bovine Serum Albumin), 2nd edition.
ASC-2	Standards for Calibration, Reference and Control Materials in Clinical Chemistry.
TSC-3	Reagent Water Specification and Test Methods for Water Use in the Clinical Laboratory.
TSC-5	Methodological Principles for Establishing Principal Assigned Values to Calibrators.
PSC-7	Guidelines for Kinetic Analysis of Enzyme Reaction.
PPC-10	Enzyme Assay Conditions: Choice of Standard Reaction Temperature for Assaying Activity Values of Enzymes of Human Sera.
PPC-11	Quantities and Units: SI.
PSC-12	Definition of Quantities and Conventions Related to Blood pH and Gas Analysis.
PSC-15	Guidelines for the Development of Definitive Methods for Use in Clinical Chemistry for the National Reference System in Clinical Chemistry.
PSC-16	Guidelines for the Development of Reference Methods in Cl. Chem. for the National Reference System in Clinical Chemistry.

Evaluation Protocols

PSEP-2	Protocol for Establishing Performance Claims for Clinical Chemical Methods. Introduction and Performance Check Experiment.
PSEP-3	Protocol for Establishing Performance Claims for Clinical Chemical Methods: II. Replication Experiment.
PSEP-4	Protocol for Establishing Performance Claims for Clinical Chemical Methods, Comparison of Methods Experiment.

Hematology

ASH-1	Standard for Evacuated Tubes for Blood Specimen Collection.
ASH-2	Standard Method for the Human Erythrocyte Sedimentation Rate (E.S.R.) Test.
TSH-3	Standard Procedures for the Collection of Diagnostic Blood Specimens by Venipuncture.
TSH-4	Standard Procedures for the Collection of Diagnostic Blood Specimens by Skin Puncture.
TSH-5	Standard Procedures for the Handling and Transport of Diagnostic Medical Specimens and Etiologic Agents.
PSH-6	Standard Assay for Determination of Erythropoietin Activity in Body Fluids.
PSH-7	Standard Procedure for Determination of Packed Cell Volume by the Microhematocrit Method.
PSH-8	Abnormal Hemoglobin Detection by Cellulose Acetate Electrophoresis.
PSH-9	Standard for the Chromatographic (Microcolumn) Determination of Hemoglobin A_2.

PSH-10	Standardized Method for Determination of Sickling Hemoglobins in Whole Blood.
PSH-12	Standard for Screening for Red Blood Cell Glucose-6 Phosphate Dehydrogenase Activity.
PSH-15	Reference Procedure for the Quantitative Determination of Hemoglobin in Blood .

Immunohematology/Blood Banking

| ASI/BB-1 | Specifications for Standard Isotonic Sodium Chloride Solution for Immunohematologic Testing |

Instrumentation

ASI-1	Preparation of Manuals for Installation, Operation, and Repair of Laboratory Instruments 2nd edition.
ASI-2	Standard for Temperature Calibration of Water Baths, Instruments, and Temperature Sensors.
TSI-3	Standard for Determining Spectrophotometric Performanc Criteria.
PSI-4	Guideline to the Selection of Accuracy Class of Thermistor.
TSI-5	Power Requirements for Clinical Laboratory Instruments and for Laboratory Power Lines.
TSI-6	Guidelines for Service of Clinical Laboratory Instruments.

Labeling

| ASL-1 | Labeling of Clinical Laboratory Materials, 2nd edition. |

Ligand Assay

| PSLA-1 | Guidelines for Standards to Assess the Quality Radioimmune Systems. |

Microbiology

ASM-1	Standard Test for Labeling Efficiency of FITC.
ASM-2	Performance Standards for Antimicrobial Disc Susceptibility Tests, 2nd edition.
TSM-4	Determination of Fluorescein/Protein Ratios in Fluorescein Isothiocynate Labeled Protein.
ASM-5	Standard for Rubella Hemagglutination-Inhibition (HAI), Reagents and Test Procedure.
PSM-11	Reference Method for Anaerobic Bacteria Susceptibility Testing.

Toxicology/Drug Monitoring

| PST/DM-1 | Standard for Development of Requisition Forms for Therapeutic Drug Monitoring and/or Overdose Toxicology. |

National Council on Radiation Protection and Measurements (NCRP)
NCRP, Publications Department, 7910 Woodmont Avenue, Suite 1016, Washington, D.C. 20014.

8	Control and Removal of Radioactive Contaminations in Laboratories, 1951.
22	Maximum Permissible Body Burdens and Maximum Permissible Concentrations of Radionuclides in Air and Water for Occupational Exposure, 1959.
30	Safe Handling of Radioactive Materials, 1964.
48	A Handbook of Radioactivity Measurements Procedures, 1978.

National Fire Protection Association (NFPA) NFPA, Publication Service Department, 470 Atlantic Avenue, Boston, Mass. 02210

10	Portable Fire Extinguishers, 1978
13	Installation of Sprinkler Systems, 1978
30	Flammable and Combustible Liquids Code, 1977
45	Fire Protection for Laboratories Using Chemicals
49	Hazardous Chemicals Data, 1975
56C	Laboratories in Health-Related Institutions, 1973
70B	Recommended Practice for Electrical Equipment Maintenance, 1977
70E	Electrical Safety Requirements for Employee Work Places, 1979
76A	Essential Electrical Systems for Health Care Facilities, 1977
77	Recommended Practice on Static Electricity, 1977
80	Fire Doors and Windows, 1979
321	Basic Classification of Flammable and Combustible Liquids, 1976
325M	Fire Hazard Properties of Flammable Liquids, Gases, and Volatile Solids, 1977.
491M	Manual of Hazardous Chemical Reactions, 1975

National Sanitation Foundation (NSF)
NSF Building, Ann Arbor, Mich. 48105

30 Cabinetry and Laboratory Furniture for Hospitals, 1973
49 Class II (Laminar Flow) Biohazard Cabinetry, 1976

Nuclear Regulatory Commission (NRC)
NRC, Washington, D.C. 20555

The U.S. Nuclear Regulatory Commission was established by authority set forth in Section 201(a) of the Energy Reorganization Act of 1974. Functions of the Atomic Energy Commission were transferred to the NRC.

The NRC Office of Nuclear Material Safety and Safeguards licenses the use of radioisotopes in industry, research, and medicine where State authority under agreements with the NRC is absent. The NRC, Office of Standards Development develops regulations, criteria, guides, standards, and codes for nuclear reactors, medical, academis, and industrial facilities, and transportation.

A. Regulations

The NRC publishes its regulations and proposed changes in the *Federal Register*. Subscription includes the basic manual and supplements for an indefinite period. Write: Superintendent of Documents, Government Printing Office, Washington, D.C. 20402.

10 CFR, chapter 1:	Part 19	Notices, Instructions, and Reports to Workers; Inspections
	Part 20	Standards for Protection Against Radiation
	Part 30	Rules of General Applicability to Licensing of Byproduct Material
	Part 35	Human Uses of Byproduct Material
	Part 71	Packaging of Radioactive Material for transport and Transportation of Radioactive Material Under Certain Conditions

B. Regulatory Guides

Regulatory Guides provide guidance to applicants. Regulatory Guides do not substitute for regulations. Compliance with them is not required. Methods and solutions other than those in the guides are acceptable if they satisfy the regulations.

Medical institution guides:
Division	6	-	Products
	7	-	Transportation
	8	-	Occupational Health
	10	-	General submitting of license applications. Licensing information is available from the License Management Branch (396-SS), U.S. NRC, Washington, D.C. 20555.

The following Radiation Regulatory Guides are relevant to clinical chemistry:

R.G. 8.13	Instruction Concerning Prenatal Radiation Exposure, 1975
R.G. 8.18	Information Relevant to Ensuring That Occupational Radiation Exposures at Medical Institutions Will Be As Low As Reasonably Achievable, 1977.
R.G. 8.20	Applications of Bioassay of I-125 and I-131 (For comment; issued 4.78)
R.G. 8.23	Radiation Safety Surveys at Medical Institutions (For comment; issued 2/79)
R.G. 10.5	Guide for the preparation of Applications for Type A Licenses of Broad Scope for Byproduct Material (For comment; issued 9/76) (only applicable to medical institutions with research programs utilizing radioactive materials)

Medical institution applying for an NRC license should write for current guides to:

License Management Branch (396-SS)
Division of Fuel Cycle and Materials Safety
Office of Nuclear Materials Safety and Safeguards U.S. NRC,
Washington, D.C. 20555

Occupational Safety and Health Administration (OSHA)
OSHA, U.S. Department of Labor, Washington, D.C. 20210.

OSHA sets standards for the protection of employees in places of employment rather than consumer safety.

Code of Federal Regulations, Title 29 - Labor, Chapter XVII, Part 1910, "Occupational Safety and Health Standards"

Code of Federal Regulations, Title 29 - Labor, Chapter XVII, Part 1926, "Safety & Health Regulations for Construction."

(See especially Sections: 1926.11, 12 (bX4), - (13), - (14), and 1926.15)

Code of Federal Regulations, Title 41 - Pulic Contracts, Property Management, Chapter 50, Public Contracts, Dept. of Labor, Part 50-204, "Safety & Health Standards for Federal Supply Contracts."

OSHA Subscription Service

OSHA standards are published in the *Federal Register*. A subscription service is available. Included are: Vol. I, General & Industry Standards and Interpretations; Vol. II, Maritime Standards; Vol. III, Construction Standards & Interpretations; Vol. IV, Other Regulations and Procedures; Vol. V, Field Operations Manual. Contact: Superintendent of Documents, U.S. Government Printing Office, Washington, D.C. 20402.

Underwriters Laboratories, Inc. (UL)
1285 Walt Whitman Road, Melville, N.Y. 11746
207 E. Ohio Street, Chicago, Ill. 60611
333 Pfingsten Road, Northbrook, Ill. 60062
1655 Scott Boulevard, Santa Clara, Va. 95050

Underwriters Laboratories, Inc., was founded in 1894 to ensure "safe" devices, systems, and materials,

407	Manifolds for Compressed Gas, 1978
416	Refrigerated Medical Equipment, 1978
467	Electrical Ground and Bonding Equipment, 1978
886	Electrical Outlet Boxes and Fittings for Use in Hazardous Location (ANSI/UL), 1976
894	Switches for Use in Hazardous Locations (ANSI/UL), 1977
913	Intrinsically Safe Electrical Circuits and Equipment for Use in Hazardous Locations and Its Associated Apparatus, 1976
943	Ground-Fault Circuit Interrupters, 1977
1022	Line Isolation Monitors, 1973
1053	Ground-Fault Sensing and Relaying Equipment, 1976
1262	Standard for Laboratory Equipment, 1976
1209	(Proposed) Centrifuges for Use in Hazardous Locations, February 1973

2. INTERNATIONAL STANDARDS

Commission on World Standards/
 World Association of Societies of Pathology
A. C. Ritchie, M.D., Banting Institute, 100 College Street, Toronto, Ontario M5G 1L5, CANADA

International Commission on Radiation Units and Measurements (ICRU)
7910 Woodmont Avenue, Suite 1016, Washington, D.C. 20014

International Commission on Radiological Protection (ICRP)
Clifton Avenue, Sutton, Surrey, England SM2 5PU

International Commission on Rules for the Approval of Electrical Equipment (CEE)

IEC Publications are available from: U.S. National Committee of the IEC, USNC/ANSI, 1430 Broadway, New York, N.Y. 10018

International Federation of Clinical Chemistry
Dr. Robert Zender, Laboratoire Hopital de la Chaux-de-Fonds, CH-2301 La Chaux-de-Fonds, SWITZERLAND

International Organization for Standardization (ISO) ANSI is the American Member Body of ISO which includes 68 national standards bodies and 17 correspondent members. ISO Central Secretariat, 1 Rue de Varembe, 1211 Geneva 20, Switzerland.

ISO 4071-1978   Exposure meters and dosimeters - General methods for testing

International Union of Pure and Applied Chemistry
Dr. R. Grasbeck, Minerva Foundation Institute for Medical Research, POB 819, SF-0010 Helsinki, FINLAND

3. NATIONAL STANDARDS

Australia

Standards Association of Australia, Standards House, 80 Arthur St., North Sydney, N.S.W., 2060 Australia. Mailing Address: PO Box 458

Austria

Oesterreichisches Normungsinstitut, Leopoldsgasse 4. A-1020 Wien, Austria (Member body of ISO)

Oesterreichischer Verband fur Elektrotechnik (OEVE), Eschenbachgasse 9, A-1010 Wien, Austria

Belgium

Institut Belge de Normalisation, Avenue de la Brabanconne 29, 1040 Bruxelles, Belge or: Comite Electrotechnique Belge Galerie Ravenstein 3, Boite 11, 1000 Bruxelles, Belgium

Canada

Bureau of Medical Devices, Environmental Health Centre, Tunney's Pasture, Ottawa, Ontario, Canada, KIA OL2

Czechoslovakia

Office for Standards and Measurements, Vaclavske nam. 19, 113 47 Praha 1, Czechoslovakia

Danish Standards Association, Post Box 77, Aurehojvej 12, DK-2900, Hellerup, Denmark

European Committee for Clinical Laboratory Standards
Dr. D. Laue, Deutsch Gesellschaft der Klinische Chemie, Neumarkt 1c, 5000 Koln 1, GERMANY

Finland

Suomen Standardisoimisliitto, PO Box 205, SF-00121 Helsinki 12, Finland

France

Association Francaise de Normalisation (AFNOR) Tour Europe, Cedex 7, 92080 Paris - la Defense

Germany

Normung Burggrafen Strasse 4-7, Postfach 1107, 1000 Berlin 30, Germany

Society of Clinical Chemistry, Germany
Prof. Dr. D. Stamm, Max-Planck-Institut fur Psychiatrie Klinik, Kraepelinstrasse 10, 8000 Munchen 40, W. GERMANY

India

Indian Standards Institution, Manak Bhavan, 9 Bahadur Shah Zafar Marg, New Delhi, 110001, India

Ireland

The Standards Division, Institute for Industrial Research and Standards, Ballymum Road, Dublin 9, Ireland. U.S. resident: Orders should be sent to: American National Standards Institute, 1430 Broadway, New York, N.Y. 10018

Japan

Japanese Standards Association, 1-24, Akaska 4, Minato-ku, Tokyo 107, Japan

Institute of Electrical Engineers of Japan (IEE), 1-3 Yuraku-cho, Chiyoda-ku, Tokyo, Japan

Japan Society of Mechanical Engineers (JSME), Sanshin-Hokusei-Building, 4-9, Yoyoki 2, Shibuya-ku, Tokyo, Japan

Society of Instrumentation and Control Engineers (SICE) c/o Keisokukaikan, 20 Kotochiracho, Minato-ku, Tokyo, Japan.

Japanese Industrial Standards Committee, Agency of Industrial Science & Technology, 33th, Mori-Building, 8-21, Toranomon 3, Minato-ku, Tokyo, Japan

Netherlands

Nederlands Normalisatie-Instituut, Polakweg 5, Riswijk, ZH-2106, Nederland. Mailing: Postbus 5810, 2280 HV Rijswijk (ZH), Nederland

New Zealand

Standards Association of New Zealand, Private Bag, Wellington, New Zealand

Norway

Norges Standardiseringsforbund, Haakon VII's gt. 2, N-Oslo 1, Norway

South Africa

South African Bureau of Standards (SABS), X191 Private Bag, Pretoria 0001, Republic of South Africa

Spain

Instituto Nacional de Racionalizacion y Normalizacion; Serrano, 150; Madrid-6, Spain
Member of ISO, CEN, FID, and CEE.

Sweden

The Swedish Planning and Rationalization Institute of the Health and Social Services (SPRI), FACK, S-10250 Stockholm, Sweden.

Switzerland

Schweizerischer Elektrotechnischer Verein (SEV), Seefeldstrasse 301, 8008 Zurich, Switzerland

United Kingdom

British Standards Institution (BSI), Sales Department, 101 Pentonville Road, London N1 9ND. Headquarters: 2 Park Street, London WIA 2BS, UK

Hospital Equipment (SL 18)
BSI Yearbook Subject index and abstract of British Standards.
BSI News Current standards and Amendments
"Yearbook Supplement" Compares International Standards to British standards

The American College of Nuclear Physicians has developed standards for quality assurance in nuclear medical practices and is using these as a basis for voluntary inspection and accreditation. The standards cover all aspects of nuclear medicine including radioassay proccedures that come under the definition of clinical laboratory tests. The program is designed to be fully compatible with that of the College of American Pathologists and the relevant standards published by NCCLS. Copies of these standards may be obtained by writing the American College of Nuclear Physicians, Suite 700, 1101 Connecticut Avenue, N.W., Washington, D.C. 20036.

REFERENCES

1. Tietz, N. W., *Fundamentals of Clinical Chemistry,* 2nd ed., W.B. Saunders, Philadelphia, 1976, 1.
2. Anon. Sales Catalog, Fisher Scientific, Pittsburgh, 1979, 938.
3. Anon., Sales Catalog, Fisher Scientific, Pittsburgh, 1979, 939.
4. Werner, M., *Microtechniques for the Clinical Laboratory: Concepts and Applications,* John Wiley & Sons, New York, 1976, 38.
5. Tietz, N. W., Fundamentals of Clinical Chemistry, 2nd ed., W.B. Saunders, Philadelphia, 1976, 27.
6. Gordon, A. J. and Ford, R. A., *The Chemist's Companion: A Handbook of Practical Data, Techniques and References,* John Wiley & Sons, New York, 1972, 429.
7. Tietz, N. W., *Fundamentals of Clinical Chemistry,* 2nd ed., W.B. Saunders, Philadelphia, 1976, 31.
8. Anon., Sales Catalog, Scientific Products, McGaw Park, Ill., 1978, 392.
9. Anon., Sales Catalog, Scientific Products, McGaw Park, Ill., 1975, 394.
10. Anon., Sales Catalog, Fisher Scientific, Pittsburgh, 1979, 332.
11. Anon., Catalog # MC 179/V, Millipore Corporation, Bedford, Mass., 1979, 13.
12. Anon., Catalog # 447D, Amicon Corporation, Lexington, Mass., 1979.
13. Anon., Catalog Price List F, Bio-Rad Laboratories, Richmond, Calif. 1980, 33.
14. Tietz, N. W., *Fundamentals of Clinical Chemistry,* 2nd ed., W.B. Saunders, Philadelphia, 1976, 32.
15. Sober, H. A., *Handbook of Biochemistry,* 2nd ed., CRC Press, Boca Raton, Fla., 1970, J-245.
16. Weast, R. C., Ed., *Handbook of Chemistry and Physics,* 61st ed., CRC Press, Boca Raton, Fla., 1980, D-116.
17. Weast, R. C., Ed., *Handbook of Chemistry and Physics,* 61st ed., CRC Press, Boca Raton, Fla., 1980, D-117.
18. Long, C., *Biochemists Handbook,* D Van Nostrand, New York, 1968, 1064.
19. Dawson, R. M. C., Elliott, D. C., Elliott, W. H., and Jones, K. M., *Data for Biochemical Research,* 2nd ed., Oxford University Press, New York, 1969, 615.
20. Gordon, A. J. and Ford, R. A., *The Chemist's Companion: A Handbook of Practical Data, Techniques and References,* John Wiley & Sons, New York, 1972, 442.
21. Tietz, N. W., *Fundamentals of Clinical Chemistry,* 2nd ed., W.B. Saunders, Philadelphia, 1976, 47.
22. The International Federation of Clinical Chemistry Newsletter, La Chaux-de-Fonds, Switzerland, #14, June 1976.
23. The International Federation of Clinical Chemistry Provisional Recommendation, reproduced in *Clin. Chem.,* 22, 532, 1976.
24. Calibration Reference Materials and Control Materials In Clinical Chemistry - Approved Standard ASC-2, National Committee for Clinical Laboratory Standards, Villanova, Pa., 1975.
25. Cali, J. P., LaFleur, P. D., Reeder, D. J., Schaffer, R., Uriano, G. A., and Velapoldi, R. A., NBS Position Paper: A National Understanding for the Development of Reference Materials, and Methods for Clinical Chemistry — Proceedings of a Conference, American Association for Clinical Chemistry, Washington, D.C., 1978.
26. Kombli, V. B. and Barnett, R. K., Control of Accuracy of Clinical Standards Solutions, *Am. J. Clin. Pathol.,* 61, 912, 1974.
27. Hanson, D. J., The Relationship of CAP Clinical Standard Solution to NBS/SRMs, *Am. J. Clin. Pathol.,* 61, 916, 1974.

28. " A National Understanding For the Development of Reference Materials And Methods For Clinical Chemistry" Proceeding of a Conference, published by AACC, Boutwell, Joseph H., Ed., American Association for Clinical Chemistry, Washington, D.C., 1978.
29. **The Internal Federation of Clinical Chemistry, Committee on Standards and Expert Panel on Nomenclature and Principles of Quality Control in Clinical Chemistry,** Provisional recommendation on quality control in clinical chemistry. I. General principles and terminology, *Z. Klin. Chem. Klin. Biochem.*, 13, 523, 1975; *Clin. Chem.*, 22, 532, 1976.
30. **Pacela, A. F. and Sloan, A. B.,** *The Guide to Biomedical Standards,* 7th ed., Quest Publishing, Brea, Calif., 1980.

Quality Assurance

INTRODUCTION

Guy T. Haven and Noel S. Lawson

Quality control techniques provide an estimate of accuracy of laboratory analyses. Continued application of these techniques allows monitoring of laboratory performance and aids in the maintenance of high quality laboratory results.

Major applications currently in use include internal and external quality control procedures. Internal quality control uses samples of known analyte content generally with each analytic run. Internal quality control procedures provide predominantly precision statistics. However, enhancement of internal quality control techniques through participation in a Regional Quality Control Program provides bias estimates, although these are based upon relatively limited numbers (hundreds) of laboratories. External quality control (proficiency testing programs) provide periodic unknown samples to thousands of laboratories. Compilation of the data from these programs provide periodic bench mark accuracy or bias estimates to individual laboratories; both types of programs use similar control materials and statistical approaches. An underlying assumption has been that these programs allow estimation of imprecision of laboratory measurements analogous to the imprecision affecting the analysis of patient samples. Equivalence of this degree of imprecision would allow comparison of control precision with medical usefulness criteria and analytic goals in order to assess the clinical adequacy of individual laboratory performance.

It must be realized, however, that factors exist which specifically lead to imprecision of patient or of control sample results. Internal and external quality control programs are limited to the control of the analytic process, not being affected by patient drug ingestion or a variety of physiologic and pathologic factors affecting patient samples. A major effort of laboratorians who have now improved the control of analytic imprecision must be to decrease the effect of preanalytic factors by the choice of more specific analytic methods, as well as by careful attention to patient preparation and sample preparation and storage in the laboratory.

Conversely, a number of factors affect lyophilized samples specifically. These include vial-to-vial variation due to inherent inaccuracies during manufacture and reconstitution as well as analyte instability prior to and after reconstitution in the laboratory. These are also partly under laboratory control by virtue of the users' careful choice of quality control material, care in control material reconstitution, and careful storage of the material before and after reconstitution.

Regardless of the difficulties in assessment of variation affecting patient samples using the control sample technique, there has been documented and continued improvement in proficiency testing results since initiation of these programs. This doubtless indicates continued improvement in analytical performance. As yet, however, the state-of-the-art does not allow direct translation of the precision of quality control results to that of patient sample measurement. The estimates provided by these programs must, therefore, be considered in conjunction with local clinical needs by the clinical pathologist of laboratory director in the choice of methods of assessment of the adequacy of their performance.

REFERENCES

1. **Barnet, R. N.,** *Clinical Laboratory Statistics,* 2nd ed. Little, Brown, Boston, 1979.
2. **Duncan, A. J.,** *Quality Control and Industrial Statistics,* Richard D. Irwin, Homewood, Ill., 1965.
3. **Inhorn, S. L.,** *Quality Assurance Practices for Health Laboratories,* American Public Health Association, Washington, D.C., 1978.
4. **Whitehead, T. P.,** *Quality Control in Clinical Chemistry,* John Wiley & Sons, New York, 1977.
5. **Young, D. S., Pestaner, L. C., and Gibberman, V.,** Effects of drugs on clinical laboratory tests, *Clin. Chem.,* 21, 1D, 1975.

ERRONEOUS OR MISLEADING RESULTS DUE TO PREANALYTIC FACTORS

William Beautyman

INTRODUCTION

Routine quality assurance, that is, the use of standards and control specimens, proficiency testing, and other aspects of process control, will not avoid errors in analysis due to something in the specimen (a so-called matrix factor) nor will such control measures prevent errors of interpretation if some unsuspected physiologic or genetic factor, a drug, a dietary component, or something in the environment, causes an unexpected result.

In 1975, Young et al.[1] published a massive compilation with 2246 references of preanalytic factors that may cause errors in laboratory results or in their interpretation. Their work was entitled "Effects of Drugs on Clinical Laboratory Tests," but the compilation also considered the effect of the following factors not usually considered to be drugs or therapeutic or toxic agents:

Aging, albumin, altitude, bananas, bedrest, blindness, blood groups, body habitus, coca cola, cocoa, coffee, contact with clot, cork stoppers, coughing, dehydration, detergents, diurnal variation, driving, eggplant, electrocautery, electroshock, erect posture, fasting, fatty foods, fever, filter paper, fish, foods, fruit, garlic, gasoline, glass containers, glassware, hair treatments, heat, height, hemodialysis, hemoglobin, hemolysis, hyperventilation, hypoxia, jaundice, ketoacids, ketones, ketosis, kwashiorkor, lactation, leucine, light, lipemia, lipochrome, lipuria, mayo enema, meals, meat, meat extract, muscle massage, muscular exercise, myoglobin, noise, obesity, oranges, phosphosoda, physical stress, physical training, plastic tubing, plums, protein, reducing diet, room temperature, sauerkraut, season, sex difference, sexual activity, sleep, sleep deprivation, smog, smoking, soft water areas, spermatozoa, spinach, standing of sample, starvation, storage of sample, stress, stress cold exposure, suntan oil, surgery, syringes, tea, tobacco smoke, tourniquet, transport of specimen, twin pregnancy, tyramine, tyrosine, kerosene, ultra violet light, water, urea, uremia, uric acid, urine flow, urine pH, urine turbidity, urine volume, valine, vanilla, vegetables, vegetarian diet, venipuncture, walnuts, water, waterload, weight, yeast, and zero gravity.

It would serve no purpose to duplicate this work but, instead, an attempt will be made to classify these phenomena, to stress their importance, to list additional reports that have appeared since the paper of Young et al., and to suggest an approach to their detection and avoidance.

Erroneous or misleading results due to preanalytic factors may be classified as follows (Table 1).

Result False Because of In Vivo Matrix (Tables 2 and 3)

It is a fundamental scientific principle that if the effect of A on system B is to be measured then only A must change.

This principle is violated every day in the clinical laboratory when a fixed analytical procedure is applied to clinical material. The material and the procedure constitute

Table 1
ERRONEOUS OR MISLEADING RESULTS DUE TO PREANALYTIC FACTORS

Result false	Matrix effect originating in vivo	Drug	Table 2
		Endogenous metabolite	Table 3
	Matrix effect originating in vitro	Substance ordinarily used as preservative, anticoagulant, etc.	Table 4
		Presence of extraneous substances	Table 5
Result correct but misleading	Does not represent in vivo status	In vitro consumption or generation of analyte	Table 6
		Subtraction or addition of analyte	Table 7
	Represents in vivo status	Matrix true but misleading	Table 8
		Pathologic, pharmacologic, toxic, genetic or physiologic change	Table 9

system B, but the analyte is contained in a matrix that varies constantly in composition in an unknown way. The reaction mixture must, therefore, vary similarly and must differ from the control specimens even when these are serum or urine based.

Matrix interferences have been long recognized, but in these days of polypharmacy and heroic measures to keep patients alive with gross metabolic disorders, the time has come to consider these matrix interferences the rule rather than the exception. The problem is to assess the degree of these interferences.

Result False Because of In Vitro Matrix (Tables 4 and 5)

Similar errors may be caused by material added to the specimen after it has been taken. The material may be added inadvertently or deliberately as a preservative or anticoagulant or with the intent to deceive.

Result Correct But Does Not Represent In Vivo Status (Tables 6 and 7)

Correct but misleading results will occur if analyte is added to or subtracted from the specimen or if the analyte is spontaneously generated or consumed in vitro after the specimen has been taken. In the case of blood, the analyte may be added to or removed from serum or plasma by the cells.

Result Correct and Represents In Vivo Status of Specimen Removed But is Misleading (Tables 8 and 9)

Results may mislead if the specimen removed from the body is not what it is thought to be, e.g., if what is considered to be capillary blood is actually mixed with interstitial fluid or what is thought to be venous blood is mixed with fluid being administered intravenously or results on blood obtained from an unusual site are interpreted as though it was from an antecubital vein. The largest group of misleading results in this category occurs because of changes produced physiologically or pharmacologically that are not recognized as such. An abnormal result may be misconstrued as reflecting disease, or a pathologic condition may be missed, because the pathologic and pharmacologic or physiologic effects act in opposite directions to produce a normal value.

DETECTION AND AVOIDANCE OF MATRIX ERRORS

The first essential is to be aware of the problem. Surprisingly few laboratory workers pay much attention to it and even fewer clinicians are aware of its existence. All laboratory workers should be trained to have a high index of suspicion for these effects

Table 2
FALSE RESULTS DUE TO DRUG EFFECTS

Analyte	Method	Interferent	Comment	Ref.
Analyte Decreased				
Acetaminophen-U	Simpson and Stewart	Ascorbic acid	Indophenols	6
Anion gap-S	Calculation	Iodide	See under chloride	7
Aspartate amino-transferase-S	SMA 12/60	Metronidazole	Absorbance at 340 nm	8
Cholesterol-S	Enzymic centrifichem	Ascorbic acid	10% ascorbic causes decrease 23 mg/%	9
Estriol-U	Kober	Hydrochlorthiazide	Formaldehyde from HCT forms polymers with estriol	10
Estriol-U	Kober	Hexamine hippurate	Formaldehyde from hexamine hippurate (see above)	11
Glucose-S	Glucose oxidase	Ascorbic acid	25 mg/dℓ causes decrease 20 mg/dℓ	12
Glucose-S	Glucose oxidase peroxidase	Acetaminophen	75% loss at 15 mg%	13
Phenobarbital-S	Gas liquid chromatography	Theophylline	Coelutes with internal standard 5-ethyl-5 p-tolyl barbituric acid	14
Analyte Increased				
Alcohol-S	Enzymatic	Isopropanol	Check osmolality	15
Alkaline phosphatase-S	Electrophoresis	Bilirubin	Apparent zone in albumin	16
Bilirubin	Jendrassik and Grof SMA 12/60	Propranolol and uremia	Retention of metabolite?	17
Catecholamines-U	Colorimetric	Labetalol	Unnecessary operation in one case	18,19
Chloride-S	Technicon® Stat-Ion	Iodide	Low iodide levels cause disproportionately high chloride levels	7
Creatinine-S	Jaffe reaction in autoanalyzer	Trimethoprim, sulfamethoxazole	Increased up to 10%	20
Diazepam-S	Spectrophotometric	N-Desmethyl metabolite	Error avoided by gas chromatography	21
Glutethimide-U	Gas chromatography	2-ethyl-2-phenyl glutaconimide (metabolite)		22
Organic acids-U	Gas chromatography	Calcium levulinate metabolites	Mimics beta keto-thiolase deficiency	23
Protein-U	Meola, Vargas and Brown	Aminoglyosides	Gentamicin Amikacin Tobramycin	24
Protein-U	Trichloroacetic acid precipitation	Penicillins	Methicillin Benzylpenicillin Cloxacillin	25
Protein-U	Biuret	Penicillins	Penicillin-G Mezlocillin Azlocillin	26
Protein-U	Sulfosalicylic acid	Nafcillin	Turbidity	27
Protein-U	Sulfosalicylic acid	Tolmetin	Turbidity	28
Porphobilinogen-U	Watson-Schwartz	Methyldopa	Indole metabolites	29
Salicylates-U	Phenistix test strip	Desferrioxamine	Red-Brown color	30

Note: S = serum and U = urine.

Table 3
FALSE RESULTS DUE TO ENDOGENOUS METABOLITES

Analyte Decreased

Analyte	Method	Interferent	Comment	Ref.
Alkaline phosphatase-S	p-Nitrophenyl	Albumin	Slight effect	32
All	RIA-liquid scintillation	Hemoglobin	Quenching	33
All	RIA-liquid scintillation	Bilirubin	Quenching	33
Asparatate aminotransferase-S	Various	Uremia	Fall most marked in SMA-12/60	34
Cholesterol-S	Enzymic centrifichem	Bilirubin	At 20 mg% decreased by 50 mg%.	9
Cholesterol-S	Enzymic Various kits	Bilirubin	Ca. 70% decreased at 20 mg%	35
CPK-S	Rosalki	Cystine	600 antiunits/mg	36
CPK-S	Rosalki	Uric acid	45 antiunits/mg	36
CPK-S	Oliver	Glutathione reductase	If glutathione in reagents	37
CPK-S	Oliver	Creatine phosphate	By inhibition of adenylate kinase	38
CPK-S	Worthington Calbiochem	Phosphohexose isomerase?	In liver disease	39—42
Creatinine-S	Kinetic Jaffe	Bilirubin	87% decrease at 24 mg%	43
Digoxin-S	Radioimmunoassay	Low albumin	Up to 3-fold decrease	44
Estrogens-U	Cohen	Urobilinogen	Up to 40% lower	45
Folate-S	Competitive protein binding	Serum protein	Serum blank partially corrects	46
Gamma glutamyltranspeptidase-S	Kinetic	Lipemia	About 50% decrease	47
Glucose-S	Glucose oxidase	Gentisic acid	25% at 80 mg%	48
Glucose-S	Glucose oxidase	Glutathione-reduced	30% at 100 mg%	48
Glucose	Reagent strip	Uric acid	20% at 14 mg%	49
Glucose-S	Glucose oxidase	Proteins — mol wt 40,000 and 500,000	Up to 20% decrease	50
Glycolic acid-U	Chromotropic acid colorimetric	Glyceric acid	Positive and negative interference noted	51
LDH-S	SMAC	Lipemia	Up to 100%	52
MCHC	Coulter counter	Hypernatremia	Up to 20% error	53
Myoglobin-S	Complement fixation	Serum binding factor	Mol wt ca. 100,000	54
Osmolality-S	Dew point osmometry	Alcohol and other volatiles	Freezing point depression not affected	55
pK'-B	Calculated from pH and pCO$_2$	Multifactorial	Up to 50% error in calculated bicarbonate	55a
Sodium-S	Flame emission	Lipemia	Water displaced by fat	56
Steroids-S	Radioimmunoassay	Nonesterified fatty acids	Steroids sequestered in fatty acid micelles	57

Analyte Increased

Analyte	Method	Interferent	Comment	Ref.
Alanineamino-transferase-S	Henry et. al.	Endogenous LDH substrate	If preincubation inadequate	58

Table 3 (continued)
FALSE RESULTS DUE TO ENDOGENOUS METABOLITES

Analyte Increased

Analyte	Method	Interferent	Comment	Ref.
Alanineamino-trans-ferase-S	Wroblewski and Ladue	Glutathione reductase	If glutathione in reagents	37
Acidity-U	Titration	Protein	Albumin effect greater than globulin	59
Ammonium-U	Titration	Protein	Albumin effect greater than globulin	59
Aspartate amino-transferase-S	Amador and Wacker	Endogenous LDH	If preincubation inadequate	58
Aspartate amino-transferase-S	Babson	Protein	Decarboxylation of oxalo acetic acid	60
Aspartate amino-transferase-S	Karmen	Glutathione reductase	If glutathione in reagents	37
Calcium-S	ACA-DuPont	Hemoglobin	575 nm absorption	61
Cholesterol-S	All methods	Related sterols	Error varies with method	62
CPK-S	Autoanalyzer II	Creatine	No blank	63
CPK-S	Modified Oliver	Adenylate kinase	Adenylate kinase increased with hemolysis	64
CPK-S	Boehringer and Mannheim	Macroisoenzyme BB	Complexed with IgG	65
CPK-S	Colorimetric	Guanidines	Up to 24% error	66
CPK-BB-S	Fluorometric	Uremia	False zone in albumin region	67
CPK-MB-S	Column chromatography	B-lipoprotein	Binds with MM to give apparent MB	68
CPK-MM-S	Anion exchange chromatography	Adenylate kinase	Coelutes with MM	69
Glucose-S	Glucose oxidase-peroxidase	Maltose	Maltase in glucose oxidase	48
Glucose-S	Folin and Wu	Saccharoids	Increased in diabetics	70
Iron-S	Modified ICSH	Hemoglobin	Marked after 96-hr storage at 4%C	71
MCHC	Coulter	Hyponatremia	Up to 20%	53
pK'-B	Calculated from pH and pCO_2	Multifactorial	Up to 60% error in calculated bicarbonate	55a
Urate-S	Phosphotungstate	Paraprotein	Due to turbidity	31
Urine oxalate-U	Precipitation and colorimetry	Nonoxalate compounds	Ca. 50% higher than gas chromatography	72
Uric acid-U	Acid ferric-reduction	Catecholamine metabolites	Up to 50% increase	73
Urine VMA-U	Diazotised p-nitroaniline	Hydroxymethylguiacocols	6 non-VMA compounds positive	74

Note: S = serum; U = urine; and B = whole blood.

and should be instructed to inform their superiors immediately if they suspect that such an error may exist.

A careful system or specimen collection and handling will avoid most of the prob-

Table 4

FALSE RESULTS DUE TO SUBSTANCES ORDINARILY USED AS
PRESERVATIVE, ANTICOAGULANT, ETC.

Analyte Decreased

Analyte	Method	Interferent	Comment	Ref.
Albumin-P	Bromcresol green-manual	Heparin	Dye precipitated and removed by centrifugation	75
Creatinine-S	Creatinine-amido-hydrolase	Oxalate	At 1 g/ℓ	76
Creatinine-S	Creatinine-amidohydrolase	EDTA	At 1 g/ℓ	76
Lead-E	Atomic absorption	Heparin	At 400 mg/ℓ	77
Post heparin Lipolysis-P	Enzymatic radioassay	Heparin as anticoagulant	Ca. 25% decrease	78

Analyte Increased

Analyte	Method	Interferent	Comment	Ref.
Albumin-P	Bromcresol green continuous flow	Heparin	Turbidity causes increased absorption	75
Angiotensin I-P	RIA	EDTA	Depresses antigen antibody binding	79
LDH-S	SMAC and SMA 12/60	Serum separators	5% Increase	80
Protein-U	Coumassie brilliant blue G250	Thymol	20% Increase	81
Uric acid-S	Acid ferric reduction	Azide	Up to 2-fold	82

Note: S = serum; U = urine; P = plasma; and E = erythrocytes.

Table 5

FALSE RESULTS DUE TO THE PRESENCE OF EXTRANEOUS SUBSTANCES

Analyte Decreased

Analyte	Method	Interferent	Comment	Ref.
Barbitol-U	EMIT	Salt	Added by patient	83
Copper	Atomic absorption	Silicone-treated stopper	Interferes with extraction	84
Ethadone-U	EMIT	Salt	Added by patient	83
Morphine-U	EMIT	Salt	Added by patient	83

Analyte Increased

Analyte	Method	Interferent	Comment	Ref.
Dilantin	Gas liquid chromatography	Tributoxyethyl phosphate	Leached from vacutainer stopper	85
Drug	Gas liquid chromatography	Unidentified	Leached from vacutainer stopper	86
Methyprylon	Gas liquid chromatography	Unidentifed	Serum separators	87
Methadone	Gas liquid chromatography	N-isopropyl N'-phenyl-p-phenylene-diamine	From rubber stoppers	88
Vitamin E	Fluorometric	Unknown	Rubber stoppers especially purple	89

Note: U = urine.

Table 6
CORRECT BUT MISLEADING RESULTS DUE TO IN VITRO
CONSUMPTION OR GENERATION OF ANALYTE

Analyte Decreased

Analyte	Effector	Comment	Ref.
Calcium ionized-S	Exposure to air	Due to increased pH	97
CPK	Blue light	Especially if frozen	98
Glutathione reductase-S	Storage	Slowed by dilution	99

Analyte Increased

Analyte	Effector	Comment	Ref.
Alcohol-U	Yeast	Up to 649 mg/dl	90
Amylase-S	Dilution	Apparent increase in lipemia	91
Fructose-U	Alkaline reaction	By chemical inversion	92
Glutamic acid-P	Storage at 20°C	Up to 6-fold increase	93
LDH-S	Serum separator	Probably from erythrocytes	94
LDH-P	Platelets	In platelet rich plasma	100
Phosphate-P	Erythrocytes	Increased after hemodialysis	100
Potassium-S	Leukocytosis	Up to 2-fold increase in 6 hr	96

Note: S = serum; U = urine; and P = plasma.

Table 7
CORRECT BUT MISLEADING RESULTS DUE TO
SUBTRACTION OR ADDITION OF ANALYTE

Analyte Decreased

Analyte	Effector	Comment	Ref.
Calcium-S	Autoanalyzer cup	Up to 10% absorption overnight	103a
Calcium-S	Stainless steel tubing in autoanalyzer	Alkaline conditions	102

Analyte Increased

Analyte	Effector	Comment	Ref.
All	Evaporation	Especially autoanalyzer cups	101
Calcium-S	Stainless steel tubing in autoanalyzer	Acid conditions	102
Calcium-S	Orange-green top vacutainer	By leaching	86
Lignocaine-U	Catheter jelly	Contains lignocaine	103
Magnesium-S	Orange-green top vacutainer	By leaching	86
Sodium-S	Orange-green top vacutainer	By leaching	86

Note: S = serum and U = urine.

Table 8
CORRECT BUT MISLEADING RESULTS DUE TO
SAMPLING EFFECTS

Analyte	Effect	Effector	Comment	Ref.
Various	Variable	Site and technique of obtaining blood sample	Marked effect on certain analytes	107

Table 9
CORRECT BUT MISLEADING RESULTS DUE TO PATHOLOGIC,
PHARMACOLOGIC TOXIC, GENETIC, OR PHYSIOLOGIC CHANGE

Analyte	Effect	Effector	Comment	Ref.
Alanine-P	Decreased	Dichoroacetate	In diabetics	108
Alanineamino-transferase-S	Increased	Halothane	Toxic 2-fold	109
Alkaline phosphatase-S	Decreased	Clofibrate	Effect on bone or liver?	110—112
Amylase-S	Increased	Zinc	Toxic pancreatitis	113
Amylase-S	Increased	Procainamide	SLE pancreatitis?	114
Amylase-S	Increased	Oxyphenbutazone	Sialadenitis	115
Anylase-S	Increased	Bumetanide	About 2-fold	116
Androgens-P	Decreased	Dexamethasone	Decreased production	117
Arginine vasopressin-P	Increased	Carbamazepine	Water intoxication	118
Arginine vasopressin-P	Decreased	Carbamazepine	? Increased sensitivity of receptors	119
Aspartate aminotransferase-S	Increased	Halothane	Toxic 2- to 3-fold	109
Bilirubin-S	Decreased	Aspirin	Pharmacologic effect	120
Bilirubin-S	Increased	Halothane	Toxic liver ca. 2-fold	109
Bromide-S	Increased	Halothane	Decomposition	121
BUN	Increased	Aspirin	In SLE and RA	122
Calcium-S	Decreased	Aspirin	Pharmacologic effect	120
Calcium-S	Decreased	Methyldopa	Due to malabsorption	123
Calcium-U	Decreased	Somatostatin	About 30%	124
Cholesterol-S	Increased	Disulfiram	In alcoholics	125
Cholesterol-S	Decreased	Milk	Main effect on LDL	126
Cholesterol-S	Decreased	Dichloroacetate	In diabetics	108
Cholesterol-S	Increased	Chlorthalidone	About 10% increase	127
Choline-E	Increased	Lithium	10-fold increase	128
Cholinesterase-E	Increased	Oral contraceptives	6% increase in women	129
Cholinesterase-P	Decreased	Oral contraceptives	15% decrease in women	129
CPK-S	Increased	Halothane	Ca. 20% increase	109
CPK-B-S	Increased	Abnormal BB	Total CK normal	130
Creatinine-S	Increased	Aspirin	In SLE and RA	122
Cyclic AMP-P	6 a.m. high 10 p.m. low	Diurnal variation	50% Increase	131
Digoxin-S	Increased	Quinidine	About 2-fold	132
Epinephrine-S	Increased	Caffeine	Ca. 3-fold	133

Table 9 (continued)
CORRECT BUT MISLEADING RESULTS DUE TO PATHOLOGIC, PHARMACOLOGIC TOXIC, GENETIC, OR PHYSIOLOGIC CHANGE

Analyte	Effect	Effector	Comment	Ref.
Estriol-U	Decreased	Ampicillin	Decreased synthesis in placenta; decreased permeability of placenta	134
Estrogen-S	Increased	Digoxin	Up to 2-fold	135
Fluoride-S	Increased	Halothane	Decomposition	121
FT3-S	Decreased	Phenytoin	Ca. 20% decrease	136
FT3-S	Decreased	Phenobarbitol	Ca. 10% decrease	136
FT4-S	Decreased	Phenytoin	Ca. 40% decrease	136
FT4-S	Decreased	Phenobarbitol	Ca. 20% decrease	136
Fibrinogen-P	Decreased	Sodium valproate	Ca. 30%	137
Folate-S	Decreased	Methyldopa	Due to malabsorption	122
Free fatty acids-S	Decreased	Acebutol	Ca. 30%	138
Free fatty acids-S	Decreased	Propranolol	Ca. 30%	138
Frusemide-S	Decreased	Phenytoin	Ca. 50% malabsorption	139
Gamma glutamyl-transferase-U	Increase	Tobramycin	Up to 4-fold increase in one case	140
Gamma glutamyl-transpeptidase-S	Decrease	Clofibrate	Ca. 50% after 2 weeks	110
Gastrin-S	Increase	Glucocorticoids	Ca. 2-fold	141
Glucose-S	Decrease	Methyldopa	Due to malabsorption	123
Glucose-S	Decrease	Dichloroacetate	In diabetes	108
Glucose-S	Decrease	Glipizide	Overcomes insulin release inhibition of diazoxide	142
HDL-S	Increase	Ethanol	5—6 oz/week gives 10% increase	143
HDL-S	Increase	Phenytoin	Up to 30%	144,145
HDL cholesterol-S	Decrease	Oral contraceptives	Especially in smokers	146
HDL cholesterol-S	Decrease	Antihypertensive drugs	Associated with increased triglycerides	147
HDL cholesterol-S	Decrease	Androgens	Marked fall in one case	148
HDL cholesterol-S	Increase	Estrogens	In women	149
HDL cholesterol-S	Decrease	Progesterones	In women	149
Homocystinuria	Increase	Oral contraceptives	Causes decreased cystathione synthase activity	150
Iron-S	Decrease	Methyldopa	Due to malabsorption	123
Ketones-P	Increase	Dichloroacetate	In diabetics	108
Lactate-S	Decrease	Dichloroacetate	In diabetes	108
LDH-S	13 Zones	Variant M polypeptide	Agarose electrophoresis	151
LDH-S	7 Zones	IgA and Beta lipoprotein	Complexed with LD2 and LD3	152
Lipids-S	Increase	Oxymetholone	Long-term hemodialysis	153
Luteinizing hormone-S	Decrease	Digoxin	Ca. 50% decrease	154

Table 9 (continued)
CORRECT BUT MISLEADING RESULTS DUE TO PATHOLOGIC, PHARMACOLOGIC TOXIC, GENETIC, OR PHYSIOLOGIC CHANGE

Analyte	Effect	Effector	Comment	Ref.
Magnesium-E	Decrease	Diuretics	2 out of 5 patients	154
Magnesium-S	Decrease	Methyldopa	Due to malabsorption	23
Metanephrine-U	Increase	Caffeine	2-fold increase	132
Norepinephrine-S	Increase	Caffeine	75% Increase	132
Normetanephrine-U	Increase	Caffeine	52% Increase	132
Plasma volume	Decrease	Carbon monoxide	Increased vascular permeability	155
Potassium-S	Decrease	Licorice	Causing cardiac arrest	156
Potassium-S	Increase	Glucose	Paradoxically in insulin dependent diabetics with hyperglycemia and ketosis	157
Prolactin-S	Decrease	Apomorphine	Ca. 30%	158
Prolactin-S	Decrease	Ergot alkaloids	See also bromocriptine	159
Prolactin-S	Increase	Estrogen	Potentiates phenothiazine effect	160
Prolactin-S	Increase	Cyproheptadine	Potentiated by TRH	161
Prolactin-S	Increase	Haloperidol	Up to 3-fold increase	162
Prolactin-S	Increase	Phenothiazines	Up to 3-fold increase	162
Prolactin-S	Increse	Reserpine	Up to 3-fold, increase	163
Prolactin-S	Increase	Monoamineoxidase inhibitors	2- to 3-fold	164
Prolactin-S	Increase	Metroclopramide	6-fold increase	165
Prolactin-S	Decrease	Bromocriptine	In hypertension	166
Prolactin-S	Decrease	Meatless diet	Ca. 50% decrease	167
Prolactin-S	Decrease	Levodopa	Especially manic depressives	168
Prolactin-S	Increase	Methyldopa	Over 4-fold increase	169
Prolactin-S	Increase	Carbidopa	Up to 50%	170
Prolactin-S	Increase	Oral contraceptives	3 Cases	171
Prolactin-S	Increase	Cimetidine	Up to 2-fold	172
Prolactin-S	Increase	Methadone	Ca. 2-fold increase	173
Prolactin-S	Decrease	Chlormethiazole	About 50%	174
Prolactin-S	Increase	Licorice	118 $\mu g/l$ in one case	175
Protein-S	Decrease	Aspirin	Pharmacological effect	120
Renin substrate-P	Increase	Oral contraceptives	Up to 4-fold increase	176
Renin-P	Increase	Caffeine	57% Increase	133
RT3-S	Decrease	Phenytoin plus phenobarbital	Ca. 30%	136
Reverse T3-S	Increase	Propranolol	Ca. 15%	177
Sodium-S	Decrease	Oxytocin	Transplacental	178
T3-S	Decrease	Propranolol	Ca. 20%	179
T3-S	Decrease	Dexamethasone	Ca. 25%	180
T3-S	Increase	Halothane	Ca. 15%	109
T4-S	Decrease	Phenytoin	Ca. 25%	136
T4-S	Decrease	Phenytoin plus phenobarbitol	Ca. 40%	136
T4-S	Decrease	Phenytoin plus carbamazepine	Ca. 25%	136

Table 9 (continued)
CORRECT BUT MISLEADING RESULTS DUE TO PATHOLOGIC, PHARMACOLOGIC TOXIC, GENETIC, OR PHYSIOLOGIC CHANGE

Analyte	Effect	Effector	Comment	Ref.
T4-S	Decrease	Danazol	Up to 40%	189
T4-S	Increase	Propranolol	Total T4 16%; free T4 18%	182
T4-S	Decrease	Dexamethasone	Ca. 10%	180
T4-S	Increase	Halothane	Slightly decreased followed by slight increased	109
TBG-S	Decrease	Danazol	Up to 29% after continued high doses	181
Testosterone binding globulin-S	Increase	Phenytoin	In women	183
Testosterone-S	Decrease	Oral contraceptives	In men	184
Thyroxin-S	Increase	Iodide	Continued small doses	185
Thyroxin-S	Decrease	Iodide	Continued high doses	185
Tryptamine-U	Decrease	Carbidopa	Up to 60%	170
Transaminase-S	Increase	Heparin	ALT and AST 3-1 to 4-fold increase	186
Triglycerides-S	Increase	Oral contraceptives	Ca. 30%	187
Triglycerides-P	Increase	Atenolol	Ca. 70%	188
Triglycerides-P	Increase	Pindolol	Ca. 30%	188
Triglycerides-P	Increase	Propranolol	Ca. 35%	188
Triglycerides-S	Increase	Atenolol	Ca. 32%	188
Triglycerides-S	Increase	Propranolol	2-fold	188
Triglycerides-S	Decrease	Dichloroacetate	In diabetics	108
Triglyceride lipase-S (hepatic)	Decrease	Uremia	Ca. 50%	189
Tryptophan-S	Decrease	Chronic hemodialysis	About 20%	190
TSH	Increase	Metoclopramide	In hypothyroidism	191
Urate-S	6 a.m. maximum 10 p.m. minimum	Diurnal variation	In pregnancy	192
Urate-S	Decrease	*o, p*-DDD	75%	193
Urate-S	Increase	Dichloroacetate	In diabetics	108
Urate-U	Variable	Beta blockers (timolol)	Small doses cause increase; high doses cause decrease	194
Various	Variable	Age	—	107
Various	Variable	Sex	—	107
Various	Variable	Posture	—	107
Various	Variable	Exercise	—	107
Various	Variable	Circadian	—	195
Various	Variable	Race	—	107
Various	Variable	Pregnancy	—	107
Various	Usually increased	Menopause	—	107
Various	Variable	Interindividual normal range greatly exceeds intraindividual range	Pathologic change may be interpreted as normal	107
Xylose absorption	Decrease	Methyldopa	Due to malabsorption	123

Note: S = serum; U = urine; P = plasma; and E = erythrocytes.

lems that are not inherent in the specimen itself. Every clinical laboratory should have a system, however elementary, designed to detect these matrix errors. This could consist, at a minimum, of a list of the most frequently used drugs and a list of the procedures in use in the laboratory with which each might be expected to interfere. Conversely, the methods book should list the matrix errors that may be associated with each method.

Levels of endogenous metabolites or drugs may exceed those reported in the literature as having no effect on a particular procedure. These unusually high levels should, therefore, always raise the possibility of matrix interference. Caution should also be exercised before accepting the statement that particular drugs do not interfere. As a rule, tests have only been carried out with the drug itself and the possible interference of drug metabolites has not been checked.

The clinical pathologist who is familiar with both the patient's pathophysiology, as well as the details of the laboratory methodology, is in the best position to anticipate and avoid or to detect these errors.

When a matrix problem is suspected, the patient's serum may be added to a serum of known reactivity and the apparent level of the analyte determined in the mixture. Examination of the urine may also be useful in investigating possible matrix errors since drugs and metabolites are often present in the urine in greater concentration than in the serum.

With the increasing use of computers in hospital and clinical laboratories, there are opportunities for correlating cumulative laboratory reports with the patient's drug profile and, in some institutions, this correlation has also been computerized[2] so that possible interferences of drugs with laboratory tests are automatically brought to the attention of the laboratory and clinical staff. Some workers are already including such information routinely in their laboratory reports.[3]

A distribution of survey material that contains interfering substances is a valuable exercise and this has already been started in Canada.[4,5]

REFERENCES

1. **Young, D. S. et al.**, Effects of drugs on clinical laboratory tests, *Clin. Chem.*, 21, 10, 1975.
2. **Friedman, R. B. et al.**, Automated monitoring of drug-test interactions, *Clin. Pharmacol. Ther.*, 24, 16, 1978.
3. **Salway, J. G.**, Drug interference with laboratory investigations, *Lancet*, 1, 483, 1977.
4. **Letellier, G. and Vinet, B.**, Etude in vitro de l'influence des drogues sur les analyses: serum de controle vs. serum frais, *5ieme Journees Internationales de Biologie Clinique*, 1976.
5. **Letellier, G. and Gagne-Billon, M.**, Study of drug interferences with biochemical tests in an interlaboratory quality control program, *Lancet*, 1, 981, 1977.
6. **Swale, J.**, Elimination of interference due to ascorbic acid when detecting paracetamol in urine, *Lancet*, 981, 1977.
7. **Fischman, R. A. et al.**, Iodide and negative anion gap, *N. Eng. J. Med.*, 1035, 298, 1978.
8. **Tighe, P. and Jones, B.**, Metronidazole interference with continuous-flow spectrophotometry of aspartate aminotransferase, *Clin. Chem.*, 25, 2057, 1979.
9. **Pesce, M. A. and Bodourian, S. H.**, Enzymic measurement of cholesterol in serum with the centrifichem centrifugal analyzer, *Clin. Chem.*, 23, 280, 1977.
10. **Watanabe, F. et al.**, Mechanism for interference of hydrochlorothiazide with the Kober reaction for determination of urinary estriol in pregnancy, *Clin. Chim. Acta*, 88, 21, 1978.

11. **Kivinen, S. and Tuimala, R.,** Decreased urinary estriol concentrations in pregnant women during hexamine hippurate treatment, *Br. Med. J.* , 2, 682, 1977.
12. **Carey, R. N., Feldbruegge, D., and Westgard, J. O.,** Evaluation of the adaptation of the glucose oxidase/peroxidase-3-methyl-2-benzothiazolinone hydrazone-n,n-dimethylaniline procedure to the technicon "SMA 12/60", and comparison with other automated methods for glucose, *Clin. Chem.,* 20, 595, 1974.
13. **Kaufmann-Raab, I., Jonen, H. G., Jahnchen, E., Kahl, G. F., and Groth, U,,** Interference by acetaminophen in the glucose oxidase-peroxidase method for blood glucose determination, *Clin. Chem.,* 22, 1729, 1976.
14. **Dechtiaruk, W., Crawford, R., and Frye, R.,** Theophylline interference in phenobarbital quantitation, *Clin. Chem.,* 25, 2055, 1979.
15. **Vasiliades, J. et al.,** Pitfalls of the alcohol dehydrogenase procedure for the emergency assay of alcohol: a case study of isopropanol overdose, *Clin. Chem.,* 24, 383, 1978.
16. **Hardin, E. et al.,** Artifactual alkaline phosphatase isoenzyme band, caused by bilirubin, on cellulose acetate electropherograms, *Clin. Chem.,* 24, 178, 1978.
17. **Stone, W. J., McKinney, T. D., and Warnock, L. G.,** Spurious hyperbilirubinemia in uremic patients on propranolol therapy, *Clin. Chem.,* 25, 1761, 1979.
18. **Harris, D. and Richards, D. A.,** Labetalol and urinary catecholamines, *Br. Med. J.,* 2, 1673, 1977.
19. **Hamilton, C. A., Jones, D. M., Darbie, M. J., and Reid, J. L.,** Does labetalol increase excretion of urinary catecholamines, *Br. Med. J.,* 2, 800, 1978.
20. **Bye, A.,** Drug interference with creatinine assay, *Clin. Chem.,* 22, 283, 1976.
21. **Rejent, R. A. and Wahl, K. C.,** Diazepam abuse: incidence, rapid screening, and confirming methods, *Clin. Chem.,* 22, 889, 1976.
22. **Fischer, L. J. and Ambre, J. J.,** Possible interference by a metabolite in gas chromatographic assay for glutethimide which employ certain nonselective liquid phases, *J. Chromatogr.,* 87, 379, 1973.
23. **Kovraa, S., Gregersen, N., Christensen, E., and Gron, I.,** Calcium levulinate medications, a pitfall in the diagnosis of organic acidurias, *Clin. Chim. Acta,* 77, 197, 1977.
24. **Meola, J. M. and Brown, H. H.,** Aminoglycoside antibiotic interference with a urinary protein method, *Clin. Chem.,* 25, 1180, 1979.
25. **Muir, A. and Hensley, W. J.,** Pseudoproteinuria due to penicillins, in the turbidometric measurement of proteins with trichloroacetic acid, *Clin. Chem.,* 25, 1662, 1979.
26. **Andrassy, K. et al.,** Pseudoproteinuria in patients taking penicillins, *Lancet,* 2, 154, 1978.
27. **Line, D. E. et al.,** Massive pseudoproteinuria caused by nafcillin, *JAMA,* 235, 1259, 1976.
28. **Ehrlich, G. E. and Worthan, G. F.,** Pseudoproteinuria in tolmetin-treated patients, *Clin. Pharmacol. Ther.,* 17, 467, 1975.
29. **Pierach, C. A., Cardinal, R. A., Petryka, Z. J., and Watson, C. J.,** Unusual Watson-Schwartz test from methyldopa, *New Engl. J. Med.,* 296, 577, 1971.
30. **Finlay, H. V. L.,** Phenistix urine test-strip and desferrioxamine, *Br. Med. J.,* 2, 356, 1978.
31. **Morris, J. and Collins, P.,** Interference with uric acid measurement (phosphoungstate method, KDA) by paraprotein, *Clin. Chem.,* 24, 1289, 1978.
32. **Foster, R. L. and Bannister, A.,** Inhibition of alkaline phosphatase activity by serum albumin, *Clin. Chem.,* 22, 1751, 1976.
33. **Helman, E. Z ., Spiehler, V., and Holland, S.,** Elimination of error caused by hemolysis and bilirubin-induced color quenching in clinical radioimmunoassays, *Clin. Chem.,* 20, 1187, 1974.
34. **Warnock, L. G., Stone, W. J., and Wagner, C.,** Decreased asparate aminotransferase ("SGOT") activity in serum of uremic patients, *Clin. Chem.,* 20, 1213, 1974,
35. **Pesce, M. A. and Bodourian, S. H.,** Interference with the enzymic measurement of cholesterol in serum by use of five reagent kits, *Clin. Chem.,* 23, 757, 1977.
36. **Jacobs, H. K., Phillipp, K. H., Sundmark, V,, and Weselake, R. J.,** Isolation and identification of a new potent inhibitor of creatine kinase from human serum, *Clin. Chim, Acta,* 85, 299, 1978.
37. **Klotzsch, S. G.,** Glutathione reductase as a possible source of error in common laboratory tests, *Clin. Chem.,* 22, 1236, 1976.
38. **Balkcom, R. M. and Amano, E.,** Adenylate kinase inhibition by creatine phosphate in CK assay, *Clin. Chem.,* 24, 1289, 1978.
39. **Mueller, R. G. et al.,** Depressed apparent creatine kinase activity in sera with abnormally high alkaline phosphatase activity, *Clin. Chem.,* 21, 268, 1975.
40. **Wearne, J. et al.,** Does alkaline phosphatase affect serum creatine kinase values?, *Clin. Chem.,* 21, 1343, 1975.
41. **Tsung, S. H.,** Relationship between alkaline phosphatase and creatine kinase activity, *Clin. Chem.,* 22, 116, 1976.
42. **Bruns, D. E. et al.,** Low apparent creatine kinase activity and prolonged lag phases in serum of patients with metastatic disease: elimination by treatment of sera with sulfhydryl agents, *Clin. Chem.,* 22, 1889, 1976.

43. **Daugherty, N. A., Hammond, K. B., and Osberg, I. M.,** Bilirubin interference with the kinetic jaffe method for serum creatinine, *Clin. Chem.,* 24, 392, 1978.

44. **Holtzman, J. L. et al.,** Methodological causes of discrepancies in radioimmunoassay for digoxin in human serum, *Clin. Chem.,* 20, 1194, 1974.

45. **Heiberg, S., Campbell, D. J., and Frohlich, J.,** Effect of urobilinogen on urinary estrogen determination, *Clin. Chem.,* 22, 124, 1976.

46. **Zettner, A. and Duly, P. E.,** New evidence for a binding principle specific for folates as a normal constituent of human serum, *Clin. Chem.,* 20, 1313, 1974.

47. **Martin, J. V., Gray, P. B., and Goldberg, D. M.,** Pitfalls in asssay and interpretation of serum gammaglutamyl transpeptidase activity, *Clin. Chim. Acta,* 61, 99, 1975.

48. **Lott, J. A. and Turner, K.,** Evaluation of Trinder's glucose oxidase method for measuring glucose in serum and urine, *Clin. Chem.,* 21, 1754, 1975.

49. **Stewart, T, C.,** Evaluation of a reagent-strip method for glucose in whole blood, as compared with a hexokinase method, *Clin. Chem.,* 22, 74, 1976.

50. **Blaedel, W. J. and Uhl, J. M.,** Nature of materials in serum that interfere in the glucose oxidase-peroxidase-o-dianisidine method for glucose and their mode of action, *Clin. Chem.,* 21, 119, 1975.

51. **Niederwieser, A., Matasovic, A., and Leumann, E. P.,** Glycolic acid in urine. A colorimetric method with values in normal adult controls and in patients with primary hyperoxaluria, *Clin. Chim. Acta,* 89, 13, 1978.

52. **Hendriks, F. R. and Groen, A.,** Pitfalls of use of lipemic serum with the technicon SMAC and DuPont ACA, *Clin. Chem.,* 24, 2062, 1978.

53. **Beautyman, W. and Bills, T.,** Hematocrit unchanged by hemodilution, *New Engl. J. Med.,* 293, 45, 1975.

54. **Kagen, L. J. and Butt, A.,** Myoglobin binding in human serum, *Clin. Chem.,* 23, 1813, 1977.

55. **Rocco, R. M.,** Volatiles and osmometry, *Clin. Chem.,* 22, 399, 1976.

55a. **Natelson, S. and Nobel, D.,** Effect of the variation of pK' of the Henderson-Hasselbalch equation on values obtained for total CO_2 calculated from pCO_2 and pH values, *Clin. Chem.,* 23, 767, 1977.

56. **Dunne, M. J. et al.,** Misleading hyponatraemia in acute pancreatitis with hyperlipaemia, *Lancet,* 1, 211, 1979.

57. **Rash, J. M., Jerkunica, I., and Sgoutas, D.,** Mechanisms of interference of nonesterified fatty acids in radioimmunoassays of steroids, *Clin. Chim. Acta,* 93, 283, 1979.

58. **Rodgerson, D. O. and Osberg, I. M.,** Sources of error in spectrophotometric measurement of aspartate aminotransferase and alanine aminotransferase activities in serum, *Clin. Chem.,* 20, 43, 1974.

59. **Litkowski, L. J. and Wilson, T. L.,** Effect of protein on titrimetry of bicarbonate, titratable acid, and ammonium in urine, *Clin. Chem.,* 25, 362, 1979.

60. **Rej, R. and Vanderlinde, R. E.,** Assay of asparatate aminotransferase activity: effects of serum and serum proteins on oxalacetate decarboxylation and dialysis, *Clin. Chem.,* 20, 454, 1974.

61. **Porter, W. H., Carroll, J. R., and Roberts, R. E.,** Hemoglobin interference with the Dupont automatic clinical analyzer procedure for calcium, *Clin. Chem.,* 23, 2145, 1977.

62. **Munster, D. J., Lever, M., and Carrell, R. W.,** Contributions of other sterols to the estimation of cholesterol, *Clin. Chim. Acta,* 68, 167, 1976.

63. **Lent, R. W., Goldschmidt, H. M. J., and Adler, D. J.,** Interference by creatine with determination of serum creatine kinase, *Clin. Chem.,* 22, 1741, 1976.

64. **Szasz, G., Gerhardt, W., Gruber, W., and Bernt, E.,** Creatine kinase in serum. II. interference by adenylate kinase with the assay, *Clin. Chem.,* 22, 1806, 1976.

65. **Stein, W. and Bohner, J.,** Immunoglobulin-bound creatine kinase BB ("macro CK") in three patients with different diseases, *Clin. Chem.,* 25, 1513, 1979.

66. **Siest, G. et al.,** Guanidine interference in determination of creatine kinase, *Clin. Chem.,* 21, 168, 1975.

67. **Aleyassine, H. and Tonks, D. B.,** Albumin-bound fluorescence: a potential source of error in fluorometric assay of creatine kinase BB isoenzyme, *Clin. Chem.,* 24, 1849, 1978.

68. **Velletri, K. et al.,** Abnormal electrophoretic mobility of a creatine kinase MM isoenzyme, *Clin. Chem.,* 21, 1837, 1975.

69. **Klein, B. and Jeunelot, C. L.,** Anion-exchange chromatography of erythrocytic and muscle adenylate kinase and its effect on the serum creatine kinase isoenzyme assays, *Clin. Chem.,* 24, 2168, 1978.

70. **Khanam, A. and Rahman, M. A.,** Studies on blood saccharoid fraction in normal and diabetic subjects, *Clin. Chim. Acta,* 71, 389, 1976.

71. **Brittenham, G.,** Spectrophotometric plasma iron determination from finger-puncture specimens, *Clin. Chim. Acta,* 91, 203, 1979.

72. **Farrington, C. J. and Chalmers, A. H.,** Gas-chromatographic estimation of urinary oxalate and its comparison with a colorimetric method, *Clin. Chem.,* 25, 1993, 1979.

73. **Liu, T. Z. and Khayam-Bashi, H.,** Positive interference of catecholamine metabolites with quantitation of urinary uric acid by the direct acid ferric-reduction method, *Clin. Chem.,* 25, 788, 1979.

74. **Knight, J. A., Fronk, S., and Haymond, R. E.,** Chemical basis and specificity of chemical screening tests for urinary vanilmandelic acid, *Clin. Chem.,* 21, 130, 1975.
75. **Bonvicini, P., Ceriotti, G., Plebani, M., and Volpe, G.,** Heparin interferes with albumin determination by dye-binding methods, *Clin. Chem.,* 25, 1459, 1979.
76. **Moss, G. A. et al.,** Kinetic enzymatic method for determining serum creatinine, *Clin. Chem.,* 21, 1422, 1975.
77. **Evenson, M. A. and Pendergast, D. D.,** Rapid ultramicro direction determination of erythrocyte lead concentration by atomic absorption spectrophotometry, with use of a graphite-tube furnace, *Clin. Chem.,* 20, 163, 1974.
78. **Elkeles, R. S. and Hambley, J.,** The effect of heparin in vitro on human plasma post-heparin lipolytic activity, *Clin. Chim. Acta,* 65, 135, 1975.
79. **Morris, R. J.,** Mechanism of interference by chelating agents and sucrose in radioimmunoassay of angiotensin I, *Clin. Chim. Acta,* 75, 503, 1977.
80. **Laessig, R. H. et al.,** Assessment of a serum separator device for obtaining serum specimens suitable for clinical analyses, *Clin. Chem.,* 22, 235, 1976.
81. **McIntosh, J. C.,** Application of a dye-binding method to the determination of protein in urine and cerebrospinal fluid, *Clin. Chem.,* 23, 1939, 1977.
82. **Liu, T. Z. and Khayam-Bashi, H. K.,** Interference of sodium azide with measurement of serum uric acid by the direct acid ferric reduction procedure, *Clin. Chem.,* 23, 581, 1977.
83. **Kim, H. J. and Cereceo, E.,** Interference by NaCl with the EMIT method of analysis for drugs of abuse, *Clin. Chem.,* 22, 1935, 1976.
84. **Healy, P. J., Turvey, W. S., and Willats, H. G.,** Interference in estimation of serum copper concentration resulting from use of silicone-coated tubes for collection of blood, *Clin. Chim. Acta,* 88, 573, 1978.
85. **Dusci, L. J. and Hackett, L. P.,** Interference in dilantin assays, *Clin. Chem.,* 22, 1236, 1976.
86. **Pragay, D. A. et al.,** Vacutainer contaminations revisited, *Clin. Chem.,* 25, 2058, 1979.
87. **Brown, H. H., Vanko, J., and Meola, J. M.,** Interference from serum separators in drug screening by gas chromtgtography, *Clin. Chem.,* 20, 919, 1974.
88. **Missen, A. W. and Gwyn, S. A.,** Another source of contamination from blood-sample containers, *Clin. Chem.,* 24, 2063, 1978.
89. **Sinclair, A. J. and Slattery, W.,** Blood-collecting tube as a contamination source in vitamin E fluorometry, *Clin. Chem.,* 24, 2073, 1978.
90. **Ball, W. et al.,** Ethanol production in infected urine, *N. Engl. J. Med.,* 301, 614, 1979.
91. **Gitlitz, P. H. and Frings, C. S.,** Interferences with the starch-iodine assay for serum amylase activity, and effects of hyperlipemi, *Clin. Chem.,* 22, 2006, 1976.
92. **Henry, R. J. et al.,** *Clinical Chemistry, Principles & Technics,* 2nd ed., Harper & Row, Hagerstown, Md., 1974, 1307.
93. **Dickinson, J. C. et. al.,** Ion exchange chromatography of the free amino acids in the plasma of the newborn infant, *Pediatrics,* 36, 2, 1965.
94. **Laessig, R. H. et al.,** Assessment of a serum separator device for obtaining serum specimens suitable for clinical analyses, *Clin. Chem.,* 22, 235, 1976.
95. **Rothwell, D. J. et al.,** Lactate dehydrogenase activities in serum and plasma, *Clin. Chem.,* 22, 1024, 1976.
96. **Bellevue, R. et al.,** Pseudohyperkalemia and extreme leukocytosis, *J. Lab. Clin. Med.,* 85, 660, 1975.
97. **Moore, E. W.,** Ionized calcium in normal serum, ultra filtrates, and whole blood determined by ion-exchange electrodes, *J. Clin. Invest.,* 49, 318, 1970.
98. **Thomson, W. H. S.,** An investigation of physical factors influencing the behaviour in vitro of serum creatine phosphokinase and other enzymes, *Clin. Chim. Acta,* 23, 105, 1969.
99. **Spooner, R. J. et al.,** Anomalous behaviour of glutathione reductase on dilution, *Clin. Chem.,* 22, 1005, 1976.
100. **Yatzidis, H. et al.,** Unexpected effect of hemodialysis on plasma phosphate changes occurring during incubation of whole blood, *N. Engl. J. Med.,* 57, 297, 1977.
101. **Burtis, C. A. et al.,** Factor influencing evaporation from sample cups and assessment of their effect on analytical error, *Clin. Chem.,* 21, 1907, 1975.
102. **Gosling, P. and Sammons, H. G.,** An interfering factor in the automated analysis of calcium, *J. Clin. Pathol.,* 32, 113, 1979.
103. **Hackett, L. P. and Dusci, L. J.,** Interference in drug screening assays, *Clin. Chem.,* 22, 933, 1976.
104. **Hall, R. A. and Whitehead, T. P.,** Adsorption of serum calcium by plastic sample cups, *J. Clin. Pathol.,* 23, 323, 1970.
105. **Husdan, H. et al.,** Effect of venous occlusion of the arm on the concentration of calcium in serum, and methods of its compensation, *Clin. Chem.,* 20, 529, 1974.

106. **Blumenfeld, T. A. et al.**, Simultaneously obtained skin-puncture serum, skin-puncture plasma, and venous serum compared, and effects of warming the skin before puncture, *Clin. Chem,*, 23, 1705, 1977.

107. **Statland, B. E.**, Fundamental issues in clinical chemistry, *Am. J. Pathol.*, 95, 243, 1979.

108. **Stacpoole, P. W, et al.**, Metabolic effects of dichloroacetate in patients with diabetes mellitus and hyperlipoproteinemia, *N. Engl. J. Med.*, 526, 298.

109. **Johnstone, R. E., Kennell, E. M., Brummond, W. Jr., Shaw, L. M., and Ebersole, R. C.**, Effect of halothane anesthesia on muscle, liver, thyroid and adrenal-function tests in man, *Clin. Chem.*, 22, 217, 1976,

110. **Ferrari, C. et al.**, Reduction of serum alkaline phosphatase and gamma-glutamyl transpeptidase activities by short term clofibrate, *N. Engl. J. Med.*, 295, 449, 1976.

111. **Kijima, Y. et al.**, Clofibrate effect on alkaline phosphatase in renal failure, *N. Engl. J. Med.*, 297, 113, 1977.

112. **Schade, R. W. B. et al.**, Clofibrate effect on alkaline phosphatase: bone or liver fraction, *N. Engl. J. Med.*, 669, 297.

113. **Murphy, J. V.**, Intoxication following ingestion of elemental zinc, *JAMA*, 212, 2119, 1970.

114. **Falko, J. M., Thomas, F. B.**, Acute pancreatitis due to procainamide-induced lupus erythematosus, *Ann. Int. Med.*, 83, 832, 1975.

115. **Chen, J. H. et al.**, Oxyphenbutazone-induced sialadenitis, *JAMA*, 238, 1399, 1977.

116. **Lynggaard, F. and Bjorndal, N.**, Bumetanide-induced hyperamylasaemia in patients with renal insufficiency, *Lancet*, 2, 1355, 1977.

117. **Kim, M. K., Rosenfield, R. L., and Dupon C.**, The effects of dexamethasone on plasma free androgens during the normal menstrual cycle, *Am. J. Obstet. Gynecol.*, 126, 982, 1976.

118. **Ferrannini, E., Pilo, A., Buzzigoli, G., Boni, C., and Tuoni, M.**, Raised plasma arginine vasopressin concentration in carbamazepine-induced water intoxication, *Br. Med. J.*, 2, 804, 1977.

119. **Stephens, W. P., Coe, J. Y., and Baylis, P. H.**, Plasma arginine vasopressin concentrations and antidiuretic action of carbamazepine, *Br. Med. J.*, 1, 1444, 1978.

120. **Routh, J. I. and Paul, W. D.**, Assessment of interference by aspirin with some assays commonly done in the clinical laboratory, *Clin. Chem.*, 22, 837, 1976.

121. **Mazze, R. I. et al.**, Inorganic fluoride nephrotoxicity: prolonged enflurane and halothane anesthesia in volunteers, *Anesthesiology*, 46, 276, 1977,

122. **Editorial**, Aspirin and renal function, *Lancet*, 1, 942, 1977.

123. **Shneerson, J. M. and Gazzard, B. G.**, Reversible malabsorption caused by methyldopa, *Br. Med. J.*, 2, 1456, 1977.

124. **Lins, P. E., Efendic, S., and Low, H.**, Somatostatin decreases urinary calcium excretion, *Lancet*, 2, 687, 1978.

125. **Major, L. F. and Goyer, P. F.**, Effects of disulfiram and pyridoxone on serum cholesterol, *Ann. Int. Med.*, 88, 53, 1978.

126. **Howard, A. N. and Marks, J.**, Hypocholesterolaemic effect of milk, *Lancet*, 2, 255, 1977.

127. **Yiamouyiannis, J.**, Chlorthalidone and serum cholesterol, *Lancet*, 2, 295, 1977.

128. **Jope, R. S.**, Choline accumulates in erythrocytes during lithium therapy, *N. Engl. J. Med.*, 833, 299.

129. **Sidell, F. R. and Kaminskis, A.**, Influence of age, sex and oral contraceptives on human blood cholinesterase activity, *Clin. Chem.*, 21, 1393, 1975.

130. **Ljungdahl, L. and Gerhardt, W.**, Creatine kinase isoenzyme variants in human serum, *Clin. Chem.*, 24, 832, 1978.

131. **Pujol-Amat, P., Davi, E., Martin-Comin, J., and Monfort, R.**, Diurnal variation in cyclic A.M.P. in plasma, *Lancet*, 1, 489, 1977.

132. **Hooymans, P. M. and Merkus, F. W. H. M.**, Effect of quinidine on plasma concentration of digoxin, *Br. Med. J.*, 2, 1022, 1978.

133. **Robertson, D. et al.**, Effects of caffeine on plasma renin activity, catecholamines and blood pressure, *N. Engl. J. Med.*, 181, 298, 1978.

134. **Sybulski, S. and Maughan, G. B.**, Effect of ampicillin administration on estradiol, estriol and cortisol levels in maternal plasma and on estriol levels in urine, *Am. J. Obstet. Gynecol.*, 124, 379, 1976.

135. **Stoffer, S. S., Hynes, K. M., Jiang, N., and Ryan, R. J.**, Digoxin and abnormal serum hormone levels, *JAMA*, 225, 1643, 1973.

136. **Yeo, P. P. B., Bates, D., Howe, J. G., Ratcliffe, W. A., Schardt, C. W., Heath, A., and Evered, D. C.**, Anticonvulsants and thyroid function, *Br. Med. J.*, 1, 1581, 1978.

137. **Dale, B. M.**, Fibrinogen depletion with sodium valproate, *Lancet*, 1, 1316, 1978.

138. **Newman, R. J.**, Comparison of the antilipolytic effect of metoprolol, acebutolol, and propranolol in man, *Br. Med. J.*, 2, 601, 1977.

139. **Fine, A., et al.**, Malabsorption of frusemide caused by phenytoin, *Br. Med. J.*, 1, 1061, 1977.

140. Beck, P. R., M.Sc., M.B., Ch.B., and Chaudrhuri, A. K. R., Effect of tobramycin on urinary γ-glutamyltransferase activity: studies in a case of renal carcinoma, *Clin. Chem.*, 22, 528, 1976.

141. Seino, S., Seino, Y., Matsukura, S., Kurahachi, H., Ikeda, M., Yawata, M., and Imura, H., Effect of glucocorticoids on gastrin secretion in man, *Gut*, 19, 10, 1978.

142. Greenwood, R. H., Mahler, R. F., and Hales, C. N., Diazoxide, glipizide, and hypoglycaemia, *Lancet*, 1, 307, 1977.

143. Castelli, W. P. et al., Alcohol and blood lipids. The cooperative lipoprotein pheno-typing study, *Lancet*, 2, 153, 1977.

144. Pelkonen, R. et al., Increase in serum cholesterol during phenytoin treatment, *Br. Med. J.*, 4, 85, 1975.

145. Sworn, M. J., Buchanan, R., and Moynihan, F. J., Increase of serum high-density lipoprotein in phenytoin users, *Br. Med. J.*, 2, 99, 1978.

146. Arntzenius, A. C. et al., Reduced high-density lipoprotein in women aged 40-41 using oral contraceptives, *Lancet*, 1, 1221, 1978.

147. Bird, H. A. and Wright, V., High-density lipoprotein cholesterol and antihypertensive drugs: the Oslo study, *Br. Med. J.*, 2, 403, 1978.

148. Masarei, J. R. and Lynch, W. J., Lowering of H.D.L.-cholesterol by androgens, *Lancet*, 2, 827, 1977.

149. Bradley, D. D. et al., Serum high-density-lipoprotein cholesterol in women using oral contraceptives, estrogens and progestins, *N. Engl. J. Med.*, 299, 17, 1978.

150. Grobe, H., Homocystinuria and oral contraceptives, *Lancet*, 1, 158, 1978.

151. Buchholz, D. H. and Donabedian, R. K., Unusual variant of lactate dehydrogenase isoenzymes, *Clin. Chem.*, 21, 162, 1975.

152. Trocha, P. J., Lactate dehydrogenase isoenzymes linked to beta-lipoproteins and immunoglobulin A., *Clin. Chem.*, 23, 1780, 1977.

153. Reeves, R. D. et al., Hyperlipidemia due to oxymetholone therapy. Occurrence in a long-term hemodialysis patient, *JAMA*, 236, 469, 1976.

154. Lim, P. and Jacob, E., Magnesium deficiency in patients on long-term diuretic therapy for heart failure, *Br. Med. J.*, 3, 620, 1972.

155. Stonesifer, L. D. et al., How carbon monoxide reduces plasma volume, *N. Engl. J. Med.*, 311, 299, 1978.

156. Pearl, K. N. and Chambers, T. L., Cardiac arrest due to liquorice-induced hypokalaemia, *Br. Med. J.*, 738, 1977.

157. Viberti, G. D., Glucose-induced hyperkalaemia: a hazard for diabetics?, *Lancet*, 1, 690, 1978.

158. Martin, J. B. et al., Inhibition by apomorphine of prolactin secretion in patients with elevated serum prolactin, *J. Clin. Endocrinol. Metab.*, 39, 180, 1974.

159. Floss, H. G. et al., Influence of ergot alkaloids on pituitary prolactin and prolactin-dependent processes, *J. Pharm. Sci.*, 62, 699, 1973.

160. Buckman, M. T. and Peake G. T., Estrogen potentiation of phenothiazine-induced prolactin *J. Clin. Endocrinol. Metab.*, 37, 977, 1973.

161. Egge, A. C. et al., Effect of cyproheptadine on TRH-stimulated prolactin and TSH release in man, *J. Clin. Endocrinol. Metab.*, 44, 210, 1977.

162. Langer, G. et al., The polactin response to neuroleptic drugs. A test of dopaminergic blockade: neuroendocrine studies in normal men, *J. Clin. Endocrinol. Metab.*, 45, 996, 1977.

163. Lee, D. A. et al., Increased prolactin levels during reserpine treatment of hypertensive patients, *JAMA*, 235, 2316, 1976.

164. Mendlewicz, J. and Youdim, M. B. H., Monoamine-oxidase inhibitors and prolactin secretion, *Lancet*, 2, 507, 1977.

165. McCallum, R. W. et al., Metoclopramide stimulates prolactin secretion in man, *J. Clin. Endocrinol. Metab.*, 42, 1148, 1976.

166. Stumpe, K. O. et al., Hyperprolactinaemia and antihypertensive effect of bromocriptine in essential hypertension, *Lancet*, 2, 211, 1977.

167. Hill, P. and Wynder, F., Diet and prolactin release, *Lancet*, 2, 806, 1976.

168. Gold, P. W. et al., Growth hormone and prolactin response to levodopa in affective illness, *Lancet*, 2, 1308, 1976.

169. Horwitz, D. et al., Effects of methyldopa in fifty hypertensive patients, *Clin. Pharmacol. Ther.*, 8, 224, 1967.

170. Brown, G. M. et al., Effect of carbidopa on prolactin, growth hormone and cortisol secretion in man, *J. Clin. Endocrinol. Metab.*, 43, 236, 1976.

171. Dericks, J. S. E. and Taubert, H. D., Elevation of serum prolactin during application of oral contraceptives, *Contraception*, 14, 1, 1976.

172. Delle Fave, G. F. et al., Gynaecomastia with cimetidine, *Lancet*, 1, 1319, 1977.

173. Gold, M. S. et al., Antipsychotic effect of opiate agonists, *Lancet,* 2, 398, 1977.
174. Cavill, I., Ricketts, C., Griffiths, G. J., Elder, G. H., and Trevett, D., Effect of chlormethiazole on serum prolactin, *Br. Med. J.,* 2, 1266, 1978.
175. Werner, S., Brismar, K., and Olsson, S., Hyperprolactinaemia and liquorice, *Lancet,* 1, 319, 1979.
176. Laragh, J. H., Oral contraceptive-induced hypertension — nine years later, *Am. J. Obstet. Gynecol.,* 126, 141, 1976.
177. Theilade, P. et al., Propranolol influences serum T3 and reverse T3 in hyperthyroidism, *Lancet,* 2, 363, 1977.
178. Chanmugam, D., Machado, V., and Mihindukulasuriya, J. C. L., Transplacental hyponatraemia due to oxytocin, *Br. Med. J.,* 1, 152, 1978.
179. Lotti, G. et al., Reduction of plasma triiodothyronine (T3) induced by propranolol, *Clin. Endocrinol.,* 6, 405, 1977.
180. Degrot, L. J. and Hoye, K., Dexamethasone suppression of serum T3 and T4, *J. Clin. Endocrinol. Metab.,* 42, 976, 1976.
181. Panaall, P. R. and Maas, D. A., Danazol and thyroid function tests, *Lancet,* 1, 102, 1977.
182. Kristensen, B. O. and Weeke, J., Propranolol-induced increments in total and free serum thyroxine in patients with essential hypertension, *Clin. Pharmacol. Ther.,* 22, 864, 1977.
183. Brock, D. J. H. and Gosden, C., Induction of sex hormone binding globulin by phenytoin, *Br. Med. J.,* 2, 934, 1977.
184. DuBois, R. M. and Freedman, S., Suppression of serum testosterone concentrations in men by an oral contraceptive preparation, *Br. Med. J.,* 2, 1261, 1977.
185. Herxheimer, H., Effect of iodide treatment on thyroid function, *N. Engl. J. Med.,* 297, 171, 1977.
186. Sonnenblick, M., Oren, A., and Jacobsohn, W., Hypertransaminasaemia with heparin therapy, *Br. Med. J.,* 2, 77, 1975.
187. Martin, J. V. et al., Enzyme induction as a possible cause of increased serum-triglycerides after oral contraceptives, *Lancet,* 1, 1107, 1976.
188. Shaw, J., England, J. D. F., and Hua, A. S. P., Beta blockers and plasma triglycerides, *Br. Med. J.,* 1, 986, 1978.
189. Mordasini, R. et al., Selective deficiency of hepatic triglyceride lipase in uremic patients, *N. Engl. J. Med.,* 297, 1362, 1977.
190. Unge, G., Lins, L., and Hultman, E., Tryptophan in patients on chronic haemodialysis, *Lancet,* 2, 937, 1977.
191. Scanlon, M. F. et al., Evidence for dopaminergic control of thyrotrotrophin secretion in man, *Lancet,* 2, 421, 1977.
192. Chamberlain, D. A. and Clark, A. N. G., Diurnal variation of serum urate in pregnancy, *Br. Med. J.,* 2, 1520, 1977.
193. Reach, G. et al., Increased urate excretion after p,p'-DDD, *Lancet,* 2, 1269, 1978.
194. Lederballe Pedersen, O. and Mikkelsen, E., Beta-blockers and uric-acid excretion, *Lancet,* 2, 1160, 1978.
195. McPherson, K. et al., The effect of age, sex and other factors on blood chemistry in health, *Clin. Chim. Acta,* 84, 373, 1978.
196. Panek, E. and Steinmetz, J., The effect of sex, deviation from ideal weight and sampling time on blood constituents in presumably healthy subjects, *Clin. Chim. Acta,* 92, 343, 1979.

CONTROL MATERIALS AND CALIBRATION STANDARDS

John W. Ross

DEFINITIONS

Control materials are analyzed solely for quality control purposes and not for calibration of procedures. Calibrators are materials with which the sample is compared in order to determine the concentration or other quantity.[1] A single material cannot fulfill both of these distinct functions for an analytic procedure.

APPLICATION OF CALIBRATORS AND CONTROL MATERIALS TO ANALYTIC METHODS

Calibrators and control materials are useful only if applicable to the analytical methods used. Methods known to have acceptably small inaccuracy due to physical or chemical matrix effects may be calibrated with primary standard solutions. However, procedural and technical constraints and complex matrix effects of biologic samples make the use of primary standards impossible in many routine methods. The calibrators and control materials selected for such methods must simulate, in all important respects, the physical and chemical properties of patient samples in order to compensate for matrix effects during analysis and to be sensitive to important changes in analytic error conditions.

The intended use of calibrators or control materials may be to monitor accuracy in external quality control, to monitor analytic imprecision in internal quality control, to measure achievement of medical goals, or to serve as a calibration standard. The suitability of the materials for these purposes is determined by their stability, homogeneity, and the degree to which their matrix effects in the analytic system are comparable to those of the biologic specimen matrix. The homogeneity of the materials determine their sensitivity to analytic imprecision and the limit to which the uncertainty interval of the assigned value of a calibrator can be reduced. Their homogeneity and stability determine the capability of the control system to detect changes in analytic error conditions. Their similarity to the biologic specimen matrix determines their ability to measure the clinical performance characteristics of the method.

TYPES OF CONTROL MATERIALS AND THEIR PRODUCTION

Although controls may be prepared from excess laboratory serum or plasma,[2,3] commercially prepared control materials are most widely used. Appropriate base materials for commercial general clinical chemistry controls are human plasma or urine. Animal serum, PVP, or protein bases such as albumin may be used as a matrix for calibrators for some analytes. However, base materials made from pooled plasma or serum of animal origin are not satisfactory for control materials for enzyme procedures[4-6] and albumin.[7,8]

The plasmas used should be fresh and are obtained best by plasmapheresis. Collection of outdated laboratory serum or plasma or blood bank plasma for commercially produced control materials leads to problems in precise definition of the base material and in reproducing the degree of degradation of the base material. Moreover, dialysis is necessary to remove the excess aqueous phase and other materials added to blood bank plasma with anticoagulants. The type of anticoagulant added to the plasmas must be defined. Excessive lipemia, hemolysis, and bacterial contamination must be avoided. Subpools of defined size, the original units of plasma, or the donor population must be screened for hepatitis B surface antigen by a third-generation test. The base material must be frozen shortly after collection and kept frozen until shortly before the processing of a lot to prevent degradation of the biologic matrix.

Control materials prepared at a single operation are assigned a unique lot number. Lots of chemistry control materials may be as large as 1700 ℓ. The initial processing of plasmas for a new lot includes: thawing and defibrination (for example, with protamine); depth and screen filtration (for example, to a particle size of 0.5 μm); initial pooling and mixing; analyte depletion by dilution or dialysis as necessary; analyte concentration adjustment to desired target ranges with defined chemical additives; and mixing of the final base pool to liquid homogeneity in a single container. Chemical contamination is avoided by careful selection of additives, and microbiologic contamination is controlled by relatively aseptic manufacturing process design.

Techniques such as selected analyte depletion by enzyme treatment, addition of antibiotics rather than filtration for sterilization, or addition of detergents to help dissolve certain additives may result in unexpected behavior of the control material in some methods and should be avoided where possible. Partially clotted sera must be avoided, and defibrination must be complete. Ketoacids and other interfering substances may be added to the pools with impure chemical additives, especially enzyme preparations.

Plug lyophilized materials are dispensed while liquid into vials under carefully controlled conditions by automated fillers which are capable of vial fill coefficients of variation (CVs) of 0.2% or less at fill volumes of 10 mℓ or greater. The vials must be free of chemical and microbiologic contamination. After the lot is dispensed, all vials are placed on trays and taken to large lyophilizers where they are frozen at approximately the same time by coolant (for example, Freon®) circulated in coils under each tray. A high vacuum is created within the chambers, and the ice is sublimated. Lyophilizer temperature is slowly raised to a final temperature which ranges from room temperature to 37°C. Each vial is securely stoppered, either under vacuum or after dry nitrogen is admitted to the lyophilizer. The type of lubricant on the stopper must be selected with care, and the stopper must be carefully sized for proper fit in the vial neck. Postlyophilization stabilization of the plugs of matrix material requires days to weeks under defined temperature conditions. Conditions of temperature, light, evaporation, and microbiologic contamination affecting the pool must be uniform during the entire vial filling and lyophilization process, whether the material remains in the original mixing container or has been dispensed into vials. Otherwise, differing treatment of vials leads to inhomogeneity of the pool.

Plug lyophilized materials have been the mainstay for internal and external laboratory quality control for many years. Two new commercial products have been introduced recently: "bead" or droplet lyophilized plasma; and plasma stabilized in liquid form. Bead lyophilized plasmas are very rapidly frozen by spraying fine droplets directly through coolant. After lyophilization, the stable beads are remixed in a single container and are dispensed with vial fill CVs which are targeted at values similar to those achieved by liquid dispensing.

The aqueous phase of stabilized liquid control materials is partially replaced by polyhydric alcohols (for example, ethylene glycol). Analyte concentrations are adjusted, and the materials are dispensed into small vials which are stored under conditions which assure defined stability.

The manufacturer's stability for liquid control materials is at least 1 year from the date of manufacture at −15 to −20°C. Lyophilized materials are stable at least 2 years from the date of manufacture at 4°C. Long-term stability data within the time of manufacturer certified usefulness are discussed by Lawson and Haven elsewhere in this chapter.

Large-scale production of control materials and calibrators uses bulk processing equipment which has practical limitations. Every lot is affected by some degree of matrix alteration, inhomogeneity, and analyte instability. Standards have been proposed which limit specimen differences in a lot of calibration or control material due to vial fill variation and other sources of inhomogeneity to ± 0.5%[9] (or ± 1%[10] of the mean value in 95% or more of the vials, corresponding to an interval CV of 0.25% (or 0.5%). Competent manufacturers should maintain in all batch records information regarding the source of base materials, anticoagulants used, hepatitis and other screening procedures employed, stabilizers added, type of chemical additives used, results of physical and chemical checks on homogeneity during processing, microbiologic content and, in the case of lyophilized materials, the residual moisture content and vial fill CVs. Any such information which is not proprietary should be made available on request.

Some useful guidelines for capable manufacture of importance to homogeneity and stability characteristics of plug lyophilized materials are

1. Vial fill variation of less than 0.2% CV for 10 ml size or larger (measured by liquid fill weight
2. Residual moisture content of less than 1% in 95% of vials (measured by drying of lyophilized plugs over P_2O_5
3. Bacterial colony count of vial 1 hr after reconstitution with sterile diluent of less than 100 cfu/ml.

Bead lyophilized materials should have similar characteristics. Vials capped under vacuum should have an audible influx of air when opened. Vials under nitrogen are less likely to leak but cannot be checked easily for leakage. The lyophilized plug should be at the bottom of the vial and retracted from the sides without layering, which may indicate increased residual moisture content.

HANDLING AND USE OF CONTROL MATERIALS IN THE LABORATORY

Long-term storage of lyophilized materials should be at 4°C with protection from light. Stabilized liquid controls require long-term storage at −15 to −20°C and protection from light.

Reconstitution of the matrix with an aqueous phase is necessary for lyophilized material. This requires an accurate volumetric delivery. While a potential source of variation, care can render this factor negligible.[11] The temperature of the reconstituting fluid is important for creatine kinase stability.[12,13] Complete solution of some analytes in lyophilized materials (magnesium, calcium, certain enzymes) may require 1 to 2 hr after reconstitution.[9]

Stability after reconstitution should be characterized and varies with material and analyte. Alkaline phosphatase increases;[14,15] creatine kinase decreases due to light and temperature effects;[12,16] and bilirubin decreases due to light effect.[17] Stability of glucose is related to the degree of bacterial contamination.[9,18] Acid phosphatase activity will decline unless the pool is acidified. Evaporation may proceed at a significant rate in open sample containers.[19-21] These effects can be managed by protecting control materials from light, refrigeration of control materials, recapping of control vials, acidification of an aliquot, and limiting exposure in open sample cups. Performing analysis of unstable enzymes at approximately the same time interval after reconstituting a vial may be helpful. Many of these effects of temperature, light evaporation, and time of analysis after reconstitution are similar to preanalytic variables in patients. (Preanalytic variation is discussed by Beautyman under "Erroneous or Misleading Results due to Preanalytic Factors" in this Volume.) One should keep in mind that these analytes in patient sera are not absolutely stable, and patient sera require similar protective measures. Recognition of the need to properly preserve control material can be the basis of a blind sample quality control program which may control preanalytic sources of variation affecting patient specimens.[22] Although stability of a calibration standard is desirable, use of a control material with analytes stable well beyond the stabilities observed in patient specimens may be injudicious.

Choice of diluent for lyophilized materials is important. Buffers used in the past include tromethamine (tris) buffers which interfered with ammonia determinations by the Berthelot reaction, ion specific electrode methods for calcium and potassium, and total protein determinations.[23] Ammonium bicarbonate interfered with protein determination and, because of the large concentration of ammonium ion present, obviated the use of the material as a control for ammonia analysis and methods for urea nitrogen which used the Berthelot or urease reaction.[23] Presently used buffers are tetramethyl ammonium bicarbonate and sodium bicarbonate. The latter buffer is physiologic and, therefore, offers no methodologic interferences (except, theoretically, on the basis of pH). At this time, there are no apparent methodologic interferences of the organic ammonium ion buffer.

Pooling vials of reconstituted lyophilized materials may reduce interval differences[9] and freezing of aliquots of reconstituted controls may be useful to conserve expensive control serum in small laboratories. However, owing to variable effects on enzymatic and lipoprotein analytes, which are difficult to control, this should be done with caution. Isoenzymes of creatine kinase vary in lability to freeze-thaw cycles,[16] as do those of lactate dehydrogenase.[24,25] Freezing of aliquots of a vial or aliquots of pooled vials should follow a carefully defined protocol to minimize interaliquot differences, particularly those created during the thawing process.[16]

Control specimens should be run routinely at different concentrations near medical decision points. Decision points include the intersections of reference intervals of defined medical importance in the normal and upper and lower pathologic ranges. Control procedures should also be used which periodically test the upper and lower limits of linearity of the analytic range of the undiluted samples.

Control materials are usually pools of plasmas from large numbers of individuals. It is impossible to be sure that such materials cannot transmit hepatitis, and these materials should be regarded at all times as hazardous and potential sources of hepatitis.

LIMITATIONS OF CONTROL MATERIALS

Plug lyophilized control materials are inhomogeneous to some extent due to limitations in the precision of the vial filling process and to post-vial fill variation in treatment of vials. Glick[26] has found that the vial-to-vial CVs of the contents of plug lyophilized serum with 10-mℓ fills range from 0.31 to 0.56%. Vial fill CVs probably vary inversely with the size of vial fill. Adams[27] noted an interval CV of 0.18% for a 10-mℓ vial fill product and vial fill CVs ranging from 0.33 to 0.91% for several 5-mℓ vial fill products. These vial-to-vial differences can contribute significantly to observed overall test imprecision if the method CV is 0.5% or less.

Protein changes occur in plug lyophilized materials that produce some cloudiness.

This turbidity is probably dependent upon post-vial fill factors such as denaturation of lipoproteins during lyophilization. The resultant turbidity varies from vial to vial with interval CVs of absorbance ranging from 6 to 16%.[28] Variation in such insoluble denatured particles could produce variation in unblanked procedures, where the contribution of blank is important. The particles are irregular and show significant light scattering in certain methods, such as certain kinetic enzyme analyzers, which may increase the observed imprecision beyond that observed on the average with native patient sera.[29]

Precision estimates may be affected by inhomogeneity of certain analytes in an otherwise homogeneous material. The apparent inhomogeneity may be method related. Significant interval variability in folic acid levels in otherwise apparently homogenous control material was reported by Rhoads.[30] Jansen et al.[31] noted inhomogeneity in several lots of control material. In one example, the inhomogeneity was limited to creatinine and glucose and, in a second example, to urea. A third case had inhomogeneity involving multiple analytes. Menson et al.[11] found that interval variation accounted for 50 to 85% of the observed variability in three plug lyophilized controls in the cases of acid phosphatase, alkaline phosphatase, and creatine kinase. Less than 1% of the observed variability of alanine aminotransferase, aspartate aminotransferase, and lactate dehydrogenase was due to interval differences in the same controls. This type of inhomogeneity is due to factors such as variation in treatment of vials or trays of vials during the lyophilization process, use of sporadically contaminated vials, bacterial conntamination, or excessive residual moisture in some vials.

Plug lyophilized materials may exhibit systematic effects upon results revealed as method related bias. Hearne and Fraser[32] noted that certain quality control materials did not give results consistent with assigned creatinine values with an enzymatic Jaffe assay due to exhaustion of NADH by competing secondary reactions. Sheehan et al.[33] reported excessive sensitivity of lyophilized controls for iron-binding capacity to changes in assay conditions which did not affect patients. This effect did not occur with frozen pooled controls or liquid controls. Handschuch and Donovan[34] noted intermethod bias in creatine kinase measurement due to anomalies of a control material used in an external quality control program. Van Helden et al.[8] in a more general report, found that only 6 of 59 lots of lyophilized control materials were characterized by intermethod analytic bias comparable to that of patient sera for ten common analytes. Grannis and Lott[35] found that control materials, while sensitive to the presence of analytic bias, did not always indicate the type or magnitude of bias. Factors in control materials producing method-related bias do not necessarily interfere with their use in monitoring precision. However, they are important considerations when interpreting evidence of changes in the error conditions of the analytic system and in assessing method accuracy.

Bead lyophilized control materials have irreducible interval differences due to vial fill variation. If variation due to reconstitition is to remain negligible, precautions must be observed. Particles of bead lyophilized material may be lost from vials during the reconstitution process since their static charge may cause the particles to cling to the stopper and sides of the vials. Excessive foaming may occur during reconstitution. Careful instructions are given by the manufacturer to avoid these sources of increased interval variation.

Liquid control materials presently available have added a high percentage of ethylene glycol. This alters the matrix in terms of specific gravity, osmolality, viscosity, and freezing point. Ethylene glycol in 34% aqueous Solution has a freezing point of $-18°C$. The osmolality is approximately 9700 mosm/kg. The viscosity is approximately 2.4 relative to water (serum averages 1.6 relative to water).[36,37] Pope et al.[38] have reported bias related to methods using dialysis when ethylene glycol is added to control serum.

There is a bias of 6 mmol/ℓ of sodium in a regional quality control program using liquid control material between ion specific electrode methods and flame photometry methods independent of the use of dialysis.[39] Systematic bias does not limit the use of the material as a precision control but reduces the interlab comparability of regional quality control group data..

Commercially available liquid materials are not useful as controls for some analytes usually controlled with lyophilized materials (for example, acid phosphatase), and they cannot be used to control osmolality.

Polyhydric alcohols protect enzyme activity in serum[40,41] and protect enzymes against cold inactivation.[42,43] While smaller CVs for enzymes would be anticipated with highly stable materials in internal quality control programs, extraordinary stability well beyond that of patient sera could be a mixed blessing. Control of preanalytic variation is an important part of overall quality control.[44] The report of Glenn and Hathaway[22] emphasizes the usefulness of quality control specimens in control of preanalytic variation affecting patient specimens. Stability most similar to the majority of patient sera is most desirable, and judgment should be exercised in selecting the available control product most suitable to the particular laboratory's need.

RELATIVE ADVANTAGES OF AVAILABLE TYPES OF CONTROL MATERIALS

The limitations in achievable homogeneity and the occurrence of unacceptable inter-vial variation in some lots of plug lyophilized control materials, particularly of labile analytes, has been discussed in the previous section. Adams et al.[27] noted improvement in interval variability of creatine kinase and lactage dehydrogenase in bead lyophilized materials compared to plug lyophilized materials. The bead lyophilization process allows stabilization of the material while the material is in a sufficiently dispersed form to permit remixing after lyophilization. The lyophilized beads are reblended to homogeneity, theoretically in the same way the original liquid pool was rendered homogeneous by stirring. This removes most of the effects of inhomogeneity due to post vial fill factors such as the lyophilization process and variations in treatments of individual vials. It does not remove variation due to vial fill or reconstitution.

The rapid freezing of very fine droplets of serum as they are sprayed through coolant provides an additional benefit. The freezing is far more rapid than the freezing of a 10-mℓ plug with coolant circulated beneath a tray upon which vials stand. The degree of protein denaturation appears to be less, and the clarity of the final product is improved. There is less interval variation of turbidity, and turbidity is only slightly above the average noted for native serum.[28]

The extent of the practical advantages of these improvements remains to be completely evaluated. The College of American Pathologists (CAP) Chemistry Survey began to use a bead lyophilized product in 1979. Comparison of Survey precision in 1978 (plug lyophilized product) to 1979 Survey precision reveals similar estimates of Survey precision for calcium, potassium, creatinine, and glucose.[45] For these analytes, representative of inorganic and more stable organic analytes, there are no important differences in overall homogeneity between plug and bead lyophilized materials. Advantages may well accrue, however, in the case of the more labile enzymes. There is a potential advantage of low levels of particulate matter in certain kinetic analyzers. There is a potential advantage in unblanked procedures in which varying serum blanks could be important.

Stabilized liquid control material avoids sources of inhomogeneity due to vial fill variation, to post-vial fill factors, and to reconstitution. The concept is, theoretically, the most ideal. Analysis of the precision of highly precise methods (those with CVs

Table 1
COMPARISON OF STATE OF THE ART AVERAGE CV TO ANALYTIC CV
EXPERIENCED WITH A LIQUID CONTROL MATERIAL

Analyte	Units	Concentration	Predicted average CV[a]	Laboratories using liquid controls[b]		
				Liquid control average CV	Number of files	Number of results
Albumin	g/dl	4.1	3.1	4.7	7	868
Alkaline phosphatase	IU/l	Normal level	6.3[d]	4.7	9	833
Biliribin	mg/dl	1.3	6.2	6.3	10	990
Calcium	mg/dl	11.5	2.1	2.6	12	1331
Chloride	mmol/l	117	1.4	1.7	11	1379
Cholesterol	mg/dl	172	4.2	4.5	10	1031
Creatine kinase	IU/l	Normal level	11.2[d]	7.7	9	1319
Creatinine	mg/dl	1.4	5.7	6.9	13	2028
Glucose	mg/dl	98	3.6	3.2	19	2665
Lactate dehydrogenase	IU/l	Normal level	6.3[d]	5.0	13	1333
Phosphorus	mg/dl	4.8	3.4	4.1	11	1125
Potassium	mmol/l	5.2	1.9	2.2	16	1741
Total protein	g/dl	6.4	2.2	2.4	12	1273
Aspartate aminotransferase	IU/l	Normal level	8.7[d]	13.1[e]	13	1144
Sodium	mmol/l	146	1.1	1.1	16	1774
Urea nitrogen	mg/dl	17	4.7	5.2	17	2746
Uric acid	mg/dl	5.3	2.9	3.2	11	1178
Totals					209	24758

Note: Average number of results in a liquid pool file = 118;
Average number of files per analyte = 12.3.

[a] Predicted average cumulative CV over 6 months for automated methods. Data is based on plug lyophilized control material. From Reference 47.
[b] Beckman "Pace" quality control program. From Reference 39.
[c] Manual methods and a few files with very high CV's which were thought to be outliers, were excluded.
[d] Cumulative average CV over 6 months for automated enzyme methods. From Reference 46.
[e] AST level in liquid control was at the mid-normal range. The comparison plug lyophilized materials are typically at the upper limits of normal.

less than 0.5%) is possible. Waste is minimal, since the excess left in a vial need not be discarded at the end of the day.

The practical benefits of liquid control materials must be more completely evaluated in view of their limitations, as defined in the previous section, and in view of their limitations, as defined in the previous section, and in view of the relatively high price of the product. Data from a small regional quality control program using stabilized liquid control materials[39] are summarized in Table 1. There is no apparent difference in experienced CVs between those predicted by the use of plug lyophilized materials[46,47] and the liquid material for the inorganic and more stable organic analytes. Three of the four enzymes have slightly lower average CVs with liquid controls. This may be significant, since in the fourth case, there is an analyte concentration discrepancy which may account for the higher CV noted with liquid material. Hearn et al.,[48] using another type of serum-based liquid control material, noted the advantage of liquid material for external quality control for analytes and methods which usually have high precision (sodium, potassium, and chloride), and the advatage of plug lyophilized materials for inorganic phosphorus and triglycerides. They reported no important differences for other surveyed analytes.

Inhomogeneity of plug lyophilized control materials is caused mainly by vial fill variation and differing vial treatments in post-vial fill processing. All analytes are affected by vial fill variation. Only less stable analytes (excluding contaminated vials and glass) are affected by post vial fill factors. Bead lyophilized materials avoid post vial fill variation, and liquid controls avoid both effects; but each of these new materials brings its own unique set of factors influencing their choice as control materials.

The practical effects of inhomogeneity are relatively unimportant for most common analytes, and all of the types of control material are useful in monitoring precision and marking achievement of medical goals. Comparison of precision data from plug lyophilized materials to similar data from bead lyophilized materials and liquid control materials reveals no important effects of inhomogeneity except for certain unstable analytes such as creatine kinase, lactate dehydrogenase, and alkaline phosphatase. Technical improvement is necessary for this category of analytes. The desirable degree of stability after reconstitution of controls is a matter of informed judgment.

Technical improvement in stability and homogeneity of control materials is unlikely to affect conclusions regarding the medical usefulness of laboratory analytic capability in general. This is due to the large analytic CV relative to inhomogeneity for most analytes and to the large biologic variation of many of the unstable analytes. However, intraindividual alkaline phosphatase and lactate dehydrogenase biologic variation is about the same size as present estimates of average analytic precision and may be exceptions where medical conclusions may change with technologic improvement in controls. (See discussion of components of precision performance standards following.)

Although there is a technical advantage of liquid controls for analytes for which there is a high degree of precision (sodium and chloride),[47] present medical conclusions are not likely to be changed, since pragmatic medical usefulness criteria find no usefulness for analytic CVs of 0.5% or less for sodium and chloride. If calcium analysis becomes more precise in the future and approaches CVs of 0.5%, a homogeneous material would be required to monitor achievement of intraindividual calcium precision goals.

ASSIGNMENT OF CALIBRATOR AND CONTROL MATERIAL VALUES

Assay values assigned to calibrators must be sufficiently accurate for the intended use. The uncertainty interval for the assigned concentration value must be small compared to the analytic precision of the method to be calibrated. A general protocol for calculation of desirable confidence intervals for calibration standards has been proposed[10] and questioned.[9] The proposed method for assigning uncertainty intervals requires replicate testing of the material in multiple laboratories. In general, the ageement between laboratory means must be less than a maximum allowable uncertainty interval, D, which is equal to 8% of the 95% normal range. For assigned values outside the normal range, D is changed multiplicatively by the ratio of the assigned value to the midpoint of the normal range.

The proposed maximum allowable uncertainty interval for control materials is 20% of the 95% normal range with proportional adjustment for pathologic values by the ratio of the midpoint of the normal range. Assigned values for control material do not require confirmation in other laboratories. The general validity of this approach of assigning calibratory and control values has been discussed.[49] Methodologic principles for assigning values to calibrators and control materials in clinical chemistry have been proposed.[50]

Efforts are presently underway to establish a national reference system in clinical chemistry.[51] The aim of the reference system is to establish a heirarchy of definitive methods and reference methods and of certified reference materials and control and testing materials by which the accuracy of definitive methods can be transferred to other methods or the degree of bias present in reference and field methods characterized. The findings of this conference, when published, may develop new approaches in assignment of calibrator values to reduce interlaboratory bias.

REFERENCES

1. **Buttner, J., Borth, R., Boutwell, J. H., and Broughton, P. M. G.**, Provisional recommendation on quality control in clinical chemistry. I. General principles and terminology, International Federation of Clinical Chemistry, *Clin. Chem.*, 22, 532, 1976.
2. **Bowers, G. N., Burnett, R. W., and McComb, R. B.**, Selected method: preparation and use of human serum control materials for monitoring precision in clinical chemistry, *Clin. Chem.*, 21, 1830, 1975.
3. **Tietz, N. W.**, Ed., *Fundamentals of Clinical Chemistry*, W. B. Saunders, Philadelphia, 1976, 92.
4. **Bretaudiere, J. and Bailey, M.**, Use and limitations of "controls" in enzyme activity measurements, in *Proc. 2nd Int. Symp. Clin. Enzymol.*, Tietz, N., Weinstock, A., and Rodgerson, D., Eds., American Association of Clinical Chemistry, Washington, D.C., 1976, 227.
5. **Fasce, C. F., Rej., R., Copeland, W. H., and Vanderlinde, R. E.**, A discussion of enzyme reference materials: applications and specifications, *Clin. Chem.*, 19, 5, 1973.
6. **Rej., R., Fasce, C. F., and Vanderlinde, R. E.**, Interlaboratory proficiency intermethod comparison, and calibration suitability in assay of serum aspartate aminotransferase activity, *Clin. Chem.*, 21, 1141, 1979.
7. **Pastewka, J. V. and Ness, A. T.**, The suitability of various serum albumin products as standards for the quantitative analysis of total protein and albumin in human body fluids, *Clin. Chim. Acta*, 12, 423, 1965.
8. **Van Helden, W. C. H., Visser, R. W. F., Van den Bergh, F. A., J.-T. M., and Souverijn, J. H. M.**, Comparison of intermethod analytic variability of patient sera and commercial quality control sera, *Clin. Chim. Acta*, 93, 335, 1979.
9. **Buttner, J., Borth, R., Boutwell, J. H., Broughton, P. M. G., and Bowyer, R. C.**, Provisional recommendation on quality control in clinical chemistry. III. Calibration and control materials, International Federation of Clinical Chemistry, *Clin. Chem.*, 23, 1784, 1977.
10. National Committee for Clinical Laboratory Standards, Calibration Reference Materials and Control Materials in Clinical Chemistry, Villanova, Pa., ASC-2, 1975.
11. **Menson, R. C., Adams, T. H., and Sanford, R. L.**, Determination of true vial-to-vial constituent variation by statistical analysis, *Clin. Chem.*, 23, 1120, 1977.
12. **Perry, B., Doumas, B., and Jendrzejczak, B.**, Effect of light and temperature on the stability of creatine kinase in human sera and controls, *Clin. Chem.*, 25, 625, 1979.
13. **Feld, R. W., Brown, L. F., Neri, B. P., and Witte, D. L.**, Effect of diluent temperature on creatine kinase values found for lyophilized controls and reference sera, *Clin. Chem.*, 24, 2039, 1978.
14. **Massion, C. G. and Frankenfeld, J. K.**, Alkaline phosphatase: lability in fresh and frozen human serum and in lyophilized control material, *Clin. Chem.*, 18, 366, 1972.
15. **Smith, A. F. and Fogg, B. A.**, Possible mechanisms for the increase in alkaline phosphatase activity of lyophilized control material, *Clin. Chem.*, 18, 1972.
16. **Nelson, D. A. and Henderson, A. R.**, Stability of commonly used thiols an of human creatine kinase isoenzymes during storage at various temperatures in various media, *Clin. Chem.*, 23, 816, 1977.
17. **Gambino, S. R.**, *Bilirubin Assay:* American Society of Clinical Pathologists, Chicago, 1968, 12.
18. **Hanok, A. and Kuo, J.**, The stability of a reconstituted serum for the assay of fifteen chemical constituents, *Clin. Chem.*, 14, 58, 1968.
19. **Glenn, G. C. and Hathaway, T. K.**, Effects of specimen evaporation on quality control, *Am. J. Clin, Pathol.*, 66, 645, 1976.
20. **Goldberg, M. and Sardi, A.**, Study of evaporation losses from test tubes and plastic sample cups, *Clin. Chem.*, 19, 662, 1973.

21. Burtis, C. A., Bequovich, J. M., and Watson, J. S., Factors influencing evaporation from sample cups, an assessment of their effect on analytic error, *Clin. Chem.*, 21, 1907, 1975.
22. Glenn, G. C. and Hathaway, T. K., Quality control by blind sample analysis, *Am. J. Clin. Pathol.*, 72, 156, 1979.
23. Young, D. S., Perstaner, L. C., and Gibberman, V., Effects of drugs on clinical laboratory tests, *Clin. Chem.*, 21(5), 1975.
24. Kreutzer, H. H. and Fennis, W. H., Lactate dehydrogenase isoenzymes on blood serum after storage at different temperatures, *Clin. Chim. Acta*, 9, 64, 1964.
25. Wilkinson, J. H., *Isoenzymes*, 2nd ed., Lippincott, Philadelphia, 1970, 163.
26. Glick, J. H., Jr., Osmotic estimation of vial-to-vial variation in contents of lyophilized sera, *Clin. Chem.*, 23, 781, 1977.
27. Adams, T. H., Menson, R. C., and Caputo, M. J., Comparison of intervial variations of Omega with four commercial control sera: Implications for improved quality control, Technical Discussion, 43, Hyland Diagnostics, Division of Travenol Laboratories, February 1979.
28. Adams, T. H., Menson, R. C., Caputo, M. J., and Doumas, B. T., The effect of control sera turbidity on intervial precision and instrument performance. Technical Discussion, No. 44, Division of Travenol Laboratories, Hyland Diagnostics, Deerfield, Ill., February 1979.
29. Atwood, J. G., Boyd, L. L., and Dowdy, A. B., Repeatability with 10 freeze-dried control materials on the KA-150, Laboratory medicine application study No. 75, Instrument Division, The Perkin-Elmer Corporation, Norwalk, Conn., 1975.
30. Rhoads, D. G. et al., Intra-lot variations in folate quality control sera, *Clin. Chem.*, 23, 910, 1977.
31. Jansen, A. P. Van Kampen, E. J., Meijers, C. A. M., Van Munster, P. J. J., and Boerma, G. J. M., Quality control and the quality of commercial test sera, *Clin. Chim. Acta*, 84, 255, 1978.
32. Hearne, C. R. and Fraser, C. G., Enzymatic vs. Jaffe (continuous-flow) assay of ceratinine in serum, *Clin. Chem.*, 25, 1665, 1979.
33. Sheehan, M., Salmon, J., and Haythorn, P., Quality control measurements of total iron binding capacity, *Clin. Chem.*, 25, 1335, 1979.
34. Handschuch, G. H. and Donovan, K. L., Creatine kinase measurement in the 1977 CAP Enzymology Survey with anomalies explained, *Clin. Chem.*, 25, 2003, 1979.
35. Grannis, G. F. and Lott, J. A., An interlaboratory comparison of analysis of clinical specimens, *Am. J. Clin. Pathol.*, 70, 567, 1978.
36. Stecher, P. G., Ed., *The Merck Index*, 8th ed., Merck & Company, Rahway, N. J., 1968, 434.
37. Weast, R. C., Ed., *CRC Handbook of Chemistry and Physics*, 58th ed., CRC Press, Boca Raton, Fla., 1977, D-228.
38. Pope, W. T., Caragher, T. E., and Grannis, G. F., An evaluation of ethylene glycol based liquid specimens for use in quality control, *Clin. Chem.*, 25, 413, 1979.
38. Pope, W. T., Caragher, T. E., and Grannis, G. F., An evaluation of ethylene glycol based liquid specimens for use in quality control, *Clin. Chem.*, 25, 413, 1979.
39. Ross, J. W., Beckman "Pace" Quality Control Program, Group Summary Report, College of American Pathologists Quality Assurance Service, Travere City, Michigan, January 1980.
40. Frajola, W. J. and Maurukas, J., A stable liquid human reference serum, *Clin Chem.*, 18, 694, 1972.
41. Tanishima, K., Minamikawa, Y., Yokogawa, N., and Takeshita, M., Protective effect of glycerol against the increase in alkaline phosphatase activity of lyophilized quality control serum, *Clin. Chem.*, 23, 1873, 1977.
42. Bradbury, S. E. and Jakoby, W. B., Glycerol as an enzyme-stabilizing agent. Effects on aldehyde dehydrogenase, *Proc. Natl. Acad. Sci. U.S.A.*, 69, 2373, 1972.
43. Ruwart, M. J. and Suelter, C. H., Activation of yeast pyruvate kinase by natural and artificial cryoprotectants, *J. Biol. Chem.*, 249, 5990, 1971.
44. Elevitch, F. R., Analytical goals in clinical chemistry: their relationship to medical care, proceedings of the Subcommittee on Analytical Goals in Clinical Chemistry, World Association of Societies of Pathology, *Am. J. Clin. Pathol.*, 71, 624, 1979.
45. Batsakis, J. G., "Omega"-New survey material, *Summing Up '79*, 9(2), 1979.
46. Lawson, N. S. and Ross, J. W., Long-term precison for selected clinical chemistry analytes as determined by data from regional quality control programs, *1976 Aspen Conference*, Elevitch, F. R., Ed., College of American Pathologists, Chicago, 1977.
47. Ross, J. W. and Fraser, M. D., Analytical clinical laboratory precision: state of the art for twenty-nine analytes, *Am. J, Clin. Pathol.*, 72, 265, 1979.
48. Hearn, T., Boone, J., Caudill, S., and Lewis, S., Comparison of results for liquid and lyophilized samples used in clinical chemistry proficiency testing, *Clin. Chem.*, 25, 1120, 1979.

49. Buttner, J., Borth, J., Boutwell, J. H., Broughton, P. M. G., and Bowyer, R. C., Provisional recommendation of quality control in clinical chemistry. VI. Quality Requirements from the point of view of health care, International Federation of Clinical Chemistry, *Clin. Chem.*, 23, 1066, 1977.
50. National Committee for Clinical Laboratory Standards, Methodological Principles for Establishing Principal Assigned Values to Calibrators, Tentative Standard TSL-5, Villanova, Pa., 1977.
51. Boutwell, J. H., Ed., A National Understanding for the Development of Reference Materials and Methods for Clinical Chemistry, Proceedings of a Conference, American Association for Clinical Chemistry, Washington, D.C., 1978.

ANALYTE STABILITY IN CONTROL AND REFERENCE MATERIALS

Noel S. Lawson and Guy T. Haven

INTRODUCTION

Commercially manufactured serum-based products are used for the purpose of calibration and quality control testing, integral to quantitation of chemical analytes in biologic fluids. These products are predominantly "plug" lyophilized,[1,2] and also comprise frozen,[3,4] lyophilized dry filled, and liquid[5] materials. Although their production may be undertaken on a small scale,[4,6] the vast majority is commercially produced in large quantities.

Quality control materials are predominantly produced from human serum, and are intended to mimic, insofar as possible, features of patient biologic fluids being analyzed. The use of quality control pools analyzed along with patient specimens allows the analyst to obtain objective data to assess the degree with which a given analytic run is representative of previous runs. This decision is based on statistics derived from mathematical reduction of results obtained by analysis of control materials on previous runs.

The commonest control material in use in the clinical chemistry laboratory is lyophilized serum. Given lots of quality control serum are generally in use for approximately 1 year,[1] and in some instances, as long as 2 years. It is desirable that analytes in quality control serum be absolutely stable, insignificantly unstable, or predictably unstable. This information assists the analyst in differentiating between shifts and/or trends in quality control data attributable to the analytic method and those caused by changes in analyte concentration in the control product itself. A wide variety of interpretations of the term *stability* have been applied over the years in numerous publications dealing with this subject. There is no uniform definition quantitating adequate stability widely applied in literature dealing with quality control materials in clinical chemistry.

In this review, we shall:

1. Summarize various strategies that have been employed to study analyte stability and quantitate instability in clinical chemistry quality control materials
2. Review published data regarding stability of numerous analytes in quality control pools, both commercially manufactured in large lots and produced noncommercially in small quantities
3. Comment on stability-related testing procedures used by manufacturers of quality control materials

The terms *control serum, control, quality control serum,* and *control materials* have been widely employed to describe products, usually lyophilized, but also including frozen and liquid materials, which are used both for quantitating analytic variability and detecting aberrations in analytic procedures. The terms *reference serum, calibration serum, calibration reference material,* and *calibrator* have been employed to describe the above materials when analyte concentrations have been established with sufficient degree of accuracy that they are used as points of reference in establishing standard curves for analytic assays.[7] Literature dealing with the stability of both types of products essentially describes an intrinsic property of the analyte-matrix combination, irrespective of this bench application. Due to mixed terminology and because these materials, whether used for quality control or calibration purposes, do not differ greatly in their biochemical properties except for analyte concentrations, we have not attempted to distinguish between reported stability characteristics of such materials according to intended use.

The use of given lots of serum over long periods of time, e.g., 12 to 24 months, necessitates maximum analyte stability in order that changes in concentration be correctly attributed to the analytic process. Purposeful manipulation of the environment of analytes in quality control materials, (e.g., by freezing, drying, and/or adding stabilizers) has not eliminated instability. Unfortunately, absolute stability cannot be guaranteed for many commonly measured substances in biologic matrix, due to such factors as light, heat, bacterial contamination, residual moisture, analyte-analyte interaction, and interaction of specimens with containers, Thus, Klugerman and Boutwell commented in 1961, "Regardless of the steps taken to preserve the components of a mixture, no such sample can be absolutely stable. This applies to control serum, whether lyophilized or in solution. Even under the most ideal conditions, deterioration of some of the constituents will eventually occur."[8]

DEFINITIONS

For the Purpose of the present discussion, the following definitions will be employed:

Stability — The state or quality of an analyte characterized by not changing detectably in measured quantity.

Instability — The state or quality of an analyte characterized by the demonstrated lack of stability. (Stability and instability refer to the measured quantity in a control product and must be interpreted in light of the analytic system employed. In absolute terms, measured concentration may change due to factors such as contamination from or absorbtion onto vials or stoppers, or due to evaporation, in addition to specific quantitative deterioration or incrementation of an analyte).

Control material — Materials, simulating analyte samples, used in an analytic system or program to afford a means both for estimating precision and for detecting systematic analytic deviations.[7]

Calibration reference materials — Control materials whose constituents levels are established with sufficient accuracy and precision that they may be used in an analytic system to establish the point(s) of reference in an assay so that patient samples may be determined by interpolated comparison.[7]

Pre-reconstitution stability — Stability of a lyophilized quality control product between the time of manufacture and reconstitution with water or specified diluent.

Post-reconstitution stability — Stability of a lyophilized quality control product after reconstitution with water or specified diluent, until the conclusion of the defined period of study.

REVIEW OF PUBLISHED REPORTS OF ANALYTE STABILITY

Tables 1 to 4 summarize stability information for 36 clinical chemistry analytes from 28 separate reports, published between 1958 and 1979.[5,6,9-34] A wide variety of data gathering techniques and definitions of instability have been employed by workers evaluating analyte stability in control materials.

Three broad approaches to the question of stability are presented. Control materials have been studied in one or more individual laboratories, primarily for the purpose of stability assessment, or secondarily, in the course of a related study. Regional Quality Control Program data from large numbers of laboratories may be probed to detect trends in monthly group mean values indicative of analyte instability. Data from inter-

Table 1
INORGANIC ANALYTE STABILITY IN CONTROL MATERIAL

Material	Calcium	Chloride	Iron	Lithium	Magnesium	Phosphorus (inorganic)	Potassium	Sodium	Ref.
Lyophilized, pre-reconstitution; 2—5°C; 2 years	Stable	Stable							9
Lyophilized pre-reconstitution, 5—8°C, 6 months	Stable	Stable				Stable	Stable	Stable	10
Pre-reconstitution, room temp., 6 months	Stable	Stable				Stable	Stable	Stable	
Post-reconstitution, 5—8°C, 48 hr	Stable	Stable				Stable	Stable	Stable	
Post-reconstitution, room temp., 48 hr	Stable	Stable				Stable	Stable	Stable	
Lyophilized, pre-reconstitution, 6—18 months, Regional Quality Control Program Data	Stable	Increased av 0.055%/month in 7 of 23 pools				Decreased av 0.18%/month in 13 of 24 pools	Increased av 0.055%/month in 5 of 21 pools	Increased av 0.034%/month in 14 of 27 pools	12
Lyophilized, pre-reconstitution, 13—33 weeks survey data	Stable	Stable	Stable	Stable	Stable	Av 0.0318 mg/dℓ decrease	Stable	Stable	15
Lyophilized delipidized serum, post-reconstitution, 5°C, 12 days	Stable					Stable			6
Lyophilized delipidized serum, enriched with bovine lipoprotein, post-reconstitution, 5°C, 10 days	Stable	Stable				Stable	Stable	Stable	22
Lyophilized, post-reconstitution −15°C, 22 days	Stable	Stable				Stable	Stable	Slight decrease	26
10°C, 22 days	Stable 18 days	Stable				Stable 7 days then decreased	Stable	Slight decrease	
Liquid, with ethylene glycol, 30% (V/V)									5
−20°C, 568 days	Stable	Stable				Stable	Stable	Stable	
4°C, 568 days	Stable	Stable				Increased	Stable	Stable	

Table 1 (continued)
INORGANIC ANALYTE STABILITY IN CONTROL MATERIAL

Material	Calcium	Chloride	Iron	Lithium	Magnesium	Phosphorus (inorganic)	Potassium	Sodium	Ref.
25°C, 568 days	Decreased	Stable				Stable 8—16 days, then increased	Stable	Stable	
Frozen, pooled serum −10°C, 42 days	Stable 33 days	Stable				Stable 33 days	Stable	Stable	33
Patient serum, −20°C, 15—17 weeks	Increased, av 1.9% in 9 of 11 patients				Increased, av 6% in 11 of 11 patients	Increased, av 1.4% in 10 of 11 patients			34
Pooled serum, −20°C, 15—17 weeks					Increasing trend				

Table 2
ORGANIC ANALYTE STABILITY IN CONTROL MATERIALS

Material	Albumin	Bilirubin (total)	Bilirubin (direct)	Cholesterol	CO₂	Creatinine	Glucose	Lipids (total)	Protein (total)	Triglyceride	Urea	Uric Acid	Ref.
Lyophilized, pre-reconstitution, 2-5°C				Stable 2 years			Stable 4 yr		Several lots stable 2 yr; 1 lot 9% increase, 2 yr; 1 lot 7% decrease, 2 yr		Stable 4 yr		9
Lyophilized pre-reconstitution, 5—8°C, 6 months						Stable	Stable		Stable		Stable		10
Pre-reconstitution, room temp, 6 months						Stable	70% Decrease		Stable		Stable		

							Ref.
Post-reconstitution, 5—8°C, 48 hr			Stable	30% Decrease		Stable	
Post-reconstitution, room temp., 48 hr			Stable	67% Decrease		Stable	
Lyophilized, pre-reconstitution, 8—17 months Regional Quality Control Program Data						Stable	12
Lyophilized, pre-reconstitution, 6—24 months Regional Quality Control Program Data				Tendency for stable or increasing values by mild methods; stable or decreasing values by rigorous methods			13
Lyophilized, pre-reconstitution, 6—21 months Regional Quality Control Program Data	Method and manufacturer related changes in 13 of 32 pools	Changes in 15 of 45 pools; calibration value changes suspected	Changes in 9 of 42 pools; calibration value changes suspected	Stable		Changes in 10 of 47 pools; decrease dominant; method and manufacturer related	Changes in 21 of 44 pools; calibration value changes suspected	14
Lyophilized, pre-reconstitution, 13—33 weeks Survey Data		Stable	Stable	Stable	Av 1.11 mg/dl decrease	Stable	Av 0.0288 mg/dl decrease	15
Lyophilized, pre-reconstitution, 6 weeks Survey data			Stable	Stable		Stable	17
Lyophilized delipidized serum; post-reconstitution 5°C, 12 days	Stable	Stable	Stable	Stable		Stable	6
Lyophilized delipidized serum, enriched with bovine lipoproteins post-reconstitution 5°C, 10 days	Stable	Stable	Stable	Stable		Stable	22
Lyophilized delipidized control, lipoprotein-containing diluent Pre-reconstitution 37°C, 2 weeks	Stable	Stable	Stable		Stable		23

Table 2 (continued)
ORGANIC ANALYTE STABILITY IN CONTROL MATERIALS

Material	Albumin	Bilirubin (total)	Bilirubin (direct)	Cholesterol	CO₂	Creatinine	Glucose	Lipids (total)	Protein (total)	Triglyceride	Urea	Uric Acid	Ref.
Post-reconstitution, 5°C, 6 months				Stable				Stable		Stable			
Lyophilized, 37°, 2 weeks during manufacture							30% Decrease						25
Lyophilized, post-reconstitution −15°C, 22 days	Stable	Stable			Slight decrease day 15—22	Stable	Stable		Stable		Stable	Stable	26
10°C, 22 days	Stable 5 days; increase days 8—22	Stable 7 days; increase days 8—22			Stable 6 days; decrease days 7—13; increase days 14—22	Stable 9 days; interval of increased concentration, days 10—15	Stable 5 days; 85% decreased, days 6—12		Stable 7 days; increased days 8—22		Stable 10 days; decreased days 11—22	Stable 10 days; decreased days 11—22	
													5
Liquid, with ethylene glycol 30% (V/V) −20°C, 568 days	Stable	Stable		Stable	Stable	Stable	Stable		Stable		Stable	Stable	
4°C, 568 days	Stable	Decrease		Stable	Decrease	Stable	Stable		Stable		Stable	Decrease	
25°C, 568 days	Stable 8—16 days, then decreased	Stable 8—16 days, then decreased		Stable	Decrease	Stable, 93 days, then decreased	Stable 93 days, then decreased		Stable		Stable	Decrease	
Frozen, pooled serum, −10°C, 42 days	Stable	Stable	Stable	Stable	Stable	Stable	10% Decrease		Stable		Stable	Stable	33
Patient serum, pooled serum, −20°C, 15—17 weeks	Stable			Stable					Stable		Stable	Stable	34

Table 3
ENZYME STABILITY IN CONTROL MATERIALS

Material	Alk. Phos.	CPK	GGTP	GDH	HBDH	LDH	SGOT (AST)	SGPT (ALT)	Ref.
Lyophilized animal and human enzymes in BSA, pre-reconstitution 33°C, 3 weeks		Stable	Stable	Stable	Stable	Stable	Stable	Stable	11
Lyophilized animal and human enzymes in human serum, pre-reconstitution 33°C, 3 weeks		Stable	Stable	10—20% decreased activity	Stable	Stable	10—20% Decreased activity	10—20% Decreased activity	
Lyophilized animal and human enzymes in BSA, post-reconstitution, 4°C, 48 hr		Stable	Stable	Stable	Stable	Stable	Stable	Stable	
Lyophilized animal and human enzymes in human serum, post-reconstitution, 4°C, 48 hr		25% Decreased activity	Stable	25% Decreased activity	Stable	Stable	Stable	Stable	
Lyophilized animal and human enzymes in BSA, post-reconstitution, 25°C, 6 hr		Stable	Stable	Stable	Stable	Stable	Stable	Stable	
Lyophilized animal and human enzymes in human serum, post-reconstitution, 25°C, 6 hr		Up to 30% decreased activity	Stable	Up to 30% decreased activity	Stable	Stable	Stable	Stable	

Table 3 (continued)
ENZYME STABILITY IN CONTROL MATERIALS

Material	Alk. Phos.	CPK	GGTP	GDH	HBDH	LDH	SGOT (AST)	SGPT (ALT)	Ref.
Lyophilized animal and human enzymes in BSA, post-reconstitution, −20°C, 4 weeks		Stable	Stable	Stable	Stable	Stable	Stable	Stable	
Lyophilized animal and human enzymes in human serum, post-reconstitution, −20°C, 4 weeks		15% Decreased activity	Stable	Stable	Stable	Stable	15% decreased activity	Stable	
Lyophilized, pre-reconstitution, 6—17 months Regional Quality Control Program Data	8 of 24 Pools, increasing activity, average 3.5%/yr	4 of 14 Pools, decreasing activity, average 17.6%/yr				No dominant directional trend in 18 pools	No dominant directional trend in 24 pools		12
Lyophilized, pre-reconstitution, 13—33 weeks Survey Data							Stable		15
Lyophilized, pre-reconstitution, approx. 12 months Regional Quality Control Program Data		3 of 4 Pools, decreased activity, av 6.5%/year							16
Lyophilized, pre-reconstitution, 6 weeks Survey Data	Stable	Stable				Stable	Stable	7.8% Decreased activity	17

Condition					Ref.
Lyophilized, pre-reconstitution, 6 months Survey Data	13—15% Increased activity	4.7% Decreased activity	Stable	Stable	8.7% Decreased activity	18
Lyophilized, pre-reconstitution, 8 months Survey Data	10% Increased activity	Stable	Stable	Stable	5% Decreased activity	19
Lyophilized, pre-reconstitution, 37°C, 1 month				4% Decreased activity	25% Decreased activity	21
Lyophilized, pre-reconstitution, 4°C, 4 ½ months				4 of 5 Pools stable	
Lyophilized, pre-reconstitution, mail shipment several days				Human and animal enzyme in BSA stable; enzyme in human serum, slight increase	
				Reaction without pyridoxal-5-PO$_4$; stable several hrs, then gradual increase	
Lyophilized, post-reconstitution, 7 days					
Lyophilized delipidized serum, post-reconstitution, 5°C, 12 days		Stable	Stable 24 hr then 25% decrease at 12 days	Stable	6
Lyophilized delipidized serum, enriched with bovine lipoprotein; post-reconstitution, 5°C, 10 days		Stable	50% Decreased activity	25% Decreased activity	22

Table 3 (continued)
ENZYME STABILITY IN CONTROL MATERIALS

Material	Alk. Phos.	CPK	GGTP	GDH	HBDH	LDH	SGOT (AST)	SGPT (ALT)	Ref.
Lyophilized, pre-reconstitution, 5°C several months	Stable								24
Lyophilized, pre-reconstitution, -15°C, 22 days	Increased activity days 1—8						Stable		26
10°C, 22 days	Increased activity						Decreased activity, approx. 20%		
Lyophilized, post-reconstitution	Increased activity 3—21% at 6 hr; 3—27% at 24 hr								27
Frozen, post-thaw	Increased activity over 4 days								
Lyophilized, post-reconstitution, 3 days	Increased activity, greater at higher temperature								28
Lyophilized, post-reconstitution 48 hr	Increased activity, greater at higher temperature								29
Frozen, Post-thaw 48 hr	Increased activity, greater at higher temperature								
Lyophilized, reconstituted room	Stable	50,60% Decreased			Stable	Stable	Stable (1 pool); 17%	20,25% decreased	30

Condition						Ref.
temp. 24 hr	activity (2 pools)			decreased activity (1 pool)	activity (2 pools)	
−20°C, 5 days	10% Increased activity in 4 days	Stable	Stable	Stable	Stable	
	Increased activity in 4 days	Stable				
4°C, 5 days	4% Increased activity	Stable	4,15% Decreased activity (2 pools)	Stable (1 pool) 25% decreased activity (1 pool)	Stable (1 pool) 15% decreased activity (1 pool)	
	Stable	Stable (1 pool) 15% decreased activity (1 pool)			15% decreased activity (1 pool)	
Lyophilized, enzymes added to bovine plasma, post-reconstitution several hours	Stable	Stable	Stable	Stable	Stable	31
Liquid, with ethylene glycol 30% (V/V)						5
−20°C, 568 days	Stable	Stable	Stable	Stable		
4°C, 568 days	Decreased activity	Decreased activity	Stable	Decreased activity		
25°C	Decreased activity, stable 8-16 days	Decreased activity, stable 5 days	Decreased activity, stable 5 days	Decreased activity		
Liquid, with ethylene glycol 30% (V/V); −10°C, 5 mo	Stable	Stable	Stable	Stable	Stable	32
Frozen, pooled serum; −10°C, 42 days	Approx. 15% decreased activity		Stable	Stable	Stable	33
Patient serum, −20°C, 15—17 weeks	Decreased activity, 11.5%		Slightly increased activity	Decreased activity 11.3%		34
Pooled serum, −20°C, 15—17 weeks	Decreased activity		Increased activity			

Table 4
STABILITY OF ANALYTES COMMONLY MEASURED BY LIGAND ASSAY

Analyte	Comment	Ref.
Thyroxine	One of three separately mailed pairs of lyophilized serum with decreased (0.25 mcg/dl) values in second sample, in CAP Radioligand Survey (3—6 month interval)	20
Thyroxine	No demonstrable difference in values between paired vials of lyophilized serum, in CAP Chemistry Survey mailings	15
T3 Uptake	One of three separately mailed pairs of lyophilized serum with decreased (0.57%) values in second sample, in CAP Radioligand Survey, (same pair which was characterized by decreased thyroxine values)	20
Cortisol	No instability in paired Radioligand Survey specimens	20
Digoxin	No instability in paired Radioligand Survey specimens	20
Folic acid	No instability in paired Radioligand Survey specimens	20
Vitamin B12	No instability in paired Radioligand Survey specimens	20
Insulin	No instability in paired Radioligand Survey specimens	20
UIBC	No instability in paired Radioligand Survey specimens	20

laboratory surveys utilizing coded duplicate pairs of vials may be analyzed to detect instability of control products during the time interval between separate mailings.

A wide variety of approaches have been used to assess stability or instability from the generated data. Various statistical approaches to the question of analyte stability are summarized in Table 5. We have employed the technique of trend analysis in monthly group data from Regional Quality Control Programs to evaluate analytic stability in lyophilized serum control products, used in conjunction with these programs.[12-14] Dominant directional instability was noted with 7 of 17 analytes (Table 6). Additionally, in the case of glucose, increasing analyte concentration was noted, when mild methods were studied (e.g., manual and automated glucose oxidase, automated neocuproine) and decreasing concentrations were detected, studying rigorous methods (automated ferricyanide, manual ortho-toluidine).

We considered changes in measured concentrations of inorganic analytes to be most likely due to interacton of given ions in the lyophilized "plug" with the containers or stoppers used for storage. Increasing glucose detected by mild methods is likely due to the ability of milder methods, which measure free glucose, to detect slow release of glucose from protein during shelf storage. Decrease in glucose, detected by rigorous methods measuring total glucose is considered to reflect glycolysis in the affected pools.

Stability data for frozen serum are taken both from studies where pooled serum was used as a long-term control[34] and from a study evaluating patient specimens stored over a period of seven weeks.[33] The data from "in-house" pools provide useful information in areas where there is paucity of published data derived from commercial frozen pools.

The class of analytes for which instability is most frequently reported is enzymes. Commonly reported changes include increased alkaline phosphatase activity in lyophlized material following reconstitution, and various decreasing activities in frozen and liquid controls. CPK is frequently reported to decrease in lyophilized material, both pre- and post-reconstitution. LDH was predominately stable. Changes, when reported, tended to be in a negative direction. When changes in SGOT (AST) were noted, they were seen in lyophilized, liquid, and frozen materials, and tended to be in a negative direction. Decreasing SGPT (ALT) was reported in lyophilized lots, both pre- and post-reconstitution.

Table 5

PUBLISHED CRITERIA EMPLOYED TO
DESCRIBE INSTABILITY OF CONTROL
SERUM

Criteria	Ref.
Observation of graphed data for trends	21
Observation of tabulated data for trends	10,30
Results beyond 1 SD from mean	33
Results beyond 2 SD from mean	26
Results beyond 10% of mean	11
Trends in average slopes of ratioed enzyme values in an interlaboratory survey	18
Results beyond ± 1% of average slopes of ratioed enzyme values in an interlaboratory survey	19
Trends in values by regression analysis	34
Trends in grouped monthly mean values by linear regression analysis with slope ≠ 0, 0.95 probability, Regional Quality Control Program Data	12—14
Analysis of value ranks, Friedman nonparametric method for randomized blocks	16,34
Paired survey mailings, *t*-test, difference in means	15,20
Paired specimens, mailed and refrigerated (4°), t-test, differences in means $p < 0.05$	21
Comparison of precision of test material against that of other control material	5

Organic analytes were generally stable. Changes in urea and uric acid, when reported, were in a negative direction. Glucose was the organic analyte most frequently reported to be unstable. Decreasing glucose was reported in lyophilized material, both pre- and post-reconstitution, as well as in frozen material and in liquid controls maintained at room temperature. Decreasing glucose in control specimens is considered to be due to glycolysis. As noted above, mild methods have detected increasing glucose in some lots of control serum.[13]

Inorganic analytes are usually reported to be stable in control materials. Data from Regional Quality Control Programs has indicated minimally changing concentrations of inorganic analytes in a proportion of pools studied, possibly due to interaction with containers and/or stoppers.

STATISTICAL APPROACHES TO STABILITY

In contemplating the correct statistical approach to be used in examining quality control serum for analyte instability, it is necessary to realize that the user and manufacturer are dealing with an extraordinarily complex group of events. Dozens of analytes in control serum exist in a highly complex matrix which, as yet, is incompletely characterized. These analytes are subject to analyte-analyte interaction, external physical factors from the environment, bacterial growth of variable quality and quantity, and are exposed to vials and stoppers during their pre-reconstitution phase.

In the post-reconstitution phase, one generally assumes that lyophilized material behaves in a fashion similar to fresh human serum. This may be the case for many analytes, but is not for many others; therefore, the fundamental assumptions that underlie

Table 6
ANALYTES WITH DOMINANT DIRECTIONAL INSTABILITY

Analyte	Number of pools	Number increased	Number decreased	Number stable	Dominant direction	Av|Coeff|/SD	Av % change/ month (year)
Potassium	21	5	1	15	Increase	0.029	0.055 (0.66)
Sodium	27	14	1	12	Increase	0.032	0.034 (0.41)
Chloride	23	7	2	14	Increase	0.033	0.055 (0.66)
Phosphorus	24	1	13	10	Decrease	0.072	−0.18 (−2.1)
Alkaline phosphatase	24	8	4	12	Increase	0.077	0.29 (3.5)
Creatine phosphokinase	14	1	4	9	Decrease	0.128	−1.47 (−17.6)
Urea N	47	2	8	37	Decrease	0.043	−0.13 (−1.55)

Note: Directional characteristics of monthly grouped means for pool-analyte combinations in Regional Quality Control Programs; coeff. is average linear regression coefficient of pools with instability in dominant direction; SD is in meq/l (potassium, sodium, chloride), mg/dl (phosphorus, urea N), and IU/l (alkaline phosphatase and creatine phosphokinase); coefficient of linear regression is given in the same units/month.

From Lawson, N. S., Haven, G. T., and Moore, T. D., *Am. J. Clin. Pathol.*, 68, 117, 1977; and Haven, G. T., Lawson, N. S., and Moore, T. D., *Am. J. Clin. Pathol.*, 72, 274, 1979. With permission.

classic stability studies may not apply in given analyte-pool combinations. An excellent example of stability characteristics of a defined analyte under defined conditions of physical interaction is given by Takemura Hocman, and Sterling[35] who examined the thermal stability of thyroxine binding globulin (TBG) and thyroxine binding pre-albumin (TBPA) when exposed to elevated temperatures over varying periods of time. The TBPA is relatively stable in its thyroid-binding ability. TBG, however, exhibits striking loss in thyroxine binding capacity, as a result of heating.

The data quantitating this loss was reduced to a series of logarithmic decay curves for the given exposure temperatures. The authors surmised that the denaturation was taking place in the TBG, rendering it less capable of binding thyroxine. The extraordinary complexity of the analyte stability question is underscored, here, in that for this very simple experiment involving a well-defined set of exposure-time-analyte conditions, one must realize that had the thyroxine-binding globulin been measured as a molecule, rather than as its ability to bind thyroxine, and had the antigenic site or sites chosen to prepare suitable antiserum been altered to a different extent by the heat, one very likely would have found the conditions of heat either producing no significant change in analyte concentration or change of a different degree. Enzyme measurement as molecular concentration, rather than activity, yields an entirely different set of stability data than that obtained by activity assay. Indeed, work with acid phosphatase measured chemically vs. measured by RIA indicates great differences in stability characteristics, with the molecular based quantitative assays detecting less instability.[36]

We have seen a wide variety of criteria applied for the determination of the threshold of instability. Commercially manufactured lyophilized control products are subjected to repeated testing by manufacturers, during the life of the pool, to detect analyte instability. Uniform guidelines for interpretation of this data are not widely applied throughout the field. Most analytes are sufficiently stable under strict adherence to manufacturers' storage specifications to allow their use as acceptable quality control products and calibrators.

The repeated testing of packaged control products over time supplements stability stress testing of manufactured products, prior to release. Pre-release testing is meant to reproduce environmental conditions (e.g., temperature, humidity, light) to which the products would logically be expected to be exposed, during time spent at the manufacturer, distributor (when applicable), and laboratory. Among other conditions, testing by manufacturers involves exposure of samples to heat for varying times in order to establish a decay-temperature-time relationship which will generate sufficient data to establish Arrhenius relationships for each analyte.[37] Using the Arrhenius approach, it is not necessary to expose samples to a large number of different temperatures, but rather exposure to limited numbers of temperatures will generate sufficient data to solve the equations.

Exposing chemistry control products to environmental stress forms a parallel to drug stability testing; however, the situations are not totally comparable. Stress testing of a drug involves a pure or relatively pure product in a known matrix with a single or limited number of breakdown products, the changes of which can be traceable to the specific environmental stress. Control serum is an infinitely more complex product whose numerous analytes are exposed to multiple environmental factors over a relatively long time continuum. They exist with minimal, yet definite, residual moisture, and may undergo analyte-analyte interaction, analyte matrix interaction, analyte container interaction, and analyte environment interaction. A still further complicating factor is bacterial growth. Although bacterial contamination, which had been a major problem in previous years,[38] has been less of a problem recently, the lyophilized quality control products available to laboratories are not sterile, containing minimal numbers of microorganisms. The number and population of microorganisms, their utilization of constituents within a control material for metabolism, and the variety of their metabolic products is variable from lot to lot. Additionally, their numbers vary in vials within given lots based on conditions of storage prior to and following reconstitution.

With respect to stability of drugs, the following definitions have been employed:[39]

Expiration date — Point at which 95% confidence line of individual assays cuts the 90% concentration line.

Shelf life — Point at which stability line cuts 100% label claim.

Outdate — Point at which line describing 95% confidence limit of individual assays about the least square fit line cuts the 100% concentration label claim.

In drug manufacture, deterioration is quantitated, predicted, and included in the strategy for manufacture. Thus, based on anticipated decay, the manufacturer may include a small excess in each dose. Thus, it is considered acceptable to anticipate a decline of this order of magnitude during the time between manufacture and the time given as shelf life. In the case of calibrators or control sera used in the analytic laboratory, such a procedure would result in allowable deterioration of the calibrators and control products which would adversely effect the quality of clinical chemistry analyses, and is not acceptable.

Calibration reference materials and control materials in clinical chemistry produced in the U.S. are manufactured under the approved standard: ASC2 of the National Committee for Clinical Laboratory Standards (1975).[7] The stability section of the standard is as follows:

> 5.1 Estimate of stability. The manufacturer shall perform appropriate tests to estimate the stability of the product. This estimate should be based upon experimental studies of representative samples of the same batch (or prior batches prepared in the same manner) in the final container, simulating with respect to time, temperature, light exposure, etc., conditions to which the product might reasonably be expected to be exposed after manufacturing, including storage, distribution, and laboratory operations, when handled according to the manufacturer's recommendations. The stability tests shall be performed on both the distributed form of the product (liquid or lyophilized) and on reconstitued samples of lyophilized products,
>
> 5.2 Labeling. The manufacturer shall make results of the stability testing available in the package insert. The stability shall be expressed by the change in concentration or activity units per unit time (e.g,, mg/dl/week or U/liter/day). When appropriate, changes may be stated as a percentage of the assigned value per unit time. Alternatively, as in the case of an extremely stable constituent or where only limited stability dates are available, an upper limit for the instability may be given (e.g., less than 1 mmol/liter/month). Estimates or upper limits of instability may be given for individual constituents or for groups of constituents.

Although no specific guidelines for acceptable degrees of instability of control products have been uniformly adopted, elements have been proffered. Thiers et al. have considered the problem of stability of individual specimens with respect to their suitability for clinical laboratory analysis.[40] The authors propose that a constituent in body fluid be considered stable under specified conditions of storage, when its mean concentration is shown to change by less than an amount equal to 1 standard deviation (SD), when a 5% risk of error is allowed for this decision. Linking the stability decision to the precision with which an analyte is measured is an important concept. A 1% deterioration in concentration might be acceptable for an analyte in a control product with inherent measurement coefficient of variation (CV) of 5 to 10%, whereas a 1% deterioration would not be acceptable for an analyte customarily measured with 1% CV.

The procedure detailed by these authors for determining acceptable conditions of storage for incoming patient specimens is not totally applicable to the question of acceptable degrees of stability of quality control serum. When considering a patient specimen, one is defining the point(s) in time at which a single analysis will yield a result no greater than 1 SD from the correct value. In the quality control serum question, one could arrive at such a decision for a single measurement or a limited number of measurements and find that the serum had, in fact, developed unpredictable increasing or decreasing concentration of one or more given analytes at a later date. Secondly, whereas 1 SD may be an acceptable risk for bias of a result from a single patient specimen, we feel that a bias of this magnitude is not acceptable for control materials or calibration reference materials, which have a multiplier effect on patient results. Thus, a 1 SD change in analyte concentration due to instability within a calibrator would be translated to a comparable bias in all specimens on a run. In a similar fashion, a deteriorating control would not be optimally effective in assisting in the correct diagnosis of out of control results. Of particular concern would be a situation wherein a laboratory was using given lots of material, both for calibration and control, exhibiting the same stability characteristics. Under this set of circumstances, the calibrator would be producing a biased assay, the control would appear to be within control, and the patients' incorrect results would be undetected.

We have used 0.5 SD as a suggested instability threshold, in studying stability of lyophilized serum used with Regional Quality Control Programs.[12] Most pools of lyophilized serum, in these programs, are used for approximately 12 to 15 months in given laboratories. The analytes shown to have dominant directional instability have been divided into those where the 0.5 SD change is calculated to occur in affected pools in less than 1 year (CPK, AP, inorganic phosphorus, glucose) and those where this change occurs over a period greater than 1 year (sodium, potassium, chloride). The rate of change in pools with decreasing urea nitrogen concentration approximated 0.5 SD/year.

Fasce, Copeland, and Vanderlinde have discussed application and specifications of enzyme reference materials.[41] They described four uses of enzyme reference materials, i.e., standardization, intra- and inter-method comparison, and precision control. With respect to stability, the authors indicate that for normal use, an enzyme material must be adequately stable. They recommended a 2-month minimum stability for statistical purposes, admitting that it is difficult to fix stability specifications.

It is clear, then, at this point in time, that lacking a clearly agreed upon and applied definition of instability, the user of control materials must rely on a variety of resources to assess adequately analyte stability in currently employed lots of control or calibration reference materials. Strict adherance to manufacturers' recommended conditions of storage is mandatory. One should be familiar with instabilities noted on the package insert. Expiration dates should be respected. The analyst should be alerted to those analytes with a known propensity for instability in control products, particularly glucose and enzymes. When observing drifts or shifts in control results obtained on these analytes, the possibility of altered analyte concentration in controls must be considered as a cause, in addition to changes in analytic conditions.

A user's reaction to an unstable analyte in a control or in reference materials would vary from no response to minimal change, to readjustment of run acceptance limits because of slight or moderate change in a control product, to changing lots of material when a change in an analyte concentration in calibration reference material has occurred or when a marked change in an analyte in control material has taken place.

SUMMARY

Serum, altered to reduce change in analyte concentration during storage, is used nearly universally for daily control of clinical chemistry analytic testing, and is widely used for calibration purposes. Although absolute stability of all analytes is desired, numerous studies have demonstrated instability of various analytes in control materials. The degree of instability detected varies from minimal to major. Commercial control products, when stored and used properly, are necessary and invaluable participants in the analytic sequence. Users should be aware that certain analytes may change concentration in some lots of material during the period of use. Tabulated summaries of published analyte stability data in control products are presented. The detected degree of change may differ as a function of the analytic method employed. Analytes most consistently reported to change concentration are glucose and enzymes. When changes in glucose and enzymes are detected, they tend to be at a rate greater than 0.5 of the analytic standard deviation per year.

ACKNOWLEDGMENTS

The authors thank Joseph Giegel, Ph.D., Dade Division, American Hospital Supply Corporation; Huey Auger, Ph,D., General Diagnostics Division, Warner Lambert Company; and Ms. Sandy Krishnamurthy, Beckman Instruments, Inc. for providing information pertaining to industrial stability testing.

REFERENCES

1. Lawson, N. S. and Haven, G. T., The Role of Regional Quality Control Programs in the practice of laboratory medicine in the United States, *Am. J. Clin. Pathol.*, 66, 268, 1976.
2. Anido, G., Preparation of quality control materials in clinical chemistry and hematology, *Proc. R. Soc. Med.*, 68, 624, 1975.
3. Hohnadel, D. C., Sunderman, F. W., Jr., Terhune, P., Reid, F. H., and Pomper, I. H., Comparisons of the precision of replicate analyses of frozen and lyophilized quality control serums, *Ann. Clin. Lab. Sci.*, 3, 335, 1973.
4. Bowers, G. N., Burnett, R. W., and McComb, R. B., Preparation and use of human serum control materials for monitoring precision in clinical chemistry, *Clin. Chem.*, 21, 1830, 1975.
5. Frajola, W. J. and Maurukas, J., A stable liquid human reference serum, *Health Lab. Sci.*, 13, 25, 1976.
6. Proksch, G. J. and Bonderman, D. P., Preparation of optically clear lyophilized human serum for use in preparing control material, *Clin. Chem.*, 22, 456, 1976.
7. Approved Standard: ASC-2, *Calibration Reference Materials and Control Materials in Clinical Chemistry*, National Committee for Clinical Laboratory Standards, Villanova, Pa., 1975.
8. Klugerman, M. R. and Boutwell, J. H., Commercial control sera in the clinical chemistry laboratory, *Clin. Chem.*, 7, 185, 1961.
9. Logan, J. E. and Allen, R. H., Control serum preparations, *Clin. Chem.*, 14, 437, 1968.
10. Klein, B. and Weissman, M., Study of dialyzed reconstituted dried serum as a clinical chemistry standard, *Clin. Chem.*, 4, 194, 1958.
11. Gruber, W., Hundt, D., Klarwein, M., and Möllering, H., Comparison of control materials containing animal and human enzymes, *J. Clin. Chem. Clin. Biochem.*, 15, 579, 1977.
12. Lawson, N, S., Haven, G. T., and Moore, T. D,, Long-term stability of enzymes, total protein, and inorganic analytes in lyophilized quality control serum, *Am. J. Clin. Pathol.*, 68, 117, 1977.
13. Lawson, N. S., Haven, G. T., and Moore, T. D., Long-term stability of glucose in lyophilized quality control serum, *Am. J. Clin. Pathol.*, 70, 523, 1978.
14. Haven, G. T., Lawson, N. S., and Moore, T. D., Stability of mean values of organic analytes in lyophilized quality control serum, *Am. J. Clin. Pathol.*, 72, 274, 1979.
15. Gilbert, R. K., A comparison of participant mean values of duplicate specimens in the CAP Chemistry Survey Program, *Am. J. Clin. Pathol.*, 66, 184, 1976.
16. DiSilvio, T. V., An examination of statistical strategies for evaluation of stability of lyophilized quality control serum under actual conditions of use, *Clin. Chem.*, 25, 1086, 1979.

17. Grannis, George F., Interlaboratory survey of enzymatic analyses. II. Intermediate studies, *Am. J. Clin. Pathol.*, 68, 142, 1977.

18. Grannis, George F. and Massion, C. G., The 1977 College of American Pathologists Enzymology Survey, *Am. J. Clin. Pathol.*, 70, 487, 1978.

19. Grannis, G. F., Massion, C. G., and Batsakis, J. G., The 1978 College of American Pathologists Survey of Analyses of Five Serum Enzymes by 450 Laboratories, *Am. J. Clin. Pathol.*, 72, 285, 1979.

20. Haven, G. T., Hansell, J. R., and Haven, M. C., Reproducibility of mean values of duplicate specimens in the Basic Ligand Assay Survey, *Am. J. Clin. Pathol.*, 70, 532, 1978.

21. Burtis, C. A., Sampson, E. J., Bayse, D. D., McKneally, S. S., and Whitner, V. S., An interlaboratory study of measurement of aspartate aminotransferase activity with use of purified enzyme materials, *Clin. Chem.*, 24, 916, 1978.

22. Proksch, G. J. and Bonderman, D. P., Use of a cholesterol-rich bovine lipoprotein to enhance cholesterol concentrations in the preparation of serum control materials, *Clin. Chem.*, 22, 1302, 1976.

23. Proksch, G. J. and Bonderman, D. P., Development of a stable lipoprotein diluent for use in reconstituting lyophilized human serum for the preparation of clear, hyperlipidemic quality-control materials, *Clin. Chem.*, 25, 1377, 1979.

24. Proksch, G. J., Bonderman, D. P., and Griep, J. A., Autoanalyzer assay for serum alkaline phosphatase activity, with sodium thymolphthalein monophosphate as substrate, *Clin. Chem.*, 19, 103, 1973.

25. Romero, P., Schneider, A., and Matthews, H., The stability of glucose in lyophilized control sera at elevated temperatures, *Clin. Chem.*, 23, 1140, 1977.

26. Hanok, A. and Kuo, J., The stability of a reconstituted serum for the assay of fifteen chemical constituents, *Clin. Chem.*, 14, 58, 1968.

27. Massion, C. G. and Frankenfeld, J. K., Alkaline phosphatase: lability in fresh and frozen human serum and in lyophilized control material, *Clin. Chem.*, 18, 366, 1972.

28. Smith, A. F. and Fogg, B. A., Possible mechanisms for the increase in alkaline phosphatase activity of lyophilized control material, *Clin. Chem.*, 18, 1518, 1972.

29. Brojer, B. and Moss, D. W., Changes in the alkaline phosphatase activity of serum samples after thawing and after reconstitution from the lyophilized state, *Clin. Chim. Acta*, 35, 511, 1971.

30. Szasz, G., Die Qualitätskontrolle von Enzymaktivitätsbestimmungen im Serum, *Z. Klin. Chem. Klin. Biochem.*, 8, 212, 1970.

31. Schneider, A. L., Edwards, G. C., and Romero, P., Preparation of a stable enzyme calibrator for clinical enzyme analyses, *Clin. Chem.*, 21, 963, 1975.

32. Pope, W. T., Caragher, T. E., and Grannis, G. F., The use of ethylene glycol based quality control materials, *Clin. Chem.*, 24, 1050, 1978.

33. Wilson, S. S., Guillan, R. A., and Hocker, E. V., Studies of the stability of 18 chemical constituents of human serum, *Clin. Chem.*, 18, 1498, 1972.

34. Williams, G. Z., Harris, E. K., and Widdowson, G. M., Comparison of estimates of long-term analytical variation derived from subject samples and control serum, *Clin. Chem.*, 23, 100, 1977.

35. Takemura, Y., Hocman, G., and Sterling, K., Thermal stability of serum thyroxine-binding proteins, *J. Clin. Endocrinol. Metab.*, 32, 222, 1971.

36. Foti, A. G., Herschman, H., and Cooper, J. F., Comparison of Human prostatic acid phosphatase by measurement of enzymatic activity and by radioimmunoassay, *Clin. Chem.*, 23, 95, 1977.

37. Garrett, E. R., Prediction of stability in pharmaceutical preparations, *J. Am. Pharm. Assoc. Sci. Ed.*, 45, 171, 1956.

38. Beeler, M. F., Samuels, M. S., Carrera, A. E., Hood, M. W., and Dickinson, C. S., Bacteriologic survey of lyophilized chemical quality control materials, *Am. J. Clin. Pathol.*, 56, 676, 1971.

39. Carstensen, J. T. and Nelson, E., Terminology regarding labeled and contained amounts in dosage forms, *J. Pharm. Sci.*, 65, 311, 1976.

40. Thiers, R. E., Wu, G. T., Reed, A. H., and Oliver, L. K., Sample stability: a suggested definition and method of determination, *Clin. Chem.*, 22, 176, 1976.

41. Fasce, C. F., Rej, R., Copeland, W. H., and Vanderlinde, R. E., A discussion of enzyme reference materials: applications and specifications, *Clin. Chem.*, 19, 5, 1973.

EVALUATION OF PRECISION

John W. Ross

INTRODUCTION

Clinical chemistry methods are usually controlled by inclusion of preanalyzed sera in each analytic run. Reduction of data generated by repeated analysis of these sera gives standards by which the validity of individual runs is determined. While an important part of internal process control, statistical standards, even when based on technologic excellence, do not identify areas of medical need where improvement is necessary or areas where present levels of performance are adequate. Campbell and Owen[1] recognized the importance of relating analytic precision to the medical context by comparing it to the component of physiologic variation relevant to the clinical application. The proceedings of the Subcommittee on Analytical Goals in Clinical Chemistry of the World Association of Societies of Pathology state, in part, that performance in a quality control program should measure the laboratory's ability to meet its medical needs.[55] The role of a quality control program, therefore, is to indicate when analytic goals are achieved and to monitor their maintenance.

EFFECT OF CONTROL SPECIMENS UPON PRECISION ESTIMATES

Control materials are the tools with which compliance to precision performance standards is measured. Control procedures are equal to this task only if the control material homogeneity, stabilty, and matrix characteristics are comparable to those of the patient specimen.

The comparability of the precision estimates by plug lyophilized sera to those by duplicates prepared by splitting fresh human sera has been the subject of several studies,[2-4] which are summarized in Table 1. The data from paired sera include a certain amount of preanalytic variation due to storage effects, evaporation, etc. These factors are considered to be in the range of usual preanalytic effects and are from sources comparable to those affecting the lyophilized control materials. For the analytes studied, there is close correspondence between precision estimates using fresh duplicate sera and plug lyophilized materials. Precision estimates for similar analytes with bead lyophilized materials and liquid materials have been compared to plug lyophilized material in a previous section. The effects of inhomogeneity due to vial fill and post-vial fill factors, stability in the time interval studied, and matrix alteration do not appear to be important for the analytes and conclusions to be discussed in the following sections with the reservations expressed in the previous section.

ANALYST BIAS IN PRECISION ESTIMATES

The possibility of distortion of precision estimates by analyst knowledge of the expected value in preanalyzed sera was studied by Weinberg and Barnett,[5] who did not find evidence of such an effect. Subsequent authors commented upon the presence of analyst bias affecting estimates of imprecision in internal quality control.[6-8] More recently, Steele et al. noted the absence of analyst bias in estimates of precision in an interlaboratory proficiency testing program using parallel known and masked survey specimens. Glenn and Hathaway have found blind sample internal quality control to be a useful tool in examining preanalytic variables affecting patient as well as con-

Table 1

COMPARISON OF ANALYTIC PRECISION ESTIMATES MADE WITH LYOPHILIZED CONTROL SERA AND PAIRED HUMAN SERUM DUPLICATES

Analyte	Williams et al., 1977[a]			Bokelund et al., 1974[b]			Van Steirteghem et al., 1978[c]		
	Normal population mean concentration	CV, paired sera from subjects	CV, plug lyophilized control	Normal population mean concentration	CV, paired sera from subjects	CV, plug lyophilized control	Normal population mean concentration	CV, paired sera from subjects	CV, plug lyophilized control
Sodium (mmol/l)				140	1.5	1.2	140	2.5	1.6
Calcium (mg/dl)	9.6	2.2	2.8	9.9	2.2	2.2	9.3	3.5	2.2
Magnesium (mg/dl)	2.1	3.6	4.5[d]						
Potassium (mmol/l)				4.3	2.1	2.2	4.3	3.2	1.9
Chloride (mmol/l)				104	2.5	2.1	105	2.0	1.6
CO$_2$ (mmol/l)							20	9.1	2.9
Glucose (mg/dl)	89	1.8	2.2						
Urea nitrogen (mg/dl)	16	5.8	4.1	14	2.4	2.2	14	5.3	3.9
Creatinine (mg/dl)				.98	3.8	1.5[e]	.97	4.6	4.3
Phosphorus (mg/dl)	3.4	1.1	2.0	3.6	3.1	2.5	3.4	4.1	2.1
Uric acid (mg/dl)				5.4	2.4	3.6	5.7	4.4	2.5
Bilirubin (mg/dl)				.45	4.4	3.1			
Iron (mcg/dl)				94	3.0	1.6			
Cholesterol (mg/dl)	215	3.3	2.7	229	1.9	3.0	198	3.6	2.6
Triglyceride (mg/dl)							114	4.8	2.9
Protein (g/dl)	7.1	2.9	2.2	6.9	1.5	1.1	6.8	2.1	1.5
Albumin (g/dl)	4.1	3.4	3.3	4.1	3.2	3.2	3.9	2.7	1.8
Alkaline phosphatase (IU/l)	40	9.0	8.3[d]	91	2.2	3.4	33	6.3	14
Acid phosphatase (IU/l)				2.4	5.4	2.3[f]			
Aspartate amino transferase (IU/l)	11	5.3	8.1[g]	8.9	5.2	4.1	17	24	8.5*

Alanine amino transferase (IU/ℓ)	50		9.8	9.6	4.3[i]	87	13	13[j]
Lactate dehydrogenase (IU/ℓ)	4.8	8.2	175	2.8	3.1	134	10	4.5
Creatine kinase (IU/ℓ)						87	4.6	4.3

Note: Analyte concentrations in the control sera correspond closely to the average concentration of the respective analytes in patient sera unless otherwise stated below.

[a]　Reference 4. Data are from a variety of automated methods. Some preinstrumental variation is included in the paired sera CVs.

[b]　Reference 2. Data from "true" serum duplicates presented. Analysis by AutoChemist. No preinstrumental variation.

[c]　Reference 3. Only data for SMAC presented. Some preinstrumental variation is included in the paired sera CVs.

[d]　After correction for linear trend.

[e]　Control concentration = 1.5 mg/dℓ.

[f]　Control concentration = 12 IU/ℓ.

[g]　Control concentration = 7 IU/ℓ.

[h]　Control concentration = 26 IU/ℓ.

[i]　Control concentration = 43 IU/ℓ.

[j]　Control concentration = 47 IU/ℓ.

Table 2
COMPARISON OF ANALYTIC PRECISION ESTIMATES BY USE OF KNOWN AND BLIND SAMPLE QUALITY CONTROL SPECIMENS

| | | Glenn and Hathaway, 1979[a] | | | | Weinberg and Barnett, 1962[b] | |
| | | 1975 Known QC specimens | | Post 1976 blind QC specimens | | | |
Analyte	Units	CV	Concentration	CV	Concentration	Known QC specimens CV	Blind QC specimens CV
Sodium	mmol/l	1.4	141	1.2	144	1.6	2.5
Calcium	mg/dl	1.7	9.4	2.1	9.2		
Potassium	mmol/l	1.9	4.2	2.5	3.9	3.7	6.7
Chloride	mmol/l	2.9	105	1.3	106	2.9	2.5
Glucose	mg/dl	3.0	87	4.2	98		
Urea nitrogen	mg/dl	5.0	13	4.4	11	16	11
Creatinine	mg/dl	7.7	0.9	11	0.9		
Phosphorus	mg/dl	2.7	4.2	2.7	3.7		
Uric acid	mg/dl	1.9	6.1	2.0	5.0	7.7	4.2
Bilirubin	mg/dl	6.0	1.3	—	0.4		
Cholesterol	mg/dl	4.5	163	6.9	164	6.0	7.2
Protein	g/dl	1.9	6.5	2.8	6.8	4.6	6.7
Alkaline phosphatase	IU/l	5.9	109	8.7[c]	47		
Aspartate aminotransferase	IU/l	6.5	62	7.9[c]	28	20	14
Lactate dehydrogenase	IU/l	6.9	235	5.9	123		

[a]   Reference 56. SMA 6-60 and SMA 12-60 used for all methods. Yearly cumulative CVs.
[b]   Reference 5. All methods were manual. CVs are cumulative over about 2 months.
[c]   The post 1976 pools have a lower analyte concentration. If the comparison was at a higher concentration noted in the 1975 known QC specimen baseline, the blind QC specimen CVs would probably compare favorably.

trols.[56] Problems with variation due to preanalytic effects such as light exposure, evaporation, and times between submission of sera and analysis were identified and controlled. Blind sample internal quality control precision is compared in Table 2 to known sample internal quality control precision, after measures were taken to limit preanalytic variation. Despite exposure of blind sample quality control specimens to routine preanalytic laboratory conditions before technical performance of the test, the average known quality control specimen CV was 3.9% with a range from 1.4 to 7.7% compared to an average blind specimen CV of 4.5% with a range from 1.2 to 11.0%. The blind samples had increased CVs for seven analytes, decreased CVs for four analytes, and no change for three analytes. The data of Weinberg and Barnett[5] are also presented. The average known sample CV was 7.8% with a range from 1.6 to 20.0%, and the average blind sample CV was 6.9% with a range from 2.5 to 14.0%. Rosenbaum[10] reviewed data from a large Regional Quality Control Program and was unable to find a consistent decrease in average CVs after the first month of each of three consecutive new regional pool lot numbers. (During the first month, the target values of the unassayed product were unknown to the analyst except in general terms.) Sax et al.[11] noted no evidence of conscious or unconscious analyst bias, either with specimens known to be quality control specimens but whose values were unknown, or with entirely blind quality control specimens.

While analyst bias has affected imprecision estimates in some circumstances, it is certainly feasible and desirable to conduct quality control programs and Regional Quality Control Programs without significant average effect of analyst bias. Analyst bias probably does not affect the main conclusions concerning the average state of the art performance.[12] This influence, when present, frustrates the need to relate quality control program results to analytic goals and medical needs. Barnett has stated, "If conscious falsification occurs, the whole quality control program is worthless. Such a phenomenon would indicate poor indoctrination of the analyst in the whole theory of the analytic method...."[13]

COMPONENTS OF ANALYTIC PRECISION

Background

Analytic precision depends upon the sum of the effects of a number of separate identifiable and measurable components of variation. This concept is essential for relating estimates of analytic imprecision to the requirements for good medical care. Cotlove[14] and Campbell and Owen[1] summarized the then scant available data regarding association of analytic precision with time span of replicate analysis. Broughton and Annan[8] delineated the three fundamental components of within laboratory analytic precision: effect of the time span of replicate analysis; effect of analyte concentration; and effect of the specimen matrix. Estimates of analytic imprecision must be known to contain the correct factor from each of these components of imprecision when comparing method to method, laboratory to laboratory, or method or laboratory to a precision performance standard.

A fourth component of precision, the effect of interlaboratory variation, is added when comparisons involve group performance.

All imprecision estimates in internal quality control programs must be made under conditions which allow as many factors as possible influencing patient results to influence the quality control specimen.

Effect of Time upon Analytic Precision

The need to study analytic variation as a function of time was anticipated some years ago by Campbell and Owen,[1] who also foresaw the importance of relating analytic variation to the relative components of physiologic variation. Cotlove et al.[14] noted that within-day imprecision was less than between-day imprecision. Broughton and Annan[8] believed that cumulative imprecision due to random analytic variation increased rapidly to a near maximum by 1 to 2 months. Longer-term imprecision was believed to be due to some extent to variation in within-day precision, but more importantly to the effect of changes in bias.

Imprecision due to time effects may be divided into within-day, short-term between-day, and long-term between-day factors. The World Association of Societies of Pathology Subcommittee on Analytic Goals in Clinical Chemistry[55] has emphasized the need for additional studies of short-term vs. long-term analytic precision.

Within-run factors involve sources of variation which are difficult to attribute and may be considered irreducible.[15] Irreducible error of an instrument may be allocated to various sources in an "error budget" in manufacturing design.[16] These sources include performance limits assigned to volumetric and gravimetric devices used, instrument sensitivity, baseline and sensitivity drift, sample interaction, calibrator inhomogeneity, operator scale reading and manual dexterity, or tolerances built into microprocessor control. Within-run precision estimates the ideal capability of the chemical method, the technology employed, and the skill with which the technology is applied to the method. Gilbert[17,18] has noted the dependence of within-run precision upon method and the degree of laboratory improvement possible by selection of the more precise of the available methods.

Longer-term precision depends upon preventive maintenance effectiveness, instrument durability, method susceptibility to variation in analyst technique, variation in reagent lots, reagent stability, limits within which calibration and instrument standardization can be maintained, and the homogeneity and stability of calibrators. Long-term precision includes "random inaccuracy" as well as random imprecision. The presence of an effect due to "random inaccuracy" has been commented upon by Kurtz,[19] Sage,[20] Gilbert,[17] and Harris,[21] in addition to Broughton and Annan,[8] and may be defined as long-term fluctuation in bias which over an extended period of time will average zero.[17] Random inaccuracy is produced by transient effects in the analytic environment such as changing analysts, calibrators, reagent vials or lots, temperature, etc., on a day-to-day or even within-day basis. Methods by which short-term and long-term precision can be evaluated by analysis of variance and applied within a quality control system have been discussed.[22,23]

Data relating analytic precision to time are given for several analytes in Table 3. The analysis of variance performed with CAP Chemistry Survey data[24] assumes that all specimens received in a given Survey shipment are analyzed within a day or usually within the same run. Longer-term laboratory Survey precision is estimated by shipments of serum duplicates 4 to 6 months apart. The CAP Survey precision estimates are not level related within the concentration intervals given, but the within-run CV may vary with concentration in proportion to the level-related variation in long-term CV, so that the ratio of short-term to long-term CV may be comparable at different concentrations. Regional quality control data[25] presented in Table 3 give average within-month CVs after 6 months and after 12 months average use of the control pools and the corresponding cumulative between-month CV. Calculations from the CAP Survey data reveal that long-term variance averages 62% of total variance and within-run variance averages 38% of total variance, measured over 4 to 6 months. Long-term variance in the regional quality control group is 36.4% of the total variance at 6 months and 38.3% of the total variance at 12 months. By combining these results and assuming that 100% of the variance has been accumulated by the 12th month, one can calculate that approximately 40% of total variance is noted within run, 62% by 1 month, and 97% by 6 months. The respective cumulative CV is 63% within run, 79% by 1 month, and 99% by 6 months. A graphic approximation of the average rate of increase of analytic variation and CV with time is illustrated in Figure 1.

Average CV will vary from lab to lab and within lab from time to time.[1,8] Statistics describing this distribution about the average figures graphed have not been calculated.

The CAP Survey long-term (4 to 6 months) CVs are uniformly larger than those noted in the Regional Quality Control Program. This is due to a larger percent of manual methods in the Survey. In the 1975 Survey, 52% of the methods were manual. The corresponding figure in the regional group program in 1979 was 28%. The effect of this variable upon the ratio of within-run CV to long-term CV was examined by calculating the ratio separately for all manual methods and all automated methods for each of the Survey analytes. The ratio is 0.53 and 0.55 for manual and automated methods, respectively. This somewhat surprising constancy permits the use of all method groupings in approximating the average increase in rate of analytic variation with time.

Effect of Concentration upon Analytic Precision

The tendency of standard deviation to increase with analyte concentration has been described by several authors.[2,8,26] A parallel tendency toward decreasing CV was also

Table 3

INCREASE IN ANALYTIC VARIATION AS A FUNCTION OF TIME

Analyte	Units	CAP Chemistry Survey 1975[a] (all methods combined)				Nine-state regional quality control program[b] 250 Laboratories, June and December 1979 (all methods combined)						
		Concentration	Av within-day CV	Av between-month CV over 3—6 months	Long-term variance as % of total	Concentration	Av within-month CV		Av between-month CV over 6 and 12 months		Component of long-term variance as % of total	
							6/79	12/79	6/79	12/79	6/79	12/79
Sodium	mmol/l	137—148	1.1	1.6	53	147	0.9	0.9	1.1	1.1	33	33
Potassium	mmol/l	4.0—4.5	1.9	2.9	57	2.9	2.4	2.3	2.9	2.8	32	33
Chloride	mmol/l	99—113	1.7	2.6	57	114	1.4	1.4	1.7	1.7	32	32
CO$_2$ Content	mmol/l	22				22	5.3	5.5	6.3	6.5	29	28
Urea nitrogen	mg/dl	13—16	7.9	9.1	25	18	4.5	4.3	5.4	5.5	31	39
Creatinine	mg/dl	2.3—2.5	5.5	12.8	82	1.7	5.0	5.0	5.9	6.0	28	31
Uric acid	mg/dl	3.9—7.4	4.2	6.9	63	5.6	3.0	2.9	3.8	3.8	38	42
Phosphorus	mg/dl	2.8—3.5	3.4	6.7	74	3.4	3.2	3.2	4.1	4.3	43	45
Iron	mcg/dl	57—87	7.3	11.6	60	110	4.4	5.2	6.2	6.3	50	32
Bilirubin	mg/dl	0.33—0.59	22	33	56	1.7	5.5	5.4	7.0	7.1	38	42
Glucose	mg/dl	68—97	3.1	5.5	68	107	3.4	3.1	3.8	3.9	20	37
Calcium	mg/dl	9.7—12.1	2.8	4.3	58	11.3	1.8	1.8	2.4	2.3	44	39
Magnesium	mg/dl	1.6—2.4	6.3	12.0	72	1.4	7.0	6.5	9.2	9.5	42	53
Osmolality	mosm/kg	268—304	1.4	2.2	60	306	1.1	1.3	1.5	1.7	46	42
Protein	g/dl	4.8—6.2	2.1	4.0	72	5.2	2.0	2.1	2.8	2.8	49	44
Cholesterol	mg/dl	137—140	3.6	6.0	64	132	3.2	3.6	4.1	4.2	39	27
Thyroxine	mcg/dl	7.7—10.0	7.4	13.4	70	6.1	6.8	6.4	7.8	8.2	24	39
Lithium	mmol/l	1.2—1.3	4.4	8.1	70	1.5	3.6	3.2	4.6	4.4	39	47
Albumin	g/dl					4.8	2.7	2.4	3.4	3.3	37	47
Triglyceride	mg/dl					89	5.7	5.3	7.3	7.3	39	47
Digoxin	ng/ml					2.2	6.4	6.0	6.9	7.0	13	27
Iron binding	mcg/dl					295	5.1	5.4	7.0	7.1	53	42

Table 3 (continued)
INCREASE IN ANALYTIC VARIATION AS A FUNCTION OF TIME

Analyte	Units	CAP Chemistry Survey 1975[a] (all methods combined)				Nine-state regional quality control program[b] 250 Laboratories, June and December 1979 (all methods combined)					
		Concentration	Av within-day CV	Av between-month CV over 3—6 months	Long-term variance as % of total	Concentration	Av within-month CV 6/79	Av within-month CV 12/79	Av between-month CV over 6 and 12 months 6/79	Av between-month CV over 6 and 12 months 12/79	Component of long-term variance as % of total 6/79	Component of long-term variance as % of total 12/79
Alkaline phosphatase	IU/l					Slight increase	3.7	3.7	5.0	4.9	45	43
Acid phosphatase	IU/l					2× increase	7.7	7.1	10.4	9.8	45	48
Creatine kinase	IU/l					Slight increase	7.3	7.7	8.9	9.0	33	27
Lactate dehydrogenase	IU/l					High normal	3.5	3.9	4.9	5.0	49	39
Aspartate aminotransferase	IU/l					High normal	6.9	6.4	8.1	8.2	27	39
Alanine aminotransferase	IU/l					High normal	8.3	7.6	9.5	9.1	24	30
Av % variance as long-term variance					62.4						36.5	38.4

[a] Reference 24. CVs are average all-method within-lab CVs over the concentration ranges given. The calculation of within-run CV assumes that all specimens of a given survey were analyzed on the same day.

[b] Reference 25. Pool use began approximately 12/78. Within-month CVs are average all-method within-month CVs calculated after 6 and 12 months average use of pool. Between-month CVs are the corresponding all-method cumulative CVs.

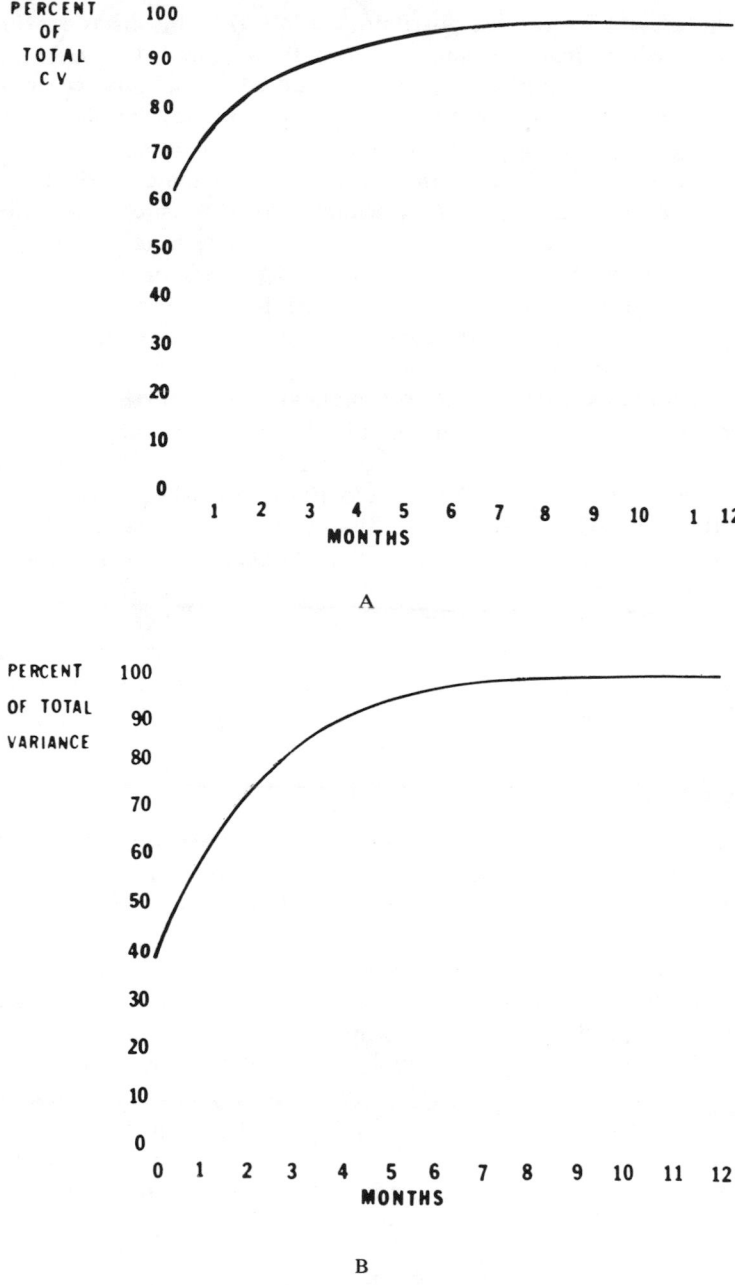

FIGURE 1.   Increase in cumulative variance and cumulative CV as a function
of time. The variance noted at 12 months experience with analytic system is set
as 100%.

described. Several studies of laboratory internal quality control data consolidated from
several large regional group programs have allowed detailed quantitative studies of the
relation of analytic CV to analyte concentration.[12,27,28] Data consolidated from multiple Regional Quality Control Programs, each using two or more analyte concentrations, are sufficiently continuous in the analyte ranges of clinical interest to estimate

state of the art mean CVs and the distribution about their means as continuous variables. The published results are estimates of long-term analytic CVs over an average of 6 months' experience. Logrithmic transformation of the CVs is required to convert the distribution of CVs to Gaussian form. Calculations can be made with the fitted polynomial coefficients in Table 4 by multiplying each coefficient by the corresponding power of a selected analyte concentration. The sum of these products estimates the natural logarithm of the average CV associated with manual or automated methods. Antilog transformation gives an estimate of the average state of the art CV at any selected concentration within the stated analyte range used for the regression analysis. Extrapolation outside of this concentration range is inappropriate.

The relationship between analyte concentration and experienced mean CVs in the concentration ranges studied is best described by polynomials varying from a constant in the case of analytes such as cholesterol, manual calcium methods, and manual cortisol methods, to a fifth-degree polynomial in the case of bilirubin. Examples of four general types of relationships are illustrated in Figures 2 to 5. The few analytes with no effect of concentration upon CV have a flat straight-line graph. Many analytes show a relationship best described by a first-order polynomial with a steady linear decrease in CV with increase in concentration. Others are best described by second-order or higher-order polynomials in which an initial phase of rapid decrease in CV is noted at lower concentrations, followed by a plateau or a less rapid decrease in CV. The fourth type is a dish-shaped second- or third-degree polynomial with improving CV followed by loss of precision in higher analyte ranges. It is probable that if the range of analyte concentration were extended sufficiently, all analytes would follow the fourth pattern, since methodologies and instrumentation are developed such that a zone of best precision and sensitivity is noted in a clinically relevant area with inevitable loss of precision above and below this zone.

Effect of Quality Control Specimens upon Precision Estimates

The effects of the quality control specimen matrix, homogeneity and stability, and the potential effect of analyst bias in precision estimates derived from internal quality control programs have been described in previous sections. Use of the available plug lyophilized materials in internal process control has provided the data upon which most conclusions regarding analytic precision and its application to relevant medical contexts are based. These conclusions apply to the most frequently analyzed chemistry analytes and appear to be medically relevant and not affected by quality control specimen characteristics with the exception of a few analytes. Technical improvement in control materials may be desirable in the cases of certain relatively unstable enzyme analytes, such as creatine kinase and lactate dehydrogenase, and analytes whose average imprecision is small, such as sodium and chloride.

DEFINITION OF ANALYTICAL GOALS

Background

The role of internal quality control programs is to monitor precision and to indicate when analytic goals are achieved. Medical goals were defined years ago in pragmatic terms by Barnett.[29] Recently, a general method by which analytical goals can be derived from the requirements of the medical care application of the test and from statistical theory has been proposed.[55] The first step in formulation of analytic goals is definition of the medical context:

1.   The subject is in apparent health and is being tested for (a) latent disease, (b) iatrogenic insult, or (c) biochemical classification

Table 4

FITTED POLYNOMIAL COEFFICIENTS RELATING ln CV TO ANALYTE CONCENTRATION[24]

Analyte	Analyte range	Units	Method	Degree (p)	b_0	b_1	b_2	b_3	b_4	b_5
Albumin	2.4—5.0	g/d*l*	Manual	1	1.89729	-0.094182				
	2.0—5.1	g/d*l*	Auto	1	1.69908	-0.135338				
Bilirubin	0.5—11.0	mg/d*l*	Manual	5	3.69413	-1.643847	0.5998153	-0.1054610	0.008497	-0.0002544
	0.4—7.9	mg/d*l*	Auto	4	3.59050	-1.936984	0.6502717	-0.0938297	0.004822	
Direct bilirubin	0.1—1.2	mg/d*l*	Manual	1	3.51181	-1.024469				
	0.1—2.7	mg/d*l*	Auto	2	3.60751	-1.614496	0.3277364			
Calcium	7.5—12.7	mg/d*l*	Manual	1	1.52195	-0.037225				
	7.0—13.4	mg/d*l*	Auto	2	2.53186	-0.314824	0.0138656			
Chloride	84—120	mmol/*l*	Manual	1	1.25632	-0.005622				
	84—122	mmol/*l*	Auto	1	1.08066	-0.006090				
Cholesterol	96—300	mg/d*l*	Manual	0	1.60412					
	71—308	mg/d*l*	Auto	1	1.45356	-0.000422				
Creatinine	0.9—6.0	mg/d*l*	Manual	1	2.22503	-0.127368				
	0.8—7.9	mg/d*l*	Auto	2	2.32256	-0.447215	0.0380555			
Glucose	74—273	mg/d*l*	Manual	2	1.79535	-0.006539	0.0000165			
	66—345	mg/d*l*	Auto	2	1.79781	-0.007112	0.0000147			
Magnesium	1.2—5.2	mg/d*l*	Manual	1	2.06700	-0.116154				
	1.1—5.5	mg/d*l*	Auto	2	3.14823	-0.697839	0.0650207			
Phosphorus	2.3—8.5	mg/d*l*	Manual	1	1.85945	-0.050849				
	2.2—8.9	mg/d*l*	Auto	2	2.16327	-0.368433	0.0287137			
Potassium	2.9—7.2	mmol/*l*	Manual	1	1.15560	-0.105518				
	2.5—7.5	mmol/*l*	Auto	2	1.89074	-0.411854	0.0308042			
Sodium	118—155	mmol/*l*	Manual	1	0.82323	-0.005187				
	118—159	mmol/*l*	Auto	1	0.71107	-0.004395				
Total protein	4.3—7.4	g/d*l*	Manual	1	1.82295	-0.156232				
	3.6—7.5	g/d*l*	Auto	1	1.31613	-0.080883				
Triglyceride	63—209	mg/d*l*	Manual	1	2.38250	-0.002882				
	59—230	mg/d*l*	Auto	2	3.02012	-0.016804	0.0000492			
Urea nitrogen	7—63	mg/d*l*	Manual	2	2.42551	-0.031732	0.0003409			
	6—62	mg/d*l*	Auto	3	2.87641	-0.102835	0.0019335	-0.0000122		

Table 4 (continued)

FITTED POLYNOMIAL COEFFICIENTS RELATING ln CV TO ANALYTE CONCENTRATION[24]

Analyte	Analyte range	Units	Method	Degree (p)	b_0	b_1	b_2	b_3	b_4	b_5
Thyroxine	2.0—17.5	mcg/dl	Manual	2	2.73234	-0.149152	0.0061842			
	2.2—16.4	mcg/dl	Auto	1	2.58242	-0.059734				
Uric acid	3.8—12.0	mg/dl	Manual	2	2.25957	-0.163792	0.0092117			
	3.7—11.3	mg/dl	Auto	2	2.38831	-0.346830	0.0204503			
Osmolality	239—344	mosm/kg/H_2O	Manual	1	1.39286	-0.003200				
Digoxin	0.5—4.0	ng/ml	Manual	3	3.54906	-1.793957	0.6399344	-0.0742913		
Cortisol	6.0—45.7	mcg/dl	Manual	0	2.31535					
Salicylate	9.0—50.9	mg/dl	Manual	0	1.74345					
	9.0—44.0	mg/dl	Auto	1	1.35191	-0.023256				
Iron	67—202	mcg/dl	Manual	1	2.75797	-0.006229				
	67—225	mcg/dl	Auto	1	1.93516	-0.005160				
Iron binding	181—379	mcg/dl	Manual	1	2.69772	-0.002446				
	200—397	mcg/dl	Auto	0	1.76052					
Lithium	0.45—2.45	mmol/l	Manual	2	2.37181	-0.931630	0.1872491			
	0.49—2.44	mmol/l	Auto	2	2.84126	-1.509160	0.3785900			

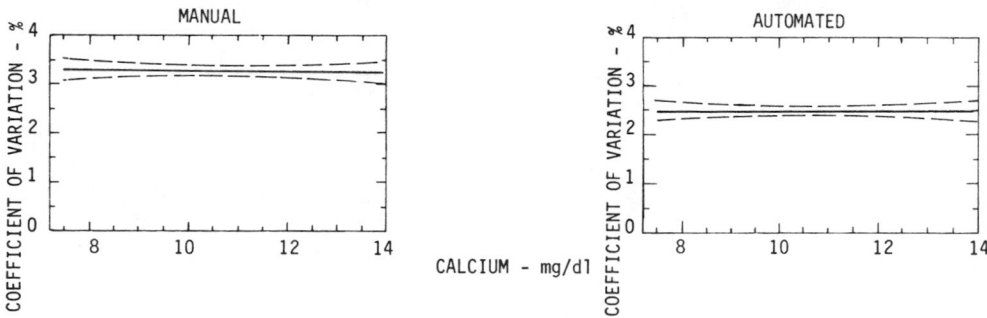

FIGURE 2.   Relationship of long-term within-laboratory calcium CV to calcium concentration.

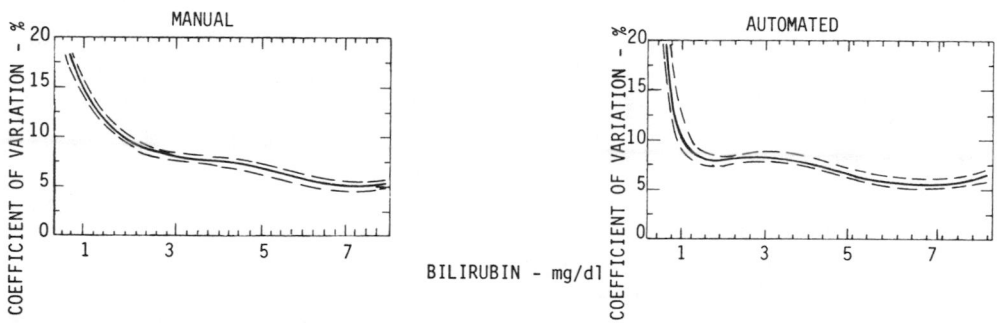

FIGURE 3.   Relationship of long-term within-laboratory bilirubin CV to bilirubin concentration.

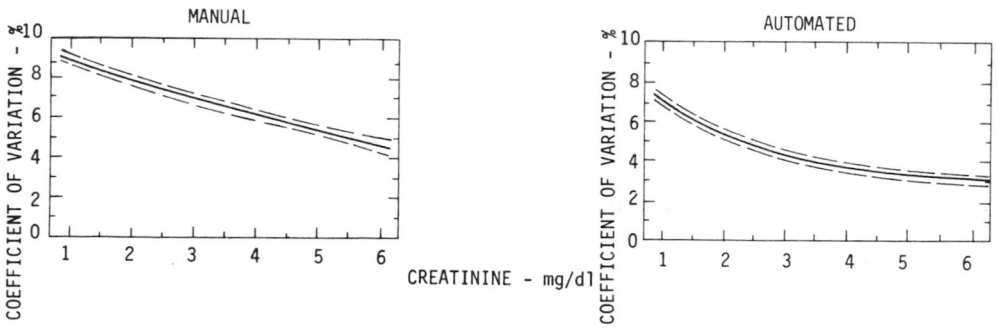

FIGURE 4.   Relationship of long-term within-laboratory creatinine CV to creatinine concentration.

2.   The patient is tested for diagnosis of disease, for (a) differential diagnosis, (b) organ system function evaluation, or (c) disease etiology
3.   The patient's disease process requires monitoring
4.   The patient's therapy requires monitoring
5.   Prior test results require confirmation

The second step is selection of an appropriate statistical model. A considerable recent literature describing the statistical techniques for analysis of physiologic variance[30-36] has been paralleled by publication of experimental data quantitating the

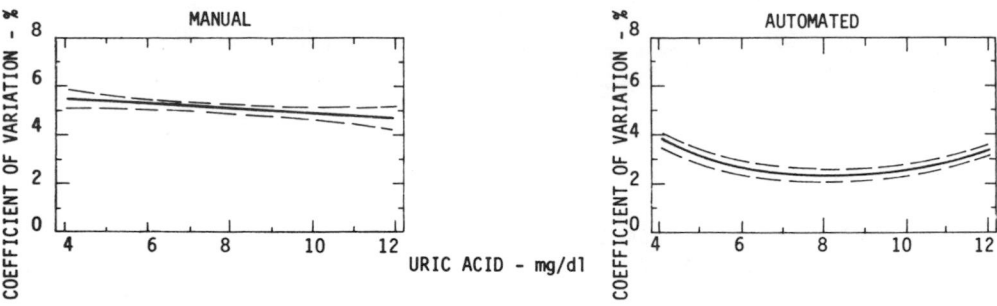

FIGURE 5.   Relationship of long-term within-laboratory uric acid CV to uric acid concentration.

group and individual components of biologic variance and the short-term and long-term components of individual biologic variance.[2,37-47] Three statistical models have been proposed based upon group testing, individual single-sample testing, and individual serial-sample testing medical applications.[21] In the medical contexts listed above, for example, screening for latent disease would require either a group statistical model or sufficient data from an individual subject to permit use of an individual based reference range.[36] Organ system evaluation and test panels for differential diagnosis require group statistical models. Monitoring of disease in an individual requires an individual subject, serial-sample statistical model, whereas biochemical classification of disease requires an individual subject, single-sample statistical model. Estimates of the components of biologic variation and analytic variation appropriate to the medical context may be used in each model to calculate the effect that analytic variation has upon test performance characteristics such as sensitivity and specificity.[21]

Within the two individual subject statistical models, intraindividual biologic variation may be best described in terms of a stationary (homeostatic) or nonstationary (random walk) statistical model.[34,36] The ranges forecast for future results vary depending on the model which best describes the individual subject's biologic variation. Predicted improvement in test performance characteristics (sensitivity and specificity) when based upon individual serial-sample and single-sample models also may vary with selection of a stationary or nonstationary statistical model.[21]

Precision Performance Standards

Precision performance standards define the level of precision which is proper and adequate in a given context. There are currently three types of precision performance standards:

1.   Statistics describing the state of the art, which is subject to improvement without stated limits
2.   Statistics describing the pragmatic usefulness of the test in medical applications
3.   Statistics describing imprecision as an allowable fraction of biologic variation.

Vanko[48] described the state of the art as the best of the available methods. Kurtz et al.[19] considered the state of the art to be the most common range of precision, "...a realistic estimate of the optimum range of instrument and method precision at the time of the study." In the latter sense, Kurtz et al.[19] and Ross and Fraser[12,27,28] have described the state-of-the-art using data based on internal quality control statistics. Estimated average CVs for several analytes at certain concentrations are listed in Table 5. Estimated mean CVs can be calculated at any concentration by interpolation by use

Table 5

COMPARISON OF STATE-OF-THE-ART CV WITH MEDICALLY
SIGNIFICANT CV BY 95% CONFIDENCE INTERVALS[28]

Analyte	Level	Units	Critical CV[a]	State-of-the-art estimated mean CV with 95% confidence intervals[b]	
				Manual	Automated
Albumin	3.5	g/dl	7.14	4.79 ± 0.20	3.41 ± 0.07
Bilirubin	1.0	mg/dl	20.00	12.84 ± 0.57	9.16 ± 0.29
	6.0	mg/dl	—	5.35 ± 0.24	3.89 ± 0.14
Direct bilirubin	0.3	mg/dl	—	24.64 ± 1.88	23.40 ± 1.75
	1.0	mg/dl	—	12.03 ± 1.42	10.20 ± 0.82
Calcium	8.0	mg/dl	—	3.40 ± 0.21	2.46 ± 0.07
	11.0	mg/dl	2.27	3.04 ± 0.13	2.11 ± 0.06
Chloride	90.0	mmol/l	2.22	2.12 ± 0.06	1.70 ± 0.05
	110.0	mmol/l	1.82	1.89 ± 0.05	1.51 ± 0.04
Cholesterol	250.0	mg/dl	8.00	4.97 ± 0.21	3.85 ± 0.12
Creatinine	1.5	mg/dl	—	7.64 ± 0.31	5.68 ± 0.16
	5.0	mg/dl	—	4.89 ± 0.20	2.82 ± 0.09
Glucose	75.0	mg/dl	—	4.05 ± 0.21	3.85 ± 0.10
	100.0	mg/dl	5.00	3.69 ± 0.11	3.43 ± 0.06
	120.0	mg/dl	4.17	3.49 ± 0.12	3.18 ± 0.05
	160.0	mg/dl	—	3.23 ± 0.15	2.82 ± 0.05
Magnesium	1.5	mg/dl	—	6.64 ± 0.66	9.47 ± 0.66
	3.5	mg/dl	—	5.26 ± 0.35	4.49 ± 0.30
Phosphorus	4.5	mg/dl	5.56	5.11 ± 0.22	2.96 ± 0.10
	7.0	mg/dl	—	4.50 ± 0.23	2.69 ± 0.09
Potassium	3.0	mmol/l	8.33	2.31 ± 0.08	2.54 ± 0.06
	6.0	mmol/l	4.17	1.69 ± 0.05	1.70 ± 0.03
Sodium	130.0	mmol/l	1.54	1.16 ± 0.03	1.15 ± 0.02
	150.0	mmol/l	1.33	1.05 ± 0.04	1.05 ± 0.03
Total protein	6.0	g/dl	—	2.42 ± 0.13	2.30 ± 0.04
	7.0	g/dl	4.28	2.07 ± 0.15	2.12 ± 0.06
Triglyceride	130.0	mg/dl	—	7.45 ± 0.36	5.30 ± 0.23
	200.0	mg/dl	—	6.09 ± 0.65	5.09 ± 0.35
Urea nitrogen	27.0	mg/dl	7.41	6.15 ± 0.31	3.56 ± 0.10
	50.0	mg/dl	—	5.43 ± 0.23	2.83 ± 0.08
Thyroxine	6.0	mcg/dl	—	7.84 ± 0.29	9.24 ± 1.01
	12.0	mcg/dl	—	6.25 ± 0.26	6.46 ± 0.71
Uric acid	6.0	mg/dl	8.33	4.99 ± 0.25	2.84 ± 0.09
	9.0	mg/dl	—	4.62 ± 0.23	2.52 ± 0.07
Osmolality	270	mosm/kg/H$_2$O	1.7	1.70 ± 0.14	—
Digoxin	1.0	ng/ml	—	10.18 ± 0.46	—
	2.5	ng/ml	—	6.70 ± 0.28	—
Cortisol	20.0	mcg/dl	—	10.13 ± 0.48	—
Salicylate	15.0	mg/dl	—	5.72 ± 0.74	2.73 ± 0.35
	40.0	mg/dl	—	5.72 ± 0.74	1.52 ± 0.41
Iron	80.0	mcg/dl	—	9.58 ± 0.58	4.58 ± 0.31
	150.0	mcg/dl	—	6.19 ± 0.28	3.19 ± 0.15
Iron binding	250.0	mcg/dl	—	8.05 ± 0.36	5.82 ± 0.31
	400.0	mcg/dl	—	5.58 ± 0.67	5.82 ± 0.31
Lithium	0.5	mmol/l	—	7.05 ± 0.55	8.86 ± 0.42
	1.5	mmol/l	—	4.04 ± 0.29	4.18 ± 0.22

[a] Barnett: medically significant CV.[29]
[b] There is slight asymmetry in the upper and lower limits due to the log$_e$ conversion. The greatest error in the limits at the points given is 0.02%, and the bounds are presented as above for the sake of clarity.

of the fitted polynomial coefficients in Table 4 as previously described. When participating in a regional quality control group, one may also obtain level-related and, in this case, specimen-related average precision data from the interlaboratory comparisons developed by computerized data reduction such as that provided by the CAP's Quality Assurance Service (QAS).

The distribution of CVs about the mean of each analyte concentration is similar to that of a log normal distribution with a skewed, elongated upper tail.[24] The upper 90% tolerance limits for the 50th, 75th, 90th, and 95th percentiles of the upper tail of the distributions at certain concentrations of medical decision importance are illustrated in Table 6. The heading of each column under tolerance limits gives the percent of laboratories with CVs less than the CV listed in the column for all manual or all automated methods. For example, in the case of albumin, 90% of the laboratories using a manual method have CVs of less than 7.9%, and 90% of the laboratories using automated methods have CVs less than 5.7%. (Population tolerance limits are calculated so that the stated percentage of the unknown population is contained within the limit with the stated degree of certainty associated with the tolerance limit each time the entire experiment is performed.) These data and other state of the art precision performance standards provide the analyst with a useful point of reference in technically achievable terms.

Medical usefulness precision standards based on pragmatic clinical use of test values have been discussed by Barnett[29,49] and Gilbert.[17] Elion-Gerritzen[50] has discussed the perception of analytic and physiologic variability by physicians and the effect of this perception upon the analytic values recognized as clinical action levels. The inverted tolerance limits for critical CV in Table 6 apply pragmatic medical usefulness criteria to estimates of long-term state of the art precision at stated concentrations. The inverted tolerance limits give the percent of laboratories whose analytic CV is less than the medically required CV cited by Barnett.[29] For example, the medically important CV for urea nitrogen at the level of 27 mg/dℓ is 7.41%. Of laboratories using manual methods 66% have CVs less than 7.41%, as do 96% of laboratories using automated methods.

If compliance to the medical usefulness criteria by 90% or more of laboratories is considered to be adequate overall clinical laboratory performance, automated analysis of 7 of 12 clinical chemistry analytes is satisfactory. Automated analysis for an additional analyte is very close to this standard and slight improvement, perhaps by increased skill of application of a method currently in use, may be all that is required. Four analytes, calcium, chloride, glucose, and sodium, have a large percentage of laboratories failing to meet medical usefulness goals. Greater skill of application of existing methods is not likely to produce adequate performance for all these analytes. Selection of inherently more precise methods by a substantial percentage of laboratories is probably necessary. Adequate automated methods appear to be available since the average CV meets medical goals.

Manual analysis is satisfactory in 90% or more of the laboratories for 1 of 13 analytes. Five additional analytes have manual performance close to this standard, suggesting that increased skill of application of existing methods presently in use may be sufficient to achieve satisfactory perfomance. Improved method selection of better methods may be necessary in manual analysis of five analytes. In the case of calcium and osmolality, adequate manual methods may not be available, since substantially less than 50% of laboratories meet medical needs.

Medical usefulness standards have also been developed which are based upon analytic precision as an allowable percent of biologic variation. At this time, a generally stated precision performance standard limits analytic imprecision allowable in a medi-

Table 6

TOLERANCE LIMITS FOR COEFFICIENTS OF VARIATION (CV) AT SELECTED ANALYTE CONCENTRATIONS WITH INVERTED TOLERANCE LIMITS GIVING PERCENT OF LABS WITH AVERAGE CV BELOW MEDICALLY IMPORTANT CVs[28]

Analyte	Analyte level	Unit	Tolerance limits 95% Man	95% Auto	90% Man	90% Auto	75% Man	75% Auto	50% Man	50% Auto	Critical CV[a]	Inverted tolerance limits for critical CV Man	Auto
Albumin	3.5	g/dl	9.0	6.6	7.9	5.7	6.3	4.5	4.9	3.5	7.14	85	97
Bilirubin	1.0	mg/dl	24.9	17.7	21.6	15.4	17.1	12.1	13.2	9.3	20.00	86	97
	6.0	mg/dl	10.4	7.5	9.0	6.5	7.1	5.2	5.5	4.0	—	—	—
Direct bilirubin	0.3	mg/dl	40.9	47.6	36.9	41.1	31.1	32.1	25.9	24.6	—	—	—
	1.0	mg/dl	20.4	20.8	18.4	17.9	15.6	14.0	13.0	10.7	—	—	—
Calcium	8.0	mg/dl	7.4	4.7	6.2	4.1	4.8	3.2	3.5	2.5	—	—	—
	11.0	mg/dl	6.5	4.0	5.5	3.5	4.2	2.8	3.1	2.1	2.27	23	56
Chloride	90.0	mmol/l	3.9	3.2	3.4	2.8	2.7	2.2	2.2	1.7	2.22	53	75
	110.0	mmol/l	3.4	2.8	3.0	2.4	2.4	2.0	1.9	1.5	1.82	44	68
Cholesterol	250	mg/dl	10.3	7.6	8.8	6.5	6.8	5.1	5.1	3.9	8.00	85	96
Creatinine	1.5	mg/dl	14.9	11.7	12.9	10.0	10.2	7.7	7.8	5.8	—	—	—
	5.0	mg/dl	9.6	5.8	8.3	5.0	6.5	3.8	5.0	2.9	—	—	—
Glucose	75	mg/dl	7.6	6.8	6.6	6.0	5.3	4.9	4.2	3.9	5.00	79	86
	100	mg/dl	6.8	6.0	6.0	5.3	4.8	4.4	3.8	3.5	4.17	67	78
	120	mg/dl	6.4	5.6	5.6	4.9	4.5	4.0	3.6	3.2	—	—	—
	160	mg/dl	6.0	5.0	5.3	4.4	4.2	3.6	3.3	2.9	—	—	—
Magnesium	1.5	mg/dl	13.7	16.6	11.8	14.8	9.2	12.2	7.1	9.9	—	—	—
	3.5	mg/dl	10.7	7.9	9.2	7.0	7.2	5.8	5.5	4.7	—	—	—
Phosphorus	4.5	mg/dl	10.1	6.2	8.7	5.3	6.8	4.1	5.3	3.0	5.56	56	92
	7.0	mg/dl	8.9	5.7	7.7	4.8	6.0	3.7	4.6	2.8	—	—	—
Potassium	3.0	mmol/l	4.1	4.6	3.6	4.0	2.9	3.3	2.4	2.6	8.33	99	99
	6.0	mmol/l	3.0	3.0	2.6	2.7	2.1	2.2	1.7	1.7	4.17	99	99
Sodium	130	mmol/l	2.1	2.1	1.9	1.8	1.5	1.5	1.2	1.2	1.54	77	79
	150	mmol/l	1.9	1.9	1.7	1.7	1.4	1.4	1.	1.1	1.33	72	73

Table 6 (continued)

TOLERANCE LIMITS FOR COEFFICIENTS OF VARIATION (CV) AT SELECTED ANALYTE CONCENTRATIONS WITH INVERTED TOLERANCE LIMITS GIVING PERCENT OF LABS WITH AVERAGE CV BELOW MEDICALLY IMPORTANT CVs[28]

Analyte	Analyte level	Unit	Tolerance limits 95% Man	95% Auto	90% Man	90% Auto	75% Man	75% Auto	50% Man	50% Auto	Critical CV[a]	Inverted tolerance limits for critical CV Man	Auto
Total protein	6.0	g/dl	6.5	4.3	5.2	3.7	3.7	3.0	2.5	2.3	—	—	—
	7.0	g/dl	5.6	4.0	4.5	3.5	3.2	2.8	2.2	2.2	4.28	88	97
Triglyceride	130	mg/dl	14.0	10.7	12.2	9.2	9.8	7.2	7.7	5.4	—	—	—
	200	mg/dl	11.8	10.4	10.3	9.0	8.3	7.0	6.5	5.3	—	—	—
Urea nitrogen	27	mg/dl	11.6	7.1	10.1	6.1	8.1	4.8	6.4	3.6	7.41	66	96
	50	mg/dl	10.2	5.7	8.9	4.9	7.1	3.8	5.6	2.9	—	—	—
Thyroxine	6.0	mcg/dl	14.9	16.1	13.0	14.4	10.3	12.0	8.0	9.9	—	—	—
	12.0	mcg/dl	11.9	11.2	10.4	1.01	8.3	8.4	6.4	6.9	—	—	—
Uric acid	6.0	mg/dl	10.4	6.3	8.9	5.3	6.9	4.0	5.2	2.9	8.33	87	99
	9.0	mg/dl	9.7	5.6	8.3	4.7	6.4	3.5	4.8	2.6	—	—	—
Osmolality	270	mosm/kg/H_2O	4.4	—	3.6	—	2.6	—	1.8	—	1.7	46	—
Digoxin	1.0	ng/ml	17.9	—	15.9	—	13.0	—	10.5	—	—	—	—
	2.5	ng/ml	11.8	—	10.4	—	8.6	—	6.9	—	—	—	—
Cortisol	20.0	mcg/dl	18.3	—	16.1	—	13.1	—	10.4	—	—	—	—
Salicylate	15.0	mg/dl	13.9	6.5	11.5	5.4	8.6	4.1	6.2	3.0	—	—	—
	40.0	mg/dl	13.9	3.9	11.5	3.3	8.6	2.5	6.2	1.8	—	—	—
Iron	80	mcg/dl	19.1	10.2	16.5	8.6	13.0	6.5	10.0	4.8	—	—	—
	150	mcg/dl	12.2	7.0	10.6	5.9	8.3	4.5	6.4	3.3	—	—	—
Iron binding	250	mcg/dl	14.9	12.2	13.0	10.4	10.5	8.0	8.3	6.0	—	—	—
	400	mcg/dl	10.7	12.2	9.5	10.4	7.6	8.0	6.0	6.0	—	—	—
Lithium	0.5	mmol/l	15.6	16.8	13.2	14.7	10.0	11.7	7.4	9.1	—	—	—
	1.5	mmol/l	8.9	8.0	7.5	6.9	5.7	5.5	4.2	4.3	—	—	—

a   Barnett: medically significant CV.[29]

cal context to less than one half of the biologic variation relevant to the medical context.[51] When this standard is met, observed variation will be no more than 11.8% larger than the biologic variation. Since, on the average, intraindividual biologic variation is less than group biologic variation and since intraindividual biologic variation increases with time, the requirement for best precision occurs in medical contexts in which an individual is studied over a short period of time.

Although this performance standard defines a generally acceptable level of analytic imprecision, for some applications the clinical context and statistical model used may be such that the appropriate standard should be defined in terms other than biologic variation. For example, in intraindividual testing during a short time span, biologic variation may be very small. The precision requirement may be best calculated with an individual series-sample statistical model in terms of the degree of certainty with which a trend of a given magnitude is to be detected in a test series of given length.[21]

For some analytes, such as sodium, the group and individual biologic variation is so small that precision performance standards expressed as a fraction of any component of biologic variation have no practical application and may be supplanted by pragmatic medical usefulness criteria. Pragmatic medical precision standards also remain useful since they often reflect physician action levels. Some knowledge of clinical action levels is required to determine the proper definitions of sensitivity and specificity to be used with statistical models which calculate the impact of given reductions in analytic imprecision upon medical test performance.[21]

Components of Precision Performance Standards

Precision performance standards depend on the sum of the effects of a number of identifiable separate components of variation which can be related to components of analytic variation. These include time, analyte concentration, and group and individual components of biologic variation.

State of the art precision performance standards are technical standards and do not contain inherent limits upon the degree of improvement which may be sought. However, they may be made quantitative and exact[28] and are useful as an assessment of the practical technical capability of laboratories and methods. The components of state of the art imprecision estimates which must be specified are time, analyte concentration, and specimen type. Time should be divided into at least three categories: within-run, short-term between-day, and long-term between-day periods. Concentration ranges should be selected within which concentration effects do not influence the conclusions drawn. Specimen effects may be important for some analytes with new quality control products. The matrix of the specimen used for precision estimates should always be relevant to the biologic matrix being analyzed. Data pertaining to long-term state-of-the-art precision are presented in Tables 3 and 5 and Figures 1 to 5. More detailed data are needed on concentration-related analytic variation for short time intervals.

Pragmatic medical usefulness[17,29,49,50] precision standards involve a general time range (such as time span of hospitalization, time between office visits, etc.) and are generally stated at a concentration range (such as normal, low, or elevated). Application of pragmatic medical usefulness standards to state of the art precision was discussed in the previous section and data are presented in Table 6.

Precision standards for physiologic analytes calculated from models applicable to medical contexts usually depend upon components of biologic variation. Total biologic variation is the sum of intraindividual variation over time and interindividual variation. Biologic variation may also have disease-dependent components, but sufficient data are available only for biologic variation in health at this time.

Estimates of average interindividual biologic variation and intraindividual biologic variation over time are listed in Table 7. Intraindividual CVs for many analytes have a steady increase with the time interval of the study, implying an effect of between-day factors. There is evidence of a dominant between-day effect for uric acid, cholesterol, and aspartate aminotransferase. An effect of between-day variation is present for chloride, bilirubin, albumin, and lactate dehydrogenase; there are insufficient data regarding possible within-day contribution for these analytes. Both within-day and between-day factors are important for urea nitrogen, creatinine, iron, total protein, alkaline phosphatase, acid phosphatase, and alanine aminotransferase. There is a resemblance in the increase of observed intraindividual variation to the approximated average increase in analytic variability with time span of replicate analysis (Figure 1) in the case of nonhomeostatic analytes; however, it is not apparent that these increases are parallel. Other analytes have CVs which reach a maximum within 1 day, implying a dominant effect of within-day factors. This group includes homeostatic analytes such as sodium and calcium and metabolically active analytes such as potassium and inorganic phosphorus.

Morrison et al.[39] found that zinc, in addition to phosphorus and potassium, had a predominant effect of within-day variation. The within-day effect of potassium and phosphorus had a characteristic diurnal pattern. Sodium, chloride, carbon dioxide, calcium, uric acid, aspartate aminotransferase, alanine aminotransferase, and cholesterol remained predictably constant within day (or varied slightly relative to analytic variation). A single sample, independent of time, may adequately estimate an average level of one of these analytes in a subject within the limits of analytic imprecision. Urea nitrogen, creatinine, total protein, albumin, bilirubin, and triglyceride had measureable within-day and between-day variation with characteristic diurnal patterns during a day. Control of collection time would limit the within-day component of such analytes and reduce observed biologic variation. The other analytes studied, such as iron, had fluctuations within-day and between-day which were large relative to analytic precision, and which were unpredictable so that control of collection time would not be effective in limiting observed biologic variation, and one sample may not adequately predict a subject's average level.

The estimates of intraindividual and interindividual variation are average figures. The possibiity of large individual deviations from these average figures is indicated by the wide observed range of intraindividual CVs noted by Aguzzi et al.[37] and Winkel et al.[46] and the large interval between the upper 95% intraindividual CV percentile range and the average intraindividual CVs noted by Harris et al.[31]

The ratio of intraindividual to interindividual biologic variance is an index of the biochemical individuality of the analyte. A low ratio, less than 0.6, implies usefulness of individual-related reference ranges, whereas a higher ratio, greater than 1.6, implies usefulness of population-based reference intervals, provided the individual under consideration has biologic variation near the average.[33,35] In the cases of analytes with a low ratio, at the time that 5% of the subject test values were falling outside of the 95% reference interval, a much higher percentage would be outside the individual-based reference interval. Individual-based reference ranges appear to be useful for magnesium, creatinine, cholesterol, uric acid, thyroxine, triglyceride, lactate dehydrogenase, alkaline phosphatase, gamma glutamyl transferase, and aspartate aminotransferase.

COMBINING ANALYTIC VARIATION AND BIOLOGIC VARIATION: EXPECTED VARIATION

Ideally, the only observed variation should be biologic variation. However, this

Table 7
ESTIMATES OF THE AVERAGE COMPONENTS OF BIOLOGIC VARIATION IN HEALTHY POPULATIONS

Analyte	Time span of study	Av intraindividual CV	Av interindividual CV	Intra/inter ratio	Av total biologic CV	Upper 95% intra CV range	Observed intraindividual CV range	Ref.
Sodium	22 weeks	0.5	0.6	0.8	1.2			33
	16 weeks	0	0.8	0	0.8	1.1		31
	15 weeks	0.6	0.5	1.2	0.8		0.79—1.7	46
	12 weeks	1.4	1.2	1.2	1.8			47
								35
	4 weeks	0.9	0	—	0.9			3
	2 weeks	0.7	0.5	1.4	0.9			45
	8 hr	0.5	—	—	—			46
	0.5 hr	0.6	—	—	—			45
Calcium	22 weeks	1.6	1.5	1.1	2.2			33
	16 weeks	1.7	2.8	0.6	3.3	3.8		31
	15 weeks	1.0	1.4	0.7	1.7			46
	12 weeks	1.6	3.4	0.5	3.8		0—4.0	47
	8.5 weeks[a]	2.1	2.7	0.8	3.4			35
	4 weeks	1.9	3.0	0.6	3.6			3
	2 weeks	1.7	0.7	2.4	1.8			45
	8 hr	0.4	—	—	—			46
	0.5 hr	1.2	—	—	—			45
Magnesium	16 weeks	1.6	6.4	0.2	6.6	5.3		31
	12 weeks	2.5	5.9	0.4	6.4		0—4.9	47
	8.5 weeks[a]	2.4	5.5	0.4	6.0			35
Potassium	22 weeks	6.2	4.7	1.3	7.9			33
	16 weeks	5.0	4.3	1.2	6.6	8.8		31
	15 weeks	6.3	1.8	3.5	6.6			46
	12 weeks	4.6	5.9	0.8	7.5		2.4—7.0	47
	4 weeks	4.4	5.6	0.8	7.1			3
	2 weeks	4.3	3.8	1.1	5.6			45
	8 hr	5.8	—	—	—			46
	0.5 hr	2.4	—	—	—			45

Table 7 (continued)
ESTIMATES OF THE AVERAGE COMPONENTS OF BIOLOGIC VARIATION IN HEALTHY POPULATIONS

Analyte	Time span of study	Av intraindividual CV	Av interindividual CV	Intra/inter ratio	Av total biologic CV	Upper 95% intra CV range	Observed intraindividual CV range	Ref.
Chloride	16 weeks	1.4	1.2	1.2	1.8	2.2		31
	15 weeks	1.2	1.6	0.8	2.0			46
	12 weeks	2.1	0.8	2.6	2.2		1.1—2.9	47
	4 weeks	1.5	1.7	0.9	2.3			3
	2 weeks	2.1	1.9	1.1	2.8			45
	0.5 hr	0.2	—	—	—			45
CO_2	16 weeks	4.2	4.4	1.0	6.1	7.2		31
	12 weeks	5.3	4.7	1.1	7.1		2.0—10	47
	4 weeks	4.4	0	—	4.4			3
Glucose	16 weeks	5.6	7.8	0.7	9.6	9.3		31
	12 weeks	6.5	2.7	2.4	7.0		4.1—11	47
	8.5 weeks[a]	6.5	8.7	0.8	10.8			35
Urea nitrogen	22 weeks	11.1	13.8	0.8	17.7			33
	16 weeks	11.9	18.3	0.7	22.0	19.4		31
	15 weeks	15.7	16.0	1.0	22.4			46
	12 weeks	13.6	13.5	1.0	19.2		8.5—18	47
	8.5 weeks[a]	16.4	21.7	0.8	26.4			35
	4 weeks	14.3	18.6	0.8	23.5			3
	2 weeks	12.3	16.4	0.8	20.5			45
	8 hr	6.0	—	—	—			46
	0.5 hr	2.2	—	—	—			45
Creatinine	22 weeks	6.3[b]	10.1	0.6	11.9			33
	15 weeks	6.0	11.6	0.5	13.1			46
	12 weeks	4.4	7.0	0.6	8.3		2.8—5.9	47
	8.5 weeks[a]	3.1	14.7	0.2	15.0			35
	4 weeks	3.4	10.4	0.3	10.9			3
	2 weeks	4.3	9.5	0.5	10.4			45
	8 hr	4.0	—	—	—			46
	0.5 hr	2.2	—	—	—			45

Phosphorus	22 weeks	6.8	10.8	0.6	12.8			33
	16 weeks	7.5	10.9	0.7	13.2	12.5		31
	15 weeks	9.5	4.2	2.3	10.4			46
	12 weeks	9.6	11.5	0.8	15.0		5.3—15	47
	8.5 weeks[a]	8.2	11.8	0.7	14.4			35
	4 weeks	7.4	13.0	0.6	15.0			3
	2 weeks	5.8	3.4	1.7	6.7			45
	8 hr	7.3	—	—	—			46
	0.5 hr	2.6	—	—	—			45
Uric acid	16 weeks	11.9[c]	13.5	0.9	18.0	16.6		31
		8.6[d]	17.2	0.5	19.2			
	15 weeks	6.9	8.8	0.8	11.2		3.8—12	46
	12 weeks	8.5	9.7	0.9	12.9			47
	8.5 weeks[a]	8.8	20.1	0.4	22.4			35
	4 weeks	8.6	17.6	0.5	19.6			3
	2 weeks	7.3	6.9	1.1	10.0			45
	8 hr	2.5	—	—	—			46
	0.5 hr	1.2	—	—	—			45
Bilirubin	22 weeks	26.0	42.6	0.6	49.7			33
	4 weeks	25.6	45.3	0.6	52.0			3
	2 weeks	23.9	17.7	1.4	29.7			45
	0.5 hr	12.1	—	—	—			45
Iron	16 weeks	29.3[d]	—	—	—			41
	15 weeks	28.8	22.3	1.3	36.4			46
	2 weeks	26.6	24.2	1.1	36.0			45
	2 weeks	22.6[c]	24.0	0.9	33.0			43
	8 hr	12.9[c]	—	—	—			46
	0.5 hr	7.3	—	—	—			45
Cholesterol	16 weeks	6.5[e]	13.6	0.5	15.1	10.5		31
		5.9[f]	18.3	0.3	19.2			
	15 weeks	4.1	13.0	0.3	13.6		1.3—6.6	46
	12 weeks	4.2	13.9	0.3	14.5			47
	8.5 weeks[a]	7.9	17.9	0.4	19.5			35
	5 weeks[g]						3.3—10.9	37
	4 weeks	4.4	15.2	0.3	15.8			3
	2 weeks	5.3	11.4	0.5	12.3			45
	1.4 weeks	4.8	19.5	0.3	20.1			44
	8 hr	3.5	—	—	—			46
	0.5 hr	3.4	—	—	—			45

Table 7 (continued)
ESTIMATES OF THE AVERAGE COMPONENTS OF BIOLOGIC VARIATION IN HEALTHY POPULATIONS

Analyte	Time span of study	Av intraindividual CV	Av interindividual CV	Intra/inter ratio	Av total biologic CV	Upper 95% intra CV range	Observed intraindividual CV range	Ref.
Triglycerides	5 weeks[g]	27.2	57.0	0.5	63.2		16.3—74.4	37
	4 weeks	27.3[a]	47.3	0.6	54.6			3
	8.5 weeks	25.0	69.8	0.4	74.1			35
	1.4 weeks							44
HDL cholesterol	5 weeks[g]						4.3—14.0	37
Total protein	22 weeks	3.0	3.6	0.8	4.7			33
	16 weeks	2.8	5.7	0.5	6.4	4.8		31
	15 weeks	2.5	3.7	0.7	4.5			46
	12 weeks	2.2	2.7	1.2	3.5		1.2—3.2	47
	8.5 weeks[a]	3.3	4.7	0.7	5.7			35
	4 weeks	2.5	2.8	09	3.8			3
	2 weeks	2.9	2.7	1.1	4.0			45
	8 hr	1.7	—	—	—			46
	0.5 hr	1.5	—	—	—			45
Albumin	22 weeks	3.2	3.9	0.8	5.0			33
	16 weeks	3.9	6.3	0.6	7.4	6.5		31
	15 weeks	3.1	2.3	1.3	3.9			46
	12 weeks	3.0	2.2	1.4	3.7		2.3—3.8	47
	4 weeks	2.3	4.8	0.5	5.3			3
	2 weeks	2.8	0.5	5.6	2.8			45
	8 hr	2.1	—	—	—			46
Transferrin	1.4 weeks	2.5	9.5	0.3	9.8			44
Iron binding	16 weeks	8.8[d]	—	—	—			31
IgG	1.4 weeks	2.7	18.1	0.2	18.3			44
IgA	1.4 weeks	3.4	41.1	0.1	41.2			44
IgM	1.4 weeks	3.1	54.0	0.1	54.1			44
Alpha-1-antitrypsin	1.4 weeks	2.9	15.7	0.2	16.0			44
Alpha-2-macroglobulin	1.4 weeks	3.1	16.6	0.2	16.9			44
Haptoglobin	1.4 weeks	8.8	70.5	0.1	71.0			44
Orosomucoid	1.4 weeks	11.1	43.1	0.3	44.5			44

Complement, C_3	1.4 weeks	3.8	19.7	0.2	20.1			44
Complement, C_4	1.4 weeks	5.9	35.6	0.2	36.1			44
Alkaline phosphatase	22 weeks	6.4	21.5	0.3	22.4			33
	15 weeks	8.6	17.8	0.5	19.8			46
	12 weeks	5.7	22.3	0.3	23.0		1.8—7.6	47
	8.5 weeks[a]	8.5	29.3	0.3	30.5			35
	4 weeks	6.3	42.2	0.1	42.7			3
	2 weeks	4.8	19.4	0.2	20.0			45
	1.4 weeks	3.5	24.8	0.1	25.0			44
	8 hr	3.3	—	—	—			46
	0.5 hr	2.0	—	—	—			45
Acid phosphatase	15 weeks	7.9	10.0	0.8	12.7			46
	2 weeks	9.9	5.9	1.5	11.5			45
	8 hr	6.7	—	—	—			46
	0.5 hr	3.3	—	—	—			45
Aspartate aminotransferase	16 weeks	11.5[c]	14.5	0.8	18.5	20.3		31
		8.9[d]	24.8	0.4	26.4			
	15 weeks	20.5	13.4	1.5	24.5			46
	12 weeks	15.1	13.6	1.1	20.3		7.0—25	47
	8.5 weeks[a]	19.3	30.8	0.6	36.3			35
	4 weeks	14.8	24.2	0.6	28.4			3
	2 weeks	24.2	21.3	1.1	32.2			45
	1.4 weeks	10.9	12.1	0.9	16.3			44
	8 hr	3.1	—	—	—			46
	0.5 hr	3.2	—	—	—			45
Alanine aminotransferase	15 weeks	33.9	41.1	0.8	53.3			46
	4 weeks	57.9	72.4	0.8	92.7			3
	2 weeks	26.4	44.5	0.6	51.7			45
	1.4 weeks	13.2	29.6	0.5	32.4			44
	8 hr	11.7	—	—	—			46
	0.5 hr	5.7	—	—	—			45
Lactate dehydrogenase	16 weeks	9.0	15.7	0.6	18.1	21.7		31
	15 weeks	7.4	7.7	0.9	10.7			46
	12 weeks	7.3	9.5	0.8	12.0			47
	8.5 weeks[a]	8.0	14.7	0.5	16.7		2.1—15	35
	4 weeks	6.2	26.1	0.2	26.8			3
	1.4 weeks	5.5	12.6	0.4	13.8			44
	8 hr	4.2	—	—	—			46

Table 7 (continued)
ESTIMATES OF THE AVERAGE COMPONENTS OF BIOLOGIC VARIATION IN HEALTHY POPULATIONS

Analyte	Time span of study	Av intraindividual CV	Av interindividual CV	Intra/inter ratio	Av total biologic CV	Upper 95% intra CV range	Observed intraindividual CV range	Ref.
Gamma glutamyl transpeptidase	8.5 weeks[a]	34.7	86.7	0.4	93.3			35
	1.4 weeks	3.9	23.8	0.2	24.1			44
Creatine kinase	8.5 weeks[a]	60.3	52.6	1.1	80.0			35
	4 weeks	82.8	82.8	1.0	117.1			3
	1.4 weeks	25.7	46.0	0.6	52.7			44
Thyroxine	8.5 weeks[a]	10.1	18.0	0.6	20.1			35
	1.4 weeks	7.5	11.9	0.6	14.1			44
Cortisol	1.4 weeks	26.6	53.1	0.5	59.4			44

Note: Within-run analytic variance is not included unless so stated. In some cases this component has been subtracted from the data tabulated in the original reference, where sufficient information was given to form a reasonable estimate of within-run variation.

[a] Range of study time interval for individual subjects was 5—12 weeks.
[b] Analytic variation estimated to be 8% = 1 CV at 103 $\mu M/\ell$.
[c] Males.
[d] Females.
[e] Males and females less than 30 years of age.
[f] Females over 30 years of age.
[g] Includes analytic variation.

would require no analytic imprecision. Predicted average, long-term, between-day analytic CVs and approximated short-term, average, analytic between-day CVs are listed in Table 8. When the short-term analytic CV is one half or less than the short-term intraindividual biologic CV, the most rigorous presently stated general analytic goal for precision for the analyte and concentration is met.

Comparable time components of intraindividual variation and of analytic variation are summed to produce an estimate of observed intraindividual CV within-day and between-days. On the average, a percent deviation from the previous result of greater than 2.7 times the calculated observed CV will occur by chance no more than 5% of the time. A percent deviation greater than 2 times the calculated observed CV from a true value would occur by chance less than 5% of the time. These are average figures and are derived from approximations in the case of short-term observed CVs. Individual laboratories should calculate corresponding figures based on their own estimates of components of analytic precision. The large range of individual variation about average figures of biologic variation should be kept in mind.

Biologic variation was estimated in healthy subjects, and application of these estimates to sick and bedridden patients has not been defined. Data for calcium and parathormone in subjects with primary hyperparathyroidism suggest that, while diurnal variation of the analytes is preserved, the variability about the individual subject's mean value is increased.[52] Biologic variability in serum bile acid concentration is increased in patients with liver disease.[53] On the other hand, Morrison et al.[39] mention cases of apparent decrease in variability of serum iron in patients with iron deficiency. There is a need to study disease-related biologic variation and reference intervals.

Statistical models for setting precision goals applicable to therapeutic drugs, as well as physiologic analytes, has been proposed by Glick.[54] The precision goals are expressed in terms of a percentage of the range of clinical interest. Glick suggests that when the purpose of a test is to distinguish an upper medically defined limit from a lower medically defined limit, the maximum random error should be 60% of the difference between the limits. He suggests an upper limit for random error equal to 20% of the difference between the range limits if the purpose is to monitor values in the neighborhood of a normal range or therapeutic range.

Although data accumulated through the operations of internal quality control programs using one of the several available types of control materials are generally suitable for documenting achievement of medical goals, all internal quality control data are insensitive to the effect of nonspecificity of analytic method upon method imprecision. Gowenlock[15] described this component of analytic imprecision years ago. Method nonspecificity may give rise to constant bias or bias which varies with random variation of interfering analytes in the biologic matrix. These effects are not observed in the constant matrix and analyte levels of control materials. They are detected only by an unexpectedly large biologic variation noted with the nonspecific method compared to the biologic variation observed with specific methods.

Table 8
ESTIMATES OF OBSERVED VARIATION EXPECTED WITHIN A HEALTHY INDIVIDUAL SUBJECT FOR CERTAIN ANALYTES RELATED TO CONCENTRATION, TIME AND AUTOMATED METHODS

Analyte	Units	Concentration	Average automated CV_L^a	Approximated Average CV_S^b	CV_D^c	CV_B^d	$CV_D'^e$	$CV_B'^f$
Sodium	mmol/l	140	1.10	0.70	0.74	0.74	1.0	1.4
Potassium	mmol/l	4.5	1.93	1.22	4.1	5.1	4.6	6.0
Chloride	mmol/l	105	1.55	0.98	0.2	1.7	1.0	2.3
Calcium	mg/dl	10.0	2.16	1.37	1.4	1.4	2.0	2.6
Magnesium	mg/dl	2.5	6.10	—	—	—	—	—
Urea nitrogen	mg/dl	20	4.46	2.78	4.1	13	5.0	14
Creatinine	mg/dl	1.0	6.78	4.29	3.1	4.6	5.5	8.0
Glucose	mg/dl	100	3.43	—	—	6.2	—	7.0
Phosphorus	mg/dl	4.5	2.96	1.87	5.0	7.8	5.5	8.5
Uric acid	mg/dl	6.0	2.84	1.80	3.4	7.8	3.8	8.5
Bilirubin	mg/dl	1.0	9.16	5.79	12	25	24	27
Iron	mcg/dl	80	4.58	2.90	10	27	11	27
Cholesterol	mg/dl	250	3.85	2.44	3.5	5.3	4.3	6.5
Triglyceride	mg/dl	130	5.30	—	—	27	—	27
Total protein	gm/dl	7.0	2.30	1.45	1.6	2.7	2.2	3.6
Albumin	gm/dl	4.5	2.97	1.88	2.1	3.1	2.8	4.3
Thyroxine	mcg/dl	9.0	7.70	—	—	8.8	—	12
Alkaline phosphatase	IU/l	Normal	6.3	—	—	6.3	—	9.0
Aspartate aminotransferase	IU/l	Normal	8.7	—	—	16	—	18
Lactate dehydrogenase	IU/l	Normal	6.3	—	—	7.2	—	—
Creatine kinase	IU/l	Normal	12	—	—	56	—	58

Note: For the average laboratory and average subject, results deviating from a prior value by an amount greater than 0.027 CV_D' (Xp) within day and 0.027 CV_B' (Xp) between months are due to analytic variation and biologic variation alone in less than 5% of the cases, where Xp is the prior result (assuming that biologic variation increases or remains the same in disease).

[a] CV_L — Predicted average long-term CV for automated methods at the stated concentration,[28] except the thyroxine CV which is for manual methods. Enzyme CVs are taken from Reference 57.

[b] CV_S — Approximation of short-term CV for automated methods. See discussion of data in Table 4 and Figure 1. Calculated by $CV_S = 0.63\ CV_L$.

[c] CV_D — Average within-day physiologic variation for healthy subjects calculated by averaging time intervals of 8 hr or less from Table 7.

[d] CV_B — Average between-day physiologic variation for healthy subjects calculated by averaging time intervals of 1.4 weeks or greater from Table 7.

[e] CV_D' — Average observed CV expected within a day for a healthy individual subject. Includes CVs and CV_D.

[f] CV_B' — Average observed CV expected over weeks to months for a healthy individual subject. Includes CV_L and CV_B

REFERENCES

1. **Campbell, D. G. and Owen, J. A.,** Clinical laboratory error in perspective, *Clin. Biochem.,* 1, 3, 1967.
2. **Bokelund, H., Winkel, P., and Statland, B.,** Factors contributing to intraindividual variation of serum constituents. III. Use of randomized duplicate serum specimens to evaluate sources of analytic error, *Clin. Chem.,* 20, 1507, 1974.
3. **Van Steirteghem, A. C., Robertson, E. A., and Young, D. S.,** Variance components of serum constituents in healthy individuals, *Clin. Chem.,* 24, 212, 1978.
4. **Williams, G. Z., Harris, E. K., and Widdowson, G. M.,** Comparison of estimates of long-term analytic variation derived from subject samples and control serum, *Clin. Chem.,* 23, 100, 1977.
5. **Weinberg, M. S. and Barnett, R. N.,** Absence of analytic bias in a quality control program, *Am. J. Clin. Pathol.,* 38, 468, 1968.
6. **Allen, J. R., Earp, R., Farrell, E. C., Jr. et al.,** Analytical bias in a quality control scheme, *Clin. Chem.,* 15, 1039, 1969.
7. **Laessig, R. H., Schwartz, T. N., Paskey, T.,** Wisconsin state quality assurance program for multi-channel chemical analyzers, in *Advances in Automated Analysis,* Edrich, M., Ed., Mediad, Tarry-town, N.Y., 1972, 223.
8. **Broughton, P. M. G. and Annan, W.,** Measurement of analytical precision in clinical chemistry, *Clin. Chim. Acta,* 32, 433, 1971.
9. **Steele, B. W., Schauble, M. K., Becklel, J. M., and Beariman, J. E.,** Evaluation of clinical chemistry laboratory performance in 20 VA hospitals, *Am. J. Clin. Pathol.,* 67, 594, 1977.
10. **Rosenbaum, J. M.,** Massachusetts Society of Pathologists Quality Control Program, personal communication, 1973.
11. **Sax, S. M., Dorman, L., Libenson, D. D., et al.,** Design and operation of an expanded system of quality control, *Clin. Chem.,* 13, 825, 1967.
12. **Ross, J. W. and Fraser, M. D.,** The effect of analyte and analyte concentration upon precision estimates in clinical chemistry, *Am. J. Clin. Pathol.,* 66, 193, 1976.
13. **Barnett, R. N.,** in *Quality Control in Clinical Chemistry,* Copeland, B. E., Ed., American Society of Clinical Pathologists, Chicago, 1963, 87.
14. **Cotlove, E., Benson, E. S., and Strandjord, P. E., Eds.,** *Multiple Laboratory Screening,* Academic Press, New York, 1969, 207.
15. **Gowenlock, A. H.,** The influence of accuracy and precision on the normal range, *Ann. Clin. Biochem.,* 6, 3, 1969.
16. **Maclin, E.,** Considerations of analytical goals for the clinical laboratory. An industrial perspective and a systems view, *1976 Aspen Conference,* Elevitch, F. R., Ed., College of American Pathologists, Chicago, 1977, 36.
17. **Gilbert, R. K.,** Progress and analytic goals in clinical chemistry, *Am. J. Clin. Pathol.,* 63, 960, 1975.
18. **Gilbert, R. K.,** CAP Interlaboratory Survey data and analytic goals, *1976 Aspen Conference,* Elev-itch, F. R., Ed., College of American Pathologists, Chicago, 1977, 63.
19. **Kurtz, S. R., Copeland, B. E., and Straumfjord, J. V.,** Guidelines for clinical chemistry quality control based on the long-term experience of sixty-one university and tertiary care referral hospitals. A reappraisal, *Am. J. Clin. Pathol.,* 68, 463, 1977.
20. **Sage, G. W.,** Quality control considerations with the Dupont ACA, *Clin. Chem.,* 20, 698, 1974.
21. **Harris, E. K.,** Statistical principles underlying analytic goal setting in clinical chemistry, *Am. J. Clin. Pathol.,* 72, 374, 1979.
22. **Russell, C. D., LeBlanc, J. H., Jr., and Wagner, H. N., Jr.,** Components of variance in laboratory quality control, *Johns Hopkins Med. J.,* 135, 344, 1974.
23. **Westgard, J. O., Falk, H., and Groth, T.,** Influence of a between-run component of variation, choice of control limits, and shape of error distribution on the performance characteristics of rules for internal quality control, *Clin. Chem.,* 25, 394, 1978.
24. **Gilbert, R. K.,** Survey data, *Chemistry,* College of American Pathologists, Chicago, 1975.
25. **Ross, J. W.,** Great Lakes-Southeastern Regional Quality Control Program, Group Summary Reports, College of American Pathologists Quality Assurance Service, Traverse City, Michigan, June and December 1979.
26. **Vikelsoe, J., Bechgaard, E., and Magid, E.,** A procedure for the evaluation of precision and accuracy of analytical methods, *Scand. J. Clin. Lab. Invest.,* 34, 149, 1974.
27. **Ross, J. W. and Fraser, M. D.,** Analytical clinical chemistry precision: state of the art for fourteen analytes, *Am. J. Clin. Pathol.,* 68, 130, 1977.
28. **Ross, J. W., Fraser, M. D., and Moore, T. D.,** Analytical clinical laboratory precision: state of the art for thirty-one analytes, *Am. J. Clin. Pathol.,* 74, 521 (Supplement), 1980.

29. Barnett, R. N., Medical significance of laboratory results, *Am. J. Clin. Pathol.,* 50, 671, 1968.
30. Harris, E. K., Distinguishing physiologic variation from analytic variation, *J. Chron. Dis.,* 23, 469, 1970.
31. Harris, E. K., Kanofsky, P., Sharkarji, G., and Cotlove, E., Biologic and analytic components of variation in long-term studies of serum constituents in normal subjects. II. Estimating biologic components of variation, *Clin. Chem.,* 16, 1022, 1970.
32. Harris, E. K. and DeMets, D. L., Effects of intra- and inter-individual variation on distributions of single measurements, *Clin. Chem.,* 18, 244, 1972.
33. Pickup, J. F., Harris, E. K., Kearns, M., and Brown, S., Intra-individual variation of some serum constituents and its relevance to population-based reference ranges, *Clin. Chem.,* 23, 842, 1977.
34. Harris, E. K., Some theory of reference values. II. Comparison of some statistical models of intraindividual variation in blood constituents, *Clin. Chem.,* 22, 1343, 1976.
35. Williams, G. Z., Widdowson, G. M., and Penton, J., Individual character of variation in time-series studies of healthy people. II. Differences in values for clinical chemical analytes in serum among demographic groups. by age and sex, *Clin. Chem.,* 24, 313, 1978.
36. Harris, E. K., Cooil, B. K., Shakarji, G., and Williams, On the use of statistical models of within-person variation in long-term studies of healthy individuals, *Clin. Chem.,* 26, 383, 1972.
37. Aguzzi, F., Poggi, N., and Maggi, M., Some data on intra-individual variability in high-density-lipoprotein cholesterol, *Clin. Chem.,* 24, 1671, 1970.
38. Cotlove, E., Harris, E. K., and Williams, Z., Biologic and analytic components of variation in long-term studies of serum constituents in normal subjects. III. Physiological and medical implications, *Clin. Chem.,* 16, 1028, 1970.
39. Morrison, B., Shenkin, A., McLelland, A., Robertson, D. A., Barrowman, M., Graham, S., Wuga, G., and Cunningham, K. J. M., Intra-individual variation in commonly analyzed serum constituents, *Clin. Chem.,* 25, 1799, 1979.
40. Statland, B. E., Bokelund, H., and Winkel, P., Factors contributing to intra-individual variation of serum constituents. IV. Effects of posture and tourniquet application on variation of serum constituents in healthy subjects, *Clin. Chem.,* 20, 1513, 1974.
41. Statland, B. E. and Winkel, P., Relationship of day-to-day variation of serum iron concentration to iron-binding capacity in healthy young women, *Am. J. Clin. Pathol.,* 67, 84, 1977.
42. Statland, B. E., Winkel, P., and Bokelund, H., Factors contributing to intraindividual variation of serum constituents. I. Within-day variation of serum constituents in healthy subjects, *Clin. Chem.,* 19, 1374, 1973.
43. Statland, B. E., Winkel, P., and Bokelund, H., Variation of serum iron concentration in young men: within-day and day-to-day changes, *Clin. Biochem.,* 9, 26, 1976.
44. Statland, B. E., Winkel, P., and Killingsworth, L. M., Factors contributing to intra-individual variation of serum constituents. VI. Physiological day-to-day variation in concentrations of 10 specific proteins in sera of healthy subjects, *Clin. Chem.,* 22, 1635, 1976.
45. Winkel, P., Statland, B. E., and Bokelund, H., Factors contributing to intraindividual variation of serum constituents. V. Short-term day-to-day and within-hour variation of serum constituents in healthy subjects, *Clin. Chem.,* 20, 1520, 1974.
46. Winkel, P., Statland, B. E., and Bokelund, H., The effects of time of venipuncture on variation of serum constituents, *Am. J. Clin. Pathol.,* 64, 433, 1975.
47. Young, D. S., Harris, E. K., and Cotlove, E., Biological and analytic components of variation in long-term studies of serum constituents in normal subjects. IV. Results of a study designed to eliminate long-term analytic deviations, *Clin. Chem.,* 17, 403, 1971.
48. Vanko, M., Selected factors which influence a quality control program, in *Advances in Automated Analysis,* Edrich, M., Ed., Mediad, Tarrytown, New York, 1971, 159.
49. Barnett, R. N., Analytic goals in clinical chemistry: the pathologist's viewpoint, *1976 Aspen Conference,* Elevitch, F. R., Ed., College of American Pathologists, Chicago, 1977, 20.
50. Elion-Gerritzen, W. E., Analytic precision in clinical chemistry and medical decisions, *Am. J. Clin. Pathol.,* 73, 183, 1980.
51. Elevitch, F. R., Ed., *1976 Aspen Conference, Conference Report,* College of American Pathologists, Chicago, 1977, 1.
52. Sinha, T. K., Miller, S., Fleming, J., et al., Demonstration of a diurnal variation in serum parathyroid hormone in primary and secondary hyperparathyroidism, *J. Clin. Endocrinol. Metab.,* 41, 1009, 1975.
53. Engelking, L. R., Dasher, C. A., and Hirschowitz, B. I., Within-day fluctuations in serum bile acid concentrations among normal control subjects and patients with hepatic disease, *Am. J. Clin. Pathol.,* 73, 196, 1980.
54. Glick, J. H., Jr., Expression of random analytic error as a percentage of the range of clinical interest, *Clin. Chem.,* 22, 475, 1976.

55. Elevitch, F. R., Analytical goals in clinical chemistry: their relationship to medical care, proceedings of the Subcommittee on Analytical Goals in Clinical Chemistry, World Association of Societies of Pathology, *Am. J. Clin. Pathol.,* 71, 624, 1979.

56. Glenn, G. C. and Hathaway, T. K., Quality control by blind sample analysis, *Am. J. Clin. Pathol.,* 72, 156, 1979.

57. Lawson, N. S. and Ross, J. W., Long term precision for selected clinical chemistry analytes as determined by data from regional quality control programs, *1976 Aspen Conference,* Elevitch, F. R., Ed., College of American Pathologists, Chicago, 1977.

IMPLEMENTATION OF AN INTERNAL QUALITY CONTROL PROGRAM

Ernest J. Kiser

INTRODUCTION

An internal quality control program is established by performing repetitive analyses of quality control materials in order to define criteria for the acceptability of the analytic runs and to quantitate analytic variability. The quality control material consisting usually of either frozen or lyophilized serum pools should generally be analyzed with each analytical run. An acceptable example of using controls less frequently is in the quality control of the DuPont ACA® where reagent and instrument stability are well-established. Two serum pools having analyte levels at clinically relevant levels should be analyzed. The most prevalent technique utilized for the interpretation of the control data obtained from an analysis is the formation of a Shewhart[1,2] type control chart. Other quality control analysis techniques, including cumulative sum (cusum) charts[2,3] average of patient normals,[4,5] duplicate analysis,[4] and techniques for nonlinear assay curves,[6,7] will be discussed. The purpose of this section is to discuss and outline the methods for establishing an internal quality control program and for utilizing one or a combination of the above interpretation techniques. Since the Shewhart type quality control chart is the most commonly used technique in the clinical laboratory, considerable emphasis will be devoted to the application of this technique.

ESTABLISHING A TARGET MEAN (X) AND STANDARD DEVIATION(S)

To establish a target mean and standard deviation for pool material, the following criteria should be observed:[8]

1.   The pool should have the same matrix as that of patient samples.
2.   If lyophilized material is used, the manufacturer should have data showing that the interval variability is acceptable.
3.   For liquid and lyophilized pools, pool origin must be traceable to the same homogeneous lot of material.
4.   Operator bias should be reduced by analysis of multiple pools in differing sequence.

The minimum requirement for establishing a target mean for an analyte in a stable control materal is that the control material be analyzed at least 20 times on 20 different days.[4] The estimate of the mean is determined by standard statistical methods. Also, the estimate of the standard deviation is determined by the standard equation. The target mean for this baseline period may be better determined by analyzing a minimum of two replicates of each control material dispersed within an analytical run for a period of at least 25 days.[6] This subgroup approach allows for a mean and a range to be determined for each run. The use and selection of grouped data is discussed by Duncan[2] and Grant and Leavenworth.[9] The target mean for this group approach is calculated as the grand mean of the group means. From this data, the estimate of the standard deviation within run (s_w) and between run (s_b) may be calculated. For this grouped data, the total standard deviation is estimated as the square root of the sum

Table 1
PERCENT ERROR (μX100)
ASSOCIATED WITH AN
ESTIMATED STANDARD
DEVIATION

Degrees of freedom	Confidence coefficient (%)		
	0.90	0.95	0.99
10	38	45	60
20	26	31	41
30	21	25	33
40	18	22	28
60	15	18	23
80	13	15	20
100	11	14	18
200	8	10	13
400	6	7	9

of S_w^2 and S_b^2. The existence of s_b can cause difficulty in determining analytical disturbances or errors; therefore, s_b should ideally be reduced.[6,10] Some data also indicates that use of s_w with chi square control rules improves the detection of random error.[10] The sample size is an important factor in determining the reliability of the mean and standard deviation for a baseline data period. The magnitude of the error (μ) in the estimate of the standard deviation(s) is given by the inequality[8]

$$(1 - \mu)\,\sigma < s < (1 + \mu)\,\sigma$$

where σ is the true standard deviation at a specified confidence level. Table 1 illustrates the value of μ for various sample sizes and confidence levels. The data is this table is also graphically shown in Natrella.[11] An example of the interpretation of the data in Table 1 is that for a group of 20 samples there is a 95% probability that σ is within 31% of s.

Any data that is widely discrepant and attributable to some documentable error (outliers) should be eliminated from the data before the determination of the target mean and estimate of the standard deviation. Segregation of these outlier values is important, since the precision of any procedure should include the inherent precision of the measurement technique and the outlier frequency. An outlier is defined as a result, X_o, which lies further than some multiple, m, of standard deviations from the mean. The value of m is related to the number of samples, n, at a 95% confidence level as listed in Table 2.

The following segregation of outliers is then accomplished by completing the following steps:[8]

1. The mean and standard deviation are calculated including all results.
2. Results more than m standard deviations from the mean are segregated.
3. A new mean and standard deviation are calculated from the remaining results.
4. Results more than m times the new standard deviation away from the new mean are segregated.
5. This process is repeated until no more outliers are found.
6. The outlier frequency is determined by dividing the total number of outliers by the total number of results.

Table 2
CRITERIA FOR OUTLIER IDENTIFICATION FOR VARIOUS SAMPLE SIZES, WITH USE OF THE DEFINITION $\overline{X}$ + ms <Xo <X−ms AND A 95% CONFIDENCE LEVEL

n	m
10	2.80
20	3.02
30	3.14
40	3.22
60	3.33
80	3.41
100	3.47
120	3.52
150	3.58
200	3.66
300	3.76
400	3.83

To verify the accuracy of the established mean for an analyte, it is useful to perform multiple analyses of both the control product and Survey Validated Serum available from the College of American Pathologists Survey Program graphically comparing the determined survey serum value to the established survey mean in order to estimate the accurate mean for the analyte in the control product.[12] The control limits for ungrouped or grouped data are generally established as the estimate of the mean plus or minus three times the standard deviation $(X \pm 3s)$.[2,6] If the data is normally distributed, this range covers 99.73% of the distribution. From Tchebycheff's Inequality, this range covers at least $1-(1/K^2)$ for the interval (mean $\pm$ Ks).[6] Since $K = 3$ in our case, the coverage will be approximately 90% of the distribution for any type of distribution (discrete or continuous, symmetric or asymmetric, unimodal or bimodal, etc). From these relationships, it may be ascertained that the three standard deviation limits encompass between 90 and 99.8% of the data for any type of distribution. From the Central Limit Theorem, it may be ascertained that the larger the sample size, the closer the mean of the sample distribution will be to the true population mean. In addition to control limits, warning limits are usually established for a distribution. The warning limits are generally defined as the estimate of the mean plus or minus two standard deviations from the estimated mean.[2,6] For a normal distribution, this range covers 95.45% distribution. Both the three sigma and two sigma limits are used in the formation and interpretation of the Shewhart-type quality control chart.[2,6] The Levey-Jennings quality control program[13] utilizes a Shewhart control chart for grouped data with both mean and range charts being used in the clinical laboratory. A discussion of control charts for range, number defective, and number of defects per unit is given by Duncan.[2]

CONTROL CHART FORMATION

After collection of the base period data and determination of its mean and standard deviation, a working control chart for the estimate of the mean is created. Figure 1

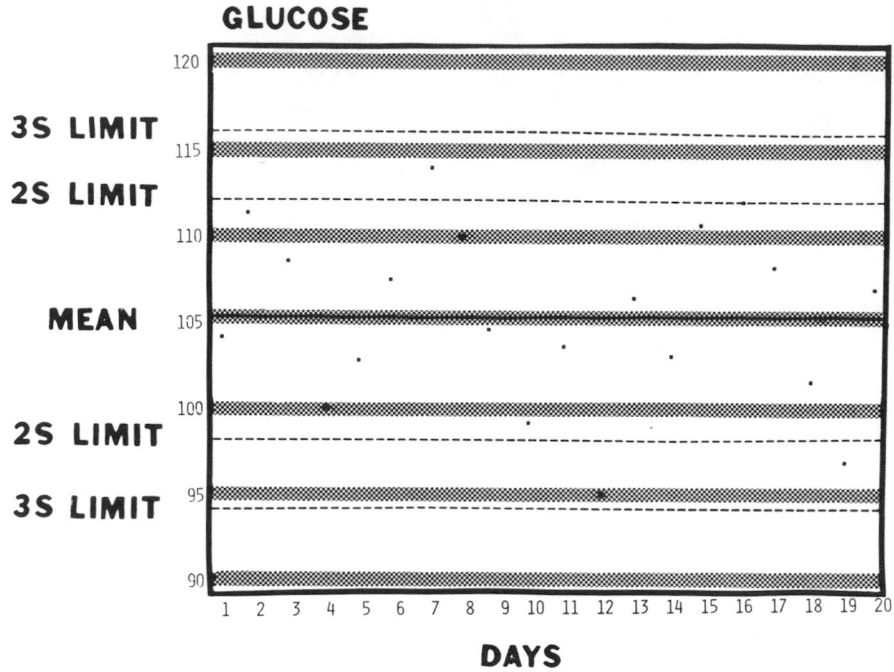

FIGURE 1.   Shewhart quality control chart for glucose with analysis under control.

shows a typical 20-day control chart for glucose with appropriate control limits diagrammed. As more data is collected for this analyte, it is plotted on the control chart and analyzed as to its acceptability. Figure 2 illustrates a continuation of the initial control chart for glucose with out-of-control runs to demonstrate the following situations in which the analysis procedure should be investigated and the results of patient samples retained:[2,14]

1.   One or more points outside the three standard deviation limit
2.   A run of seven or more increasing (or decreasing) points
3.   Two points exceeding the two standard deviation warning limit

A study utilizing computer simulation for testing the application of control rules singularly or in combination to optimize error detection (p_{ed}), while minimizing false rejections (p_{fr}), has been described by Westgard et al.[15] The decision which must be determined for a laboratory is what value of p_{fr} is tolerable. A conservative value of 0.01 is appropriate to minimize the number of false rejections in a clinical laboratory. For collection of only one or two observations per run, a combination of one control exceeding the 3s limit and the cumulative sum limit (see cumulative sum decision limit in this section) appear to provide good performance for this small number of samples. One control value exceeding the 3s limit is responsive to random and systematic error, while the cumulative sum limit is responsive to systematic errors. A control exceeding the 2s warning limit should ideally cause results to be held until another run is performed. If on the second run the control exceeds 2s, the process should be stopped and investigated. If the control of the second run does not exceed 2s, the results for both runs may be released. Westgard and Groth[16] also provide insight into the use of various control limits singularly, or in combination for detection of systemic or ran-

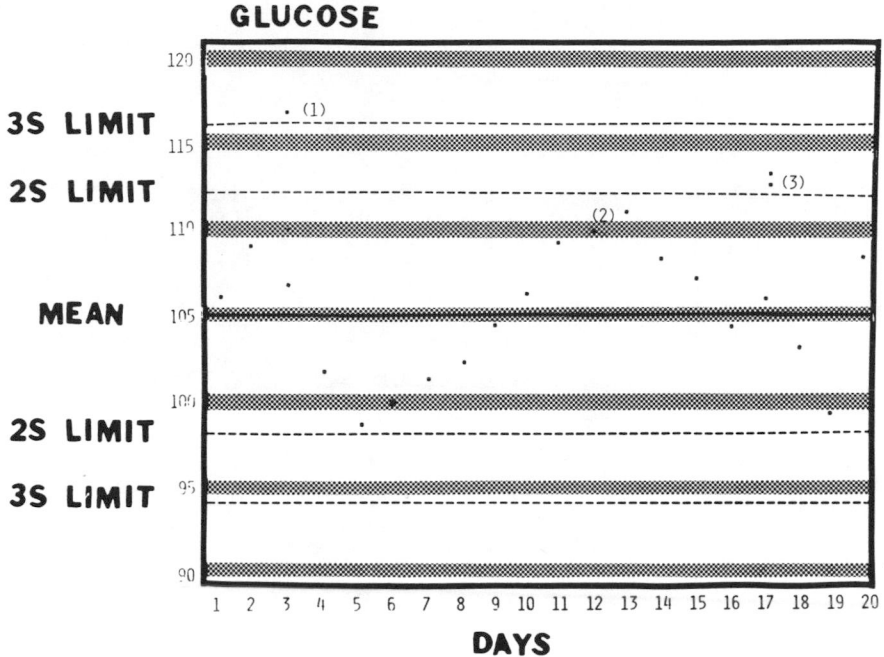

FIGURE 2. Shewhart quality control chart for glucose with example of out-of-control runs.

dom error, by utilizing power functions of control rules together with computer simulation. These functions demonstrate the relationship between the probability that a control rule will generate a rejection signal and the size of analytical errors that are to be detected. The availability of a real time computer system would be most beneficial for the implementation of this type of quality control.

CUMULATIVE SUM (CUSUM)

Background

Another quality control system which does not require the determination of a standard deviation is the cusum technique.[2,3] This is a useful technique when prior data are not available for an assay, since it may be initiated immediately. The method involves subtraction of a target value from the day's observed value. This difference obtained daily is algebraically added to the previous day's difference and plotted as the cumulative difference from the target value. As in the case of Shewhart charts, cusum charts may be constructed for means, ranges, variances, or number defective.[2] Traditionally, interpretation of an out-of-control value on a cusum chart is determined by using a V-shaped mask. The construction of such a mask for the mean is discussed in Duncan[2] and in this section. On a practical basis, six successive points increasing or decreasing from the center line are considered indicative of an uncontrolled system.

Cumulative Sum Control Charts (CSCC)

The mean of the ith sample of size n is denoted by $\bar{x}_i$. We plot on the control chart points with the coordinates (m, Ym) where m is the sample number and

$$Ym = \sigma_{\bar{x}}^{-1} \sum_i (\bar{x}_i - \mu_o)$$

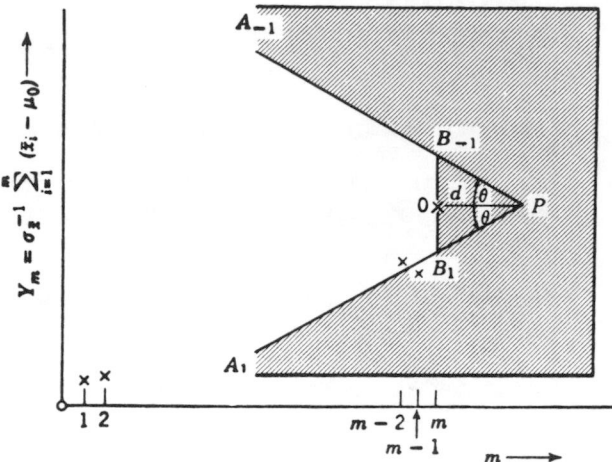

FIGURE 3. Mask for cumulative sum control chart for the mean.

with μ_o equal to the proposed "target mean". The chart is interpreted by placing a mask, which is the shaded area of Figure 3, over the chart with the point O over the last point plotted on the chart, with the line OP horizontal. The process is said to be out of control if any points of the CSCC are covered by the mask. If any points lie below the straight line A_1B_1, is regarded as an indication of increase in the process average; if any points lie above $A_{-1}B_{-1}$, a decrease is indicated. (For each chart, an appropriate scaling factor is utilized. This will be considered later in the section).

In order to determine the dimensions of the mask, we must calculate the size 2θ of the angle $B_{-1}PB_1$, and the length, d, from O to the vertex, P, of this angle in Figure 3. To do so, we first decide on the size of errors of the first and second kind. (As long as β is very small, less than 0.01, say, this can be deleted from the exact formula.) The approximate formulas for θ and d are

$$\theta = \text{arc tan} (\delta/2)$$

$$d = 2\delta^{-2} \ln \alpha \tag{1}$$

where $\delta = D/\sigma_{\bar{x}}$ and "D" is the least size shift (in either direction) which it is desired to detect with fair certainty $(1 - \beta$, or power). The probability of error of the first kind for a two-sided test is taken here as 2α and for a one-sided test as α.

The procedure for a CSCC on means is as follows:

1. Decided on α (or 2α) and D
2. Determine θ and d from Equation 1
3. Plot the sequential points Y, moving the mask with each successive point
4. At each step make a decision to (a) continue the test; (b) accept a shift of $\mu_o + D$; or (c) accept a shift of $\mu_o - D$.

If σ (and hence $\sigma_{\bar{x}}$) is not known or assumed before the test, this can be estimated by

$$s_p = \sqrt{\Sigma s_i^2 / m}$$

Table 3
CUMULATIVE SUM CONTROL LIMITS FOR SAMPLE MEANS

Values of d

δ	θ	$2\alpha = 0.10$ $\alpha = 0.05$	$2\alpha = 0.05$ $\alpha = 0.025$	$2\alpha = 0.02$ $\alpha = 0.01$	$2\alpha = 0.01$ $\alpha = 0.005$	$2\alpha = 0.0027$ $\alpha = 0.00135^c$	$2\alpha = 0.002$ $\alpha = 0.001$	$2\alpha = 0.001$ $\alpha = 0.0005$
0.2	5° 43′	149.8	184.4	230.6	264.9	330.4	345.4	380.0
0.4	11°19′	37.4	46.1	57.6	66.2	82.6	86.3	95.0
0.6	16°42′	16.6	20.5	25.6	29.4	36.7	38.4	42.2
0.8	21°48′	9.36	11.5	14.4	16.6	20.6	21.6	23.8
1.0	26°34′	5.99	7.38	9.21	10.6	13.2	13.8	15.2
1.2	30°58′	4.16	5.12	6.40	7.36	9.18	9.59	10.6
1.4	35°0′	3.06	7.76	4.70	5.41	6.74	7.05	7.76
1.6	38°40′	2.34	2.88	3.60	4.14	5.16	5.40	5.94
1.8	41°59′	1.85	2.28	2.84	3.27	4.08	4.26	4.69
2.0	45°41′	1.50	1.84	2.30	2.65	3.30	3.45	3.80
2.2	47°44′	1.24	1.52	1.90	2.19	2.74	2.85	3.14
2.4	50°12′	1.04	1.28	1.69	1.84	2.29	2.40	2.64
2.6	52°26′	0.89	1.09	1.36	1.57	1.95	2.04	2.25
2.8	54°28′	0.76	0.94	1.17	1.35	1.69	1.76	1.94
3.0	56°19′	0.67	0.82	1.02	1.18	1.47	1.54	1.69

[a] Two-sided test.
[b] One-sided test.
[c] These are comparable to the "3α" limits used in the Shewhart chart, i.e., $2\alpha = 0.0027$ and $\alpha = 0.00135$.

where s_i^2 is the variance from sample i, and m is the number of samples. As long as the number of degrees of freedom $\mu = m(n - 1)$ is greater tha 30 or 40, this estimate by s_p is quite reliable.

An alternative to calculating the values of θ and d is to make use of Table 3. For selected values of δ and α, the values of θ and d can be taken directly from this table.

To attempt to calculate the dimensions of the mask is somewhat meaningless unless the appropriate scaling factor is brought into these dimensions. This is different from the Shewhart chart where only the vertical scale is important. Let k units for the ordinate be equal to one unit for the abscissa. Then the equations for θ^* and d^* become

$$\theta^* = \text{arc tan } (D/2k)$$

$$d^* = -2\delta^{-2} \ln \alpha = d$$

Note that if Table 3 is to be used for the determination of the dimensions of the mask for θ^* and d^*, it must be entered twice as follows:

1. To select θ^*, choose the row for D/k rather than δ
2. To select d^* (for a designated α or 2α) choose the appropriate row for δ and appropriate column for α or 2α

Another approach for interpretation of cusum control chart system is the decision limit method.[3,17] For this case, the cusum is compared with a numerical limit rather than utilizing a V-shaped mask. The cusum is not initiated until a reference point (k) determined half way between the mean for the process in control and the mean when the process is considered out of control, is exceeded. After exceeding the reference point, the cusum is calculated as the difference between the exceeding observation and

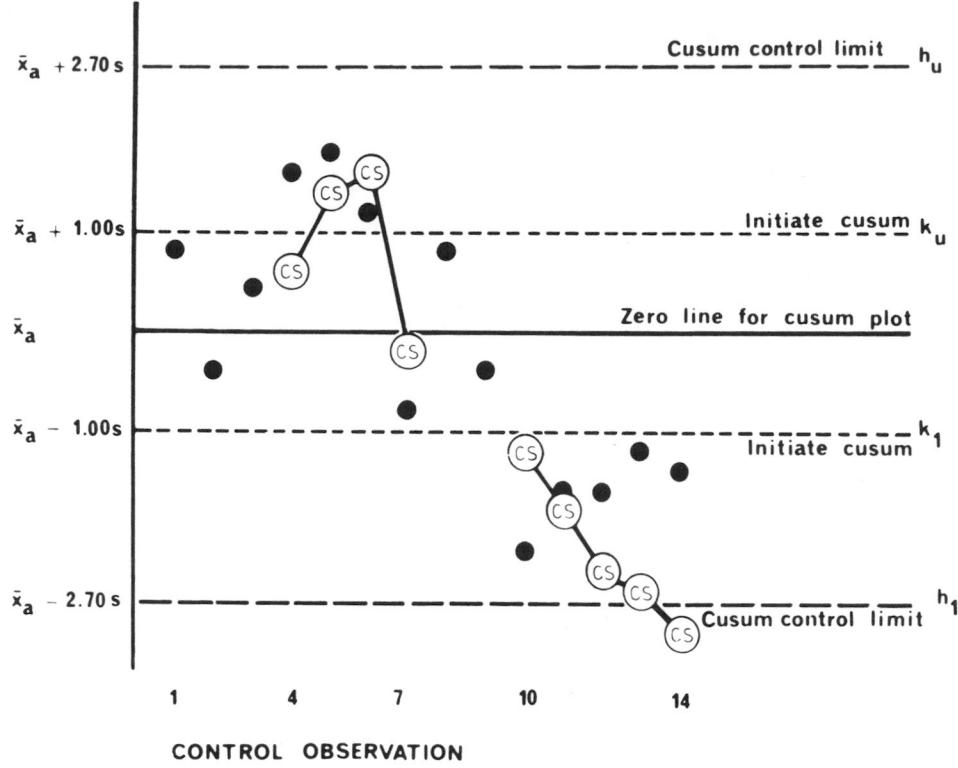

FIGURE 4.   Decision limit cusum chart for which a tabular data record is not required.

the reference point. This process continues until the cusum changes sign (process is in control) or until the cusum exceeds a control limit (h) (process is out of control). Figure 4 shows an example of this approach using the estimate of the mean $\overline{X}\alpha$ as the zeroline, $(\overline{X}\alpha \pm 1.05s)$ as the reference points, and $\overline{X}\alpha \pm 2.70s)$ as the control limits. This decision limit cusum chart method may be simply combined with the Shewhart chart by eliminating the 2s Shewhart warning limit and by combining the 3s Shewhart control limit with the cusum control limit as demonstrated by Westgard et al.[17] From computer simulation studies, it appears that this combined Shewhart cusum control chart improves the detection of systemic errors and aids in distinguishing between random and systemic errors in solving out-of-control problems.[17]

AVERAGE OF PATIENT NORMALS

This technique of quality control involves determining the normal range in an unselected population for an analyte as described by Hoffman and Waid.[5] The normal range for a population is generally considered to be $\overline{X} \pm 2s$, for a normally distributed analyte. To apply this method, one calculates the mean of the normal patient values obtained each day and plots them on a chart constructed with the mean of the normal range as the center line and plus or minus two times the standard error (SE) of the average for groups of normal values as the control limits. The standard error of the average for groups of normal values is derived as $SE = s/\sqrt{n}$ where n is the usual number of normal values per day. The effectiveness of error detection for this method increases as the number of normal daily patient values increases, with at least 16 being

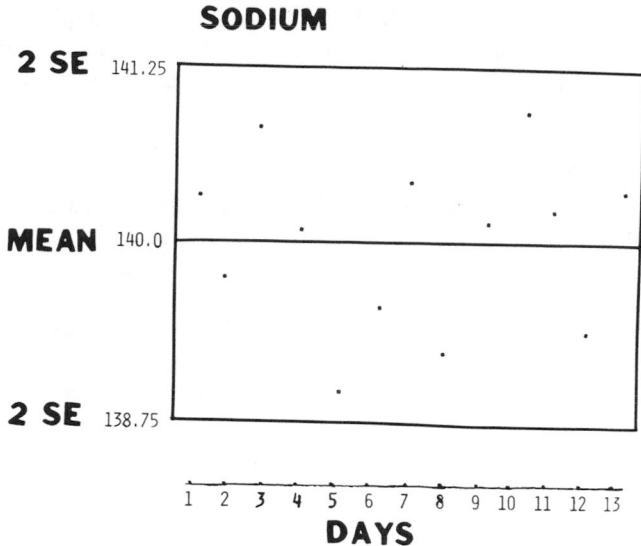

FIGURE 5. Average of normals control chart for sodium with a normal range of 135 to 145 meq/ℓ and 16 normal patient values per day.

required, which yields a 90% probability of detecting a 0.72s shift in the mean.[4,18] Using sodium as an example of a normally distributed analyte with a normal range of 135 to 145 meq/ℓ and an n equal 16, Figure 5 shows a control chart created for use with this method. The average of normals approach has been criticized for lack of sensitivity and on a statistical basis.[2] To improve this technique, it has been recommended that the same number of patient normals be used daily to calculate the mean and that the control limits should be computed without reference to a range estimate. The greatest merit of the use of the average of normals concept is that it attempts to control the whole analytic process from sample collection to the final result.

DUPLICATE MEASUREMENTS[4]

Duplicate analysis was once considered the best method for assuring accuracy of results. For some procedures where significant analyte loss is possible, e.g., extraction, duplicate analysis, especially "blind duplicate analysis", is useful for assessing precision only if each sample is treated independently and is dispersed throughout a run. Where identical samples are analyzed at the same time, the samples value may be biased by foreknowledge of the expected results. If the measurement is biased from the true mean, duplicate samples provide little information regarding the accuracy of the measurement.

IMMUNOASSAYS

Besides using control material together with the previously discussed interpretation techniques for immunoassay results, different criteria are required to evaluate the quality of nonlinear standard curves. Rodbard outlines a fairly extensive procedure for evaluating radioimmunoassay standard curves with the aid of computer curve analysis techniques.[7] Quality control of immunoassay may be approached similarly to other assays with utilization of other control techniques including duplication and dilution. From a practical standpoint for laboratories without computer capabilities, it is recommended that at least one or more of the assays parameters, response for zero dose (B_o), bound to total ratio for zero dose $(B/T)_o$ dose resulting in 50% displacement of the labeled antigen bound for zero dose (0.5 B/Bo), be recorded and observed.[6,7,20] The parameter, $(B/T)_o$ is generally considered to be the most useful parameter to monitor, since it is sensitive to antibody affinity, damage to labeled antigen, and changes in assay conditions. For laboratories with computerized data reduction capabilities, the value of the estimate of residual variance of the standard errors of the fitted parameters and of the confidence limits for the curve should be determined and observed by charting for each analyte.[21] Another approach of internal quality control of radioimmunoassay is the use of a modified cusum technique which is similar to other cusum techniques previously discussed.[22] With the advance of various types of immunoassay, further definition of control parameters for nonlinear standard curves will evolve.

REFERENCES

1. **Shewhart, W. A.,** *Economic Control of Quality Manufactured Product,* Van Nostrand, New York, 1931.
2. **Duncan, A. J.,** *Quality Control and Industrial Statistics,* Richard D. Irwin, Homewood, Ill., 1965.
3. **Davies, U. L. and Goldsmith, P. L.,** *Statistical Methods in Research and Production,* 4th ed., Hafner Publishing, New York, 1972.
4. **Barnett, R. N.,** *Clinical Laboratory Statistics,* 2nd ed., Little, Brown, Boston, 1979.
5. **Hoffman, R. G. and Waid, M. E.,** The "average of normals" method of quality control, *Am. J. Clin. Pathol.,* 43, 134, 1965.
6. **Inhorn, S.L.,** *Quality Assurance Practices for Health Laboratories,* American Public Health Association, Washington, D.C., 1978.
7. **Rodbard, D.,** Statistical quality control and routine data processing for radioimmunoassays and immunoradiometric assays, *Clin. Chem.,* 20, 1255, 1974.
8. **Burnett, R. W.,** Accurate estimation of standard deviations of quantitative methods used in clinical chemistry, *Clin. Chem.,* 21, 1935, 1975.
9. **Grant, E. L. and Leavenworth, R. S.,** *Statistical Quality Control,* McGraw-Hill, New York, 1972.
10. **Westgard, J. O., Falk, H., and Groth, T.,** Influence of a between-run component of variation, choice of control limits, and shape of error distribution on the performance characteristics of rules for internal quality control, *Clin. Chem.,* 25, 394, 1979.
11. **Natrella, M. G.,** Experimental Statistics, National Bureau of Standards Handbook 91, U.S. Government Printing Office, Washington, D.C., 1963.
12. **Grannis, G. F.,** Use of survey-validated reference materials (survey serum) to establish target values of quality control pools, *Am. J. Clin. Pathol.,* 70, 580, 1978.
13. **Levey, S. and Jennings, E. R.,** The use of control charts in the clinical laboratory, *Am. J. Clin. Pathol.,* 20, 1059, 1950.
14. **Haven, G. T.,** Outline for quality control decisions, *Pathologist,* 28, 373, 1974.
15. **Westgard, J. O., Groth, T., Aronsson, T., Falk, H., and de Verdier, C. H.,** Performance characteristics of rules for internal quality control: probabilities for false rejection and error detection, *Clin. Chem.,* 23, 1857, 1977.
16. **Westgard, J. O. and Groth, T.,** Power functions for statistical control rules, *Clin. Chem.,* 25, 863, 1979.
17. **Westgard, J. O., Groth, T., Aronsson, T., and de Verdier, C. H.,** Combined Shewhart-Cusum control chart for improved quality control in clinical chemistry, *Clin. Chem.,* 23, 1881, 1977.
18. **Hoffman, R. G.,** Patients' tests for quality control, *Clin. Chem.,* 15, 533, 1969.
19. **Reed, A. H.,** Use of patient data for quality control of clinical laboratory tests, *Clin. Chem.,* 16, 129, 1970.
20. **Walker, W. H. C.,** An approach to immunoassay, *Clin. Chem.,* 23, 384, 1977.
21. **Rodbard, D., Lenox, R. H., Wray, H. L., and Ramseth, D.,** Statistical characterization of the random errors in the radioimmunoassay dose-response variable, *Clin. Chem.,* 22, 350, 1976.
22. **Kemp, K. W., Nix, A. B. J., Wilson, D. W., and Griffiths, K.,** Internal quality control of radioimmunoassays, *J. Endocrinol.,* 26, 203, 1978.

REGIONAL QUALITY CONTROL PROGRAMS

Jerald M. Rosenbaum

INTRODUCTION AND HISTORY

A Regional Quality Control Program (RQCP) is composed of a group of laboratories, organized to purchase quality control materials and data processing services. Since the first RQCPs were initiated in 1967 in Colorado and then Massachusetts, programs of this type have grown dramatically in size. Initially targeted at clinical chemistry laboratories, these programs have subsequently diversified to include specialized disciplines within the broad scope of clinical chemistry, as well as other laboratory disciplines.

REGULATORY REQUIREMENTS FOR DAILY QUALITY CONTROL

The "Partnership For Health Amendments of 1967", better known as the Clinical Laboratories Improvement Act of 1967, formalized for the first time the federal government's role to ensure that clinical laboratories, to be licensed, must establish and maintain a quality control program adequate to assure the "accuracy of laboratory procedures and services". This requirement must be met before a federal license can be issued to laboratories affected by that Act, which applied by various individual states to laboratories within their borders and, in 1978, similar standards were established by the Health Care Finance Agency for the Medicare Program. Since most laboratories accept Medicare patients and Medicare funds, requirements for maintenance of a quality control program apply in a uniform fashion to most of the nation's clinical laboratories. These federal licensure standards mandate an internal quality control program which includes daily monitoring of all laboratory tests, documentation that these tests have been routinely monitored, and guidelines or rules that personnel must follow when the quality control program indicates that a test or tests is out-of-control. This entire program is termed internal process control. Replicate testing of commercially supplied lyophilized quality control pools is the most prevalent form of internal process control.

REPLICATE TESTING PROGRAMS AS THE BASIS OF REGIONAL CONTROL PROGRAMS

Since quality control samples in a replicate testing program are part of a single pool of material, it is advantageous to develop large pools of control materials for use by many laboratories. Each laboratory in these programs establishes a target mean value for each measured analyte, as well as limits of acceptability for that procedure, generally based upon a multiple of the observed standard deviation.

Introduction of high speed data processing by computers in the last decade has promoted, usually on a shared basis, the appropriate data processing needed to round out the internal process control program, i.e., regularly scheduled printouts for examination and documentation.

ORGANIZATION OF THE TYPICAL RQCP

The RQCP provides control material and data processing resources to its participating members. To accomplish this, a professional society, such as a state or regional pathologists' group, typically appoints a steering committee with an executive director or regional coordinator, to negotiate with potential vendors of control materials and data processing services. Furthermore, the executive director, on behalf of the steering committee, organizes the regional group of consumer laboratories and operates the program on a daily basis. The RQCP may also provide educational seminars on quality control topics for its members.

The typical RQCP is organized on a regional or statewide, rather than national basis, since each RQCP in clinical chemistry already obtains the largest serum pools produced by commercial vendors. At the present time, maximum pool size is approximately 1500 ℓ per pool for lyophilized products. Thus, a relatively small group of laboratories, approximately 200 to 400, will consume a pool of this size in approximately 12 to 18 months. While pool-producing capacity limits the size of the clinical chemistry regional quality control group, it does not seriously compromise its usefulness.

REGIONAL QUALITY CONTROL PRODUCTS

Purchasing power of hundreds of laboratories is greater than that of a single laboratory, providing incentive for commercial suppliers to provide better quality control materials at lower prices than can be otherwise obtained. Further, by using unassayed control pools, more than half the cost of such commercial materials can be saved. Target values for individual analytes are subsequently defined by participant use.

The manufacturer provides sufficiently large quantities of control products to meet the needs of the participating laboratories for a year or more. In addition, the program can specify and customize the number and concentration of analytes in the control product. Regional specifications and the resulting bidding process by commercial vendors has led to significant improvement in products provided by industry. Advantages of long lasting quality control pools include smoother operation of the entire program and less frequent need for individual laboratories to establish new target values and confidence limits.

COMPUTERIZED DATA PROCESSING IN REGIONAL PROGRAMS

The data processing service of the regional quality control program provides on a regular basis, usually monthly, precision statistics and control limits to participating laboratories. These data are presented in a computerized report along with the laboratory's previous statistics for the immediate past few months and cumulative quality control data for the life of the present control pool(s) (Figure 1). The computer printout provides documentation for the laboratory and for any third party that an ongoing internal quality control program is operating in the laboratory.

The computer's access to the precision data for many laboratories allows for the first time the collection of data from hundreds or thousands of laboratories regarding the day-to-day reproducibility of laboratory procedures. These precision statistics indicate the state-of-the-art precision of laboratory tests and they are updated with each monthly print-out. They provide laboratory directors and supervisors, as well as private third-party or government overseers of laboratory quality, with useful bench mark estimates of test precision on a day to day and month to month level, by method, for many analytes.

Since 100 to 400 laboratories in a RQCP are utilizing the same serum pool, it is possible for the computer collating the information it receives from participating laboratories to formulate pool analyte target values (mean values) in addition to precision

YOUR LABORATORY

FILE NO.	CONSTITUENT GENERAL METHOD SPECIFIC METHOD UNIT OF MEASURE LOT DESCRIPTION	LINE TITLE	LOT THRU FEB.	THIS MONTH FEB.	YOUR MONTHLY STATISTICS PREVIOUS MONTHS		
					FIRST JAN.	SECOND DEC.	THIRD NOV.
001	CALCIUM	MEAN	8.6	8.6	8.6	8.7	8.6
	CRSLPH CPLXN AUT	S.D.	.12	.12	.11	.13	.12
	ACME INSTRUMENT	C.V.	1.4	1.4	1.3	1.5	1.4
	MG/DL	RSULTS	102	30	31	22	21
	GL/SE/SPXL376DD	S.D.I.	- .22	- .21	- .54	-.25	-.40
002	CALCIUM	MEAN	10.7	10.7	NO	NO	10.7
	CRSLPH CPLXN AUT	S.D.	.19	.16	DATA	DATA	.14
	ACME INSTRUMENT	C.V.	1.8	1.5	RECD	RECD	1.3
LOW	MG/DL	RSULTS	55	31			24
	GL/SE/SPXP9574DD	S.D.I.	- .20	- .13			-.15
003	GLUCOSE	MEAN	208	208	NO	NO	NO
	HEXOKINASE AUT	S.D.	7.5	7.5	DATA	DATA	DATA
	EXCLO INSTRUMENT	C.V.	3.6	3.6	RECD	RECD	RECD
	MG/DL	RSULTS	16	16	0	0	0
	GL/SE/SPXP9574DD	S.D.I.	.14	.13			
003 OLD	GLUCOSE	MEAN	229	228	231	235	227
	HEXOKINASE AUT	S.D.	9.8	5.9	9.5	6.3	7.9
	EXCLO INSTRUMENT	C.V.	4.3	2.6	4.1	2.7	3.5
	MG/DL	RSULTS	386	12	28	29	25
	S.E. CHEM 76A DD	S.D.I.	1.22	.87	1.50	2.22	.72

FIGURE 1.   Individual laboratory statistics from the CAP-QAS data summary.

data for the various reagent/instrument combinations used by particpants. For instance, the condensed group summary report of the CAP-QAS program provides summarized information by general methods (defined as reagent system or instrument class), specific methods (specific instrument or kit manufacturer), and all methods for the analytes in the RQCP data base (Figure 2). Similar to the assay sheet provided with commercially available assayed quality control products, this report facilitates method comparisons with reference to method bias. It surpasses the assay sheet in usefulness by also offering average experienced standard deviations and coefficients of variation for all reported method combinations, thereby, providing both useful accuracy and precision data for use when laboratories select instruments or reagent kits. The quality of RQCP data is superior to assay sheets because of the large number of data points contributing to the RQCP statistics.

ANALYTICAL ACCURACY

The condensed group summary report (Figure 2) provides participant-generated target values for all analytes by method. In addition to this, there has been preliminary effort of some regional control programs to obtain definitive method or reference method analysis of serum pools for selected analytes.[1] While definitive or reference methodology provides numbers of potential high validity and low bias, these services are not obtainable on a regular basis and may be very costly when available. They are, thus, likely to remain of secondary importance to participant results as a source of target values for interlaboratory quality control programs.[1]

GROUP SUMMARY REPORT

		COLLEGE OF AMERICAN PATHOLOGISTS									PAGE 1
LOT TO DATE		QUALITY ASSURANCE SERVICE									
THRU JAN 1977		GROUP SUMMARY REPORT									
		ANY SOC OF PATH REGIONAL Q/C PROGRAM								COPYRIGHT 1977 C.A.P.	
CONSTITUENT	::	CHEMISTRY POOL 1-77					::	CHEMISTRY POOL 11-77			::
UNIT OF MEASURE											
GENERAL METHOD	AVG.	S.D. OF	AVG.	AVG.	NUMBER	NUMBER	AVG.	S.D. OF	AVG.	AVG.	NUMBER NUMBER
SPECIFIC METHOD	MEAN	MEANS	S.D.	C.V.	VALUES	FILES	MEAN	MEANS	S.D.	C.V.	VALUES FILES
CALCIUM											
MG/DL											
CRESOLPHCPLX AUT											
PROGCHEM	9.8	.06	.27	2.8	1994	7	11.8	.09	.29	2.5	2085	7
DPNT ACA	9.6	.08	.25	2.6	2875	10	12.0	.11	.46	3.8	2681	9
HYC MK X	9.5	.06	.26	2.7	2631	9	12.1	.08	.39	3.2	2733	9
HYCEL 17	9.5	.07	.24	2.5	2945	10	11.8	.12	.41	3.5	2541	9
HYC SP17	9.7	.05	.18	1.9	2410	8	12.1	.09	.36	3.0	2384	8
TECH AAI	9.7	.07	.19	2.0	2290	8	11.7	.13	.43	3.7	2136	7
TECHAAII	9.6	.08	.20	2.1	2007	7	12.0	.08	.38	3.2	2008	7
TECH SMA	9.4	.09	.34	3.6	2647	9	11.9	.13	.42	3.5	2773	9
TECHSMAC	9.7	.09	.31	3.2	2232	7	11.6	.21	.34	2.9	2103	7
ALL SPEC. METHODS	9.6	.14	.25	2.5	22031	75	11.9	.20	.39	3.2	21444	72
ALKALINE HPE											
AMERICAN MONITOR	9.5	.35	.35	3.6	686	5	11.6	.44	.44	3.8	670	5
ALL SPEC. METHODS	9.5	.35	.35	3.6	686	5	11.6	.44	.44	3.8	670	5

FIGURE 2.   Condensed group summary report from the CAP-QAS program.

Table 1
DAILY DECISIONS — TWO POOL PROGRAMS

I.   Accept run if
    A.   Both controls read within ± 2 SD of your established mean
    B.   One control reads within ± 2 SD; other control between 2 to 3 SD (once only)

II.   Reject run if
    A.   One control is greater than ± 3 SD from the mean
    B.   Both controls are greater than ± 2 SD from the mean
    C.   One control is between 2 to 3 SD on two successive runs

III. If the run is rejected
    A.   Hold patient results
    B.   Reconstitute fresh control specimen
    C.   Repeat control determinations using both old and new control pools and up to five patient specimens from the run.
        1.   If the new (freshly reconstituted) control specimen is now "in control" (within ± 2 SD of the mean value for that constituent as determined in your laboratory), compare the patients' repeated values with their original values; if the two values from each patient's specimen show a percentage difference less than twice your laboratory's CV for that constituent on control specimens, consult with supervisor to release patient results; if the patient values have shown a greater percentage change, repeat the entire group of patient samples
        2.   If control values remain unacceptable, troubleshoot procedures in conjunction with supervisor
        3.   Discard any control vials which have erroneous values for any constituent

IV. Examine procedure
    A.   If seven successive control samples (either level) are consistently above or below the mean, even if runs are in control on all occasions

USING A RQCP IN THE LABORATORY

Using the resources of a RQCP, a working system for daily quality control may be established in the laboratory as follows:

1.   Obtain sufficient quality control material for 12 months minimum of daily analyses. At the same time, submit descriptions of your test methods to the data processing service.

2.   Establish the means and control limits either by in-house calculation or by submitting results of the daily quality control analyses to the data processing service.

3.   Levey-Jennings wall charts displaying the laboratory's selected confidence limits or control limits are also provided. Using these charts, plot subsequent control values daily. Observe each day whether consistent drifts, shifts, or erratic results are occurring and require study. Apply a consistent decision scheme[2] to control results (Table 1).

4.   Using the RQCP monthly print-out, compare your CV with the average CV of all laboratories using methods similar to yours. If your CV is substantially higher, method reevaluation and/or method change may be indicated.

5.   To determine the bias of your method, compare the mean results of your method to that of the appropriate comparative group, be it all the methods mean or a selected comparative method mean. These comparisons are readily made on the CAP-QAS print-out as Standard Deviation Interval (SDI) (Figure 1). The SDI is defined as laboratory mean minus mean of laboratory means divided by the standard deviation of laboratory means. In general, any SDI values greater than ± 2 require explanation and possible corrective action. While method bias evaluation by SDI comparison in RQCPs is useful, a better understanding of method

bias is obtained when these results are combined with the SDI results from proficiency test surveys, whose SDI are generally more reliable than those in RQCPs.

Quality control systems fail because of inattention, inadequate training, undefined decision schemes, failure to follow decisions schemes, and the pressure of the workload superseding quality control efforts. Regional quality control programs, emphasizing a systems approach, eliminate much of the time-consuming quality control calculations and, through regional education, foster an appreciation for the benefits of quality control in the clinical laboratory.

REFERENCES

1. **Gilbert, R. K. and Rosenbaum, J. M.**, Accuracy in interlaboratory quality control programs, *Am. J. Clin. Pathol.*, 72, (Suppl. 260), 1979.
2. **Haven, Guy T.**, Outline for quality control decisions, *Pathologist*, 38, 373, 1974.

EXTERNAL QUALITY CONTROL PROGRAMS

Guy T. Haven

DEFINITION

These programs, also termed proficiency testing or survey programs, entail periodic provision of a series of unknown samples to hundreds or thousands or laboratories, typically for analysis of multiple analytes. After each laboratory completes the analyses, the results are returned to a central computer facility for compilation and comparison of each laboratory's results to those obtained by peer or reference laboratories.[1]

PURPOSES

These programs assess the "state-of-the-art" for specific analyses, the standard deviation of all results reflecting interlaboratory precision.[2] If the true content of an analyte is known, comparison with participant results reflects "state-of-the-art" accuracy.

The individual laboratory's results obtained on these unknown specimens can be compared with peer laboratories, reference laboratories, or defined performance criteria. The programs supplement internal quality control procedures in the laboratory and serve as an educational stimulus to improve laboratory performance.

Many of the programs allow detailed study of the performance of individual methods and instruments. The programs also yield consensus values for specimens which can subsequently be provided to laboratories for use in the evaluation of analytic methods and the classification of laboratory errors.

SOURCES

Proficiency testing programs in chemistry are provided by private associations, the Center for Disease Control, states, and private industry. The most widely accepted comprehensive programs are provided by the College of American Pathologists (CAP), the Center for Disease Control (CDC), the American Association of Bioanalysts (AAB), and the American Society for Internal Medicine (ASIM).

CAP surveys are available in most major areas of clinical chemistry and are accepted in lieu of the CDC proficiency testing program and state programs in 35 states. The CDC proficiency testing program is accepted in lieu of state programs by 31 states. The AAB program is accepted in lieu of state proficiency testing programs by 29 states, while the ASIM program, directed at clinicians' office laboratories, meets proficiency testing requirements in 22 states.

Programs provided in specific limited chemistry disciplines include the American College of Nuclear Physicians Radioassay Program, the California Thoracic Society program in blood gas analysis, the American Association for Clinical Chemistry's (AACC) therapeutic drug monitoring program, the Proficiency Testing Service (PTS), and several programs provided by reagent and instrument manufacturers. Programs provided by the AACC, PTS, and commercial concerns are specifically directed toward education and laboratory improvement, not being designed to fulfill federal or state regulatory requirements. An international program directed at laboratory improvement is also provided by Burroughs Wellcome, primarily to laboratories outside the U.S.

Proficiency testing programs in chemistry are also provided by 15 states. In 8 of these states, participation is mandatory, while in 7 states, other federal or private programs are accepted in lieu of the state program.

Generally, of greatest benefit to the laboratory and regulatory agencies are the CDC, program as well as the larger programs provided by private agencies which develop large data bases for interlaboratory comparison.

THE SAMPLE

Lyophilized serum is most widely used because of its homogeneity, stability, and large lot size. In specialized areas, e.g., blood gas analysis, other homogeneous and stable samples are utilized. The unknown samples are forwarded to all participating laboratories simultaneously.[3]

The volume of the sample is generally limited both to control program costs and to discourage replicate analyses in the laboratory. It is important that clear instructions for preparation of the material for analysis and description of its stability under specified storage conditions accompany the specimen. Analyte concentrations generally span the analytic range, since both precision and accuracy may be affected by concentration.[1]

LABORATORY ANALYSIS

After sample receipt, the laboratory should promptly analyze the specimens, handling them in a fashion identical to patient samples. This is difficult to accomplish in many laboratories, since most control sera differ in appearance from patient samples, alerting the analyst that a special sample is being processed. This potentially allows special attention towards the proficiency testing sample.

Some laboratories, in fear of regulatory agencies, resort to special handling of proficiency testing samples in order to obtain passing scores. They may perform replicate analyses, place the samples after standards or controls in continuous-flow systems, analyze the samples by selected technologists, review the results for an analyte determined by several analytic methods choosing the most visually appealing, or even compare results with another laboratory participating in the program. These practices have deleterious effects on the morale of the analyst, are contrary to the intent of proficiency testing programs, and have undefined effects on the performance of the individual laboratory.

Current proficiency testing programs are limited in the number of analytes. They also are generally limited by allowing submission of only one result for each analyte, while multiple procedures, e.g., stat, batch, profile, may be in use. The data input forms provided by the program provide for accurate description of the method used. However, any proficiency testing input form is unlike familiar patient reporting forms and, therefore, offers extra potential for clerical error.

COMPUTER SUMMARIZATION

Data from all laboratories should ideally be summarized and returned promptly to describe the performance of each laboratory, as well as the performance of each method, for a given analyte.[3] This requires definition of a target value and a performance standard in order to relate them to individual results and grouped results (e.g., by method).

Target values can be established by use of a definitive method, as the mean of a reference laboratory group, as the mean of results obtained by participants using a reference method, as a weighed-in value for some analytes, e.g., drugs, as the grand mean after removal of outliers, or by combinations of these techniques.[3]

Gilbert[4] has studied the comparability of grand mean values (Table 1), as well as target values obtained by participants using a well-accepted method (Table 2) vs. results obtained at the National Bureau of Standards (NBS) using definitive methods. This study demonstrates that the grand mean and the target values provided by the participants generally are very close to the accurate value determined by the NBS. Thus, laboratories can generally accept target values based upon participants' results to reflect, with minimum bias, the accurate content of analyte in the sample.

Outliers, i.e., data differing greatly from the majority of results, are frequently due to arithmetic or clerical errors.[3] They are most commonly removed by eliminating all results lying outside the mean ± 3 SD, followed by subsequent recalculation of mean and SD for comparison and evaluation. Although this assumes Gaussian distribution, proficiency testing results at least approximate this. Other techniques, including visual review of all the data, use of percentiles, or arbitrary truncation have also been used.[3] After outliers are excluded, the results are assumed to reflect normal laboratory performance.

Performance standards for individual results may be based on a multiple of the SD of laboratories using the same method or of reference laboratory results. They may also be based on medical usefulness criteria or historically achieved performance.

Grading systems allow both laboratorians and regulatory agencies to assess laboratory performance. Systems available include depiction as a positive or negative difference in units from the target value, or in relation to the SD of all results, the SD of laboratories using the same method, or a reference method. In addition, the use of histograms or mathematical schemes to provide a laboratory performance score have been developed.[3]

LABORATORY UTILIZATION OF SURVEY DATA

Proficiency testing results provide primarily retrospective studies of accuracy for each analyte. They represent a supplement to an ongoing internal quality control program which primarily provides precision statistics.

Comparison of average long-term bias estimates for individual analytes studied in a proficiency testing program and in regional quality control programs yields a significant albeit imperfect correlation.[6] Thus, between proficiency testing intervals, laboratories can correct bias by careful attention to internal quality control results.

When the proficiency testing program depicts a statistically significant discrepancy from the target value for any analyte, the type of error should be classified by reviewing records for a clerical or calculation error, checking for specimen misidentification, and checking whether the control specimen was unusual, e.g., contaminated, turbid, or whether laboratory environment, e.g., temperature or humidity, was abnormal. The internal quality control results on the day of sample analysis should also be reviewed at this time for bias or excessive variability.

An aliquot of the proficiency testing sample should be saved frozen or extra samples of the same material, e.g., CAP Survey Serum, should be again assayed for those analytes which yielded erroneous results. If the results remain similar and in error, this confirms the presence of a long-term bias. If results are variable in magnitude or sign but still in error, long-term precision is questionable. In either case, a major study of the method is indicated.

Such study should include careful review of the written procedure in use, documen-

Table 1
CALCULATED ACCURACY OF CAP ALL RESULT MEAN VALUES[a]

Analyte	Units	Regression		Calculated accuracy						Av coefficient of variation
		Slope	Intercept	Low value		High value		Selected value		
Calcium	mg/dl	0.9280	0.6093	9.30	−0.6%	12.27	−2.2%	10.0	−1.1%	4.2%
	mmol/l		0.1506	2.33		3.07		2.50		
Chloride	mmol/l	0.9269	6.7974	101.4	−0.6%	115.1	−1.4%	105	−0.8%	2.4%
Iron	µg/dl	0.7890	14.0657	56.6	+3.7%	158.5	−12.2%	80	−3.5%	12.7%
	mmol/l		2.5230	10.1		28.4		14		
Lithium	mmol/l	0.9730	0.0295	0.41	+4.5%	1.30	−0.4%	1.0	+0.2%	13.6%
Magnesium	mg/dl	0.8384	0.3098	1.58	+3.4%	2.47	−3.6%	2.0	−0.7%	12.9%
	mmol/l		0.1142	0.65		1.01		0.8		
Potassium	mmol/l	0.9854	0.0977	3.75	+1.1%	4.67	+0.6%	4.0	+1.0%	2.4%
Sodium	mmol/l	0.9254	10.0305	137.4	−0.2%	151.9	−0.9%	145	−0.5%	1.5%

[a] Accuracy is calculated as a percentage difference from the NBS value at the upper and lower end of the concentration range included in the series and at a selected value.

From Gilbert, R. K., Accuracy of clinical laboratories studied by comparison with definitive methods, *Am. J. Clin. Pathol.*, 70, 450, 1978. With permission.

tation that procedural details are being followed, documentation that maintenance and monitoring procedures are being followed, and if consistent bias is detected, special attention should be paid to the standards of calibrators in use.[3]

In the event that repeat samples yield satisfactory results, it is likely that the proficiency testing results represented random error or transient bias on the day of testing. Attention to ongoing internal quality control results, comparing the observed coefficient of variation to other laboratories using the same method and alternate methods in a Regional Quality Control Program, may suggest whether a good method is being performed poorly (high CV vs. other laboratories using the same method) or whether better methods are available.

When selecting new methods, assessment of relative method performance can be aided by reviewing summarized proficiency testing data, e.g., recent Survey Summary: Calcium (Table 3). Both relative precision (SD and CV) and accuracy (nearness to the target value) can be readily reviewed for a variety of methods which are in active use in the field. Longitudinal studies of method performance are also available.[4] Thus, proficiency testing programs can provide a useful educational tool to the individual laboratory in its search for laboratory improvement.

REGULATORY USE OF PROFICIENCY TESTING DATA

Just as a laboratorian can evaluate one's own laboratory's performance using proficiency testing data, so can a regulatory agency evaluate laboratory performance. This premise was incorporated into the 1965 Medicare Act and the 1967 Clinical Laboratory Improvement Act. Regulatory review was further revised in 1974 for independent laboratories under Medicare.[3] Participation in proficiency testing programs is also required by many states, as well as by the accreditation programs of JCAH, CAP, and AIHA.

The degree of intervention by regulatory agencies is variable. Some agencies require only participation in a proficiency testing program. More frequently, the agencies require that the laboratories provide them copies of proficiency testing results, either by referring copies to the agency or having the reports available during an on-site inspection. In the most widely used approach, the governmental agency designates approved proficiency testing programs, allowing the laboratory to select from the approved list.[3] The agency then receives reports for its accumulation and evaluation. This latter approach is used by the Veterans Administration and military hospitals, as well as 42 states.

In some cases, the regulatory agency both operates the proficiency testing program, as well as evaluates the results. Such a program is provided by the CDC, some states, and New York City.[3]

While the grading schemes of regulatory agencies differ, it is important that the evaluation criteria are reasonable and widely accepted. If unrealistic performance criteria are employed, the laboratories are indirectly encouraged to handle proficiency testing samples much differently from patient samples. When laboratories process proficiency testing samples in a nonroutine fashion, they are likely to overlook laboratory errors which affect routine results. Thus, overemphasis on proficiency testing may actually be counterproductive to good laboratory performance on patient samples. All parties must, therefore, recognize that the educational and laboratory improvement aspects of proficiency testing should be encouraged in order to benefit the ultimate consumer of laboratory services, the patient.

Table 2
CALCULATED ACCURACY OF CAP TARGET VALUES[a]

		Regression		Calculated accuracy						Average coefficient of variation
Analyte	Units	Slope	Intercept	Low value		High value		Selected value		
Calcium	mg/dl	0.9685	0.2326	9.30	+0.2%	12.27	-1.2%	10.0	-0.8%	3.4%
	mmol/l		-0.0049	2.33		3.07		2.50		
Chloride	mmol/l	0.9740	1.3599	101.4	-1.3%	115.1	-1.4%	105	-1.3%	2.1%
Iron	µg/dl	0.7019	19.6680	56.	+19.7%	158.5	-17.4%	80	-5.2%	13.9%
	mmol/l		3.5354	10.1		28.4		14		
Lithium	mmol/l	0.9739	0.0353	0.41	+6.0%	1.30	+0.1%	1.0	+0.9%	14.6%
Magnesium	mg/dl	0.9364	0.0986	1.58	-0.1%	2.47	-2.4%	2.0	-1.4%	9.4%
	mmol/l		0.0381	0.65		1.01		0.8		
Potassium	mmol/l	0.9816	0.1112	3.75	+1.1%	4.67	+0.5%	4.0	+0.9%	2.6%
Sodium	mmol/l	0.8934	14.3634	137.4	-0.2%	151.9	-1.2%	145	-0.6%	1.5%

[a] Accuracy is calculated as a percentage difference from the NBS value at the upper and lower end of the concentration range included in the series and at a selected value.

From Gilbert, R. K., Accuracy of clinical laboratories studied by comparison with definitive methods, *Am. J. Clin. Pathol.*, 70, 450, 1978. With permission.

Table 3
SURVEY SUMMARY: CALCIUM

1979 Set C-B

Constituent/method and system	Specimen C-6				Specimen C-7				Specimen C-8			
	No. labs	Mean	SD	CV	No. labs	Mean	SD	CV	No. labs	Mean	SD	CV
Alizarin												
All results	150	9.47	0.65	6.9	149	9.63	0.68	7.0	58	10.01	0.70	7.0
Centrifugal analyzers	136	9.44	0.65	6.9	135	9.61	0.67	7.0	57	9.99	0.69	6.9
Union Carbide Centrifichem	136	9.44	0.65	6.9	135	9.61	0.67	7.0	57	9.99	0.69	6.9
All multi. analyzers	143	9.45	0.65	6.9	142	9.62	0.67	6.9	57	9.99	0.69	6.9
Atomic absorption spectrophotometry												
All results	111	9.85	0.39	3.9	111	10.37	0.37	3.6	99	11.19	0.43	3.8
Perkin Elmer	64	9.85	0.40	4.0	64	10.36	0.33	3.2	60	11.17	0.43	3.9
All atomic absorption spectrophotometry[a]	102	9.85	0.39	3.9	102	10.37	0.34	3.2	93	11.19	0.44	3.9
Calcein — fluorometric												
All results	190	9.37	0.55	5.9	190	9.78	0.67	6.9	78	10.40	0.82	7.8
Oxford (titrator) kits	47	9.15	0.69	7.5	47	9.19	0.92	10.0	—	—	—	—
All manual procedures	61	9.15	0.69	7.6	61	9.29	0.93	10.0	—	—	—	—
Corning titrator	90	9.47	0.32	3.4	90	9.90	0.43	4.4	38	10.46	0.57	5.5
Precision Systems titrator	24	9.54	0.61	6.4	25	10.07	0.86	8.5	—	—	—	—
All titrators	124	9.49	0.38	4.0	124	9.93	0.51	5.1	58	10.52	0.77	7.3
Colcein—nonfluormetric												
All results	21	9.04	1.54	17.1	21	9.51	1.65	17.3	—	—	—	—
Cresolphthalein complexone												
All results	4375	9.56	0.49	5.1	4378	10.04	0.55	5.5	2112	11.01	0.48	4.3
Manual, in-house reagent	33	9.51	0.43	4.5	34	9.98	0.55	5.5	—	—	—	—
American monitor kits	79	9.28	0.54	5.8	80	9.66	0.80	8.3	—	—	—	—
Bio-Dynamics kits	25	9.21	0.78	8.4	24	9.42	0.89	9.4	—	—	—	—
Data Medical Association kits	53	9.32	0.71	7.7	53	10.03	0.76	7.6	—	—	—	—
Dow kits	343	9.29	0.63	6.8	344	9.68	0.66	6.8	47	10.39	0.50	4.8
Hycel kits	31	9.73	0.47	4.8	32	10.26	0.49	4.8	—	—	—	—
Worthington kits	92	9.53	0.55	5.7	91	9.93	0.70	7.1	—	—	—	—

Table 3 (continued)
SURVEY SUMMARY: CALCIUM

1979 Set C-B

Constituent/method and system	Specimen C-6				Specimen C-7				Specimen C-8			
	No. labs	Mean	SD	CV	No. labs	Mean	SD	CV	No. labs	Mean	SD	CV
Other, manual	22	9.31	0.75	8.0	21	9.81	0.83	8.4	—	—	—	—
All manual	787	9.38	0.63	6.7	784	9.81	0.73	7.4	99	10.59	0.61	5.8
Abbott ABA 50	93	9.38	0.59	6.3	93	9.67	0.71	7.3	—	—	—	—
Abbott ABA 100	192	9.28	0.56	6.0	195	9.57	0.66	6.9	63	10.00	0.69	6.9
American Monitor programachem	43	9.48	0.62	6.5	42	9.78	0.56	5.7	—	—	—	—
American Monitor KDA	189	9.36	0.32	3.4	189	9.92	0.36	3.6	124	10.58	0.44	4.1
Centrifugal analyzers	220	9.03	0.72	8.0	218	9.43	0.79	8.4	71	9.94	0.82	8.2
Chemetrics	84	9.10	0.76	8.4	84	9.44	0.79	8.3	—	—	—	—
Coulter 22	94	9.80	0.29	2.9	93	10.41	0.30	2.8	63	11.27	0.31	2.8
Dupont ACA	752	9.53	0.28	2.9	760	10.16	0.25	2.5	420	11.03	0.25	2.3
ElectroNucleonics GemENI	130	8.68	0.65	7.5	128	9.07	0.72	8.0	34	9.40	0.94	9.9
Gilford 3500	153	9.28	0.54	5.8	153	9.66	0.57	5.9	47	10.11	0.52	5.1
Gilford System 4	41	9.18	0.42	4.6	42	9.63	0.57	5.9	—	—	—	—
Gilford System 5	61	9.28	0.43	4.7	61	9.55	0.64	6.7	—	—	—	—
Hycel HMA 1600	45	9.61	0.70	7.3	44	10.28	0.87	8.4	—	—	—	—
Hycel 17	45	9.84	0.28	2.8	46	10.43	0.27	2.5	30	11.25	0.29	2.6
Hycel Super 17	162	9.95	0.22	2.2	163	10.52	0.21	2.0	102	11.28	0.21	1.8
Technicon AAII	48	9.80	0.39	3.9	48	10.29	0.33	3.2	38	11.18	0.37	3.3
Technicon AAI and AAII	65	9.80	0.36	3.6	65	10.33	0.33	3.2	52	11.17	0.36	3.2
Technicon SMA 6/60	24	9.82	0.20	2.0	25	10.18	0.23	2.3	23	11.10	0.25	2.2
Technicon SMA II	129	9.74	0.24	2.5	131	10.17	0.26	2.6	105	11.11	0.25	2.3
Technicon SMA 12/60	735	9.80	0.27	2.7	737	10.22	0.29	2.8	569	11.15	0.31	2.8
All Technicon SMA's	909	9.79	0.26	2.7	913	10.21	0.28	2.8	704	11.14	0.30	2.7
Technicon SMAC	295	9.91	0.25	2.6	295	10.42	0.23	2.2	255	11.35	0.25	2.2
Union Carbide centrifichem	44	9.52	0.64	6.7	44	9.87	0.67	6.8	—	—	—	—
Other, automated	110	9.49	0.69	7.2	109	9.95	0.78	7.8	40	10.86	0.79	7.3
All multi. analyzers	3578	9.60	0.44	4.6	3582	10.09	0.49	4.9	2005	11.04	0.45	4.1

	No.	Mean	SD	CV%	No.	Mean	SD	CV%	No.	Mean	SD	CV%
EDTA Titration — other												
All Results	98	9.33	0.77	8.2	98	9.74	1.04	10.7	—	—	—	—
Oxford titrator (manual) kits	45	9.18	0.89	9.7	45	9.58	1.05	10.9	—	—	—	—
All manual procedures	77	9.33	0.77	8.2	78	9.71	1.06	10.9	—	—	—	—
All titrators	20	9.32	0.78	8.3	20	9.86	0.97	9.8	—	—	—	—
HPE (monitor)												
All results	72	9.61	0.55	5.7	72	10.08	0.61	6.0	—	—	—	—
American Monitor kits	61	9.61	0.58	6.0	61	10.11	0.63	6.2	—	—	—	—
All manual procedures	67	9.62	0.56	5.8	67	10.13	0.60	5.9	—	—	—	—
Calcium — mg/dℓ Methylthymol blue												
All results	306	9.45	0.59	6.2	306	9.72	0.61	6.3	72	10.29	0.65	6.3
Pierce (rapid stat) kits	153	9.44	0.52	5.5	153	9.76	0.59	6.0	33	10.28	0.69	6.7
All manual procedures	166	9.45	0.53	5.6	166	9.79	0.61	6.2	35	10.28	0.67	6.5
Centrifugal analyzers	96	9.45	0.74	7.8	95	9.67	0.66	6.8	28	10.38	0.60	5.8
ElectroNucleonics GemENI	34	8.97	0.75	8.4	34	9.29	0.62	6.7	—	—	—	—
I.L. Multistat III	42	9.83	0.49	5.0	42	9.92	0.48	4.8	—	—	—	—
All multi. analyzers	139	9.41	0.67	7.1	138	9.64	0.61	6.3	37	10.30	0.62	6.1
Chlorophosphonaze III												
All results	433	10.04	0.60	5.9	430	10.69	0.62	5.8	34	11.48	0.63	5.5
Harleco kits	163	10.12	0.55	5.4	163	10.71	0.56	5.3	—	—	—	—
All manual procedures	176	10.12	0.57	5.7	177	10.71	0.59	5.5	—	—	—	—
Harleco/IL Clinicard	226	10.03	0.59	5.9	224	10.67	0.67	6.3	—	—	—	—
All multi. analyzers	254	10.01	0.59	5.9	253	10.65	0.68	6.4	—	—	—	—
All procedures/all results	5867	9.58	0.55	5.7	5856	10.04	0.63	6.3	2534	10.93	0.59	5.4
Outliers excluded	5.4%				4.8%				4.2%			

a   Comparative method.

From Gilbert, R. K., Accuracy of clinical laboratories studied by comparison with definitive methods, *Am. J. Clin. Pathol.*, 70, 450, 1978. With permission.

REFERENCES

1. **Gilbert, R. K.,** CAP interlaboratory survey data and analytical goals, in *Proc. 1976 Aspen Conf. Anal. Goals Clin. Chem.,* Elevitch, F. R., Ed., College of American Pathologists, Skokie, Ill., 1977, 63.
2. **Buttner, J., Borth, R., Boutwell, J. H., Broughton, P. M. G., and Bowyer, R. C.,** Provisional recommendation on quality control in clinical chemistry, *Clin. Chem.,* 24, 1213, 1978.
3. **Forney, F. E., Blumberg, J. M., Brooke, M. M., Eavensen, E., Gilbert, R. K., and Kaufmann, W.,** Laboratory evaluation and certification, in *Quality Assurance Practices for Health Laboratories,* Inhorn, S. L., Ed., American Public Health Association, Washington, D.C., 1978, chap. 2.
4. **Gilbert, R. K.,** Accuracy of clinical laboratories studied by comparison with definitive methods, *Am. J. Clin. Pathol.,* 70, 450, 1978.
5. **Schork, M. A., Greenhouse, J. B., and Williams, G. W.,** Normality and other statistical considerations as they relate to selected quantitative measures in the CAP survey, *Am. J. Clin. Pathol.,* 68, 112, 1977.
6. **Haven, G. T.,** A cumulative report of survey and QAS activities: a management tool for the pathologist, *Pathologist,* 32, 170, 1978.

VENDORS

Guy T. Haven

CHEMISTRY CONTROL PRODUCTS

Beckman Instrument Inc.
2500 Harbor Boulevard
Fullerton, Calif. 92634

Dade
P.O. Box 520672
Miami, Fla. 33152

Fisher Diagnostics
Fisher Scientific Co.
711 Forbes Avenue
Pittsburgh, Pa. 15219

General Diagnostics
201 Tabor Road
Morris Plains, N.J. 07950

Hyland Diagnostics
Bannockburn Executive Plaza
2275 Half Day Road
Deerfield, Ill. 60015

Ortho Diagnostics
Raritan, N.J. 08869

Technicon
511 Benedict Avenue
Tarrytown, N.Y. 10591

Wellcome Reagents Division
Burroughs Wellcome Co.
3030 Cornwallis Road
Research Triangle Park, N.C. 27709

PROFICIENCY TESTING PROGRAMS

American Association of Bioanalysts
205 W. Levee Street
Brownville, Tex. 78520

American Association for Clinical Chemistry
1725 K Street N.W.
Washington, D.C. 20006

American College of Nuclear Physicians
Suite 700
1101 Connecticut Avenue, N.W.
Washington, D.C. 20036

American Society of Internal Medicine
2250 M. Street, N.W.
Suite 620
Washington, D.C. 20037

California Thoracic Society
424 Pendleton Way
Oakland, Calif. 94621

Center for Disease Control
Atlanta, Ga. 30333

College of American Pathologists
7400 N. Skokie Boulevard
Skokie, Ill. 60077

Proficiency Test Service
c/o Institute of Clinical Science
1833 Delancey Place
Philadelphia, Pa. 19103

Wellcome Reagents Division
Burroughs Wellcome Co.
3030 Cornwallis Road
Research Triangle Park, N.C. 27709

DAILY QUALITY CONTROL DATA PROCESSORS

College of American Pathologists
7400 N. Skokie Boulevard
Skokie, Ill. 60077

Dade
P.O. Box 520672
Miami, Fla. 33152

General Diagnostics
201 Tabor Road
Morris Plains, N.J. 07950

Hyland Diagnostics
Bannockburn Executive Plaza
2275 Half Day Road
Deerfield, Ill. 60015

Ortho Diagnostics
Raritan, N.J. 08869

Wellcome Reagents Div.
Burroughs Wellcome Co.
3030 Cornwallis Road
Research Triangle Park, N.C. 27709

Apparatus and Instrumentation

INTRODUCTION

William H. C. Walker

Clinical chemistry employs a wide variety of instrumentation and apparatus, some developed specifically for clinical samples, some of general usage in analytical chemistry, and some highly specialized such as mass spectrometry. This area deals with instrumentation specialized for and dedicated to clinical chemistry. Discussions of instrumentation of general applicability can be found elsewhere.[1,2]

Clinical chemistry has to meet constraints of high-sample throughput, rapid turnaround, and long-term precision. In addition, multiple tests often must be performed on samples of limited volume. These may be in fixed groupings, termed profiles, or their combination may vary from sample to sample, termed selective, or discretionary testing. The variety of multichannel analyzers that has been developed to meet such demands is exceeded only by the diversity of needs encountered in different settings which range from a slow turnaround reference laboratory to a round-the-clock intensive care service for premature infants. Considerations of microsamples, rapid response, and guaranteed continuty of service are dominant in one setting, but absent in the other, where the capability to handle a large workload at a low cost usually is crucial. The choice of appropriate instrumentation has to take into account not only such patterns of workload, but also of overall laboratory management constraints. Are labor costs prohibitive, thereby justifying capital intensive investment? What technical skills are available? Does technical knowhow related to one type of instrumentation represent a major investment? What are the skills available for out-of-hours service?

In general, it is desirable to restrict the variety of different instrumentation to the necessary minimum, but unavoidably, a number of special apparatus dedicated to a single task such as the assay of osmolality may be necessary. The final choice of equipment is never easy; hard information on performance never is complete and some arbitrary judgements are inevitable. It is a tribute to the versatility of much instrumentation and to the application of their operators that so many laboratories using quite different equipment can provide similar services, all apparently satisfactorily.

REFERENCES

1. Ackerman, P. G., Eds., *Electronic Instrumentation in the Clinical Laboratory*, Little, Brown, Boston, 1972.
2. Hicks, R., Schenken, J. R., and Steinrauf, M. A., Eds., *Laboratory Instrumentation*, Harper & Row, New York, 1974.
3. Werner, M., Ed., *Microtechniques for the Clinical Laboratory*, John Wiley & Sons, New York, 1976.
4. Haeckel, R., Mitchell, F. L., Büttner, H., Hjelm, M., Geary, T. D., Sandbland, B., and Craig, T. M., Decision criteria for the selection of analytical instrumentation used in clinical chemistry, *J. Autom. Chem.*, 2, 22, 1980.

ANALYTICAL BALANCES

Albert F. Plant

INTRODUCTION

Most analyses carried out today in the clinical laboratory utilize volumetric procedures, often referred to as "wet chemistry". Unfortunately, while easy and convenient, these procedures are not the most accurate available. Much greater accuracy is obtainable through gravimetric or weight procedures.

A primary reason for the preference given to volumetric procedures was the inconvenience of the double-pan balance which was the only readily available weighing device until the 1950s. To obtain accurate weighings required not only patience, but an almost artistic level of skill, and the balances seemed to require constant recalibration, particularly if they were used by more than one individual. A second reason was the introduction of mass-produced calibrated pipettes, mechanized liquid dispensers, and a variety of other low-cost glassware, which allowed continuing simplification of volumetric methods. Further, the reliability of measurements of volume in common with those of time and electrical current, has improved by several orders of magnitude within the last few decades. Despite these advances, volumetric measurement of liquids is still imprecise when compared to gravimetry. Since 1900, the reliability of balances has been about 1 ppm, which even now, stands three orders of magnitude better than volumetry. Few other measurements approach the reliability of mass measurements, and there is little need for better performance because other procedural steps invariably make a greater contribution to overall error.

The single most important factor in the resurgence of gravimetric analysis, was the development of the single-pan analytical balance (Figure 1). This allows gravimetry to be performed rapidly and with minimal dexterity. Even microbalances can be operated reliably with a modicum of training, and the direct weighing of microsamples eliminates much of the effort of older methods, where the analyst had to first weigh a sample in bulk, then prepare dilutions, and finally aliquots for analysis. Microgravimetry eliminates most of the error associated with dilutions and, at the same time, requires only a fraction of the original sample material, which is important when limited sample is available. Gravimetry and microgravimetry have the further advantage that they permit the analysis of solid samples in their natural state and with reduced contamination from solvents and other reagents.

Among laboratory balances, the following classes are generally recognized:

Analytical	100—200-g capacity	0.1-mg repeatability
Semi-micro	75—100-g capacity	0.01-mg repeatability
Micro	1—30-g capacity	0.001-mg repeatability
Ultra-micro	less than 5-g capacity	0.1-μg repeatability

MASS MEASUREMENT

Mass cannot be measured directly, but must be measured through some related physical property, such as gravity. At any particular location, mass is subject to the local effect of gravity and, thus, has weight, which is the force exerted by a mass under the influence of gravity.

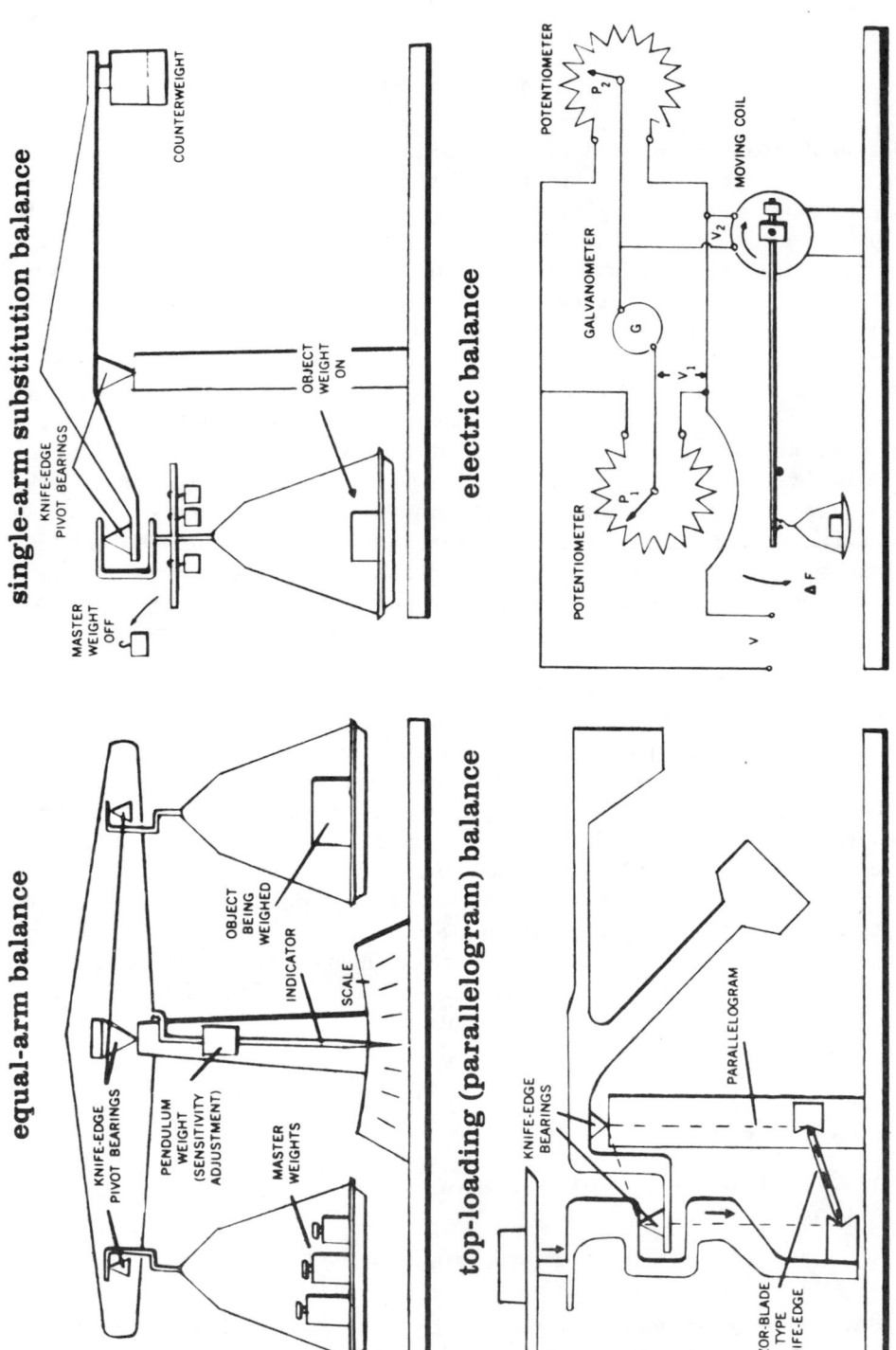

FIGURE 1.   Four basic types of beam balances. (Courtesy of Industrial Research Magazine.)

$$m = \frac{W}{g}$$

Unfortunately, the term "weighing" is applied equally to weight, as well as mass determinations, although the same mass at different locations can have different weights. This usage comes about because the parameters are usually interchangeable. Weight is the more conveniently measured physical property, and in a single locality, mass and weight remain proportional.

FEATURES FOR THE EVALUATION OF BALANCES

The most important feature of any balance is its repeatability or precision. Repeatability is the ability to achieve the same value for a given sample time after time, and it is a direct function of balance quality. In comparing various balances, this specification is the major consideration. The repeatability is a statistical value expressed in mass units and defined as the standard deviation (σ) of repeated weighings of the same mass.

$$\sigma^2 = \pm \frac{\Sigma(x-\bar{x})^2}{n-1}$$

where $\times$ = an individual weighing value; $\bar{x}$ = the test group's average value; and n = the number of weighings made (at least 10).

Random errors cause individual weights to be dispersed with a Gaussian distribution. Two standard deviations on either side of the mean value will include about 95% of all weighings, while three standard deviations will include 99.5%.

The accuracy of a balance defines how well the balance reading matches the true value of measured mass. In practice, accuracy is determined by averaging ten weighings made under identical conditions. A balance may be highly repeatable, but inaccurate. This is often of little importance, since absolute weights are rarely needed. More frequently, the difference between two weighings is required and only this difference need be repeatable and accurate.

A third important characteristic, particularly in microbalances is the sensitivity of the balance. This parameter defines the minimum detectable mass change that will produce a measurable deflection in the weighing system. Sensitivity should always be viewed in conjunction with balance repeatability, since it is possible to have great sensitivity with poor repeatability. Torsional suspension beam microbalances generally have a load-precision ratio exceeding 10^7 and the load capacity can exceed four million times the sensitivity.

BALANCE DESIGN

The classical analytical balance is the two-pan balance (Figure 1), in which an unknown mass is suspended from one arm of a beam supported on a knife edge, and compared to a known mass suspended from the opposite beam arm. At present, most precision balances are of the single-pan type (Figure 1). An unknown mass is compared to a fixed, known mass on the other side of the fulcrum. Moveable or substitution weights suspended from the sample side of the beam are removed to compensate for the addition of the unknown sample. Usually the weight adjustment is carried out to the nearest 0.1 g. This control, which removes the substitution weights, indicates their mass value (and, thus, the mass value of the unknown) by a visual display. The mag-

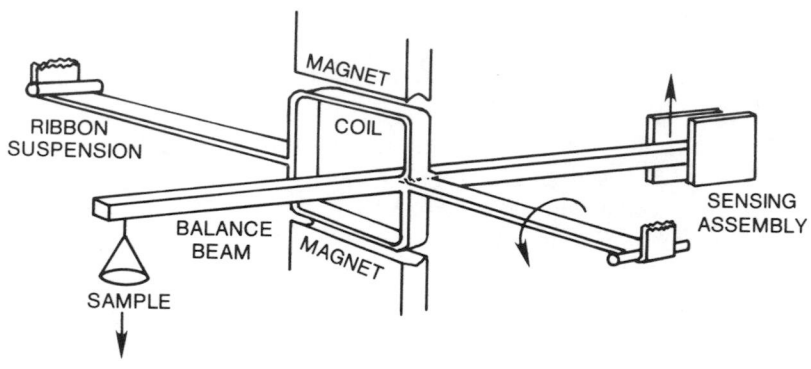

FIGURE 2.   Electromagnetic balance with a torsional ribbon suspension.

nified projection of a pointer attached to the beam and viewed on a calibrated fixed scale indicates the beam deflection due to the remainder of the weight. In some newer balances, an electrical circuit replaces the optical system. The magnitude of the electrical current required in the circuit to bring the beam to an equilirium position is proportional to the mass being measured, and calibration defines this proportionality. Some balances include more advanced electronics, such as microprocessors, allowing live, moving animals to be reliably weighed by automatically averaging hundreds of readings made over several minutes.

Further refinement is used in torsion suspension balances. A taut suspension band supports the beam carrying the sample by means of a tensed metal ribbon. This eliminates the need for a knife-edge fulcrum. The first instruments to combine the stability of electronic circuitry and the ruggedness of taut band suspension were microbalances. At present, the knife-edge construction has been replaced in many top-loading and analytical laboratory balances. Top-loading balances employ a sample pan atop the beam and a fulcrum mechanism. Most beam microbalances use torsional suspension. In some, a suspension fiber (the fulcrum from which the balance beam is suspended) is rigidly attached to the balance beam tilts, it twists the suspension fiber longitudinally, creating a restoring torque in the fiber.

The quartz fiber torsion balance developed by Nernst[1] in 1903 remains unchallenged among mechanical balances. In electromagnetic balances, the beam is attached to a coil placed in a magnetic field to balance the load. As the mass placed on the beam causes it to drop below the horizontal position, an appropriate current is passed through the coil to bring the beam back to the horizontal position by creating a restoring torque in the coil. Since the beam is always rebalanced to the same horizontal position, beam deflection does not affect sensitivity, and it is not necessary to read the final significant figures of weight on a projected optical scale. Recently, a fully electronic top-loading analytical balance has become available, which rounds off the last visible digit and also monitors its own reliability.

Electromagnetic balances (Figure 1) have been improved considerably in ease of operation, reliability, and range. Over the last decade, typical commercial units have increased their measuring range over 20-fold and their sensitivity over 10-fold. Most of these torsional suspension balances now incorporate a ribbon suspension (Figure 2), rather than the earlier torsion fiber, but their operation is essentially unchanged. The advantage of the ribbon is that it defines the axis of rotation and is selfcentering.

APPLICATIONS

For most applications, the use of the analytical balance is simply to determine the sample mass at various points in an analytical procedure. There is, however, a growing usage of the balance as a primary analytical tool, ranging from the calibration of pipettes and glassware to true analysis.[2]

Current balance technology has allowed the revival of the Pregl electrogravimetric technique[3] which was developed in 1911 but fell into disuse because it required demanding weighing operations. However, even where solutions are involved, samples can be handled most reliably gravimetrically rather than volumetrically. In laser Raman spectroscopy, for example, microbalances are used to sample both solid specimens (less than 1 μg) and liquid specmens (less than 1 μl). Other applications in biology include the weighing of fresh sections of frozen tissue in a cryostat,[4] the determination of freeze-drying weights of specimens,[5] measurement of the freezing of small tissue samples and thin frozen sections,[6-9] determination of lactate and phosphate concentrations in human working muscles,[10] determinations of the hepatic and serum vitamin B12 content in liver disease,[11] measurements of the sorption of water by spores, heat-killed spores, and vegetable cells,[12] and other uses.[13-14]

However, gravimetry and microgravimetry are generally not used by themselves as isolated analytical techniques. Rather, they are ideal adjuncts to other methods because of their sensitivity and accuracy. For instance, microgravimetry is very useful in the preparation of submicrogram quantities of drug standards prior to gas chromatographic analysis,[15] and in the analysis of hair and fingernail samples for toxic substances, such as lead and arsenic.[16,17]

In 1926, McBain and Bakr[18] described the gravimetric adsorption technique to estimate the adsorption of gases onto charcoal or crystalline surfaces, but the lack of appropriate balances curtailed its use. A modular gravimetric adsorption system developed by Williams[19] incorporates an electrobalance and is particularly suited to biological specimens (Figure 3). To measure initial weight, the sample can be decontaminated *in situ,* and adsorbates then can be added easily. This technique is now used in measuring the water vapor sorption properties of dried milks and wheys[20] and milk powder components,[21,22] and in the determination of sorption isotherms in foods.[23,24]

Thermogravimetric analysis (TGA), or thermogravimetry (TG), uses heat like a reagent and continually records the weight and temperature of a test sample as it is heated (Figure 4). This technique and related procedures, such as Differential Thermal Analysis (DTA) (Figure 5), Differential Scanning Colorimetry (DSC), and Evolved Gas Analysis (EGA), have found widespread use. Of particular interest to biology is the study of samples in different gaseous environments. West[25], for example, identified the release of carbon dioxide and water from cortical bone under various gases. Depending on bond strength, water requires different temperatures for release. Pfiel[30] measuring free and bound water in burn tissue, also found that normal tissue contains two types of bound water in addition to free water. Similarly, Heydinger and associates[27] used TGA for edema measurements following burns. Further, TGA has been used to determine amino acid compositions in human plasma prothrombin,[28] in the mass determination of lipids,[29] and in other applications.[30]

Surface tension measurements are another field of application for analytical balances. Surface tension can be sensed by the pull or force that a liquid exerts on an object in contact with it. A static Wilhelmy plate, a dynamic Wilhelmy plate, or a duNouy ring[31] can be used as the contact device, and this is connected to the weighing arm of the balance. Electronic microbalances are particularly suited for such measurements, since their mode of operation is to restore the torque of the balancing arm,

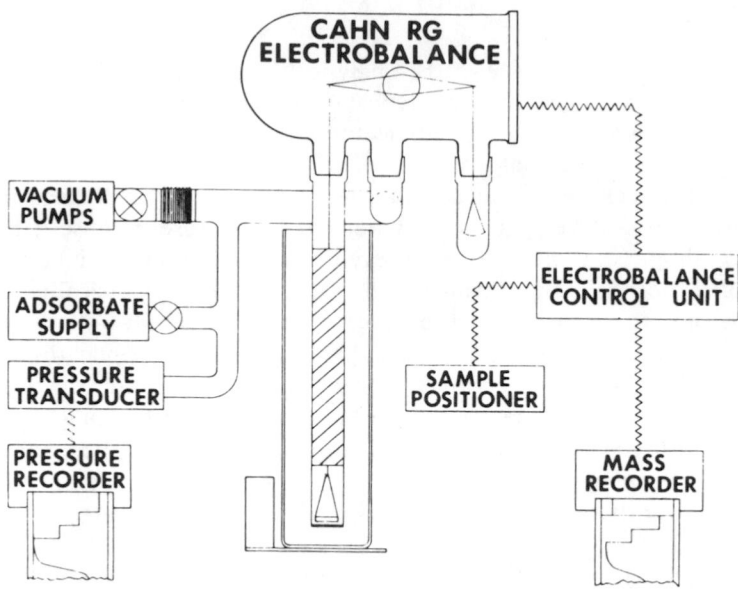

FIGURE 3.   Modular gravimetric adsorption system according to Williams.

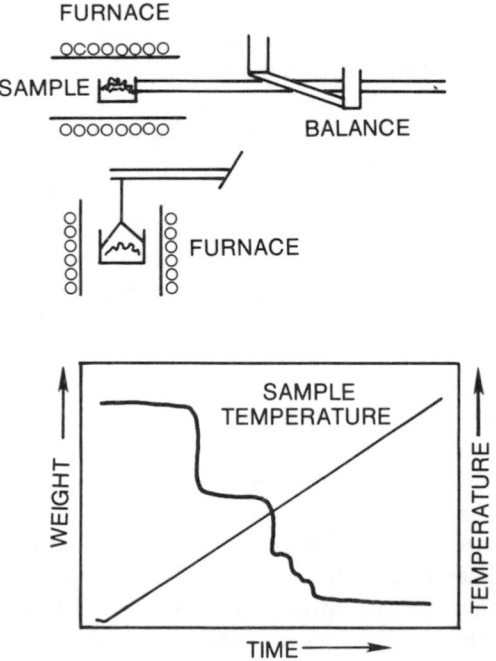

FIGURE 4.   Thermogravimetry or thermal gravimetric analysis.

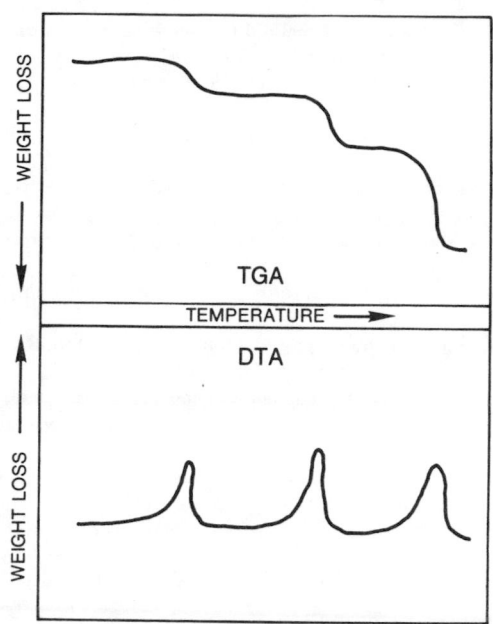

FIGURE 5. Thermal analysis curves.

which indicates the surface tension. Such measurements are important in studying the continual loss and replacement of the alveolar surfactant film in the lungs during the breathing cycle. Mendenhall and Mendenhall[32-35] applied this technique in their investigation of potential toxicity on pulmonary mechanics and fluid dynamics. Similarly, Cook and Webb[36] used it to study lung surfactants in chronic smokers. Finally, the same technique has been applied in kinetic studies of enzyme reactions[37] and in studying various proteins.

REFERENCES

1. **Nernst, W.**, Uber molekulargewichts-bestimmung bein sehr hohen temperaturen, *Z. Elektrochem.*, 9, 622, 1903.
2. **Plant, A. F.**, Microbalances, in *Microtechniques for the Clinical Laboratory,* Werner, M., Ed., John Wiley & Sons, New York, 1976, 17.
3. **Szabadviary, F.,** *History of Analytical Chemistry,* Pergamon Press, Oxford, 1966, 302.
4. Weighing fresh frozen tissue sections in a cryostat at −20 C, Cahn Techniques Bulletin No. 1, Cahn Division of Ventron Instruments Corporation, Paramount, Calif.
5. **MacKenzie, A. P. and Luyet, B. J.**, Apparatus for the automatic recording of freeze-drying rates at controlled specimen temperatures, *Biocynamica,* 9, 192, 1964.
6. **MacKenzie, A. P., et al.**, Apparatus for the automatic recording of freeze-drying rates at controlled specimen temperatures, *Biocynamica,* 9, 193, 1964.
7. **Stumpf, W. E., et al.**, Freeze drying of small tissue samples and thin frozen sections below −60 C: a simple method for cryosorption pumping, *J. Histochem. Cytochem.,* 15, 243, 1967.
8. Techniques #1, Weighing fresh frozen tissue sections in a cryostat at −20 C, Cahn Instruments, Paramount, Calif.

9. Atherton, J. C. et al., Evaluation of a method for weighing small tissue samples: investigations into freezing and evaporation, *Pflügers Arch.*, 309, 203, 1969.
10. Karlsson, J., Lactate and phosphagen concentrations in working muscle of man, *Acta Phys. Scand. Suppl.*, 358, 1971.
11. Nelson, R. S. et al., Hespatic and serum vitamin B12 content in liver disease, *Gastroenterology*, 38, 188, 1960.
12. Greca, N. et al., Sorption of water by spores, heat-killed spores, and vegetable cells, *Can. J. Microbiol.*, 16, 573, 1970.
13. Williams, C. J. et al., Monitoring employee exposure to hardous airborne dusts, *Pollut. Eng.*, 38, 1974.
14. Williams, C. J. et al., An industrial hazard — silica dust, *Am. Lab.*, 17, 1975; *Intern. Lab.*, 67, 1975.
15. Street, H. V., Gas-liquid chromatography of submicrogram amounts of drugs. III. Analysis of alkaloids in biological media, *J. Chromatogr.*, 29, 68, 1967.
16. Shapiro, H. A., Arsenic content of human hair and nails, *Zacchia*, 3, 436, 1966.
17. Shapiro, H. A., Arsenic content of human hair and nails. Its interpretation, *J. Forenic Med.*, 14, 65, 1967.
18. McBain, J. W. and Bakr, A. M., A new sorption balance, *J. Am. Chem. Soc.*, 48, 690, 1926.
19. Williams, C. J., Gravimetric adsorption, *Am. Lab.*, 40, 1969.
20. Berlin, E. et al., Water vapor sorption properties of various dried milks and wheys, *J. Dairy Sci.*, 51, 1912, 1968.
21. Berlin, E. et al., Comparison of water vapor sorption by milk powder components, *J. Dairy Sci.*, 51, 1912, 1968.
22. Berlin, E. et al., Effect of temperature on water vapor sorption by dried milk powders, *J. Dairy Sci.*, 53, 146, 1970.
23. Elvanides, S. M., An Analysis and Interpretation of Thermodynamic Properties of Foods as Determined From Moisture Sorption Data, Ph.D. thesis, Michigan State University, East Lansing, 1973.
24. Gal, S., Recent advances in techniques for the determination of sorption isotherms, Proc. Intern. Symp. Water Relation Foods, *Recent Adv. Food Sci.*, 5, 1975.
25. West, A. O., Determination of the activation energy change of calcium carbonate under influence of various environmental gases, in *Proc. 2nd Int. Conf. Thermal Analysis*, Vol. 2. Schwenker, R. F., Jr., Ed., Academic Press, New York, 1969, 1095.
26. Heydinger, D. K. et al., The measurement of edema following burns, *J. Lab. Clin. Med.*, 77, 451, 1974.
27. Heydinger, D. K. et al., The measurement of edema following burns, *J. Lab. Clin. Med.*, 77, 451, 1971.
28. Lanchantin, G. F. et al., Amino acid composition of human plasma prothrombin, *J. Biol. Chem.*, 243, 5479, 1968.
29. Leerkamp, J. et al., Mass determination of lipids with automated thermogravimetric analytical procedure, *Lipids*, 9, 415, 1974.
30. Pfeil, R. W., Biological application of thermal analysis, in *Proc. 3rd Toronto Symp. Thermal Analysis*, McAdie, H. G., Ed., Toronto Section, Chemical Institute of Canada, February 1969.
31. Gast, T., *In Ultra Micro Weight Determination on Controlled Environmenta*, Wolsky, S. P. and Zdanuk, E. J., Eds., Interscience, New York, 1969, 444.
32. Mendenhall, R. H., Pulmonary mechanics, *Arch. Environ. Health*, 6, 74, 1963.
33. Mendenahll, R. M. and Mendenhall, A. L., Jr., Surface balance for production of rapid changes of surface area with continuous measurement of surface tension, *Rev. Sci. Instrum.*, 34, 1350, 1963.
34. Mendenahll, R. M. et al., A study of some biological surfactants, *Ann. N.Y. Acad. Sci.*, 902.
35. Mendenahll, R. M., A surface balance for the study of surfactant movement, *Rev. Sci. Instrum.*, 42, 878, 1971.
36. Cook, W. A. and Webb, W. R., Surfactants in chronic smokers, *Ann. Thorac. Surg.*, 2, 327, 1966.
37. Lagocki, J. W. et al., The kinetic study of enzyme action on substrate monolayers, *J. Biol. Chem.*, 248, 580, 1973.

CENTRIFUGES AND CENTRIFUGATION

Robert D. Strickland

GENERAL DESCRIPTION

Centrifugation is the physical means of separating solid-liquid or liquid-liquid mixtures under the influence of a centrifugal force. A centrifuge is a device for the practical application of this principle. Since centrifugation results in forces many times that of gravity, separations overcome diffusion effects and result in more compacting of separated zones. There are consequent large reductions in sedimentation time. Routine applications in clinical chemistry are the sedimentation of solids from liquids (e.g., cells from plasma or protein precipitate separation) or separation of immiscible liquids (e.g., solvent extraction). Centrifugation is particularly useful for separating materials susceptible to excessive air exposure, for breaking liquid emulsions, sedimentation of very fine or gelatinous material, and with appropriate equipment, separating temperature labile particles.

PHYSICAL PRINCIPLES

Separation by centrifugation is achieved by rotating the suspension in a tube at the periphery of a rotor (radius r) at a constant angular velocity (ω) about a central spindle. Although the rotational speed of the rotor remains constant, the direction of the velocity of the mass in the tube is always changing. At any instant, the direction of velocity is tangential to the circular path of the mass and results in a centrifugal acceleration of the mass directed perpendicular to the tangent, i.e., towards the axis of the rotor along its radius. The acceleration (a) is defined by

$$a = \omega^2 r \tag{1}$$

The force acting on the mass (m) moving with this acceleration is the centripetal force (F_c) directed inwardly towards the center of rotation.

$$F_c = m \omega^2 r \tag{2}$$

The equal and opposite reaction is an outward force on the mass. This is the centrifugal force effecting sedimentation. More precisely, mass (m) is the effective mass of the sedimenting particle under consideration (actual mass minus a buoyancy correction to allow for solvent mass displaced). When considering centrifugal forces, it is more useful for comparison purposes to measure the relative centrifugal force (RCF), which is the ratio of the centrifugal acceleration to gravitational acceleration ($\omega^2 r/g$). Speeds of rotation are measured in revolutions per minute (r/min).

For any centrifuge,

$$RCF = k \cdot r \, (r/min)^2 \tag{3}$$

where r is the distance from the axis to the point of consideration (usually tip of tube). If r is in inches, k is 0.0000284. If r is in centimeters, k is 0.00001118.

A particle's sedimentation rate under the influence of a centrifugal force is affected by the characteristics of the liquid medium, as well as the physical properties of the particle. If the particle is assumed to be a sphere, as its sedimentation velocity reaches a constant value, the net force causing sedimentation equals the force resisting its motion through the liquid (friction or drag force). The particle's sedimentation rate (v) is given by

$$v = \frac{d^2 [\rho_p - \rho_l] \cdot g'}{18 \mu} \qquad (4)$$

where d is the particle diameter, ρ_p is the particle density, ρ_l is the liquid density, μ is the viscosity of the medium and g' is the gravitational field.[1] This equation shows that the sedimentation rate of a given particle increases with its size, its density, and the centrifugal field. The rate decreases with increased density and viscosity of the liquid medium. If the density of the liquid exceeds that of the particle, then the latter migrates away from the direction of gravitation (i.e., floats). When considering the sedimentation of particles at very high speeds (ultracentrifugation), the *sedimentation coefficient (s)* is a useful index for characterization. This is defined as the sedimentation velocity per unit of centrifugal force and has the dimension of time.

$$s = \frac{1}{\omega^2 r} \cdot \frac{dr}{dt} \qquad (5)$$

Sedimentation coefficients are more conveniently expressed in Svedbergs (S) where one S is 10^{-13} sec.

METHODS OF SEPARATION

Sedimentation rate increases with particle size and centrifugal force. Common laboratory separations involve relatively large particles and the need for complete separation of solid components from suspension. This *nonselective centrifugation*, therefore, needs only low centrifugation speeds for short time spans. For separating and characterizing complex mixtures of cellular, subcellular, or macromolecular sized particles, the following techniques are used in conjunction with high speed or ultracentrifuges:

1. Differential centrifugation (pelleting)
2. Adjusted density centrifugation
3. Rate zonal density gradient centrifugation
4. Isopycnic density gradient centrifugation (equilibrium banding)

For details of these procedures, see *Separation Methods in the Centrifuge* by Charles H. Chervenka in a future volume. A comprehensive review of gradient density techniques is also given by Griffith.[1]

CHOICE OF CENTRIFUGE AND ACCESSORIES

The basic structure of a centrifuge is a motor-driven vertical spindle on which various rotors or heads may be mounted. Samples are contained in fitted tubes in the periphery of the rotor except in special flow-through type of rotors. The unit is surrounded by a protective bowl and casing with a hinged or sliding lid. Design and materials are such that maximum stability with resistance to excessive stress is achieved,

Table 1
LABORATORY CENTRIFUGES

Type	Maximum rotor speed (r/min)	Vacuum/ refrigeration	Rotor types available
Bench-top			
Micro	15,000	No/no	Built-in
Standard	6,000	No/yes	1, 4, 5
Ultra	100,000	No/no	1, 6
Floor model			
Low speed	6,000	No/yes	1, 4, 5
High speed	21,000	Yes/yes	1, 2, 4, 5, 6, 7, 8
Ultra	80,000	High/yes	1, 2, 3, 6, 7, 8
Analytical	75,000	High/yes	9

Note: See Table 2 for rotor types available.

Courtesy of C. H. Chervenka, Beckman Instruments, Palo Alto, Calif.

especially in high speed units. Variable speed controls, timers, tachometers, and braking systems are optional accessories in smaller models, but standard in larger units. Some models have safety features, such as automatic lid locks during rotation or flexible spindle shafts, which stop the motor under serious imbalance.

A laboratory's choice of centrifuge and accessories should be made after consideration of the number and volume of samples to be separated during a run and at what RCF, the needs for fixed angle or horizontal separation, and timing or refrigeration conditions. Centrifuges fall into three main classes, general purpose generating maximum speeds of approximately 6000 r/min, high speed up to 21,000 r/min, and ultracentrifuges up to 80,000 r/min. (Table 1). (Hematocrit and blood bank systems are not considered, although some general purpose centrifuges have applications here.) General purpose models may be bench type, varying in size and capacity, or floor models which are usually equipped with more control options and, because of their dimensions, tend to have a greater capacity and ability to generate greater force. Refrigerated models are available in both types. The smaller bench models offer economy and convenience for occasional, low volume applications. For situations demanding more versatility in sample size and number and greater forces, the larger bench or floor models should be considered. Most radioimmunoassays require a refrigerated unit. High-speed centrifuges incorporate temperature, speed, timing, and braking controls and are capable of forces up to 50,000 g. These instruments will duplicate, at a lesser cost, some functions of preparative ultracentrifuges when separation of particles greater than 0.2 μm diameter is involved. They have density gradient capabilities and, with suitable rotors, are used for sedimentary concentration through a continuous-flow process. High speed microcentrifuges generate up to 15,000 r/min. Due to the small rotor radius, maximum forces are only up to 12,000 g, but this is suitable for the rapid centrifugation of 2 ml or less sample volume. Applications are in microchemistry systems and radioimmunoassay procedures not affected by temperature.

Ultracentrifuges provide a high and precisely controlled rotor speed at uniform temperature. High-strength casing is necessary and the centrifuge bowl is evacuated to reduce heating from air friction. Sophisticated electronic control of sequencing along with safeguards for emergency shutdown contribute to the high cost of modern units. They are still mainly research instruments, but clinical applications such as in lipoprotein fractionation and hormone receptor isolation, indicate their potential in the larger

Table 2
TYPICAL ROTORS AVAILABLE

Type	Type number	Maximum centrifugal force × g	Number of samples	Sample volume (ml)	Usual applications
Fixed angle	1	500,000	4—100	17—3000	A, B, D
Vertical tube	2	510,000	8	40—300	A, D
Swinging bucket					
High speed	3	480,000	3—6	15—230	D, G
Low speed	4	6,800	4—6	200—6000	B, F
Rack	5	2,600	to 336	—	C
Batch	6	100,000	1	2—1670	B
Zonal	7	250,000	1	10—200	B, D
Continuous-flow	8	100,000	1	9 l/hr	B, E
Analytical	9	370,000	1—15	—	D, G

Note: A — simple pelleting; B — large-scale pelleting; C — large number samples; D — density gradient methods; E — continuous-flow separations; F — blood processing; and G — measure physical properties.

Courtesy of C. H. Chervenka, Beckman Instruments, Palo Alto, Calif.

service laboratory. An innovative version of the ultracentrifuge is the Airfuge® (Beckman Instruments), which uses a small turbine rotor driven by air pressure. This small bench top unit (23 lb) is capable of 100,000 r/min and 178,000 g for small volume separations.

Manufacturers' literature shows the wide range of rotors or carriers available for many specific applications (Table 2). Most separations are effected in angle head rotors in which the suspension is held at a fixed angle, or swinging bucket systems in which vertical tubes swing out horizontally during centrifugation. Speeds and rates of sedimentation are higher with the former and their angle is optimized to reduce sedimentation distance. However, they are less suitable for gradient density work. Glass tubes may be centrifuged at low speeds (below 5000 r/min), but stresses at higher speeds dictate the need for polymer or metal tubes chosen according to their chemical resistance (Table 3).

MAINTENANCE AND SAFETY CONSIDERATIONS

Modern laboratory centrifuges should give trouble-free service when given regular preventative maintenance and operated according to the manufacturers' recommendations. Due to the high forces generated in centrifugation, stress resulting from imbalance of the rotating parts may cause mechanical damage and present a hazard to the operator. Secure anchorage (brakes on wheels, rubber cushions) where vibrations will not cause a hazard, use only of rotors or accessories recommended for the instrument with frequent checks for signs of corrosion or hair cracks, and correct balancing and seating of buckets with symmetrical loading around the head are important. Tubes should be of material able to stand the stress of the speed chosen with any breakages or spillage being promptly removed from the centrifuge bowls. Factors such as frequency of use, excessive exposure to maximum forces, frequent autoclaving, and mechanical damage such as etching or abrasion in glass will contribute to tube failures under stress. High-speed rotors should be stored away from exposure to corrosive agents or mechanical damage.

Table 3
MATERIALS SELECTION GUIDE FOR SAMPLE CONTAINERS IN CENTRIFUGATION

	Stainless steel	Corex®	Pyrex®[a]	Cellulose acetate butyrate	Nylon	Poly-allomer	Poly-carbonate	Poly-ethylene	Poly-propylene	Poly-sulfone
Transparent		•	•	•			•			•
Translucent		•	•		•	•	•	•	•	
Temperature limits										
Below 0°C	•	•	•		•	•	•	•	•	•
0°C to 60°C	•	•	•	•	•	•	•	•	•	•
Up to 120°C	•	•	•		•	•	•	•	•	•
Up to 180°C	•	•	•							
Maximum "Speed"[b]										
3000 r/min	•	•	•	•	•	•	•	•	•	•
5000 r/min	•	•	1	•	•	•	•	•	•	•
Rotor "speed" limit	•			•	•	•	•		•	•
Chemical resistance[c]										
Weak acids	•	•	•	•	•	•	•	•	•	•
Acids		2			3	4		4	4	•
Weak bases	•	•	•	•	•	•		•	•	•
Alkalies	•	•	•		3	4		4	4	•
Organic solvents	•	•	•		3	4			•	•
Salt solutions, includes chlorides and hypochlorites	•	•		•	•	•	•	•	•	•

Note: 1 = only tubes smaller than 50 mℓ; 2 = not hydrofluoric or phosphoric acids; 3 = except formic acid, phenol, meta cresol, cresylic acid, xylenol, oxidizing agents, and mineral acids; and 4 = below 60°C.

[a] Registered trademark of Corning Glass Works.

[b] When supported in recommended adapter, refer to manufacturer's literature.

[c] Guide only, when in doubt pretest sample lots.

Courtesy of Du Pont Biomedical Products Division, Newtown, Conn.

Table 4
MANUFACTURERS OF CENTRIFUGES WITH
CLINICAL CHEMISTRY APPLICATIONS

Manufacturer	Centrifuge type
Beckman Instruments Inc., Spinco Division, 1117 California Avenue, Palo Alto, Calif. 94304	A,B,D,E,F,G
Bio-Dynamics/bmc, 9115 Hague Road, Indianapolis, Ind. 46250	A
Brinkman Instruments Inc., Cantiague Road, Westbury, N.Y. 11590	E
Clay Adams, Division of Becton Dickinson and Co., Parsippany, N.J. 07054	A
Du Pont Company, Instrument Products, Biomedical Division, Newtown, Conn. 06470	A,D,F,G
Fisher Scientific Co., 711 Forbes Avenue, Pittsburg, Pa. 15219	A,E
GCA/Precision Scientific, 3737 West Cortland Street, Chicago, Ill. 60647	A
International Equipment Company, Division of Damon, 300 Second Avenue, Needham Heights, Mass. 02194	A,B,C,D,F,G
Lab-Line Instruments Inc., 15th and Bloomingdale Avenues, Melrose Park, Ill. 60160	A
MSE Scientific Instruments. Distributor: Johns Scientific, 219 Broadview Avenue, Toronto, Ontario M4M 2G4	A,B,C,D,F,G
Sargent-Welch Scientific Company, 7300 North Linder Avenue, Skokie, Ill. 60076	A
The Drucker Company, 3240 West 16th Avenue, Hialeah, Fla. 33012	A,C,D,E

Note: A — bench; B — bench, refrigerated; C — floor; D — floor, refrigerated; E — high speed micro; F — high speed; and G — preparative ultracentrifuge.

Aerosol formation from the centrifuging of infectious biological material is a particular problem. Exposure may be minimized by centrifuging in an area of negative pressure ventilation, having stoppered sample tubes, and using sealed rotor lids where feasible. When breakage occurs, components (rotors, buckets, etc.) should be disinfected with 2% glutaraldehyde or autoclaved. Bowls should be cleaned with disinfectant.

INFORMATION SOURCES

Descriptive brochures of centrifuge systems, as well as technical and application bulletins produced by manufacturers are valuable sources of information (Table 4). Hinton[2] and also Ambler and Keith[3] have more extensive reviews of the topic.

REFERENCES

1. **Griffith, O. M.,** *Techniques of Preparative, Zonal, and Continuous Flow Ultracentrifugation,* 2nd ed., Spinco Division, Beckman Instruments, Palo Alto, Calif., 1976.
2. **Hinton, R. H.,** Centrifugation, in *Scientific Foundations of Clinical Biochemistry,* Vol. 1, Williams, D. L., Nunn, R. F., and Marks, V., Eds., Heinemann, London, 1978, chap. 22.
3. **Ambler, C. M. and Keith, F. W., Jr.,** Centrifuging, in *Techniques of Chemistry,* Vol. 12, 3rd ed., Perry, E. S. and Weissberger, A., Eds., John Wiley & Sons, New York, 1978, chap. 6.

TEMPERATURE, THERMOMETERS, AND THE GALLIUM STANDARD

George N. Bowers, Jr.

INTRODUCTION

Clinical laboratory personnel utilize predominately two types of thermometers in their daily tasks requiring temperature measurements and/or control. The *first,* the mercury-in-glass thermometer, while familiar to everyone, is more often misused then most of us appreciate. This deceptively simple instrument actually *multiplies* rather than reduces the chance of errors in making accurate temperature measurements when it is employed without proper knowledge and careful handling. Many readily accessible publications[1-5] deal fully with the selection, calibration, and correct use of these exceedingly fragile, albeit ubiquitous, measurement tools.

The *second* temperature measurement tool which is becoming ever more common in the clinical laboratory today is the electronic thermometer utilizing predominately small, rapidly responding thermistors.[6-8] Over narrow temperature ranges near ambient, the semiconductor materials of the thermistor show marked changes in electrical resistance as a result of small changes in temperature. Thus, electrical resistance of the thermistor, rather than a visual analog due to contraction or expansion of mercury, is the property which is measured. Fortunately, simple electronic circuits which are stable can be employed and their output is easily coupled to any number of recording devices such as meters, chart recorders, printers, data-loggers, controllers, microprocessors, and/or computers. In the manufacture of thermistor resistance thermometers, sources of error due to nonlinearity, interchangeability, and long-term instability are systematically reduced to such a level that specifications for repeatability and accuracy can be shown to be valid.[9]

INTERNATIONAL PRACTICAL TEMPERATURE SCALE (IPTS)

No matter what the type of thermometer, its scale must be appropriately calibrated to some scale to be of utility. There have been many temperature scales based on the properties of any number of different materials. These scales were so numerous in the past that one glass thermometer made in Europe about 1750 had 18 scales attached to it.[10] Fortunately, today there is only *one* temperature scale used in science. This is the *International Practical Temperature Scale (IPTS)* as defined by the equilibrium states of the materials listed in Table 1 — Defining Fixed Points, or in Table 2 — Secondary Reference Points.[11-13] The IPTS scale is rigorously maintained and carefully disseminated by the ten national laboratories given in Table 3. In the U.S., the agency responsible for those standardization functions is the Heat Division, Institute of Basic Standards at the National Bureau of Standards (NBS). The responsible agent in other countries is also given in Table 3.

Dissemination of the scale into field laboratories for everyday usage is accomplished primarily through calibrations. Thermometers of several types which meet prescribed specifications are calibrated either directly at thermometric fixed points of the IPTS or indirectly by comparison to a Standard Platinum Resistance Thermometer (SPRT), the designated interpolating instrument for temperatures within the range of interest for clinical laboratories.[14] A review of the detailed metrological procedures involved in scaling the SPRT to the IPTSs fixed points and its subsequent use to calibrate other thermometers can be found in several NBS publications.[1,12,14-16] Until the recent availability of the gallium melting point cells from NBS, the only fixed point that clinical

Table 1
DEFINING FIXED POINTS OF THE IPTS-68[a]

Fixed points	Assigned value of International Practical Temperature	
	$T_{68}(K)$	$t_{68}(°C)$
Triple point of equilibrium hydrogen[b]	13.81	−259.34
Boiling point of equilibrium hydrogen at a pressure of 33 330.6 Pa(25/76 standard atmosphere)[b,c]	17.042	−256.108
Boiling point of equilibrium hydrogen[b,c]	20.28	−252.87
Boiling point of neon[c]	27.102	−246.048
Triple point of oxygen	54.361	−218.789
Triple point of argon[d]	83.798	−189.352
Condensation point of oxygen[c,d]	90.188	−182.962
Triple point of water	273.16	0.01
Boiling point of water[e]	373.15	100
Freezing point of tin[e]	505.1181	231.9681
Freezing point of zinc	692.73	419.58
Freezing point of silver	1235.08	961.93
Freezing point of gold	1337.58	1064.43

[a]   Except for the triple points and the equilibrium hydrogen point at 17.042 K, the assigned values of temperature are for equilibrium states at a pressure of 101 325 Pa (1 standard atmosphere). If differing isotopic abundances could significantly affect the fixed point temperatures, the abundances are specified.

[b]   Equilibrium hydrogen means that the hydrogen has its equilibrium ortho-para composition at the relevant temperature. "Ortho" and "para" are the designations for the molecular configurations (nuclear spin arrangements) of hydrogen.

[c]   Fractionation of isotopes or impurities dictate the use of boiling points (vanishingly small vapor fractions) for hydrogen and neon and condensation point (vanishingly small liquid fraction) for oxygen.

[d]   The triple point of argon may be used as an alternative to the condensation point of oxygen.

[e]   The freezing point of tin may be used as an alternative to the boiling point of water.

From Mangum, B. W., *Clin. Chem.,* 23, 713, 1977.

laboratory personnel could easily work with was that of the ice point at 0°C. Reproducibility of the ice point value was the practical means of checking to see that a prior NBS scale calibration was still valid and, thus, your measurement accurate to the IPTS. As Ween[4] points out, ice point determinations are complex, not simple measurements. Thus, if one does not faithfully reproduce the conditions under which the ice point value was obtained at NBS, subsequent determination may lead to confusion and ambiguity. Fortunately, gallium gives us yet another thermometric fixed point at the center of our range of clinical interest (i.e., 20 to 40°C), with which to check the accuracy of our temperature measurements to the IPTS.

THE GALLIUM MELTING-POINT STANDARD

Thornton has described the unique melting properties of pure gallium which make

Table 2
SECONDARY REFERENCE POINTS

Equilibrium state	International Practical Temperature	
	$T_{68}(K)$	$t_{68}(°C)$
Triple point of normal hydrogen[a]	13.956	−259.194
Boiling point of normal hydrogen[a]	20.397	−252.753
Triple point of neon	24.561	−248.589
Triple point of nitrogen	63.146	−210.004
Boiling point of nitrogen	77.344	−195.806
Boiling point of argon	87.294	−185.856
Sublimation point of carbon dioxide	194.674	−78.476
Freezing point of mercury	234.314	−38.836
Ice point[b]	273.15	0
Triple point of phenoxybenzene (diphenyl ether)	300.02	26.87
Triple point of benzoic acid	395.52	122.37
Freezing point of indium	429.784	156.634
Freezing point of bismuth	544.592	271.442
Freezing point of cadmium	594.258	321.108
Freezing point of lead	600.652	327.502
Boiling point of mercury	629.81	356.66
Boiling point of sulfur	717.824	444.74
Melting point of the copper-aluminium eutectic	821.41	548.26
Freezing point of antimony	903.905	630.755
Freezing point of aluminum	933.61	660.46
Freezing point of copper	1358.03	1084.88
Freezing point of nickel	1728	1455
Freezing point of cobalt	1768	1495
Freezing point of palladium	1827	1554
Freezing point of platinum	2042	1769
Freezing point of rhodium	2236	1963
Melting point of aluminum oxide	2327	2054
Freezing point of iridium	2720	2447
Melting point of niobium	2750	2477
Melting point of molybdenum	2896	2623
Melting point of lungsten	3695	3422

[a]   Normal hydrogen is a mixture of 75% orthohydrogen and 25% parahydrogen.

[b]   The ice point is a very close approximation to the temperature defined as being 0.01 K below the triple point of water.

From Mangum, B. W., *Clin. Chem.*, 23, 715, 1977. With permission.

it so useful as a thermometric fixed point.[15] Figure 1, taken from Thornton's paper, shows how increasing purity of gallium is associated with a sharper transition to a prolonged plateau at 29.77 + °C. Magnum and Thornton's subsequent detailed work with high-purity gallium on the triple point measured in cells admitting SPRTs directly gives a value of 29.77406 ± 0.001 °C with an overall systematic uncertainty to IPTS of approximately ± 1 mC.[16]* Since the gallium melting point is pressure dependent (dt/DP = −2.01 mC/std. atm), the atmospheric melting point of the SRM 1968 cells,[19] is

* Another triple point, that of ultra pure mercury at −38.8417°C has recently been described,[17] and its freezing point at atmospheric conditions −38.836°C may offer help to those few clinical laboratories requiring a fixed point below the ice point, but above the sublimation point of carbon dioxide at −78.476°C.

Table 3

NATIONAL STANDARDS LABORATORIES THAT CALIBRATE THERMOMETERS

Country	National Standards Laboratory
Australia	National Measurement Laboratory (NML)
Canada	National Research Council (NRC)
Federal Republic of Germany	Physikalisch Technische Bundesanstalt (PTB)
France	Institute National de Metrologie (INM)
German Democratic Republic	Deutsches Amt für Mass und Gewicht (DAMG)
Italy	Istituto di Metrologia "G. Colonnetti" (IMGC)
Japan	Ministry of International Trade and Industry's National Research Laboratory of Metrology (NRLM)
Russia	Physico-Technical and Radio-Technical Measurement Institute (PRMI) Mendeleev Institute (VNIIM)
U.K.	National Physical Laboratory (NPL)
U.S.	National Bureau of Standards (NBS)

From Mangum, B. W., *Clin. Chem.,* 23, 715, 1977. With permission.

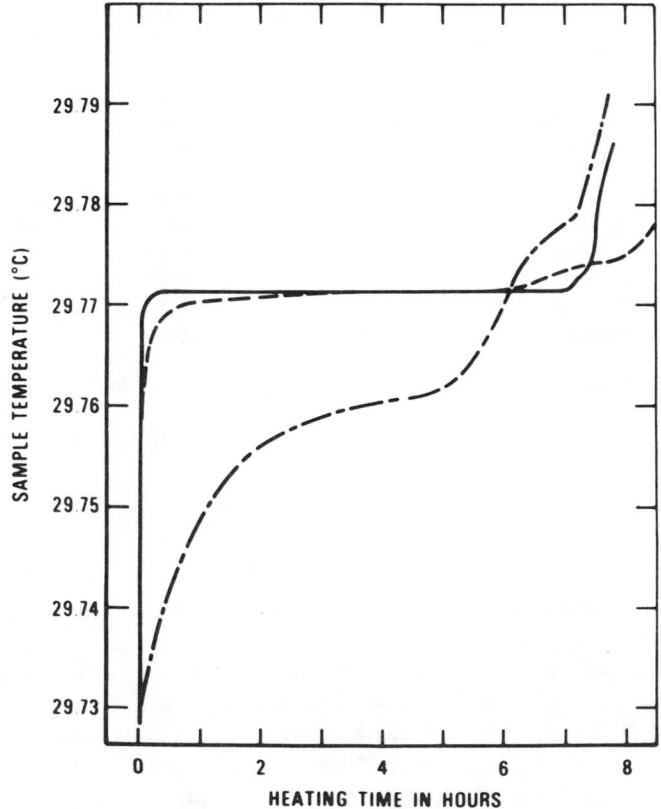

FIGURE 1. Melting curves of 99.99999% ("7N's"), 99.999% ("5N's"), and 99.9% ("3N's") pure gallium for about the same gradient in each case is shown by——, –––, and– –, respectively.

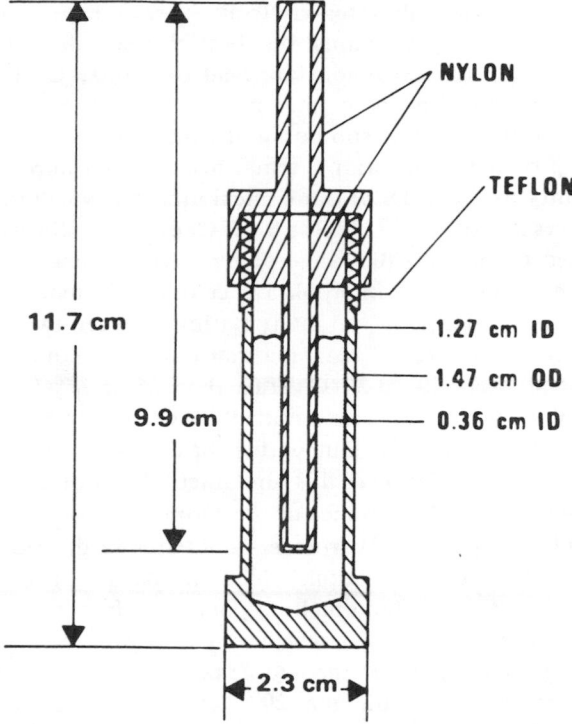

FIGURE 2. A cross-sectional drawing of a SRM No. 1968 gallium melting-point cell.

29.772°C. The Office of Standard Reference Materials at NBS now issues a small gallium melting point cell as SRM No. 1968. The design of SRM 1968 is given in Figure 2 and shows the central nylon well in which a small thermistor resting at the bottom is surrounded by an environment of melting-gallium.

Since obtaining a gallium cell from NBS in September 1976, we at Hartford have used measurements at the gallium melting point, rather than the ice point measurements as the primary means of checking the calibration of thermometer scales. The small probes of electronic thermometers are placed directly into the well of the gallium cell and calibration is very easy, requiring only a few minutes. We have found these gallium calibrated electronic thermometer system so useful, stable, and accurate that we now employ two of them* in our service laboratory.[18] Checking the calibration of mercury-in-glass thermometers at other than the ice point requires a carefully regulated comparison type water bath. At national measurement laboratories and in industrial heat laboratories specially constructed comparison baths are used for thermometer calibrations. Such comparison baths are exceedingly stable between 20 to 40°C and the uncertainty of bath temperatures to the IPTS via calibrated SPRTS is said to be 0.0001°C.[9,12,15] Using clinical laboratory water baths to check thermometer calibrations requires that they be carefully stabilized. In baths in which temperature fluctuates randomly or cyclically, the sensitive small thermister probes will follow the changes, while the massive more slowly responding mercury-in-glass thermometer will integrate them, thus, temporal differences in the indications of these two types of thermometers

---

* Model 45CV Cuvette Thermometer with Bead Thermister Probe No. 4501 from Yellow Springs Inst. Co., Oberland, Ohio.

is observed.[9] Using a gallium calibrated electronic thermometer with bead thermistor as the comparison instrument, we adjust our best clinical laboratory water bath* to cycle around 29.77°C so that excursions are held to ± 0.02°C. The indication on a mercury-in-glass thermometer properly immersed in such a bath even when the readability of the scale is to 0.01°C may still be significantly in error due to such effects as parallex, air cooling of the stem, prior thermal history, and improper storage conditions. The uncertainty to the IPTS of our clinical laboratory calibrations of mercury-in-glass thermometers is probably + 0.03 to 0.05°C, even with the use of gallium.

The closer a fixed point is to the range of temperature one uses each day (20 to 40°C) the less chance there is of interpolative errors. If a temperature of daily use coincides exactly with a fixed point, no interpolation between two fixed points is necessary and there is only the error of realizing that one fixed point. For years at Hartford, we have arbitrarily employed 30°C rather than 25 or 37°C as the reaction temperature for clinical enzymology assays. With the availability of the gallium melting point standard from NBS, this assay temperature has been switched to 29.77°C so that we can realize the ease of calibration to this fundamental fact of nature.[18]

This review emphasizes the importance of thermometric fixed points in determining the IPTS. *The Scale* is disseminated from national standardization laboratories into working laboratories through calibrations which can be traced to the fixed points of IPTS. The availability of the gallium melting point standard from NBS as SRM No. 1968 should improve the accuracy with which we can realize the IPTS in clinical laboratories by providing a fixed point in the middle of our most useful range of temperature. Since this melting point of gallium at 29.772°C can now be easily and accurately realized in working laboratories anywhere, this temperature has been suggested as *the* reaction temperature for reference methods in clinical enzymology both nationally and internationally.[12,18]

---

* Model TE45/High, Tampson (Holland), marketed by Nexlab Instruments, Inc., Portsmouth, N.H.

REFERENCES

1. Swindells, J. F., Calibration of liquid-in-glass thermometers, in National Bureau of Standards Monograph 90, Superintendent of Documents, U.S. Government Printing Office, Washington, D.C., 1965.
2. Wise, J. A., Liquid-in-glass thermometers, in National Bureau of Standards Monograph 150, Superintendent of Documents, U.S. Government Printing Office, Washington, D.C., 1976.
3. Mangum, B. W. and Wise, J. A., Description and use of precision thermometers for the clinical laboratory, SRM 993 and SRM 934, in NBS Special Publication 260-18, National Bureau of Standards, Washington, D.C., 1974; The National Bureau of Standards precision thermometers for the clinical laboratory, *Clin. Chem.,* 20, 670, 1974.
4. Ween, S., Correct application of liquid-in-glass thermometers for accurate temperature measurements in the clinical laboratory, *Clin. Chem.,* 22, 1112, 1976.
5. NCCLS Standard for Temperature Calibration of Water Baths, Instruments and Temperature Sensors, Approved Standard ASI-2, National Committee for Clinical Laboratory Standards, Villanova, Pa., 1977.
6. Trolander, H. W., Current state of electrical thermometry for biological applications, in *Temperature, Its Measurement and Control in Science and Industry,* Vol. 4, Plumb, H. H., Ed., Instrument Society of America, Pittsburgh, Pa., 1972, 2035.
7. Sapoff, M., Thermistors for biomedical use, in *Temperature, Its Measurement and Control in Science and Industry,* Vol. 4, Plumb, H. H., Ed., Instrument Society of America, Pittsburgh, Pa., 1972, 2035.
8. Sachse, H. B., *Semiconducting Temperature Sensors and Their Applications,* John Wiley & Sons, New York, 1975.
9. Sostman, H. E., The gallium melting-point standard. Its role in manufacture and quality control of electronic thermometers for the clinical laboratory, *Clin. Chem.,* 23, 729, 1977.
10. Middleton, W. E. K., *A History of the Thermometer and Its Use in Meterology,* John-Hopkins Press, Baltimore, 1966.
11. Comptes rendus des séances de la Trèizieme Conference Génèrales des Poids et Mesures, Annexe 2, 1967 to 1968, p. A1; Comité Consultatif de Thermométrie, 8e Session, 1967, Annexe 18; *Metrologia,* 5, 35, 1969; Comptes rendus des séances de la Quinzieme Confèrence Génèrale des Poids et Mesures, Resolution 7, Annexe 2, 1975, P. A1; *Metrologia,* 12, 7, 1976.
12. Mangum, B. W., The gallium melting-point standard. Its role in our temperature measuring system, *Clin. Chem.,* 23, 771, 1977.
13. Preston-Thomas, H., The International Practical Temperature Scale of 1968, *Metrologia,* 12, 7, 1976.
14. Riddle, J. L., Furukawa, G. T., and Plumb, H. H., Platinum resistance thermometry, in National Bureau of Standards, Monograph 126, Superintendent of Documents, U.S. Government Printing Office, Washington, D.C., 1973.
15. Thornton, D. D., The gallium melting-point standard. A determination of the liquid-solid equilibrium temperature of pure gallium on the International Practical Temperature Scale of 1968, *Clin. Chem.,* 23, 719, 1977.
16. Mangum, B. W. and Thornton, D. D., Determination of the triple-point temperature of Gallium, *Metrologia,* 15, 201, 1979.
17. Furukawa, G. T. and Bigge, W. R., The Triple-Point of Mercury as a Thermometric Standard, Proces Verbaux du CCT, Document No. CCT/1976-18 June 1976 (Proc. 11th Session).
18. Bowers, G. N., Jr. and Inman, S. R., The gallium melting-point standard: its application and evaluation for temperature measurements in the clinical laboratory, *Clin. Chem.,* 23, 733, 1977.

AUTOMATED ANALYZERS

William H. C. Walker

Analyses of simple solutes in plasma have increased exponentially over the past 30 years and this has only been possible because of advances in instrumentation. Photoelectric colorimeters came into general clinical laboratory use in the 50s. By the late 50s, the chemical methods for most current high-throughput analytes were established except for those using enzymes as reagents, but output was limited due to slow manipulation of pipettes, tubes, and cuvettes by hand.

At that time, a major instrumental innovation occurred with the introduction of continuous-flow analysis in the Technicon® AutoAnalyser. Proportioning pumps, dialysers, reaction in moving air-segmented liquid streams, and flow-through colorimeters replaced the labor intensiveness of manual procedures and simultaneously reduced the volume of sample needed. An analysis rate of one sample per minute with the ability to link several channels in parallel led to rapid growth of multiple analyses on individual samples, a procedure at first termed "screening" and, more recently, "profiling," especially when the tests have some functional relationship. The Technicon® SMA6/60 and 12/60 analysed 6 and 12 analytes simultaneously at 60 samples per hour.

This equipment was revolutionary, not just in its hydrodynamic concepts, but because it heralded the first comprehensive manufacturer support for the analytical process rather than for the instrument alone. User education, company supported method development, marketing of reagents, field repair service, and provision of loan units ensured reliable function of instrumentation of intimidating complexity. Such comprehensive support for the user has become the model for subsequent manufacturers of Clinical Laboratory Analysers.

Continuing development of continuous-flow technology during the past 25 years, embracing solid state circuitry, computers, and miniaturization has maintained the leadership role of this approach. The Vickers® SP120 provides computer control of Technicon® 12/60 and 6/60 multichannel analysers utilizing a DEC PDP-8/A, with doubling of sampling rate to 120/hr and adds positive sample identification, carryover correction, and output formatting. The Technicon® SMAII uses a similar computer controlled approach, and the SMAC is a computerized self-monitoring machine with miniaturized flow-system which analyzes 20 tests at 150 samples per hour with further reduction in the sample volume required.

The second instrumental innovation, Centrifugal Fast Analysis, was developed at Oak Ridge National Laboratories and became widely available in the early 1970s. The use of centrifugal force to synchronously add 30 samples to reagents and transfer them to individual cuvettes at the periphery of a rotor was combined with sophisticated optics and electronics to permit repeated absorbance readings on all the samples while they continued to rotate. Early reaction rates could be measured on large numbers of samples with precision, rendering this instrumentation applicable to almost all routine assays and particularly to enzyme assays. The small volumes of plasma (5 to 10 $\mu\ell$) and of reagents used are of advantage in pediatric work and when using expensive reagents.

Over the past 20 years, there has been a great variety of instruments designed to mechanize the steps used in manual assays. Since individual samples with their reagents stay separate from each other in their own tubes throughout the assay, these mechanized instruments have collectively been termed discrete analysers, in contradistinction to continuous-flow analysers. Electrical, mechanical, and pneumatic devices have been used to regulate, incubate, and measure responses.

Initial removal of protein by precipitation and centrifugation often required in manual assays was replaced by dialysis in continuous-flow analysers. Centrifugation is not readily mechanized and discrete anlysers have found no ready substitute. In consequence, special methods had to be developed for analytes influenced by protein matrix effects. Persistent matrix effects invalidated earlier discrete analyser methods, but these problems have been gradually overcome and adequate methodology now exists for all important analytes. Further, improvements in mechanical design and the introduction of microprocessors for control, timing, and function monitoring have led to development of mechanized discrete analysers of reliability and good performance.

Whereas continuous-flow analysers always have a dedicated channel for each analyte with up to 20 channels operating in parallel, the majority of discrete analysers have a single, mechanized track which handles the selected tests in series. Cost containment programs, rational or not, increasingly demand that unrequested analyses are not provided, however inexpensive they may be. Parallel analysers meet this requirement by reading only the analyses that are requested; serial analysers load tubes for the analyses appropriate to each sample. The resultant ability to analyse only for those tests requested is termed "selective" analysis.

In the Greiner® Selective Analyser, GSA 2D, analyses are performed serially at a rate of 300 tests per hour on any of 30 tests. A smaller version, the G300, provides 200 tests per hour out of 20 available tests. Each analysis is carried out on a transport line that allows up to four reagents to be added with stirring and incubation for adjustable times up to a total of almost 20 min. Aliquots are transferred automatically to a photometer with appropriate selection of wavelength. Start-up is rapid and stats are readily accommodated.

The American Monitor KDA is a similar serial machine, with computer control of its analytical steps. Six different reagents may be dispensed during any procedure with up to four incubation periods and a choice of three temperatures. Reagents are dispensed by a hydrostatic system based on precise timing of a fixed-flow rate. Both end-point and kinetic assays are available with a selection of 26 tests. In the American Monitor "Parallel", up to 30 tests may be performed simultaneously on each sample at a sampling rate 240/hr. Analysis time is about 12 min, with sample volumes of 5 to 50 $\mu\ell$ depending on the test. The computer controls standardization, start-up and shutdown, and there is positive sample identification.

The Hycel series of computer controlled instruments are SKS60 with 15 tests, Super 17 with 18 tests, and "M" with 26 tests. The first two sample at 60 samples per hour and the last at 120 samples per hour with capability to run any number of tests on each sample. The "M" series has positive sample identification and provides monochromatic, bichromatic, and kinetic modes of analysis.

The Abbott ABA100 is a versatile single-channel discrete analyser dedicated at any time to a single analysis, with rapid changeover to other analyses. It has end-point or kinetic capability using bichromatic spectrophotometry and generates results at up to 150 tests per hour on 5 to 25 $\mu\ell$ serum. Forty-six chemistries are available. In the Abbott VP Bichromatic Analyser, extensive computer based self-monitoring capability has been added and test throughput is increased to up to 450 tests per hour.

There has been a continuing trend to transfer from the laboratory bench to the factory production line those repetitive tasks that are unrelated directly to patient sample handling. Control sera, standards, and reagents have long been factory prepared and available in bulk or as kits suitable for small batches.

The transfer of identical aliquots of reagent into tubes prior to addition of sample is a task still often performed in the laboratory manually or by machine, but this is changing. There are kits in which the tubes already contain an appropriate amount of

Table 1
CLASSIFICATION OF ANALYTICAL
SYSTEMS

Reagent dispensing	Incubation Mode	System
Pipettes	Tubes	Manual
Separate dispenser	Tubes	Work simplification, kits
Separate dispenser	Centrifugal disc	Centrifugal fast analyser
Integral dispenser	Continuous-flow	Continuous-flow analyser
Integral dispenser	Mechanized	Discrete analysers
By manufacturer	Tubes	Kits
By manufacturer	Mechanized	Dupont® ACA
By manufacturer	Thin film	Ektachem®

reagents dispensed in the factory and lyophilized. Anticoagulated blood collection tubes and urine testing dip sticks are examples of predispensed reagents.

This approach is applied to an automated discrete analyser in the Dupont® Automatic Clinical Analyser, ACA. A heat-sealed plastic pack is provided for each test. The pack contains up to seven preweighed reagents in separate compartments and contains its own integral cuvette. It carries binary coded instructions to the instrument leading to appropriate injections of sample and diluents, aspirations, and other manipulations including choice of wavelength for colorimetry. In operation, a sample cup with patient identification card is loaded, along with appropriate reagent packs for each test required. Mechanical transfer occurs from filling station to processing, heating, delay, and reading stations. Thirty-five tests are available, singly or in any combination. There is positive sample identification with the first test available after about 7 min and further tests at a rate of 97 per hour. Such instrumentation greatly reduces the skills required of the operator while maintaining analytical quality. Sample volume is 20 to 500 $\mu\ell$ depending on the test.

The third conceptual innovation in clinical chemical analysis has come from marrying custom dispensing of reagents to the multilayer technology familiar to all users of instant photography. The Ektachem system under development by Kodak provides a plastic backing disc on which are layered in sequence reagents, diffusion barriers, and other elements. A measured drop of plasma (5 $\mu\ell$) is applied to the surface of the film and proceeds in sequence through the layers, propelled by capillary force. The endproduct is measured *in situ* by reflectance photometry or by other transducers. Electrolytes, for instance, have a pontentiometric reading taken direct from the film which is formulated as a disposable specific-ion electrode. Such integral factory packaging of reagents, flow paths, and cuvette greaty reduces the need both for complex laboratory-based instrumentation and for sophisticated operator skills. Multilayer chemistries are already well-developed and tested for a wide range of common analytes, and it is certain that this technology will gain widespread acceptance in the 1980s. A broad classification of analytical systems is shown in Table 1.

Emergency medicine and intensive care require a guarantee of rapid analytical results. Complex analytical machines will fail eventually and back-up instrumentation must be available, capable of providing at least a partial service at short notice. Contingency planning is the responsibility of the user, but manufacturers have a major role in easing this burden. Modularity of design with the ability to interchange identical units often allows the more important tests to proceed. Automatic shutdown when instrument malfunction occurs prevents loss of samples and erroneous results, but it should be capable of manual override to maintain limited function when no physical hazard exists. Good preventative maintenance routines, operator training in trouble-shooting and service, and adequate service documentation are important and sometimes neglected.

Frequently, back-up instrumentation is itself not maintained in good order and operators lack familiarity with its use. Good management requires such instrumentation to be kept in regular low volume routine use, with methodological differences minimized to ensure consistency of results. Where continuity of service is essential, it is sound practice to have two identical instruments operating together to provide the routine service and either one alone able to provide the same service, albeit at a slower rate during periods of preventative maintenance and unscheduled downtime.

There are a variety of smaller discrete analysers dedicated to a limited group of analytes, for example glucose, BUN, and electrolytes. These instruments generally have the same parallel arrangement of channels as is seen in continuous-flow analysis. The Beckman Glucose and BUN Analyser measures glucose by an oxygen electrode monitoring oxygen consumption during the reaction of glucose with glucose oxidase. Urea is measured by the conductivity change resulting from cleavage of urea by urease to form ammonium and bicarbonate ions. Analyses take less than 1 min and require 28 $\mu\ell$ serum.

The Beckman Astra 8 provides glucose, BUN, creatinine, sodium, potassium, chloride, and bicarbonate simultaneously at 70 samples per hour with a sample consumption of 175 $\mu\ell$. The Astra 4 provides any four of these tests. Both instruments have specific-ion electrodes for sodium and potassium estimation, and the same principle is used in the Nova Sodium-Potassium Analysers, which can provide results on whole blood. Glucose, BUN, and creatinine are analysed together in the 1L919 using kinetic spectrophotometric methods.

A valuable document entitled, "Characteristics and Attributes of Instruments Intended for Automated Analysis in Clinical Chemistry" has been produced by the International Union of Pure and Applied Chemistry. It provides a glossary of terms and sets out the desirable attributes of automated instrument design. Automated instruments are defined as those that are self-monitoring and self-adjusting, with automatic detection of malfunction and its description by informative error messages. Instruments should be compact with minimal special demands on ventilation, drainage, or gases. They should have provision for positive specimen identification, with linkage to the output. Reporting should be both human and machine readable with a standard interface for computer linkage. The sample should be screened for inadequacy, held under conditions that ensure stability and analysed without interaction with other samples and with minimal drift and imprecision. The volume of sample required should be compatible with pediatric use and there should be provision for insertion of an emergency sample ahead of the run.

No instrument possesses all these desirable attributes and indeed they are not equally important in all settings. The field is, however, highly competitive and established manufacturers frequently update their major instruments. Table 2 sets out the salient characteristics of some of the instruments available. It is important that updated performance claims be obtained direct from the manufacturer and that the operating experience of users in the field be solicited.

Table 2
AUTOMATED ANALYSERS

Instrument	Manufacturer	Performance claim
SMA II	Technicon Instrument Company Tarrytown, N.Y.	18 chemistries 90 samples/hr 1620 tests/hr
SMAC	Technicon Instrument Company Tarrytown, N.Y.	20 chemistries 150 samples/hr Selective test output
SP120	Vickers America Medical Corporation, P.O. Box 101 Whitehouse Station, N.J.	18 chemistries 120 samples/hr when used with Technicon SMA 6/60 and 12/60
AKEA	Datex Instrumentation Box 531, 00101 Helsinki Finland	60—80 samples/hr Continuous flow
CM2/4	Breda Scientific Box 3336 Breda	40—80 samples/hr Continuous flow
Gemsaec	Electronucleonics Fairfield, N.J.	16 cuvette rotor
Gemini	Electronucleonics Fairfield, N.J.	20 cuvette rotor
Centrifichem	Union Carbide New York, N.Y.	30 cuvette rotor
Rotochem IIa	American Instruments Co. Silver Springs, Md.	36 cuvette rotor
Multistate III	Instrumentation Laboratory Lexington, Mass.	20 cuvette rotor
ABA100	Abbott Laboratories South Pasadena, Calif.	46 chemistries 150 tests/hr
ABA-VP	Abbott Laboratories South Pasadena, Calif.	300—450 tests/hr
ACA III	Dupont Instruments Wilmington, Del.	35 chemistries 100 tests/hr
SKS60	Hycel Inc. Houston, Tx.	15 chemistries 60 samples/hr 900 tests/hr
Super 17	Hycel Inc. Houston, Tx.	18 chemistries 60 samples/hr 1020 tests/hr
M	Hycel Inc., Houston, Tx.	26 chemistries 120 samples/hr 3120 tests/hr
ASTRA 8	Beckman Instruments Fullerton, Calif.	7 chemistries 70 samples/hr 490 tests/hr
ASTRA 4	Beckman Instruments Fullerton, Calif.	4 chemistries 70 samples/hr 280 tests/hr
Seven Channel Analyser	Instrumentation Laboratories Lexington, Mass.	7 chemistries 60 samples/hr 420 tests/hr
GSA2D	Greiner Instruments Langenthal, Switzerland	30 chemistries 300 tests/hr
G300	GreinerInstruments Langenthal, Switzerland	20 chemistries 200 tests/hr
KDA	American Monitor P.O. Box 68505 Indianapolis, Ind.	26 chemistries 150 tests/hr

Table 2 (continued)
AUTOMATED ANALYSERS

Instrument	Manufacturer	Performance claim
Parallel	American Monitor P.O. Box 68505 Indianapolis, Ind.	30 chemistries 240 samples/hr 7200 tests/hr
Kem-O-Lab	Coulter Electronics Hialeah, Fla.	40 samples/hr 240 tests/hr
PRISMA	LKB Produkter Stockholm, Sweden	20 chemistries 150—300 samples/hr

REFERENCES

1. Schwartz, M. K., Multiple Analysers in clinical chemistry, in *Recent Advances in Clinical Biochemistry,* Vol. 1, Alberti, K. G. M. M., Ed., Churchill Livingstone, London, 1978.
2. Characteristics and attributes of instruments intended for automated analyses in clinical chemistry, *IUPAC Inf. Bull.,* 3, 233, 1978.
3. Evanston, M. A., Criteria for analytical instrumentation in a clinical analysis laboratory, *Anal. Chem.,* 51, 1411A, 1979.

CONTINUOUS-FLOW ANALYZERS

Jacob B. Levine

INTRODUCTION

Continuous-flow technology using air-segmented fluid streams involves the use of a peristaltic pump which delivers air or fluid at a rate proportional to the internal cross sectional area of the pump tubing, a timing device to introduce air bubbles at regular intervals, a dialyser to separate low molecular ratio solutes from protein, and a series of delay coils and incubation steps to allow reaction to proceed. Detection employs a variety of transducers: photometer, fluorometer, nephelometer, flame photometer, or ion-selective electrode with or without prior removal of the air bubbles. General reviews have been provided by Schwartz[1] and Snyder et al.[2]

Detailed characteristics of these units are described by Furman[3] in an extensive review of the literature, with operating characteristics and user experience. Along with analytical systems developed and supported by Technicon®, there have been many specialized applications developed by users, many of which have been described in the series, "Advances in Automated Analyses" (Mediaid Inc., New York). Microtechniques, some permitting analysis on less than 1 $\mu\ell$ serum per test, are described by Werner[4] and Walker.[5]

Theoretical aspects of continuous-flow have been reviewed by Thiers[6] and Walker,[7] and a model of dispersion in segmented streams has been developed by Snyder and Adler.[8,9]

In the last 10 years, development of continuous-flow technology for clinical instrumentation has been concentrated in three areas:

1. Development of computer-controlled, multichannel clinical analyzers such as the Technicon SMAC® and SMA® II systems
2. Development of stable reagent systems utilizing immobilized enzymes
3. Development of ion-selective electrodes for the routine measurement of sodium and potassium.

COMPUTER-CONTROLLED ANALYZERS

In the development of the Technicon SMAC® system, a major technical change occurred in the design of continuous-flow analyzers involving continuous monitoring of output curves to control system performance (Figure 1). In previous SMA systems, control functions such as calibration, phasing, and baseline adjustments were left in the hands of the operator. In addition, the evaluation of the analog signals produced was left to the judgement of the operator. Daily phasing manually of all analytical channels to the slowest test in the analyzer's profile array was required to maintain a fixed time relationship between the "steady-state" segments of all the tests in the profile. In practice, this meant that only a small portion of the analog information in the test curve was being used. In a 12-test Technicon SMA® system operating at an analytical rate of 60 samples per hour, only 5 sec out of the available 60 sec of analog signal (1/12) was used.

Once the decision was made to operate the Technicon SMAC® system at a rate of 150 samples per hour, it was necessary to redefine the control function of the system.

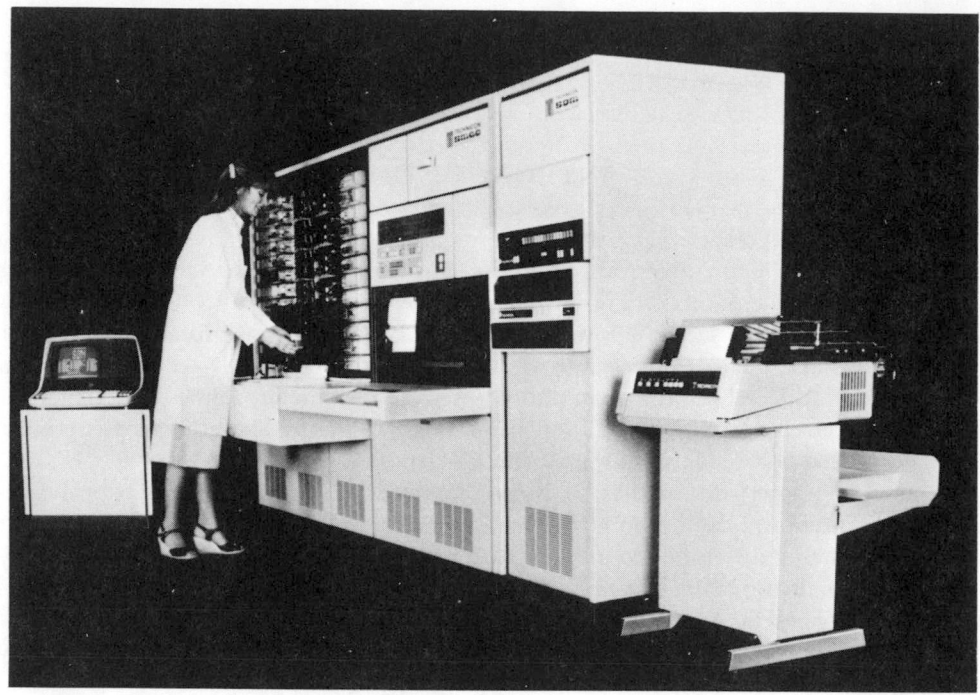

FIGURE 1.   Technicon SMAC® Analyzer.

The 24 sec of analog signal available per sample, if shared between 20 tests, would have limited the system to 1.2 sec of analog signal per test. A computer-controlled monitoring scheme was developed which viewed the entire test curve from its initial rise to "wash out". This strategy was employed first on the Technicon SMAC® system introduced in 1972 and the Technicon SMA® II system, introduced in 1976.

In the Technicon SMAC system, each analog curve produced by a sample signal produces approximately 24 sec of analog information which is measured in a chopping photometer which scans the channel every 240 msec. This signal includes the contribution of intra-sample air bubbles which are injected into the flow stream every 667 msec. In this system, unlike earlier systems, air bubbles are not removed from the analytical stream prior to passage through the flowcell. Air segments and liquid sample-reagent mixtures pass sequentially through the flowcells on the individual cartridges. Fiber optics connect the flowcells to the chopping photometer which contains 26 visible and 13 UV channels. The analog signals are then converted into digital form and the response due to the air bubble passing through the flowcell is filtered out. Response from the ion-selective electrode detectors for sodium and potassium is handled in a similar manner. The curves are then compared to the pattern expected for that particular chemistry, more than 5000 curves being monitored each hour. When curve quality does not meet preselected criteria, the computer flags the assay for review. At the same time, the computer interprets curve abnormalities in terms of specific machine malfunctions, and alerts the operator to the need for corrective action.

The Technicon SMA® II system uses a similar curve monitoring approach, but retains conventional hydraulics and debubbling of the flow stream prior to the flowcell. Data points are monitored once every second with a sampling rate of 90 per hour.

The utilization of the process control computer on the Technicon SMAC® and SMA® II systems also provides opportunity for handling of patient data. Those tests which are specifically requested may have their analytical curves converted to reportable concentration units. The system can also be programmed for institution-specific reference intervals (normal ranges). The latest versions of both systems incorporate patient demographics and calculation of test ratios. In addition, on-line monitoring of quality control materials is possible to assist the operator in reviewing the quality of the test data prior to release. Both the Technicon SMAC® and the Technicon SMA® II system are available with Laboratory Information System (LIS) output boards that are compatible with industry specifications.

STABLE REAGENT SYSTEMS UTILIZING IMMOBILIZED ENZYMES

In parallel with the development of high-capacity continuous-flow analyzers, there has been an increasing concern with the analytical accuracy of these systems as compared with "reference methods". For many analytes, these reference methods involve use of highly specific enzyme reagents. However, the high cost of purified enzyme reagents encourages some scheme for their recycling. Recent development of physical or chemical binding techniques which coat an enzyme on the walls of a tube which can be placed in a flow stream offers such an opportunity. Successful applications of this technique include the measurement of glucose by glucose oxidase or hexokinase coated coils (Figure 2). Recently, a more specific uric acid assay was developed by immobilizing uricase and coupling it with the measurement of the peroxide liberated by treatment of the sample as it passes through the coil.

Both the hexokinase and uricase coils have demonstrated life in excess of 5 weeks when used in routine operation on both the Technicon SMAC® system and SMA systems. As many as 30,000 assays have been performed using an individual coil before it required replacement. Not only does the laboratory benefit from improved accuracy, but there is the benefit of no longer having to prepare reagents on a daily basis. Currently, over half of the Technicon SMAC® systems in the U.S. are utilizing the hexokinase procedure routinely.

ION-SELECTIVE ELECTRODES FOR SODIUM AND POTASSIUM

In the early development of the Technicon SMAC®, analysis of sodium and potassium by a flame photometry was found to be a rate limiting step. A possible analytical alternative was the development of ion-selective electrodes for the measurement of these cations. This approach proved to be feasible with the development of a glass electrode for the measurement of sodium and an electrode which incorporated valinomycin in a PVC matrix for the measurement of potassium. A sampling rate of 150 per hour was attainable with these electrodes.

This electrode technology has also been applied to stand alone, bench-top analyzers for the analysis of sodium, potassium, chloride, and CO_2. The Technicon STAT/ LYTE® system, introduced in 1976, incorporated the hydraulic technology and electrodes used on the Technicon SMAC® system. Chloride measurements used a Ag/ AgCl electrode and CO_2 was determined by monitoring the pH change of a solution into which the dialyzed gas passed after acidification of the sample. The system was designed for 24-hr operation and incorporated an automated two-point recalibration every 30 min and a series of diagnostic routines to assist the operator in maintenance and troubleshooting. It operates at a rate of 36 samples per hour, producing results in 2.5 min on 200 μl serum or plasma.

In 1979, an improved version of this four-channel electrolyte analyser, the C800®, was introduced (Figure 3). This system operates at 72 samples per hour on less than

FIGURE 2.   Hexokinase coated reaction coils.

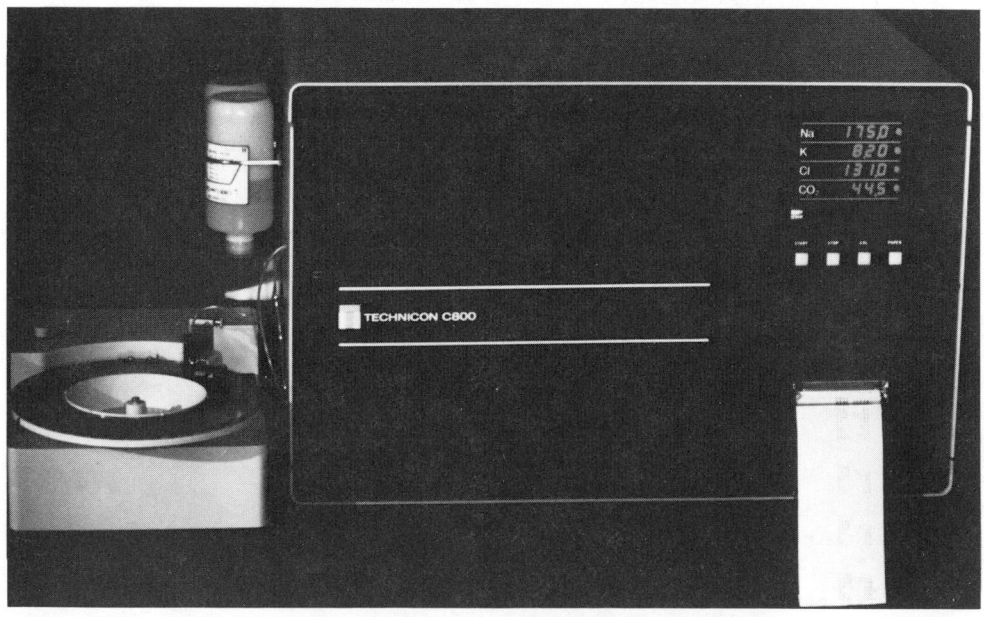

FIGURE 3.   Technicon® 4-channel electrolyte analyzer, C800®.

100 $\mu\ell$ serum or plasma. It incorporates an improved chloride electrode system and utilizes "curve regeneration" techniques to improve throughput. Among its operating features is a stand-by mode which primes the system every half hour and maintains stability of calibration around the clock making it available for STAT analysis at all times.

REFERENCES

1. **Schwartz, M. K.,** Continuous flow analysis, *Anal. Chem.,* 45, 739A, 1973.
2. **Snyder, L., Levine, J., Stoy, R., and Conetta, A.,** Automated chemical analysis, update on continuous-flow approach, *Anal. Chem.,* 48, 942A, 1976.
3. **Furman, W. B.,** *Continuous Flow Analysis,* Marcel Dekker, New York, 1976.
4. **Werner, M.,** Microtechniques for the Clinical Laboratory, John Wiley & Sons, New York, 1976, Chaps. 19—21.
5. **Walker, W. H. C.,** Practical aspects of continuous flow analysis, in *Advances in Automated Analysis,* Mediaid, New York, 1977, 82.
6. **Thiers, R. E.,** in *Clinical Chemistry,* Henry, R. J., Cannon, D. C., and Winkelman, J. W., Eds., Harper & Row, Hagerstown, Md., 1974, Chap. 10.
7. **Walker, W. H. C.,** in *Continuous Flow Analysis,* Furman, W. B., Ed., Marcel Dekker, New York, 1976, 207.
8. **Snyder, L. R. and Adler, H. J.,** Dispersion in segmented flow through glass tubing in continuous flow analysis, the ideal model, *Anal. Chem.,* 48, 1017, 1976.
9. **Snyder, L. R. and Adler, H. J.,** Dispersion in segmented flow through glass tubing in continuous flow analysis, the non-ideal model, *Anal. Chem.,* 48, 1022, 1976.

CENTRIFUGAL ANALYZERS

Thomas O. Tiffany

INTRODUCTION

The centrifugal analyzer was developed by Anderson.[1-6] The concept evolved from a device which used small chambers to measure precise fluid volumes in centrifugal fields and used centrifugal force to fill and empty them completely.[7] Through the unique coupling of centrifugal sample/reagent addition to the dynamic photometric monitoring of multiple reaction mixtures, a new concept in discrete sample analysis for the clinical laboratory was developed. In this new analytical system, centrifugal force was used to add aliquots of a reagent discretely to individual samples and to propel these reaction mixtures radially into separate equally spaced cuvettes located on the outer edge of a rotor. The absorbance of these reaction mixtures was measured simultaneously in the same time temperature space. The combination of the simultaneous photometric measurements with the numerical power of a computer permitted the application of a variety of equilibrium and kinetic methods.[8,9]

The early development of centrifugal analyzers can best be viewed from Anderson's own reflections — "The Birth and Early Childhood of Centrifugal Analyzers".[10] The subject of the centrifugal analyzer and its use in the clinical laboratory has been reviewed by Burtis et al.,[11] Cross,[12] and Tiffany.[13,14] These reviews provide more detailed discussions of the definitions, concepts, and applications contained in this chapter. Finally, the methods applied to centrifugal analyzers are discussed in volumes edited by Savory and Cross[15] and Price and Spencer.[16]

This section is a reference guide to centrifugal analyzers. The object is to present appropriate definitions, concepts, and applications related to centrifugal analyzers and to provide references. The intended result is a resource document which provides pointers from various centrifugal analyzer topics to the literature.

The subject of centrifugal analyzers is divided into three parts: (1) instrumentation; (2) concepts; and (3) applications. Instrumentation is presented as a series of definitions. Concepts which relate to the distinguishing features of centrifugal analyzers are presented. The applications of these concepts and methods are categorized in the third section.

INSTRUMENTATION

The centrifugal analyzer consists of three major elements: (1) a centrifugal analytic module; (2) a computer; and (3) reagent and sample metering (discrete loading and dynamic loading). These major elements or modules are listed with their component parts in Table 1. This section will define these modules and elements.

Centrifugal Analytical Module

The centrifugal analytical module consists of a transfer disc, a cuvette rotor, a photometric system, a device for synchronizing transmission signals, and a unit for control and measurement of temperature.

Transfer Disc

The transfer disc is an inert machined or molded circular device, usually constructed from Teflon®, which contains concentrically located chambers for sample and reagent.

Table 1
CENTRIFUGAL ANALYZER
COMPONENTS

I.   Centrifugal Analytical Module
    A.   Transfer Disc
        1.   Sample Chamber
        2.   Reagent Chamber
    B.   Cuvette Rotor
        1.   Optical Windows
        2.   Slotted Spacer Annulus
        3.   Cuvette Chambers
        4.   Syphons
        5.   Rotor Assembly
        6.   Rotor Drive
    C.   Combined Transfer Disc/Cuvette Rotors
        1.   Three Piece Rotor
        2.   Flexible Disposable Rotor
        3.   Rigid Fixed Pathlength Disposable Rotor
        4.   Variable Pathlength Disposable Rotors
        5.   Rotor Holder
        6.   Rotor Drive
        7.   Rotor Brake
    D.   Photometric System
        1.   Absorbance Photometers
            a.   Light Source
            b.   Monochromator
            c.   Detector
        2.   Fluorescence/Light Scatter Photometer
            a.   Light Source
            b.   Monochromator
            c.   Detector
        3.   Luminescence Measurements
    E.   Transmission Signal Synchronization
    F.   Temperature Measurement and Control

II. Data Processor Module
    A.   Computer
        1.   Central Processing Unit
        2.   Memory
        3.   Input Output Ports
    B.   Data Storage Device
        1.   Cassette
        2.   Flexible Disc
        3.   Fixed Disc
    C.   Peripheral Data Display Devices
        1.   Analog Display Oscilliscope
        2.   Programmable Storage Scope
        3.   X-Y Plotter
        4.   Printer
        5.   Video Display Unit
    D.   Software
        1.   High Level Language
        2.   Assembler Language
        3.   Algorithms

III. Reagent and Sample Metering
    A.   Discrete Sample/Reagent Loading
        1.   Manual Pipetting Devices
        2.   Mechanical Loading Devices
    B.   Dynamic Loading

Sample chambers — A set of concentrically located holes or steps in a transfer disc designed to hold an aliquot of sample separate from reagent. The sample chambers are designed to be angled radially such that the sample aliquot remains in the sample chamber at low centrifugal velocity, but transfers to radially located reagent or cuvette chambers upon attaining higher centrifugal speed.

Reagent chambers — The reagent chambers are likewise located concentrically in the transfer disc and, generally, they are positioned radially in front of the sample chamber. They are also designed to retain their aliquoted reagent volume at low centrifugal speed. The optimal design of the transfer disc provides the reagent chamber located before the sample chamber. This permits the mixing and washing out of sample with reagent when the transfer disc/cuvette rotor has attained sufficient centrifugal speed.

Cuvette Rotor

The cuvette rotor consists of a slotted spacer annulus compressed between two optical windows. These three components are housed into a rotor assembly. The rotor assembly is attached to a rotor drive. The assembled cuvette rotor receives the radially driven reagent sample mixtures into slotted cuvette chambers. These mixtures are further mixed and eventually emptied from the rotor by syphons located in the slotted spacer annulus.

Optical windows — A set of windows consisting of a thin quartz or Pyrex® glass annulus (top window) and a glass disc (bottom window) which form the optical windows of the rotor.

Slotted spacer annulus — The slotted spacer annulus defines the pathlength btween the top window annulus and the bottom rotor disc. Each slot in the annulus defines a cuvette chamber. The annulus must be inert to chemical reactions. Thus, it is generally made from Teflon® or Teflon®-coated metals.

Cuvette chamber — A radially located slot in the spacer annulus which defines one of several cuvette chambers in the rotor.

Syphon — An inverted u-shaped slot (hole) in the spacer annulus extending from the bottom of the cuvette chamber and radially along the edge of the cuvette chamber about two thirds up the length of the chamber. The slot then turns 180° and extends to the outer edge of spacer annulus. The syphon defines the maximum fluid volume which can be contained in the sealed rotor cuvette. The syphon also provides a conduit, whereby air can be pulled into the cuvette by applying a negative pressure to the center of the rotor and waste liquid can be emptied by applying a positive pressure at the center of the rotor.

Rotor assembly — The rotor assembly consists of a top rotor housing, a top window annulus, the spacer annulus, the bottom window disc, and the lower rotor housing bolted together.

Rotor drive—The rotor assembly is attached to a drive motor to form the rotor drive. The assembled rotor and drive attain speeds of 2000 r/min during acceleration, but operate during run at rotors speeds of 500 to 1000 r/min.

The above definitions of the transfer disc and cuvette rotor describe first generation centrifugal analyzer technology as exemplified by Anderson's original centrifugal analyzer and the first commercial centrifugal analyzers. The following section defines the rotor technology used in the first NASA miniature centrifugal analyzer and the newer commercial second-generation centrifugal analyzers.

Combined Transfer Disc/Cuvette Rotors

The transfer disc and the cuvette rotor can be combined into an integrated analytical rotor. The integrated analytical rotor was first developed for the miniature centrifugal analyzer.[17,18] The combined rotor concept permits these types of rotors to be fabricated from optical grade plastic or molded from thermal plastics. Thus, the combined rotor becomes smaller, requires less sample and reagent volume, and is removable from its rotor drive. The separation of the analytical rotor from the drive provides for more flexibility in analytical applications. As an example, rotors have been designed for fluorescence and absorbance photometric measurements.[19]

Three-piece rotor — The miniature centrifugal analyzer rotor was a three-piece rotor. It was constructed by combining an ultraviolet transmitting (UVT) top window disc, a black acrylic center disc, and a bottom UVT acrylic window disc. The black center disc contained the reagent chambers, the sample chambers, and the cuvette chambers. It served also as a spacer to maintain the optical pathlength. Many types of these rotors were designed and fabricated for various analytical applications.[20,24]

Flexible disposable rotor — A flexible plastic two-piece rotor has been designed for the Electronucleonics Gemini Centrifugal Analyzer.

Rigid fixed pathlength disposable rotor — A rigid, UVT acrylic plastic rotor has been developed for the Instrumentation Laboratory, Inc. Micro Centrifugal Analyzer. This rotor provides both fluorescence and absorbance optical windows.[14,25]

Variable pathlength disposable rotors — A new acrylic rotor has been developed by the Hoffman-La Roche & Co. Ltd. for their Cobas Centrifugal Analyzer. This rotor is designed for absorbance measurements in cuvettes lying longitudinal to the light beam.[26] This minimizes volume errors due to reagent pipetting.

Rotor holder — The rotor holder provides the indexing, transport, and temperature sensors required to support the removable analytical rotor.

Rotor drive — The rotor holder is coupled to a high speed DC motor to form the rotor drive.

Fluorescence/Light Scattering Photometer

The fluorescence/light scattering photometer can be designed to use a right-angle optical configuration,[19,24,27] a front-surface optical configuration,[28] or a low-angle optical configuration.[29] The light source can be a quartz halogen lamp, a xenon arc lamp, or a laser.[24,28,29] The monochromator device used for excitation wavelength selection can be an interference filter or a grating monochromator. The detector is generally a sensitive photomultiplier tube.

Luminescence measurements — The measurement of photons emitted from chemical or biochemical reactions is called chemiluminescence or bioluminescence.[30] Burtis et al. described the chemiluminescence measurement of cholesterol using luminol/hydrogen peroxide with a miniature centrifugal analyzer.[22]

Transmission Signal Synchronization

A rotating analytical rotor will produce a continuous series of transmission signals. To synchronize these transmission signals with the computer, a rotor pulse and individual cuvette pulses are generated. This is accomplished by shining light through holes located in the rotor, the rotor holder, or on the rotor drive onto an appropriate light sensitive detector.

Temperature Measurement and Control

Temperature measurement within centrifugal analyzers is generally accomplished by use of calibrated thermistors. The thermistor is located within a cuvette or in an indexing pin next to the cuvette. Temperature control involves feedback from the temperature sensor to a control circuit. Temperature is adjusted in the rotor by means of electrically heating, by radiant heating, or by forced-hot air heating of the analytical rotor. Other temperature sensing media have been employed. An attractive idea for dynamic temperature measurement is the use of the optical thermometer where absorbance change of temperature sensitive dyes is related to temperature.[31-34]

The Data Processor Module

The data processor module consists of a computer, a data storage device, one or more peripheral data display devices, and the software to make it operate.

Computer

The computers used in most centrifugal analyzers are either minicomputers or microcomputers. Regardless of their size, these computers consist of three basic elements: (1) the central processing unit; (2) the memory hierarchy; and (3) the input/output ports.[35]

The central processing unit — The central processing unit (CPU) consists of a control logic unit and an arithmetic and logic unit. The CPU controls the operation of the entire computer. It reads information from memory, interprets instructions, performs operations on the data according to the instructions, writes the results back into memory, and moves information between memory levels or through the input/output ports.

The memory hierarchy — Stores the instructions and the data in the system so they can be retrived quickly on demand by the CPU. The input and output ports (I/O) are the paths whereby information is fed to the computer or taken out by means of magnetic tape, terminals, etc.

Data Storage Device

Media used to store analytical results from centrifugal analyzer runs or software programs are cassettes, flexible discs, or fixed discs units.

The cassette tape — Is capable of storage of 256 K words. It is a reliable storage media, but requires several seconds to retrieve data.

The flexible disc — Comes in mini and standard disc sizes. This device is capable of storing more than one million words of data. It requires milliseconds to 1 sec for data retrieval.

The fixed disc — Is a rapid data retrieval device. It is capable of storing several million words of data.

Peripheral Data Display Devices

Centrifugal analyzers use different types of data display, depending upon the applications. These data display units include: (1) analog display oscilliscope; (2) a programmable storage scope; (3) an X-Y plotter, (4) printers; and (5) video display unit.

The analog data display oscilloscope — Shows transmission signal vs. cuvette position. This is a visual qualitative means of assessing reactions in progress. It has limited quantitative use.[36]

The programmable storage scope — Used to provide graphical representation of centrifugal analyzer data. It is used primarily as a research tool.[37]

The X-Y plotter — Provides hard copy of the graphical centrifugal analyzer results; it also is a research tool.

Printer — A variety of printer terminals are available. These printers provide final copy of centrifugal analyzer results.

Video display unit (VDU) — A more representative name for the cathode ray tube (CRT). This device serves as a programming module, a rapid visual data display module, and a communication module.

Software

The data processing module of the centrifugal analyzer requires software programs to operate. The analysis programs are generally written in high-level language such as Fortran, Basic, or Focal®. The input/output routines are generally written in assembly language. The logical organization of program steps to solve a numerical problem is called an algorithm. The versatility of computer controlled instruments using high-level program language is that a number of algorithms can be written to accommodate different types of reactions. A variety of analytical software has been written for centrifugal analyzers.[8,9,37-40] The flexibility of the centrifugal analyzer is demonstrated by these software applications.

Reagent and Sample Metering

The centrifugal analyzer requires a device to load sample and reagent. Traditionally, this is accomplished by loading discrete sample and reagent aliquots. Discrete loading can be done by hand. The first centrifugal analyzer was a hand pipetted device. Mechanized loaders were developed to improve sample throughput. These mechanical loaders have employed fixed volume syringes,[41] mechanically controlled digital syringes,[20,42] and electronically controlled digital stepper driven syringes.[43]

Dynamic loading is a process whereby a liquid stream is proportioned evenly in all cuvettes of a spinning rotor.[44,45] Burtis et al. demonstrated several applications of this technique of loading sample reagent or diluent into a spinning analytical rotor.[46] The techniques have also been used to load antibody into a rotor to check for antigen excess.[19]

CONCEPTS

General Conceptional Requirements

The requirements for a new class of analytical systems called "fast analytical systems" were defined by Anderson to be simultaneity, scaling, engagement, micro sample volumes, rapid sample to reagent addition, visible reaction monitoring, and rapid data reduction. These requirements form the basic concepts of centrifugal analysis.

Simultaneity — Is the concept of all reactions starting together and proceeding in parallel. This is important not only in time-dependent color development in colorimetry, but also in kinetic methods of analysis using two-point fixed time absorbance measurements of continuous reactions and in multipoint methods using regression methods to determine enzyme activity.

Scaling — Is an important concept which suggests that effective automated analytical systems should be able to analyze larger numbers of samples in batch or few samples with equal ease and economy. According to this concept, large and small-scale analyses are performed in the same manner. One technologist handles each type of analyses, and the accuracy and precision are comparable for both. Further, similar instrumentation and procedures can be used in large hospitals or in small hospitals or clinics in the city, as well as in rural communities.

Engagement — Refers to the fact that all of the analytical system functions must be sufficiently fast to permit the technologist to interact, respond, and control. The use of *micro volumes* of sample and reagent by the analytical system will lead to economy (smaller amounts of expensive reagents will be used) and convenience (the sample volume required from an individual patient will be minimized). *Rapid reagent-to-sample addition* is a basic concept of centrifugal analysis. Through very rapid acceleration of the analytical rotor, sample and reagent are admixed. Simultaneity is the obvious result of rapid sample-to-reagent addition. This concept also enables early reaction absorbance measurements, which is essential to some kinetic methods of analysis.[9,37]

Visual reaction monitoring — By observation of the transmission signals displayed on an oscilloscope CRT is provided for the technologist to monitor all reactions in real time, troubleshoot, and rapidly repeat an analysis when needed. The trend in newer microprocessor analyzers is to develop error checking algorithms which minimize the need of the CRT, but maintain the concept of monitoring reactions in real time for substrate exhaustion, nonlinear kinetics, inappropriate samples, and other reaction qualities which would indicate the need to flag or repeat an analysis.

Data reduction — Must be rapid and capable of being performed during the time required for the analysis or directly thereafter. It should be flexible enough to allow many different analytical approaches to be taken, including enzyme activity assays, equilibrium methods, and kinetic methods.

Integration and Application of General Concepts

The centrifugal analyzer was the result of the combination of the analytical systems requirements. There are three concepts that are unique to centrifugal analyzers; dynamic referencing, centrifugal discrete reagent/sample addition, and dynamic reagent/sample addition.

Dynamic referencing refers to the measurement of dark current, reference cuvette transmittance, and sample cuvettes transmittances during each rotor revolution. One rotor revolution occurs approximately every 50 msec. The logarithm of the ratio of the dark current corrected reference transmittance to the dark current corrected sample transmittance provides an absorbance measurement similar to a double-beam photometer.[5] Therefore, long-term source lamp drift is minimized as an error component. A similar concept exists for fluorescence relative intensity measurements. Thus, during each rotor revolution, the relative intensity of a standard reference cuvette, a blank cuvette, and the unknown sample cuvettes are measured.[27] This provides a ratio referencing of fluorescence measurements.

The concept of the separate or self-contained sample/reagent transfer disc with its radially located sample/reagent chambers provides the vehicle that causes rapid discrete sample/reagent addition upon acceleration of the analytical rotor.[1]

Dynamic liquid addition to the rotor is a concept unique to centrifugal analyzers.[45] By controlled injection of a liquid stream onto a series of identical splitting vanes located around the inner circumference of the rotor, reagent or sample can be aliquoted into each rotor cuvette.[46] This concept has been used in the development of a centrifugal radioimmunoassay system.[47,48]

Analytical Concepts

The centrifugal analyzer is a stationary photometer or spectrophotometer with multiple reaction cuvettes rotating past the detector. Thus, most colorimetric methods adaptable to a double-beam photometer can be applied to the centrifugal analyzer.[11-13] This extends likewise to fluorometric and light scattering types of measurements.[49] This provides a *flexibility* concept in methods development.

Kinetic Methods

The classification of kinetic methods of analysis used in clinical chemistry has been presented by Pardue.[50] Physicochemical methods of analysis such as photometric methods are divided into equilibrium and kinetic methods. Kinetic methods are classed as variable sensor-signal methods (centrifugal analyzer absorbance measurements) and fixed sensor signal methods. The variable sensor methods are further catagorized as direct response two-point and multiple-point methods. The concept of classification of kinetic methods of analyses is useful for better communication of analyses schemes.

Table 2

ENZYME ASSAYS APPLIED TO
CENTRIFUGAL ANALYZERS

	Ref.
Routine clinical enzymology	
Aspartate aminotransferase (AST)	53
Alanine aminotransferase (ALT)	53
Lactate dehydrogenase (LD)	53
Creatine kinase (CK)	53
α-Hydroxy Butyrate Dehydrogenase (HBD)	53
Alkaline phosphatase (AP)	53
Acid phosphatase	71
Gamma glutomyl transferase (GGT)	72
5' nucleotidase	73
Amylase	74
Special clinical enzymology	
Sorbitol dehydrogenase	75
Glucose-6-phosphate dehydrogenase in erythrocytes	76, 77
Serum alkaline phosphatase	78
Isoenzymes	
Adenylate kinase	77
Galactose-1-phosphate uridyltransferase	77
Glycerol dehyde-3-phosphate dehydrogenase	77
Glucosephosphate isomerase	77
Glutathione peroxidase	77
Glutathione reductase	77
Hexokinase	77

The two kinetic methods most used on centrifugal analyzers are the fixed-time and multipoint regressions methods. The former method is useful for substrate or analyte measurements, while the latter is useful for enzyme activity measurements. Pardue provides some useful concepts of multiple-point regression analysis which are directly applicable to the centrifugal analyzer.

Equilibrium Methods

An equilibrium method is one in which the forward reaction rate is equal to the reverse reaction rate. There is no further apparent change in measured analyte concentration. A centrifugal analyzer can be programmed to do equilibrium methods.[8,9] The advantage provided by the centrifugal analyzer is sample blank correction, either through early absorbance measurements or by use of bichromatic absorbance measurements.

APPLICATIONS TO CLINICAL ANALYSES

The first analysis on centrifugal analyzers was total protein by a modified Biuret reaction. However, the initial interest in applications for the centrifugal analyzers was toward enzyme activity assays.[51] Several algorithms have been developed to monitor and measure enzyme activity.[11,20,38,50,52] The maximum utilization of the centrifugal analyzer parallel analysis feature and computerization was in the development of algorithms to calculate enzyme kinetic parameters. The subject of enzyme and isoenzyme measurement on the centrifugal analyzer is reviewed by Skillen.[53] The enzyme assays applied to centrifugal analyzers are provided in Table 2.

The centrifugal analyzer became locked in a role in the early 1970s as an enzyme analyzer. Ingle and Crouch published an excellent paper on analytical rate measurements.[54] They demonstrated the relationship between linear absorbance change and concentration within a fixed time interval. The concepts of rate analysis were extended to centrifugal analyzers in a paper on the application of enzymatic kinetic rate and end-point analyses of substrate to centrifugal analyzers.[9] This paper was an attempt to demonstrate that centrifugal analyzers could be used for both enzyme activity and substrate analysis, thus, making it a clinical chemistry analyzer. The kinetic substrate theory has been extended to optimize the amount of enzyme used in the reaction.[55-57] The subject of enzyme mediated analyte measurement and its application to the centrifugal analyzer is reviewed by Ziegenhorn.[58]

Interest in the use of kinetic measurement has spread beyond the study of enzyme mediated reactions. Kinetic assays have been used for the measurement of other analytes, such as creatinine, by the use of nonenzyme mediated reactions.[59,60] There are several advantages to this approach which is reviewed by Ertingshausen.[61] When the speed of the reaction is such that collection of kinetic data is not possible or when sample blanking is required in equilibrium methods, then bichromatic absorbance measurement can be used. This facility can also be used to correct for sample interference. This topic is discussed by Price and Spencer.[62] Equilibrium and kinetic enzyme substrate and analyte methods applied to centrifugal analyzers are listed in Table 3.

The centrifugal analyzer has been used for enzyme immunoassays. The proper design of the analytical system makes it possible to automate these procedures. Johnson has developed a robust four-parameter nonlinear least squares routine which is capable of fitting EIA data in the general case.[40]

Other centrifugal analyzer applications have included turbidimetric protein methods,[63,64] coagulation methods,[65,66] protein measurements by use of light scattering measurements,[67-69] and hemagglutination measurements.[24,70]

The present and future applications of the centrifugal analyzer have been well-reviewed.[11-16,25] These demonstrate that the concept is being applied to the clinical laboratory.

FUTURE TRENDS

The future trend for centrifugal analysis will be the integration of multiple units into single-reporting stations through the use of computers. These computers will serve as data collation systems in the smallest sense and as data management systems in the larger applications.

The analyzers will also have expanded analytical functions through application of fluorescence and other types of optical systems.

These analyzers should be able to accommodate "stats", small individual workloads, or larger workloads. Whether these predictions come about depends upon the continued acceptance of the present centrifugal analyzers in the clinical laboratory.

Table 3
CLINICAL LABORATORY ANALYTE METHODS APPLIED TO CENTRIFUGAL ANALYZERS

	Ref.
Metabolites and inorganic ions	
Glucose	9
Urea	9
Creatinine	59
Uric acid	79
Cholesterol	80
HDL-cholesterol	81
Triglycerides	80, 82
Bilirubin	83, 84
Iron/binding	85
Inorganic phosphorous	86, 87
Ammonia	88
Lactate	89, 92
Galactose	90
2-3, Diphosphoglycerate	91
β-Hydroxybutyrate	92
Acetoacetate	92
Ethanol	93
Calcium	27
Proteins	
Total protein	51, 94
Albumin	95—97
Fibrinogen	98
Transferrin	99
Immunoglobulins	100
Therapeutic drug analyses	
Anti-epileptic drugs	101
Theophylline	102
Other applications	
Urine drugs of abuse analyses	103

REFERENCES

1. **Anderson, N. G.**, Analytical techniques for cell fractions. XII. A multiple cuvet-rotor for a new analytical system, *Anal. Biochem.*, 28, 545, 1969.
2. **Anderson, N. G.**, Computer interfaced fast analyzers, *Science*, 166, 317, 1969.
3. **Anderson, N. G.**, Analytical techniques for cell fractions. XIV. Use of drainage syphons in a fast-analyzer cuvet-rotor, *Anal. Biochem.*, 32, 59, 1969.
4. **Anderson, N. G.**, Analytical techniques for cell fractions. XVI. Preparation of protein-free supernatants with a "Z" path rotor, *Anal. Biochem.*, 31, 272, 1969.
5. **Anderson, N. G.**, Basic principles of fast analyzers, *Am. J. Clin. Pathol.*, 53, 778, 1970.
6. **Anderson, N. G.**, The development of fast analyzers, *Z. Anal. Chem.*, 261, 257, 1972.
7. **Anderson, N. G.**, Analytical techniques for cell fractions. IX. Measurement and transfer of small fluid volumes, *Anal. Biochem.*, 23, 207, 1968.

8. Kelley, M. T. and Janse, J. M., Programming concepts for the GeMSAEC rapid photometric analyzer, *Clin. Chem.*, 17, 701, 1971.
9. Tiffany, T. O., Jansen, J. M., Burtis, C. A., Overton, J. B., and Scott, C. D., Enzymatic kinetic rate and end-point analyses of substrate by use of a GeMSAEC fast analyzer, *Clin. Chem.*, 18, 829, 1972.
10. Anderson, N. G., The birth and early childhood of centrifugal analyzers, in *Methods for the Centrifugal Analyzer,* Savory, J. and Cross, R. E., Eds., American Association for Clinical Chemistry, Washington, D.C., 1978, xii.
11. Burtis, C. A., Tiffany, T. O., and Mrocheck, J. E., The centrifugal analyzer and its use in clinical analysis, in *Methods for the Centrifugal Analyzer,* Savory, J. and Cross, R. E., Eds., American Association for Clinical Chemistry, Washington, D.C., 1978, 3.
12. Cross, R. E., Centrifugal analyzers, *Lab. Med.,* 9, 8, 1978.
13. Tiffany, T. O., Centrifugal fast analyzers in clinical laboratory analysis, *Crit. Rev. Clin. Lab. Sci.,* 5, 129, 1974.
14. Tiffany, T. O., The concepts of centrifugal analyzers, in *Centrifugal Analyzers in Clinical Chemistry,* Price, C. P. and Spencer, K., Eds., Praeger Publishers, East Sussex, U.K., 1980, chap. 1.
15. Savory, J. and Cross, R. E., Eds., *Methods for the Centrifugal Analyzer,* American Association for Clinical Chemistry, Washington, D.C., 1978, 287.
16. Price, C. P. and Spencer, K., Eds., *Centrifugal Analyzers in Clinical Chemistry,* Praeger Publishers, Eastbourne, East Sussex, U.K., 1980.
17. Anderson, N. G., Burtis, C. A., Mailen, J. C., Scott, C. D., and Willis, D. D., Feasibility of miniaturization of a fast analyzer, *Anal. Lett.,* 5, 153, 1972.
18. Burtis, C. A., Mailen, J. C., Johnson, W. F., Scott, C. D., and Tiffany, T. O., Development of a miniature fast analyzer, *Clin. Chem.,* 18, 753, 1972.
19. Tiffany, T. O., Parrela, J. M., and Burtis, C. A., Specific protein analysis by light scatter measurement with a miniature centrifugal fast analyzer, *Clin. Chem.,* 20, 1055, 1974.
20. Burtis, C. A., Johnson, W. F., Mailen, J. C., Overton, J. B., Tiffany, T. O., and Watsky, M. B., Development of an analytical system based around a miniature fast analyzer, *Clin. Chem.,* 19, 895, 1973.
21. Burtis, C. A., Johnson, W. F., and Tiffany, T. O., The development of rotors having separate samples and reagent transfer channels for use with centrifugal analyzers, *Anal. Lett.,* 7, 591, 1974.
22. Burtis, C. A., Bostick, W. D., and Johnson, W. F., Development of a multipurpose optical system for use with a centrifugal analyzer, *Clin. Chem.,* 21, 1225, 1975.
23. Mrocheck, J. E., Burtis, C. A., Johnson, W. F., Bauer, M. L., et al., A new portable centrifugal analyzer with expanded versatility, *Clin. Chem.,* 23, 1416, 1977.
24. Tiffany, T. O., Parella, J. M., Burtis, C. A., Johnson, W. F., and Scott, C. D., Blood grouping with a miniature centrifugal analyzer, *Clin. Chem.,* 20, 1043, 1974.
25. Tiffany, T. O., Manning, G. B., Hills, L. P., and Frankart, M. L., The application of fluorescence measurements to centrifugal analyzers, in *Centrifugal Analyzers in Clinical Chemistry,* Price, C. P. and Spencer, K., Eds., Praeger Publishers, East Sussex, U.K., 1980, chap. 21.
26. Eisenwiener, H. G. and Keller, M., Absorbance measurements in cuvettes lying longitudinal to the light beam, *Clin. Chem.,* 25, 117, 1979.
27. Tiffany, T. O., Watsky, M. B., Burtis, C. A., and Thacker, L. H., Fluorometric fast analyzer: some applications to fluorescence measurements in clinical chemistry, *Clin. Chem.,* 19, 871, 1973.
28. Tiffany, T. O., Burtis, C. A., Mailen, J. C., and Thacker, L. H., Dynamic multicuvette fluorometer-spectrophotometer based on the GeMSAEC fast analyzer principle, *Anal. Chem.,* 45, 1716, 1973.
29. Buffone, G. J., Cross, R. E., Savory, J., and Soodak, C., Measurement of laser induced near front surface light scattering with a parallel fast analyzer system, *Anal. Chem.,* 46, 2047, 1974.
30. Whitehead, T. P., Kricha, L. J., Carter, T. J. N., and Thorpe, G. H. G., Analytical luminescence: its potential in the clinical laboratory, *Clin. Chem.,* 25, 1531, 1979.
31. Hatcher, D. W. and Anderson, N. G., An optical density thermometer, *Anal. Lett.,* 2, 373, 1969.
32. Blume, P., Experience with the GeMSAEC fast analyzer at the University of Minnesota, in *Enzymology in Medicine,* Blume, P., Ed., Academic Press, New York, 1974, 229.
33. Oliver, R. W. A. and Stoot, A., An optical thermometer: the design, construction, and calibration of a photometric-temperature scale for use with a thermometric solution, *J. Phys. E-J,* 7, 275, 1974.
34. Bowie, L., Esters, F., Bolin, J., and Gochman, N., Development of an aqueous temperature indicating thermometer and its application to clinical laboratory instrumentation, *Clin. Chem.,* 22, 449, 1976.
35. Terman, L. M., The role of microelectronics in data processing, *Sci. Am.,* 237, 163, 1977.
36. Mashburn, D. N., Stevens, R. H., Willis, D. D., Elrod, L. H., and Anderson, N. G., Analytical technique for cell fractions. XVII. The G-11C fast analyzer system, *Anal. Biochem.,* 35, 98, 1970.

37. **Tiffany, T. O., Chilcote, D. D., and Burtis, C. A.,** Evaluation of kinetic enzyme parameters by use of a small computer interfaced "fast analyzer" — An addition to automated clinical enzymology, *Clin. Chem.,* 19, 909, 1973.
38. **Statland, B. E. and Louderback, A. L.,** Nonlinear regression analysis approach for determining "true" lactate dehydrogenase activity in serum with the centrifugal analyzer ("Rotochem"), *Clin. Chem.,* 18, 845, 1972.
39. **Tiffany, T. O., Hills, L. P., Thayer, P. C., Frankart, M., and Jurczyk, C.,** The automatic measurement of SYVA EMIT® chemistries, *Clin. Chem.,* 25, 1095, 1979 A.
40. **Johnson, G. F., Tiffany, T. O., and Kiser, E. J.,** A robust nonlinear least squares (NLLS) algorithm for fitting emit® standard curve data, *Clin. Chem.,* 25, 1079, 1979 A.
41. **Burtis, C. A., Johnson, W. F., Mailen, J. C., and Attrill, J. E.,** Automated sample-reagent loader for use with the GeMSAEC fast analyzer, *Clin. Chem.,* 18, 433, 1972.
42. **Skinner, A. G., Holder, R., Northam, B. E., and Whitehead, T. P.,** Suitability of the micromedic automatic pipette and hamilton precision liquid dispenser and automatic serum and reagent addition to the GeMSAEC fast analyzer, *Scand. J. Clin. Lab. Invest.,* 29, 16.5, 1972.
43. **Nishi, H. H.,** Mechanized, direct digital-control pipette for automation of ultramicro chemical analyses, *Clin. Chem.,* 18, 771, 1972.
44. **Candler, E. L., Nunley, C. E., and Anderson, N. G.,** Analytical techniques for cell fractions. VI. Multiple gradient-distributor rotor (BXXI), *Anal. Biochem.,* 21, 253, 1967.
45. **Scott, C. D. and Mailen, J. C.,** Dynamic introduction of wholeblood samples into fast analyzers, *Clin. Chem.,* 18, 749, 1972.
46. **Burtis, C. A., Johnson, W. F., Overton, J. B., Tiffany, T. O., and Mailen, J. C.,** Parametric optimization and analytical application of the technique of dynamic introduction of liquids into centrifugal fast analyzers, *Clin. Chem.,* 20, 932, 1974.
47. **Ertingshausen, G., Shapiro, S. I., Green, G., and Zborowski, G.,** Adaptation of a T_3-uptake test and of radioimmunoassays for serum digoxin, thyroxine, and triiodothyronine to an automated radioimmunoassay system — "Centria", *Clin. Chem.,* 21, 1305, 1975.
48. **Meriadec, B., Jean-Pierre, J., and Robert, H.,** A new and universal free/bound separation technique for the "Centria" automated radioimmunoassay system, *Clin. Chem.,* 1596, 1979.
49. **Tiffany, T. O.,** Applications of centrifugal analyzers to fluorescence and chemiluminescence analyses, in *Modern Fluorescence Spectroscopy,* Vol. 2, Wehry, E. L., Ed., Plenum Press, New York, 1976, 1.
50. **Pardue, H. L.,** A comprehensive classification of kinetic methods of analysis used in clinical chemistry, *Clin. Chem.,* 23, 2189, 1977.
51. **Hatcher, D. W. and Anderson, N. G.,** GeMSAEC: a new tool for clinical chemistry total serum protein with the biruet reaction, *Am. J. Clin. Pathol.,* 52, 645, 1969.
52. **Maclin, E., Saylor, R., and Fennell, R.,** An improved algorithm for kinetic analysis for a new chemical analyzer, *Clin. Chem.,* 21, 1004, 1975.
53. **Sckillen, A. W.,** Enzyme and isoenzyme analysis in centrifugal analyzers, in *Clinical chemistry,* Price, C. P. and Spencer, K., Eds., W. B. Saunders, Philadelphia, 1980, chap. 12.
54. **Ingle, J. D., Jr. and Crouch, S. R.,** Theoretical and experimental factors influencing the accuracy of analytical rate measurements, *Anal. Chem.,* 43, 697, 1971.
55. **Attwood, J. G. and Cesare, D.,** *Clin. Chem.,* 21, 1263, 1975.
56. **Davies, J. E. and Renoe, B.,** Optimized wide interval rate measurements of substrate, *Anal. Chem.,* 51, 526, 1979.
57. **Davis, J. E. and Pevnick, J.,** Optimization of the coupled enzymatic measurement of substrate, *Anal. Chem.,* 51, 529, 1979.
58. **Ziegenhorn, J.,** Enzymatic substrate analysis by use of centrifugal analyzers in centrifugal analyzers, in *Clinical Chemistry,* Price, C. P. and Spencer, K. W., Eds., W. B. Saunders, Philadelphia, 1980, chap. 13.
59. **Fabiny, D. L. and Ertingshausen, G.,** Automated reaction-rate method for determination of serum creatinine with the centrifiChem, *Clin. Chem.,* 17, 696, 1971.
60. **Heerspink, W. and Eisenweiner, H. G.,** Comparative studies of the initial rate method (modified centrifiChem method) and the Fran Konit Method to assay creatinine in serum and urine, *Clin. Chem. Acta,* 63, 317, 1975.
61. **Ertingshausen, G.,** Non-enzymatic substrate analysis on centrifugal analyzers in centrifugal analyzers, in *Clinical Chemistry,* Price, C. P. and Spencer, K., Eds., W. B. Saunders, Philadelphia, 1980, chap. 11.
62. **Price, C. P. and Spencer, K.,** Multiple wavelength spectrophotometry and the centrifugal analyzer in centrifugal analyzers, in *Clinical Chemistry,* Price, C. P. and Spencer, K., Eds., W. B. Saunders, Philadelphia, 1980, chap. 14.
63. **Blom, M. and Hjorne, N.,** Profile analysis of blood proteins with a centrifugal analyzer, *Clin. Chem.,* 22, 657, 1976.

64. Finley, P. R., Williams, J. R., Lichiti, D. A., Griffith, F., and Thies, C. A., Immunochemical determination of human immunoglobulins: use of kinetic turbidimetry and a 36-place centrifugal analyzer, *Clin. Chem.*, 25, 526, 1979.

65. Bostick, W. D., Bauer, M. L., Morton, J. M., and Burtis, C. A., Coagulation-time determination with automatic multivariable analysis by use of a miniature centrifugal fast analyzer, *Clin. Chem.*, 21, 1288, 1975.

66. Dunikoski, L. K., Simultaneous coagulation testing: prothombin prothrombin time determinations with a centrifugal analyzer, *Clin. Chem.*, 24, 1053, 1978.

67. Buffone, G. J., Savory, J., and Cross, R. E., Use of a laser equipped centrifugal analyzer for kinetic measurement of serum IgG, *Clin. Chem.*, 20, 1320, 1974.

68. Buffone, G. J., Savory, J., Cross, R. E., and Hammond, J. E., Evaluation of kinetic light scattering as an approach to the measurement of specific proteins with the centrifugal analyzer. I. Methadology, *Clin. Chem.*, 21, 1731, 1975.

69. Buffone, G. J., Savory, J., and Hermans, J., Evaluation of kinetic light scattering as an approach to the measurement of specific proteins with the centrifugal analyzer, *Clin. Chem.*, 21, 1735, 1975.

70. Wenz, B., Feng, C. S., and Karmen, A., Improved method for detecting hemagglutination by centrifugal analysis, *Clin. Chem.*, 25, 1613, 1979.

71. Savory, J., Cross, R. W., and Hill, K. E., Measurement of acid phosphatase with a centrifugal analyzer in *Methods for Centrifugal Analyzer*, Savory, J. and Cross, R. E., Eds., American Association for Clinical Chemistry, Washington, D.C., 1978, 75.

72. Cross, R. E., Savory, J., and Hill, K. E., Adaptation to the centrifugal analyzer of an optimized assay for gamma glutamyltransferase, in *Methods for Centrifugal Analyzer*, Savory, J. and Cross, R. E., Eds., American Association for Clinical Chemistry, Washington, D.C., 1978, 79.

73. Harvey, M. S., VanderStoel, A. G., and Egbert, T. B., Determination of serum 5' nucleotidase with a centrifugal analyzer, *Clin. Chem.*, 25, 918, 1979.

74. Whitlow, K. J., Gochman, N., Forrester, R. L., and Wataji, L. J., Maltotetrose as a substrate for enzyme coupled assay of amylase activity in serum and urine, *Clin. Chem.*, 25, 481, 1979.

75. Dooley, J. F., Turnquick, L. J., and Racich, L., Kinetic determination of serum sorbital dehydrogenase activity with a centrifugal analyzer, *Clin. Chem.*, 25, 2026, 1979.

76. Catalano, E. W., Johnson, G. F., and Solomon, H. M., Measurement of erychrocyte glucose-6-phosphate dehydrogenase activity with a centrifugal analyzer, *Clin. Chem.*, 21, 134, 1975.

77. Fielek, S. and Mohrenweiser, H. W., Erythrocyte enzyme deficiencies assessed with a miniature centrifugal analyzer, *Clin. Chem.*, 25, 384, 1979.

78. Statland, B. E., Nishi, H. H., and Young, D. S., Serum alkaline phosphatase: total activity and isoenzyme determinations made by use of the centrifugal fast analyzer, *Clin. Chem.*, 18, 1468, 1972.

79. Bartl, K., Brandhuber, M., and Ziegenhorn, J., Improved automated kinetic determination of uric acid by use of uricase/catalase/aldehyde dehydrogenase, *Clin. Chem.*, 25, 619, 1979.

80. Wentz, P. W., Cross, R. E., and Savory, J., An integrated approach to lipid profiling enzymatic determination of cholesterol and triglycerides with a centrifugal analyzer, *Clin. Chem.*, 22, 188, 1976.

81. Ash, O. K. and Hentschel, W. M., High density lipoproteins estimated by an enzymatic cholesterol procedure with a centrifugal analyzer, *Clin. Chem.*, 24, 2180, 1978.

82. Tiffany, T. O., Morton, J. M., Hall, E. M., and Garrett, A. S., Jr., Kinetic enzyme fixed-time and integral analysis of serum triglycerides a clinical evaluation, *Clin. Chem.*, 20, 476, 1974.

83. Ertingshausen, G., Byrd, D. L. F., Tiffany, T. O., and Casey, S. J., Single reagent method for rapid determination of total bilirubin with the centrifiChem analyzer, *Clin. Chem.*, 19, 1366, 1973.

84. Brody, J. P., Valdes, R., and Savory, J., Centrifugal analyzer method for total bilirubin in serum by use of diazotized-2-chloroaniline-5-sulfonic acid, *Clin. Chem.*, 25, 2068, 1979.

85. Sharpe, L. A., Delaney, K. K., Breakell, G. J., Richor, W. J., and Siegel, L., Measurement of total iron-binding capacity with a centrifugal analyzer, *Clin. Chem.*, 25, 640, 1979.

86. Daly, J. A. and Ertingshausen, G., Direct method for determining inorganic phosphate in serum with the centrifiChem, *Clin. Chem.*, 18, 263, 1972.

87. Pesce, M. A., Bodourian, S. H., and Nicholson, J. F., Enzymatic method for determination of inorganic phosphate in serum and urine with a centrifugal analyzer, *Clin. Chem.*, 20, 332, 1974.

88. Li, P. K. and Shull, B., Fixed-time kinetic assay of plasma ammonia with NADPH as co-factor with a centrifugal analyzer, *Clin. Chem.*, 25, 611, 1979.

89. Pesce, M. A., Boudorin, S. H., and Nicholson, J. F., Rapid kinetic measurement of lactate in plasma with a centrifugal analyzer, *Clin. Chem.*, 21, 1932, 1975.

90. Gabrielli, M., Serum galactose determination with centrifugal analyzers, *Clin. Chem.*, 24, 1990, 1978.

91. Hickey, T. M., Uddin, D. E., and Lutz, A. K., Measurement of erythrocyte 2, 3, diphosphoglycerate with a centrifugal analyzer, *Clin. Chem.*, 25, 1314, 1979.

92. Hansen, J. L. and Freier, E., Direct assays of lactate, pyruvate, β-hydroxybutyrate, and acetoacetate with a centrifugal analyzer, *Clin. Chem.*, 24, 475, 1978.

93. **Jung, G. and Ferard, G.,** Enzyme coupled measurement of ethanol in whole blood and plasma with a centrifugal analyzer, *Clin. Chem.,* 24, 873, 1978.
94. **Hanson, N. and Freier, E.,** An improved method for determination of total serum proteins on the fast analyzer, *Am. J. Med. Technol.,* 39, 229, 1973.
95. **Blom, M. and Hjorne, N.,** Immunochemical determination of serum albumin with a centrifugal analyzer, *Clin. Chem.,* 21, 195, 1975.
96. **Savory, J., Heintges, G., Sonowane, M., and Cross, R. E.,** Measurement of total protein and albumin in serum with a centrifugal analyzer, *Clin. Chem.,* 22, 1102, 1976.
97. **Haythorn, P. and Sheehan, M.,** Improved centrifugal analyzer assay of albumin, *Clin. Chem.,* 25, 194, 1979.
98. **Brunelle, P. and Basuyau, J. P.,** The determination of plasma fibrinogen using a centrifugal analyzer, in *Centrifugal Analyzers in Clinical Chemistry,* Price, C. P. and Spencer, K., Eds., Praeger Publishers, East Sussex, U.K., 1980, chap. 25.
99. **Spencer, K. and Price, C. P.,** The use of the IL Multistat III centrifugal analyzer for kinetic immunoturbidimetry: the measurement of transferrin in centrifugal analyzers, in *Clinical Chemistry,* Price, C. P. and Spencer, K., Eds., Praeger Publishers, East Sussix, U.K., 1980, chap. 31.
100. **Severiratne, C. J. and Moores, S.,** Kinetic turbidimetric determination of serum immunoglobulins using a Multistat III centrifugal analyzer, in *Centrifugal Analyzer in Clinical Chemistry,* Price, C. P. and Spencer, K., Eds., Praeger Publishers, East Sussix, U.K., 1980, chap. 30.
101. **Tracy, R. P., Ebnet, L. E., and Moyer, T. P.,** Use of decreased reagent volumes in enzyme immunoassays, *Clin. Chem.,* 25, 1868, 1979.
102. **Henry, V., Dentsch, J., and Lum, G.,** Enzyme immunoassay of theophylline with a centrifugal analyzer and comparison with an ultraviolet method, *Clin. Chem.,* 24, 514, 1978.
103. **Hills, L. P., Tiffany, T. O., Huey, E., and Jurczyk, C.,** Adaptation of emit drugs of abuse assays to the centrifugal analyzer, *Clin. Chem.,* 25, 1094, 1979.

CALORIMETRY

Nadja N. Rehak

ANALYTICAL CALORIMETRY

Calorimetry is a physicochemical technique that is used for measurement of heat changes which occur in an in vitro system or in an intact organism due to physical or chemical processes. The instrument designed to measure such heat changes is called a "calorimeter".

Calorimetric measurements are used extensively in thermochemistry for determinations of thermodynamic data. Calorimetry has also found application in biology, biomedicine, and microbiology where it is used mainly as a research tool in studies of bacterial growth,[1-2] cell metabolism,[3-4] antibiotic sensitivity,[5] etc. The use of calorimetry in analytical work has been hindered because of instrumentation. However, with the development of calorimeters that are more sensitive, have fast response, and do not require long equilibration time and large sample volume, the potential for calorimetry to become a competitive analytical technique has greatly increased.

All chemical reactions, in which the total energy of the initial reactants differs from the total energy of the final products, are accompanied by an absorption or a release of energy in the form of heat. The energy difference between the initial and final components of a reaction system, and thus, the amount of heat change during this reaction, is independent of the reaction path and of the number or nature of intermediate products.

Analytical calorimetry is based on the direct proportionality between the amount of heat change during the observed chemical reaction and the quantity of the reacting substances. The total heat released (exothermic process) or absorbed (endothermic process) during a reaction at constant pressure is

$$Q = \Delta H \times c \times v \tag{1}$$

where Q is the total heat in calories,* ΔH is the enthalpy change (or heat content) in calories per mole, c is the concentration of the reactant in mole per liter, and v is the volume of the reactant in liter. The enthalpy change is the heat that is released (ΔH value negative) or absorbed (ΔH value positive) during the reaction of one mole of a substance, i.e., $c \times v = 1$ mol. The value of ΔH is a constant for a given chemical reaction that is carried out under defined conditions (pH, solvent, temperature, etc.). The ΔH for any chemical reaction can be determined experimentally by calorimetric measurements or can be calculated from the heats of formation[6-7] of reactants, ΔHf_R°, and of products, ΔHf_P°.

$$\Delta H = \Delta Hf_P^\circ - \Delta Hf_R^\circ \tag{2}$$

Table 1 shows examples of ΔH values for enzyme catalyzed reactions that have been determined by calorimetric measurements or by calculation.

It can be seen from Equation 1 that, if the value of ΔH is known and the volume of the reactant is constant, the heat of reaction is a direct measure of the concentration, c. However, it must be remembered that, in contrast with other commonly used ana-

---

* The SI unit is the joule (J); 1 calorie (cal) = 4.1868 J.

Table 1
SELECTED ΔH VALUES FOR ENZYME CATALYZED REACTIONS

Reaction	Buffer/pH/temperature	ΔH kcal/mole	Ref.
Glucose + ATP $\xrightarrow{\text{HK}}$ Glucose-6-P + ADP	TRIS/7.6/30.8°C	−14.67	14
Pyruvate + NADH + H⁺ $\xrightarrow{\text{LD}}$ L-lactate + NAD⁺	Phosphate/7.4/30°C	−11.31	13
	TRIS/7.4/30°C	−3.67	13
Urate + 2 H_2O + O_2 $\xrightarrow{\text{URC}}$ Allantoin + CO_2 + H_2O_2	TRIS/9.0/30°C	−35.76	17
H_2O_2 $\xrightarrow{\text{CTL}}$ H_2O + ½ O_2	Calculated	−35.16	17
2 H_2O_2 $\xrightarrow{\text{CTL}}$ 2 H_2O + O_2	Saline/25°C	−24.00	23
	Calculated	−24.03	6
Cholesterol + O_2 $\xrightarrow{\text{CHO}}$ Cholestenone + H_2O_2	Phosphate/6.9/30°C	−36.67	13
H_2O_2 $\xrightarrow{\text{CTL}}$ H_2O + ½ O_2			
BAEE $\xrightarrow{\text{TRP}}$ BA + ethanol	TRIS/7.8/30°C	−9.82	13

Note: ADP: adenosine 5'-diphosphate; ATP: adenosine 5'-triphosphate; BA: N-benzoyl-l-arginine; BAEE: N-benzoyl-1-arginine ethyl ester; CHO: cholesterol oxidase (1.1.3.6); CTL: catalase (1.11.1.6); LD: lactate dehydrogenase (1.1.1.27); NAD: nicotinamide-adenine dinucleotide; NADH: nicotinamide-adenine dinucleotide, reduced form; TRIS: tris (hydroxymethyl) aminomethane hydrochloride; TRP: trypsin (3.4.21.4); and URC: uricase (3.5.1.5).

lytical techniques, calorimetry measures an overall chemical process and not only a single reaction component. Since the heat effects are additive, the measured heat change for a process that consists of more than one reaction, is equal to the sum of the individual reaction heats

$$Q_M = Q_1 + Q_2 + \dots Q_n \tag{3}$$

where n is the number of reactions. Thus, coupling of reactions results in an increase of the value of Q_M, and therefore, in an increased sensitivity of calorimetric measurements. However, this also means that any side reaction, which is more likely to occur in a complex matrix system, can contribute to the overall measured heat and cause interference. The extent of this "interfering" heat contribution, Q_I, depends on the concentration of an interfering reactant, c_I, and on the ΔH_I value of the interfering reaction. The interference is negligible if $Q_M \gg Q_I$, i.e., if $c_M \gg c_I$ and/or $\Delta H_M \gg \Delta H_I$.

The interference due to a side reaction can be altered by changing the reaction conditions to minimize the value of ΔH_I. Otherwise, corrections must be made by measuring the "blank heat", which includes Q_I and any other heat effects due to dilutions, evaporation, friction of flowing solutions, etc.

Even though this nonspecificity of heat source can be regarded as a disadvantage of calorimetric technique, it makes calorimetry a general analytical tool. Therefore, the calorimetric technique can be applied to the measurement of various compounds even in the most complex systems, such as biological specimen, cell suspensions, etc. However, the specificity of the measured process must be assured, for example, with the use of specific enzymes.*

INSTRUMENTATION

It is beyond the scope of this chapter to review all different types of calorimeters that are used in analytical studies. The following discussion will be concerned only with the basic instrumental designs and measuring principles. The reader is referred to references for a more detailed description of analytical calorimeters.[8-10]

Apparatus

The basic unit of a calorimeter consists of a reaction vessel and a heat measuring device. The unit is enclosed in a container which is maintained at a constant temperature ($\pm 0.001°C$). The temperature controlled environment is provided either by a thermostated water bath or a metal block or shield.

The reaction vessel is essentially a small container in which the process to be measured is carried out. The material that the vessel is made of should be nonreactive with the content of vessel and have a low heat capacity value (glass, plastic, metal). The design of a reaction vessel varies with the type of calorimeter and will be described later.

The primary heat measuring devices are thermocouples, thermopiles, and thermis-

---

* It is interesting to compare the nonspecific calorimetric technique with the commonly used spectrophotometric method. The spectrophotometric analysis is based on the direct proportionality between the absorbance, A, and the concentration, c, of a reactant, which is expressed as $A = a \times c \times b$ (compare to Equation 1), where a is molar absorptivity (compare to ΔH) and b is the light path (compare to v in Equation 1). The absorbance is measured at a wavelength that is specific for a given compound. However, in a complex matrix, several compounds can absorb light of the same wavelength and cause interferences (compare to "interfering" heat contribution). In the case of biological fluids, the accuracy of the spectrophotometric measurement is enhanced with the use of enzymes.

tors which operate on the basic principles of electrical thermometry. A thermocouple consists of two wires of different metals that are connected at both ends to form a closed circuit. The reference junction of a thermocouple is maintained at constant temperature. The test junction is in a thermal contact with the system whose temperature is to be measured. The thermal electromotive force (EMF) that is produced due to a temperature difference between the junctions, is a function of the temperature of the test junction and can be measured as a voltage. The output of a thermocouple depends on the type of metals in the circuit, but for most thermocouples the sensitivity is a few millivolts per degree Celsius. A thermopile is made of several thermocouples that are connected in series. Again, the reference junctions are maintained at a constant temperature and the test junctions are exposed to the heat source. The thermal EMF of a thermopile is the sum of the individual EMFs of all the thermocouples, so that the sensitivity is increased. A thermistor is a resistive element which consists of a sintered mixture of metal oxides. The resistance of a thermistor is a function of temperature and can be measured. Thermistors have a high negative coefficient of temperature and, therefore, high sensitivity.

The voltage output of a thermopile or a thermocouple is amplified and measured. The resistance of a thermistor is measured with a resistor bridge system with a null balance detector. The output is measured as an unbalance potential in volts.

The described heat measuring devices are also used as temperature control sensors for maintenance of a constant temperature environment in a calorimeter.

Design of Calorimeters

All calorimeters are designed on the basic principles of the adiabatic and/or the heat conduction type of measurements.

In the ideal adiabatic calorimeter, the reaction vessel is thermally insulated so that no exchange of heat occurs between the vessel and the surrounding. A change of heat is measured quantitatively as the change in temperature, ΔT, inside the vessel

$$Q = h \times \Delta T \tag{4}$$

where h is the heat capacity of the vessel and its content. In the ideal heat conduction calorimeter, the heat is transferred quantitatively between the reaction vessel and the surrounding. The flow of heat, F, which is proportional to the temperature difference between the vessel and the surrounding, is measured. The total heat change is proportional to the time integral of the heat flow

$$Q = p \int F \, dt \tag{5}$$

where p is the proportionality constant.

In practice, all calorimeters operate with heat losses for which corrections are made usually by "calibration". This involves the measurement of heat changes of processes with a well established ΔH or ΔT, such as chemical reactions or electrical heat sources. The amount of heat, or the change in temperature, that is measured is

$$Q_M (\Delta T_M) = k \times Q_T \tag{6}$$

where Q_T is the theoretical heat change and k is the calibration constant of a calorimeter.

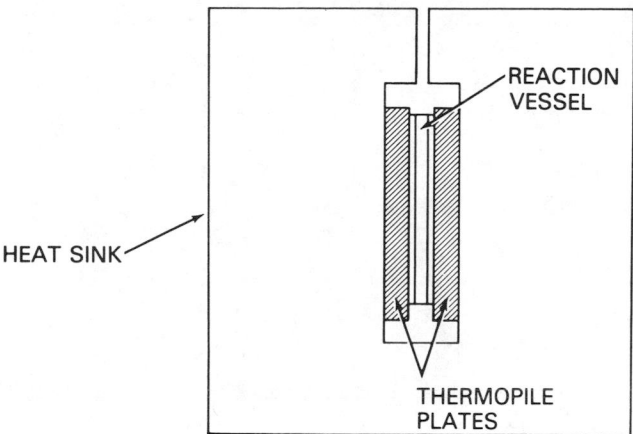

FIGURE 1. Schematic representation of a cross-section through a basic unit of a thermopile conduction calorimeter. (From Rehak, N. N. and Young, D. S., *Clin. Chem.*, 24, 1414, 1978. With permission.)

Generally, in the adiabatic type of calorimeters, the transfer of heat is prevented by enclosing the reaction vessel in a vacuum chamber (such as in case of dewar) or by keeping the temperature of the vessel and the surrounding equal during the entire reaction process. This is achieved by appropriate cooling or heating of either the surrounding or the vessel. The extent of this required heat effects is the measure of ΔT inside the vessel. The temperature change is measured with a thermistor or a thermocouple. In the heat flow type of calorimeters, the reaction vessel is placed between thermopile plates and seated in an aluminum block of a comparatively large mass, called heat sink (Figure 1). The heat flow from and to the vessel is transferred through the thermopile. The time integral of the measured voltage output is proportional to the heat change inside the reaction vessel

$$Q = k \int V \, dt \tag{7}$$

where k is the calibration constant (see Equation 6).

Mode of Calorimetric Measurements

Calorimetric measurements can be carried out in a batch or a flow mode. In batch experiments, the reaction vessel must be washed, refilled with reagent(s), and brought to a temperature equilibrium inside the calorimeter after each measurement. A single-compartment reaction vessel (adiabatic and some heat conduction designs) is filled with one of the reagent and the reaction is initiated by addition of a second reagent by means of pipetting, injecting, titration, or breaking a glass capsule, that contains the reagent, inside the vessel. Two-compartment vessels (Figure 2) are used extensively in heat conduction calorimeters; both compartments are filled with the appropriate reagents at the same time. The mixing is accomplished by rotating the heat sink of the calorimeter, which contains the vessel.

Flow calorimeters (adiabatic and heat conduction designs) have a reaction vessel through which reagents are pumped at a constant speed. The flow can be either continuous or aliquoted (stop-flow). Before mixing, the reagents flow through a tubing which is in thermal equilibrium inside of the calorimeter. The mixing is accomplished by

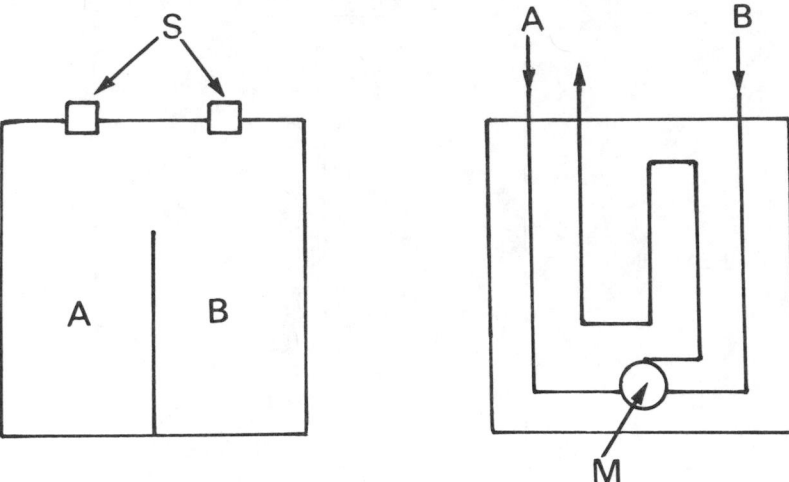

FIGURE 2.   Schematic representation of a batch type (left) and a flow-type (right) reaction vessels of a thermopile conduction calorimeter. A and B are the appropriate reagent chambers/flow tubes, S are sealing devices, and M is the mixing chamber.

joining the two streams in a Y junction or a mixing chamber of the vessel (Figure 2). In stop-flow measurements, the reaction is initiated inside the vessel and the combined volume of the two reagents should not exceed the total volume capacity of the vessel, i.e., batch mode operation. In continuous-flow measurements, the mixing of reagents can occur before the entry to the vessel. However, the process to be measured by continuous-flow technique must be either fast or slow enough to be detected in the flow-through vessel (i.e., should not start after the combined stream leaves the vessel). Both batch and flow calorimeters can be constructed as twin instruments with two basic units (two reaction vessels and heat measuring devices). The differential output of a twin calorimeter allows automatic "blank heat" corrections.

The most important characteristics of a calorimeter are the response time and the sensitivity. In analytical work, where throughput of samples is an important factor, instruments with fast response time and high sensitivity are preferred. Modern calorimeters that can measure heat changes taking place in solutions of micromolar concentrations (i.e., sensitivity of 10^{-6} °C or 10^{-6} cal) are called microcalorimeters. The volume of sample required for microcalorimetric measurement is also small and can be as low as 20 $\mu\ell$. The time constant of heat conduction microcalorimeters can range from 35 to 400 sec, while adiabatic calorimeters can have time constant as small as 6 sec or less.

Calculations

The output signal of a calorimeter is usually measured continuously as a function of time. The data can be recorded in a digital form suitable for processing by a computer, or graphically, on a strip-chart recorder, as a voltage vs. time curve (thermogram or enthalpogram). The recorder chart ordinate can be calibrated in temperature units (°C/cm) so that the temperature change, ΔT, can be determined by the height of the thermogram. The total heat, Q, is calculated by integrating the area under thermogram.

The shape of a thermogram depends on the rate of the measured process and the response time of the calorimeter. If both are fast, the initial slope of a thermogram is

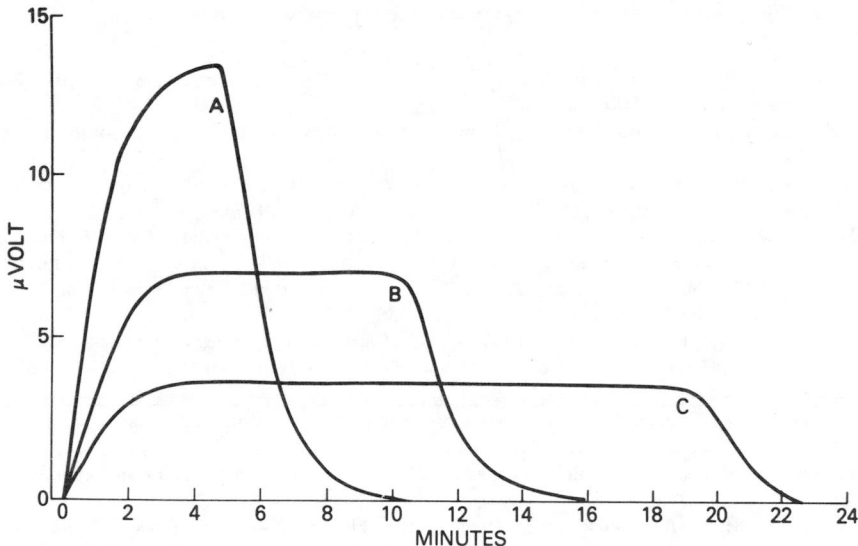

FIGURE 3. Effect of the rate of enzymatic reaction on the shape of thermogram; A, initial slope 7.2 μV/min, Q = 15 mcal, enzyme activity 5000 U/ℓ; B, initial slope 3.7 μV/min, Q = 17 mcal, enzyme activity 2500 U/ℓ; and C, initial slope 1.9 μV/min, Q = 16 mcal, enzyme activity 1250 U/ℓ. (From Rehak, N. N. and Young, D. S., *Clin. Chem.*, 24, 1414, 1978. With permission.)

steep and the time of measurement is short. Figure 3 shows a series of thermograms that were obtained by changing the rate of the measured reaction. It can be seen that, while the total heat of reaction remains the same, the peak height and the initial slope of the thermogram depends on the reaction rate that can be measured as the change of heat per unit of time.[11-13]

APPLICATION IN CLINICAL CHEMISTRY

Although calorimetry is not a routine analytical technique in clinical laboratories, it has been used successfully for quantitative and kinetic measurements of body fluid components.[12-19] In all assays, specific enzyme catalyzed reactions were employed, using either soluble or immobilized enzymes.[18-21] Even though heat conduction microcalorimeters are generally considered more suitable for analytical studies because of their long-term heat stability, flow-type adiabatic microcalorimeters are, at the present time, more adaptable to automation[22] and, because of the fast response time, the throughput of samples is also much faster. (About 37 per hour as compared to about 8 per hour with a stop-flow heat conduction calorimeter). A detailed description of microcalorimeters, measurement techniques, and data processing applicable to clinical assays has been published.[24]

REFERENCES

1. **Forrest, W. W.**, Bacterial calorimetry, in *Biochemical Microcalorimetry,* Brown, H. D., Ed., Academic Press, New York, 1969, 165.
2. **Delin, S., Monk, P., and Wadso, I.,** Flow microcalorimetry as an analytical tool in microbiology, *Sci. Tools,* 16, 22, 1969.
3. **Levin, K.,** Determination of heat production from erythrocytes in normal man and in anemic patients with flow microcalorimetry, *Scand. J. Clin. Lab. Invest.,* 32, 55, 1973.
4. **Ross, P. D., Fletcher, A. P., and Jamieson, G. A.,** Microcalorimetric study of isolated blood platelets in the presence of thrombin and other aggregating agents, *Biochem. Biophys. Acta,* 313, 106, 1973.
5. **Binford, J. S., Binford, L. F., and Adler, P.,** A semi-automated microcalorimetric method of antibiotic sensitivity testing, *Am. J. Clin. Pathol.,* 59, 86, 1973.
6. **Wagman, D. D., Evans, W. H., Parker, V. B., et al.,** Selected values of chemical thermodynamic properties, in NBS Technical Note 270-3, 270-4, 270-6, 270-7, National Bureau of Standards, Washington, D.C.
7. **Domalski, E. S.,** Selected values of heats of combustion and heats of formation of organic compounds containing the elements C, H, N, O, P, and S, *J. Phys. Chem. Ref. Data,* 1, 221, 1972.
8. **Spink, C. and Wadso, I.,** Calorimetry as an analytical tool in biochemistry and biology, *Methods Biochem. Anal.,* 23, 1, 1976.
9. **Wadso, I.,** Microcalorimetry and its application in biological sciences, in *New Techniques in Biophysics and Cell Biology,* Vol. 2, Pain, R. and Smith, B., Eds., John Wiley & Sons, New York, 1974, 85.
10. **Sturtevant, J. M.,** Calorimetry, *Methods Enzymol.,* 26, 227, 1972.
11. **Berger, R. L., Davids, N., and Panek, E.,** The design and development of a stopped-flow microcalorimeter for the study of enzyme kinetics, *J. Calorimetric Anal. Therm.,* 6, 1, 1975.
12. **Grime, J. K., Tan, B., and Jordan, J.,** The determination of serum cholinesterase activity by kinetic direct injection enthalpimetry, *Anal. Chim. Acta,* 109, 393, 1979.
13. **Rehak, N. N. and Young, D. S.,** Prospective applications of calorimetry in the clinical laboratory, *Clin. Chem.,* 24, 1414, 1978.
14. **Goldberg, R. N., Prosen, E. J., Staples, B. R., et al.,** Heat measurements applied to biochemical analysis; glucose determination by hexokinase catalyzed phosphorylation, *Anal. Chem.,* 47, 786, 1975.
15. **McGlothlin, S. D. and Jordan, J.,** Thermochemical determination of glucose in serum, plasma, and whole blood without deproteinization, *Clin. Chem.,* 21, 741, 1975.
16. **McGlothlin, C. D. and Jordan, J.,** Enzymatic enthalpimetry, a new approach to clinical analysis; glucose determination by hexokinase catalyzed phosphorylation, *Anal. Chem.,* 47, 786, 1975.
17. **Rehak, N. N., Janes, G., and Young, D. S.,** Calorimetric enzymic measurement of uric acid in serum, *Clin. Chem.* 23, 195, 1977.
18. **Bowers, L. D., Canning, L. M., Jr., Sayers, C. N., and Carr, P. W.,** Rapid-flow enthalpimetric determination of urea in serum, with use of an immobilized urease reactor, *Clin. Chem.,* 22, 1314, 1976.
19. **Bowers, L. D. and Carr, P. W.,** An immobilized-enzyme flow-enthalpimetric analyzer: application to glucose determination by direct phosphorylation catalyzed by hexokinase, *Clin. Chem.,* 22, 1427, 1976.
20. **Danielsson, B., Gadd, K., Mattiasson, B., and Mosbach, K.,** Enzyme thermistor determination of glucose in serum using immobilized glucose oxidase, *Clin. Chim. Acta,* 81, 163, 1977.
21. **Forrester, L. J., Yourtee, D. M., and Brown, H. D.,** Immobilized enzyme modules for use in analytical microcalorimetry, *Anal. Lett.,* 7—9, 599, 1974.
22. **Picker, P.,** New concepts in design and applications of flow microcalorimetry, *Can. Res. Dev.,* 11, 1974.
23. **Nelson, D. P. and Kiesow, L. A.,** Enthalpy of decomposition of hydrogen peroxide by catalase at 25°C, *Anal. Biochem.,* 49, 474, 1972.
24. **Martin, C. J. and Marini, M. A.,** Microcalorimetry in biochemical analysis, *Crit. Rev. Anal. Chem.,* 8, 221, 1979.

MASS SPECTROMETRY

Robert E. Hill

INTRODUCTION

The clinical laboratory is faced with a diverse array of analytical problems and their solution has required the adoption of many techniques using a wide range of instrumentation. One of these, the mass spectrometer, is a recent addition. This review highlights some of the features of the instrumentation and techniques which are relevant in the clinical laboratory and to the clinical chemist.

A mass spectrometer produces ions from the molecules under investigation, separates these ions according to their mass/charge (m/e) ratios and measures their relative abundance. These data, which may be graphical or tabular, constitute the mass-spectrum (Figure 1).

The technique is attractive as an analytical procedure because the mass-spectrum reflects the composition of the material under investigation and, therefore, serves as a source of qualitative information and also because estimation of the number of ions produced, either total, or of selected m/e value, permits quantitative analysis. Finally, these data are usually acquired from very small amounts of substance (nanogram range).

INSTRUMENTATION

Each mass spectrometer is a system produced by the careful matching of separate components. These are an inlet system, which determines the mode of entry of the sample; an ion source, which is responsible for the ionization of the sample molecules; a mass analyser, which separates the ions according to their m/e ratio; a detection system into which the separated ions are directed; and finally, since the ionization, analysis, and detection require low pressure environments, each instrument must be equipped with an appropriate vacuum system (Figure 2).

The designs and properties of the components determine the characteristics and analytical capabilities of the whole instrument and it is, therefore, important to assess these properties and to match them to the analytical task at hand.

The majority of the applications in clinical chemistry are met by mass spectrometers which have inlet systems interfaced to a gas chromatograph (GC/MS).[1-3] More recently, liquid chromatographs have also been interfaced.[2,4,5] This interface technology has allowed the mass spectrometer to serve as a sensitive, selective, and quantitative detector for any substance which can be isolated by these forms of chromatography.

The major problem in this area is the pressure differential between the chromatograph and mass spectrometer and the fact that the interface design must allow efficient sample transfer into the ion source over a wide range of sample molecular weight and stability.

In GC/MS, to control the pressure differential one must control the quantity of gas entering the mass spectrometer. If one considers packed and capillary columns, the practical range of carrier gas flow is 1 to 25 ml/min. However, for maintenance of the high vacuum which is required in the mass analyser to limit ion molecule collision, the total gas flow into an electron impact source should not exceed 2 ml/min, and the actual limit will be determined by the pumping capacity of the vacuum system, i.e., the greater this capacity, the higher the gas flow which can be tolerated without ex-

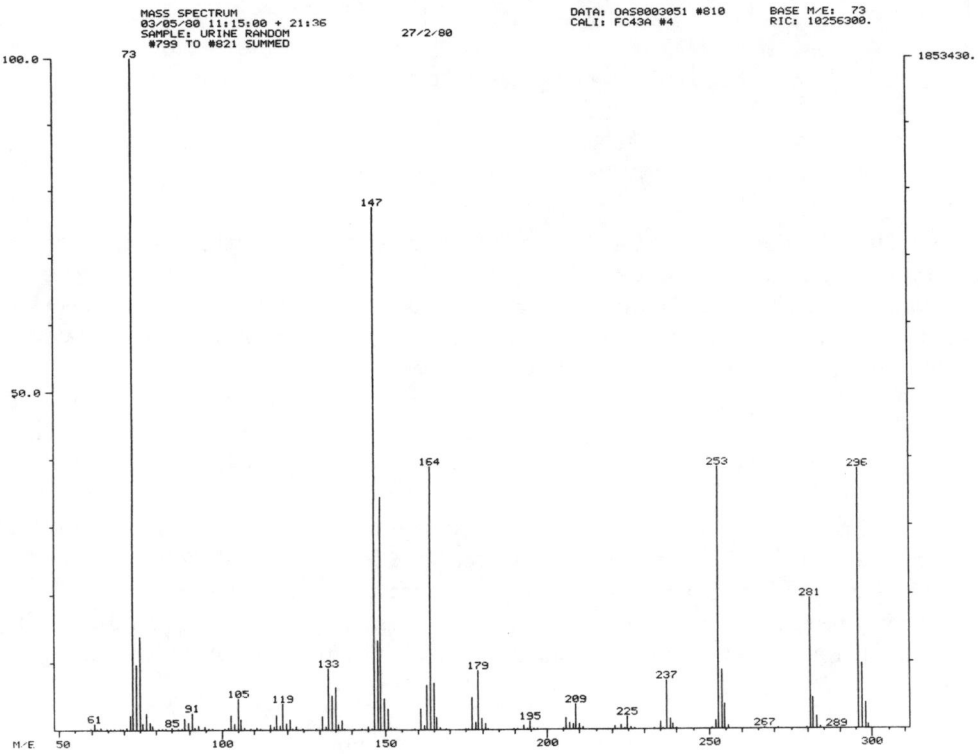

FIGURE 1.   Typical mass spectrum, di-TMS-derivative of o-hydroxyphenyl acetic acid from a patient with uncontrolled phenylketonuria.

SCHEMATIC OF GC/MS/DS

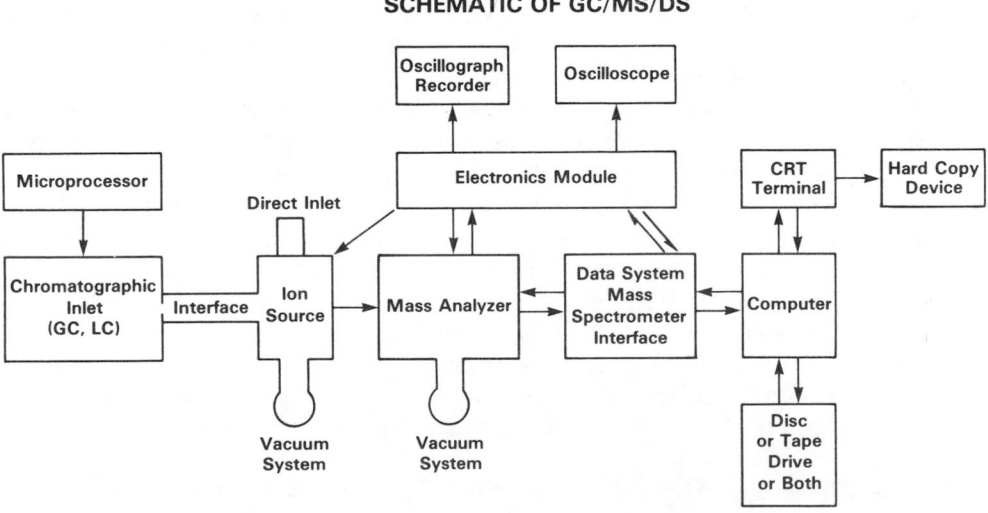

FIGURE 2.   Block diagram of chromatograph mass spectrometer-data system.

ceeding operating pressures. It is evident then that interface design must at least account for chromatographic mode (packed, capillary), and the sample type and its stability, and that ion source design and the vacuum system are also critically important to optimum operation of the interface.

The simplest interface is a direct line which contains a restrictor to produce the required pressure drop. This is convenient, but not versatile, since optimum performance occurs only at one flow rate depending upon the dimensions of the restrictor. Alternatively, a variable needle valve can be used, but one should be aware of dead volume effects and also of the difficulty of designing a needle valve using inert materials. Direct interfaces are suitable for the connection of capillary columns to electron impact sources and both packed and capillary columns to chemical ionization sources. The latter operate at higher pressures, allowing gas flows of 10 to 15 ml/min into the source.

Connection of a packed column (flow 10 to 25 ml/min) to an electron impact source (allowed flow 0.1 to 2 ml/min) requires an enrichment device to increase the partial pressure of the sample in the carrier gas because the source can only accept a small percentage, e.g., 10%, of the GC eluent. There have been several successful designs. Today, with the advent of efficient vacuum systems, the "single stage" jet separator is the most commonly available in commercial instruments. Briefly, effluent from the gas chromatograph passes through a fine jet into a small vacuum chamber which is maintained by an independent mechanical pump. Directly opposed and at a small distance away, a second jet leads to the ion source. As the eluent traverses this distance, the lighter carrier gas, usually helium, undergoes greater diffusion and can be pumped away leaving a stream enriched with the heavier sample components. This simple device allows adequate enrichment and gives good yields from gas flow of 15 to 25 ml/min. When constructed of glass, it is well suited to labile compounds and it can be heated enough to avoid sample condensation. Its problems are that it may become plugged, it is less efficient for lower molecular weight compounds, and it has only one optimum flow rate.

Progress in the interfacing of liquid chromatographs has been much slower and commercial systems are only recently available. There are two major types, the direct liquid introduction, which is limited to chemical ionization using the chromatographic solvent as the reagent gas (*vide infra*); and the "moving band" interface, which can be utilized with electron impact and chemical ionization with no restrictions on the reagent gas. The fact that the moving band interface also transfers up to 40% of sample, whereas the direct introduction transfers at best 1%, would seem to make the former a better choice at this time. However, it is rather early to decide which will evolve as the optimum technique.

Finally, a commonly available inlet system worth consideration is the solid probe. This mode is suitable for compounds which cannot be handled by chromatography and, since it allows introduction directly into the source, it provides maximum sensitivity. The sample is placed at the probe tip which is introduced into the source through a vacuum lock. It should be possible to heat the tip using a temperature program. Although it is unsuitable for the analysis of mixtures, it is desirable as a "backup" to chromatographic interfaces.

Following their introduction and prior to analysis by the mass analyser, the sample molecules must undergo ionization in the ion source.[6] There are at least ten types of ion source in current use. The present discussion is limited to the electron impact and chemical ionization sources,[7] since they are routinely available as well-tested commercial designs and have found wide application in clinical chemistry.

The electron impact (EI) source is the most common and can be viewed as the "standard" general purpose design for organic analysis. This source effects ionization and fragmentation by impact between sample molecules and electrons having energies in excess of the sample ionization potential. Conventionally, 70 eV electrons are used, although any instrument should allow other values to be chosen. Some practical features in a good EI source are a separate pumping system, i.e., the source and analyser system of the mass spectrometer are "differentially pumped". This feature also optimizes the analyser's performance. Independent heating of the source in excess of 300°C should be available to avoid sample condensation and to facilitate overnight "bakeout". The source should be easily removable for cleaning and should be mounted on a self-centering device for easy installation. There should be isolation valves so that diffusion vacuum pumps can remain in operation after venting and, in magnetic sector instruments, it is also desirable to be able to isolate the mass analyser. Alternatively, one might consider an instrument with turbomolecular pumps which have a rapid pump down rate, thus, obviating the need for isolation valves.

In the chemical ionization source, the sequence of events is ionization of a reagent gas (e.g., methane) by a high energy electron beam (200 to 500 eV), followed by reactions between reagent gas ions and sample molecules to produce sample ions. Construction is similar to the EI source, except that it requires a "tight" design to maintain the relatively high pressure of reagent gas required for the ionization process. Unlike the EI source, the CI source operates with gas flows of about 10 to 15 mℓ/min which means that, in CI/GC/MS, eluent from a packed column can be channeled directly into the source (*vide infra*). To maintain operating vacuum in the mass analyzer, it is essential that the CI source be differentially pumped by a large capacity vacuum system. A well-designed source should also permit easy selection of reagent gases, for instance up to three gases may be permanently plumbed in and selected by push button or similar operation. Since they have similar construction, and since CI spectra often augment EI data, it has become popular to fit mass spectrometers with combination EI/CI sources which can be switched from one mode to another by a simple one-step operation. All the major manufacturers offer this option and the flexibility is useful for a clinical laboratory.

There are three major types of mass analysers, the magnetic sector field, the quadrupole mass filter, and the time-of-flight analyser. The first two are more common in instruments that have been used for clinical analyses. The characteristics of the mass analyser are the major determinants of the resolution (*vide infra*) of a mass spectrometer, i.e., the ability of the instrument to differentiate ions of different m/e ratios. In turn, different analytical problems require different levels of resolution and it follows that the choice of a particular mass analyser should be made with the analytical problems in mind.

Magnetic instruments are available with either single or double focusing analysers. In single focusing instruments, positive ions are accelerated from the source by a potential V, such that their exiting kinetic energy is

$$\frac{1}{2} mv^2 = EV \tag{1}$$

As the ions enter the magnetic field of the analyser, they experience a force HeV which causes them to adopt a curved trajectory, resulting in an equal centripetal force, i.e.,

$$\frac{mv^2}{r} = HeV \tag{2}$$

(H = magnetic field strength, r = radius of curvature). Combination of 1 and 2 gives the basic "mass spectrometer" equation:

$$\frac{m}{e} = \frac{H^2 r^2}{2V} \tag{3}$$

which demonstrates that the values H and V determine the m/e value of those ions following trajectory r and arriving at a collector slit placed on this path. It also follows that a mass spectrum can be "scanned" either by changing V or H. The latter is the more common mode and Equation 3 predicts that a linear scan of the magnetic field will produce a nonlinear mass scan. A linear scan can be produced by scanning the accelerating voltage which has the advantage of being relatively easy to control by a digital computer, and scanning speeds can also be increased since limitations associated with magnetic field scanning are avoided. On the other hand, wide variation of the large accelerating voltages (kV range) used in magnetic instruments can lead to major problems in the transmission of ions from the source through the analyser.

Since ions are formed in different positions in the source and possess different "natural" or initial kinetic energies, the resultant beam at any particular m/e value, will, after acceleration by V, have an energy divergence, i.e., ions of a given mass will move at different velocities. Equation 2 can be rewritten:

$$r = \frac{mv}{He} \tag{4}$$

This equation demonstrates that the radial path is a function of momentum and that any velocity divergence will cause a spread in the value of r, thereby broadening each individual peak and limiting resolution. A double focusing instrument eliminates this "energy/velocity" divergence before the ions enter the magnetic field. The ions are first subjected to a radial electrostatic field E, which causes them to adopt a circular path, such that:

$$mv2/r = eE \tag{5}$$

By adjusting E, one can select ions of the same mass but different velocities, which eliminates the velocity focusing aberration in the ensuing magnetic field and results in a significant increase in available resolution.

In recent years, the quadrupole mass filter has become popular, especially in instruments designed for GC/MS work.[8,9] The construction and theoretical principles are entirely different from the magnetic sector. As the name implies, the mass filter is produced by four parallel rods held in a square array. The theory of the instrument dictates that for best performance these rods should have hyperbolic cross sections. However, instruments with circular rods equipped with appropriate ion lens systems to direct the ions from the source into the "mass filter" provide overall acceptable performance.

The rods of the quadrupole (circular or hyperbolic) are diagonally paired. A positive DC voltage and a radio frequency voltage is applied to one pair, while the other receives an equal but negative DC voltage and a radio frequency voltage 180 degrees out of phase with the first. This arrangement creates an oscillating electrostatic field within the quadrupole, and it is this field which acts as the mass filter. In any given instrument, the value of the applied voltages determine the nature of the field. Charged particles enter the field from the ion source under a small accelerating voltage (~30 V vs. kV values in magnetic instruments) and, since there is no force along the logitudinal axis of the filter, they travel at a constant velocity. However, the oscillating field causes them to adopt an oscillating trajectory. This trajectory is defined by complicated equations of motion which relate the mass of the ions to the magnitude of the applied voltages and to the dimensional constants of the quadrupole. It can be shown that for given operating conditions, i.e., applied voltages, only ions of a certain m/e value will traverse the field. Others, at lower and higher mass, will adopt unstable trajectories which cause them to strike the quadrupole rods. Thus, by varying the value of the voltages, it is possible to scan the mass range of ions which will traverse the filter. Normally, this is done by varying both voltages while keeping the ratio constant. This mode leads to a linear scan function which facilitates mass calibration.

Electron multipliers are commonly used as detector devices on mass spectrometers applied to organic analysis. The current associated with an ion per second is in the order of 10^{-19} A and modern components are capable of counting and amplifying such low values. The typical gain of a good secondary electron multiplier should be about 10^{+6}. In quadrupoles, the multiplier should be mounted in an off-axis position to avoid photons and neutral particles which may pass through the filter. Charged ions are then attracted to multiplier by the potential of the first dynode. This arrangement reduces noise but increases the likelihood of "mass discrimination", because ion beams of increasing mass undergo increasing divergence as they pass through the mass filter. Deflection of this dispersed beam into the offset dynode is less efficient and there is a concurrent loss of sensitivity as mass increases. This phenomenon is termed "mass discrimination".

The output signal of the electron multiplier is either further amplified by an electrometer to drive a recording device or to input an analogue to digital converter, or alternatively, it can be digitized directly by a pulse counting system. Digitization is a prerequisite for on-line automatic data processing.

The introduction of on-line dedicated mini-computers[10] has coincided with the rapid growth of GC/MS, and now it is common to refer to these analytical systems as GC/MS/DS, i.e., gas chromatography-mass spectrometer-data systems. In modern instruments, the data system is usually made up of a computer (16 bit, 32 K machines are common) with a data storage device, either magnetic tape or more commonly a dedicated disc system. Dual disc systems are popular with a fixed disc containing the operating software and removable discs for data storage. Communication between the operator and the instruments is usually via a CRT terminal and, finally, there must be a hard copy output device, e.g., printer/plotter. The other major piece of hardware is the interface system between the mass spectrometer and the computer. Dedicated data systems of this type perform three basic functions. First, they must acquire the data; secondly, they may control the operating parameters of the mass spectrometer; and finally, they must contain interactive software allowing data processing. Modern instruments should allow processing of current or historic files during an acquisition. Consideration of the problems associated with data acquisition and mass spectrometer control by computer are beyond the scope of this article.

The use of data systems (in GC/MS) is attractive because they capture large amounts of useful data that would otherwise be lost. For instance, in a 20 min GC run, a 1 sec scan time of a mass range of 500 amu would produce 600,000 data points. Needless to say, in the normal run, not all of these will be associated with chromatographic peaks. However, in modern instruments, particularly quadrupoles where the scanning voltage can easily be brought under digital control, it is not uncommon to use such fast scan rates, especially for capillary GC runs. By cyclic scanning in this mode, all the data is collected and there is no chance that data will be lost by the operator failing to record a spectrum of interest.

Presentation of the data is a function of the expertise of the programmers and commercial systems compete with each other in the flexibility and novelty of their output software. The versatility of some of these systems is truly amazing and it can be a major selling feature of an instrument. However, prospective purchasers should always remember that these data processing "marvels" operate on information *acquired* from the mass spectrometer and, in the final analysis, no amount of sophisticated software will substitute for good component design in the mass spectrometer.

Software packages frequently include identification routines[11] which, in dedicated mini-computer systems, go under the title of "library search". Most, if not all, commercial GC/MS/DS systems incorporate a disc-based mass spectral library together with the operating software required to "search" this data base. In the traditional type of search, the system compares the spectrum of interest with each spectrum in the library and selects and displays those which most closely resemble the unknown. The display also usually includes a "match coefficient" to indicate the closeness of "fit" between the unknown and the library spectrum. Modern software also often incorporates interactive routines which allow the operator to modify the search procedures, e.g., to search for compounds of given molecular weight. Another approach, particularly useful in GC/MS where the analyst may be looking for particular compounds in a class mixture, is the reverse library search. Here, the program takes the spectrum of interest from the library data base and searches data from the unknown for the presence of this compound. The accuracy of this procedure can be augmented by including in the library a chromatographic parameter such as the retention index. In this way, compounds are identified by their mass spectral and chromatographic correlations. The accuracy of all library search procedures depends upon the content and integrity of the data base and on the algorithms of the search procedure. Moreover, commercial data bases are generally compiled from several sources, using different types of mass spectrometer and sets of operating parameters. Needless to say, the accuracy of the search match increases if the unknown and library spectrum are collected under similar conditions. For this reason, it is often advisable for users to construct their own libraries made up of compounds pertinent to particular applications. This is especially relevant in the reverse search mode in which a dedicated library containing relatively few compounds can be used to search the GC/MS run for the presence of these compounds.

RESOLUTION

The ability of a mass spectrometer to separate ions of different m/e ratio is defined by the resolution and the type of mass analyser is its main determinant. Increasing resolution adds to the capital cost and to the difficulty of operating an instrument, and it is important to consider carefully the resolution required for the analytical problems at hand. For the magnetic instruments, resolution is most often quoted according to the "valley definition", which gives a constant value over the entire mass range. The 10% valley definition is common and states that resolution (R)

$$R = M1/(M1 - M2)$$

where M1 and M2 are approximately equally intense and adjacent mass peaks separated so that the height of the valley between them is 10% of the peak height. The choice of the 10% valley is arbitrary and other values may be referred to. Furthermore, it should be understood that a quoted figure for resolution is meaningless unless accompanied by a definition of the way it was determined. Using the 10% valley definition, instruments with less than R3000 are usually termed "low resolution"; with R3000 to 8000, "medium resolution"; and above R8000, "high resolution". Single focusing instruments can deliver low and medium resolution data, whereas double focusing instruments usually operate at a resolution greater than R12000.

In quadrupole instruments, resolution, is usually given by $M/\Delta M$ where M is the mass and ∆M the peak width at half the height. This definition is applicable because it is usual to operate quadrupoles using their "natural", scan function where the peak width is constant over the total mass range. It is important to realize that these conditions produce a different value for resolution over the entire mass range and that to be meaningful, any value quoted must be accompanied by the mass and the mass range of the instrument. Since the resolution is controlled electronically, in theory it can be set to very high values. However, as resolution increases, so does mass discrimination. In practice, good commercial instruments should have at least "unit resolution" (peak width at base 1 amu) with a mass range of 2 to about 1200; and should be able to deliver nondiscriminated spectra up to the region of about 500 amu. These specifications would be suitable for many applications in clinical chemistry, but any analyst contemplating the analysis of higher masses (e.g., prostaglandin derivatives) should critically assess mass discrimination data.

SENSITIVITY

The sensitivity of an analytical method is a particularly important consideration in the organic analyses of clinical chemistry, where the compounds of interest may be present in the biological matrix only at micromolar and nanomolar levels. The sensitivity of a mass spectrometer refers to the total charge collected at a specified mass during the consumption of a known amount of sample in the source. To compare instruments, an accurate definition should state the chemical substance, the instrument resolution, and the electron generating filament current. Most manufacturers of GC/MS systems do not write sensitivity specifications in this way. A more popular method is to quote the signal to noise ratio achieved at a particular mass following the injection of a given amount of compounds onto the GC column. To be useful, however, this specification must also include a full description of the operating conditions. These are the sample identity, all chromatographic parameters, (carrier flow, column temperature, retention time, and peak width), source filament current, mass spectrometer resolution, and the mass peak being measured. In addition, if sensitivity in the scanning mode is being quoted, the scan speed and the mass range must be defined. Instruments that are equipped with facilities for capillary columns, chemical ionization, and selected ion monitoring should include a separate set of specifications for each mode.

In modern GC/MS and LC/MS, it is background noise rather than instrumental hardware design that generally sets the limit of sensitivity. There are, however, two operating parameters that can significantly influence sensitivity — resolution and scan rate.

Resolution and sensitivity vary inversely. In magnetic instruments, resolution is increased by reducing the widths of source and collector slits producing a corresponding decrease in beam intensity. In quadrupoles, there are no variable apertures and resolution is determined electronically, but higher resolution dictates that a larger proportion of ions will oscillate into the rods which again reduces the total number reaching the detector. This is heightened by the mass discrimination properties of the quadrupole, which can serve to reduce sensitivity even more at higher masses. This problem can be alleviated by an ion optic system which can be adjusted so that the ions enter the quadrupole as a narrow beam; by reducing the fringing field of the mass filter, and by using hyperbolic rods to maximize the transmission radius of the filter.

In general, the greater the scan rate, the lower the sensitivity. An ion beam of about 6000 ions per second would produce a current of about 10^{-15} amp which is several orders of magnitude above the limit of the detectors but in a scanning mode, the rate of scan will determine the time for which this beam is focused on the detector. As the scan rate increases, the number of ions will decrease to a limiting value that can no longer be statistically distinguished from random noise. Thus, sensitivity is impaired to the extent that small peaks in the spectrum will not be counted and an incomplete record will result. This outcome illustrates an important aspect of the meaning of sensitivity in mass spectrometry. Identification is usually the goal and this requires a statistically valid sample of ions from the least intense peak of the spectrum that is required for unequivocal identification.

CHANGING SENSITIVITY

Several different approaches may be taken in any mass spectrometry problem and one must match the approach to the analytical question. The diversity of questions stemming from clinical chemistry problems means that the analyst should be fully aware of the different ways one can use a mass spectrometer in order to improve both sensitivity and the spectral data in general.

Photoplate recording is a detection method that amounts to integrating the ion counts over several scans. It is used almost exclusively in high resolution work and has found little practical application in clinical chemistry.

Another approach is *selected ion monitoring* which, in contrast, has proved extremely useful and its application has been the topic of two recent and excellent reviews.[12,13] In this mode, the mass spectrometer is used to detect preselected ions which are characteristic of the analyte and which are preferably of high abundance. In essence, one is making use of the relationship between scan rate and sensitivity, i.e., sensitivity is increased because the time spent on each peak is significantly greater than when the mass range is scanned. As a result, a far greater portion of the ion beam associated with the selected mass reaches the detector with an increase in the signal-to-noise ratio. The technique is well suited to quantitative and qualitative analysis and there are many clinical chemistry applications reported. A drawback is that the analyst must have prior knowledge of the mass spectrum.

The voltage controlled mass selection of a quadrupole mass filter allows virtually instantaneous peak switching over a wide mass range and the quadrupole is, therefore, highly suitable for selective ion monitoring. This property has been further facilitated by controlling voltage switching with a digital computer, and commercial systems are available which allow the simultaneous monitoring of up to 25 ions over any mass range within the total mass range of the instrument.

Traditionally, in magnetic instruments, ions are selected by attenuating the accelerating voltage. However, lower voltages influence ion optics and the selected ions' abundance which, in turn, affects sensitivity and limits the extent of the mass difference between selected ions, thereby restricting the usefulness of the technique. This situation has been improved by computer controlled techniques which allow rapid switching of the magnetic field strength, as well as the accelerating voltage.

Sensitivity can be selectively increased by modifying ionization conditions. In the electron impact mode, the electron energy is traditionally set at 70 eV. This energy is a major determinant of the total positive ion current and also the resultant fragmentation pattern. In some cases, the fragmentation at 70 EV will be excessive, and the percentage ion current at the molecular ion and larger fragment ions will be small. Adjustment of the electron energy to a lower value still above the ionization potential, may produce a different spectral pattern in which the molecular ion intensity is increased. If this ion is being used for quantitation or identification, an apparent increase in sensitivity may be realized. However, the overall intensity of the total ion current will generally fall as the electron energy decreases. Also, EI mass spectra are traditionally catalogued using 70 eV data, and it is difficult to compare spectra obtained with different values.

The chemical ionization source can be used to produce mass spectra in which the relative abundance of the "quasi molecular" ion is large without concomittant loss of overall mass spectral sensitivity. Since the number of reagent gas molecules is high, the probability of reaction with sample molecules is also high and overall sensitivity is maintained. However, the total ion current will now be shared between fewer ions than in the EI mode. The simpler mass spectrum which is produced contains a few ions of high intensity and also is influenced by the reagent gas used. Such spectra are often easier to interpret and for some compounds, represent a significant increase in analytical sensitivity.

Chemical ionization also forms negative ions with some compounds.[14] Moreover, with those compounds that possess positive electron affinity and a large cross section for electron capture, either naturally or following derivitization, the negative ion current can be two to three orders of magnitude greater than the positive. This greatly enhanced signal brings the mass spectrometer into new realms of sensitivity which have already found application in clinical chemistry.[15] Derivatization to a volatile product is often a necessary prerequisite to the gas chromatography in GC/MS, and selection of a suitable derivatizing reagent can also permit negative ion production. Instruments with the hardware necessary for negative ion detection are now commercially available, and quadrupole instruments are especially relevant since their mass filter allows equal passage of both positive and negative ions of the same m/e value. By pulsing the polarity of the ion source potential and the focusing lens potentials at a rate of about 20 kHz, beams of positive and negative ions leave the source and enter the mass filter in rapid succession, and incorporation of dual ion detection systems of opposite polarities allows the acquisition of both positive and negative ion spectra of a sample. Instruments offering this "pulsed positive negative ion chemical ionization" flexibility are now readily available.

APPLICATIONS

Mass spectrometry has been used for some time in special areas such as blood gas analysis;[16] however, it is fair to say that its recent growth in clinical chemistry has centered around organic analysis using combined gas chromatography/mass spectrometry.[17] (A notable exception is the development of the definitive method for calcium using isotope dilution techniques[18]) Gas chromatography/mass spectrometry has been applied to two major analytical problems of organic analysis in clinical chemistry. These are the analysis of complex mixtures termed "metabolic profiling",[19] and the analysis of specific compounds where the mass spectrometer can be viewed as a selective and sensitive gas chromatographic detector.

Traditionally, profile analysis has been largely qualitative, but recent computer-based techniques have permitted quantitative work.[20,21] The aim of specific component analysis is almost entirely quantitative. Interest in metabolic profiling using gas chromatograph/mass spectrometer/data systems is growing. Its efficiency has been greatly increased by the use of open tubular glass capillary columns, which are now generally available, and any instrument should be designed to accept these columns if profiling is the major application. The technique has been applied to almost all of the body fluids, particularly urine, and to practically all the major compound groups, e.g., steroids, organic acids, volatiles, carbohydrates, amino acids, and bile acids. Excellent reviews on the scope of the approach have appeared recently.[22,23]

For the quantitative analysis of specific compounds, selected ion monitoring is the method of choice and, in modern instruments, computers again are playing a major and growing role. The technique has been applied, either in the electron impact or chemical ionization mode, to many different compounds and its special application to clinical chemistry has been reviewed recently.[12,13] It has found particular application in the development of reference methods for many compounds using stable isotope labelled internal standards, for example, calcium, uric acid, cholesterol, cortisol, and glucose.[24] Many other reports describe quantitation of a wide variety of analytes by this technique and, currently, it seems that the rate limiting step in the growth of this work is not mass spectrometry technique, but the availability of pure internal standards. Use of negative ion chemical ionization selected ion monitoring has enabled new levels of sensitivity to be attained and, no doubt, will permit the accurate analysis of materials hitherto beyond the limits of measurement.

CONCLUSION

This brief review is deliberately broad and lacking in specifics. More detailed information is available in many excellent reviews and books. The list of references is comprehensive and, hopefully, will prove useful to the clinical chemist who is interested in using mass spectrometry. Some topics such as field desorption and field ionization which should find application in clinical problems have been omitted, since well-tested commercial instruments are not yet available.

Modern mass spectrometers, particularly those with chromatographic interfaces and with data systems, are extremely sophisticated and complicated instruments, bringing together many techniques in which the chemist often has little or no formal training. However, it is imperative that analysts gain as full an understanding as possible of all the features of these instruments, since only with this information will they be able to choose an appropriate instrument and be able to use it to its fullest potential.

SUPPLIERS

1. Finnigan Instruments, 845 W. Maude Avenue, Sunnyvale, Calif., 94086
2. Hewlett Packard, 1501 Page Mill Road, Palo Alto, Calif., 94304.
3. Kratos Instruments, 24 Booker Street, Westwood, N.J., 07675
4. LKB Instruments, 122221 Parklawn Drive, Rockville, Md., 20852
5. Ribermag GC/MS, 49 Quai Du Halage, 92502 Rueil-Malmaison, France
6. VG Micromass Ltd., Ion Path, Road 3, Winsford Ches, England CW7 3BX

REFERENCES

1. **McFadden, W. H.,** *Techniques of Combined Gas Chromatography/Mass Spectrometry: Applications in Organic Analysis,* John Wiley & Sons, New York, 1973.
2. **McFadden, W. H.,** *J. Chromatogr. Sci.,* 17, 2, 1979.
3. **Fenselau, C.,** *Anal. Chem.,* 49, 563A, 1977.
4. **Guichon, G. and Aprino, P. J.,** *Anal. Chem.,* 51, 682A, 1979.
5. **Aprino, P. J., Dawkins, B. G., and McClafferty, F. W.,** *J. Chromatogr. Sci.,* 12, 574, 1974.
6. **Milberg, R. M. and Carter Cook, J., Jr.,** *J. Chromatogr. Sci.,* 17, 17, 1979.
7. **Munson, B.,** *Anal. Chem.,* 49, 772A, 1977.
8. **Feser, K. and Kogler, W.,** *J. Chromatogr. Sci.,* 17, 57, 1979.
9. **Dawson, P. H.,** *Adv. Mass. Spectrom.,* 1, 905, 1978.
10. **Chapman, J. R.,** *Computers in Mass Spectrometry,* Academic Press, New York, 1978.
11. **McClafferty, F. W. and Venkatarachan, R.,** *J. Chromatogr. Sci.,* 17, 24, 1979.
12. **Falkner, F. C., Sweetman, B. J., and Throck Watson, J.,** *Appl. Spectrosc. Rev.,* 10(1), 51, 1975.
13. **Bjorkhem, I.,** *Crit. Rev. Clin. Lab. Sci.,* 11, 53, 1979.
14. **Hunt, D. F. and Sethi, S. K., A.C.S.** *High Performance Mass Spectrometry: Chemical Applications, (Symposium Series, No. 79),* Gross, M. L., Ed., American Chemical Society, Washington, D.C., 1978.
15. **Hunt, D. F. and Crow, F. W.,** *Anal. Chem.,* 50, 1781, 1978.
16. **Roboz, J.,** Mass spectrometry in clinical chemistry in *Advances in Clinical Chemistry,* Bodansky, O. and Latner, A. L., Eds., Academic Press, New York, 1975, 67, 132.
17. **Lawson, A. M.,** *Clin. Chem.,* 21, 803, 1975.
18. **Moore, L. J. and Machlan, L. A.,** *Anal. Chem.,* 44, 2291, 1972.
19. **Horning, E. and Horning, M.,** *Clin. Chem.,* 17, 802, 1971.
20. **Gates, S. C., Smisko, M. J., Ashendel, C. L., Young, N. D., Holland, J. F., and Sweeley, C. C.,** *Anal. Chem.,* 50, 433, 1978.
21. **Smith, D. H., Achenbach, M., Yeager, W. J., Anderson, P. J., Fitch, W. L., and Rindfleisch, T. C.,** *Anal. Chem.,* 49, 1623, 1977.
22. **Jellum, E.,** *J. Chromatogr. Biomed. Appl.,* 143, 427, 1977.
23. **Gates, S. C. and Sweeley, C. C.,** *Clin. Chem.,* 24, 1663, 1978.
24. **Bjorkhem, I., Blomstrand, R., Lantto, O., Svensson, L., and Ohman, G.,** *Clin. Chem.,* 22, 1789, 1976.

OSMOMETRY

M. J. McQueen

INTRODUCTION

Osmometry is the technique for measuring the molar concentration of solute particles in a solvent.[1-3] In biological systems the solvent is water. The addition of solute produces changes in the physical properties of the solvent: increases in the boiling point and osmotic pressure; decreases in the freezing point and vapor pressure. Such changes are common to the addition of any solute. Since they are mathematically interconvertible and are proportional to solute concentration, they are called *colligative properties.*

The extent of the changes relates to the number, not the mass or charge, of particles dissolved in the solvent. One mole of any substance contains the same *number* of molecules. In the ideal case where the solute does not dissociate in solution, 1 mol dissolved in 1 kg of water contains 6.023×10^{23} molecules *(Avogadro's number).* However, if each molecule does dissociate into two or three particles, the total number of particles will increase by the appropriate factor and the osmolality will double or triple. The changes in the colligative properties of 1 kg of pure water when it contains 1 mol of nondissociated solute are given in Table 1. When each molecule of the solute dissociates into two or three particles each of these colligative properties is changed by a factor of two or three. Since the change in these properties relates to the number of particles, the measurement of any one property allows calculation of the other three and of the concentration of dissolved particles.

It is readily appreciated that biological systems do not behave in a mathematically *ideal* way. There may be association between solute molecules when more than one solute is present. An electrolyte solution may not completely dissociate in plasma, e.g., sodium chloride (NaCl) is 93% dissociated at the concentration usually found in plasma. Solute molecules can associate with solvent. When effective concentrations of solutes are lower than the molar concentrations, a correction factor called the *activity coefficient* can be applied to correct for the deviation from the *ideal.*

The activity coefficient for NaCl is 0.93 at the concentrations usually found in plasma, for urea it is 0.94 and for glucose 1.00:

$$\text{Osmolality} = \phi \cdot n \cdot C \tag{1}$$

where ϕ = activity coefficient; n = number of particles dissolved in the solution; and C = concentration in mole per kilogram water.

MEASUREMENT

Despite the problems of nonideal solutions, osmolality can be measured, the two colligative properties most frequently assessed being depression of the freezing point and depression of the vapor pressure. The greatest contribution to the plasma osmolality is made by the electrolyte ions, sodium, potassium, chloride, and bicarbonate, since they are present in relatively high concentrations. Glucose and urea normally make only a small contribution to plasma osmolality, but by virtue of their low molecular weight, they can contribute a large number of molecules when their plasma concentrations are increased in disease. The plasma proteins make a small contribution

Table 1

CHANGES IN THE COLLIGATIVE
PROPERTIES OF 1 KG OF PURE WATER
WHEN IT CONTAINS 1 MOL OF
NONDISSOCIATED SOLUTE

Boiling point	Increases by 0.52°C
Osmotic pressure	Increases to 17,000 mm Hg (2267 kPa)
Freezing point	Decreases by 1.858°C
Vapor pressure	Falls by 0.3 mm Hg (0.04 kPa)

to the total osmotic pressure, since although they have a large relative molecular mass, they contribute only a small *number* of molecules.

The freezing point depression and vapor pressure depression methods measure the total osmolality due to the small ions or molecules (*crystalloids*) plus the high molecular weight solute particles (*colloids*). The colloid osmotic pressure can be measured in the clinical laboratory and is important in the regulation of water entering and leaving capillaries.[4,5]

Freezing Point Depression

The fluid to be tested is placed in a glass cuvette and then immersed in a bath containing a solution such as ethylene glycol maintained at −7°C (Centigrade or Celsius) by utilizing the Peltier effect. A thermistor is lowered into the sample to measure the temperature. The sample is stirred continuously and is rapidly cooled to −7°C. The sample does not solidify, but remains in a liquid state, as may all liquids when cooled below their freezing points. They are then said to be *super-cooled*. The sample is then raised above the liquid in the cooling bath and a physical stimulus such as vigorous vibration of the stirring rod is applied and the solvent crystallizes. As a result of this fusion, heat is released which then warms the solution to its freezing point. This is an equilibrium temperature at which both freezing and thawing of the solution is occurring. The equilibrium temperature may be maintained for up to 2 to 3 min until all the released heat is absorbed by the cooling bath, when the temperature begins to fall again and the sample freezes.

The temperature of the sample is measured by connecting the thermistor, the resistance of which increases with decrease in temperature, to a Wheatstone bridge circuit. The depression of the freezing point produced by 1000 mmol of solute per kilogram of water is 1.86°C. A galvanometer is used to null the current in the Wheatstone circuit and a measuring potentiometer is used to balance the circuit. The change of temperature from −7°C to that at which there is equilibrium in freezing and thawing of the sample, produces a change in the resistance of the thermistor which is reflected in a change in current in the Wheatstone circuit. The balancing potentiometer is then adjusted to bring the galvanometer to the null position. Solutions of known osmolality are used to calibrate the balancing potentiometer. Since temperature changes are linearly related to osmolality, the freezing point depression of the sample as represented by the changes in the readings of the balancing potentiometer can be expressed directly in mosmol per kilogram of water (or millimole per kilogram H_2O, when using SI units).

Vapor Pressure Depression

This has been identified as one of the four colligative properties of a solution and within the range of interest to clinical work is linearly related to the osmolality. Vapor pressure osmometers have been based on condensation, evaporation, or equilibrium between condensation and evaporation, of the solvent. When equilibrium between condensation and evaporation is involved, the measurement chamber is at ambient temperature, which is essentially room temperature. The instrument should not be located where rapid temperature changes could be caused by direct sunlight, by heating or cooling systems, or where any thermal gradient could occur. Within the chamber there is a sensitive thermocouple hygrometer and its resting voltage is used to establish a null or zero reference point. The sample is then put into the chamber, sealed, and given time for thermal and vapor equilibration. The thermocouple junction is cooled to a temperature below the dew or condensation point by passing a current through it (the Peltier effect). A thin film of water forms on the junction surface by condensing from the air in the chamber. As a result of this condensation, heat is released and raises the temperature of the junction asymptotically toward the dew point. No more water condenses when the junction temperature reaches the dew point.

The relationship between vapor pressure depression and dew point depression is given by:

$$\Delta T = \frac{\Delta e}{S} \qquad (2)$$

where ΔT = dew point depression in °C (Centigrade or Celsius); Δe = the difference between saturation and chamber vapor pressure; and S = slope of the vapor pressure temperature function at ambient temperature.

The slope, S, is given by the Claussius-Clopeyron equation:

$$S = \frac{e_o \lambda}{RT^2} \qquad (3)$$

where T = temperature (Centigrade or Celsius); e_o = saturation vapor pressure; λ = latent heat of vaporization; and R = the universal gas constant.

The dew point depression, ΔT, is obtained from the measurement of the voltage of the thermocouple. This voltage is equal to ΔT multiplied by the thermocouple calibrator, which is 61 μVolts/°C.

The reading on the output meter is proportional to the temperature depression of the thermocouple junction, which is equivalent to the dew point temperature depression in the chamber and can, in turn, be related to vapor pressure depression.

Both measurement systems so far described have a precision of ± 2 mmol/kg of water. The vapor pressure depression osmometers can function with sample volumes of 8 $\mu\ell$ and freezing point depression osmometers are available which use 20 $\mu\ell$ or 60 $\mu\ell$, but most require 200 $\mu\ell$ or 2.0 mℓ.

Colloid Osmotic Pressure

A sample chamber and reference chamber are separated by a semipermeable membrane which prevents passage from the sample chamber of solutes with a relative molecular mass of 30,000 or higher. The reference chamber is filled with a 150 mmol/ℓ sodium chloride solution in direct contact with a sensitive pressure transducer. When saline is in the sample chamber, there is no pressure gradient across the membrane and this serves as a baseline. When serum or plasma is introduced into the sample compartment, the protein molecules which cannot cross the membrane cause water molecules to diffuse across the membrane from the reference chamber into the sample chamber. This creates a negative pressure in the reference chamber which is equal to the colloid osmotic pressure and which, in turn, is converted by the pressure transducer into electric signals. These are expressed as positive pressures (mmHg) which equal the colloid osmotic pressure. Standardization can be achieved by using an albumin solution of known concentration.

CLINICAL APPLICATIONS

Serum osmolality ranges (reference range 270 to 295 mmol/kg H_2O) are used to identify characteristic clinical syndromes of hyper- and hypo-osmolar states.[6,7] The hyperosmolar states may result from a deficiency of water. In the presence of normal renal function, this can arise from reduced intake, an excess loss in sweat, burns, or gastrointestinal fluids. The concentrating ability of the kidney can be impaired in diabetes insipidus, head injury, tumor, and fluoride toxicity.

Solute accumulation such as alcohol, glucose in diabetes mellitus, and urea in renal failure, can also lead to hyperosmolality. It should be remembered that while an increased plasma osmolality should serve to arouse clinical suspicion of increased urea or alcohol in a patient, both of these solutes readily diffuse across cell membranes and are not *active* osmotically, since they do not contribute to any osmotic gradient across cell membranes.[6]

The hypo-osmolar states may arise from an accumulation of water and, in the presence of normal renal function, require the ingestion of more than 15 ℓ/day. More commonly, failure to excrete water is due to impaired perfusion, reduced glomerular filtration, or increased arginine vasopressin (antidiuretic hormone) secretion. There can be a deficit in solute due to the sick-cell syndrome or to decreased intake in cachexia or chronic alcoholism. Diuretics and Addison's disease can also produce increased loss of solute and, hence, a solute deficit.

Urine osmolality has been used to assess the concentrating ability and solute output of the kidney. In the presence of normal renal function, the ratio of urine osmolality to plasma osmolality can be used as an estimate of the state of body hydration: oliguria of dehydration and that of renal impairment can be differentiated.[8]

Any difference between measured and calculated osmolality,[9] delta osmolality (Δosm), suggests the presence of osmotically measureable particles such as ethanol, isopropanol, methanol, and mannitol.[10] In the presence of a volatile substance such as ethanol, a difference can be detected only if osmolality is measured using a freezing point osmometer.[11,12] As the amount of ethanol increases in the solution, the increase in the solution vapor pressure results in a vapor pressure osmometer giving an apparent lowering of osmolality. Conversely, freezing point osmometers may give falsely high readings in the presence of hyperlipidemia, since the viscosity of the material may hinder uniform freezing.

Measurement of sweat osmolality has been reported as a satisfactory alternative to the measurement of sweat chloride in the detection of cystic fibrosis.[13]

The gradient between colloid osmotic pressure and pulmonary artery wedge pressure gives a good correlation with pulmonary edema.[14] The reduction in colloid osmotic pressure arising from a fall in plasma protein concentration produces a reduced gradient and correlates well with the development of pulmonary edema.

REFERENCES

1. Freier, E. F., Osmometry, in *Fundamentals of Clinical Chemistry,* Tietz, N. W., Ed., W. B. Saunders, Philadelphia, 1976, 153.
2. Bevan, D. R., Osmometry. I. Terminology and principles of measurement, *Anaesthesia,* 33, 794, 1978.
3. McSwiney, R. R., Osmosis and osmometry, in *Scientific Foundations of Clinical Biochemistry,* Williams, D. L., Nunn, F. R., and Marks, V., Eds., William Heinemann, London, 1978, 388.
4. Gyton, A. C. and Lindsey, A. W., Effect of left atrial pressure and decreased plasma protein concentration on the development of pulmonary edema, *Circ. Res.,* 7, 649, 1959.
5. Morissette, M. P., Colloid osmotic pressure: its measurement and clinical value, *Can. Med. Assoc. J.,* 116, 897, 1977.
6. Loeb, J. N., The hyperosmolar state, *N. Engl. J. Med.,* 240, 1184, 1974.
7. Bevan, D. R., Osmometry. III. Clinical applications, *Anaesthesia,* 33, 809, 1978.
8. Coran, A. G., Das, J. B., and Evaklis, A. J., Use of osmometry in the pre-operative and post-operative management of the newborn, *J. Pediatr. Surg.,* 6, 529, 1975.
9. Dorwart, W. V. and Chalmers, L., Comparison of methods for calculating serum osmolality from chemical concentrations, and the prognostic value of such calculations, *Clin. Chem.,* 21, 190, 1975.
10. Glasser, L., Sternglanz, P. D., Combie, J., and Robinson, A., *Am. J. Clin. Pathol.,* 60, 605, 1973.
11. Champion, H. R., Baker, S. P., Benner, C., Fisher, R., Capham, Y. H., Long, W. B., Adams Cowley, R., and Gill, W., Alcohol intoxication and serum osmolality, *Lancet,* 1, 1402, 1975.
12. Barlow, W. K., Volatiles and osmometry (cont.), *Clin. Chem.,* 22, 1230, 1976.
13. Webster, L. and Lochlin, H., Cystic fibrosis screening by sweat analysis, *Med. J. Aust.,* 1, 923, 1977.
14. Rackow, E. C., Fein, I. A., and Leppo, J., Colloid osmotic pressure as a prognostic indicator of pulmonary edema and mortality in the critically ill, *Chest,* 72, 709, 1977.

SCANNING MICRODENSITOMETRY

J. Chayen

GENERAL BACKGROUND

The Function of Scanning Microdensitometry
Definition

Microdensitometry or microspectrophotometry, is concerned with spectrophotometric measurements of chromophores, whether natural or experimentally induced, in individual cells or parts of a cell. It is analogous to conventional spectrophotometry in which the biological cell, suitably magnified microscopically, replaces the cuvette of the conventional spectrophotometer. Although, whereas the chromophore, measured spectrophotometrically, is in solution, that measured by microdensitometry is frequently inhomogeneously precipitated. Consequently, a scanning system has to be incorporated into the microdensitometer to allow precise measurement of such optically inhomogeneously distributed chromophores.

The Use of Microdensitometry

Modern quantitative cytochemistry[1-3] is now a precise, quantitative form of cellular biochemistry that can yield assays of several enzymes from a single needle biopsy. It is much used in exfoliative cytology[4-6] and in hematology.[7,8] It is the basis of the cytochemical bioassays of polypeptide hormones that, while assaying bioactive hormone, are about 1000 times as sensitive as the equivalent radioimmunoassays.[1,2,9,10]

Since a sample of tissue will rarely contain a single cell-population, it is necessary to relate biochemical activity to histology.[1] In this way, the biochemical activity of the target-cells can be measured irrespective of any contribution from surrounding, unaffected cells. The ability to relate biochemical activity to histology requires a microscope as well as a spectrophotometer; the precise measurement of an optically inhomogeneous colored precipitate necessitates the use of a scanning system.

Physical Theory of Microdensitometry
Basic Considerations

The fundamental law, both of spectrophotometry and of microdensitometry, is the Beer-Lambert Law which relates the concentration (c) of the chromophore to the absorbance (A: also called "extinction") of the solution;

$$A = \log_{10} (I_0/I) = kcl \tag{1}$$

where I and I_0 are the intensities of the transmitted and of the incident light respectively; l is the path-length of the light through the solution of the chromophore; and k is the absorptivity (or extinction coefficient; also called the coefficient of absorbance) of that chromophore at the particular wavelength used for the measurement.

In microdensitometry, it is required to measure the mass (M), not the concentration, of the chromophore. Since M = c × volume (v), and v = l multiplied by the projected surface area (P), c is equal to M/Pl. Consequently, Equation 1 becomes:

$$A = k (M/Pl)l = kM/P$$

$$\text{or} \quad M = A \cdot P/k \tag{2}$$

This relationship is satisfactory provided that the chromophore is in true solution or, otherwise, evenly distributed throughout the area (P) to be measured. However, in cells, this rarely pertains to either natural, or experimentally induced, chromophores. The problem posed by an inhomogeneously distributed chromophore is overcome by measuring the absorbance of each minute and optically homogeneous fraction (p) of the total projected area (P) and summating all the measurements.

Consequently, Equation 2 becomes:

$$M = (A_1 p + A_2 p + A_3 p + \ldots + A_n p)/k$$

$$\text{or } M = (A_1 + A_2 + A_3 + \ldots + A_n) \, p/k \tag{3}$$

In practice, this is achieved by sending a flying-spot of light (normally of the wavelength at which the chromophore absorbs maximally) across the area to be measured and summating (or integrating) the absorptions recorded at every position in the scan. The size of this flying-spot is critical, as will be discussed later. However, it should be noted that precise measurements can be made with certainty if the flying-spot is so small that the region encompassed by it appears optically homogeneous. This is ensured by recourse to the fact that the resolution of the microscope (d) is defined by:

$$d = 0.6\lambda/N.A.$$

where N.A. is the numerical aperture of the objective, and λ is the wavelength of light used. Consequently, with light of 550 nm wavelength, and an oil-immersion objective of N.A. 1.3, the resolution (d) will be $(0.6 \times 550.10^{-6})/1.3$ mm, or 0.25 μm. Thus, if the diameter of the flying-spot is 0.25 μm (or 0.2 μm for shorter wavelengths), the area within that spot will be optically homogeneous; if it is greater, the microscope is capable of resolving optical inhomogeneities within the area (p) covered by the spot, and this can cause distributional error in the measurement of an inhomogeneously distributed chromophore.

Problems in Microdensitometry

Various optical effects can provide potential sources of error (Table 1). These potential problems have been discussed exhaustively[11-16,18-22] and have been taken into account in defining correct procedures of microdensitometry. Provided these are followed, precise measurement will result. For example, spectrophotometric and microdensitometric measurements of the concentration of cytochrome P-450 in isolated microsomes, measured by the absorption of this chromophore at 450 nm, yielded identical results.[23]

Distributional Error

If no scanning system is employed, it is well recognized that the inhomogeneous distribution of a chromophore can cause "distributional error", which is the most serious of all potential errors in microdensitometry.[11,12,16,18,21,22,24,25] It remains a potential hazard when inappropriate sizes of the scanning-spot are used (as emphasized by[14]) and, therefore, merits further discussion.

Suppose a cell contains its chromophore in discrete granules, each of which has an absorbance of 0.7 (and, therefore, they transmit 20% of the light incident on them). Suppose these granules occupy one fifth of the projected area of the cell and light from the area of the whole cell is transmitted to the photomultiplier. There will be four regions with 100% transmission and one with 20% transmission. Thus, the mean

Table 1

THEORETICALLY POTENTIAL ERRORS IN MICRODENSITOMETRY

Possible error	Conditions favoring the error	Comment
Glare	High absorbance values in the specimen;[16] exacerbated by over-large field diaphragm	Caused by reflections within the microscope-system; up to 3% of incident light may be reflected inside the microscope;[13] this will not influence measurements as long as absorbance is kept low by using thin specimens, or short cytochemical reaction-times
Out of focus effects	Very thick specimens	Probably exaggerated by some microscopists,[12,14,15] provided that specimens are not unduly thick
Dichroism	Chromophore in oriented structures	Test with polarized light; although theoretically a potential error, it does not seem so in general practice[16]
Convergent illumination through microscope lenses	Normal microscopy	Error likely to be negligible at high refractive index[16] (e.g., use suitable mountant)
Scatter	Reflections by small particles of very different refractive index from rest of cell	Overcome by suitable refractive index of mountant;[14,16] also by measuring at absorption maximum (A_{max}) and absorption minimum (A_{min}); then $A = A_{max} - A_{min}$
Diffraction	Accentuated by "stopped-down" condenser and by high absorbance in specimen	Insignificant if microscope is used correctly, with condenser at full numerical aperture, and if absorbance of each point in the specimen is below 0.7 (preferably 0.5)
Distributional error	Chromophore inhomogeneously distributed, e.g., in small particles	Will not operate if the scanning-spot is 0.25 μm in diameter[12,14]
Absorption coefficient of solid chromophore	Precipitated chromophore may have different absorption-characteristics from that in solution	The absorption coefficient of the chromophore must be determined under the conditions in which it is measured (as done by Reference 18)

transmitted light will be 420/5 or 84%. From Equation 1, substituting 100 for I_o and 84 for I, it follows that the absorbance recorded for this cell will be 0.076. Yet, if this chromophore were to be distributed throughout the cell, the extinction of the cell would be 0.7 divided by 5, or 0.14 (assuming a very flat cell, where "volume" is approximately synonymous with "projected area"). Consequently, distributional error would have given a value only 54% of the true value. Hence, the need to scan and to measure each region separately, integrating all the readings, as is done in scanning microdensitometry.

By similar reasoning,[22] if the chromophore filled half the field, giving an absorbance of 0.7 for each point in that half-field, the absorbance registered for the whole field would be 0.22, whereas if the chromophore had been dispersed evenly across the whole field, the absorbance would have been 0.35. This is the type of error that can occur when one uses a scanning-spot of twice the area of the granules of the chromophore. Thus, the size of the scanning-spot is also critical.

Under ideal conditions, the diameter of the scanning-spot is 0.2 μm so that the sample of the specimen within the spot, for each measurement, is homogeneous. Frequently, however, it is not practicable to use an oil-immersion objective, as is necessary to achieve this size of scanning-spot. Sufficiently accurate measurements can sometimes be made with a scanning-spot having a diameter just smaller than the smallest optical inhomogeneity within the specimen. This particularly applies[11] when the maximum absorbance of each region occupied by the chromophore is small, namely between 0.1 and 0.3. (However, it should be noted[24] that this does not apply to absorbances of less than 0.1.) This dispensation is much used in the cytochemical bioassays where the need for speedy measurement often precludes the use of oil-immersion optics; but, in the assays of thyroid stimulators[26] in which the chromophore is deposited in small granules that may be between 0.3 and 0.5 μm in diameter, oil-immersion optics, with a scanning-spot of 0.2 μm diameter, are essential; scanning-spots of greater diameter completely fail to measure the response.[14,27] Indeed, the use of a scanning-spot of twice the area of the colored optical inhomogeneity yields the same error, in each measurement, as occurs when the whole, uniformly colored cell is measured by a photomultiplier with a sensing area twice that of the cell (as discussed above).

The distributional error is related to both the absorbance of the particles and their area relative to that of the scanning spot.[16]

INSTRUMENTATION

The Instrument

The requirements of a scanning and integrating microdensitometer are

1. That it should have a fully variable monochromator, not merely a system of filters.
2. That the scanning spot or diaphragm, at a diameter of 0.2 μm, should give an ample signal-to-noise ratio.
3. That the scan should cover most, or all, of the selected area, so avoiding sampling error.
4. That the whole operation of selection and measurement should be rapid.
5. That the instrument should be robust.

Of the various commercial microdensitometers, only the Vickers M85 (Figure 1) has been proven to meet all these requirements.

FIGURE 1. The Vickers M85 scanning and integrating microdensitometer. This is a conventional microscope with condenser (a), objectives (b), and eyepieces (c) for inspecting the histology of the specimen. The spectrophotometer component comprises controls for adjusting the light-source (d) of the spectrophotometer; for regulating the band-width of the monochromator (e) and selecting the wavelength (f). There is also a control for adjusting the size of the scanning-spot (g). The "relative absorption" is shown on the digital meter (h) and the area occupied by the chromophore can be derived from another meter (i). The limit of the field to be measured is monitored by an auxiliary photomultiplier (housed inside j) which also controls the "gating" level. The measuring photomultiplier is housed below the microscope (k). To the left is shown an automatic camera that can be used to record the field that is being examined; this attaches opposite the monitoring auxiliary photomultiplier. (Photograph courtesy of Vickers Instruments.)

Procedure

The specimen is first examined by conventional microscopy to determine its histology. The target-cell, or region to be measured, suitably magnified, is placed under an optical, illuminated mask which is viewed by an auxillary photomultiplier. This mask instructs the instrument to measure only what is within the mask.

The monochromator and controls for the scanning flying-spot are housed above the microscope, so the light used for measurement traverses the opposite path to that used for inspection microscopy. The required wavelength and the size of the scanning-spot are selected. Then, with the light passing only through clear field (to compensate for contributions from the mountant, the slide, and coverslip), the instrument is set for 90 to 95% transmission under the selected conditions. This "blank" reading is recorded and subtracted from all subsequent readings.

The specimen is then brought back into the field, and the first region to be measured is placed under the mask and measured. Each measurement takes about 3 sec. The specimen is again examined, and the next region to be measured is selected. With a relatively homogeneous population, it may suffice to measure 20 such regions.

Mean Absorbance

The measurements are recorded as "relative absorption" and will vary from one instrument to another, depending on the characteristics of the measuring photomultiplier and on the voltage applied to it. Such measurements can be converted to units of absolute absorbance by means of a calibration graph. This is constructed, for each instrument, by recording the relative absorption of fields composed of filters of known absorbance. From such graphs, the mean absorbance (or mean integrated absorbance, or extinction) of fields corresponding to masks of known sizes can be obtained. Since the mean absorbance will be less than unity, it is customary to multiply the value by 100 to give "the mean integrated absorbance (or extinction) × 100".

The "relative absorption" is a function of the sum of all the absorbances and, therefore, will increase as the projected area of the colored region increases. Thus, if we have two nuclei, each containing the same mass of a chromophore (e.g., after the Feulgen reaction for DNA), but one being twice the area of the other, they will have the same relative absorption; however (assuming them to be flat discs), one will have half the mean absorbance of the other. Hence, if the mask is not kept of constant diameter or is not completely filled by the colored structure, to obtain the total mass of that chromophore it is necessary to multiply the mean absorbance by the area occupied by the chromophore; the M85 has a facility for recording the area of the chromophore. Such corrections are not required if the results are reported as "relative absorption" which includes a factor for area, as well as for absorbance.

Pathlength

When recording the mass of chromophore in each nucleus, or each cell, the pathlength is immaterial: it does not matter if one cell is flattened and another is tall in the light-path. However, pathlength is critical when comparing the mass of a chromophore in sections, where the pathlength will vary with the thickness of the sections. However, sections of uniform thickness (within ± 5%) can be cut on a cryostat that is fitted with an automatic drive[1,2,22] that ensures a constant speed of cutting.[28]

Absorbance of Discrete Regions

A microscope field of mean absorbance of 0.5, in which the chromophore is unevenly distributed, will have regions of 100% transmission (in between regions containing the chromophore). Consequently, the absorbance of each particulate mass of chromophore will be higher than 0.5; it may be unacceptably high and exacerbate errors shown in Table 1. Consequently, it is good practice to check the absorbance of the most dense chromophore-regions by immobilizing the scanning-spot over each such region and checking its absorbance on the control meter that directly records absorbance.

REFERENCES

1. **Chayen, J.**, The cytochemical approach to hormone assay, *Int. Rev. Cytol.*, 53, 333, 1978.
2. **Chayen, J.**, *The Cytochemical Bioassay of Polypeptide Hormones,* Springer-Verlag Berlin, 1980.
3. **Pattison, J. R., Bitensky, L., and Chayen, J.**, Eds., *Quantitative Cytochemistry and its Applications,* Academic Press, London, 1979.
4. **Sandritter, W.**, A review of nucleic acid cytophotometry in general pathology, in *Quantitative Cytochemistry and its Applications,* Pattison, J. R., Bitensky, L., and Chayen, J., Eds., Academic Press, London, 1979, 1.
5. **Millett, J. A. and Husain, O. A. N.**, Analysis of chromatin in carcinoma-*in-situ,* in *Quantitative Cytochemistry and its Applications,* Pattison, J. R., Bitensky, L., and Chayen, J., Eds., Academic Press, London, 1979, 37.
6. **Husain, O. A. N. and Millett, J. A.**, The detection of malignancy in the cervix, in *Quantitative Cytochemistry and its Applications,* Pattison, J. R., Bitensky, L., and Chayen, J., Eds., Academic Press, London, 1979, 231.
7. **Wickramasinghe, S. N.**, Studies on the cell cycle in human bone marrow, in *Quantitative Cytochemistry and its Applications,* Pattison, J. R., Bitensky, L., and Chayen, J., Eds., Academic Press, London, 1979, 9.
8. **Stuart, J.**, Quantitative enzyme cytochemistry in acute leukaemia, in *Quantitative Cytochemistry and its Applications,* Pattison, J. R., Bitensky, L., and Chayen, J., Eds., Academic Press, London, 1979, 113.
9. **Chayen, J., Daly, J. R., Loveridge, N., and Bitensky, L.**, The cytochemical bioassay of hormones, *Recent Progr. Horm. Res.,* 32, 33, 1976.
10. **Bitensky, L. and Chayen, J.**, Quantitative cytochemistry — the basis of sensitive bioassays, for comparison of bio- and immuno-reactive hormone values, *Clin. Chem.,* 24, 1399, 1978.
11. **Chayen, J. and Denby, E. F.,** *Biophysical Technique as Applied to Cell Biology,* Methuen, London, 1968.
12. **Goldstein, D. J.**, Aspects of scanning microdensitometry. II. Spot size, focus and resolution, *J. Microsc.,* 93, 15, 1971.
13. **Goldstein, D. J.**, Aspects of scanning microdensitometry. I. Stray light (glare), *J. Microsc.,* 92, 1, 1970.
14. **Bitensky, L.**, Microdensitometry, in *Assessment of Quantitative Histochemical Techniques,* Ciba Symp., Excerpta Medica, 1980.
15. **Deeley, E. M.**, An integrating microdensitometer for biological cells, *J. Sci. Instrum.,* 32, 263, 1955.
16. **Walker, P. M. B.**, Ultraviolet microspectrophotometry, in *General Cytochemical Methods,* Vol. 1, Danielli, J. F., Ed., Academic Press, New York, 1958, 163.
17. **Butcher, R. G.**, Precise cytochemical measurement of neotetrazolium formazan by scanning and integrating microdensitometry, *Histochemie,* 32, 171, 1972.
18. **Swift, H. H. and Rasch, E.**, Microphotometry with visible light, in *Physical Techniques in Biological Research,* Vol. 3, Oster, G. and Pollister, A. W., Eds., Academic Press, New York, 1956, 353.
19. **Caspersson, T.**, Methods for the determination of the absorption spectra of cell structures, *J. Roy. Micros. Soc.,* 60, 8, 1940.
20. **Walker, P. M. B. and Richards, B. M.**, Quantitative microscopical techniques for single cells, in *The Cell,* Vol. 1, Brachet, J. and Mirsky, A. E., Eds., Academic Press, New York, 1959, 91.
21. **Fukuda, M., Böhm, N., and Fujita, S.**, Cytophotometry and its biological application, *Progr. Histochem. Cytochem.,* 11, 1, 1978.
22. **Chayen, J.**, Microdensitometry, in *Biochemical Mechanisms of Liver Injury,* Slater, T. F., Ed., Academic Press, London, 1978, 257.
23. **Gooding, P. E., Chayen, J., Sawyer, B., and Slater, T. F.**, Cytochrome P-450 distribution in rat liver and the effect of sodium phenobarbitone administration, *Chem. Biol. Interactions,* 20, 299, 1978.
24. **Glick, D., Engström, A., and Malmstrom, B. G.**, A critical evaluation of quantitative histo- and cytochemical microscopic techniques, *Science,* 114, 253, 1951.
25. **Gomori, G.,** *Microscopic Histochemistry,* University of Chicago Press, Illinois, 1952.
26. **Bitensky, L., Alaghband-Zadeh, J., and Chayen, J.**, Studies on thyroid stimulating hormone and the long-acting thyroid stimulating hormone, *Clin. Endocrinol.,* 3, 363, 1974.
27. **Bitensky, L., Butcher, R. G., and Chayen, J.**, Quantitative cytochemistry in the study of lysosomal function, in *Lysosomes in Biology and Pathology,* Vol. 3, Dingle, J. T., Ed., North-Holland, Amsterdam, 1973, 465.
28. **Butcher, R. G.**, The chemical determination of section thickness, *Histochemie,* 28, 131, 1971.

STANDARDS FOR INSTRUMENT SERVICE

William H. C. Walker

In the tentative standard TS1-6 of the National Committee for Clinical Laboratory Standards (NCCLS) (771E Lancaster Avenue, Villanova, Pa.), an interdisciplinary committee has set out "Guidelines for Service of Clinical Laboratory Instruments" and this document is summarized here. The standard covers operator training, installation, preventive maintenance, and repair requirements for the proper service of clinical laboratory instruments. It defines the several and the mutual responsibilities of the user and the provider or manufacturer of an instrument. Computer systems are not considered, except insofar as they are an integral part of the instrument.

The manufacturer should provide a statement of performance characteristics including warranty provisions, details of service plans, training programs, operating procedures, including calibration and quality control, troubleshooting, preventive maintenance, and environmental impact, including installation needs, operational hazards, and waste disposal.

The user should provide operators of sufficient skill to become proficient in the use of the instrument. Training needs are defined at four levels: no additional training (e.g., for use of waterbaths); self-training; training by manufacturer in user's laboratory; and training by manufacturer off-site. The manufacturer should provide a basic installation-operation-repair manual which conforms to NCCLS standard ASI-1 together with automatic updates of the manual for the market-life of the instrument. Continuing education of users and user groups, when appropriate, should be supported by the manufacturer.

Subject to defined prerequisites the manufacturer should provide users at their expense, with field-service training and a field repair manual.

Specifications are set out for description of test equipment and for the parts lists required for preventive maintenance and repair. The manufacturer should provide replacement parts during the period of production of the instrument and for at least 5 years thereafter.

Five levels of service are defined according to the urgency of return to operation: 8 hr; 24 hr; 1 week; indefinite; and repair not economical. For the first four, telephone consultation should be available, for the first three, field service should be available, and for the first two, necessary replacement parts should be available in the user's laboratory, and test backup should be provided by the user. If the instrument is out of service for longer than the time specified in the level of service, a loan instrument or component should be available from the manufacturer. The user has a responsibility to provide backup procedures that will allow continued operation on a short-term basis.

The manufacturer should provide service during the period of production of an instrument and for at least 5 years thereafter.

This summary indicates the major points of the NCCLS document, but the full text with its detailed apportionment of responsibilities and obligations between user and manufacturer is an invaluable bench mark for any prospective purchaser of major equipment. Many manufacturers and users fall far short of the expectations listed and all have some deficiencies. Field service within 1 week is the hallmark of only a few of the industry leaders. Many users are neglectful of preventive maintenance. A frequent failing is the lack of any clear definition of responsibilities for support and maintenance negotiated at the time of purchase. Provision to suitably qualified users of field-service training might go a long way to improve the service delays which are often an inevitable consequence of geographic separation of user and manufacturer. Advances in the self-diagnostic capabilities of microprocessors and the ability of instrument microprocessors to communicate by telephone with manufacturers' computer diagnostic facilities may be expected greatly to enhance the service potential of suitably trained users equipped with a stock of replacement parts.

REFERENCE

1. Preparation of Manuals for Installation, Operation and Repair of Laboratory Instruments, ASI-1, National Committee for Clinical Laboratory Standards, Villanova, Pa.

Index

INDEX

A

AAB, see American Association of Bioanalysts
AACC, see American Association of Clinical
    Chemists
AAMI, see Association for the Advancement of
    Medical Instrumentation
Abfarad, 75
Absolute weights, 457
Absorbed dose, 241—242
    radiation field and, 246—247
    units for, 9, 16
Absorbed dose rate, 241—242
    units for, 16
Absorption
    carbonyl, 228—232
    discrete regions, 534
    gravimetric, 459
    mean, 534
    relative, 534
    values of, 195—197
Absorption frequencies
    cumulated double bonds, 227—228
    double bond region, 228—236
    single bonds to hydrogen, 223—226
    triple bonds, 227—228
Abvolt, 76
Acceptable Common Names, 62
Accreditation of laboratories, 324
Accuracy, see also Precision, 339, 359
    analytical, 435
    balance, 457
    retrospective studies of, 441
Acetate buffer, 111
Acid-base ratios at given pH, 101
Acid phosphatase, 385
Acids, see also specific acids
    aldaric, 176
    aldonic, 176
    concentration of, 309
    ionization constants for, 101
    pH values of, 100
Aconitate buffer, 110
Acoustical phenomena, see also Sound, 18—20
Acre-foot, 71
ACS, see American Chemical Society
Activated carbon, 299
Addition of analyte, 347
AIHA, 443
Air-displacement piston pipettes, 295—296
Alcohols, see also specific alcohols, 225
    polyhydric, 364
Aldaric acid, 176
Alditols, 174—175
Aldonic acid, 176
Aldoses, 169—172
Aliquoting adaptors with microsyringes, 296
Alkaline phosphatase, 382
Alkenes, 224, 233
Allowable percent of biologic variation, 406

Alphabets, 69
ALT, 382
American Association of Blood Banks, 324
American Association of Bioanalysts (AAB), 439,
    449
American Association of Clinical Chemists
    (AACC), 322, 439, 449
American Association of Equine Practitioners, 37
American Chemical Society (ACS), 59, 287
    Committee on Analytical Reagents of, 299
American College of Nuclear Physicians, 334, 449
    Radioassay Program of, 439
American National Metric Council, 3
American National Standards Institute (ANSI),
    61, 325
American Society for Quality Control, 325
American Society for Testing and Materials
    (ASTM), 325—326
American Society of Internal Medicine (ASIM),
    439, 449
American Society of Textile Chemists and
    Colorists, 61
Amide, 225
Amine buffers for biological research, 117,
    118—119
Amino acids
    chemical properties of, 177—183
    ionization constants of, 148, 149
    pH values of, 148
    physical properties of, 177—183
    solubilities of, 144—147
Amino sugars, 173
Ammonium, 225
Amne, 225
Ampere, 6, 11
    defined, 10
Analog data display oscilloscope, 493
Analysis, see also Assays; Testing; specific tests
    accuracy of, 435
    centrifugal, 489—494
    duplicate, 423
    error in, 292
    goals of, see Analytic goals
    gravimetric, 455
    kinetic methods of, 495—496
    thermogravimetric, 459
    variability in, 423
Analyst bias in precision estimates, 391—395
Analytes
    addition of, 347
    concentration of, 395, 409
    consumption of, 347
    dominant directional instability, 384
    generation of, 347
    inorganic, 383
    organic, 383
    stability of, 371—389
    subtraction of, 347
Analytic balance, see Balances
Analytic calorimetry, 503—505

Analytic centrifugal module, 489
Analytic goals, 339, 400
  for precision, 417
Analytic components of, 395—400
  concentration and, 396—400
  time and, 395—396
Analytic reagent grade (AR), 299
Analytic runs, 423
Analytic systems, 494
Analytic weight tolerances, 265
Analyzers
  automated, 477—482
  centrifugal, 489—502
  computer-controlled, 483—485
  continuous-flow, 477, 478, 483—487
  time-of-flight, 514
Anemia, 37
Angstrom, 15, 72, 74, 75
Angular measurement, see also specific units of
      measurements, 15
Animals
  domestic, 39—40
  laboratory, 23, 25—26
Annulus, 491
ANSI, see American National Standards Institute
Anticoagulants, 342, 346
Antigens, 360
Apothecaries' fluid measure, 77
Applications
  analytical balance, 459—461
  mass spectrometry, 520—521
  medical, 404
  practical, 14—20
Aqueous solutions, 89—92, 275, 276
  dilute, 273
  pH of, 94
  solvents for extraction of, 313
AR, see Analytical reagent grade
Are, 15
Area, 77
  conversion factors for units of, 80
  units of, 78, 80
Aromatic C-H, 224
Aromatic compounds, 233
Artificial radioisotope dosimetry, 255
ASIM, see American Society for Internal
      Medicine
Aspen Conference on Analytical Goals in Clinical
      Chemistry, 320
Assays
  enzyme, 496
  immuno-, 431—432
  ligand, 382
  quality control of, 431
  values in, 366
Association for the Advancement of Medical
      Instrumentation (AAMI), 326
AST, 382
ASTM, see American Society for Testing and
      Materials
Atmosphere, 15, 69—73, 75, 76
Automated analyzers, 477—482
Average CV, 396

Average of patient normals, 423, 430—431
Avogram, 72
Avoidance of matrix errors, 342—352
Azo compounds, 232

B

Bacterial growth, 385
Balances, 455—462
  accuracy of, 457
  applications of, 459—461
  design of, 457—458
  double-pan, 455
  electromagnetic, 458
  evaluation of, 457
  quartz fiber torsion, 458
  sensitivity of, 457
  single-pan, 455
  torsion suspension, 458
  two-pan, 457
Bar, 15, 20, 69, 71, 75, 76
Barbitol buffer, 118
Barn, 15
Barometer
  Fortin type, 266
  mercurial, 266, 268
Barometric corrections, 266—272
Barometric latitude-gravity, 269—270
Barometry, 266—272
Barrel, 71
Barye, 76
Base materials, 359
Base quantities, 8—9
Bases
  concentration of, 309
  ionization constants for, 101
  pH values of, 100
Base SI units, defined, 9—12
Bead lyophilized materials, 364
Bead lyophilized plasma, 360
Beckman Instrument, Inc., 449
Beer-Lambert Law, 529
Bending of N-H, 226
Benzene ring substitution patterns, 234
Bequerel, see also Curie, 16
Bias, 364
  analyst, 391—395
  method, 363, 437
Biochemical individuality, 410
Biochemical names, 61
Biological research amine buffers, 117—119
Biologic material pH values, 100
Biologic variation, 404
  allowable percent of, 406
  group components of, 409
  healthy populations, 411—416
  individual components of, 409
  interindividual, 410
  intraindividual, 410
  total, 409
Biomedical standards, 325
BIPM, see Bureau International des Poids and
      Mesures

Blind sample, 394
Blood, see also Serum
  collection of, 310—314
  samples of, 37
Blood chemistry values, 38, 39—40
BMD, see Bureau of Medical Devices
Board foot, 70
Boiling points
  organic solvents, 164—166
  water, 120
Bonds
  double, 227—236
  single, 223—226
  triple, 227—228
Borax-NaOH buffer, 115
Boric acid-borax buffer, 115
Boron-free glass, 289
BRH, see Bureau of Radiological Health
British Association for the Advancement of
    Science, 5
British (Imperial) capacity measure conversion
    factors, 84
British standards, 99
Btu, 71, 73
  per degree Fahrenheit, 73
Bucket, 70
Buffers, 309, 362
  acetate, 111
  aconitate, 110
  amine, 117—119
  barbitol, 118
  borax-NaOH, 115
  boric acid-borax, 115
  cacodylate, 118
  carbonate-bicarbonate, 114
  citrate, 111
  citrate-phosphate, 115
  enzyme studies and, 109—116
  glycine-HCl, 110
  glycine-NaOH, 116
  hydrochloric acid-potassium chloride, 110
  maleate, 117
  pH of, 96—98
  phosphate, 116
  phthalate-hydrochloric acid, 110
  phthalate-sodium hydroxide, 117
  reference pH, 95
  succinate, 117
  tris(hydroxymethyl)-aminomethane (tris), 114
  tris(hydroxymethyl)-aminomethane-maleate
      (tris-maleate), 117
Bureau International des Poids et Mesures
    (BIPM), 5, 6
Bureau of Medical Devices (BMD), 322, 326—327
Bureau of Radiological Health (BRH), 327—328
Burets, 266
Burroughs Wellcome Co., 439, 449, 450
Bushel, 70, 73

C

Cable length, 74

Cacodylate buffer, 118
Calcium, 445—447
Calibrated capillary tubes, 294
Calibration
  reference materials for, 371, 372
  serum for, 371
  spectroscope, 278
  standards for, 359—369
  thermometer, 272—273, 472, 473
  volumetric glassware, 264
Calibrators, 314—315
  assignment of values for, 366—367
  defined, 359, 371
California Thoracic Society, 439, 449
Callibration and Test Material (CTM), 318
Calorie, see also Joule, 18, 73
  per degree centigrade, 73
  thermochemical, 17
Calorie-gram, 71
Calorimetry, 503—510
Candela, 11
  defined, 10
  per square centimeter, 71, 73
  per square foot, 71
  per square inch, 73
  per square meter, 71, 73
Candle power, 74
CAP, see College of American Pathologists
Capacitance units, 9
Capacity, 77
  British (Imperial), 84
  conversion factors for units of, 81
  dry, 82
  liquid, 83
Capillary containers, 293—294
Carat, 72, 75
Carbohydrates, 169—176
Carbon, 299
Carbonate-bicarbonate buffer, 114
Carbonyl absorption, 228—231
  position of, 232
Carbonyl band intensities, 232
CAS, see Chemical Abstracts Service
CBN, see Council on Biochemical Nomenclature
CDC, see Center for Disease Control
CEE, see International Commission on Rules for
    the Approval of Electrical Equipment
Cellulose membranes, 300
Cellulose powders, 308
  for column chromatography, 306
Celsius degrees, 87
Center for Disease Control (CDC), 320, 323, 324,
    439, 449
Centigrade heat unit, 73
Centimeter, 71, 72, 75, 76
  of mercury, 69, 76
  of water, 71, 76
  per degree centigrade, 73
Centimeter water, 18
Central processing unit (CPU), 493
Centrifugal Fast Analysis, 477
Centrifugation, 262, 463—468, 478, 489—502
  separation by, 463

Centrifuges, 463—468
  choice of, 464—466
Certified Reference Material (CRM), 318
CGPM, see Conference Generale des Poids et
  Mesures
Chain, 71, 72, 74, 76
Chambers, 491
Change
  genetic, 348—351
  pathologic, 348—351
  pharmacologic toxic, 348—351
  physiologic, 348—351
Charge, 9
CHEMDEX, System Development Corporation,
  60
Chemical Abstracts Service (CAS), 59—62
  registry system of, 60, 62
Chemical conversion factors, 93
Chemical ionization, 520
Chemically pure grade (CP), 299
Chemical nomenclature, 59—63
  systematic, 59—60
Chemical properties
  amino acids, 177—183
  lipids, 184—189
  steroid hormones, 190—192
Chemical resistance chart of Nalgene labware,
  257—259
Chemical resistance of resins, 260—261
Chemicals, 299
  plastics and, 260
  standard names of, 60—62
Chemical solution correction factors, 263
Chemical Substance Index, 59, 60
Chemical units of concentration, 308—309
Chemistry
  blood, 23, 38—40
  wet, 455
Chemistry control products, 449
CHEMLINE, 60
CHEMNAME, Lockheed, 60
Chromatography
  column, 306
  liquid, 513
CI, see Colour Index
CIPM, see Comite International des Poids et
  Mesures
Circular mil, 76
Circular millimeter, 76
Citrate buffer, 111
Citrate-phosphate buffer, 115
Classification, 321
Class I medical device, 321
Class II medical device, 321
Class III medical device, 321
Cleaning solutions for glassware, 289
Clerical error, 440
Clinical chemistry device performance standards
  development, 321—322
Clinical enzymology reference methods, 474
Clinical Laboratories Improvement Act of 1967,
  321, 433, 443
Clinical laboratory standards, 324—334

CODATA, see Committee on Data for Science
  and Technology
Code of Federal Regulations, 61
Coefficients of variation (CV), 394, 399, 400,
  404, 443
  average, 396
  distribution of, 406
  interindividual, 410
  intraindividual, 410, 417
  tolerance limits for, 407—408
Coherence, defined, 9
Coherent system, 9
Collection of specimens, 44, 345
College of American Pathologists (CAP), 299,
  320—322, 326, 334, 364, 396, 439, 443,
  449, 450
Commission on Laboratory Inspection and
  Accreditation of, 324
  standard solutions of, 318
Colloid osmotic pressure, 525—526
Colored filters, 198—204
Color names, 61
Colour Index (CI), 61
Column chromatography cellulose powders, 306
Combustible mixtures, 163
Comite International des Poids et Mesures
  (CIPM), 5, 6
Commercial grade, 299
Commission on Laboratory Inspection and
  Accreditation of CAP, 324
Commission on Quantities and Units in Clinical
  Chemistry, 6
Commission on Symbols, Units, and
  Nomenclature (SUN Commission), 6
Commission on World Standards, World
  Association of Societies of Pathology, 331
Committee on Analytical Reagents, American
  Chemical Society, 299
Committee on Data for Science and Technology
  (CODATA), 6, 21
Committee on Teaching of Chemistry, 8
Common laboratory animals, 23
  names for, 25—33
Common laboratory tests, 93
Common reagent solution pH values, 99
Computers, 434—435, 450, 493
  analyzers controlled by, 483—485
  summarization by, 440—441
Concentrations
  acid, 309
  analyte, 395, 409
  analytic precision and, 396—400
  base, 309
  chemical unit expression of, 308—309
  dilute solution, 273
  glucose, 276
  hydrogen ion, 287
  measures of, 273—274
  salt, 276
  solution, 273, 306—308
  units for, 88, 89
Condensed Chemical Dictionary, 62
Conductance units, 9

Conference Generale des Poids et Mesures
  (CGPM), 5, 6, 10, 12, 15
Consensus laboratory standards, 287
Constriction pipettes, 293
Consumption of analyte, 347
Containers, 293—294
Continuous-flow analyzers, 477, 478, 483—487
Control
  defined, 371
  internal process, 433
  limits on, 425
  products of, 449
Control charts
  cumulative, 423
  cumulative sum, 427—430
  formation of, 425—427
  Shewhart, 423
Control materials, 314, 319, 359—369
  advantages of, 364—366
  analyte stability in, 371—389
  assignment of values of, 366—367
  defined, 359, 371, 372
  enzyme stability in, 377—381
  folic acid levels in, 363
  handling of, 361—362
  homogeneity of, 364
  inorganic analyte stability in, 373—374
  interval differences in, 363
  interval variability in folic acid levels in, 363
  limitations of, 362—364
  liquid, 363—365
  lot number for, 360
  maximum allowable uncertainty interval for,
    366
  organic analyte stability in, 374—376
  plug lyophilized, 362, 366
  precision estimates and, 391
  production of, 359—361
  quality, 371
  turbidity of, 363
  types of, 359—361
  use of, 361—362
  vial-to-vial differences in, 362
Control pools, 434
Control serum
  defined, 371
  instability of, 383
  lyophilized, 392—393
Conventional metric units, 23
Conversion
  area units, 80
  British (Imperial) capacity measures, 84
  capacity units, 81, 84
  chemical, 93
  concentration units, 88
  dose quantities, 244—246
  electrolytes, 90—92
  length units, 79
  mass units, 85
  roentgen dose rates, 252
  SI, 43
  solutions, 87
  thermometer scales, 87

troy weight units, 86
units, 69—77
U.S. dry capacity measures, 82
U.S. liquid capacity measure, 83
volume units, 81
Corrections
  barometric, 266—272
  for chemical solutions, 263
  for water, 263
  temperature, 268
Corvac, 314
Cosmetic ingredient names, 61
Cosmetics, Toiletries and Fragrance Association,
  61
Coulomb per kilogram, see also Roentgen, 16
Coulomb per kilogram second, see also Roentgen
  per second, 16
Council on Biochemical Nomenclature (CBN), 6
CP, see Chemically pure grade
CPK, 382
CPU, see Central processing unit
CQUCC, see Commission on Quantities and
  Units in Clinical Chemistry
CRM, see Certified Reference Material
CSCC, see Cumulative sum control charts
CTM, see Callibration and Test Material
Cubic centimeter, 70, 71, 73, 75
Cubic foot, 70, 71, 73
  per hour, 74
  per second, 74
Cubic foot-atmosphere, 71, 73
Cubic inch, 70, 71, 73, 75
Cubic meter, 70, 71, 73
Cubic yard, 70, 72, 74
Cumulated double bond absorption frequencies,
  227—228
Cumulative sum (cusum), 426—430
Cumulative sum control charts (CSCC), 423,
  427—430
Curie, see also Bequerel, 16, 70
Curve monitoring, 484
Cusum, see Cumulative sum
Cuvette chamber, 491
Cuvette rotors, 491
  transfer disc and, 491—492
CV, see Coefficients of variation

D

Dade, 449, 450
Daily quality control
  data processors in, 450
  regulatory requirements for, 433
Dalton, 70, 74
Data display devices, 493
Data processing, see Computers
Data reduction, 495
Data storage devices for computer, 493
Day, 15
Deci-normal solutions
  oxidation reagents, 312
  reagents, 311, 312

reduction reagents, 312
  salts, 311
Decision limit method, 429
Decision points, 362
Definitive method, 319, 435
Degrees
  angular, 15
  Celsius, 70, 87
  Fahrenheit, 70, 87
  Rankine, 87
  Reaumur, 87
Dehydrating reagents, 313
Deionized water, 299, 300
Delivery error, 292
Density
  ethyl alcohol, 125
  liquid, 125
  magnetic flux, 9
  mercury, 124
  water, 120
Depression
  dew point, 525
  freezing-point, 273—275, 524
  vapor pressure, 524—525
Design
  balance, 457—458
  calorimeter, 506—507
Detection of matrix errors, 342—352
Deterioration of drugs, 386
Deviation, 423—425
Dew point depression, 525
Diagnostic devices, 321, 323
Dialysis, 478
Dictionaries, 60
Dieletric constant of water, 121
Differential Scanning Colorimetry (DSC), 459
Differential Thermal Analysis (DTA), 459
Diluent for lyophilized materials, 362
Dilute solutions, 273, 306
  concentration of, 273
Dilutors, 291—298
Directional instability, 384
Discrete region absorbance, 534
Disintegration
  law of, 237—238
  per second, 70
Disposable rotors, 492
Distilled water, 299, 300
Distribution
  CV, 406
  error in, 530—532
  normal, 425
Diuretic agents, 37
Division, 12
Domestic animal blood chemistry values, 39—40
Dominant directional instability, 384
Dose
  absorbed, 9, 16, 241—242, 246—247
  conversion of, 244—246
  gamma-emitting isotope, 256
  ion, 243
  roentgen, 252
Dose equivalent, 248

units of, 9, 16
Dosimetry
  artificial radioisotope, 255
  radiation, 239—240
Double bonds, 227—228
  absorption frequencies of, 228—236
Double-pan analytical balance, 455
Double usage of prefixes, 14
Drachm, 70
Dram, 70, 72, 74, 75
Droplet lyophilized plasma, 360
Drugs, 341
  deterioration of, 386
  effects of, 343, 352
  expiration date of, 386
  names of, 61
  outdate of, 386
  shelf life of, 386
  stability of, 385, 386
Dry capacity, 82
Drying agents, 310
Dry measure, 77
DSC, see Differential Scanning Colorimetry
DTA, see Differential Thermal Analysis
Duplicates
  analysis of, 423
  measurements of, 431
  paired human serum, 392—393
Dye names, 61
Dyne, 72, 73, 75
  per centimeter, 70
  per square centimeter, 69, 71—73, 75, 76

E

EEC, see European Economic Community
EGA, see Evolved Gas Analysis
Electrical units, 6, 9
Electrodes, 485—487
Electrogravimetric technique, 459
Electrolyte conversion factors, 90—92
Electromagnetic balances, 458
Electronic thermometer, 469
Electron multipliers, 516
Electronvolt, 71, 74
Elements, 68
Emission spectra, 220—222
Endogenous metabolites, 344—345, 352
Energy units, 9, 18
Engagement, 494
Environmental Protection Agency (EPA), 62
Enzymes, 382
  activities of in serum, 32—33, 364
  assays of, 496
  buffers for, 109—116
  catalyzed reactions of, 504
  immobilized, 485
  names of, 61
  reference materials for, 387
  stability of in control materials, 377—381
Enzymology, 474
Equilibrium methods, 496

Equines, see Horses
Erg, 71, 73, 74
  per square centimeter, 70
Erroneous results, 341—358
Errors, 424
  analytical, 292
  clerical, 440
  delivery, 292
  distributional, 530—532
  interpolative, 474
  interpretation, 341
  matrix, 342—352
  microdensitometry, 531
  random, 457
  standard, 430
  systematic, 426
Ethers, 235
Ethyl alcohol density, 125
Ethylene glycol, 363
European Economic Community (EEC), 7
Evacuated tubes, 313, 314
Evolved Gas Analysis (EGA), 459
Expiration data of drugs, 386
Exponents, 12
Exposure, 242—244
  units of, 16
Exposure rate, 242—244
  units of, 16
Extensive quantity, 9
External quality control, 339, 359, 439—448
Extraction solvents, 313
Extraneous substances, 346
Extrapolation among species, 38

F

Fahrenheit degrees, 87
Farad, 75
Fast analytical systems, 494
Fathom, 71, 74, 76
FDA, see Food and Drug Administration
Filter paper, 300
  Schleicher and Schuell, 302—303
  Whatman, 301
Filters
  colored, 198—204
  glass fiber, 300
  membrane, 300, 305—306
  microfiber, 304
  Millipore membrane, 300
  quadrupole mass, 514, 515
  Whatman glass microfiber 304
  Wratten, 205—219
Fisher Diagnostics, 449
Fixed pathlength disposable rotor, 492
Fixed points, 469
Flame emission spectra, 220—222
Flasks, 289
Flexible disposable rotor, 492
Flint glass, 289
Flow calorimeters, 507
Fluid measure, 77

Fluorescence-light scattering photometer, 492
Fluorescent indicators, 102, 105—106
Flux, 9
Flux density, 9
Folic acid, 363
Food and Drug Administration (FDA), 326
  standards program of, 322
Food, Drug and Cosmetic Act, 61
Foot, 71, 72, 74, 76
Foot-candle, 71
Foot-lambert, 71, 73
Foot-pound, 71, 73
Foot-poundal, 73
Foot of water, 69, 72, 75, 76
Force
  centrifugal, 262
  units of, 9, 18
Fortin type barometer, 266
Freezing-point depression, 273—275, 524
Frequency units, 9
Fresh human sera, 391
Furlong, 71, 74, 76

G

Gal, 15
Gallium melting-point standard, 470—474
Gallium standard, 469—475
Gallon, 70—72, 74, 75
  per second, 74
Gamma-emitting isotope dose rates, 256
Gamma ray constant, 249
Gardner's Handbook of Chemical Synonyms and Trade Names, 62
Gas chromatography-mass spectrometry (GC-MS), 511, 515, 516, 518, 520
Gas mixtures, 271—272
Gauss, 72
Gauss-square centimeter, 74
General Diagnostics, 449, 450
General Subject Index, 60
Generation of analyte, 347
Genetically stable breeds of horses, 38
Genetic change, 348—351
Geometric rule of similarity, 291
Gill, 70, 72, 74, 76
Giorgi System, 6
Glass, 289
Glass fiber filters, 300
Glass microfiber filters, 304
Glass volumetric apparatus, 263
Glassware, 289—290
  cleaning solutions for, 289
  types of, 289
  volumetric, 264
Globulin, 385
Glucose, 382, 387
  concentrations of, 276
Glycine-HCl buffer, 110
Glycine-NaOH buffer, 116
GMP, see Good manufacturing practices
Goals

analytic, 339, 400, 417
    precision, 417
Good manufacturing practice (GMP) for medical
        devices, 322
"Good Laboratory Practice for Nonclinical
        Laboratory Studies," 5
Governmental regulation, see Regulatory
        agencies; Regulatory requirements
Grades, 299
Grading systems, 441
Graduated pipettes, 289
Grain, 70, 72, 75
    per gallon, 76
Gram, 70, 72, 73
    per square centimeter, 69, 73, 75
Gram-centimeter, 73
Gram-force per centimeter, 76
Gram-force per square centimeter, 71
Gravimetry, 455, 459
Gray, see also Rad, 16
Great Britain's Society of Dyers and Colourists,
        61
Greek alphabet, 69
Grotthus-Draper law, 239
Group components of biologic variation, 409
Group testing, 404
Growth of bacteria, 385
Grunbaum pipette, 294
Guidelines for acceptable degrees of instability,
        386

H

Hackh's Chemical Dictionary, 62
Halogen compounds, 236
Hand, 71
*Handbook of Chemical Synonyms and Trade
        Names*, 62
Handling
    control materials, 361—362
    specimens, 345
Healthy population biologic variations, 411—416
Hearing, 18—20
Heat units, 18
Hectare, 15
Hematologic reference values, 23
    horses, 41
    laboratory animals, 25—26
    leukocytes, 26—27
Hemolysis, 44
Hepatitis B surface antigen, 360
Hertz, 20
Highest purity, see Chemically pure grade
High silica glass, 289
Homogeneity
    control materials, 364
    plug lyophilized materials, 361
Hormones, 190—192
Horse industry, 37
Horsepower-hour, 73
Horses, 37—42
    blood chemistry values in, 38

genetically stable breeds of, 38
    hematologic values for, 41
    normal, 38
    racing, 37
Hour, 15
Human laboratory data, 43—57
Human serum, 391—393
Hydrochloric acid-potassium chloride buffer, 110
Hydrogen, 223—226
Hydrogen ion concentration, 287
tris(Hydroxymethyl)-aminomethane (tris) buffer,
        114
tris(Hydroxymethyl)-aminomethane-maleate (tris-
        maleate) buffer, 117
Hyland Diagnostics, 449, 450
Hyperosmolality, 526
Hypoplastic anemia, 37

I

ICRP, see International Commission on
        Radiological Protection
ICRU, see International Commission on
        Radiation Units and Measurements
ICSH, see International Committee for
        Standardization in Hematology
IEC, see International Electrotechnical
        Commission
IFCC, see International Federation of Clinical
        Chemistry
Illuminance units, 9
Imines, 225, 232
Immobilized enzymes, 485
Immunoassays, 431—432
    quality control of, 431
Imperial see British
Imprecision, 396
Inaccuracy, 396
Inch, 71, 72, 74—76
    per degree Fahrenheit, 73
Inch of mercury, 69, 71—73, 76
Inch of water, 71, 73, 76
Index Guide, 60
Individual components of biologic variation, 409
Individual for reference, 23
Individuality, 410
Individual-related reference ranges, 410
Inductance units, 9
Infrared spectroscopy, 223—236
Inhomogeneity of plug lyophilized control
        materials, 366
Inorganic analytes, 383
    stability of in control materials, 373—374
Inorganic compound solubility, 126—138
Inorganic ions, 236
Inorganic phosphorus, 387
Inosamines, 174—175
Inositols, 174—175
Inososes, 174—175
Inspection of laboratories, 324
Instability
    acceptable degrees of, 386

control serum, 383
defined, 372
dominant directional, 384
Instrumentation
calorimetry, 505
centrifugal analysis, 489—494
mass spectrometry, 511—517
microdensitometry, 532—534
standards for service of, 537—538
Intensities of carbonyl bands, 232
Intensive quantity, 9
Interdepartmental Committee on Pest Control, 61
Interindividual biologic variation, 410
Interindividual CVs, 410
Interlaboratory surveys, 372—382
Interlaboratory variation, 395
Internal quality control, 339, 359, 391, 395, 400,
   423—433
precision of, 394
International Commission on Radiation Units
   and Measurements (ICRU), 6, 16, 239,
   248, 331
International Commission on Radiological
   Protection (ICRP), 331
International Commission on Rules for the
   Approval of Electrical Equipment (CEE),
   331
International Committee for Standardization in
   Hematology (ICSH), 6, 318
International Council of Scientific Unions, 6
International Electrotechnical Commission (IEC),
   6
International Federation of Clinical Chemistry
   (IFCC), 6, 315, 331
International meter, 5
International Practical Temperature Scale (IPTS),
   469, 473
thermometric fixed points of, 469
International Society of Hematology, 318
International standards, 331—332
International Standards Organization (ISO), 6,
   12, 315, 331
International System of Units (SI), 3, 5, 8—14,
   23, 43, 239
non-SI units for use with, 13, 15
International Union of Biochemistry (IUB), 6, 61
International Union of Geodesy and Geophysics
   (IUGG), 6
International Union of Pharmacology
   (IUPHAR), 6
International Union of Physiological Sciences
   (IUPS), 6
International Union of Pure and Applied
   Chemistry (IUPAC), 6, 8, 59, 60
International Union of Pure and Applied Physics
   (IUPAP), 6
Interpolative errors, 474
Interpretation errors, 341
Interval differences, 362
in control materials, 363
Intervals of time, 15
Intraindividual biologic variation, 410
Intraindividual CV, 410, 417

In-vitro diagnostic devices, 321, 323
In-vitro matrix, 342
In-vivo matrix, 341
Ionic product constant of water, 121
Ionization, 520
Ionization constants, 310
acids, 101
amino acids, 148, 149
bases, 101
Ionizing radiations, 16
Ions
dose rate for, 243
hydrogen, 287
inorganic, 236
selected monitoring of, 519
Ion-selective electrodes, 485—487
IPTS, see International Practical Temperature
   Scale
ISO, see International Standards Organization
Isotopes, 256
IUB, see International Union of Biochemistry
IUGG, see International Union of Geodesy and
   Geophysics
IUPAC, see International Union of Pure and
   Applied Chemistry
IUPAP, see International Union of Pure and
   Applied Physics
IUPHAR, see International Union of
   Pharmacology
IUPS, see International Union of Physiological
   Sciences

J

JCAH, see Joint Commission on Accreditation of
   Hospitals
Joint Commission on Accreditation of Hospitals
   (JCAH), 443
Joule, see also Calorie; Kilopond meter; Rem, 16,
   18, 71, 73
per abcoulomb, 73
per ampere-hour, 73
per centimeter, 72, 73
per coulomb, 73
per coulomb per degree centigrade, 73
per degree centigrade, 73
per electron, 73
per meter, 73
per statcoulomb, 73
Jugular venipuncture, 37

K

Kelvin, 11, 70
defined, 10
Ketoses, 169—172
Kilocalorie per mole, 71
Kilogram, 6, 11
defined, 10
per square meter, 72, 73, 75
Kilogram-force, 18

Kilogram-force meter, 18
Kilogram-force per square meter, 69, 71
Kilogram-meter, 73
Kilojoule, 18
    per mole, 71
Kiloline, 74
Kilometer, 75
Kilopascal, see also Millimeter mercury, 18
Kilopond, see also Newton, 18
Kilopond meter, see also Joule, 18
Kilopond meter per minute, see also Watt, 18
Kilowatt-hour, 73
Kinetic methods of analysis, 495—496
Kip per square inch, 76
Knot, 15

L

Labeling, 323
Laboratories
    accreditation of, 324
    clinical standards for, 324—334
    common tests in, 93
    consensus standards for, 287
    inspection of, 324
    medical, 324
    national standards, 472, 474
    peer, 439
    performance, 321, 441
    reference, 439
Laboratory Animal Data Bank (LADB), 24
Laboratory animals
    common, 23
    common names for, 25—33
    hematologic reference values for, 25—26
    scientific names for, 25—33
Laboratory data on humans, 43—57
Laboratory Product Problem Reporting (LPPR),
        323
Labware, 257—261
LADB, see Laboratory Animal Data Bank
Lambert, 71, 73, 74
Lambert law, 17
Lasix, 37
LDH, 382
Legislation, see also specific legislation, 5—8
Length, 77
    units of, 78, 79
Leukocyte hematologic reference values, 26—27
Levey-Jennings quality control program, 425
Levey-Jennings wall charts, 437
Ligand assay, 382
Limits
    control, 425
    control material, 362—364
    reference, 23
    tolerance, 407—408
    warning, 425, 430
Link, 71, 72, 75, 76
Lipid properties, 184—189
Liquid capacity, 83
Liquid chromatographs, 513

Liquid control materials, 363—365
Liquid-in-glass thermometers, 273
Liquids
    density of, 125
    measure of, 77
Liter, 6, 15, 70, 72, 73, 76
    defined, 13
    per minute, 74
Liter-atmosphere, 73
Lockheed, 60
Lot number for control materials, 360
Loudness, 19—20
Low actinic glass, 289
LPPR Program, see Laboratory Product Problem
        Reporting Program
Lumen, 74
    per square centimeter, 73, 74
    per square foot, 71, 73, 74
    per square meter, 71, 74
Luminescence measurements, 9, 205, 492
Lux, 71, 74
Lyophilized materials, 392—393, 440
    bead, 360, 364
    diluent for, 362
    droplet, 360
    plug, 360—363, 366, 391
    reconstituted, 362
    storage of, 361

M

Magnetic flux units, 9
Magnetic sector field, 514
Maleate buffer, 117
Masks, 427
Mass, 77
    measurement of, 455—457
    oxygen atom, 70, 74
    units of, 71, 74, 85
Mass filters, 514, 515
Mass spectrometry (MS), see also Gas
        chromatography-mass spectrometry,
        511—522
    applications of, 520—521
    instrumentation for, 511—517
    resolution in, 517—518
    sensitivity in, 518—519
Mathematical operations, 12
Mathematical signs, 13
Matrix, 363
    in-vitro, 342
    in-vivo, 341
    specimen, 395
Matrix effects, 359, 478
Matrix errors
    avoidance of, 342—352
    detection of, 342—352
Matrix factor, 341
Maximum allowable uncertainty interval for
        control materials, 366
Maxwell, 74
Mean absorbance, 534

Measurements
  angular, 15
  British (Imperial) capacity, 84
  calorimetric, 507—508
  capacity, 82—84
  concentration, 273—274
  conversion of units of, 77
  dry, 77
  duplicate, 431
  fluid, 77
  liquid, 77
  liquid capacity, 83
  luminescence, 492
  mass, 455—457
  osmolality, 523—526
  osmotic pressure, 273—277
  pH, 94—108
  pressure, 18—20, 273—277
  surface tension, 459
  temperature, 492—493
  U.S. dry capacity, 82
  U.S. liquid capacity, 83
Media, 310
Medical applications, 404
"Medical Device and Diagnostic Products
      Standards Survey," 321
Medical devices
  classes of, 321
  GMPs for, 322
  reporting program for, 323
Medical laboratory accreditation standards, 324
Medically important compound molecular
      weights, 193—194
Medical precision standards, 409
Medical usefulness
  criteria for, 339
  precision standards for, 406
Medicare Act, 443
MEDLARS system, National Library of
      Medicine, 60
Melting-point standard, 470—474
Membranes
  cellulose, 300
  microporous polymeric, 300
  Millipore, 300
  "skinned," 301—307
  specifications for, 305—306
Merck Index, 62
Mercury, 124
Mercury barometer, 266
  temperature correction for, 268
Mercury-in-glass thermometer, 469, 474
Metabolites, 344—345, 352
Metal salt solubility, 126—138
Meter, 9, 11, 20, 71, 74—76
  defined, 10
  international, 5
  per second, 20
Meter-candle, 74
Metering, 494
Meter-lambert, 71
Method bias, 363, 437
Metre, see Meter

Metric Board, 6
Metric Conversion Act of 1975 (PL 94-168), 3, 6
Metric System of Weights and Measures, 8
Metric ton (tonne), 15
Metric units, see also specific units, 23
Microdensitometry, 529—535
Microdispensers, 291—298
Microfarad, 75
Microgram, 76
Microgravimetry, 455, 459
Micron, 75
Microporous polymeric membranes, 300
Microspectrophotometry, see Microdensitometry
Microsyringes with aliquoting adaptors, 296
Mil, 72, 75
Mile, 71, 72, 75, 76
  nautical, 15
Milligram, 72, 75
  per inch, 70
  per kilogram, 76
  per liter, 76
  per millimeter, 70
Milliliter, 70, 74
Millimeter, 72
Millimeter of mercury, see also Kilopascal, 18,
      69, 71, 75, 76
Millimicron, 75
Milliphot, 74
Millipore membrane filter, 300
Minim, 70, 72, 74, 76
Minute, 15
  per liter, 18
Miscibility of organic solvent pairs, 139—140
Misleading results, 341—358
Mixed indicators, 105
Mixtures
  combustible, 163
  gas, 271—272
MKS, 6
Module
  centrifugal analytical, 489
  data processor, 493
Molal solution, 308
Mole, 11
  defined, 10
Molecular weights of medically important
      compounds, 193—194
Monitoring
  curve, 484
  selected ion, 519
  visual reaction, 495
Monogastric species, 38
Monosaccharides, 169—173
MS, see Mass spectrometry
Multiple use of prefixes, 14
Multiplication, 12
Multipliers, 516

N

Nalgene labware, 260—261
  chemical resistance chart for, 257—259

Nanometer, 75
National Bureau of Standards (NBS), 8, 265, 319,
     321, 441, 472
  gallium melting point standard from, 474
  Standard Reference Materials of, 315
National Committee for Clinical Laboratory
     Standards (NCCLS), 6, 314, 315,
     319—321, 324, 328—329, 334, 386, 537
National Fire Protection Association (NFPA),
     329—330
National Formulary (NF) grade, 299
National legislation, 6—8
National Library of Medicine, 24
  MEDLARS system of, 60
National reference system, 367
National standards, 332—333, 472, 474
Nautical mile, 15
NCCLS, see National Committee for Clinical
     Laboratory Standards
Newton, see also Kilopond, 18, 73, 75
  per square meter, 69, 71, 75
Newton second per meter, 20
NF, see National Formulary
NFPA, see National Fire Protection Association
N-H bending, 226
Nitro, 234
Nitroso, 234
Noise levels, 19
Nomenclature, 59—63
Nonionizing radiation, 16—17
Nonproprietary names, 61
Non-SI units, 12—13
  for use with SI, 13, 15
Nonspecificity, 417
Nonstationary statistical model, 404
Normal distribution, 425
Normal horse, 38
Normal solution, 308
Normals for patients, 423, 430—431
Normal values, 23, 43
Nox, 74
NRC, see Nuclear Regulatory Commission
Nuclear Regulatory Commission (NRC),
     330—331
Nuclides, 237
Numerical prefixes, 69
Numerical value, 21
  defined, 8

O

Oak Ridge National Laboratories, 477
"Obligations of Clinical Investigators of
     Regulated Articles," 5
Occupational Safety & Health Administration
     (OSHA), 331
Octave, 20
Office of Management and Budget (OMB), 322
Office of Standard Reference Materials, 473
Ohm, 11
On-line dictionaires, 60
Optical windows, 491

Organic analytes, 383
  stability of in control materials, 374—376
Organic reagent purification, 299
Organic solvents
  boiling points of, 164—166
  miscibility of, 139—140
  physical properties of, 150—162
Ortho Diagnostics, 449, 450
Oscilloscope, 493
OSHA, see Occupational Safety & Health
     Administration
Osmolality, 274, 277, 523—527
  measurement of, 523—526
  serum, 526
  sweat, 526
  urine, 526
Osmotic coefficient, 277
Osmotic pressure, 275
  colloid, 525—526
  measurements of, 273—277
Ounce, 70, 71, 72, 74, 75
  per square inch, 73
Outdate of drugs, 386
Outliers, 424, 441
  segregation of, 424
Overflow pipettes, 293
Oxidation reagents, 312
Oxidation-reduction indicators, 107—108
Oximes, 232

P

Paired human serum duplicates, 392—393
Partial pressures of gas mixtures, 271—272
"Partnership for Health Amendments of 1967",
     321, 433, 443
Part per million, 76
Pascal, 20, 70, 72, 73, 75, 76
Pascal-second, 76
  per meter, 20
  per meter cubed, 20
Pathlength, 534
Pathologic change, 348—351
Patients
  normals for, 423, 430—431
  preparation of, 43—44
Peck, 74
Peer laboratories, 439
Pennyweight, 70, 72, 75
Performance characteristics of laboratory results,
     321, 441
Performance criteria, 439, 443
Performance standards, 323, 441
  precision, 404—410
Periodic table of elements, 68
Peripheral data display devices for computers,
     493
Pesticide ingredient names, 61—62
pH, 287
  acid-base ratios at given, 101
  acids, 100
  amino acids, 148

aqueous solutions, 94
bases, 100
biologic materials, 100
buffer solutions, 98
defined, 94
determination indicators for, 103—105
indicator methods for, 102—105
measurements of, 94—108
reagent solutions, 99
secondary British standards, 99
standard solutions for determination of, 95
Pharmacologic toxic change, 348—351
pH buffer solutions, 96, 97
reference, 95
Phenol, 225
Phenylbutazone, 37
Phosphate buffer, 116
Phosphorus
compounds of, 235
inorganic, 387
Phot, 71, 74
Photometers, 492
Photoplate recording, 519
Phthalate-hydrochloric acid buffer, 110
Phthalate-sodium hydroxide buffer, 117
Physical properties
amino acids, 177—183
lipids, 184—189
organic solvents, 150—162
resins, 260—261
steroid hormones, 190—192
Physical unit expression of solution
concentrations, 306—308
Physiologic variation, 348—351, 391, 403
Physiology units, 17—20
Pica, 72
Pied, 75
Pint, 70, 71, 72, 74, 76
Pipettes
air-displacement piston, 295—296
constriction, 293
graduated, 289
Grunbaum, 294
overflow, 293
serological, 289, 290
transfer, 289
types of, 289
volumetric, 266, 289, 290
Piston pipettes, 295—296
PL 94-168, see Metric Conversion Act of 1975
Plasma, 360
Plastics and chemicals, 260
Plasticware, 290
Plotters, Plug lyophilized materials, 360, 362,
363, 391
homogeneity of, 361
inhomogeneity of, 366
stability of, 361
Point, 72
Poise, 76
Polyhydric alcohols, 364
Polymeric membranes, 300

Pooling vials of reconstituted lyophilized
materials, 362
Pools, 434
Population-based reference intervals, 410
Populations
healthy, 411—416
reference, 23
Position of carbonyl absorption, 232
Post-reconstitution stability, 372
Post-vial fill processing, 366
Potassium, 485—487
Potential, 9
Pound, 70, 72, 73, 75
per million gallons, 76
per square foot, 71, 72, 73, 75
per square inch, 71, 72, 73, 75
Poundal, 73
per square foot, 76
Pound-force
per square foot, 70, 76
per square inch, 70, 76
Powders
cellulose, 306, 308
for column chromatography, 306
Power units, 9, 18
Practical grades, 299
Pragmatic medical precision standards,
Preanalytic factors, 339, 341—358, 364
Precision, see also Accuracy, 287
analytic, 395—400, 417
evaluation of, 391—422
goals for, 417
internal quality control, 394
medical, 409
medical usefulness, 409
state-of-the-art, 409, 434, 439
Precision estimates
analyst bias in, 391—395
control materials and, 391
quality control specimens and, 400
Precision performance standards, 404—409
components of, 409—410
Prefixes, 13—14
multiple use of, 14
numerical, 69
Pregl electrogravimetric technique, 459
Pre-reconstitution stability, 372
Preservatives, 342, 346
measurement of, 18
osmotic, 273—277, 525—526
partial, 271—272
units of, 9,
vapor, 524—525
Printers, 493
Problem reporting program, 323
Process control, 433
Production of control materials, 359—361
Proficiency testing, 324, 339, 439, 441, 449
regulatory use of data in, 443—448
Proficiency Testing Service (PTS), 439, 449
Programmable storage scope, 493
PTS, see Proficiency Testing Service

Public Health Service Code, 324
Pulmonary resistance units, 18
Pump bodies, 296—298
Pure grade, 299
Pure Primary Reference Materials, 315, 317
Purification of organic reagents, 299
Purified grade, 299
Purity, 299

Q

QAS, see Quality Assurance Service
Quadrupole mass filter, 514, 515
Quality assurance, 341
Quality Assurance Service (QAS) of CAP, 406
Quality control, 394
    daily, 433, 450
    data processors in, 450
    defined, 371
    external, 339, 359, 439—448
    immunoassay, 431
    internal, see Internal quality control
    Levey-Jennings, 425
    materials for, 371
    precision of, 394
    precision estimates and, 400
    role of, 391
    Shewhart chart of, 425
Quality factor, 248
Quantity
    base, 8—9
    defined, 8
    kind of, 21
Quart, 70, 71, 72, 74, 76
Quartz fiber torsion balance, 458

R

Racing horses, 37
Rad, see also Gray, 15, 16
Radian, 12
Radiation
    absorbed dose of, 246—247
    dosimetry of, 239—240
    ionizing, 16
    nonionizing, 16—17
    quantities of, 240
    units of, 16—17
Radioactive material specific activity, 238—239
Radioactivity, 237—238
    units of, 9, 16
Radioisotopes, 237—250
    artificial, 255
Rad per second, see also Gray per second, 16
Random errors, 396, 457
Rankine degrees, 87
RBE, see Relative biological effectiveness
Reagent chambers, 491
Reagent grade, 299
Reagents, 88, 310
    common, 99

deci-normal solutions of, 311
dehydrating, 313
metering of, 494
organic, 299
oxidation, 312
reduction, 312
single vial, 287
stable, 485
Reaumur degrees, 87
Reciprocal meter, 20
Reciprocal second, 20
Reconstituted lyophilized materials, 362
Reconstitution, 361
Recording, 519
Reduction reagents, 312
Reference individual, 23
Reference intervals, 410
Reference laboratories, 439
Reference limits, 23
Reference materials
    analyte stability in, 371—389
    calibration, 371, 372
    enzyme, 387
    hierarchy of, 316
    standard, 315
Reference methods, 319, 435
    in clinical enzymology, 474
Reference pH buffer solutions, 95
Reference points, 469, 471
Reference population, 23
Reference ranges, 410
Reference sample, 23
Reference serum, 371
Reference system, 367
Reference values, 23
    hematologic, see Hematologic reference values
Refractive index of water, 121
Regional Quality Control Program (RQCP), 339,
    372, 383, 394—396, 399, 433—438
    organization of, 433—434
Registry of Toxic Effects of Chemical Substances,
    62
Registry system of CAS, 60
Regulatory agencies, 440
    needs of, 5
Regulatory requirements, 287, 320—323
    for daily quality control, 433
    for proficiency testing data, 443—448
Relative absorption, 534
Relative biological effectiveness (RBE), 247
Rem, see also Joule per kilogram, 16
Replicate testing programs, 433
Reporting of results, 20—21
Resins, 260—261
Resistance
    electric, 9
    pulmonary, 18
    vascular, 18
Response time of calorimeters, 508
Results
    erroneous, 341—358
    laboratory, 321
    misleading, 341—358

reporting of, 20—21
Retrospective studies of accuracy, 441
R-H, 226
Rigid fixed pathlength disposable rotor, 492
Rod, 71, 72, 75, 76
Roentgen, see also Coulomb per kilogram, 15, 16, 242
Roentgen dose rate conversion, 252
Roentgen per second, see also Coulomb per kilogram second, 16
Rotor, 491
  cuvette, 491—492
  disposable, 492
  flexible disposable, 492
  three-piece, 492
RQCP, see Regional Quality Control Program
Ruminants, 38
Rydberg unit of energy, 71

S

Safety, 466—468
Salts
  concentrations of, 276
  deci-normal solutions of, 311
Samples, see also Specimens
  blind, 394
  blood, 37
  chambers for, 491
  metering of, 494
  reference, 23
  size of, 424
Sampling effects, 348
Saturated C-C, 223
Saturated C-H, 223
Saturated solutions, 306
Scaling, 494
Scanning microdensitometry, 529—535
Scientific names for laboratory animals, 25—33
Scruple, 70, 72, 75
SDC, see System Development Corporation
SDI, see Standard Deviation Interval
SE, see Standard error
Second, 6, 11, 15, 20
  defined, 10
  per liter, 18
Secondary British standards pH values, 99
Secondary reference points, 469, 471
Segregation of outliers, 424
Selected biomedical standards, 325
Selected ion monitoring, 519
Sensitivity
  balance, 457
  calorimeters, 508
  changing of, 519—520
  mass spectrometry and, 518—519
Separation
  aids for, 287, 300—306
  by centrifugation, 463
  methods of, 464
  serum, 313
Serial-sample testing, 404

Serological pipettes, 289, 290
Serum, see also Blood
  calibration, 371
  chemistry of, 23
  control, 371, 383, 392—393
  duplicates of, 392—393
  enzyme activities in, 32—33, 364
  human, 391
  lyophilized, 391, 440
  osmolality ranges of, 526
  plug lyophilized, 391
  quality control, 371
  reference, 371
  separation tubes for, 313
SGOT, 382
SGPT, 382
Shelf life of drugs, 386
Shewhart quality control charts, 423, 425, 427
SI, see International System of Units
Signs, mathematical, 13
"SI for the Health Professions, The," 16
Similarity geometric rule, 291
Simultaneity, 494
Single bonds to hydrogen, 223—226
Single-pan analytical balance, 455
Single-sample testing, 404
Single vial reagents, 287
"Skinned" membranes, 301—306, 307
Slotted spacer annulus, 491
SMA system, 484
SMAC system, 483, 485
Sodium, 485—487
Solubilities, 141—143
  amino acids, 144, 144—147
  inorganic compounds, 126—138
  metal salts, 126—138
Solutions, 306—310
  aqueous, see Aqueous solutions
  buffer, see Buffers
  conversion formulas for, 87
  deci-normal, 311, 312
  dilute, 273, 306
  glassware cleaning, 289
  molal, 308
  normal, 308
  pH buffer, 95, 96, 97
  physical unit expression of concentrations of, 306—308
  reagent, 99
  saturated, 306
  standard, 95, 166—169, 306, 318
  supersaturated, 306
  volumetric, 263
Solvents
  for aqueous solution extraction, 313
  organic, 139—140, 150—162, 164—166
Sound, see also Acoustical phenomena, 18—20
  units of, 20
Spacer annulus, 491
Species extrapolation, 38
Specific activity of radioactive materials, 238—239
Specifications of membrane filters, 305—306

Specific gamma ray constant, 249
Specific gravity, 278
Specimens, see also Samples
    collection of, 44, 345
    handling of, 345
    matrix of, 395
    quality control, 394, 400
    storage of, 44
    transportation of, 44
Spectra of flame emission, 220—222
Spectrometry, 529
    infrared, 223—236
    mass, see Mass spectrometry
Spectrophotometric Standards, 315, 317
Spectroscope calibration wave lengths, 278
SPRT, see Standard Platinum Resistance
        Thermometer
Square centimeter, 76
Square foot, 76
Square inch, 76
Square meter, 76
Square mile, 76
Square rod, 76
Square yard, 76
SRM, see Standard Reference Materials
Stability, 364, 371
    analyte, 371—389
    defined, 372
    drug, 385, 386
    enzyme, 377—381
    inorganic analyte, 373—374
    lyophilized material, 361
    organic analyte, 374—376
    plug lyophilized material, 361
    post-reconstitution, 361, 372
    pre-reconstitution, 372
    statistical approaches to, 383—387
    testing of, 385
Stable reagent systems, 485
Standard deviation, 423—425
Standard Deviation Interval (SDI), 437
Standard error (SE), 430
Standard materials, 315—319
Standard methods, 319—320
Standard names of chemical substances, 60—62
Standard Platinum Resistance Thermometer
        (SPRT), 469
Standard Reference Materials (SRM), 315, 474
Standards
    biomedical, 325
    calibration, 359—369
    clinical chemistry device performance,
        321—322
    clinical laboratory, 324—334
    FDA, 322
    gallium, 469—475
    gallium melting point, 470—474
    instrument service, 537—538
    international, 331—332
    laboratory, 287
    laboratory accreditation, 324
    laboratory inspection, 324
    medical laboratory accreditation, 324

    medical precision, 409
    medical usefulness precision, 406
    national, 332—333
    performance, 323, 404—410, 441
    pragmatic medical precision, 409
    precision, 494—410
    regulatory, 320—323
    United States, 325—331
    voluntary development of, 322
Standards for Blood Banks and Transfusion
        Services, 324
Standards laboratories, 472, 474
Standard solutions, 166—169, 306
    CAP, 318
    pH determination, 95
State-of-the-art precision, 409, 434, 439
Statfarad, 75
Stationary statistical model, 404
Statistical models, 403, 417
    nonstationary, 404
    stability and, 383—387
    stationary, 404
Statvolt, 76
Steradian, 12
Sterilization, 313
Steroid hormones, 190—192
Stock solutions, 166—169
Storage
    lyophilized material, 361
    specimen, 44
Storage scope, 493
Stress testing, 385
Studies, see also Assays; Testing
    accuracy, 441
    enzyme, 109—116
Subcommittee on Analytic Goals in Clinical
        Chemistry, WASP, 391, 395
Substitution patterns of benzene ring, 234
Subsystems, 21
Subtraction of analyte, 347
Succinate buffer, 117
Sugars, 173
Sulphur compounds, 235
SUN Commission, see Commission on Symbols,
        Units, and Nomenclature
Supersaturated solution, 306
Super-system, 21
Supplementary units, 12
Sure-Sep, 314
Surface antigens of hepatitis B, 360
Surface tension
    measurements of, 459
    of water, 121
Surveys, 439
    interlaboratory, 372—382
Survey Validated Serum, 425
Suspension of torsion, 458
Sweat osmolality, 526
Synchronization of transmission signals, 492
Syphon, 491
Systematic chemical nomenclature, 59—60
Systematic errors, 426
System Development Corporation (SDC), 60

T

Target mean deviation, 423—425
TBG, see Thyroxine binding globulin
TBPA, see Thyroxine binding pre-albumin
Technical grade, 299
Technicon, 449
  SMA system by, 484
  SMAC system by, 483, 485
Temperature, 469—475
  control of, 492—493
  measurement of, 492—493
  variation in, 89
Temperature correction
  for glass volumetric apparatus, 263
  for mercury barometer, 268
  for volumetric solutions, 263
Testing
  drug stability, 385
  group, 404
  laboratory, 93
  proficiency, 324, 339, 439, 441, 443—449
  replicate, 433
  serial-sample, 404
  single-sample, 404
  stability stress, 385
TG, see Thermogravimetry
TGA, see Thermogravimetric analysis
Therapeutic agents, 341
Thermal resistant glass, 289
Thermistors, 469, 505—506
Thermochemical calorie, 17
Thermochemistry, 503
Thermocouples, 505
Thermodynamic data, 503
Thermogravimetric analysis (TGA), 459
Thermogravimetry (TG), 459
Thermometers, 469—475
  calibration of, 272—273, 472, 473
  conversion of, 87
  electronic, 469
  liquid-in-glass, 273
  mercury-in-glass, 469, 474
  thermistor resistance, 469, 505—506
Thermometric fixed points of IPTS, 469
Thermopiles, 505
Three-piece rotor, 492
Thyroid-binding ability, 385
Thyroxine binding globulin (TBG), 385
Thyroxine binding pre-albumin (TBPA), 385
Time, 409
  analytic precision and, 395—396
  intervals of, 15
Time-of-flight analyzer, 514
Titration data, 88
Tolerances
  for analytical weights, 265
  for CVs, 407—408
  for volumetric burets, 266
  for volumetric pipets, 266
Tonne, see Ton
Ton (short), 75

  per square foot, 70
  per square inch, 70
Torr, 76
Torsion balance, 458
Total biologic variation, 409
Toxic agents, 341
Toxic change in drugs, 348—351
Toxic Substances Control Act Chemical
      Inventory, 62
TOXLINE data base, 60
Transfer disc, 489
  with cuvette rotors, 491—492
Transfer pipettes, 289
Transmission
  Corning colored filters, 198—204
  luminous, 205
  values of, 195—197
  Wratten filters, 205—219
Transmission signal synchronization, 492
Transportation of specimens, 44
Triple bond absorption frequencies, 227—228
Tris(hydroxymethyl)-aminomethane (tris) buffer,
      114
Tris(hydroxymethyl)-aminomethane-maleate (tris-
      maleate) buffer, 117
Troy weight conversion factors, 86
Tubes
  capillary, 294
  evacuated, 313
  serum separation, 313
  Vacutainer, 314
Turbidity, 363, 364
Two-barrel dilutors, 296
Two-pan balance, 457

U

UL, see Underwriters Laboratories, Inc.
Ultrafiltration, 307
Unassayed control pools, 434
Uncertainty interval for control materials, 366
Underwriters Laboratories, Inc. (UL), 331
Units, see also specific units
  absorbed dose, 9, 16
  area, 78, 80
  capacitance, 9
  capacity, 81
  concentration, 88, 89
  conductance, 9
  conventional metric, 23
  conversion of, 69—77
  defined, 8
  dose equivalent, 9, 16
  electrical, 6, 9
  energy, 9, 18
  exposure, 16
  force, 9, 18
  frequency, 9
  heat, 18
  illuminance, 9
  inductance, 9
  length, 78, 79

luminous, 9
magnetic flux, 9
mass, 85
non-SI, 12—13
    for use with SI, 13, 15
physiology, 17—20
power, 9, 18
pressure, 9, 18
pulmonary resistance, 18
radiation, 16—17
radioactivity, 9, 16
representation of, 21
SI, 5, 8—12, 43
sound, 20
supplementary, 12
troy weight, 86
vascular resistance, 18
volume, 78, 81
work, 18
Urea, 383
Uric acid, 383
Urine
constituents of, 37
osmolality of, 526
Uronic acid, 176
U.S. Adopted Names (USAN) Council, 61
U.S. dry capacity measure conversion factors, 82
U.S. liquid capacity measure conversion factors,
    83
U.S. Metric Board, 3
U.S. National Bureau of Standards, see National
    Bureau of Standards
USP. see U.S. Pharmacopeia
U.S. Pharmacopeia (USP), 61, 299, 323
U.S. Pharmacopeia (USP) grade, 299
U.S. standards, 325—331
USAN Council, see U.S. Adopted Names Council
Use of control materials, 361—362
USP, see U.S. Pharmacopeia

V

Vapor pressure depression, 524—525
Variable pathlength disposable rotors, 492
Variation
analytic, 423
biologic, see Biologic variation
interindividual biologic, 410
interlaboratory, 395
interval, 363
intraindividual biologic, 410
physiologic, 391, 403
preanalytic, 364
temperature, 89
total biologic, 409
vial fill, 366
within-run, 395
Vascular resistance units, 18
VDU, see Video display unit
Venipuncture, 37

Veterans Administration, 443
Vial fill variation, 366
Vial-to-vial differences in control materials, 362
Video display unit (VDU), 493
Viscosity, 363
of water, 121
Visual reaction monitoring, 495
Volt, 11, 76
Volume, 77
of mercury, 124
of water, 122—123
Volumetric apparatus, 263
Volumetric burets, 266
Volumetric flasks, 289
Volumetric glassware calibration, 264
Volumetric indicators, 103—105
Volumetric pipettes, 289, 290
tolerances for, 266
Volumetric procedures, 455
Volumetric solution temperature correction, 263
Volume units, 78
conversion factors for, 81
Voluntary standards development, 322
V-shaped mask, 427

W

Warning limits, 425, 430
WASP, see World Association of Societies of
    Pathology
Water
boiling points of, 120
correction factors for, 263
deionized, 299, 300
density of, 120
dieletric constant of, 121
distilled, 299, 300
ionic product constant of, 121
refractive index of, 121
surface tension of, 121
viscosity of, 121
volume properties of, 122—123
Watt, see also Kilopond meter per minute, 18, 20,
    74
per square meter, 20
Watt-hour, 73
Watt-second, 73
Wave lengths, 251
for spectroscope calibration, 278
Weber, 74
Weighing, 457
Weights, 457
absolute, 457
analytical, 265
molecular, 193—194
troy, 86
Wet chemistry, 455
Whatman filter paper, 301
Whatman glass microfiber filters, 304

WHO, see World Health Organization
Within-run variation, 395
Work units, 18
World Association of Societies of Pathology
    (WASP), 6, 331, 391, 395
World Health Assembly, 14—16
World Health Organization (WHO), 16, 23, 43
Wratten filter transmission, 205—219

X

X-Y plotter, 493

Y

Yard, 71, 72, 75, 76